T0236601

Network Information Theory

This comprehensive treatment of network information theory and its applications provides the first unified coverage of both classical and recent results. With an approach that balances the introduction of new models and new coding techniques, readers are guided through Shannon's point-to-point information theory, single-hop networks, multihop networks, and extensions to distributed computing, secrecy, wireless communication, and networking. Elementary mathematical tools and techniques are used throughout, requiring only basic knowledge of probability, whilst unified proofs of coding theorems are based on a few simple lemmas, making the text accessible to newcomers. Key topics covered include successive cancellation and superposition coding, MIMO wireless communication, network coding, and cooperative relaying. Also covered are feedback and interactive communication, capacity approximations and scaling laws, and asynchronous and random access channels. This book is ideal for use in the classroom, for self-study, and as a reference for researchers and engineers in industry and academia.

Abbas El Gamal is the Hitachi America Chaired Professor in the School of Engineering and the Director of the Information Systems Laboratory in the Department of Electrical Engineering at Stanford University. In the field of network information theory, he is best known for his seminal contributions to the relay, broadcast, and interference channels; multiple description coding; coding for noisy networks; and energy-efficient packet scheduling and throughput–delay tradeoffs in wireless networks. He is a Fellow of IEEE and the winner of the 2012 Claude E. Shannon Award, the highest honor in the field of information theory.

Young-Han Kim is an Assistant Professor in the Department of Electrical and Computer Engineering at the University of California, San Diego. His research focuses on information theory and statistical signal processing. He is a recipient of the 2008 NSF Faculty Early Career Development (CAREER) Award and the 2009 US–Israel Binational Science Foundation Bergmann Memorial Award.

NETWORK INFORMATION THEORY

Abbas El Gamal
Stanford University

Young-Han Kim
University of California, San Diego

CAMBRIDGE
UNIVERSITY PRESS

CAMBRIDGE
UNIVERSITY PRESS

University Printing House, Cambridge CB2 8BS, United Kingdom

One Liberty Plaza, 20th Floor, New York, NY 10006, USA

477 Williamstown Road, Port Melbourne, VIC 3207, Australia

314-321, 3rd Floor, Plot 3, Splendor Forum, Jasola District Centre, New Delhi - 110025, India

79 Anson Road, #06-04/06, Singapore 079906

Cambridge University Press is part of the University of Cambridge.

It furthers the University's mission by disseminating knowledge in the pursuit of education, learning and research at the highest international levels of excellence.

www.cambridge.org
Information on this title: www.cambridge.org/9781108453240

© Cambridge University Press 2011

First published 2011
3rd printing 2014
First paperback edition 2018

A catalogue record for this publication is available from the British Library

ISBN 978-1-107-00873-1 Hardback
ISBN 978-1-108-45324-0 Paperback

Additional resources for this publication at www.cambridge.org/9781107008731

*To our families
whose love and support
made this book possible*

Contents

Part II Single-Hop Networks

Appendices

Preface

Network information theory aims to establish the fundamental limits on information flow in networks and the optimal coding schemes that achieve these limits. It extends Shannon's fundamental theorems on point-to-point communication and the Ford–Fulkerson max-flow min-cut theorem for graphical unicast networks to general networks with multiple sources and destinations and shared resources. Although the theory is far from complete, many elegant results and techniques have been developed over the past forty years with potential applications in real-world networks. This book presents these results in a coherent and simplified manner that should make the subject accessible to graduate students and researchers in electrical engineering, computer science, statistics, and related fields, as well as to researchers and practitioners in industry.

The first paper on network information theory was on the two-way channel by Shannon (1961). This was followed a decade later by seminal papers on the broadcast channel by Cover (1972), the multiple access channel by Ahlswede (1971, 1974) and Liao (1972), and distributed lossless compression by Slepian and Wolf (1973a). These results spurred a flurry of research on network information theory from the mid 1970s to the early 1980s with many new results and techniques developed; see the survey papers by van der Meulen (1977) and El Gamal and Cover (1980), and the seminal book by Csiszár and Körner (1981b). However, many problems, including Shannon's two-way channel, remained open and there was little interest in these results from communication theorists or practitioners. The period from the mid 1980s to the mid 1990s represents a "lost decade" for network information theory during which very few papers were published and many researchers shifted their focus to other areas. The advent of the Internet and wireless communication, fueled by advances in semiconductor technology, compression and error correction coding, signal processing, and computer science, revived the interest in this subject and there has been an explosion of activities in the field since the mid 1990s. In addition to progress on old open problems, recent work has dealt with new network models, new approaches to coding for networks, capacity approximations and scaling laws, and topics at the intersection of networking and information theory. Some of the techniques developed in network information theory, such as successive cancellation decoding, multiple description coding, successive refinement of information, and network coding, are being implemented in real-world networks.

Development of the Book

The idea of writing this book started a long time ago when Tom Cover and the first author considered writing a monograph based on their aforementioned 1980 survey paper. The first author then put together a set of handwritten lecture notes and used them to teach a course on multiple user information theory at Stanford University from 1982 to 1984. In response to high demand from graduate students in communication and information theory, he resumed teaching the course in 2002 and updated the early lecture notes with recent results. These updated lecture notes were used also in a course at EPFL in the summer of 2003. In 2007 the second author, who was in the 2002 class, started teaching a similar course at UC San Diego and the authors decided to collaborate on expanding the lecture notes into a textbook. Various versions of the lecture notes have been used since then in courses at Stanford University, UC San Diego, the Chinese University of Hong Kong, UC Berkeley, Tsinghua University, Seoul National University, University of Notre Dame, and McGill University among others. The lecture notes were posted on arXiv in January 2010. This book is based on these notes. Although we have made an effort to provide a broad coverage of the results in the field, we do not claim to be all-inclusive. The explosion in the number of papers on the subject in recent years makes it almost impossible to provide a complete coverage in a single textbook.

Organization of the Book

We considered several high-level organizations of the material in the book, from source coding to channel coding or vise versa, from graphical networks to general networks, or along historical lines. We decided on a pedagogical approach that balances the introduction of new network models and new coding techniques. We first discuss single-hop networks and then multihop networks. Within each type of network, we first study channel coding settings, followed by their source coding counterparts, and then joint source–channel coding. There were several important topics that did not fit neatly into this organization, which we grouped under Extensions. The book deals mainly with discrete memoryless and Gaussian network models because little is known about the limits on information flow for more complex models. Focusing on these models also helps us present the coding schemes and proof techniques in their simplest possible forms.

The first chapter provides a preview of network information theory using selected examples from the book. The rest of the material is divided into four parts and a set of appendices.

Part I. Background (Chapters 2 and 3). We present the needed basic information theory background, introduce the notion of typicality and related lemmas used throughout the book, and review Shannon's point-to-point communication coding theorems.

Part II. Single-hop networks (Chapters 4 through 14). We discuss networks with single-round one-way communication. Here each node is either a sender or a receiver. The material is divided into three types of communication settings.

- *Independent messages over noisy channels (Chapters 4 through 9).* We discuss noisy single-hop network building blocks, beginning with multiple access channels (many-to-one communication) in Chapter 4, followed by broadcast channels (one-to-many communication) in Chapters 5 and 8, and interference channels (multiple one-to-one communications) in Chapter 6. We split the discussion on broadcast channels for a pedagogical reason—the study of general broadcast channels in Chapter 8 requires techniques that are introduced more simply through the discussion of channels with state in Chapter 7. In Chapter 9, we study Gaussian vector channels, which model multiple-antenna (multiple-input multiple-output/MIMO) communication systems.

- *Correlated sources over noiseless links (Chapters 10 through 13).* We discuss the source coding counterparts of the noisy single-hop network building blocks, beginning with distributed lossless compression in Chapter 10, followed by lossy compression with side information in Chapter 11, distributed lossy compression in Chapter 12, and multiple description coding in Chapter 13. Again we spread the discussion on distributed compression over three chapters to help develop new ideas gradually.

- *Correlated sources over noisy channels (Chapter 14).* We discuss the general setting of sending uncompressed sources over noisy single-hop networks.

Part III. Multihop networks (Chapters 15 through 20). We discuss networks with relaying and multiple rounds of communication. Here some of the nodes can act as both sender and receiver. In an organization parallel to Part II, the material is divided into three types of settings.

- *Independent messages over graphical networks (Chapter 15).* We discuss coding for networks modeled by graphs beyond simple routing.

- *Independent messages over noisy networks (Chapters 16 through 19).* In Chapter 16, we discuss the relay channel, which is a simple two-hop network with a sender, a receiver, and a relay. We then discuss channels with feedback and the two-way channel in Chapter 17. We extend results on the relay channel and the two-way channel to general noisy networks in Chapter 18. We further discuss approximations and scaling laws for the capacity of large wireless networks in Chapter 19.

- *Correlated sources over graphical networks (Chapter 20).* We discuss source coding counterparts of the channel coding problems in Chapters 15 through 18.

Part IV. Extensions (Chapters 21 through 24). We study extensions of the theory discussed in the first three parts of the book to communication for computing in Chapter 21, communication with secrecy constraints in Chapter 22, wireless fading channels in Chapter 23, and to problems at the intersection of networking and information theory in Chapter 24.

Appendices. To make the book as self-contained as possible, Appendices A, B, and E provide brief reviews of the necessary background on convex sets and functions, probability and estimation, and convex optimization, respectively. Appendices C and D describe techniques for bounding the cardinality of auxiliary random variables appearing in many capacity and rate region characterizations, and the Fourier–Motzkin elimination procedure, respectively.

Presentation of the Material

Each chapter typically contains both teaching material and advanced topics. Starred sections contain topics that are either too technical to be discussed in detail or are not essential to the main flow of the material. The chapter ends with a bulleted summary of key points and open problems, bibliographic notes, and problems on missing proof steps in the text followed by exercises around the key ideas. Some of the more technical and less central proofs are delegated to appendices at the end of each chapter in order to help the reader focus on the main ideas and techniques.

The book follows the adage "a picture is worth a thousand words." We use illustrations and examples to provide intuitive explanations of models and concepts. The proofs follow the principle of making everything as simple as possible but not simpler. We use elementary tools and techniques, requiring only basic knowledge of probability and some level of mathematical maturity, for example, at the level of a first course on information theory. The achievability proofs are based on joint typicality, which was introduced by Shannon in his 1948 paper and further developed in the 1970s by Forney and Cover. We take this approach one step further by developing a set of simple lemmas to reduce the repetitiveness in the proofs. We show how the proofs for discrete memoryless networks can be extended to their Gaussian counterparts by using a discretization procedure and taking appropriate limits. Some of the proofs in the book are new and most of them are simplified—and in some cases more rigorous—versions of published proofs.

Use of the Book in Courses

As mentioned earlier, the material in this book has been used in courses on network information theory at several universities over many years. We hope that the publication of the book will help make such a course more widely adopted. One of our main motivations for writing the book, however, is to broaden the audience for network information theory. Current education of communication and networking engineers encompasses primarily point-to-point communication and wired networks. At the same time, many of the innovations in modern communication and networked systems concern more efficient use of shared resources, which is the focus of network information theory. We believe that the next generation of communication and networking engineers can benefit greatly from having a working knowledge of network information theory. We have made every effort to present some of the most relevant material to this audience as simply and clearly as possible. In particular, the material on Gaussian channels, wireless fading channels, and Gaussian networks can be readily integrated into an advanced course on wireless communication.

The book can be used as a main text in a one-quarter/semester first course on information theory with emphasis on communication or a one-quarter second course on information theory, or as a supplementary text in courses on communication, networking, computer science, and statistics. Most of the teaching material in the book can be covered in a two-quarter course sequence. Slides for such courses are posted at http://arxiv.org/abs/1001.3404/.

Dependence Graphs

The following graphs depict the dependence of each chapter on its preceding chapters. Each box contains the chapter number and lighter boxes represent dependence on previous parts. Solid edges represent required reading and dashed edges represent recommended reading.

Part II.

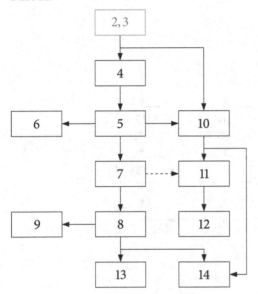

2 Information measures and typicality
3 Point-to-point information theory
4 Multiple access channels
5 Degraded broadcast channels
6 Interference channels
7 Channels with state
8 General broadcast channels
9 Gaussian vector channels
10 Distributed lossless compression
11 Lossy compression with side information
12 Distributed lossy compression
13 Multiple description coding
14 Joint source–channel coding

Part III.

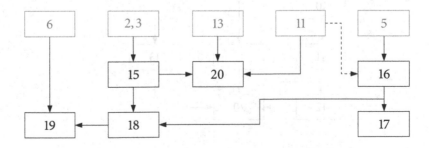

15 Graphical networks
16 Relay channels
17 Interactive channel coding
18 Discrete memoryless networks
19 Gaussian networks
20 Compression over graphical networks

Part IV.

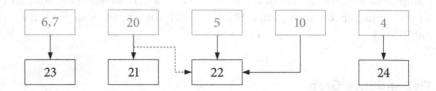

21 Communication for computing   22 Information theoretic secrecy
23 Wireless fading channels   24 Networking and information theory

In addition to the dependence graphs for each part, we provide below some interest-based dependence graphs.

Communication.

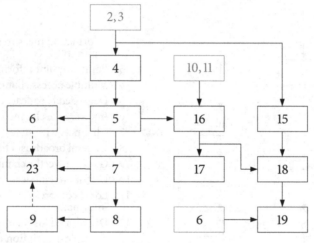

Data compression.

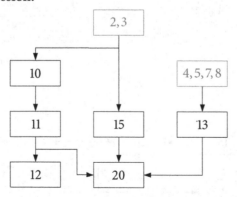

Abbas El Gamal   Palo Alto, California
Young-Han Kim   La Jolla, California
                                               July 2011

Acknowledgments

The development of this book was truly a community effort. Many colleagues, teaching assistants of our courses on network information theory, and our postdocs and PhD students provided invaluable input on the content, organization, and exposition of the book, and proofread earlier drafts.

First and foremost, we are indebted to Tom Cover. He taught us everything we know about information theory, encouraged us to write this book, and provided several insightful comments. We are also indebted to our teaching assistants—Ehsan Ardestanizadeh, Chiao-Yi Chen, Yeow-Khiang Chia, Shirin Jalali, Paolo Minero, Haim Permuter, Han-I Su, Sina Zahedi, and Lei Zhao—for their invaluable contributions to the development of this book. In particular, we thank Sina Zahedi for helping with the first set of lecture notes that ultimately led to this book. We thank Han-I Su for his contributions to the chapters on quadratic Gaussian source coding and distributed computing and his thorough proofreading of the entire draft. Yeow-Khiang Chia made invaluable contributions to the chapters on information theoretic secrecy and source coding over graphical networks, contributed several problems, and proofread many parts of the book. Paolo Minero helped with some of the material in the chapter on information theory and networking.

We are also grateful to our PhD students. Bernd Bandemer contributed to the chapter on interference channels and proofread several parts of the book. Sung Hoon Lim contributed to the chapters on discrete memoryless and Gaussian networks. James Mammen helped with the first draft of the lecture notes on scaling laws. Lele Wang and Yu Xiang also provided helpful comments on many parts of the book.

We benefited greatly from discussions with several colleagues. Chandra Nair contributed many of the results and problems in the chapters on broadcast channels. David Tse helped with the organization of the chapters on fading and interference channels. Mehdi Mohseni helped with key proofs in the chapter on Gaussian vector channels. Amin Gohari helped with the organization and several results in the chapter on information theoretic secrecy. Olivier Lévêque helped with some of the proofs in the chapter on Gaussian networks. We often resorted to John Gill for stylistic and editorial advice. Jun Chen, Sae-Young Chung, Amos Lapidoth, Prakash Narayan, Bobak Nazer, Alon Orlitsky, Ofer Shayevitz, Yossi Steinberg, Aslan Tchamkerten, Dimitris Toumpakaris, Sergio Verdú, Mai Vu, Michèle Wigger, Ram Zamir, and Ken Zeger provided helpful input during the writing of this book. We would also like to thank Venkat Anantharam, François Baccelli, Stephen Boyd, Max Costa, Paul Cuff, Suhas Diggavi, Massimo Franceschetti, Michael Gastpar,

Andrea Goldsmith, Bob Gray, Te Sun Han, Tara Javidi, Ashish Khisti, Gerhard Kramer, Mohammad Maddah-Ali, Andrea Montanari, Balaji Prabhakar, Bixio Rimoldi, Anant Sahai, Anand Sarwate, Devavrat Shah, Shlomo Shamai, Emre Telatar, Alex Vardy, Tsachy Weissman, and Lin Zhang.

This book would not have been written without the enthusiasm, inquisitiveness, and numerous contributions of the students who took our courses, some of whom we have already mentioned. In addition, we would like to acknowledge Ekine Akuiyibo, Lorenzo Coviello, Chan-Soo Hwang, Yashodhan Kanoria, Tae Min Kim, Gowtham Kumar, and Moshe Malkin for contributions to some of the material. Himanshu Asnani, Yuxin Chen, Aakanksha Chowdhery, Mohammad Naghshvar, Ryan Peng, Nish Sinha, and Hao Zou provided many corrections to earlier drafts. Several graduate students from UC Berkeley, MIT, Tsinghua, University of Maryland, Tel Aviv University, and KAIST also provided valuable feedback.

We would like to thank our editor Phil Meyler and the rest of the Cambridge staff for their exceptional support during the publication stage of this book. We also thank Kelly Yilmaz for her wonderful administrative support. Finally, we acknowledge partial support for the work in this book from the DARPA ITMANET and the National Science Foundation.

Notation

We introduce the notation and terminology used throughout the book.

Sets, Scalars, and Vectors

We use lowercase letters $x, y, \ldots$ to denote constants and values of random variables. We use $x_i^j = (x_i, x_{i+1}, \ldots, x_j)$ to denote an $(j - i + 1)$-sequence/column vector for $1 \leq i \leq j$. When $i = 1$, we always drop the subscript, i.e., $x^j = (x_1, x_2, \ldots, x_j)$. Sometimes we write $\mathbf{x}, \mathbf{y}, \ldots$ for constant vectors with specified dimensions and x_j for the j-th component of $\mathbf{x}$. Let $\mathbf{x}(i)$ be a vector indexed by time i and $x_j(i)$ be the j-th component of $\mathbf{x}(i)$. The sequence of these vectors is denoted by $\mathbf{x}^n = (\mathbf{x}(1), \mathbf{x}(2), \ldots, \mathbf{x}(n))$. An all-one column vector $(1, \ldots, 1)$ with a specified dimension is denoted by $\mathbf{1}$.

Let $\alpha, \beta \in [0, 1]$. Then $\bar{\alpha} = (1 - \alpha)$ and $\alpha * \beta = \alpha \bar{\beta} + \beta \bar{\alpha}$.

Let $x^n, y^n \in \{0, 1\}^n$ be binary n-vectors. Then $x^n \oplus y^n$ is the componentwise modulo-2 sum of the two vectors.

Calligraphic letters $\mathcal{X}, \mathcal{Y}, \ldots$ are used exclusively for finite sets and $|\mathcal{X}|$ denotes the cardinality of the set $\mathcal{X}$. The following notation is used for common sets:

- $\mathbb{R}$ is the real line and $\mathbb{R}^d$ is the d-dimensional real Euclidean space.

- $\mathbb{F}_q$ is the finite field $\mathrm{GF}(q)$ and $\mathbb{F}_q^d$ is the d-dimensional vector space over $\mathrm{GF}(q)$.

Script letters $\mathscr{C}, \mathscr{R}, \mathscr{P}, \ldots$ are used for subsets of $\mathbb{R}^d$.

For a pair of integers $i \leq j$, we define the discrete interval $[i : j] = \{i, i + 1, \ldots, j\}$. More generally, for $a \geq 0$ and integer $i \leq 2^a$, we define

- $[i : 2^a) = \{i, i + 1, \ldots, 2^{\lfloor a \rfloor}\}$, where $\lfloor a \rfloor$ is the integer part of a, and

- $[i : 2^a] = \{i, i + 1, \ldots, 2^{\lceil a \rceil}\}$, where $\lceil a \rceil$ is the smallest integer $\geq a$.

Probability and Random Variables

The probability of an event $\mathcal{A}$ is denoted by $\mathrm{P}(\mathcal{A})$ and the conditional probability of $\mathcal{A}$ given $\mathcal{B}$ is denoted by $\mathrm{P}(\mathcal{A} | \mathcal{B})$. We use uppercase letters $X, Y, \ldots$ to denote random variables. The random variables may take values from finite sets $\mathcal{X}, \mathcal{Y}, \ldots$ or from the real line $\mathbb{R}$. By convention, $X = \emptyset$ means that X is a degenerate random variable (unspecified constant) regardless of its support. The probability of the event $\{X \in \mathcal{A}\}$ is denoted by $\mathrm{P}\{X \in \mathcal{A}\}$.

In accordance with the notation for constant vectors, we use $X_i^j = (X_i, \ldots, X_j)$ to denote a $(j - i + 1)$-sequence/column vector of random variables for $1 \le i \le j$. When $i = 1$, we always drop the subscript and use $X^j = (X_1, \ldots, X_j)$.

Let $(X_1, \ldots, X_k)$ be a tuple of k random variables and $\mathcal{J} \subseteq [1:k]$. The subtuple of random variables with indices from $\mathcal{J}$ is denoted by $X(\mathcal{J}) = (X_j : j \in \mathcal{J})$. Similarly, given k random vectors $(X_1^n, \ldots, X_k^n)$,

$$X^n(\mathcal{J}) = (X_j^n : j \in \mathcal{J}) = (X_1(\mathcal{J}), \ldots, X_n(\mathcal{J})).$$

Sometimes we write $\mathbf{X}, \mathbf{Y}, \ldots$ for random (column) vectors with specified dimensions and X_j for the j-th component of $\mathbf{X}$. Let $\mathbf{X}(i)$ be a random vector indexed by time i and $X_j(i)$ be the j-th component of $\mathbf{X}(i)$. We denote the sequence of these vectors by $\mathbf{X}^n = (\mathbf{X}(1), \ldots, \mathbf{X}(n))$.

The following notation is used to specify random variables and random vectors.

- $X^n \sim p(x^n)$ means that $p(x^n)$ is the probability mass function (pmf) of the discrete random vector X^n. The function $p_{X^n}(\tilde{x}^n)$ denotes the pmf of X^n with argument $\tilde{x}^n$, i.e., $p_{X^n}(\tilde{x}^n) = \mathrm{P}\{X^n = \tilde{x}^n\}$ for all $\tilde{x}^n \in \mathcal{X}^n$. The function $p(x^n)$ without subscript is understood to be the pmf of the random vector X^n defined over $\mathcal{X}_1 \times \cdots \times \mathcal{X}_n$.

- $X^n \sim f(x^n)$ means that $f(x^n)$ is the probability density function (pdf) of the continuous random vector X^n.

- $X^n \sim F(x^n)$ means that $F(x^n)$ is the cumulative distribution function (cdf) of X^n.

- $(X^n, Y^n) \sim p(x^n, y^n)$ means that $p(x^n, y^n)$ is the joint pmf of X^n and Y^n.

- $Y^n | \{X^n \in \mathcal{A}\} \sim p(y^n | X^n \in \mathcal{A})$ means that $p(y^n | X^n \in \mathcal{A})$ is the conditional pmf of Y^n given $\{X^n \in \mathcal{A}\}$.

- $Y^n | \{X^n = x^n\} \sim p(y^n | x^n)$ means that $p(y^n | x^n)$ is the conditional pmf of Y^n given $\{X^n = x^n\}$.

- $p(y^n | x^n)$ is a collection of (conditional) pmfs on $\mathcal{Y}^n$, one for every $x^n \in \mathcal{X}^n$. $f(y^n | x^n)$ and $F(y^n | x^n)$ are similarly defined.

- $Y^n \sim p_{X^n}(y^n)$ means that Y^n has the same pmf as X^n, i.e., $p(y^n) = p_{X^n}(y^n)$. Similar notation is used for conditional probability distributions.

Given a random variable X, the expected value of its function $g(X)$ is denoted by $\mathrm{E}_X(g(X))$, or $\mathrm{E}(g(X))$ in short. The conditional expectation of X given Y is denoted by $\mathrm{E}(X|Y)$. We use $\mathrm{Var}(X) = \mathrm{E}[(X - \mathrm{E}(X))^2]$ to denote the variance of X and $\mathrm{Var}(X|Y) = \mathrm{E}[(X - \mathrm{E}(X|Y))^2 \,|\, Y]$ to denote the conditional variance of X given Y.

For random vectors $\mathbf{X} = X^n$ and $\mathbf{Y} = Y^k$, $K_{\mathbf{X}} = \mathrm{E}[(\mathbf{X} - \mathrm{E}(\mathbf{X}))(\mathbf{X} - \mathrm{E}(\mathbf{X}))^T]$ denotes the covariance matrix of $\mathbf{X}$, $K_{\mathbf{XY}} = \mathrm{E}[(\mathbf{X} - \mathrm{E}(\mathbf{X}))(\mathbf{Y} - \mathrm{E}(\mathbf{Y}))^T]$ denotes the crosscovariance matrix of $(\mathbf{X}, \mathbf{Y})$, and $K_{\mathbf{X}|\mathbf{Y}} = \mathrm{E}[(\mathbf{X} - \mathrm{E}(\mathbf{X}|\mathbf{Y}))(\mathbf{X} - \mathrm{E}(\mathbf{X}|\mathbf{Y}))^T] = K_{\mathbf{X} - \mathrm{E}(\mathbf{X}|\mathbf{Y})}$ denotes the conditional covariance matrix of $\mathbf{X}$ given $\mathbf{Y}$, that is, the covariance matrix of the minimum mean squared error (MMSE) for estimating $\mathbf{X}$ given $\mathbf{Y}$.

We use the following notation for standard random variables and random vectors:

- $X \sim \text{Bern}(p)$: X is a Bernoulli random variable with parameter $p \in [0, 1]$, i.e.,

$$X = \begin{cases} 1 & \text{with probability } p, \\ 0 & \text{with probability } 1 - p. \end{cases}$$

- $X \sim \text{Binom}(n, p)$: X is a binomial random variable with parameters $n \geq 1$ and $p \in [0, 1]$, i.e.,

$$p_X(k) = \binom{n}{k} p^k (1 - p)^{n-k}, \quad k \in [0 : n].$$

- $X \sim \text{Unif}(\mathcal{A})$: X is a discrete uniform random variable over a finite set $\mathcal{A}$.
  $X \sim \text{Unif}[i : j]$ for integers $j > i$: X is a discrete uniform random variable over $[i : j]$.

- $X \sim \text{Unif}[a, b]$ for $b > a$: X is a continuous uniform random variable over $[a, b]$.

- $X \sim \text{N}(\mu, \sigma^2)$: X is a Gaussian random variable with mean μ and variance σ^2.
  $Q(x) = \text{P}\{X > x\}$, $x \in \mathbb{R}$, where $X \sim \text{N}(0, 1)$.

- $\mathbf{X} = X^n \sim \text{N}(\boldsymbol{\mu}, K)$: $\mathbf{X}$ is a Gaussian random vector with mean vector $\boldsymbol{\mu}$ and covariance matrix K, i.e.,

$$f(\mathbf{x}) = \frac{1}{\sqrt{(2\pi)^n |K|}} e^{-\frac{1}{2}(\mathbf{x}-\boldsymbol{\mu})^T K^{-1}(\mathbf{x}-\boldsymbol{\mu})}.$$

We use the notation $\{X_i\} = (X_1, X_2, \ldots)$ to denote a discrete-time random process. The following notation is used for common random processes:

- $\{X_i\}$ is a $\text{Bern}(p)$ process means that $(X_1, X_2, \ldots)$ is a sequence of independent and identically distributed (i.i.d.) $\text{Bern}(p)$ random variables.

- $\{X_i\}$ is a $\text{WGN}(P)$ process means that $(X_1, X_2, \ldots)$ is a sequence of i.i.d. $\text{N}(0, P)$ random variables. More generally, $\{X_i, Y_i\}$ is a $2\text{-WGN}(P, \rho)$ process means that (X_1, Y_1), $(X_2, Y_2), \ldots$ are i.i.d. jointly Gaussian random variable pairs with $\text{E}(X_1) = \text{E}(Y_1) = 0$, $\text{E}(X_1^2) = \text{E}(Y_1^2) = P$, and correlation coefficient $\rho = \text{E}(X_1 Y_1)/P$.

We say that $X \to Y \to Z$ form a Markov chain if $p(x, y, z) = p(x)p(y|x)p(z|y)$. More generally, we say that $X_1 \to X_2 \to X_3 \to \cdots$ form a Markov chain if $p(x_i|x^{i-1}) = p(x_i|x_{i-1})$ for $i \geq 2$.

Common Functions

The following functions are used frequently. The logarithm function log is assumed to be base 2 unless specified otherwise.

- Binary entropy function: $H(p) = -p \log p - \bar{p} \log \bar{p}$ for $p \in [0, 1]$.

- Gaussian capacity function: $\text{C}(x) = (1/2) \log(1 + x)$ for $x \geq 0$.

- Quadratic Gaussian rate function: $\text{R}(x) = \max\{(1/2) \log x, \, 0\} = (1/2)[\log x]^+$.

ϵ–δ **Notation**

We use ϵ, $\epsilon' > 0$ exclusively to denote "small" constants such that $\epsilon' < \epsilon$. We use $\delta(\epsilon) > 0$ to denote a function of ϵ that tends to zero as $\epsilon \to 0$. When there are multiple such functions $\delta_1(\epsilon), \delta_2(\epsilon), \ldots, \delta_k(\epsilon)$, we denote them all by a generic function $\delta(\epsilon)$ that tends to zero as $\epsilon \to 0$ with the understanding that $\delta(\epsilon) = \max\{\delta_1(\epsilon), \delta_2(\epsilon), \ldots, \delta_k(\epsilon)\}$. Similarly, we use $\epsilon_n \geq 0$ to denote a generic function of n that tends to zero as $n \to \infty$.

We say that $a_n \doteq 2^{nb}$ for some constant b if there exists some $\delta(\epsilon)$ (with ϵ defined in the context) such that for n sufficiently large,

$$2^{n(b-\delta(\epsilon))} \leq a_n \leq 2^{n(b+\delta(\epsilon))}.$$

Matrices

We use uppercase letters $A, B, \ldots$ to denote matrices. The entry in the i-th row and the j-th column of a matrix A is denoted by $A(i, j)$ or A_{ij}. A transpose of a matrix A is denoted by A^T, i.e., $A^T(i, j) = A(j, i)$. We use $\mathrm{diag}(a_1, a_2, \ldots, a_d)$ to denote a $d \times d$ diagonal matrix with diagonal elements $a_1, a_2, \ldots, a_d$. The $d \times d$ identity matrix is denoted by I_d. The subscript d is omitted when it is clear from the context. For a square matrix A, $|A| = \det(A)$ denotes the determinant of A and $\mathrm{tr}(A)$ denotes its trace.

A symmetric matrix A is said to be positive definite (denoted by $A \succ 0$) if $\mathbf{x}^T A \mathbf{x} > 0$ for all $\mathbf{x} \neq 0$. If instead $\mathbf{x}^T A \mathbf{x} \geq 0$ for all $\mathbf{x} \neq 0$, then the matrix A is said to be positive semidefinite (denoted by $A \succeq 0$). For symmetric matrices A and B of the same dimension, $A \succ B$ means that $A - B \succ 0$ and $A \succeq B$ means that $A - B \succeq 0$.

A singular value decomposition of an $r \times t$ matrix G of rank d is given by $G = \Phi \Gamma \Psi^T$, where Φ is an $r \times d$ matrix with $\Phi^T \Phi = I_d$, Ψ is a $t \times d$ matrix with $\Psi^T \Psi = I_d$, and $\Gamma = \mathrm{diag}(\gamma_1, \ldots, \gamma_d)$ is a $d \times d$ positive diagonal matrix.

For a symmetric positive semidefinite matrix K with an eigenvalue decomposition $K = \Phi \Lambda \Phi^T$, we define its symmetric square root as $K^{1/2} = \Phi \Lambda^{1/2} \Phi^T$, where $\Lambda^{1/2}$ is a diagonal matrix with diagonal elements $\sqrt{\Lambda_{ii}}$. Note that $K^{1/2}$ is symmetric positive definite with $K^{1/2} K^{1/2} = K$. We define the symmetric square root inverse $K^{-1/2}$ of a symmetric positive definite matrix K as the symmetric square root of K^{-1}.

Order Notation

Let $g_1(N)$ and $g_2(N)$ be nonnegative functions on natural numbers.

- $g_1(N) = o(g_2(N))$ means that $g_1(N)/g_2(N)$ tends to zero as $N \to \infty$.

- $g_1(N) = O(g_2(N))$ means that there exist a constant a and an integer n_0 such that $g_1(N) \leq a g_2(N)$ for all $N > n_0$.

- $g_1(N) = \Omega(g_2(N))$ means that $g_2(N) = O(g_1(N))$.

- $g_1(N) = \Theta(g_2(N))$ means that $g_1(N) = O(g_2(N))$ and $g_2(N) = O(g_1(N))$.

CHAPTER 1

Introduction

We introduce the general problem of optimal information flow in networks, which is the focus of network information theory. We then give a preview of the book with pointers to where the main results can be found.

1.1 NETWORK INFORMATION FLOW PROBLEM

A networked system consists of a set of information sources and communication nodes connected by a network as depicted in Figure 1.1. Each node observes one or more sources and wishes to reconstruct other sources or to compute a function based on all the sources. To perform the required task, the nodes communicate with each other over the network.

- What is the limit on the amount of communication needed?

- How can this limit be achieved?

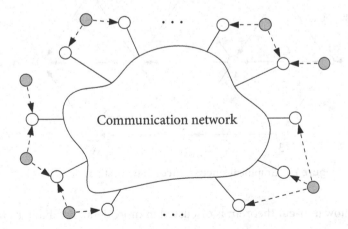

Figure 1.1. Elements of a networked system. The information sources (shaded circles) may be data, video, sensor measurements, or biochemical signals; the nodes (empty circles) may be computers, handsets, sensor nodes, or neurons; and the network may be a wired network, a wireless cellular or ad-hoc network, or a biological network.

These information flow questions have been answered satisfactorily for graphical unicast (single-source single-destination) networks and for point-to-point communication systems.

1.2 MAX-FLOW MIN-CUT THEOREM

Consider a *graphical* (wired) network, such as the Internet or a distributed storage system, modeled by a directed graph $(\mathcal{N}, \mathcal{E})$ with link capacities C_{jk} bits from node j to node k as depicted in Figure 1.2. Assume a unicast communication scenario in which source node 1 wishes to communicate an R-bit message M to destination node N. What is the network capacity C, that is, the maximum number of bits R that can be communicated reliably?

The answer is given by the *max-flow min-cut theorem* due to Ford and Fulkerson (1956) and Elias, Feinstein, and Shannon (1956). They showed that the capacity (maximum flow) is equal to the minimum cut capacity, i.e.,

$$C = \min_{\mathcal{S} \subset \mathcal{N} : 1 \in \mathcal{S}, N \in \mathcal{S}^c} C(\mathcal{S}),$$

where $C(\mathcal{S}) = \sum_{j \in \mathcal{S}, k \in \mathcal{S}^c} C_{jk}$ is the capacity of the cut $(\mathcal{S}, \mathcal{S}^c)$. They also showed that the capacity is achieved without errors using simple routing at the intermediate (relay) nodes, that is, the incoming bits at each node are forwarded over its outgoing links. Hence, under this networked system model, information can be treated as a commodity to be shipped over a transportation network or electricity to be delivered over a power grid.

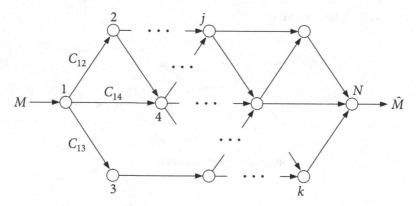

Figure 1.2. Graphical single-source single-destination network.

The max-flow min-cut theorem is discussed in more detail in Chapter 15.

1.3 POINT-TO-POINT INFORMATION THEORY

The graphical unicast network model captures the topology of a point-to-point network with idealized source and communication link models. At the other extreme, Shannon

(1948, 1959) studied communication and compression over a single link with more complex source and link (channel) models. He considered the communication system architecture depicted in Figure 1.3, where a sender wishes to communicate a k-symbol source sequence U^k to a receiver over a noisy channel. To perform this task, Shannon proposed a general *block coding* scheme, where the source sequence is mapped by an encoder into an n-symbol input sequence $X^n(U^k)$ and the received channel output sequence Y^n is mapped by a decoder into an estimate (reconstruction) sequence $\hat{U}^k(Y^n)$. He simplified the analysis of this system by proposing simple discrete memoryless models for the source and the noisy channel, and by using an *asymptotic* approach to characterize the necessary and sufficient condition for reliable communication.

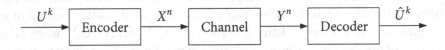

Figure 1.3. Shannon's model of a point-to-point communication system.

Shannon's ingenious formulation of the point-to-point communication problem led to the following four fundamental theorems.

Channel coding theorem. Suppose that the source is a maximally compressed k-bit message M as in the graphical network case and that the channel is discrete and memoryless with input X, output Y, and conditional probability $p(y|x)$ that specifies the probability of receiving the symbol y when x is transmitted. The decoder wishes to find an estimate $\hat{M}$ of the message such that the probability of decoding error $P\{\hat{M} \neq M\}$ does not exceed a prescribed value P_e. The general problem is to find the tradeoff between the number of bits k, the block length n, and the probability of error P_e.

This problem is intractable in general. Shannon (1948) realized that the difficulty lies in analyzing the system for any given finite block length n and reformulated the problem as one of finding the *channel capacity* C, which is the maximum communication rate $R = k/n$ in bits per channel transmissions such that the probability of error can be made arbitrarily small when the block length n is sufficiently large. He established a simple and elegant characterization of the channel capacity C in terms of the maximum of the mutual information $I(X; Y)$ between the channel input X and output Y:

$$C = \max_{p(x)} I(X; Y) \quad \text{bits/transmission.}$$

(See Section 2.3 for the definition of mutual information and its properties.) Unlike the graphical network case, however, capacity is achieved only *asymptotically* error-free and using sophisticated coding.

Lossless source coding theorem. As a "dual" to channel coding, consider the following lossless data compression setting. The sender wishes to communicate (store) a source sequence *losslessly* to a receiver over a *noiseless* binary channel (memory) with the minimum number of bits. Suppose that the source U is discrete and memoryless, that is, it generates an i.i.d. sequence U^k. The sender encodes U^k at rate $R = n/k$ bits per source symbol into an n-bit index $M(U^k)$ and sends it over the channel. Upon receiving the index M, the decoder finds an estimate $\hat{U}^k(M)$ of the source sequence such that the probability of error $\mathrm{P}\{\hat{U}^k \neq U^k\}$ is less than a prescribed value. Shannon again formulated the problem as one of finding the minimum lossless compression rate R^* when the block length is arbitrarily large, and showed that it is characterized by the entropy of U:

$$R^* = H(U) \quad \text{bits/symbol.}$$

(See Section 2.1 for the definition of entropy and its properties.)

Lossy source coding theorem. Now suppose U^k is to be sent over the noiseless binary channel such that the receiver can reconstruct it with some distortion instead of losslessly. Shannon assumed the per-letter distortion $(1/k) \sum_{i=1}^{k} \mathrm{E}(d(U_i, \hat{U}_i))$, where $d(u, \hat{u})$ is a measure of the distortion between the source symbol u and the reconstruction symbol $\hat{u}$. He characterized the *rate–distortion function* $R(D)$, which is the optimal tradeoff between the rate $R = n/k$ and the desired distortion D, as the minimum of the mutual information between U and $\hat{U}$:

$$R(D) = \min_{p(\hat{u}|u):\mathrm{E}(d(U,\hat{U}))\leq D} I(U;\hat{U}) \quad \text{bits/symbol.}$$

Source–channel separation theorem. Now we return to the general point-to-point communication system shown in Figure 1.3. Let C be the capacity of the discrete memoryless channel (DMC) and $R(D)$ be the rate–distortion function of the discrete memoryless source (DMS), and assume for simplicity that $k = n$. What is the necessary and sufficient condition for communicating the DMS over the DMC with a prescribed distortion D? Shannon (1959) showed that $R(D) \leq C$ is necessary. Since $R(D) < C$ is sufficient by the lossy source coding and channel coding theorems, *separate* source coding and channel coding achieves the fundamental limit. Although this result holds only when the code block length is unbounded, it asserts that using bits as a "universal" interface between sources and channels—the basis for digital communication—is essentially optimal.

We discuss the above results in detail in Chapter 3. Shannon's asymptotic approach to network performance analysis will be adopted throughout the book.

1.4 NETWORK INFORMATION THEORY

The max-flow min-cut theorem and Shannon's point-to-point information theory have had a major impact on communication and networking. However, the simplistic model of a networked information processing system as a single source–destination pair communicating over a noisy channel or a graphical network does not capture many important aspects of real-world networks:

• Networked systems have multiple sources and destinations.

• The task of the network is often to compute a function or to make a decision.

- Wireless communication uses a shared broadcast medium.

- Networked systems involve complex tradeoffs between competition for resources and cooperation for the common good.

- Many networks allow for feedback and interactive communication.

- Source–channel separation does not hold for networks in general.

- Network security is often a primary concern.

- Data from the sources is often bursty and network topology evolves dynamically.

Network information theory aims to answer the aforementioned information flow questions while capturing some of these aspects of real-world networks. In the following, we illustrate some of the achievements of this theory using examples from the book.

1.4.1 Multiple Sources and Destinations

Coding for networks with many sources and destinations requires techniques beyond routing and point-to-point source/channel coding. Consider the following settings.

Graphical multicast network. Suppose we wish to send a movie over the Internet to multiple destinations (multicast). Unlike the unicast case, routing is not optimal in general even if we model the Internet by a graphical network. Instead, we need to use *coding* of incoming packets at the relay nodes.

We illustrate this fact via the famous "butterfly network" shown in Figure 1.4, where source node 1 wishes to send a 2-bit message $(M_1, M_2) \in \{0,1\}^2$ to destination nodes 6 and 7. Assume link capacities $C_{jk} = 1$ for all edges (j, k). Note that using routing only, both M_1 and M_2 must be sent over the edge $(4, 5)$, and hence the message cannot be communicated to both destination nodes.

However, if we allow the nodes to perform simple modulo-2 sum operations in addition to routing, the 2-bit message can be communicated to both destinations. As illustrated in Figure 1.5, relay nodes 2, 3, and 5 forward multiple copies of their incoming bits,

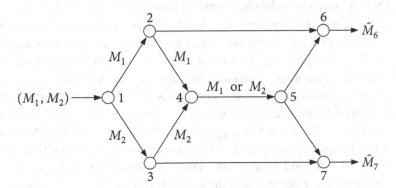

Figure 1.4. Butterfly network. The 2-bit message (M_1, M_2) cannot be sent using routing to both destination nodes 6 and 7.

and relay node 4 sends the modulo-2 sum of M_1 and M_2. Using this simple scheme, both destination nodes 6 and 7 can recover the message error-free.

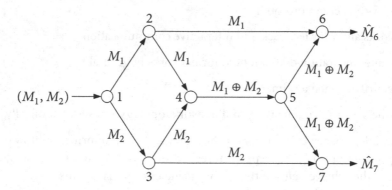

Figure 1.5. The 2-bit message can be sent to destination nodes 6 and 7 using linear network coding.

In Chapter 15, we show that linear network coding, which is a generalization of this simple scheme, achieves the capacity of an arbitrary graphical multicast network. Extensions of this multicast setting to lossy source coding are discussed in Chapters 13 and 20.

Distributed compression. Suppose that a sensor network is used to measure the temperature over a geographical area. The output from each sensor is compressed and sent to a base station. Although compression is performed separately on each sensor output, it turns out that using point-to-point compression is not optimal when the sensor outputs are *correlated*, for example, because the sensors are located close to each other.

Consider the distributed lossless compression system depicted in Figure 1.6. Two sequences X_1^n and X_2^n are drawn from correlated discrete memoryless sources $(X_1, X_2) \sim p(x_1, x_2)$ and compressed separately into an nR_1-bit index M_1 and an nR_2-bit index M_2, respectively. A receiver (base station) wishes to recover the source sequences from the index pair (M_1, M_2). What is the minimum sum-rate R_{sum}^*, that is, the minimum over $R_1 + R_2$ such that both sources can be reconstructed losslessly?

If each sender uses a point-to-point code, then by Shannon's lossless source coding theorem, the minimum lossless compression rates for the individual sources are $R_1^* = H(X_1)$ and $R_2^* = H(X_2)$, respectively; hence the resulting sum-rate is $H(X_1) + H(X_2)$. If instead the two sources are jointly encoded, then again by the lossless source coding theorem, the minimum lossless compression sum-rate is $H(X_1, X_2)$, which can be much smaller than the sum of the individual entropies. For example, let X_1 and X_2 be binary-valued sources with $p_{X_1, X_2}(0, 0) = 0.495$, $p_{X_1, X_2}(0, 1) = 0.005$, $p_{X_1, X_2}(1, 0) = 0.005$, and $p_{X_1, X_2}(1, 1) = 0.495$; hence, the sources have the same outcome 0.99 of the time. From the joint pmf, we see that X_1 and X_2 are both Bern(1/2) sources with entropy $H(X_1) = H(X_2) = 1$ bit per symbol. By comparison, their joint entropy $H(X_1, X_2) = 1.0808 \ll 2$ bits per symbol pair.

Slepian and Wolf (1973a) showed that $R_{\text{sum}}^* = H(X_1, X_2)$ and hence that the minimum

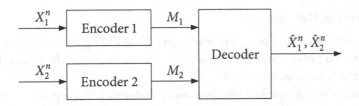

Figure 1.6. Distributed lossless compression system. Each source sequence X_j^n, $j = 1, 2$, is encoded into an index $M_j(X_j^n) \in [1 : 2^{nR_j})$, and the decoder wishes to reconstruct the sequences losslessly from (M_1, M_2).

sum-rate for distributed compression is asymptotically the same as for centralized compression! This result is discussed in Chapter 10. Generalizations to distributed lossy compression are discussed in Chapters 11 and 12.

Communication for computing. Now suppose that the base station in the temperature sensor network wishes to compute the *average* temperature over the geographical area instead of the individual temperature values. What is the amount of communication needed?

While in some cases the rate requirement for computing a function of the sources is the same as that for recovering the sources themselves, it is sometimes significantly smaller. As an example, consider an n-round online game, where in each round Alice and Bob each select one card without replacement from a virtual hat with three cards labeled 1, 2, and 3. The one with the larger number wins. Let X^n and Y^n be the sequences of numbers on Alice and Bob's cards over the n rounds, respectively. Alice encodes her sequence X^n into an index $M \in [1 : 2^{nR}]$ and sends it to Bob so that he can find out who won in each round, that is, find an estimate $\hat{Z}^n$ of the sequence $Z_i = \max\{X_i, Y_i\}$ for $i \in [1 : n]$, as shown in Figure 1.7. What is the minimum communication rate R needed?

By the aforementioned Slepian–Wolf result, the minimum rate needed for Bob to reconstruct X is the conditional entropy $H(X|Y) = H(X, Y) - H(Y) = 2/3$ bit per round. By exploiting the structure of the function $Z = \max\{X, Y\}$, however, it can be shown that only 0.5409 bit per round is needed.

This card game example as well as general results on communication for computing are discussed in Chapter 21.

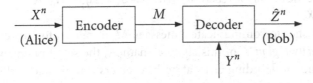

Figure 1.7. Online game setup. Alice has the card number sequence X^n and Bob has the card number sequence Y^n. Alice encodes her card number sequence into an index $M \in [1 : 2^{nR}]$ and sends it to Bob, who wishes to losslessly reconstruct the winner sequence Z^n.

1.4.2 Wireless Networks

Perhaps the most important practical motivation for studying network information theory is to deal with the special nature of wireless channels. We study models for wireless communication throughout the book.

The first and simplest wireless channel model we consider is the point-to-point Gaussian channel $Y = gX + Z$ depicted in Figure 1.8, where $Z \sim N(0, N_0/2)$ is the receiver noise and g is the channel gain. Shannon showed that the capacity of this channel under a prescribed average transmission power constraint P on X, i.e., $\sum_{i=1}^{n} X_i^2 \le nP$ for each codeword X^n, has the simple characterization

$$C = \frac{1}{2} \log(1 + S) = \mathsf{C}(S),$$

where $S = 2g^2 P/N_0$ is the received signal-to-noise ratio (SNR).

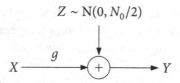

Figure 1.8. Gaussian point-to-point channel.

A wireless network can be turned into a set of separate point-to-point Gaussian channels via time or frequency division. This traditional approach to wireless communication, however, does not take full advantage of the broadcast nature of the wireless medium as illustrated in the following example.

Gaussian broadcast channel. The downlink of a wireless system is modeled by the Gaussian broadcast channel

$$Y_1 = g_1 X + Z_1,$$
$$Y_2 = g_2 X + Z_2,$$

as depicted in Figure 1.9. Here $Z_1 \sim N(0, N_0/2)$ and $Z_2 \sim N(0, N_0/2)$ are the receiver noise components, and $g_1^2 > g_2^2$, that is, the channel to receiver 1 is stronger than the channel to receiver 2. Define the SNRs for receiver $j = 1, 2$ as $S_j = 2g_j^2 P/N_0$. Assume average power constraint P on X.

The sender wishes to communicate a message M_j at rate R_j to receiver j for $j = 1, 2$. What is the *capacity region* $\mathscr{C}$ of this channel, namely, the set of rate pairs (R_1, R_2) such that the probability of decoding error at both receivers can be made arbitrarily small as the code block length becomes large?

If we send the messages M_1 and M_2 in different time intervals or frequency bands, then we can reliably communicate at rate pairs in the "time-division region" $\mathscr{R}$ shown in Figure 1.9. Cover (1972) showed that higher rates can be achieved by adding the codewords for the two messages and sending this sum over the entire transmission block. The stronger receiver 1 decodes for both codewords, while the weaker receiver 2 treats the other codeword as noise and decodes only for its own codeword. Using this *superposition coding* scheme, the sender can reliably communicate the messages at any rate pair in the capacity region $\mathscr{C}$ shown in Figure 1.9b, which is strictly larger than the time-division region $\mathscr{R}$.

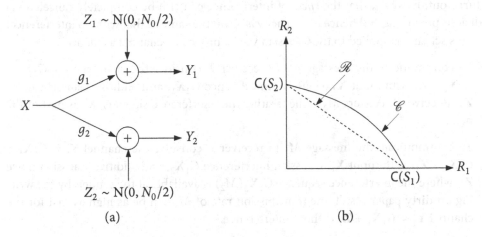

(a)   (b)

Figure 1.9. (a) Gaussian broadcast channel with SNRs $S_1 = g_1^2 P > g_2^2 P = S_2$. (b) The time-division inner bound $\mathscr{R}$ and the capacity region $\mathscr{C}$.

This superposition scheme and related results are detailed in Chapter 5. Similar improvements in rates can be achieved for the uplink (multiple access channel) and the intercell interference (interference channel), as discussed in Chapters 4 and 6, respectively.

Gaussian vector broadcast channel. Multiple transmitter and receiver antennas are commonly used to enhance the performance of wireless communication systems. Coding for these multiple-input multiple-output (MIMO) channels, however, requires techniques beyond single-antenna (scalar) channels. For example, consider the downlink of a MIMO wireless system modeled by the Gaussian vector broadcast channel

$$\mathbf{Y}_1 = G_1 \mathbf{X} + \mathbf{Z}_1,$$
$$\mathbf{Y}_2 = G_2 \mathbf{X} + \mathbf{Z}_2,$$

where G_1, G_2 are r-by-t channel gain matrices and $\mathbf{Z}_1 \sim \mathrm{N}(0, I_r)$ and $\mathbf{Z}_2 \sim \mathrm{N}(0, I_r)$ are noise components. Assume average power constraint P on $\mathbf{X}$. Note that unlike the single-antenna broadcast channel shown in Figure 1.9, in the vector case neither receiver is necessarily stronger than the other. The optimum coding scheme is based on the following *writing on dirty paper* result. Suppose we wish to communicate a message over a Gaussian vector channel,

$$\mathbf{Y} = G\mathbf{X} + \mathbf{S} + \mathbf{Z}$$

where $\mathbf{S} \sim \mathrm{N}(0, K_S)$ is an *interference* signal, which is independent of the Gaussian noise

$\mathbf{Z} \sim N(0, I_r)$. Assume average power constraint P on $\mathbf{X}$. When the interference sequence S^n is available at the receiver, it can be simply subtracted from the received sequence and hence the channel capacity is the same as when there is no interference. Now suppose that the interference sequence is available only at the sender. Because of the power constraint, it is not always possible to presubtract the interference from the transmitted codeword. It turns out, however, that the *effect* of interference can still be completely canceled via judicious precoding and hence the capacity is again the same as that with no interference! This scheme is applied to the Gaussian vector broadcast channel as follows.

- To communicate the message M_2 to receiver 2, consider the channel $\mathbf{Y}_2 = G_2 \mathbf{X}_2 + G_2 \mathbf{X}_1 + \mathbf{Z}_2$ with input $\mathbf{X}_2$, Gaussian interference $G_2 \mathbf{X}_1$, and additive Gaussian noise $\mathbf{Z}_2$. Receiver 2 recovers M_2 while treating the interference signal $G_2 \mathbf{X}_1$ as part of the noise.

- To communicate the message M_1 to receiver 1, consider the channel $\mathbf{Y}_1 = G_1 \mathbf{X}_1 + G_1 \mathbf{X}_2 + \mathbf{Z}_1$, with input $\mathbf{X}_1$, Gaussian interference $G_1 \mathbf{X}_2$, and additive Gaussian noise $\mathbf{Z}_1$, where the interference sequence $G_1 \mathbf{X}_2^n(M_2)$ is available at the sender. By the writing on dirty paper result, the transmission rate of M_1 can be as high as that for the channel $\mathbf{Y}_1' = G_1 \mathbf{X}_1 + \mathbf{Z}_1$ without interference.

The writing on dirty paper result is discussed in detail in Chapter 7. The optimality of this scheme for the Gaussian vector broadcast channel is established in Chapter 9.

Gaussian relay channel. An ad-hoc or a mesh wireless network is modeled by a Gaussian *multihop* network in which nodes can act as relays to help other nodes communicate their messages. Again reducing such a network to a set of links using time or frequency division does not take full advantage of the shared wireless medium, and the rate can be greatly increased via node cooperation.

As a canonical example, consider the 3-node relay channel depicted in Figure 1.10a. Here node 2 is located on the line between nodes 1 and 3 as shown in Figure 1.10b. We assume that the channel gain from node k to node j is $g_{jk} = r_{jk}^{-3/2}$, where r_{jk} is the distance between nodes k and j. Hence $g_{31} = r_{31}^{-3/2}$, $g_{21} = r_{21}^{-3/2}$, and $g_{32} = (r_{31} - r_{21})^{-3/2}$. Assume average power constraint P on each of X_1 and X_2.

Suppose that sender node 1 wishes to communicate a message M to receiver node 3 with the help of relay node 2. On the one extreme, the sender and the receiver can communicate *directly* without help from the relay. On the other extreme, we can use a *multihop* scheme where the relay plays a pivotal role in the communication. In this commonly used scheme, the sender transmits the message to the relay in the first hop and the relay recovers the message and transmits it to the receiver concurrently in the second hop, causing interference to the first-hop communication. If the receiver is far away from the sender, that is, the distance r_{31} is large, this scheme performs well because the interference due to the concurrent transmission is weak. However, when r_{31} is not large, the interference can adversely affect the communication of the message.

In Chapter 16, we present several coding schemes that outperform both direct transmission and multihop.

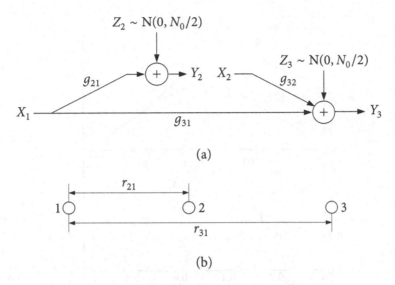

Figure 1.10. (a) Gaussian relay channel. (b) Node placements: relay node 2 is placed along the lines between sender node 1 and receiver node 2.

- *Decode–forward.* The direct transmission and multihop schemes are combined and further enhanced via coherent transmission by the sender and the relay. The receiver decodes for the signals from both hops instead of treating the transmission from the first hop as interference. Decode–forward performs well when the relay is closer to the sender, i.e., $r_{21} < (1/2)r_{31}$.

- *Compress–forward.* As an alternative to the "digital-to-digital" relay interface used in multihop and decode–forward, the compress–forward scheme uses an "analog-to-digital" interface in which the relay compresses the received signal and sends the compression index to the receiver. Compress–forward performs well when the relay is closer to the receiver.

- *Amplify–forward.* Decode–forward and compress–forward require sophisticated operations at the nodes. The amplify–forward scheme provides a much simpler "analog-to-analog" interface in which the relay scales the incoming signal and transmits it to the receiver. In spite of its simplicity, amplify–forward can outperform decode–forward when the relay is closer to the receiver.

The performance of the above relaying schemes are compared in Figure 1.11. In general, it can be shown that both decode–forward and compress–forward achieve rates within 1/2 bit of the capacity, while amplify–forward achieves rates within 1 bit of the capacity.

We extend the above coding schemes to general multihop networks in Chapters 18 and 19. In particular, we show that extending compress–forward leads to a noisy network coding scheme that includes network coding for graphical multicast networks as a special case. When applied to Gaussian multihop multicast networks, this noisy network coding scheme achieves within a constant gap of the capacity independent of network topology,

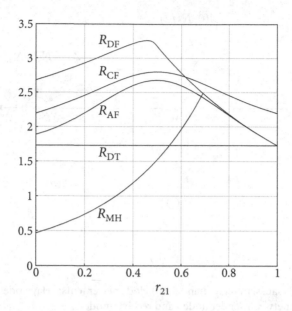

Figure 1.11. Comparison of the achievable rates for the Gaussian relay channel using direct transmission (R_{DT}), multihop (R_{MH}), decode–forward (R_{DF}), compress–forward (R_{CF}), and amplify–forward (R_{AF}) for $N_0/2 = 1$, $r_{31} = 1$ and $P = 10$.

channel parameters, and power constraints, while extensions of the other schemes do not yield such performance guarantees.

To study the effect of interference and path loss in large wireless networks, in Chapter 19 we also investigate how capacity scales with the network size. We show that relaying and spatial reuse of frequency/time can greatly increase the rates over naive direct transmission with time division.

Wireless fading channels. Wireless channels are *time varying* due to scattering of signals over multiple paths and user mobility. In Chapter 23, we study fading channel models that capture these effects by allowing the gains in the Gaussian channels to vary randomly with time. In some settings, channel capacity in the Shannon sense is not well defined. We introduce different coding approaches and corresponding performance metrics that are useful in practice.

1.4.3   Interactive Communication

Real-world networks allow for feedback and node interactions. Shannon (1956) showed that feedback does not increase the capacity of a memoryless channel. Feedback, however, can help simplify coding and improve reliability. This is illustrated in the following example.

Binary erasure channel with feedback. The binary erasure channel is a DMC with binary input $X \in \{0, 1\}$ and ternary output $Y \in \{0, 1, e\}$. Each transmitted bit (0 or 1) is erased

$(Y = e)$ with probability p. The capacity of this channel is $1 - p$ and achieving it requires sophisticated block coding. Now suppose that noiseless causal feedback from the receiver to the sender is present, that is, the sender at each time i has access to all previous received symbols Y^{i-1}. Then we can achieve the capacity simply by retransmitting each erased bit. Using this simple feedback scheme, on average $n = k/(1 - p)$ transmissions suffice to reliably communicate k bits of information.

Unlike point-to-point communication, feedback can achieve higher rates in networks with multiple senders/receivers.

Binary erasure multiple access channel with feedback. Consider the multiple access channel (MAC) with feedback depicted in Figure 1.12, where the channel inputs X_1 and X_2 are binary and the channel output $Y = X_1 + X_2$ is ternary, i.e., $Y \in \{0, 1, 2\}$. Suppose that senders 1 and 2 wish to communicate independent messages M_1 and M_2, respectively, to the receiver at the same rate R. Without feedback, the *symmetric capacity*, which is the maximum rate R, is $\max_{p(x_1)p(x_2)} H(Y) = 3/4$ bits/transmission.

Noiseless causal feedback allows the senders to *cooperate* in communicating their messages and hence to achieve higher symmetric rates than with no feedback. To illustrate such cooperation, suppose that each sender first transmits its k-bit message uncoded. On average $k/2$ bits are "erased" (that is, $Y = 0 + 1 = 1 + 0 = 1$ is received). Since the senders know through feedback the exact locations of the erasures as well as the corresponding message bits from both messages, they can cooperate to send the erased bits from the first message (which is sufficient to recover both messages). This cooperative retransmission requires $k/(2 \log 3)$ transmissions. Hence we can increase the symmetric rate to $R = k/(k + k/(2 \log 3)) = 0.7602$. This rate can be further increased to 0.7911 by using a more sophisticated coding scheme that sends new messages simultaneously with cooperative retransmissions.

In Chapter 17, we discuss the *iterative refinement* approach illustrated in the binary erasure channel example; the cooperative feedback approach for multiuser channels illustrated in the binary erasure MAC example; and the two-way channel. In Chapters 20

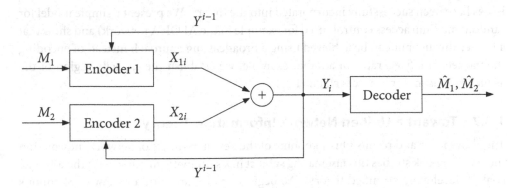

Figure 1.12. Feedback communication over a binary erasure MAC. The channel inputs X_{1i} and X_{2i} at time $i \in [1 : n]$ are functions of (M_1, Y^{i-1}) and (M_2, Y^{i-1}), respectively.

through 22, we show that interaction can also help in distributed compression, distributed computing, and secret communication.

1.4.4 Joint Source–Channel Coding

As we mentioned earlier, Shannon showed that separate source and channel coding is asymptotically optimal for point-to-point communication. It turns out that such separation does not hold in general for sending correlated sources over multiuser networks. In Chapter 14, we demonstrate this breakdown of separation for lossless communication of correlated sources over multiple access and broadcast channels. This discussion yields natural definitions of various notions of *common information* between two sources.

1.4.5 Secure Communication

Confidentiality of information is a crucial requirement in networking applications such as e-commerce. In Chapter 22, we discuss several coding schemes that allow a legitimate sender (Alice) to communicate a message reliably to a receiver (Bob) while keeping it secret (in a strong sense) from an eavesdropper (Eve). When the channel from Alice to Bob is stronger than that to Eve, a confidential message with a positive rate can be communicated reliably without a shared secret key between Alice and Bob. By contrast, when the channel from Alice to Bob is weaker than that to Eve, no confidential message can be communicated reliably. We show, however, that Alice and Bob can still agree on a secret key through interactive communication over a public (nonsecure) channel that Eve has complete access to. This key can then be used to communicate a confidential message at a nonzero rate.

1.4.6 Network Information Theory and Networking

Many aspects of real-world networks such as bursty data arrivals, random access, asynchrony, and delay constraints are not captured by the standard models of network information theory. In Chapter 24, we present several examples for which such networking issues have been successfully incorporated into the theory. We present a simple model for random medium access control (used for example in the ALOHA network) and show that a higher throughput can be achieved using a broadcasting approach instead of encoding the packets at a fixed rate. In another example, we establish the capacity region of the asynchronous multiple access channel.

1.4.7 Toward a Unified Network Information Theory

The above ideas and results illustrate some of the key ingredients of network information theory. The book studies this fascinating subject in a systematic manner, with the ultimate goal of developing a unified theory. We begin our journey with a review of Shannon's point-to-point information theory in the next two chapters.

PART I

PRELIMINARIES

CHAPTER 2

Information Measures and Typicality

We define entropy and mutual information and review their basic properties. We introduce basic inequalities involving these information measures, including Fano's inequality, Mrs. Gerber's lemma, the maximum differential entropy lemma, the entropy power inequality, the data processing inequality, and the Csiszár sum identity. We then introduce the notion of typicality adopted throughout the book. We discuss properties of typical sequences and introduce the typical average lemma, the conditional typicality lemma, and the joint typicality lemma. These lemmas as well as the aforementioned entropy and mutual information inequalities will play pivotal roles in the proofs of the coding theorems throughout the book.

2.1 ENTROPY

Let X be a discrete random variable with probability mass function (pmf) $p(x)$ (in short $X \sim p(x)$). The uncertainty about the outcome of X is measured by its *entropy* defined as

$$H(X) = -\sum_{x \in \mathcal{X}} p(x) \log p(x) = -\mathsf{E}_X(\log p(X)).$$

For example, if X is a Bernoulli random variable with parameter $p = \mathsf{P}\{X = 1\} \in [0, 1]$ (in short $X \sim \text{Bern}(p)$), the entropy of X is

$$H(X) = -p \log p - (1 - p) \log(1 - p).$$

Since the Bernoulli random variable will be frequently encountered, we denote its entropy by the *binary entropy function* $H(p)$. The entropy function $H(X)$ is a nonnegative and concave function in $p(x)$. Thus, by Jensen's inequality (see Appendix B),

$$H(X) \le \log |\mathcal{X}|,$$

that is, the uniform pmf over $\mathcal{X}$ maximizes the entropy.

Let X be a discrete random variable and $g(X)$ be a function of X. Then

$$H(g(X)) \le H(X),$$

where the inequality holds with equality if g is one-to-one over the support of X, i.e., the set $\{x \in \mathcal{X} : p(x) > 0\}$.

Conditional entropy. Let $X \sim F(x)$ be an arbitrary random variable and $Y | \{X = x\} \sim p(y|x)$ be discrete for every x. Since $p(y|x)$ is a pmf, we can define the entropy function $H(Y | X = x)$ for every x. The *conditional entropy* (or *equivocation*) $H(Y|X)$ of Y given X is the average of $H(Y | X = x)$ over X, i.e.,

$$H(Y|X) = \int H(Y | X = x) \, dF(x)$$
$$= -\mathsf{E}_{X,Y}(\log p(Y|X)).$$

Conditional entropy is a measure of the remaining uncertainty about the outcome of Y given the "observation" X. Again by Jensen's inequality,

$$H(Y|X) \le H(Y) \tag{2.1}$$

with equality if X and Y are independent.

Joint entropy. Let $(X, Y) \sim p(x, y)$ be a pair of discrete random variables. The *joint entropy* of X and Y is defined as

$$H(X, Y) = -\mathsf{E}(\log p(X, Y)).$$

Note that this is the same as the entropy of a single "large" random variable (X, Y). The chain rule for pmfs, $p(x, y) = p(x)p(y|x) = p(y)p(x|y)$, leads to a *chain rule* for joint entropy

$$H(X, Y) = H(X) + H(Y|X) = H(Y) + H(X|Y).$$

By (2.1), it follows that

$$H(X, Y) \le H(X) + H(Y) \tag{2.2}$$

with equality if X and Y are independent.

The definition of entropy extends to discrete random vectors. Let $X^n \sim p(x^n)$. Then again by the chain rule for pmfs,

$$H(X^n) = H(X_1) + H(X_2|X_1) + \cdots + H(X_n|X_1, \ldots, X_{n-1})$$
$$= \sum_{i=1}^{n} H(X_i|X_1, \ldots, X_{i-1})$$
$$= \sum_{i=1}^{n} H(X_i|X^{i-1}).$$

Using induction and inequality (2.2), it follows that $H(X^n) \le \sum_{i=1}^{n} H(X_i)$ with equality if $X_1, X_2, \ldots, X_n$ are mutually independent.

Next, we consider the following two results that will be used in the converse proofs of many coding theorems. The first result relates equivocation to the "probability of error."

Fano's Inequality. Let $(X, Y) \sim p(x, y)$ and $P_e = P\{X \neq Y\}$. Then

$$H(X|Y) \leq H(P_e) + P_e \log |\mathcal{X}| \leq 1 + P_e \log |\mathcal{X}|.$$

The second result provides a lower bound on the entropy of the modulo-2 sum of two binary random vectors.

Mrs. Gerber's Lemma (MGL). Let $H^{-1} : [0, 1] \to [0, 1/2]$ be the inverse of the binary entropy function, i.e., $H(H^{-1}(v)) = v$.

- Scalar MGL: Let X be a binary random variable and let U be an arbitrary random variable. If $Z \sim \text{Bern}(p)$ is independent of (X, U) and $Y = X \oplus Z$, then

$$H(Y|U) \geq H\big(H^{-1}(H(X|U)) * p\big).$$

- Vector MGL: Let X^n be a binary random vector and U be an arbitrary random variable. If Z^n is a vector of independent and identically distributed $\text{Bern}(p)$ random variables independent of (X^n, U) and $Y^n = X^n \oplus Z^n$, then

$$\frac{H(Y^n|U)}{n} \geq H\left(H^{-1}\left(\frac{H(X^n|U)}{n}\right) * p\right).$$

The proof of this lemma follows by the convexity of the function $H(H^{-1}(v) * p)$ in v and using induction; see Problem 2.5.

Entropy rate of a stationary random process. Let $X = \{X_i\}$ be a stationary random process with X_i taking values in a finite alphabet $\mathcal{X}$. The *entropy rate* $\overline{H}(X)$ of the process X is defined as

$$\overline{H}(X) = \lim_{n \to \infty} \frac{1}{n} H(X^n) = \lim_{n \to \infty} H(X_n|X^{n-1}).$$

2.2 DIFFERENTIAL ENTROPY

Let X be a continuous random variable with probability density function (pdf) $f(x)$ (in short $X \sim f(x)$). The *differential entropy* of X is defined as

$$h(X) = -\int f(x) \log f(x) \, dx = -\mathsf{E}_X(\log f(X)).$$

For example, if $X \sim \text{Unif}[a, b]$, then

$$h(X) = \log(b - a).$$

As another example, if $X \sim \text{N}(\mu, \sigma^2)$, then

$$h(X) = \frac{1}{2} \log(2\pi e \sigma^2).$$

The differential entropy $h(X)$ is a concave function of $f(x)$. However, unlike entropy it is not always nonnegative and hence should not be interpreted directly as a measure of information. Roughly speaking, $h(X) + n$ is the entropy of the quantized version of X using equal-size intervals of length 2^{-n} (Cover and Thomas 2006, Section 8.3).

The differential entropy is invariant under translation but not under scaling.

- Translation: For any constant a, $h(X + a) = h(X)$.

- Scaling: For any nonzero constant a, $h(aX) = h(X) + \log|a|$.

The maximum differential entropy of a continuous random variable $X \sim f(x)$ under the average power constraint $\mathsf{E}(X^2) \le P$ is

$$\max_{f(x):\mathsf{E}(X^2)\le P} h(X) = \frac{1}{2}\log(2\pi e P)$$

and is attained when X is Gaussian with zero mean and variance P, i.e., $X \sim \mathrm{N}(0, P)$; see Remark 2.1 and Problem 2.6. Thus, for any $X \sim f(x)$,

$$h(X) = h(X - \mathsf{E}(X)) \le \frac{1}{2}\log(2\pi e \, \mathrm{Var}(X)). \tag{2.3}$$

Conditional differential entropy. Let $X \sim F(x)$ be an arbitrary random variable and $Y\,|\,\{X = x\} \sim f(y|x)$ be continuous for every x. The *conditional differential entropy* $h(Y|X)$ of Y given X is defined as

$$
\begin{aligned}
h(Y|X) &= \int h(Y|X = x)\, dF(x) \\
&= -\mathsf{E}_{X,Y}(\log f(Y|X)).
\end{aligned}
$$

As for the discrete case in (2.1), conditioning reduces entropy, i.e.,

$$h(Y|X) \le h(Y) \tag{2.4}$$

with equality if X and Y are independent.

We will often be interested in the sum of two random variables $Y = X + Z$, where X is an arbitrary random variable and Z is an independent continuous random variable with bounded pdf $f(z)$, for example, a Gaussian random variable. It can be shown in this case that the sum Y is a continuous random variable with well-defined density.

Joint differential entropy. The definition of differential entropy can be extended to a continuous random vector X^n with joint pdf $f(x^n)$ as

$$h(X^n) = -\mathsf{E}(\log f(X^n)).$$

For example, if X^n is a Gaussian random vector with mean μ and covariance matrix K, i.e., $X^n \sim \mathrm{N}(\mu, K)$, then

$$h(X^n) = \frac{1}{2}\log((2\pi e)^n |K|).$$

By the chain rule for pdfs and (2.4), we have

$$h(X^n) = \sum_{i=1}^{n} h(X_i | X^{i-1}) \le \sum_{i=1}^{n} h(X_i) \tag{2.5}$$

with equality if $X_1, \ldots, X_n$ are mutually independent. The translation and scaling properties of differential entropy continue to hold for the vector case.

- Translation: For any real-valued vector a^n, $h(X^n + a^n) = h(X^n)$.

- Scaling: For any real-valued nonsingular $n \times n$ matrix A,

$$h(AX^n) = h(X^n) + \log|\det(A)|.$$

The following lemma will be used in the converse proofs of Gaussian source and channel coding theorems.

Maximum Differential Entropy Lemma. Let $\mathbf{X} \sim f(x^n)$ be a random vector with covariance matrix $K_{\mathbf{X}} = \mathsf{E}[(\mathbf{X} - \mathsf{E}(\mathbf{X}))(\mathbf{X} - \mathsf{E}(\mathbf{X}))^T] \succ 0$. Then

$$h(\mathbf{X}) \le \frac{1}{2}\log\left((2\pi e)^n |K_{\mathbf{X}}|\right) \le \frac{1}{2}\log((2\pi e)^n |\mathsf{E}(\mathbf{XX}^T)|), \tag{2.6}$$

where $\mathsf{E}(\mathbf{XX}^T)$ is the correlation matrix of X^n. The first inequality holds with equality if and only if $\mathbf{X}$ is Gaussian and the second inequality holds with equality if and only if $\mathsf{E}(\mathbf{X}) = 0$. More generally, if $(\mathbf{X}, \mathbf{Y}) = (X^n, Y^k) \sim f(x^n, y^k)$ is a pair of random vectors $K_{\mathbf{X}|\mathbf{Y}} = \mathsf{E}[(\mathbf{X} - \mathsf{E}(\mathbf{X}|\mathbf{Y}))(\mathbf{X} - \mathsf{E}(\mathbf{X}|\mathbf{Y}))^T]$ is the covariance matrix of the error vector of the minimum mean squared error (MMSE) estimate of $\mathbf{X}$ given $\mathbf{Y}$, then

$$h(\mathbf{X}|\mathbf{Y}) \le \frac{1}{2}\log\left((2\pi e)^n |K_{\mathbf{X}|\mathbf{Y}}|\right) \tag{2.7}$$

with equality if $(\mathbf{X}, \mathbf{Y})$ is jointly Gaussian.

The proof of the upper bound in (2.6) is similar to the proof for the scalar case in (2.3); see Problem 2.7. The upper bound in (2.7) follows by applying (2.6) to $h(\mathbf{X}|\mathbf{Y} = \mathbf{y})$ for each $\mathbf{y}$ and Jensen's inequality using the concavity of $\log|K|$ in K. The upper bound on differential entropy in (2.6) can be further relaxed to

$$h(X^n) \le \frac{n}{2}\log\left(2\pi e\left(\frac{1}{n}\sum_{i=1}^{n} \mathrm{Var}(X_i)\right)\right) \le \frac{n}{2}\log\left(2\pi e\left(\frac{1}{n}\sum_{i=1}^{n} \mathsf{E}(X_i^2)\right)\right). \tag{2.8}$$

These inequalities can be proved using Hadamard's inequality or more directly using (2.5), (2.3), and Jensen's inequality.

The quantity $2^{2h(X^n)/n}/(2\pi e)$ is often referred to as the *entropy power* of the random vector X^n. The inequality in (2.8) shows that the entropy power is upper bounded by the average power. The following inequality shows that the entropy power of the sum of two independent random vectors is lower bounded by the sum of their entropy powers. In a sense, this inequality is the continuous analogue of Mrs. Gerber's lemma.

Entropy Power Inequality (EPI).

- Scalar EPI: Let $X \sim f(x)$ and $Z \sim f(z)$ be independent random variables and $Y = X + Z$. Then

$$2^{2h(Y)} \geq 2^{2h(X)} + 2^{2h(Z)}$$

  with equality if both X and Z are Gaussian.

- Vector EPI: Let $X^n \sim f(x^n)$ and $Z^n \sim f(z^n)$ be independent random vectors and $Y^n = X^n + Z^n$. Then

$$2^{2h(Y^n)/n} \geq 2^{2h(X^n)/n} + 2^{2h(Z^n)/n}$$

  with equality if X^n and Z^n are Gaussian with $K_X = aK_Z$ for some scalar $a > 0$.

- Conditional EPI: Let X^n and Z^n be conditionally independent given an arbitrary random variable U, with conditional pdfs $f(x^n|u)$ and $f(z^n|u)$, and $Y^n = X^n + Z^n$. Then

$$2^{2h(Y^n|U)/n} \geq 2^{2h(X^n|U)/n} + 2^{2h(Z^n|U)/n}.$$

The scalar EPI can be proved, for example, using a sharp version of Young's inequality or de Bruijn's identity; see Bibliographic Notes. The proofs of the vector and conditional EPIs follow by the scalar EPI, the convexity of the function $\log(2^v + 2^w)$ in (v, w), and induction.

Differential entropy rate of a stationary random process. Let $X = \{X_i\}$ be a stationary continuous-valued random process. The *differential entropy rate* $\bar{h}(X)$ of the process X is defined as

$$\bar{h}(X) = \lim_{n \to \infty} \frac{1}{n} h(X^n) = \lim_{n \to \infty} h(X_n | X^{n-1}).$$

2.3 MUTUAL INFORMATION

Let $(X, Y) \sim p(x, y)$ be a pair of discrete random variables. The information about X obtained from the observation Y is measured by the *mutual information* between X and Y defined as

$$\begin{aligned}
I(X; Y) &= \sum_{(x,y) \in \mathcal{X} \times \mathcal{Y}} p(x, y) \log \frac{p(x, y)}{p(x)p(y)} \\
&= H(X) - H(X|Y) \\
&= H(Y) - H(Y|X) \\
&= H(X) + H(Y) - H(X, Y).
\end{aligned}$$

The mutual information $I(X; Y)$ is a nonnegative function of $p(x, y)$, and $I(X; Y) = 0$ if and only if (iff) X and Y are independent. It is concave in $p(x)$ for a fixed $p(y|x)$, and convex in $p(y|x)$ for a fixed $p(x)$. Mutual information can be defined also for a pair of continuous random variables $(X, Y) \sim f(x, y)$ as

$$I(X; Y) = \int f(x, y) \log \frac{f(x, y)}{f(x)f(y)} \, dx \, dy$$

$$= h(X) - h(X|Y)$$

$$= h(Y) - h(Y|X)$$

$$= h(X) + h(Y) - h(X, Y).$$

Similarly, let $X \sim p(x)$ be a discrete random variable and $Y|\{X = x\} \sim f(y|x)$ be continuous for every x. Then

$$I(X; Y) = h(Y) - h(Y|X) = H(X) - H(X|Y).$$

In general, mutual information can be defined for an *arbitrary* pair of random variables (Pinsker 1964) as

$$I(X; Y) = \int \log \frac{d\mu(x, y)}{d(\mu(x) \times \mu(y))} \, d\mu(x, y),$$

where $d\mu(x, y)/d(\mu(x) \times \mu(y))$ is the *Radon–Nikodym derivative* (see, for example, Royden 1988) of the joint probability measure $\mu(x, y)$ with respect to the product probability measure $\mu(x) \times \mu(y)$. Equivalently, it can be expressed as

$$I(X; Y) = \sup_{\hat{x}, \hat{y}} I(\hat{x}(X); \hat{y}(Y)),$$

where $\hat{x}(x)$ and $\hat{y}(y)$ are finite-valued functions, and the supremum is over all such functions. These definitions can be shown to include the above definitions of mutual information for discrete and continuous random variables as special cases (Gray 1990, Section 5) by considering

$$I(X; Y) = \lim_{j,k \to \infty} I([X]_j; [Y]_k),$$

where $[X]_j = \hat{x}_j(X)$ and $[Y]_k = \hat{y}_k(Y)$ can be any sequences of finite quantizations of X and Y, respectively, such that the quantization errors $(x - \hat{x}_j(x))$ and $(y - \hat{y}_k(y))$ tend to zero as $j, k \to \infty$ for every x, y.

Remark 2.1 (Relative entropy). Mutual information is a special case of the *relative entropy* (*Kullback–Leibler divergence*). Let P and Q be two probability measures such that P is absolutely continuous with respect to Q, then the relative entropy is defined as

$$D(P||Q) = \int \log \frac{dP}{dQ} \, dP,$$

where dP/dQ is the Radon–Nikodym derivative. Thus, mutual information $I(X; Y)$ is the relative entropy between the joint and product measures of X and Y. Note that by the convexity of $\log(1/x)$, $D(P||Q)$ is nonnegative and $D(P||Q) = 0$ iff $P = Q$.

Conditional mutual information. Let $(X, Y)|\{Z = z\} \sim F(x, y|z)$ and $Z \sim F(z)$. Denote the mutual information between X and Y given $\{Z = z\}$ by $I(X; Y|Z = z)$. Then the *conditional mutual information* $I(X; Y|Z)$ between X and Y given Z is defined as

$$I(X; Y|Z) = \int I(X; Y|Z = z)\, dF(z).$$

For $(X, Y, Z) \sim p(x, y, z)$,

$$
\begin{aligned}
I(X; Y|Z) &= H(X|Z) - H(X|Y, Z) \\
&= H(Y|Z) - H(Y|X, Z) \\
&= H(X|Z) + H(Y|Z) - H(X, Y|Z).
\end{aligned}
$$

The conditional mutual information $I(X; Y|Z)$ is nonnegative and is equal to zero iff X and Y are conditionally independent given Z, i.e., $X \to Z \to Y$ form a Markov chain. Note that unlike entropy, no general inequality relationship exists between the conditional mutual information $I(X; Y|Z)$ and the mutual information $I(X; Y)$. There are, however, two important special cases.

- Independence: If $p(x, y, z) = p(x)p(z)p(y|x, z)$, that is, if X and Z are independent, then

$$I(X; Y|Z) \geq I(X; Y).$$

This follows by the convexity of $I(X; Y)$ in $p(y|x)$ for a fixed $p(x)$.

- Conditional independence: If $Z \to X \to Y$ form a Markov chain, then

$$I(X; Y|Z) \leq I(X; Y).$$

This follows by the concavity of $I(X; Y)$ in $p(x)$ for a fixed $p(y|x)$.

The definition of mutual information can be extended to random vectors in a straightforward manner. In particular, we can establish the following useful identity.

Chain Rule for Mutual Information. Let $(X^n, Y) \sim F(x^n, y)$. Then

$$I(X^n; Y) = \sum_{i=1}^{n} I(X_i; Y|X^{i-1}).$$

The following inequality shows that processing cannot increase information.

Data Processing Inequality. If $X \to Y \to Z$ form a Markov chain, then

$$I(X; Z) \leq I(X; Y).$$

Consequently, for any function g, $I(X; g(Y)) \leq I(X; Y)$, which implies the inequality in (2.2). To prove the data processing inequality, we use the chain rule to expand $I(X; Y, Z)$ in two ways as

$$I(X; Y, Z) = I(X; Y) + I(X; Z|Y) = I(X; Y)$$
$$= I(X; Z) + I(X; Y|Z) \geq I(X; Z).$$

The chain rule can be also used to establish the following identity, which will be used in several converse proofs.

Csiszár Sum Identity. Let $(U, X^n, Y^n) \sim F(u, x^n, y^n)$. Then

$$\sum_{i=1}^{n} I(X_{i+1}^n; Y_i | Y^{i-1}, U) = \sum_{i=1}^{n} I(Y^{i-1}; X_i | X_{i+1}^n, U),$$

where $X_{n+1}^n, Y^0 = \emptyset$.

2.4 TYPICAL SEQUENCES

Let x^n be a sequence with elements drawn from a finite alphabet $\mathcal{X}$. Define the *empirical pmf* of x^n (also referred to as its *type*) as

$$\pi(x|x^n) = \frac{|\{i: x_i = x\}|}{n} \quad \text{for } x \in \mathcal{X}.$$

For example, if $x^n = (0, 1, 1, 0, 0, 1, 0)$, then

$$\pi(x|x^n) = \begin{cases} 4/7 & \text{for } x = 0, \\ 3/7 & \text{for } x = 1. \end{cases}$$

Let $X_1, X_2, \ldots$ be a sequence of independent and identically distributed (i.i.d.) random variables with $X_i \sim p_X(x_i)$. Then by the (weak) law of large numbers (LLN), for each $x \in \mathcal{X}$,

$$\pi(x|X^n) \to p(x) \quad \text{in probability.}$$

Thus, with high probability, the random empirical pmf $\pi(x|X^n)$ does not deviate much from the true pmf $p(x)$. For $X \sim p(x)$ and $\epsilon \in (0, 1)$, define the set of ϵ-typical n-sequences x^n (or the typical set in short) as

$$\mathcal{T}_\epsilon^{(n)}(X) = \left\{ x^n \colon |\pi(x|x^n) - p(x)| \leq \epsilon p(x) \text{ for all } x \in \mathcal{X} \right\}.$$

When it is clear from the context, we will use $\mathcal{T}_\epsilon^{(n)}$ instead of $\mathcal{T}_\epsilon^{(n)}(X)$. The following simple fact is a direct consequence of the definition of the typical set.

Typical Average Lemma. Let $x^n \in \mathcal{T}_\epsilon^{(n)}(X)$. Then for any nonnegative function $g(x)$ on $\mathcal{X}$,

$$(1 - \epsilon) \, \mathsf{E}(g(X)) \le \frac{1}{n} \sum_{i=1}^{n} g(x_i) \le (1 + \epsilon) \, \mathsf{E}(g(X)).$$

Typical sequences satisfy the following properties:

1. Let $p(x^n) = \prod_{i=1}^{n} p_X(x_i)$. Then for each $x^n \in \mathcal{T}_\epsilon^{(n)}(X)$,

    $$2^{-n(H(X)+\delta(\epsilon))} \le p(x^n) \le 2^{-n(H(X)-\delta(\epsilon))},$$

    where $\delta(\epsilon) = \epsilon H(X)$. This follows by the typical average lemma with $g(x) = -\log p(x)$.

2. The cardinality of the typical set is upper bounded as

    $$\left| \mathcal{T}_\epsilon^{(n)}(X) \right| \le 2^{n(H(X)+\delta(\epsilon))}.$$

    This can be shown by summing the lower bound in property 1 over the typical set.

3. If $X_1, X_2, \ldots$ are i.i.d. with $X_i \sim p_X(x_i)$, then by the LLN,

    $$\lim_{n \to \infty} \mathsf{P}\{X^n \in \mathcal{T}_\epsilon^{(n)}(X)\} = 1.$$

4. The cardinality of the typical set is lower bounded as

    $$\left| \mathcal{T}_\epsilon^{(n)}(X) \right| \ge (1 - \epsilon) 2^{n(H(X)-\delta(\epsilon))}$$

    for n sufficiently large. This follows by property 3 and the upper bound in property 1.

The above properties are illustrated in Figure 2.1.

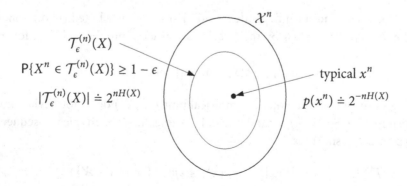

Figure 2.1. Properties of typical sequences. Here $X^n \sim \prod_{i=1}^{n} p_X(x_i)$.

2.5   JOINTLY TYPICAL SEQUENCES

The notion of typicality can be extended to multiple random variables. Let (x^n, y^n) be a pair of sequences with elements drawn from a pair of finite alphabets $(\mathcal{X}, \mathcal{Y})$. Define their joint empirical pmf (joint type) as

$$\pi(x, y | x^n, y^n) = \frac{|\{i : (x_i, y_i) = (x, y)\}|}{n} \quad \text{for } (x, y) \in \mathcal{X} \times \mathcal{Y}.$$

Let $(X, Y) \sim p(x, y)$. The set of jointly ϵ-typical n-sequences is defined as

$$\mathcal{T}_\epsilon^{(n)}(X, Y) = \{(x^n, y^n) : |\pi(x, y | x^n, y^n) - p(x, y)| \le \epsilon p(x, y) \text{ for all } (x, y) \in \mathcal{X} \times \mathcal{Y}\}.$$

Note that this is the same as the typical set for a single "large" random variable (X, Y), i.e., $\mathcal{T}_\epsilon^{(n)}(X, Y) = \mathcal{T}_\epsilon^{(n)}((X, Y))$. Also define the set of conditionally ϵ-typical n sequences as $\mathcal{T}_\epsilon^{(n)}(Y | x^n) = \{y^n : (x^n, y^n) \in \mathcal{T}_\epsilon^{(n)}(X, Y)\}$. The properties of typical sequences can be extended to jointly typical sequences as follows.

1. Let $(x^n, y^n) \in \mathcal{T}_\epsilon^{(n)}(X, Y)$ and $p(x^n, y^n) = \prod_{i=1}^n p_{X,Y}(x_i, y_i)$. Then

   (a) $x^n \in \mathcal{T}_\epsilon^{(n)}(X)$ and $y^n \in \mathcal{T}_\epsilon^{(n)}(Y)$,

   (b) $p(x^n) \doteq 2^{-nH(X)}$ and $p(y^n) \doteq 2^{-nH(Y)}$,

   (c) $p(x^n | y^n) \doteq 2^{-nH(X|Y)}$ and $p(y^n | x^n) \doteq 2^{-nH(Y|X)}$, and

   (d) $p(x^n, y^n) \doteq 2^{-nH(X,Y)}$.

2. $|\mathcal{T}_\epsilon^{(n)}(X, Y)| \doteq 2^{nH(X,Y)}$.

3. For every $x^n \in \mathcal{X}^n$,

   $$|\mathcal{T}_\epsilon^{(n)}(Y | x^n)| \le 2^{n(H(Y|X) + \delta(\epsilon))},$$

   where $\delta(\epsilon) = \epsilon H(Y | X)$.

4. Let $X \sim p(x)$ and $Y = g(X)$. Let $x^n \in \mathcal{T}_\epsilon^{(n)}(X)$. Then $y^n \in \mathcal{T}_\epsilon^{(n)}(Y | x^n)$ iff $y_i = g(x_i)$ for $i \in [1 : n]$.

The following property deserves a special attention.

Conditional Typicality Lemma. Let $(X, Y) \sim p(x, y)$. Suppose that $x^n \in \mathcal{T}_{\epsilon'}^{(n)}(X)$ and $Y^n \sim p(y^n | x^n) = \prod_{i=1}^n p_{Y|X}(y_i | x_i)$. Then, for every $\epsilon > \epsilon'$,

$$\lim_{n \to \infty} \mathsf{P}\{(x^n, Y^n) \in \mathcal{T}_\epsilon^{(n)}(X, Y)\} = 1.$$

The proof of this lemma follows by the LLN. The details are given in Appendix 2A. Note that the condition $\epsilon > \epsilon'$ is crucial to applying the LLN because x^n could otherwise be on the boundary of $\mathcal{T}_\epsilon^{(n)}(X)$; see Problem 2.17.

The conditional typicality lemma implies the following additional property of jointly typical sequences.

5. If $x^n \in \mathcal{T}_{\epsilon'}^{(n)}(X)$ and $\epsilon' < \epsilon$, then for n sufficiently large,

$$\left|\mathcal{T}_{\epsilon}^{(n)}(Y|x^n)\right| \geq (1 - \epsilon)2^{n(H(Y|X)-\delta(\epsilon))}.$$

The above properties of jointly typical sequences are illustrated in two different ways in Figure 2.2.

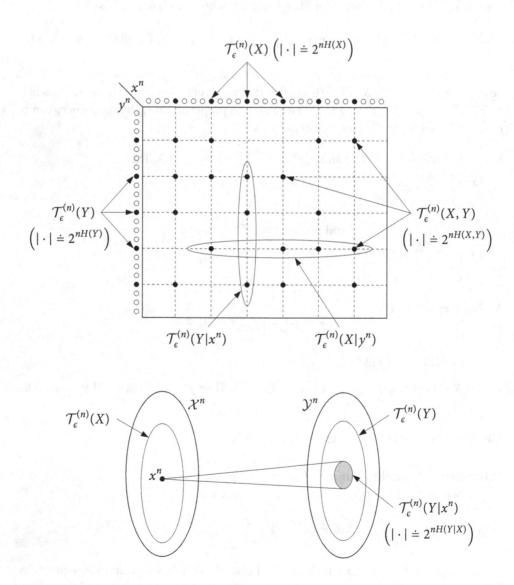

Figure 2.2. Properties of jointly typical sequences.

2.5.1   Joint Typicality for a Triple of Random Variables

Let $(X, Y, Z) \sim p(x, y, z)$. The set of jointly ϵ-typical (x^n, y^n, z^n) sequences is defined as

$$\mathcal{T}_\epsilon^{(n)}(X, Y, Z) = \{(x^n, y^n, z^n): |\pi(x, y, z|x^n, y^n, z^n) - p(x, y, z)| \leq \epsilon p(x, y, z)$$
$$\text{for all } (x, y, z) \in \mathcal{X} \times \mathcal{Y} \times \mathcal{Z}\}.$$

Since this is equivalent to the typical set for a single "large" random variable (X, Y, Z) or a pair of random variables $((X, Y), Z)$, the properties of (jointly) typical sequences continue to hold. For example, suppose that $(x^n, y^n, z^n) \in \mathcal{T}_\epsilon^{(n)}(X, Y, Z)$ and $p(x^n, y^n, z^n) = \prod_{i=1}^n p_{X,Y,Z}(x_i, y_i, z_i)$. Then

1. $x^n \in \mathcal{T}_\epsilon^{(n)}(X)$ and $(y^n, z^n) \in \mathcal{T}_\epsilon^{(n)}(Y, Z)$,

2. $p(x^n, y^n, z^n) \doteq 2^{-nH(X,Y,Z)}$,

3. $p(x^n, y^n|z^n) \doteq 2^{-nH(X,Y|Z)}$,

4. $|\mathcal{T}_\epsilon^{(n)}(X|y^n, z^n)| \leq 2^{n(H(X|Y,Z)+\delta(\epsilon))}$, and

5. if $(y^n, z^n) \in \mathcal{T}_{\epsilon'}^{(n)}(Y, Z)$ and $\epsilon' < \epsilon$, then for n sufficiently large, $|\mathcal{T}_\epsilon^{(n)}(X|y^n, z^n)| \geq 2^{n(H(X|Y,Z)-\delta(\epsilon))}$.

The following two-part lemma will be used in the achievability proofs of many coding theorems.

Joint Typicality Lemma. Let $(X, Y, Z) \sim p(x, y, z)$ and $\epsilon' < \epsilon$. Then there exists $\delta(\epsilon) > 0$ that tends to zero as $\epsilon \to 0$ such that the following statements hold:

1. If $(\tilde{x}^n, \tilde{y}^n)$ is a pair of arbitrary sequences and $\tilde{Z}^n \sim \prod_{i=1}^n p_{Z|X}(\tilde{z}_i|\tilde{x}_i)$, then

$$\mathsf{P}\{(\tilde{x}^n, \tilde{y}^n, \tilde{Z}^n) \in \mathcal{T}_\epsilon^{(n)}(X, Y, Z)\} \leq 2^{-n(I(Y;Z|X)-\delta(\epsilon))}.$$

2. If $(x^n, y^n) \in \mathcal{T}_{\epsilon'}^{(n)}$ and $\tilde{Z}^n \sim \prod_{i=1}^n p_{Z|X}(\tilde{z}_i|x_i)$, then for n sufficiently large,

$$\mathsf{P}\{(x^n, y^n, \tilde{Z}^n) \in \mathcal{T}_\epsilon^{(n)}(X, Y, Z)\} \geq 2^{-n(I(Y;Z|X)+\delta(\epsilon))}.$$

To prove the first statement, consider

$$\mathsf{P}\{(\tilde{x}^n, \tilde{y}^n, \tilde{Z}^n) \in \mathcal{T}_\epsilon^{(n)}(X, Y, Z)\} = \sum_{\tilde{z}^n \in \mathcal{T}_\epsilon^{(n)}(Z|\tilde{x}^n, \tilde{y}^n)} p(\tilde{z}^n|\tilde{x}^n)$$

$$\leq |\mathcal{T}_\epsilon^{(n)}(Z|\tilde{x}^n, \tilde{y}^n)| \cdot 2^{-n(H(Z|X)-\epsilon H(Z|X))}$$

$$\leq 2^{n(H(Z|X,Y)+\epsilon H(Z|X,Y))} 2^{-n(H(Z|X)-\epsilon H(Z|X))}$$

$$\leq 2^{-n(I(Y;Z|X)-\delta(\epsilon))}.$$

Similarly, for every n sufficiently large,

$$P\{(x^n, y^n, \tilde{Z}^n) \in \mathcal{T}_\epsilon^{(n)}(X, Y, Z)\} \geq |\mathcal{T}_\epsilon^{(n)}(Z|x^n, y^n)| \cdot 2^{-n(H(Z|X)+\epsilon H(Z|X))}$$
$$\geq (1-\epsilon)2^{n(H(Z|X,Y)-\epsilon H(Z|X,Y))}2^{-n(H(Z|X)+\epsilon H(Z|X))}$$
$$\geq 2^{-n(I(Y;Z|X)+\delta(\epsilon))},$$

which proves the second statement.

Remark 2.2. As an application of the joint typicality lemma, it can be easily shown that if $(X^n, Y^n) \sim \prod_{i=1}^n p_{X,Y}(x_i, y_i)$ and $\tilde{Z}^n|\{X^n = x^n, Y^n = y^n\} \sim \prod_{i=1}^n p_{Z|X}(\tilde{z}_i|x_i)$, then

$$P\{(X^n, Y^n, \tilde{Z}^n) \in \mathcal{T}_\epsilon^{(n)}\} \doteq 2^{-nI(Y;Z|X)}.$$

Other applications of the joint typicality lemma are given in Problem 2.14.

2.5.2 Multivariate Typical Sequences

Let $(X_1, X_2, \ldots, X_k) \sim p(x_1, x_2, \ldots, x_k)$ and $\mathcal{J}$ be a nonempty subset of $[1:k]$. Define the subtuple of random variables $X(\mathcal{J}) = (X_j : j \in \mathcal{J})$. For example, if $k = 3$ and $\mathcal{J} = \{1, 3\}$, then $X(\mathcal{J}) = (X_1, X_3)$. The set of ϵ-typical n-sequences $(x_1^n, x_2^n, \ldots, x_k^n)$ is defined as $\mathcal{T}_\epsilon^{(n)}(X_1, X_2, \ldots, X_k) = \mathcal{T}_\epsilon^{(n)}((X_1, X_2, \ldots, X_k))$, that is, as the typical set for a single random variable $(X_1, X_2, \ldots, X_k)$. We can similarly define $\mathcal{T}_\epsilon^{(n)}(X(\mathcal{J}))$ for every $\mathcal{J} \subseteq [1:k]$.

It can be easily checked that the properties of jointly typical sequences continue to hold by considering $X(\mathcal{J})$ as a single random variable. For example, if $(x_1^n, x_2^n, \ldots, x_k^n) \in \mathcal{T}_\epsilon^{(n)}(X_1, X_2, \ldots, X_k)$ and $p(x_1^n, x_2^n, \ldots, x_k^n) = \prod_{i=1}^n p_{X_1, X_2, \ldots, X_k}(x_{1i}, x_{2i}, \ldots, x_{ki})$, then for all $\mathcal{J}, \mathcal{J}' \subseteq [1:k]$,

1. $x^n(\mathcal{J}) \in \mathcal{T}_\epsilon^{(n)}(X(\mathcal{J}))$,

2. $p(x^n(\mathcal{J})|x^n(\mathcal{J}')) \doteq 2^{-nH(X(\mathcal{J})|X(\mathcal{J}'))}$,

3. $|\mathcal{T}_\epsilon^{(n)}(X(\mathcal{J})|x^n(\mathcal{J}'))| \leq 2^{n(H(X(\mathcal{J})|X(\mathcal{J}'))+\delta(\epsilon))}$, and

4. if $x^n(\mathcal{J}') \in \mathcal{T}_{\epsilon'}^{(n)}(X(\mathcal{J}'))$ and $\epsilon' < \epsilon$, then for n sufficiently large,

$$|\mathcal{T}_\epsilon^{(n)}(X(\mathcal{J})|x^n(\mathcal{J}'))| \geq 2^{n(H(X(\mathcal{J})|X(\mathcal{J}'))-\delta(\epsilon))}.$$

The conditional and joint typicality lemmas can be readily generalized to subsets $\mathcal{J}_1$, $\mathcal{J}_2$, and $\mathcal{J}_3$ and corresponding sequences $x^n(\mathcal{J}_1)$, $x^n(\mathcal{J}_2)$, and $x^n(\mathcal{J}_3)$ that satisfy similar conditions.

SUMMARY

- Entropy as a measure of information

- Mutual information as a measure of information transfer

- Key inequalities and identities:

  - Fano's inequality

  - Mrs. Gerber's lemma

  - Maximum differential entropy lemma

  - Entropy power inequality

  - Chain rules for entropy and mutual information

  - Data processing inequality

  - Csiszár sum identity

- Typical sequences:

  - Typical average lemma

  - Conditional typicality lemma

  - Joint typicality lemma

BIBLIOGRAPHIC NOTES

Shannon (1948) defined entropy and mutual information for discrete and continuous random variables, and provided justifications of these definitions in both axiomatic and operational senses. Many of the simple properties of these quantities, including the maximum entropy property of the Gaussian distribution, are also due to Shannon. Subsequently, Kolmogorov (1956) and Dobrushin (1959a) gave rigorous extensions of entropy and mutual information to abstract probability spaces.

Fano's inequality is due to Fano (1952). Mrs. Gerber's lemma is due to Wyner and Ziv (1973). Extensions of the MGL were given by Witsenhausen (1974), Witsenhausen and Wyner (1975), and Shamai and Wyner (1990).

The entropy power inequality has a longer history. It first appeared in Shannon (1948) without a proof. Full proofs were given subsequently by Stam (1959) and Blachman (1965) using de Bruijn's identity (Cover and Thomas 2006, Theorem 17.7.2). The EPI can be rewritten in the following equivalent inequality (Costa and Cover 1984). For a pair of independent random vectors $X^n \sim f(x^n)$ and $Z^n \sim f(z^n)$,

$$h(X^n + Z^n) \geq h(\tilde{X}^n + \tilde{Z}^n), \tag{2.9}$$

where $\tilde{X}^n$ and $\tilde{Z}^n$ are a pair of independent Gaussian random vectors with proportional covariance matrices, chosen so that $h(X^n) = h(\tilde{X}^n)$ and $h(Z^n) = h(\tilde{Z}^n)$. Now (2.9) can be proved by the strengthened version of Young's inequality (Beckner 1975, Brascamp and Lieb 1976); see, for example, Lieb (1978) and Gardner (2002). Recently, Verdú and Guo (2006) gave a simple proof by relating the minimum mean squared error (MMSE) and mutual information in Gaussian channels; see Madiman and Barron (2007) for a similar proof from a different angle. Extensions of the EPI are given by Costa (1985), Zamir and Feder (1993), and Artstein, Ball, Barthe, and Naor (2004).

There are several notions of typicality in the literature. Our notion of typicality is that of *robust typicality* due to Orlitsky and Roche (2001). As is evident in the typical average lemma, it is often more convenient than the more widely known notion of *strong typicality* (Berger 1978, Csiszár and Körner 1981b) defined as

$$\mathcal{A}_{\epsilon}^{*(n)} = \left\{ x^n : |\pi(x|x^n) - p(x)| \leq \frac{\epsilon}{|\mathcal{X}|} \text{ if } p(x) > 0, \ \pi(x|x^n) = 0 \text{ otherwise} \right\}.$$

Another widely used notion is *weak typicality* (Cover and Thomas 2006) defined as

$$\mathcal{A}_{\epsilon}^{(n)}(X) = \left\{ x^n : \left| -\frac{1}{n} \log p(x^n) - H(X) \right| \leq \epsilon \right\}, \tag{2.10}$$

where $p(x^n) = \prod_{i=1}^{n} p_X(x_i)$. This is a weaker notion than the one we use, since $\mathcal{T}_{\epsilon}^{(n)} \subseteq \mathcal{A}_{\delta}^{(n)}$ for $\delta = \epsilon H(X)$, while in general for some $\epsilon > 0$ there is no $\delta' > 0$ such that $\mathcal{A}_{\delta'}^{(n)} \subseteq \mathcal{T}_{\epsilon}^{(n)}$. For example, every binary n-sequence is weakly typical with respect to the Bern$(1/2)$ pmf, but not all of them are typical. Weak typicality is useful when dealing with discrete or continuous stationary ergodic processes because it is tightly coupled to the Shannon–McMillan–Breiman theorem (Shannon 1948, McMillan 1953, Breiman 1957, Barron 1985), commonly referred to as the *asymptotic equipartition property (AEP)*, which states that for a discrete stationary ergodic process $X = \{X_i\}$,

$$\lim_{n \to \infty} -\frac{1}{n} \log p(X^n) = \bar{H}(X).$$

However, we will encounter several coding schemes that require our notion of typicality. Note, for example, that the conditional typicality lemma fails to hold under weak typicality.

PROBLEMS

2.1.   Prove Fano's inequality.

2.2.   Prove the Csiszár sum identity.

2.3.   Prove the properties of jointly typical sequences with $\delta(\epsilon)$ terms explicitly specified.

2.4.   *Inequalities.* Label each of the following statements with =, ≤, or ≥. Justify each answer.

(a) $H(X|Z)$ vs. $H(X|Y) + H(Y|Z)$.

(b) $h(X + Y)$ vs. $h(X)$, if X and Y are independent continuous random variables.

(c) $h(X + aY)$ vs. $h(X + Y)$, if $Y \sim N(0, 1)$ is independent of X and $a \geq 1$.

(d) $I(X_1, X_2; Y_1, Y_2)$ vs. $I(X_1; Y_1) + I(X_2; Y_2)$, if $p(y_1, y_2|x_1, x_2) = p(y_1|x_1)p(y_2|x_2)$.

(e) $I(X_1, X_2; Y_1, Y_2)$ vs. $I(X_1; Y_1) + I(X_2; Y_2)$, if $p(x_1, x_2) = p(x_1)p(x_2)$.

(f) $I(aX + Y; bX)$ vs. $I(X + Y/a; X)$, if $a, b \neq 0$ and $Y \sim N(0, 1)$ is independent of X.

2.5. *Mrs. Gerber's lemma.* Let $H^{-1}: [0, 1] \to [0, 1/2]$ be the inverse of the binary entropy function.

(a) Show that $H(H^{-1}(v) * p)$ is convex in v for every $p \in [0, 1]$.

(b) Use part (a) to prove the scalar MGL

$$H(Y|U) \geq H(H^{-1}(H(X|U)) * p).$$

(c) Use part (b) and induction to prove the vector MGL

$$\frac{H(Y^n|U)}{n} \geq H\left(H^{-1}\left(\frac{H(X^n|U)}{n}\right) * p\right).$$

2.6. *Maximum differential entropy.* Let $X \sim f(x)$ be a zero-mean random variable with finite variance and $X^* \sim f(x^*)$ be a zero-mean Gaussian random variable with the same variance as X.

(a) Show that

$$-\int f(x) \log f_{X^*}(x)\, dx = -\int f_{X^*}(x) \log f_{X^*}(x)\, dx = h(X^*).$$

(b) Using part (a) and the nonnegativity of relative entropy (see Remark 2.1), conclude that

$$h(X) = -D(f_X \| f_{X^*}) - \int f(x) \log f_{X^*}(x)\, dx \leq h(X^*)$$

with equality iff X is Gaussian.

(c) Following similar steps, show that if $\mathbf{X} \sim f(\mathbf{x})$ is a zero-mean random vector and $\mathbf{X}^* \sim f(\mathbf{x}^*)$ is a zero-mean Gaussian random vector with the same covariance matrix, then

$$h(\mathbf{X}) \leq h(\mathbf{X}^*)$$

with equality iff $\mathbf{X}$ is Gaussian.

2.7. *Maximum conditional differential entropy.* Let $(\mathbf{X}, \mathbf{Y}) = (X^n, Y^k) \sim f(x^n, y^k)$ be a pair of random vectors with covariance matrices $K_{\mathbf{X}} = E[(\mathbf{X} - E(\mathbf{X}))(\mathbf{X} - E(\mathbf{X}))^T]$ and $K_{\mathbf{Y}} = E[(\mathbf{Y} - E(\mathbf{Y}))(\mathbf{Y} - E(\mathbf{Y}))^T]$, and crosscovariance matrix $K_{\mathbf{XY}} = E[(\mathbf{X} - E(\mathbf{X}))(\mathbf{Y} - E(\mathbf{Y}))^T] = K_{\mathbf{YX}}^T$. Show that

$$h(\mathbf{X}|\mathbf{Y}) \leq \frac{1}{2} \log((2\pi e)^n |K_{\mathbf{X}} - K_{\mathbf{XY}}K_{\mathbf{Y}}^{-1}K_{\mathbf{YX}}|)$$

with equality if $(\mathbf{X}, \mathbf{Y})$ is jointly Gaussian.

2.8. *Hadamard's inequality.* Let $Y^n \sim N(0, K)$. Use the fact that

$$h(Y^n) \le \frac{1}{2} \log\left((2\pi e)^n \prod_{i=1}^{n} K_{ii} \right)$$

to prove Hadamard's inequality

$$\det(K) \le \prod_{i=1}^{n} K_{ii}.$$

2.9. *Conditional entropy power inequality.* Let $X \sim f(x)$ and $Z \sim f(z)$ be independent random variables and $Y = X + Z$. Then by the EPI,

$$2^{2h(Y)} \ge 2^{2h(X)} + 2^{2h(Z)}$$

with equality iff both X and Z are Gaussian.

(a) Show that $\log(2^v + 2^w)$ is convex in (v, w).

(b) Let X^n and Z^n be conditionally independent given an arbitrary random variable U, with conditional densities $f(x^n|u)$ and $f(z^n|u)$, respectively. Use part (a), the scalar EPI, and induction to prove the conditional EPI

$$2^{2h(Y^n|U)/n} \ge 2^{2h(X^n|U)/n} + 2^{2h(Z^n|U)/n}.$$

2.10. *Entropy rate of a stationary source.* Let $X = \{X_i\}$ be a discrete stationary random process.

(a) Show that

$$\frac{H(X^n)}{n} \le \frac{H(X^{n-1})}{n-1} \quad \text{for } n = 2, 3, \dots.$$

(b) Conclude that the entropy rate

$$\bar{H}(X) = \lim_{n \to \infty} \frac{H(X^n)}{n}$$

is well-defined.

(c) Show that for a continuous stationary process $Y = \{Y_i\}$,

$$\frac{h(Y^n)}{n} \le \frac{h(Y^{n-1})}{n-1} \quad \text{for } n = 2, 3, \dots.$$

2.11. *Worst noise for estimation.* Let $X \sim N(0, P)$ and Z be independent of X with zero mean and variance N. Show that the minimum mean squared error (MMSE) of estimating X given $X + Z$ is upper bounded as

$$E\left[(X - E(X|X + Z))^2\right] \le \frac{PN}{P + N}$$

with equality if Z is Gaussian. Thus, Gaussian noise is the worst noise if the input to the channel is Gaussian.

2.12.   *Worst noise for information.* Let X and Z be independent, zero-mean random variables with variances P and N, respectively.

(a) Show that

$$h(X|X + Z) \le \frac{1}{2} \log\left(\frac{2\pi ePN}{P + N}\right)$$

with equality iff both X and Z are Gaussian. (Hint: Use the maximum differential entropy lemma or the EPI or Problem 2.11.)

(b) Let X^* and Z^* be independent zero-mean Gaussian random variables with variances P and N, respectively. Use part (a) to show that

$$I(X^*; X^* + Z^*) \le I(X^*; X^* + Z)$$

with equality iff Z is Gaussian. Thus, Gaussian noise is the worst noise when the input to an additive channel is Gaussian.

2.13.   *Joint typicality.* Let $(X, Y) \sim p(x, y)$ and $\epsilon > \epsilon'$. Let $X^n \sim p(x^n)$ be an arbitrary random sequence and $Y^n | \{X^n = x^n\} \sim \prod_{i=1}^n p_{Y|X}(y_i|x_i)$. Using the conditional typicality lemma, show that

$$\lim_{n \to \infty} \mathsf{P}\{(X^n, Y^n) \in \mathcal{T}_\epsilon^{(n)} \mid X^n \in \mathcal{T}_{\epsilon'}^{(n)}\} = 1.$$

2.14.   *Variations on the joint typicality lemma.* Let $(X, Y, Z) \sim p(x, y, z)$ and $0 < \epsilon' < \epsilon$. Prove the following statements.

(a) Let $(X^n, Y^n) \sim \prod_{i=1}^n p_{X,Y}(x_i, y_i)$ and $\tilde{Z}^n | \{X^n = x^n, Y^n = y^n\} \sim \prod_{i=1}^n p_{Z|X}(\tilde{z}_i|x_i)$, conditionally independent of Y^n given X^n. Then

$$\mathsf{P}\{(X^n, Y^n, \tilde{Z}^n) \in \mathcal{T}_\epsilon^{(n)}(X, Y, Z)\} \doteq 2^{-nI(Y;Z|X)}.$$

(b) Let $(x^n, y^n) \in \mathcal{T}_{\epsilon'}^{(n)}(X, Y)$ and $\tilde{Z}^n \sim \mathrm{Unif}(\mathcal{T}_\epsilon^{(n)}(Z|x^n))$. Then

$$\mathsf{P}\{(x^n, y^n, \tilde{Z}^n) \in \mathcal{T}_\epsilon^{(n)}(X, Y, Z)\} \doteq 2^{-nI(Y;Z|X)}.$$

(c) Let $x^n \in \mathcal{T}_{\epsilon'}^{(n)}(X)$, $\tilde{y}^n$ be an arbitrary sequence, and $\tilde{Z}^n \sim p(\tilde{z}^n|x^n)$, where

$$p(\tilde{z}^n|x^n) = \begin{cases} \dfrac{\prod_{i=1}^n p_{Z|X}(\tilde{z}_i|x_i)}{\sum_{z^n \in \mathcal{T}_\epsilon^{(n)}(Z|x^n)} \prod_{i=1}^n p_{Z|X}(z_i|x_i)} & \text{if } \tilde{z}^n \in \mathcal{T}_\epsilon^{(n)}(Z|x^n), \\ 0 & \text{otherwise.} \end{cases}$$

Then

$$\mathsf{P}\{(x^n, \tilde{y}^n, \tilde{Z}^n) \in \mathcal{T}_\epsilon^{(n)}(X, Y, Z)\} \le 2^{-n(I(Y;Z|X) - \delta(\epsilon))}.$$

(d) Let $(\tilde{X}^n, \tilde{Y}^n, \tilde{Z}^n) \sim \prod_{i=1}^n p_X(\tilde{x}_i) p_Y(\tilde{y}_i) p_{Z|X,Y}(\tilde{z}_i|\tilde{x}_i, \tilde{y}_i)$. Then

$$\mathsf{P}\{(\tilde{X}^n, \tilde{Y}^n, \tilde{Z}^n) \in \mathcal{T}_\epsilon^{(n)}(X, Y, Z)\} \doteq 2^{-nI(X;Y)}.$$

2.15. *Jointly typical triples.* Given $(X, Y, Z) \sim p(x, y, z)$, let

$$\mathcal{A}_n = \{(x^n, y^n, z^n): (x^n, y^n) \in \mathcal{T}_\epsilon^{(n)}(X, Y),$$
$$(y^n, z^n) \in \mathcal{T}_\epsilon^{(n)}(Y, Z), (x^n, z^n) \in \mathcal{T}_\epsilon^{(n)}(X, Z)\}.$$

(a) Show that

$$|\mathcal{A}_n| \le 2^{n(H(X,Y)+H(Y,Z)+H(X,Z)+\delta(\epsilon))/2}.$$

(Hint: First show that $|\mathcal{A}_n| \le 2^{n(H(X,Y)+H(Z|Y)+\delta(\epsilon))}$.)

(b) Does the corresponding lower bound hold in general? (Hint: Consider $X = Y = Z$.)

Remark: It can be shown that $|\mathcal{A}_n| \doteq 2^{n(\max H(\tilde{X},\tilde{Y},\tilde{Z}))}$, where the maximum is over all joint pmfs $p(\tilde{x}, \tilde{y}, \tilde{z})$ such that $p(\tilde{x}, \tilde{y}) = p_{X,Y}(\tilde{x}, \tilde{y})$, $p(\tilde{y}, \tilde{z}) = p_{Y,Z}(\tilde{y}, \tilde{z})$, and $p(\tilde{x}, \tilde{z}) = p_{X,Z}(\tilde{x}, \tilde{z})$.

2.16. *Multivariate typicality.* Let $(U, X, Y, Z) \sim p(u, x, y, z)$. Prove the following statements.

(a) If $(\tilde{U}^n, \tilde{X}^n, \tilde{Y}^n, \tilde{Z}^n) \sim \prod_{i=1}^n p_U(\tilde{u}_i)p_X(\tilde{x}_i)p_Y(\tilde{y}_i)p_Z(\tilde{z}_i)$, then

$$P\{(\tilde{U}^n, \tilde{X}^n, \tilde{Y}^n, \tilde{Z}^n) \in \mathcal{T}_\epsilon^{(n)}(U, X, Y, Z)\} \doteq 2^{-n(I(U;X)+I(U,X;Y)+I(U,X,Y;Z))}.$$

(b) If $(\tilde{U}^n, \tilde{X}^n, \tilde{Y}^n, \tilde{Z}^n) \sim \prod_{i=1}^n p_{U,X}(\tilde{u}_i, \tilde{x}_i)p_{Y|X}(\tilde{y}_i|\tilde{x}_i)p_Z(\tilde{z}_i)$, then

$$P\{(\tilde{U}^n, \tilde{X}^n, \tilde{Y}^n, \tilde{Z}^n) \in \mathcal{T}_\epsilon^{(n)}(U, X, Y, Z)\} \doteq 2^{-n(I(U;Y|X)+I(U,X,Y;Z))}.$$

2.17. *Need for both ϵ and ϵ'.* Let (X, Y) be a pair of independent Bern(1/2) random variables. Let $k = \lfloor (n/2)(1 + \epsilon) \rfloor$ and x^n be a binary sequence with k ones followed by $(n - k)$ zeros.

(a) Check that $x^n \in \mathcal{T}_\epsilon^{(n)}(X)$.

(b) Let Y^n be an i.i.d. Bern(1/2) sequence, independent of x^n. Show that

$$P\{(x^n, Y^n) \in \mathcal{T}_\epsilon^{(n)}(X, Y)\} \le P\left\{\sum_{i=1}^k Y_i < (k + 1)/2\right\},$$

which converges to 1/2 as $n \to \infty$. Thus, the fact that $x^n \in \mathcal{T}_\epsilon^{(n)}(X)$ and $Y^n \sim \prod_{i=1}^n p_{Y|X}(y_i|x_i)$ does not necessarily imply that $P\{(x^n, Y^n) \in \mathcal{T}_\epsilon^{(n)}(X, Y)\}$.

Remark: This problem illustrates that in general we need $\epsilon > \epsilon'$ in the conditional typicality lemma.

APPENDIX 2A   PROOF OF THE CONDITIONAL TYPICALITY LEMMA

We wish to show that

$$\lim_{n\to\infty} \mathsf{P}\{|\pi(x,y|x^n,Y^n) - p(x,y)| > \epsilon p(x,y) \text{ for some } (x,y) \in \mathcal{X} \times \mathcal{Y}\} = 0.$$

For $x \in \mathcal{X}$ such that $p(x) \neq 0$, consider

$$\mathsf{P}\{|\pi(x,y|x^n,Y^n) - p(x,y)| > \epsilon p(x,y)\}$$

$$= \mathsf{P}\left\{\left|\frac{\pi(x,y|x^n,Y^n)}{p(x)} - p(y|x)\right| > \epsilon p(y|x)\right\}$$

$$= \mathsf{P}\left\{\left|\frac{\pi(x,y|x^n,Y^n)\pi(x|x^n)}{p(x)\pi(x|x^n)} - p(y|x)\right| > \epsilon p(y|x)\right\}$$

$$= \mathsf{P}\left\{\left|\frac{\pi(x,y|x^n,Y^n)}{\pi(x|x^n)p(y|x)} \cdot \frac{\pi(x|x^n)}{p(x)} - 1\right| > \epsilon\right\}$$

$$\leq \mathsf{P}\left\{\frac{\pi(x,y|x^n,Y^n)}{\pi(x|x^n)p(y|x)} \cdot \frac{\pi(x|x^n)}{p(x)} > 1 + \epsilon\right\} + \mathsf{P}\left\{\frac{\pi(x,y|x^n,Y^n)}{\pi(x|x^n)p(y|x)} \cdot \frac{\pi(x|x^n)}{p(x)} < 1 - \epsilon\right\}.$$

Now, since $x^n \in \mathcal{T}_{\epsilon'}^{(n)}(X)$, $1 - \epsilon' \leq \pi(x|x^n)/p(x) \leq 1 + \epsilon'$,

$$\mathsf{P}\left\{\frac{\pi(x,y|x^n,Y^n)}{\pi(x|x^n)p(y|x)} \cdot \frac{\pi(x|x^n)}{p(x)} > 1 + \epsilon\right\} \leq \mathsf{P}\left\{\frac{\pi(x,y|x^n,Y^n)}{\pi(x|x^n)} > \frac{1+\epsilon}{1+\epsilon'}p(y|x)\right\} \qquad (2.11)$$

and

$$\mathsf{P}\left\{\frac{\pi(x,y|x^n,Y^n)}{\pi(x|x^n)p(y|x)} \cdot \frac{\pi(x|x^n)}{p(x)} < 1 - \epsilon\right\} \leq \mathsf{P}\left\{\frac{\pi(x,y|x^n,Y^n)}{\pi(x|x^n)} < \frac{1-\epsilon}{1-\epsilon'}p(y|x)\right\}. \qquad (2.12)$$

Since $\epsilon' < \epsilon$, we have $(1 + \epsilon)/(1 + \epsilon') > 1$ and $(1 - \epsilon)/(1 - \epsilon') < 1$. Furthermore, since Y^n is generated according to the correct conditional pmf, by the LLN, for every $y \in \mathcal{Y}$,

$$\frac{\pi(x,y|x^n,Y^n)}{\pi(x|x^n)} \to p(y|x) \quad \text{in probability.}$$

Hence, both upper bounds in (2.11) and (2.12) tend to zero as $n \to \infty$, which, by the union of events bound over all $(x,y) \in \mathcal{X} \times \mathcal{Y}$, completes the proof of the conditional typicality lemma.

CHAPTER 3

Point-to-Point Information Theory

We review Shannon's basic theorems for point-to-point communication. Over the course of the review, we introduce the techniques of random coding and joint typicality encoding and decoding, and develop the packing and covering lemmas. These techniques will be used in the achievability proofs for multiple sources and channels throughout the book. We rigorously show how achievability for a discrete memoryless channel or source can be extended to its Gaussian counterpart. We also show that under our definition of typicality, the lossless source coding theorem is a corollary of the lossy source coding theorem. This fact will prove useful in later chapters. Along the way, we point out some key differences between results for point-to-point communication and for the multiuser networks discussed in subsequent chapters.

3.1 CHANNEL CODING

Consider the point-to-point communication system model depicted in Figure 3.1 in which a sender wishes to reliably communicate a message M at a rate R bits per transmission to a receiver over a noisy communication channel (or a noisy storage medium). Toward this end, the sender encodes the message into a codeword X^n and transmits it over the channel in n time instances (also referred to as transmissions or channel uses). Upon receiving the noisy sequence Y^n, the receiver decodes it to obtain the estimate $\hat{M}$ of the message. The channel coding problem is to find the channel capacity, which is the highest rate R such that the probability of decoding error can be made to decay asymptotically to zero with the code block length n.

We first consider the channel coding problem for a simple *discrete memoryless channel* (DMC) model $(\mathcal{X}, p(y|x), \mathcal{Y})$ (in short $p(y|x)$) that consists of a finite input set (or alphabet) $\mathcal{X}$, a finite output set $\mathcal{Y}$, and a collection of conditional pmfs $p(y|x)$ on $\mathcal{Y}$ for every $x \in \mathcal{X}$. Thus, if an input symbol $x \in \mathcal{X}$ is transmitted, the probability of receiving an output symbol $y \in \mathcal{Y}$ is $p(y|x)$. The channel is stationary and memoryless in the

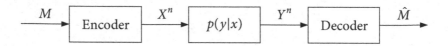

Figure 3.1. Point-to-point communication system.

sense that when it is used n times with message M drawn from an arbitrary set and input $X^n \in \mathcal{X}^n$, the output $Y_i \in \mathcal{Y}$ at time $i \in [1:n]$ given (M, X^i, Y^{i-1}) is distributed according to $p(y_i|x^i, y^{i-1}, m) = p_{Y|X}(y_i|x_i)$. Throughout the book, the phrase "discrete memoryless (DM)" will refer to "finite-alphabet and stationary memoryless".

A $(2^{nR}, n)$ code for the DMC $p(y|x)$ consists of

- a message set $[1:2^{nR}] = \{1, 2, \ldots, 2^{\lceil nR \rceil}\}$,

- an encoding function (encoder) $x^n \colon [1:2^{nR}] \to \mathcal{X}^n$ that assigns a *codeword* $x^n(m)$ to each message $m \in [1:2^{nR}]$, and

- a decoding function (decoder) $\hat{m} \colon \mathcal{Y}^n \to [1:2^{nR}] \cup \{e\}$ that assigns an estimate $\hat{m} \in [1:2^{nR}]$ or an error message e to each received sequence y^n.

Note that under the above definition of a $(2^{nR}, n)$ code, the memoryless property implies that

$$p(y^n|x^n, m) = \prod_{i=1}^{n} p_{Y|X}(y_i|x_i). \tag{3.1}$$

The set $\mathcal{C} = \{x^n(1), x^n(2), \ldots, x^n(2^{\lceil nR \rceil})\}$ is referred to as the *codebook* associated with the $(2^{nR}, n)$ code. We assume that the message is uniformly distributed over the message set, i.e., $M \sim \mathrm{Unif}[1:2^{nR}]$.

The performance of a given code is measured by the probability that the estimate of the message is different from the actual message sent. More precisely, let $\lambda_m(\mathcal{C}) = \mathrm{P}\{\hat{M} \neq m \mid M = m\}$ be the conditional probability of error given that message m is sent. Then, the *average probability of error* for a $(2^{nR}, n)$ code is defined as

$$P_e^{(n)}(\mathcal{C}) = \mathrm{P}\{\hat{M} \neq M\} = \frac{1}{2^{\lceil nR \rceil}} \sum_{m=1}^{2^{\lceil nR \rceil}} \lambda_m(\mathcal{C}).$$

A rate R is said to be *achievable* if there exists a sequence of $(2^{nR}, n)$ codes such that $\lim_{n \to \infty} P_e^{(n)}(\mathcal{C}) = 0$. The *capacity* C of a DMC is the supremum over all achievable rates.

Remark 3.1. Although the message M depends on the block length n (through the message set), we will not show this dependency explicitly. Also, from this point on, we will not explicitly show the dependency of the probability of error $P_e^{(n)}$ on the codebook $\mathcal{C}$.

3.1.1 Channel Coding Theorem

Shannon established a simple characterization of channel capacity.

Theorem 3.1 (Channel Coding Theorem). The capacity of the discrete memoryless channel $p(y|x)$ is given by the information capacity formula

$$C = \max_{p(x)} I(X; Y).$$

In the following, we evaluate the information capacity formula for several simple but important discrete memoryless channels.

Example 3.1 (Binary symmetric channel). Consider the *binary symmetric channel* with *crossover probability p* (in short BSC(p)) depicted in Figure 3.2. The channel input X and output Y are binary and each binary input symbol is flipped with probability p. Equivalently, we can specify the BSC as $Y = X \oplus Z$, where the noise $Z \sim \text{Bern}(p)$ is independent of the input X. The capacity is

$$C = \max_{p(x)} I(X; Y)$$

$$= \max_{p(x)} (H(Y) - H(Y|X))$$

$$= \max_{p(x)} (H(Y) - H(X \oplus Z|X))$$

$$= \max_{p(x)} (H(Y) - H(Z|X))$$

$$\overset{(a)}{=} \max_{p(x)} H(Y) - H(Z)$$

$$= 1 - H(p),$$

where (a) follows by the independence of X and Z. Note that the capacity is attained by $X \sim \text{Bern}(1/2)$, which, in turn, results in $Y \sim \text{Bern}(1/2)$.

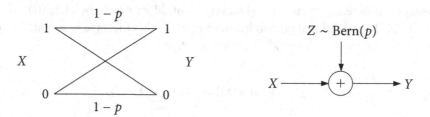

Figure 3.2. Equivalent representations of the binary symmetric channel BSC(p).

Example 3.2 (Binary erasure channel). Consider the *binary erasure channel* with *erasure probability p* (BEC(p)) depicted in Figure 3.3. The channel input X and output Y are binary and each binary input symbol is erased (mapped into an erasure symbol e) with probability p. Thus, the receiver knows which transmissions are erased, but the sender does not. The capacity is

$$C = \max_{p(x)} (H(X) - H(X|Y))$$

$$\overset{(a)}{=} \max_{p(x)} (H(X) - pH(X))$$

$$= 1 - p,$$

where (a) follows since $H(X|Y = y) = 0$ if $y = 0$ or 1, and $H(X|Y = e) = H(X)$. The capacity is again attained by $X \sim \text{Bern}(1/2)$.

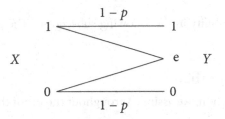

Figure 3.3. Binary erasure channel BEC(p).

Example 3.3 (Product DMC). Let $p(y_1|x_1)$ and $p(y_2|x_2)$ be two DMCs with capacities C_1 and C_2, respectively. The *product DMC* is a DMC $(\mathcal{X}_1 \times \mathcal{X}_2, p(y_1|x_1)p(y_2|x_2), \mathcal{Y}_1 \times \mathcal{Y}_2)$ in which the symbols $x_1 \in \mathcal{X}_1$ and $x_2 \in \mathcal{X}_2$ are sent simultaneously in parallel and the received outputs Y_1 and Y_2 are distributed according to $p(y_1, y_2|x_1, x_2) = p(y_1|x_1)p(y_2|x_2)$. The capacity of the product DMC is

$$
\begin{aligned}
C &= \max_{p(x_1,x_2)} I(X_1, X_2; Y_1, Y_2) \\
&= \max_{p(x_1,x_2)} \left(I(X_1, X_2; Y_1) + I(X_1, X_2; Y_2|Y_1) \right) \\
&\overset{(a)}{=} \max_{p(x_1,x_2)} \left(I(X_1; Y_1) + I(X_2; Y_2) \right) \\
&= \max_{p(x_1)} I(X_1; Y_1) + \max_{p(x_2)} I(X_2; Y_2) \\
&= C_1 + C_2,
\end{aligned}
$$

where (a) follows since $Y_1 \to X_1 \to X_2 \to Y_2$ form a Markov chain, which implies that $I(X_1, X_2; Y_1) = I(X_1; Y_1)$ and $I(X_1, X_2; Y_2|Y_1) \le I(Y_1, X_1, X_2; Y_2) = I(X_2; Y_2)$ with equality iff X_1 and X_2 are independent.

More generally, let $p(y_j|x_j)$ be a DMC with capacity C_j for $j \in [1:d]$. A product DMC consists of an input alphabet $\mathcal{X} = \times_{j=1}^{d} \mathcal{X}_j$, an output alphabet $\mathcal{Y} = \times_{j=1}^{d} \mathcal{Y}_j$, and a collection of conditional pmfs $p(y_1, \ldots, y_d|x_1, \ldots, x_d) = \prod_{j=1}^{d} p(y_j|x_j)$. The capacity of the product DMC is

$$
C = \sum_{j=1}^{d} C_j.
$$

To prove the channel coding theorem, we need to show that the *information* capacity in Theorem 3.1 is equal to the *operational* capacity defined in the channel coding setup. This involves the verification of two statements.

- **Achievability.** For every rate $R < C = \max_{p(x)} I(X; Y)$, there exists a sequence of $(2^{nR}, n)$ codes with average probability of error $P_e^{(n)}$ that tends to zero as $n \to \infty$. The proof of achievability uses random coding and joint typicality decoding.

- **Converse.** For every sequence of $(2^{nR}, n)$ codes with probability of error $P_e^{(n)}$ that tends to zero as $n \to \infty$, the rate must satisfy $R \le C = \max_{p(x)} I(X; Y)$. The proof of the converse uses Fano's inequality and basic properties of mutual information.

We first prove achievability in the following subsection. The proof of the converse is given in Section 3.1.4.

3.1.2   Proof of Achievability

For simplicity of presentation, we assume throughout the proof that nR is an integer.

Random codebook generation. We use random coding. Fix the pmf $p(x)$ that attains the information capacity C. Randomly and independently generate 2^{nR} sequences $x^n(m)$, $m \in [1 : 2^{nR}]$, each according to $p(x^n) = \prod_{i=1}^{n} p_X(x_i)$. The generated sequences constitute the codebook C. Thus

$$p(C) = \prod_{m=1}^{2^{nR}} \prod_{i=1}^{n} p_X(x_i(m)).$$

The chosen codebook C is revealed to both the encoder and the decoder before transmission commences.

Encoding. To send message $m \in [1 : 2^{nR}]$, transmit $x^n(m)$.

Decoding. We use *joint typicality decoding*. Let y^n be the received sequence. The receiver declares that $\hat{m} \in [1 : 2^{nR}]$ is sent if it is the unique message such that $(x^n(\hat{m}), y^n) \in T_\epsilon^{(n)}$; otherwise—if there is none or more than one such message—it declares an error e.

Analysis of the probability of error. Assuming that message m is sent, the decoder makes an error if $(x^n(m), y^n) \notin T_\epsilon^{(n)}$ or if there is another message $m' \neq m$ such that $(x^n(m'), y^n) \in T_\epsilon^{(n)}$. Consider the probability of error averaged over M and codebooks

$$P(\mathcal{E}) = \mathsf{E}_C\left(P_e^{(n)}\right)$$

$$= \mathsf{E}_C\left(\frac{1}{2^{nR}} \sum_{m=1}^{2^{nR}} \lambda_m(C)\right)$$

$$= \frac{1}{2^{nR}} \sum_{m=1}^{2^{nR}} \mathsf{E}_C(\lambda_m(C))$$

$$\stackrel{(a)}{=} \mathsf{E}_C(\lambda_1(C))$$

$$= P(\mathcal{E} \mid M = 1),$$

where (a) follows by the symmetry of the random codebook generation. Thus, we assume without loss of generality that $M = 1$ is sent. For brevity, we do not explicitly condition on the event $\{M = 1\}$ in probability expressions whenever it is clear from the context.

The decoder makes an error iff one or both of the following events occur:

$$\mathcal{E}_1 = \{(X^n(1), Y^n) \notin T_\epsilon^{(n)}\},$$

$$\mathcal{E}_2 = \{(X^n(m), Y^n) \in T_\epsilon^{(n)} \text{ for some } m \neq 1\}.$$

Thus, by the union of events bound,

$$P(\mathcal{E}) = P(\mathcal{E}_1 \cup \mathcal{E}_2) \leq P(\mathcal{E}_1) + P(\mathcal{E}_2).$$

We now bound each term. By the law of large numbers (LLN), the first term $P(\mathcal{E}_1)$ tends to zero as $n \to \infty$. For the second term, since for $m \ne 1$,

$$(X^n(m), X^n(1), Y^n) \sim \prod_{i=1}^{n} p_X(x_i(m))p_{X,Y}(x_i(1), y_i),$$

we have $(X^n(m), Y^n) \sim \prod_{i=1}^{n} p_X(x_i(m))p_Y(y_i)$. Thus, by the extension of the joint typicality lemma in Remark 2.2,

$$P\{(X^n(m), Y^n) \in \mathcal{T}_\epsilon^{(n)}\} \le 2^{-n(I(X;Y)-\delta(\epsilon))} = 2^{-n(C-\delta(\epsilon))}.$$

Again by the union of events bound,

$$P(\mathcal{E}_2) \le \sum_{m=2}^{2^{nR}} P\{(X^n(m), Y^n) \in \mathcal{T}_\epsilon^{(n)}\} \le \sum_{m=2}^{2^{nR}} 2^{-n(C-\delta(\epsilon))} \le 2^{-n(C-R-\delta(\epsilon))},$$

which tends to zero as $n \to \infty$ if $R < C - \delta(\epsilon)$.

Note that since the probability of error averaged over codebooks, $P(\mathcal{E})$, tends to zero as $n \to \infty$, there must exist a sequence of $(2^{nR}, n)$ codes such that $\lim_{n\to\infty} P_e^{(n)} = 0$, which proves that $R < C - \delta(\epsilon)$ is achievable. Finally, taking $\epsilon \to 0$ completes the proof.

Remark 3.2. To bound the average probability of error $P(\mathcal{E})$, we divided the error event into two events, each of which comprises events with $(X^n(m), Y^n)$ having the same joint pmf. This observation will prove useful when we analyze more complex error events in later chapters.

Remark 3.3. By the Markov inequality, the probability of error for a random codebook, that is, a codebook consisting of random sequences $X^n(m)$, $m \in [1 : 2^{nR}]$, tends to zero as $n \to \infty$ in probability. Hence, most codebooks are good in terms of the error probability.

Remark 3.4. The capacity with the *maximal* probability of error $\lambda^* = \max_m \lambda_m$ is equal to that with the average probability of error $P_e^{(n)}$. This can be shown by discarding the worst half of the codewords (in terms of error probability) from each code in the sequence of $(2^{nR}, n)$ codes with $\lim_{n\to\infty} P_e^{(n)} = 0$. The maximal probability of error for each of the codes with the remaining codewords is at most $2P_e^{(n)}$, which again tends to zero as $n \to \infty$. As we will see, the capacity with maximal probability of error is not always equal to that with average probability of error for multiuser channels.

Remark 3.5. Depending on the structure of the channel, the rate $R = C$ may or may not be achievable. We will sometimes informally say that C is achievable to mean that every $R < C$ is achievable.

3.1.3 Achievability Using Linear Codes

Recall that in the achievability proof, we used only *pairwise* independence of codewords $X^n(m)$, $m \in [1 : 2^{nR}]$, rather than mutual independence among all of them. This observation has an interesting consequence—the capacity of a BSC can be achieved using *linear* codes.

Consider a BSC(p). Let $k = \lceil nR \rceil$ and $(u_1, u_2, \ldots, u_k) \in \{0, 1\}^k$ be the binary expansion of the message $m \in [1 : 2^k - 1]$. Generate a random codebook such that each codeword $x^n(u^k)$ is a linear function of u^k (in binary field arithmetic). In particular, let

$$
\begin{bmatrix} x_1 \\ x_2 \\ \vdots \\ x_n \end{bmatrix} = \begin{bmatrix} g_{11} & g_{12} & \cdots & g_{1k} \\ g_{21} & g_{22} & \cdots & g_{2k} \\ \vdots & \vdots & \ddots & \vdots \\ g_{n1} & g_{n2} & \cdots & g_{nk} \end{bmatrix} \begin{bmatrix} u_1 \\ u_2 \\ \vdots \\ u_k \end{bmatrix},
$$

where $g_{ij} \in \{0, 1\}$, $i \in [1 : n]$, $j \in [1 : k]$, are generated i.i.d. according to Bern($1/2$).

Now we can easily check that $X_1(u^k), \ldots, X_n(u^k)$ are i.i.d. Bern($1/2$) for each $u^k \neq 0$, and $X^n(u^k)$ and $X^n(\tilde{u}^k)$ are independent for each $u^k \neq \tilde{u}^k$. Therefore, using the same steps as in the proof of achievability for the channel coding theorem, it can be shown that the error probability of joint typicality decoding tends to zero as $n \to \infty$ if $R < 1 - H(p) - \delta(\epsilon)$. This shows that for a BSC there exists not only a good sequence of codes, but also a good sequence of *linear* codes.

It can be similarly shown that random linear codes achieve the capacity of the binary erasure channel, or more generally, channels for which the input alphabet is a finite field and the information capacity is attained by the uniform pmf.

3.1.4   Proof of the Converse

We need to show that for every sequence of $(2^{nR}, n)$ codes with $\lim_{n \to \infty} P_e^{(n)} = 0$, we must have $R \leq C = \max_{p(x)} I(X; Y)$. Again for simplicity of presentation, we assume that nR is an integer. Every $(2^{nR}, n)$ code induces a joint pmf on (M, X^n, Y^n) of the form

$$
p(m, x^n, y^n) = 2^{-nR} p(x^n | m) \prod_{i=1}^{n} p_{Y|X}(y_i | x_i).
$$

By Fano's inequality,

$$
H(M | \hat{M}) \leq 1 + P_e^{(n)} nR = n\epsilon_n,
$$

where ϵ_n tends to zero as $n \to \infty$ by the assumption that $\lim_{n \to \infty} P_e^{(n)} = 0$. Thus, by the data processing inequality,

$$
H(M | Y^n) \leq H(M | \hat{M}) \leq n\epsilon_n. \tag{3.2}
$$

Now consider

$$
\begin{aligned}
nR &= H(M) \\
&= I(M; Y^n) + H(M | Y^n) \\
&\overset{(a)}{\leq} I(M; Y^n) + n\epsilon_n \\
&= \sum_{i=1}^{n} I(M; Y_i | Y^{i-1}) + n\epsilon_n
\end{aligned}
$$

$$\leq \sum_{i=1}^{n} I(M, Y^{i-1}; Y_i) + n\epsilon_n$$

$$\overset{(b)}{=} \sum_{i=1}^{n} I(X_i, M, Y^{i-1}; Y_i) + n\epsilon_n$$

$$\overset{(c)}{=} \sum_{i=1}^{n} I(X_i; Y_i) + n\epsilon_n$$

$$\leq nC + n\epsilon_n, \tag{3.3}$$

where (a) follows from (3.2), (b) follows since X_i is a function of M, and (c) follows since the channel is memoryless, which implies that $(M, Y^{i-1}) \to X_i \to Y_i$ form a Markov chain. The last inequality follows by the definition of the information capacity. Since ϵ_n tends to zero as $n \to \infty$, $R \leq C$, which completes the proof of the converse.

3.1.5 DMC with Feedback

Consider the DMC with *noiseless causal feedback* depicted in Figure 3.4. The encoder assigns a symbol $x_i(m, y^{i-1})$ to each message $m \in [1 : 2^{nR}]$ and past received output sequence $y^{i-1} \in \mathcal{Y}^{i-1}$ for $i \in [1 : n]$. Hence (3.1) does not hold in general and a $(2^{nR}, n)$ feedback code induces a joint pmf of the form

$$(M, X^n, Y^n) \sim p(m, x^n, y^n) = 2^{-nR} \prod_{i=1}^{n} p(x_i | m, y^{i-1}) p_{Y|X}(y_i | x_i).$$

Nonetheless, it can be easily shown that the chain of inequalities (3.3) continues to hold in the presence of such causal feedback. Hence, feedback *does not* increase the capacity of the DMC. In Chapter 17 we will discuss the role of feedback in communication in more detail.

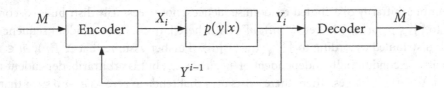

Figure 3.4. DMC with noiseless causal feedback.

3.2 PACKING LEMMA

The packing lemma generalizes the bound on the probability of the decoding error event $\mathcal{E}_2$ in the achievability proof of the channel coding theorem; see Section 3.1.2. The lemma will be used in the achievability proofs of many multiuser source and channel coding theorems.

Recall that in the bound on $P(\mathcal{E}_2)$, we had a fixed input pmf $p(x)$ and a DMC $p(y|x)$. As illustrated in Figure 3.5, we considered a set of $(2^{nR} - 1)$ i.i.d. codewords $X^n(m)$,

$m \in [2 : 2^{nR}]$, each distributed according to $\prod_{i=1}^{n} p_X(x_i)$, and an output sequence $\tilde{Y}^n \sim \prod_{i=1}^{n} p_Y(\tilde{y}_i)$ generated by the codeword $X^n(1) \sim \prod_{i=1}^{n} p_X(x_i)$, which is independent of the set of codewords. We showed that the probability that $(X^n(m), \tilde{Y}^n) \in \mathcal{T}_\epsilon^{(n)}$ for some $m \in [2 : 2^{nR}]$ tends to zero as $n \to \infty$ if $R < I(X; Y) - \delta(\epsilon)$.

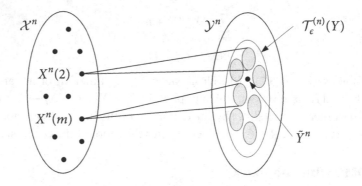

Figure 3.5. Illustration of the setup for the bound on $P(\mathcal{E}_2)$.

The following lemma extends this bound in three ways:

1. The codewords that are independent of $\tilde{Y}^n$ need not be mutually independent.

2. The sequence $\tilde{Y}^n$ can have an arbitrary pmf (not necessarily $\prod_{i=1}^{n} p_Y(\tilde{y}_i)$).

3. The sequence $\tilde{Y}^n$ and the set of codewords are conditionally independent given U^n that has a general joint pmf with $\tilde{Y}^n$.

Lemma 3.1 (Packing lemma). Let $(U, X, Y) \sim p(u, x, y)$. Let $(\tilde{U}^n, \tilde{Y}^n) \sim p(\tilde{u}^n, \tilde{y}^n)$ be a pair of arbitrarily distributed random sequences, not necessarily distributed according to $\prod_{i=1}^{n} p_{U,Y}(\tilde{u}_i, \tilde{y}_i)$. Let $X^n(m)$, $m \in \mathcal{A}$, where $|\mathcal{A}| \leq 2^{nR}$, be random sequences, each distributed according to $\prod_{i=1}^{n} p_{X|U}(x_i|\tilde{u}_i)$. Further assume that $X^n(m)$, $m \in \mathcal{A}$, is pairwise conditionally independent of $\tilde{Y}^n$ given $\tilde{U}^n$, but is arbitrarily dependent on other $X^n(m)$ sequences. Then, there exists $\delta(\epsilon)$ that tends to zero as $\epsilon \to 0$ such that

$$\lim_{n \to \infty} P\{(\tilde{U}^n, X^n(m), \tilde{Y}^n) \in \mathcal{T}_\epsilon^{(n)} \text{ for some } m \in \mathcal{A}\} = 0,$$

if $R < I(X; Y|U) - \delta(\epsilon)$.

Note that the packing lemma can be readily applied to the linear coding case where the $X^n(m)$ sequences are only pairwise independent. We will later encounter cases for which $U \neq \emptyset$ and $(\tilde{U}^n, \tilde{Y}^n)$ is not generated i.i.d.

Proof. Define the events

$$\tilde{\mathcal{E}}_m = \{(\tilde{U}^n, X^n(m), \tilde{Y}^n) \in \mathcal{T}_\epsilon^{(n)}\} \quad \text{for } m \in \mathcal{A}.$$

By the union of events bound, the probability of the event of interest can be bounded as

$$P\left(\bigcup_{m\in\mathcal{A}}\tilde{\mathcal{E}}_m\right) \le \sum_{m\in\mathcal{A}} P(\tilde{\mathcal{E}}_m).$$

Now consider

$$
\begin{aligned}
P(\tilde{\mathcal{E}}_m) &= P\{(\tilde{U}^n, X^n(m), \tilde{Y}^n) \in \mathcal{T}_\epsilon^{(n)}(U, X, Y)\}\\
&= \sum_{(\tilde{u}^n,\tilde{y}^n)\in\mathcal{T}_\epsilon^{(n)}} p(\tilde{u}^n, \tilde{y}^n)\, P\{(\tilde{u}^n, X^n(m), \tilde{y}^n) \in \mathcal{T}_\epsilon^{(n)}(U, X, Y) \mid \tilde{U}^n = \tilde{u}^n, \tilde{Y}^n = \tilde{y}^n\}\\
&\overset{(a)}{=} \sum_{(\tilde{u}^n,\tilde{y}^n)\in\mathcal{T}_\epsilon^{(n)}} p(\tilde{u}^n, \tilde{y}^n)\, P\{(\tilde{u}^n, X^n(m), \tilde{y}^n) \in \mathcal{T}_\epsilon^{(n)}(U, X, Y) \mid \tilde{U}^n = \tilde{u}^n\}\\
&\overset{(b)}{\le} \sum_{(\tilde{u}^n,\tilde{y}^n)\in\mathcal{T}_\epsilon^{(n)}} p(\tilde{u}^n, \tilde{y}^n) 2^{-n(I(X;Y|U)-\delta(\epsilon))}\\
&\le 2^{-n(I(X;Y|U)-\delta(\epsilon))},
\end{aligned}
$$

where (a) follows by the conditional independence of $X^n(m)$ and $\tilde{Y}^n$ given $\tilde{U}^n$, and (b) follows by the joint typicality lemma in Section 2.5 since $(\tilde{u}^n, \tilde{y}^n) \in \mathcal{T}_\epsilon^{(n)}$ and $X^n(m) \mid \{\tilde{U}^n = \tilde{u}^n, \tilde{Y}^n = \tilde{y}^n\} \sim \prod_{i=1}^n p_{X|U}(x_i|\tilde{u}_i)$. Hence

$$\sum_{m\in\mathcal{A}} P(\tilde{\mathcal{E}}_m) \le |\mathcal{A}| 2^{-n(I(X;Y|U)-\delta(\epsilon))} \le 2^{-n(I(X;Y|U)-R-\delta(\epsilon))},$$

which tends to zero as $n \to \infty$ if $R < I(X; Y|U) - \delta(\epsilon)$. This completes the proof of the packing lemma.

3.3 CHANNEL CODING WITH INPUT COST

Consider a DMC $p(y|x)$. Suppose that there is a nonnegative cost function $b(x)$ associated with each input symbol $x \in \mathcal{X}$. Assume without loss of generality that there exists a zero-cost symbol $x_0 \in \mathcal{X}$, i.e., $b(x_0) = 0$. We further assume an average input cost constraint

$$\sum_{i=1}^n b(x_i(m)) \le nB \quad \text{for every } m \in [1:2^{nR}],$$

(in short, average cost constraint B on X). Now, defining the channel capacity of the DMC with cost constraint B, or the *capacity–cost function*, $C(B)$ in a similar manner to capacity without cost constraint, we can establish the following extension of the channel coding theorem.

Theorem 3.2. The capacity of the DMC $p(y|x)$ with average cost constraint B on X is

$$C(B) = \max_{p(x):\, \mathrm{E}(b(X))\le B} I(X; Y).$$

Note that $C(B)$ is nondecreasing, concave, and continuous in B.

Proof of achievability. The proof involves a minor change to the proof of achievability for the case with no cost constraint in Section 3.1.2 to ensure that every codeword satisfies the cost constraint.

Fix the pmf $p(x)$ that attains $C(B/(1 + \epsilon))$. Randomly and independently generate 2^{nR} sequences $x^n(m)$, $m \in [1 : 2^{nR}]$, each according to $\prod_{i=1}^{n} p_X(x_i)$. To send message m, the encoder transmits $x^n(m)$ if $x^n(m) \in \mathcal{T}_\epsilon^{(n)}$, and consequently, by the typical average lemma in Section 2.4, the sequence satisfies the cost constraint $\sum_{i=1}^{n} b(x_i(m)) \le nB$. Otherwise, it transmits $(x_0, \ldots, x_0)$. The analysis of the average probability of error for joint typicality decoding follows similar lines to the case without cost constraint. Assume $M = 1$. For the probability of the first error event,

$$P(\mathcal{E}_1) = P\{(X^n(1), Y^n) \notin \mathcal{T}_\epsilon^{(n)}\}$$

$$= P\{X^n(1) \in \mathcal{T}_\epsilon^{(n)}, (X^n(1), Y^n) \notin \mathcal{T}_\epsilon^{(n)}\} + P\{X^n(1) \notin \mathcal{T}_\epsilon^{(n)}, (X^n(1), Y^n) \notin \mathcal{T}_\epsilon^{(n)}\}$$

$$\le \sum_{x^n \in \mathcal{T}_\epsilon^{(n)}} \prod_{i=1}^{n} p_X(x_i) \sum_{y^n \notin \mathcal{T}_\epsilon^{(n)}(Y|x^n)} \prod_{i=1}^{n} p_{Y|X}(y_i|x_i) + P\{X^n(1) \notin \mathcal{T}_\epsilon^{(n)}\}$$

$$\le \sum_{(x^n, y^n) \notin \mathcal{T}_\epsilon^{(n)}} \prod_{i=1}^{n} p_X(x_i) p_{Y|X}(y_i|x_i) + P\{X^n(1) \notin \mathcal{T}_\epsilon^{(n)}\}.$$

Thus, by the LLN for each term, $P(\mathcal{E}_1)$ tends to zero as $n \to \infty$. The probability of the second error event, $P(\mathcal{E}_2)$, is upper bounded in exactly the same manner as when there is no cost constraint. Hence, every rate $R < I(X; Y) = C(B/(1 + \epsilon))$ is achievable. Finally, by the continuity of $C(B)$ in B, $C(B/(1 + \epsilon))$ converges to $C(B)$ as $\epsilon \to 0$, which implies the achievability of every rate $R < C(B)$.

Proof of the converse. Consider a sequence of $(2^{nR}, n)$ codes with $\lim_{n \to \infty} P_e^{(n)} = 0$ such that for every n, the cost constraint $\sum_{i=1}^{n} b(x_i(m)) \le nB$ is satisfied for every $m \in [1 : 2^{nR}]$ and thus $\sum_{i=1}^{n} E[b(X_i)] = \sum_{i=1}^{n} E_M[b(x_i(M))] \le nB$. As before, by Fano's inequality and the data processing inequality,

$$nR \le \sum_{i=1}^{n} I(X_i; Y_i) + n\epsilon_n$$

$$\overset{(a)}{\le} \sum_{i=1}^{n} C(E[b(X_i)]) + n\epsilon_n$$

$$\overset{(b)}{\le} nC\left(\frac{1}{n} \sum_{i=1}^{n} E[b(X_i)]\right) + n\epsilon_n \tag{3.4}$$

$$\overset{(c)}{\le} nC(B) + n\epsilon_n,$$

where (a) follows by the definition of $C(B)$, (b) follows by the concavity of $C(B)$, and (c) follows by the monotonicity of $C(B)$. This completes the proof of Theorem 3.2.

3.4 GAUSSIAN CHANNEL

Consider the discrete-time additive white Gaussian noise channel model depicted in Figure 3.6. The channel output corresponding to the input X is

$$Y = gX + Z, \tag{3.5}$$

where g is the *channel gain*, or *path loss*, and $Z \sim N(0, N_0/2)$ is the noise. Thus, in transmission time $i \in [1 : n]$, the channel output is

$$Y_i = gX_i + Z_i,$$

where $\{Z_i\}$ is a white Gaussian noise process with average power $N_0/2$ (in short, $\{Z_i\}$ is a WGN($N_0/2$) process), independent of the channel input $X^n = x^n(M)$. We assume an *average transmission power constraint*

$$\sum_{i=1}^{n} x_i^2(m) \leq nP \quad \text{for every } m \in [1 : 2^{nR}]$$

(in short, average power constraint P on X). The Gaussian channel is quite popular because it provides a simple model for several real-world communication channels, such as wireless and digital subscriber line (DSL) channels. We will later study more sophisticated models for these channels.

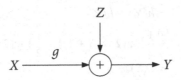

Figure 3.6. Additive white Gaussian noise channel.

We assume without loss of generality that $N_0/2 = 1$ (since one can define an equivalent Gaussian channel by dividing both sides of (3.5) by $\sqrt{N_0/2}$) and label the received power (which is now equal to the received *signal-to-noise ratio* (SNR)) g^2P as S. Note that the Gaussian channel is an example of the channel with cost discussed in the previous section, but with continuous (instead of finite) alphabets. Nonetheless, its capacity under power constraint P can be defined in the exact same manner as for the DMC with cost constraint.

Remark 3.6. If causal feedback from the receiver to the sender is present, then X_i depends only on the message M and the past received symbols Y^{i-1}. In this case X_i is *not* in general independent of the noise process. However, the message M and the noise process $\{Z_i\}$ are always assumed to be independent.

Remark 3.7. Since we discuss mainly additive white Gaussian noise channels, for brevity we will consistently use "Gaussian" in place of "additive white Gaussian noise."

3.4.1 Capacity of the Gaussian Channel

The capacity of the Gaussian channel is a simple function of the received SNR S.

Theorem 3.3. The capacity of the Gaussian channel is

$$C = \sup_{F(x):\mathsf{E}(X^2)\leq P} I(X;Y) = \mathsf{C}(S),$$

where $\mathsf{C}(x) = (1/2)\log(1 + x)$, $x \geq 0$, is the Gaussian capacity function.

For low SNR (small S), C grows linearly with S, while for high SNR, it grows logarithmically.

Proof of the converse. First note that the proof of the converse for the DMC with input cost constraint in Section 3.3 applies to arbitrary (not necessarily discrete) memoryless channels. Therefore, continuing the chain of inequalities in (3.4) with $b(x) = x^2$, we obtain

$$C \leq \sup_{F(x):\mathsf{E}(X^2)\leq P} I(X;Y).$$

Now for any $X \sim F(x)$ with $\mathsf{E}(X^2) \leq P$,

$$
\begin{aligned}
I(X;Y) &= h(Y) - h(Y|X) \\
&= h(Y) - h(Z|X) \\
&= h(Y) - h(Z) \\
&\stackrel{(a)}{\leq} \frac{1}{2}\log(2\pi e(S+1)) - \frac{1}{2}\log(2\pi e) \\
&= \mathsf{C}(S),
\end{aligned}
$$

where (a) follows by the maximum differential entropy lemma in Section 2.2 with $\mathsf{E}(Y^2) \leq g^2 P + 1 = S + 1$. Since this inequality becomes equality if $X \sim \mathrm{N}(0, P)$, we have shown that

$$C \leq \sup_{F(x):\mathsf{E}(X^2)\leq P} I(X;Y) = \mathsf{C}(S).$$

This completes the proof of the converse.

Proof of achievability. We extend the achievability proof for the DMC with cost constraint to show that $C \geq \mathsf{C}(S)$. Let $X \sim \mathrm{N}(0, P)$. Then, $I(X;Y) = \mathsf{C}(S)$. For every $j = 1, 2, \ldots$, let $[X]_j \in \{-j\Delta, -(j-1)\Delta, \ldots, -\Delta, 0, \Delta, \ldots, (j-1)\Delta, j\Delta\}$, $\Delta = 1/\sqrt{j}$, be a quantized version of X, obtained by mapping X to the closest quantization point $[X]_j = \hat{x}_j(X)$ such that $|[X]_j| \leq |X|$. Clearly, $\mathsf{E}([X]_j^2) \leq \mathsf{E}(X^2) = P$. Let $Y_j = g[X]_j + Z$ be the output corresponding to the input $[X]_j$ and let $[Y_j]_k = \hat{y}_k(Y_j)$ be a quantized version of Y_j defined in the same manner. Now, using the achievability proof for the DMC with cost constraint, we can show that for each j, k, any rate $R < I([X]_j; [Y_j]_k)$ is achievable for the channel with input $[X_j]$ and output $[Y_j]_k$ under power constraint P.

We now show that $I([X]_j; [Y_j]_k)$ can be made as close to $I(X; Y)$ as desired by taking j, k sufficiently large. First, by the data processing inequality,

$$I([X]_j; [Y_j]_k) \le I([X]_j; Y_j) = h(Y_j) - h(Z).$$

Since $\text{Var}(Y_j) \le S + 1$, $h(Y_j) \le h(Y)$ for all j. Thus, $I([X]_j; [Y_j]_k) \le I(X; Y)$. For the other direction, we have the following.

Lemma 3.2. $\liminf_{j \to \infty} \lim_{k \to \infty} I([X]_j; [Y_j]_k) \ge I(X; Y)$.

The proof of this lemma is given in Appendix 3A. Combining both bounds, we have

$$\lim_{j \to \infty} \lim_{k \to \infty} I([X]_j; [Y_j]_k) = I(X; Y),$$

which completes the proof of Theorem 3.3.

Remark 3.8. This discretization procedure shows how to extend the coding theorem for a DMC to a Gaussian or any other well-behaved continuous-alphabet channel. Similar procedures can be used to extend coding theorems for finite-alphabet multiuser channels to their Gaussian counterparts. Hence, in subsequent chapters we will not provide formal proofs of such extensions.

3.4.2 Minimum Energy Per Bit

In the discussion of the Gaussian channel, we assumed average power constraint P on each transmitted codeword and found the highest reliable transmission rate under this constraint. A "dual" formulation of this problem is to assume a given transmission rate R and determine the *minimum energy per bit* needed to achieve it. This formulation can be viewed as more natural since it leads to a fundamental limit on the energy needed to reliably communicate one bit of information over a Gaussian channel.

Consider a $(2^{nR}, n)$ code for the Gaussian channel. Define the average power for the code as

$$P = \frac{1}{2^{nR}} \sum_{m=1}^{2^{nR}} \frac{1}{n} \sum_{i=1}^{n} x_i^2(m),$$

and the average energy per bit for the code as $E = P/R$ (that is, the energy per transmission divided by bits per transmission).

Following similar steps to the converse proof for the Gaussian channel in the previous section, we can show that for every sequence of $(2^{nR}, n)$ codes with average power P and $\lim_{n \to \infty} P_e^{(n)} = 0$, we must have

$$R \le \frac{1}{2} \log(1 + g^2 P).$$

Substituting $P = ER$, we obtain the lower bound on the energy per bit $E \ge (2^{2R} - 1)/(g^2 R)$.

We also know that if the average power of the code is P, then any rate $R < C(g^2 P)$ is achievable. Therefore, reliable communication at rate R with energy per bit $E > (2^{2R} - 1)/(g^2 R)$ is possible. Hence, the *energy-per-bit–rate function*, that is, the minimum energy-per-bit needed for reliable communication at rate R, is

$$E_b(R) = \frac{1}{g^2 R}(2^{2R} - 1).$$

This is a monotonically increasing and strictly convex function of R (see Figure 3.7). As R tends to zero, $E_b(R)$ converges to $E_b^* = (1/g^2)2 \ln 2$, which is the minimum energy per bit needed for reliable communication over a Gaussian channel with noise power $N_0/2 = 1$ and gain g.

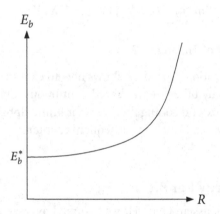

Figure 3.7. Minimum energy per bit versus transmission rate.

3.4.3   Gaussian Product Channel

The *Gaussian product channel* depicted in Figure 3.8 consists of a set of parallel Gaussian channels

$$Y_j = g_j X_j + Z_j \quad \text{for } j \in [1:d],$$

where g_j is the gain of the j-th channel component and $Z_1, Z_2, \ldots, Z_d$ are independent zero-mean Gaussian noise components with the same average power $N_0/2 = 1$. We assume an average transmission power constraint

$$\frac{1}{n} \sum_{i=1}^{n} \sum_{j=1}^{d} x_{ji}^2(m) \leq P \quad \text{for } m \in [1:2^{nR}].$$

The Gaussian product channel is a model for continuous-time (waveform) additive Gaussian noise channels; the parallel channels represent different frequency bands, time slots, or more generally, orthogonal signal dimensions.

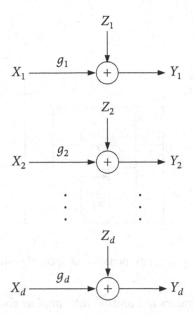

Figure 3.8. Gaussian product channel: d parallel Gaussian channels.

The capacity of the Gaussian product channel is

$$C = \max_{\substack{P_1, P_2, \ldots, P_d \\ \sum_{j=1}^{d} P_j \leq P}} \sum_{j=1}^{d} C(g_j^2 P_j). \tag{3.6}$$

The proof of the converse follows by noting that the capacity is upper bounded as

$$C \leq \sup_{F(x^d): \sum_{j=1}^{d} E(X_j^2) \leq P} I(X^d; Y^d) = \sup_{F(x^d): \sum_{j=1}^{d} E(X_j^2) \leq P} \sum_{j=1}^{d} I(X_j; Y_j)$$

and that the supremum is attained by mutually independent $X_j \sim N(0, P_j)$, $j \in [1 : d]$. For the achievability proof, note that this bound can be achieved by the discretization procedure for each component Gaussian channel. The constrained optimization problem in (3.6) is convex and can be solved by forming the Lagrangian; see Appendix E. The solution yields

$$P_j^* = \left[\lambda - \frac{1}{g_j^2} \right]^+ = \max\left\{ \lambda - \frac{1}{g_j^2}, 0 \right\},$$

where the Lagrange multiplier λ is chosen to satisfy the condition

$$\sum_{j=1}^{d} \left[\lambda - \frac{1}{g_j^2} \right]^+ = P.$$

This optimal power allocation has the *water-filling* interpretation illustrated in Figure 3.9.

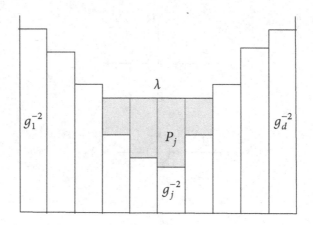

Figure 3.9. Water-filling interpretation of optimal power allocation.

Although this solution maximizes the mutual information and thus is optimal only in the asymptotic sense, it has been proven effective in practical subcarrier bit-loading algorithms for DSL and orthogonal frequency division multiplexing (OFDM) systems.

3.5 LOSSLESS SOURCE CODING

In the previous sections, we considered reliable communication of a maximally compressed information source represented by a uniformly distributed message over a noisy channel. In this section we consider the "dual" problem of communicating (or storing) an uncompressed source over a noiseless link (or in a memory) as depicted in Figure 3.10. The source sequence X^n is encoded (described or compressed) into an index M at rate R bits per source symbol, and the receiver decodes (decompresses) the index to find the estimate (reconstruction) $\hat{X}^n$ of the source sequence. The lossless source coding problem is to find the lowest compression rate in bits per source symbol such that the probability of decoding error decays asymptotically to zero with the code block length n.

We consider the lossless source coding problem for a *discrete memoryless source* (DMS) model $(\mathcal{X}, p(x))$, informally referred to as X, that consists of a finite alphabet $\mathcal{X}$ and a pmf $p(x)$ over $\mathcal{X}$. The DMS $(\mathcal{X}, p(x))$ generates an i.i.d. random process $\{X_i\}$ with $X_i \sim p_X(x_i)$. For example, the Bern(p) source X for $p \in [0, 1]$ has a binary alphabet and the Bern(p) pmf. It generates a Bern(p) random process $\{X_i\}$.

Figure 3.10. Point-to-point compression system.

A $(2^{nR}, n)$ *lossless source code* of rate R bits per source symbol consists of

- an encoding function (encoder) $m\colon \mathcal{X}^n \to [1:2^{nR}) = \{1, 2, \ldots, 2^{\lfloor nR \rfloor}\}$ that assigns an index $m(x^n)$ (a codeword of length $\lfloor nR \rfloor$ bits) to each source n-sequence x^n, and

- a decoding function (decoder) $\hat{x}^n\colon [1:2^{nR}) \to \mathcal{X}^n \cup \{e\}$ that assigns an estimate $\hat{x}^n(m) \in \mathcal{X}^n$ or an error message e to each index $m \in [1:2^{nR})$.

The probability of error for a $(2^{nR}, n)$ lossless source code is defined as $P_e^{(n)} = \mathsf{P}\{\hat{X}^n \neq X^n\}$. A rate R is said to be *achievable* if there exists a sequence of $(2^{nR}, n)$ codes such that $\lim_{n \to \infty} P_e^{(n)} = 0$ (hence the coding is required to be only *asymptotically* error-free). The *optimal rate* R^* for lossless source coding is the infimum of all achievable rates.

3.5.1 Lossless Source Coding Theorem

The optimal compression rate is characterized by the entropy of the source.

Theorem 3.4 (Lossless Source Coding Theorem). The optimal rate for lossless source coding of a discrete memoryless source X is

$$R^* = H(X).$$

For example, the optimal lossless compression rate for a $\mathrm{Bern}(p)$ source X is $R^* = H(X) = H(p)$. To prove this theorem, we again need to verify the following two statements:

- **Achievability.** For every $R > R^* = H(X)$ there exists a sequence of $(2^{nR}, n)$ codes with $\lim_{n \to \infty} P_e^{(n)} = 0$. We prove achievability using properties of typical sequences. Two alternative proofs will be given in Sections 3.6.4 and 10.3.1.

- **Converse.** For every sequence of $(2^{nR}, n)$ codes with $\lim_{n \to \infty} P_e^{(n)} = 0$, the source coding rate $R \geq R^* = H(X)$. The proof uses Fano's inequality and basic properties of entropy and mutual information.

We now prove each statement.

3.5.2 Proof of Achievability

For simplicity of presentation, assume nR is an integer. For $\epsilon > 0$, let $R = H(X) + \delta(\epsilon)$ with $\delta(\epsilon) = \epsilon H(X)$. Hence, $|\mathcal{T}_\epsilon^{(n)}| \leq 2^{n(H(X)+\delta(\epsilon))} = 2^{nR}$.

Encoding. Assign a distinct index $m(x^n)$ to each $x^n \in \mathcal{T}_\epsilon^{(n)}$. Assign $m = 1$ to all $x^n \notin \mathcal{T}_\epsilon^{(n)}$.

Decoding. Upon receiving the index m, the decoder declares $\hat{x}^n = x^n(m)$ for the unique $x^n(m) \in \mathcal{T}_\epsilon^{(n)}$.

Analysis of the probability of error. All typical sequences are recovered error-free. Thus, the probability of error is $P_e^{(n)} = \mathsf{P}\{X^n \notin \mathcal{T}_\epsilon^{(n)}\}$, which tends to zero as $n \to \infty$. This completes the proof of achievability.

3.5.3 Proof of the Converse

Given a sequence of $(2^{nR}, n)$ codes with $\lim_{n \to \infty} P_e^{(n)} = 0$, let M be the random variable corresponding to the index generated by the encoder. By Fano's inequality,

$$H(X^n | M) \leq H(X^n | \hat{X}^n) \leq 1 + nP_e^{(n)} \log |\mathcal{X}| = n\epsilon_n,$$

where ϵ_n tends to zero as $n \to \infty$ by the assumption that $\lim_{n \to \infty} P_e^{(n)} = 0$. Now consider

$$
\begin{aligned}
nR &\geq H(M) \\
&= I(X^n; M) \\
&= nH(X) - H(X^n | M) \\
&\geq nH(X) - n\epsilon_n.
\end{aligned}
$$

By taking $n \to \infty$, we conclude that $R \geq H(X)$. This completes the converse proof of the lossless source coding theorem.

3.6 LOSSY SOURCE CODING

Recall the compression system shown in Figure 3.10. Suppose that the source alphabet is continuous, for example, the source is a sensor that outputs an analog signal, then lossless reconstruction of the source sequence would require an infinite transmission rate! This motivates the lossy compression setup we study in this section, where the reconstruction is only required to be *close* to the source sequence according to some *fidelity criterion* (or distortion measure). In the scalar case, where each symbol is separately compressed, this lossy compression setup reduces to scalar quantization (analog-to-digital conversion), which often employs a mean squared error fidelity criterion. As in channel coding, however, it turns out that performing the lossy compression in blocks (vector quantization) can achieve better performance.

Unlike the lossless source coding setup where there is an optimal compression rate, the lossy source coding setup involves a tradeoff between the rate and the desired distortion. The problem is to find the limit on such tradeoff, which we refer to as the rate–distortion function. Note that this function is the source coding equivalent of the capacity–cost function in channel coding.

Although the motivation for lossy compression comes from sources with continuous alphabets, we first consider the problem for a DMS $(\mathcal{X}, p(x))$ as defined in the previous section. We assume the following per-letter distortion criterion. Let $\hat{\mathcal{X}}$ be a *reconstruction* alphabet and define a *distortion measure* as a mapping

$$d: \mathcal{X} \times \hat{\mathcal{X}} \to [0, \infty).$$

This mapping measures the cost of representing the symbol x by the symbol $\hat{x}$. The *average distortion* between x^n and $\hat{x}^n$ is defined as

$$d(x^n, \hat{x}^n) = \frac{1}{n} \sum_{i=1}^{n} d(x_i, \hat{x}_i).$$

For example, when $\mathcal{X} = \hat{\mathcal{X}}$, the *Hamming distortion measure* (loss) is the indicator for an error, i.e.,

$$d(x, \hat{x}) = \begin{cases} 1 & \text{if } x \neq \hat{x}, \\ 0 & \text{if } x = \hat{x}. \end{cases}$$

Thus, $d(\hat{x}^n, x^n)$ is the fraction of symbols in error (bit error rate for the binary alphabet).

Formally, a $(2^{nR}, n)$ *lossy source code* consists of

- an encoder that assigns an index $m(x^n) \in [1 : 2^{nR})$ to each sequence $x^n \in \mathcal{X}^n$, and
- a decoder that assigns an estimate $\hat{x}^n(m) \in \hat{\mathcal{X}}^n$ to each index $m \in [1 : 2^{nR})$.

The set $\mathcal{C} = \{\hat{x}^n(1), \ldots, \hat{x}^n(2^{\lfloor nR \rfloor})\}$ constitutes the *codebook*.

The expected distortion associated with a $(2^{nR}, n)$ lossy source code is defined as

$$E(d(X^n, \hat{X}^n)) = \sum_{x^n} p(x^n) d(x^n, \hat{x}^n(m(x^n))).$$

A rate–distortion pair (R, D) is said to be *achievable* if there exists a sequence of $(2^{nR}, n)$ codes with

$$\limsup_{n \to \infty} E(d(X^n, \hat{X}^n)) \leq D. \tag{3.7}$$

The *rate–distortion function* $R(D)$ is the infimum of rates R such that (R, D) is achievable.

3.6.1 Lossy Source Coding Theorem

Shannon showed that mutual information is again the canonical quantity that characterizes the rate–distortion function.

> **Theorem 3.5 (Lossy Source Coding Theorem).** The rate–distortion function for a DMS X and a distortion measure $d(x, \hat{x})$ is
>
> $$R(D) = \min_{p(\hat{x}|x):E(d(X,\hat{X})) \leq D} I(X; \hat{X})$$
>
> for $D \geq D_{\min} = \min_{\hat{x}(x)} E[d(X, \hat{x}(X))].$

Similar to the capacity–cost function in Section 3.3, the rate–distortion function $R(D)$ is nonincreasing, convex, and continuous in $D \geq D_{\min}$ (see Figure 3.11). Unless noted otherwise, we will assume throughout the book that $D_{\min} = 0$, that is, for every symbol $x \in \mathcal{X}$ there exists a reconstruction symbol $\hat{x} \in \hat{\mathcal{X}}$ such that $d(x, \hat{x}) = 0$.

Example 3.4 (Bernoulli source with Hamming distortion). The rate–distortion function for a $\mathrm{Bern}(p)$ source X, $p \in [0, 1/2]$, and Hamming distortion measure is

$$R(D) = \begin{cases} H(p) - H(D) & \text{for } 0 \leq D < p, \\ 0 & \text{for } D \geq p. \end{cases}$$

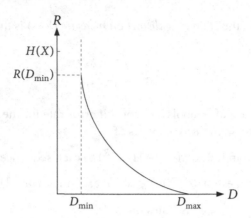

Figure 3.11. Graph of a typical rate–distortion function. Note that $R(D) = 0$ for $D \geq D_{\max} = \min_{\hat{x}} \mathsf{E}(d(X, \hat{x}))$ and $R(D_{\min}) \leq H(X)$.

To show this, recall that

$$R(D) = \min_{p(\hat{x}|x):\mathsf{E}(d(X,\hat{X}))\leq D} I(X; \hat{X}).$$

If $D \geq p$, $R(D) = 0$ by simply taking $\hat{X} = 0$. If $D < p$, we find a lower bound on $R(D)$ and then show that there exists a test channel $p(\hat{x}|x)$ that attains it. For any joint pmf that satisfies the distortion constraint $\mathsf{E}(d(X, \hat{X})) = \mathsf{P}\{X \neq \hat{X}\} \leq D$, we have

$$
\begin{aligned}
I(X; \hat{X}) &= H(X) - H(X|\hat{X}) \\
&= H(p) - H(X \oplus \hat{X}|\hat{X}) \\
&\geq H(p) - H(X \oplus \hat{X}) \\
&\overset{(a)}{\geq} H(p) - H(D),
\end{aligned}
$$

where (a) follows since $\mathsf{P}\{X \neq \hat{X}\} \leq D$. Thus

$$R(D) \geq H(p) - H(D).$$

It can be easily shown that this bound is attained by the *backward* BSC (with $\hat{X}$ and Z independent) shown in Figure 3.12, and the associated expected distortion is D.

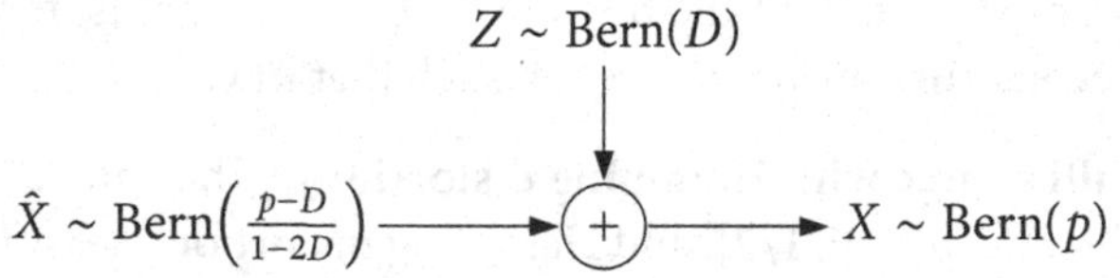

$$Z \sim \text{Bern}(D)$$

$$\hat{X} \sim \text{Bern}\left(\tfrac{p-D}{1-2D}\right) \longrightarrow \boxed{+} \longrightarrow X \sim \text{Bern}(p)$$

Figure 3.12. The backward BSC (test channel) that attains the rate–distortion function $R(D)$.

3.6.2 Proof of the Converse

The proof of the lossy source coding theorem again requires establishing achievability and the converse. We first prove the converse.

We need to show that for any sequence of $(2^{nR}, n)$ codes with

$$\limsup_{n \to \infty} \mathsf{E}(d(X^n, \hat{X}^n)) \leq D, \tag{3.8}$$

we must have $R \geq R(D)$. Consider

$$
\begin{aligned}
nR &\geq H(M) \\
&\geq I(M; X^n) \\
&\geq I(\hat{X}^n; X^n) \\
&= \sum_{i=1}^{n} I(X_i; \hat{X}^n | X^{i-1}) \\
&\stackrel{(a)}{=} \sum_{i=1}^{n} I(X_i; \hat{X}^n, X^{i-1}) \\
&\geq \sum_{i=1}^{n} I(X_i; \hat{X}_i) \\
&\stackrel{(b)}{\geq} \sum_{i=1}^{n} R(\mathsf{E}[d(X_i, \hat{X}_i)]) \\
&\stackrel{(c)}{\geq} nR(\mathsf{E}[d(X^n, \hat{X}^n)]),
\end{aligned}
$$

where (a) follows by the memoryless property of the source, (b) follows by the definition of $R(D) = \min I(X; \hat{X})$, and (c) follows by the convexity of $R(D)$. Since $R(D)$ is continuous and nonincreasing in D, it follows from the bound on distortion in (3.8) that

$$R \geq \limsup_{n \to \infty} R(\mathsf{E}[d(X^n, \hat{X}^n)]) \geq R\left(\limsup_{n \to \infty} \mathsf{E}[d(X^n, \hat{X}^n)]\right) \geq R(D).$$

This completes the proof of the converse.

3.6.3 Proof of Achievability

The proof uses random coding and joint typicality encoding. Assume that nR is an integer.

Random codebook generation. Fix the conditional pmf $p(\hat{x}|x)$ that attains $R(D/(1 + \epsilon))$, where D is the desired distortion, and let $p(\hat{x}) = \sum_x p(x)p(\hat{x}|x)$. Randomly and independently generate 2^{nR} sequences $\hat{x}^n(m)$, $m \in [1 : 2^{nR}]$, each according to $\prod_{i=1}^{n} p_{\hat{X}}(\hat{x}_i)$. These sequences constitute the codebook $\mathcal{C}$, which is revealed to the encoder and the decoder.

Encoding. We use *joint typicality encoding*. Given a sequence x^n, find an index m such that $(x^n, \hat{x}^n(m)) \in \mathcal{T}_\epsilon^{(n)}$. If there is more than one such index, choose the smallest one among them. If there is no such index, set $m = 1$.

Decoding. Upon receiving the index m, the decoder sets the reconstruction sequence $\hat{x}^n = \hat{x}^n(m)$.

Analysis of expected distortion. Let $\epsilon' < \epsilon$ and M be the index chosen by the encoder. We bound the distortion averaged over the random choice of the codebook $\mathcal{C}$. Define the "encoding error" event

$$\mathcal{E} = \{(X^n, \hat{X}^n(M)) \notin \mathcal{T}_\epsilon^{(n)}\},$$

and consider the events

$$\mathcal{E}_1 = \{X^n \notin \mathcal{T}_{\epsilon'}^{(n)}\},$$
$$\mathcal{E}_2 = \{X^n \in \mathcal{T}_{\epsilon'}^{(n)}, (X^n, \hat{X}^n(m)) \notin \mathcal{T}_\epsilon^{(n)} \text{ for all } m \in [1 : 2^{nR}]\}.$$

Then by the union of events bound,

$$P(\mathcal{E}) \le P(\mathcal{E}_1) + P(\mathcal{E}_2).$$

We bound each term. By the LLN, the first term $P(\mathcal{E}_1)$ tends to zero as $n \to \infty$. Consider the second term

$$P(\mathcal{E}_2) = \sum_{x^n \in \mathcal{T}_{\epsilon'}^{(n)}} p(x^n) \, P\{(x^n, \hat{X}^n(m)) \notin \mathcal{T}_\epsilon^{(n)} \text{ for all } m \mid X^n = x^n\}$$

$$= \sum_{x^n \in \mathcal{T}_{\epsilon'}^{(n)}} p(x^n) \prod_{m=1}^{2^{nR}} P\{(x^n, \hat{X}^n(m)) \notin \mathcal{T}_\epsilon^{(n)}\}$$

$$= \sum_{x^n \in \mathcal{T}_{\epsilon'}^{(n)}} p(x^n) (P\{(x^n, \hat{X}^n(1)) \notin \mathcal{T}_\epsilon^{(n)}\})^{2^{nR}}.$$

Since $x^n \in \mathcal{T}_{\epsilon'}^{(n)}$ and $\hat{X}^n(1) \sim \prod_{i=1}^n p_{\hat{X}}(\hat{x}_i)$, it follows by the second part of the joint typicality lemma in Section 2.5 that for n sufficiently large

$$P\{(x^n, \hat{X}^n(1)) \in \mathcal{T}_\epsilon^{(n)}\} \ge 2^{-n(I(X;\hat{X}) + \delta(\epsilon))},$$

where $\delta(\epsilon)$ tends to zero as $\epsilon \to 0$. Since $(1 - x)^k \le e^{-kx}$ for $x \in [0, 1]$ and $k \ge 0$, we have

$$\sum_{x^n \in \mathcal{T}_{\epsilon'}^{(n)}} p(x^n) (P\{(x^n, \hat{X}^n(1)) \notin \mathcal{T}_\epsilon^{(n)}\})^{2^{nR}} \le (1 - 2^{-n(I(X;\hat{X}) + \delta(\epsilon))})^{2^{nR}}$$

$$\le \exp(-2^{nR} \cdot 2^{-n(I(X;\hat{X}) + \delta(\epsilon))})$$

$$= \exp(-2^{n(R - I(X;\hat{X}) - \delta(\epsilon))}),$$

which tends to zero as $n \to \infty$ if $R > I(X; \hat{X}) + \delta(\epsilon)$.

Now, by the law of total expectation and the typical average lemma,

$$E_{\mathcal{C}, X^n}[d(X^n, \hat{X}^n(M))] = P(\mathcal{E}) \, E_{\mathcal{C}, X^n}[d(X^n, \hat{X}^n(M)) \mid \mathcal{E}] + P(\mathcal{E}^c) \, E_{\mathcal{C}, X^n}[d(X^n, \hat{X}^n(M)) \mid \mathcal{E}^c]$$

$$\le P(\mathcal{E}) d_{\max} + P(\mathcal{E}^c)(1 + \epsilon) \, E(d(X, \hat{X})),$$

where $d_{\max} = \max_{(x,\hat{x}) \in \mathcal{X} \times \hat{\mathcal{X}}} d(x, \hat{x})$. Hence, by the assumption on the conditional pmf $p(\hat{x}|x)$ that $E(d(X, \hat{X})) \leq D/(1 + \epsilon)$,

$$\limsup_{n \to \infty} E_{C, X^n}[d(X^n, \hat{X}^n(M))] \leq D$$

if $R > I(X; \hat{X}) + \delta(\epsilon) = R(D/(1 + \epsilon)) + \delta(\epsilon)$. Since the expected distortion (averaged over codebooks) is asymptotically $\leq D$, there must exist a sequence of codes with expected distortion asymptotically $\leq D$, which proves the achievability of the rate–distortion pair $(R(D/(1 + \epsilon)) + \delta(\epsilon), D)$. Finally, by the continuity of $R(D)$ in D, it follows that the achievable rate $R(D/(1 + \epsilon)) + \delta(\epsilon)$ converges to $R(D)$ as $\epsilon \to 0$, which completes the proof of achievability.

Remark 3.9. The above proof can be extended to unbounded distortion measures, provided that there exists a symbol $\hat{x}_0$ such that $d(x, \hat{x}_0) < \infty$ for every x. In this case, encoding is modified so that $\hat{x}^n = (\hat{x}_0, \ldots, \hat{x}_0)$ whenever joint typicality encoding fails. For example, for an erasure distortion measure with $\mathcal{X} = \{0, 1\}$ and $\hat{\mathcal{X}} = \{0, 1, e\}$, where $d(0, 0) = d(1, 1) = 0$, $d(0, e) = d(1, e) = 1$, and $d(0, 1) = d(1, 0) = \infty$, we have $\hat{x}_0 = e$. When $X \sim \text{Bern}(1/2)$, it can be easily shown that $R(D) = 1 - D$.

3.6.4 Lossless Source Coding Revisited

We show that the lossless source coding theorem can be viewed as a *corollary* of the lossy source coding theorem. This leads to an alternative *random coding* achievability proof of the lossless source coding theorem. Consider the lossy source coding problem for a DMS X, reconstruction alphabet $\hat{\mathcal{X}} = \mathcal{X}$, and Hamming distortion measure. Setting $D = 0$, we obtain

$$R(0) = \min_{p(\hat{x}|x):E(d(X,\hat{X}))=0} I(X; \hat{X}) = I(X; X) = H(X),$$

which is equal to the optimal lossless source coding rate R^* as we have already seen in the lossless source coding theorem.

Here we prove that operationally $R^* = R(0)$ without resorting to the fact that $R^* = H(X)$. To prove the converse ($R^* \geq R(0)$), note that the converse for the lossy source coding theorem under the above conditions implies that for any sequence of $(2^{nR}, n)$ codes if the average symbol error probability

$$\frac{1}{n} \sum_{i=1}^{n} P\{\hat{X}_i \neq X_i\}$$

tends to zero as $n \to \infty$, then $R \geq R(0)$. Since the average symbol error probability is smaller than or equal to the block error probability $P\{\hat{X}^n \neq X^n\}$, this also establishes the converse for the lossless case.

To prove achievability ($R^* \leq R(0)$), we can still use random coding and joint typicality encoding! We fix a test channel

$$p(\hat{x}|x) = \begin{cases} 1 & \text{if } x = \hat{x}, \\ 0 & \text{otherwise}, \end{cases}$$

and define $\mathcal{T}_\epsilon^{(n)}(X, \hat{X})$ in the usual way. Then, $(x^n, \hat{x}^n) \in \mathcal{T}_\epsilon^{(n)}$ implies that $x^n = \hat{x}^n$. Following the achievability proof of the lossy source coding theorem, we generate a random code $\hat{x}^n(m)$, $m \in [1 : 2^{nR}]$, and use the same encoding and decoding procedures. Then, the probability of decoding error averaged over codebooks is upper bounded as

$$P(\mathcal{E}) \le P\{(X^n, \hat{X}^n) \notin \mathcal{T}_\epsilon^{(n)}\},$$

which tends to zero as $n \to \infty$ if $R > I(X; \hat{X}) + \delta(\epsilon) = R(0) + \delta(\epsilon)$. Thus there exists a sequence of $(2^{nR}, n)$ lossless source codes with $\lim_{n\to\infty} P_e^{(n)} = 0$.

Remark 3.10. We already know how to construct a sequence of asymptotically optimal lossless source codes by uniquely labeling each typical sequence. The above proof, however, shows that random coding can be used to establish *all* point-to-point communication coding theorems. Such unification shows the power of random coding and is aesthetically pleasing. More importantly, the technique of specializing a lossy source coding theorem to the lossless case will prove crucial later in Chapters 11 and 21.

3.7 COVERING LEMMA

The covering lemma generalizes the bound on the probability of the encoding error event $\mathcal{E}$ in the achievability proof of the lossy source coding theorem. The lemma will be used in the achievability proofs of several multiuser source and channel coding theorems.

Recall that in the bound on $P(\mathcal{E})$, we had a fixed conditional pmf $p(\hat{x}|x)$ and a source $X \sim p(x)$. As illustrated in Figure 3.13, we considered a set of 2^{nR} i.i.d. reconstruction sequences $\hat{X}^n(m)$, $m \in [1 : 2^{nR}]$, each distributed according to $\prod_{i=1}^n p_{\hat{X}}(\hat{x}_i)$ and an independently generated source sequence $X^n \sim \prod_{i=1}^n p_X(x_i)$. We showed that the probability that $(X^n, \hat{X}^n(m)) \in \mathcal{T}_\epsilon^{(n)}$ for some $m \in [1 : 2^{nR}]$ tends to one as $n \to \infty$ if $R > I(X; \hat{X}) + \delta(\epsilon)$.

The following lemma extends this bound by assuming that X^n and the set of codewords are conditionally independent given a sequence U^n with the condition that U^n and X^n are jointly typical with high probability. As such, the covering lemma is a dual to the packing lemma in which we do not wish any of the untransmitted (independent) codewords to be jointly typical with the received sequence given U^n.

Lemma 3.3 (Covering Lemma). Let $(U, X, \hat{X}) \sim p(u, x, \hat{x})$ and $\epsilon' < \epsilon$. Let $(U^n, X^n) \sim p(u^n, x^n)$ be a pair of random sequences with $\lim_{n\to\infty} P\{(U^n, X^n) \in \mathcal{T}_{\epsilon'}^{(n)}(U, X)\} = 1$, and let $\hat{X}^n(m)$, $m \in \mathcal{A}$, where $|\mathcal{A}| \ge 2^{nR}$, be random sequences, conditionally independent of each other and of X^n given U^n, each distributed according to $\prod_{i=1}^n p_{\hat{X}|U}(\hat{x}_i|u_i)$. Then, there exists $\delta(\epsilon)$ that tends to zero as $\epsilon \to 0$ such that

$$\lim_{n\to\infty} P\{(U^n, X^n, \hat{X}^n(m)) \notin \mathcal{T}_\epsilon^{(n)} \text{ for all } m \in \mathcal{A}\} = 0,$$

if $R > I(X; \hat{X}|U) + \delta(\epsilon)$.

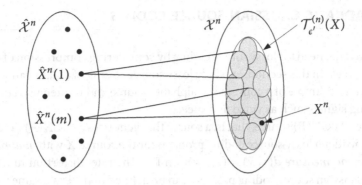

Figure 3.13. Illustration of the setup for the bound on $P(\mathcal{E})$.

Proof. Define the event

$$\mathcal{E}_0 = \{(U^n, X^n) \notin \mathcal{T}_{\epsilon'}^{(n)}\}.$$

Then, the probability of the event of interest can be upper bounded as

$$P(\mathcal{E}) \leq P(\mathcal{E}_0) + P(\mathcal{E} \cap \mathcal{E}_0^c).$$

By the condition of the lemma, $P(\mathcal{E}_0)$ tends to zero as $n \to \infty$. For the second term, recall from the joint typicality lemma that if $(u^n, x^n) \in \mathcal{T}_{\epsilon'}^{(n)}$, then for n sufficiently large,

$$P\{(u^n, x^n, \hat{X}^n(m)) \in \mathcal{T}_{\epsilon}^{(n)} \mid U^n = u^n, X^n = x^n\} = P\{(u^n, x^n, \hat{X}^n(m)) \in \mathcal{T}_{\epsilon}^{(n)} \mid U^n = u^n\}$$
$$\geq 2^{-n(I(X;\hat{X}|U)+\delta(\epsilon))}$$

for each $m \in \mathcal{A}$ for some $\delta(\epsilon)$ that tends to zero as $\epsilon \to 0$. Hence, for n sufficiently large,

$$P(\mathcal{E} \cap \mathcal{E}_0^c) = \sum_{(u^n,x^n)\in\mathcal{T}_{\epsilon'}^{(n)}} p(u^n, x^n)\, P\{(u^n, x^n, \hat{X}^n(m)) \notin \mathcal{T}_{\epsilon}^{(n)} \text{ for all } m \mid U^n = u^n, X^n = x^n\}$$

$$= \sum_{(u^n,x^n)\in\mathcal{T}_{\epsilon'}^{(n)}} p(u^n, x^n) \prod_{m\in\mathcal{A}} P\{(u^n, x^n, \hat{X}^n(m)) \notin \mathcal{T}_{\epsilon}^{(n)} \mid U^n = u^n\}$$

$$\leq \left(1 - 2^{-n(I(X;\hat{X}|U)+\delta(\epsilon))}\right)^{|\mathcal{A}|}$$

$$\leq \exp\left(-|\mathcal{A}| \cdot 2^{-n(I(X;\hat{X}|U)+\delta(\epsilon))}\right)$$

$$\leq \exp\left(-2^{n(R-I(X;\hat{X}|U)-\delta(\epsilon))}\right),$$

which tends to zero as $n \to \infty$, provided $R > I(X; \hat{X}|U) + \delta(\epsilon)$. This completes the proof.

Remark 3.11. The covering lemma continues to hold even when independence among all the sequences $\hat{X}^n(m)$, $m \in \mathcal{A}$, is replaced with pairwise independence; see the mutual covering lemma in Section 8.3.

3.8   QUADRATIC GAUSSIAN SOURCE CODING

We motivated the need for lossy source coding by considering compression of continuous-alphabet sources. In this section, we study lossy source coding of a Gaussian source, which is an important example of a continuous-alphabet source and is often used to model real-world analog signals such as video and speech.

Let X be a WGN(P) source, that is, a source that generates a WGN(P) random process $\{X_i\}$. We consider a lossy source coding problem for the source X with *quadratic (squared error) distortion measure* $d(x, \hat{x}) = (x - \hat{x})^2$ on $\mathbb{R}^2$. The rate–distortion function for this quadratic Gaussian source coding problem can be defined in the exact same manner as for the DMS case. Furthermore, Theorem 3.5 with the minimum over arbitrary test channels applies and the rate–distortion function can be expressed simply in terms of the power-to-distortion ratio.

Theorem 3.6. The rate–distortion function for a WGN(P) source with squared error distortion measure is

$$R(D) = \inf_{F(\hat{x}|x):\, \mathsf{E}((X-\hat{X})^2)\leq D} I(X; \hat{X}) = \mathsf{R}\left(\frac{P}{D}\right),$$

where $\mathsf{R}(x) = (1/2)[\log x]^+$ is the quadratic Gaussian rate function.

Proof of the converse. It is easy to see that the converse proof for the lossy source coding theorem extends to continuous sources with well-defined density such as Gaussian, and we have

$$R(D) \geq \inf_{F(\hat{x}|x):\, \mathsf{E}((X-\hat{X})^2)\leq D} I(X; \hat{X}). \tag{3.9}$$

For $D \geq P$, we set $\hat{X} = \mathsf{E}(X) = 0$; thus $R(D) = 0$. For $0 \leq D < P$, we first find a lower bound on the infimum in (3.9) and then show that there exists a test channel that attains it. Consider

$$
\begin{aligned}
I(X; \hat{X}) &= h(X) - h(X|\hat{X}) \\
&= \frac{1}{2}\log(2\pi e P) - h(X - \hat{X}|\hat{X}) \\
&\geq \frac{1}{2}\log(2\pi e P) - h(X - \hat{X}) \\
&\geq \frac{1}{2}\log(2\pi e P) - \frac{1}{2}\log(2\pi e\, \mathsf{E}[(X - \hat{X})^2]) \\
&\overset{(a)}{\geq} \frac{1}{2}\log(2\pi e P) - \frac{1}{2}\log(2\pi e D) \\
&= \frac{1}{2}\log\frac{P}{D},
\end{aligned}
$$

where (a) follows since $\mathsf{E}((X - \hat{X})^2) \leq D$. It is easy to show that this bound is attained by the backward Gaussian test channel shown in Figure 3.14 and that the associated expected distortion is D.

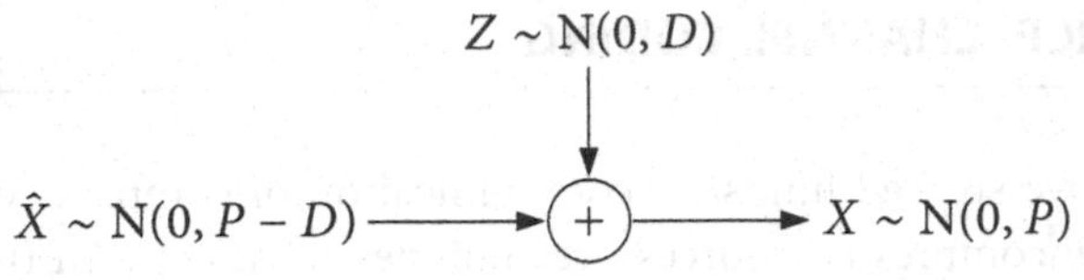

$$Z \sim \mathrm{N}(0, D)$$

$$\hat{X} \sim \mathrm{N}(0, P - D) \longrightarrow \boxed{+} \longrightarrow X \sim \mathrm{N}(0, P)$$

Figure 3.14. The backward Gaussian test channel that attains the minimum in (3.9).

Proof of achievability. We extend the achievability proof for the DMS to the case of a Gaussian source with quadratic distortion measure by using the following discretization procedure. Let D be the desired distortion and let $(X, \hat{X})$ be a pair of jointly Gaussian random variables attaining $I(X; \hat{X}) = R((1 - 2\epsilon)D)$ with distortion $\mathrm{E}((X - \hat{X})^2) = (1 - 2\epsilon)D$. Let $[X]$ and $[\hat{X}]$ be finitely quantized versions of X and $\hat{X}$, respectively, such that

$$\mathrm{E}(([X] - [\hat{X}])^2) \le (1 - \epsilon)^2 D,$$
$$\mathrm{E}((X - [X])^2) \le \epsilon^2 D. \tag{3.10}$$

Then by the data processing inequality,

$$I([X]; [\hat{X}]) \le I(X; \hat{X}) = R((1 - 2\epsilon)D).$$

Now, by the achievability proof for the DMS $[X]$ and reconstruction $[\hat{X}]$, there exists a sequence of $(2^{nR}, n)$ rate–distortion codes with asymptotic distortion

$$\limsup_{n \to \infty} \frac{1}{n} \mathrm{E}\big(d([X]^n, [\hat{X}]^n)\big) \le (1 - \epsilon)^2 D, \tag{3.11}$$

if $R > R((1 - 2\epsilon)D) \ge I([X]; [\hat{X}])$. We use this sequence of codes for the original source X by mapping each x^n to the codeword $[\hat{x}]^n$ that is assigned to $[x]^n$. Then

$$\limsup_{n \to \infty} \mathrm{E}\big(d(X^n, [\hat{X}]^n)\big) = \limsup_{n \to \infty} \frac{1}{n} \sum_{i=1}^{n} \mathrm{E}((X_i - [\hat{X}]_i)^2)$$

$$= \limsup_{n \to \infty} \frac{1}{n} \sum_{i=1}^{n} \mathrm{E}(((X_i - [X_i]) + ([X_i] - [\hat{X}]_i))^2)$$

$$\overset{(a)}{\le} \limsup_{n \to \infty} \frac{1}{n} \sum_{i=1}^{n} \Big(\mathrm{E}((X_i - [X_i])^2) + \mathrm{E}(([X_i] - [\hat{X}]_i)^2) \Big)$$

$$+ \limsup_{n \to \infty} \frac{2}{n} \sum_{i=1}^{n} \sqrt{\mathrm{E}((X_i - [X_i])^2)\, \mathrm{E}(([X_i] - [\hat{X}]_i)^2)}$$

$$\overset{(b)}{\le} \epsilon^2 D + (1 - \epsilon)^2 D + 2\epsilon(1 - \epsilon)D$$

$$= D,$$

where (a) follows by Cauchy's inequality and (b) follows by (3.10) and (3.11), and Jensen's inequality. Thus, $R > R((1 - 2\epsilon)D)$ is achievable for distortion D. Using the continuity of $R(D)$ completes the proof of achievability.

3.9   JOINT SOURCE–CHANNEL CODING

In previous sections we studied limits on communication of compressed sources over noisy channels and uncompressed sources over noiseless channels. In this section, we study the more general joint source–channel coding setup depicted in Figure 3.15. The sender wishes to communicate k symbols of an uncompressed source U over a DMC $p(y|x)$ in n transmissions so that the receiver can reconstruct the source symbols with a prescribed distortion D. A straightforward scheme would be to perform separate source and channel encoding and decoding. Is this separation scheme optimal? Can we do better by allowing more general joint source–channel encoding and decoding?

Figure 3.15. Joint source–channel coding setup.

It turns out that separate source and channel coding is asymptotically optimal for sending a DMS over a DMC, and hence the fundamental limit depends only on the rate–distortion function of the source and the capacity of the channel.

Formally, let U be a DMS and $d(u, \hat{u})$ be a distortion measure with rate–distortion function $R(D)$ and $p(y|x)$ be a DMC with capacity C. A $(|\mathcal{U}|^k, n)$ joint source–channel code of rate $r = k/n$ consists of

- an encoder that assigns a codeword $x^n(u^k) \in \mathcal{X}^n$ to each sequence $u^k \in \mathcal{U}^k$ and

- a decoder that assigns an estimate $\hat{u}^k(y^n) \in \hat{\mathcal{U}}^k$ to each sequence $y^n \in \mathcal{Y}^n$.

A rate–distortion pair (r, D) is said to be achievable if there exists a sequence of $(|\mathcal{U}|^k, n)$ joint source–channel codes of rate r such that

$$\limsup_{k \to \infty} \mathsf{E}\big[d(U^k, \hat{U}^k(Y^n)) \big] \le D.$$

Shannon established the following fundamental limit on joint source–channel coding.

Theorem 3.7 (Source–Channel Separation Theorem). Given a DMS U and a distortion measure $d(u, \hat{u})$ with rate–distortion function $R(D)$ and a DMC $p(y|x)$ with capacity C, the following statements hold:

- If $rR(D) < C$, then (r, D) is achievable.

- If (r, D) is achievable, then $rR(D) \le C$.

Proof of achievability. We use separate lossy source coding and channel coding.

- Source coding: For any $\epsilon > 0$, there exists a sequence of lossy source codes with rate

$R(D/(1 + \epsilon)) + \delta(\epsilon)$ that achieve expected distortion less than or equal to D. We treat the index for each code in the sequence as a message to be sent over the channel.

- Channel coding: The sequence of source indices can be reliably communicated over the channel if $r(R(D/(1 + \epsilon)) + \delta(\epsilon)) \leq C - \delta'(\epsilon)$.

The source decoder finds the reconstruction sequence corresponding to the received index. If the channel decoder makes an error, the distortion is upper bounded by $d_{\max}$. Because the maximal probability of error tends to zero as $n \to \infty$, the overall expected distortion is less than or equal to D.

Proof of the converse. We wish to show that if a sequence of codes achieves the rate–distortion pair (r, D), then $rR(D) \leq C$. By the converse proof of the lossy source coding theorem, we know that

$$R(D) \leq \frac{1}{k} I(U^k; \hat{U}^k).$$

Now, by the data processing inequality,

$$\frac{1}{k} I(U^k; \hat{U}^k) \leq \frac{1}{k} I(U^k; Y^n).$$

Following similar steps to the converse proof for the DMC, we have

$$\frac{1}{k} I(U^k; Y^n) \leq \frac{1}{k} \sum_{i=1}^{n} I(X_i; Y_i) \leq \frac{1}{r} C.$$

Combining the above inequalities completes the proof of the converse.

Remark 3.12. Since the converse of the channel coding theorem holds when causal feedback is present (see Section 3.1.5), the separation theorem continues to hold with feedback.

Remark 3.13. As in Remark 3.5, there are cases where $rR(D) = C$ and the rate–distortion pair (r, D) is achievable via joint source–channel coding; see Example 3.5. However, if $rR(D) > C$, the rate–distortion pair (r, D) is not achievable. Hence, we informally say that source–channel separation holds in general for sending a DMS over a DMC.

Remark 3.14. As a special case of joint source–channel coding, consider the problem of sending U over a DMC losslessly, i.e., $\lim_{k \to \infty} \mathsf{P}\{\hat{U}^k \neq U^k\} = 0$. The separation theorem holds with the requirement that $rH(U) \leq C$.

Remark 3.15. The separation theorem can be extended to sending an arbitrary stationary ergodic source over a DMC.

Remark 3.16. As we will see in Chapter 14, source–channel separation does not hold in general for communicating multiple sources over multiuser channels, that is, even in the asymptotic regime, it may be beneficial to leverage the structure of the source and channel jointly rather than separately.

3.9.1   Uncoded Transmission

Sometimes optimal joint source–channel coding is simpler than separate source and channel coding. This is illustrated in the following.

Example 3.5. Consider communicating a Bern$(1/2)$ source over a BSC(p) at rate $r = 1$ with Hamming distortion less than or equal to D. The separation theorem shows that $1 - H(D) < 1 - H(p)$, or equivalently, $D > p$, can be achieved using separate source and channel coding. More simply, we can transmit the binary sequence over the channel *without any coding* and achieve average distortion $D = p$!

Similar *uncoded transmission* is optimal also for communicating a Gaussian source over a Gaussian channel with quadratic distortion (with proper scaling to satisfy the power constraint); see Problem 3.20.

Remark 3.17. In general, we have the following condition for the optimality of uncoded transmission. A DMS U can be communicated over a DMC $p(y|x)$ uncoded if $X \sim p_U(x)$ attains the capacity $C = \max_{p(x)} I(X; Y)$ of the channel and the test channel $p_{Y|X}(\hat{u}|u)$ attains the rate–distortion function $R(D) = \min_{p(\hat{u}|u):\mathsf{E}(d(U,\hat{U}))\leq D} I(U; \hat{U})$ of the source. In this case, $C = R(D)$.

SUMMARY

- Point-to-point communication system architecture

- Discrete memoryless channel (DMC), e.g., BSC and BEC

- Coding theorem: achievability and the converse

- Channel capacity is the limit on channel coding

- Random codebook generation

- Joint typicality decoding

- Packing lemma

- Feedback does not increase the capacity of a DMC

- Capacity with input cost

- Gaussian channel:

  - Capacity with average power constraint is achieved via Gaussian codes

  - Extending the achievability proof from discrete to Gaussian

  - Minimum energy per bit

  - Water filling

- Discrete memoryless source (DMS)

- Entropy is the limit on lossless source coding
- Rate–distortion function is the limit on lossy source coding
- Joint typicality encoding
- Covering lemma
- Lossless source coding theorem is a corollary of lossy source coding theorem
- Rate–distortion function for Gaussian source with quadratic distortion
- Source–channel separation
- Uncoded transmission can be optimal

BIBLIOGRAPHIC NOTES

The channel coding theorem was first proved in Shannon (1948). There are alternative proofs of achievability for this theorem that yield stronger results, including Feinstein's (1954) maximal coding theorem and Gallager's (1965) random coding exponent technique, which yield stronger results. For example, it can be shown (Gallager 1968) that the probability of error decays exponentially fast in the block length and the random coding exponent technique gives a very good bound on the optimal error exponent (reliability function) for the DMC. These proofs, however, do not extend easily to many multiuser channel and source coding problems. In comparison, the current proof (Forney 1972, Cover 1975b), which is based on Shannon's original arguments, is much simpler and can be readily extended to more complex settings. Hence, we will adopt the random codebook generation and joint typicality decoding approach throughout.

The achievability proof of the channel coding theorem for the BSC using a random linear code is due to Elias (1955). Even though random linear codes allow for computationally efficient encoding (by simply multiplying the message by a *generator matrix G*), decoding still requires an exponential search, which limits its practical value. This problem can be mitigated by considering a linear code ensemble with special structures, such as Gallager's (1963) low density parity check (LDPC) codes, which have efficient decoding algorithms and achieve rates close to capacity (Richardson and Urbanke 2008). A more recently developed class of capacity-achieving linear codes is polar codes (Arıkan 2009), which involve an elegant information theoretic low-complexity decoding algorithm and can be applied also to lossy compression settings (Korada and Urbanke 2010). Linear codes for the BSC or BEC are examples of structured codes. Other examples include lattice codes for the Gaussian channel, which have been shown to achieve the capacity by Erez and Zamir (2004); see Zamir (2009) for a survey of recent developments.

The converse of the channel coding theorem states that if $R > C$, then $P_e^{(n)}$ is bounded away from zero as $n \to \infty$. This is commonly referred to as the *weak converse*. In comparison, the *strong converse* (Wolfowitz 1957) states that if $R > C$, then $\lim_{n \to \infty} P_e^{(n)} = 1$. A similar statement holds for the lossless source coding theorem. However, except for a few cases to be discussed later, it appears to be difficult to prove the strong converse for most multiuser settings. As such, we only present weak converse proofs in our main exposition.

The capacity formula for the Gaussian channel under average power constraint in Theorem 3.3 is due to Shannon (1948). The achievability proof using the discretization procedure follows McEliece (1977). Alternative proofs of achievability for the Gaussian channel can be found in Gallager (1968) and Cover and Thomas (2006). The discrete-time Gaussian channel is the model for a continuous-time (waveform) bandlimited Gaussian channel with bandwidth $W = 1/2$, noise power spectral density (psd) $N_0/2$, average transmission power P (area under psd of signal), and channel gain g. If the channel has bandwidth W, then it is equivalent to $2W$ parallel discrete-time Gaussian channels (per second) and the capacity (see, for example, Wyner (1966) and Slepian (1976)) is

$$C = W \log \left(1 + \frac{g^2 P}{W N_0} \right) \quad \text{bits/second.}$$

For a wideband channel, the capacity C converges to $(S/2) \ln 2$ as $W \to \infty$, where $S = 2g^2 P/N_0$. Thus the capacity grows linearly with S and can be achieved via simple binary code as shown by Golay (1949). The minimum energy per bit for the Gaussian channel also first appeared in this paper. The minimum energy per bit can be also viewed as a special case of the reciprocal of the capacity per unit cost studied by Csiszár and Körner (1981b, p. 120) and Verdú (1990). The capacity of the spectral Gaussian channel, which is the continuous counterpart of the Gaussian product channel, and its water-filling solution are due to Shannon (1949a).

The lossless source coding theorem was first proved in Shannon (1948). In many applications, one cannot afford to have any errors introduced by compression. Error-free compression ($\mathsf{P}\{X^n \neq \hat{X}^n\} = 0$) for fixed-length codes, however, requires that $R \geq \log |\mathcal{X}|$. Using *variable-length* codes, Shannon (1948) also showed that error-free compression is possible if the average rate of the code is larger than the entropy $H(X)$. Hence, the limit on the average achievable rate is the same for both lossless and error-free compression. This is not true in general for distributed coding of correlated sources; see Bibliographic Notes in Chapter 10.

The lossy source coding theorem was first proved in Shannon (1959), following an earlier result for the quadratic Gaussian case in Shannon (1948). The current achievability proof of the quadratic Gaussian lossy source coding theorem follows McEliece (1977). There are several other ways to prove achievability for continuous sources and unbounded distortion measures (Berger 1968, Dunham 1978, Bucklew 1987, Cover and Thomas 2006). As an alternative to the expected distortion criterion in (3.7), several authors have considered the stronger criterion

$$\lim_{n \to \infty} \mathsf{P}\{d(X^n, \hat{x}^n(m(X^n))) \leq D\} = 1$$

in the definition of achievability of a rate–distortion pair (R, D). The lossy source coding theorem and its achievability proof in Section 3.6 continue to hold for this alternative distortion criterion. In the other direction, a strong converse—if $R < R(D)$, then

$P\{d(X^n, \hat{X}^n) \le D\}$ tends to zero as $n \to \infty$—can be established (Csiszár and Körner 1981b, Theorem 2.3) that implies the converse for the expected distortion criterion.

The lossless source coding theorem can be extended to discrete stationary ergodic (not necessarily i.i.d.) sources (Shannon 1948). Similarly, the lossy source coding theorem can be extended to stationary ergodic sources (Gallager 1968) with the following characterization of the rate–distortion function

$$R(D) = \lim_{k \to \infty} \min_{p(\hat{x}^k | x^k) : E(d(X^k, \hat{X}^k)) \le D} \frac{1}{k} I(X^k; \hat{X}^k).$$

However, the notion of ergodicity for channels is more subtle and involved. Roughly speaking, the capacity is well-defined for discrete channels such that for every time $i \ge 1$ and shift $j \ge 1$, the conditional pmf $p(y_i^{j+i} | x_i^{j+i})$ is time invariant (that is, independent of i) and can be estimated using appropriate time averages. For example, if $Y_i = g(X_i, Z_i)$ for some stationary ergodic process $\{Z_i\}$, then it can be shown (Kim 2008b) that the capacity is

$$C = \lim_{k \to \infty} \sup_{p(x^k)} \frac{1}{k} I(X^k; Y^k).$$

The coding theorem for more general classes of channels with memory can be found in Gray (1990) and Verdú and Han (1994). As for point-to-point communication, the essence of the multiuser source and channel coding problems is captured by the memoryless case. Moreover, the multiuser problems with memory often have only uncomputable "multiletter" expressions as above. We therefore restrict our attention to discrete memoryless and white Gaussian noise sources and channels.

The source–channel separation theorem was first proved in Shannon (1959). The general condition for optimality of uncoded transmission in Remark 3.17 is given by Gastpar, Rimoldi, and Vetterli (2003).

PROBLEMS

3.1. *Memoryless property.* Show that under the given definition of a $(2^{nR}, n)$ code, the memoryless property $p(y_i | x^i, y^{i-1}, m) = p_{Y|X}(y_i | x_i)$, $i \in [1 : n]$, reduces to

$$p(y^n | x^n, m) = \prod_{i=1}^{n} p_{Y|X}(y_i | x_i).$$

3.2. *Z channel.* The Z channel has binary input and output alphabets, and conditional pmf $p(0|0) = 1$, $p(1|1) = p(0|1) = 1/2$. Find the capacity C.

3.3. *Capacity of the sum channel.* Find the capacity C of the union of two DMCs $(\mathcal{X}_1, p(y_1|x_1), \mathcal{Y}_1)$ and $(\mathcal{X}_2, p(y_2|x_2), \mathcal{Y}_2)$, where, in each transmission, one can send a symbol over channel 1 or channel 2 but not both. Assume that the output alphabets are distinct, i.e., $\mathcal{Y}_1 \cap \mathcal{Y}_2 = \emptyset$.

3.4. *Applications of the packing lemma.* Identify the random variables U, X, and Y in the packing lemma for the following scenarios, and write down the packing lemma condition on the rate R for each case.

(a) Let $(X_1, X_2, X_3) \sim p(x_1)p(x_2)p(x_3|x_1, x_2)$. Let $X_1^n(m)$, $m \in [1 : 2^{nR}]$, be each distributed according to $\prod_{i=1}^n p_{X_1}(x_{1i})$, and $(\tilde{X}_2^n, \tilde{X}_3^n) \sim \prod_{i=1}^n p_{X_2, X_3}(\tilde{x}_{2i}, \tilde{x}_{3i})$ be independent of $X_1^n(m)$ for $m \in [1 : 2^{nR}]$.

(b) Let $(X_1, X_2, X_3) \sim p(x_1, x_2)p(x_3|x_2)$ and $R = R_0 + R_1$. Let $X_1^n(m_0)$, $m_0 \in [1 : 2^{nR_0}]$, be distributed according to $\prod_{i=1}^n p_{X_1}(x_{1i})$. For each m_0, let $X_2^n(m_0, m_1)$, $m_1 \in [1 : 2^{nR_1}]$, be distributed according to $\prod_{i=1}^n p_{X_2|X_1}(x_{2i}|x_{1i}(m_0))$. Let $\tilde{X}_3^n \sim \prod_{i=1}^n p_{X_3}(\tilde{x}_{3i})$ be independent of $(X_1^n(m_0), X_2^n(m_0, m_1))$ for $m_0 \in [1 : 2^{nR_0}]$, $m_1 \in [1 : 2^{nR_1}]$.

3.5. *Maximum likelihood decoding.* The achievability proof of the channel coding theorem in Section 3.1.1 uses joint typicality decoding. This technique greatly simplifies the proof, especially for multiuser channels. However, given a codebook, the joint typicality decoding is not optimal in terms of minimizing the probability of decoding error (it is in fact surprising that such a suboptimal decoding rule can still achieve capacity).

Since the messages are equally likely, maximum likelihood decoding (MLD)

$$\hat{m} = \arg\max_m p(y^n|m) = \arg\max_m \prod_{i=1}^n p_{Y|X}(y_i|x_i(m))$$

is the optimal decoding rule (when there is a tie, choose an arbitrary index that maximizes the likelihood). Achievability proofs using MLD are more complex but provide tighter bounds on the optimal error exponent (reliability function); see, for example, Gallager (1968).

In this problem we use MLD to establish achievability of the capacity for a BSC(p), $p < 1/2$. Define the Hamming distance $d(x^n, y^n)$ between two binary sequences x^n and y^n as the number of positions where they differ, i.e., $d(x^n, y^n) = |\{i : x_i \neq y_i\}|$.

(a) Show that the MLD rule reduces to the minimum Hamming distance decoding rule—declare $\hat{m}$ is sent if $d(x^n(\hat{m}), y^n) < d(x^n(m), y^n)$ for all $m \neq \hat{m}$.

(b) Now fix $X \sim \text{Bern}(1/2)$. Using random coding and minimum distance decoding, show that for every $\epsilon > 0$, the probability of error averaged over codebooks is upper bounded as

$$\begin{aligned} P_e^{(n)} &= \mathsf{P}\{\hat{M} \neq 1 \mid M = 1\} \\ &\leq \mathsf{P}\{d(X^n(1), Y^n) > n(p + \epsilon) \mid M = 1\} \\ &\quad + (2^{nR} - 1)\, \mathsf{P}\{d(X^n(2), Y^n) \leq n(p + \epsilon) \mid M = 1\}. \end{aligned}$$

(c) Show that the first term tends to zero as $n \to \infty$. It can be shown using the

Chernoff–Hoeffding bound (Hoeffding 1963) that

$$P\{d(X^n(2), Y^n) \le n(p + \epsilon) \,|\, M = 1\} \le 2^{-n(1 - H(p + \epsilon))}.$$

Using these results, show that any $R < C = 1 - H(p)$ is achievable.

3.6. *Randomized code.* Suppose that in the definition of the $(2^{nR}, n)$ code for the DMC $p(y|x)$, we allow the encoder and the decoder to use random mappings. Specifically, let W be an arbitrary random variable independent of the message M and the channel, i.e., $p(y_i|x^i, y^{i-1}, m, w) = p_{Y|X}(y_i|x_i)$ for $i \in [1:n]$. The encoder generates a codeword $x^n(m, W)$, $m \in [1 : 2^{nR}]$, and the decoder generates an estimate $\hat{m}(y^n, W)$. Show that this randomization does not increase the capacity of the DMC.

3.7. *Nonuniform message.* Recall that a $(2^{nR}, n)$ code for the DMC $p(y|x)$ consists of an encoder $x^n = \phi_n(m)$ and a decoder $\hat{m} = \psi_n(y^n)$. Suppose that there exists a sequence of $(2^{nR}, n)$ codes such that $P_e^{(n)} = P\{M \ne \hat{M}\}$ tends to zero as $n \to \infty$, where M is uniformly distributed over $[1 : 2^{nR}]$. (In other words, the rate R is achievable.) Now suppose that we wish to communicate a message M' that is arbitrarily (not uniformly) distributed over $[1 : 2^{nR}]$.

(a) Show that there exists a sequence of $(2^{nR}, n)$ codes with encoder–decoder pairs (ϕ_n', ψ_n') such that

$$\lim_{n \to \infty} P\{M' \ne \hat{M}'\} = 0.$$

(Hint: Consider a random ensemble of codes $\Phi_n' = \phi_n \circ \sigma$ and $\Psi_n' = \sigma^{-1} \circ \psi_n$, where σ is a random permutation. Show the probability of error, averaged over M' and σ, is equal to $P_e^{(n)}$ and conclude that there exists a good permutation σ for each M'.)

(b) Does this result imply that the capacity for the maximal probability of error is equal to that for the average probability of error?

3.8. *Independently generated codebooks.* Let $(X, Y) \sim p(x, y)$, and $p(x)$ and $p(y)$ be their marginals. Consider two randomly and independently generated codebooks $C_1 = \{X^n(1), \ldots, X^n(2^{nR_1})\}$ and $C_2 = \{Y^n(1), \ldots, Y^n(2^{nR_2})\}$. The codewords of C_1 are generated independently each according to $\prod_{i=1}^n p_X(x_i)$, and the codewords for C_2 are generated independently according to $\prod_{i=1}^n p_Y(y_i)$. Define the set

$$C = \{(x^n, y^n) \in C_1 \times C_2 : (x^n, y^n) \in T_\epsilon^{(n)}(X, Y)\}.$$

Show that

$$E(|C|) \doteq 2^{n(R_1 + R_2 - I(X;Y))}.$$

3.9. *Capacity with input cost.* Consider the DMC $p(y|x)$ with cost constraint B.

(a) Using the operational definition of the capacity–cost function $C(B)$, show that it is nondecreasing and concave for $B \ge 0$.

(b) Show that the information capacity–cost function $C(B)$ is nondecreasing, concave, and continuous for $B \geq 0$.

3.10. *BSC with input cost.* Find the capacity–cost function $C(B)$ for a BSC(p) with input cost function $b(1) = 1$ and $b(0) = 0$.

3.11. *Channels with input–output cost.* Let $b(x, y)$ be a nonnegative input–output cost function on $\mathcal{X} \times \mathcal{Y}$. Consider a DMC $p(y|x)$ in which every codeword $x^n(m)$, $m \in [1 : 2^{nR}]$, must satisfy the average cost constraint

$$\mathsf{E}(b(x^n(m), Y^n)) = \frac{1}{n} \sum_{i=1}^{n} \mathsf{E}(b(x_i(m), Y_i)) \leq B,$$

where the expectation is with respect to the channel pmf $\prod_{i=1}^{n} p_{Y|X}(y_i|x_i(m))$. Show that the capacity of the DMC with cost constraint B is

$$C(B) = \max_{p(x):\mathsf{E}(b(X,Y))\leq B} I(X; Y).$$

(Hint: Consider the input-only cost function $b'(x) = \mathsf{E}(b(x, Y))$, where the expectation is taken with respect to $p(y|x)$.)

3.12. *Output scaling.* Show that the capacity of the Gaussian channel $Y = gX + Z$ remains the same if we scale the output by a nonzero constant a.

3.13. *Water-filling.* Consider the 2-component Gaussian product channel $Y_j = g_jX_j + Z_j$, $j = 1, 2$, with $g_1 < g_2$ and average power constraint P.

(a) Above what power P should we begin to use both channels?

(b) What is the energy-per-bit–rate function $E_b(R)$ needed for reliable communication at rate R over the channel? Show that $E_b(R)$ is strictly monotonically increasing and convex in R. What is the minimum energy per bit for the 2-component Gaussian product channel, i.e., $\lim_{R \to 0} E_b(R)$?

3.14. *List codes.* A $(2^{nR}, 2^{nL}, n)$ list code for a DMC $p(y|x)$ with capacity C consists of an encoder that assigns a codeword $x^n(m)$ to each message $m \in [1 : 2^{nR}]$ and a decoder that upon receiving y^n tries to finds the list of messages $\mathcal{L}(y^n) \subseteq [1 : 2^{nR}]$ of size $|\mathcal{L}| \leq 2^{nL}$ that contains the transmitted message. An error occurs if the list does not contain the transmitted message M, i.e., $P_e^{(n)} = \mathsf{P}\{M \notin \mathcal{L}(Y^n)\}$. A rate–list exponent pair (R, L) is said to be achievable if there exists a sequence of $(2^{nR}, 2^{nL}, n)$ list codes with $P_e^{(n)} \to 0$ as $n \to \infty$.

(a) Using random coding and joint typicality decoding, show that any (R, L) is achievable, provided $R < C + L$.

(b) Show that for every sequence of $(2^{nR}, 2^{nL}, n)$ list codes with $P_e^{(n)} \to 0$ as $n \to \infty$, we must have $R \leq C + L$. (Hint: You will need to develop a modified Fano's inequality.)

3.15. *Strong converse for source coding.* Given a sequence of $(2^{nR}, n)$ lossless source codes with $R < H(X)$, show that $P_e^{(n)} \to 1$ as $n \to \infty$. (Hint: A $(2^{nR}, n)$ code can represent only 2^{nR} sequences in $\mathcal{X}^n$. Using typicality, show that if $R < H(X)$, the probability of these 2^{nR} sequences converges to zero, no matter how we choose them.)

3.16. *Infinite alphabet.* Consider the lossless source coding problem for a discrete, but infinite-alphabet source X with finite entropy $H(X) < \infty$. Show that $R^* = H(X)$. (Hint: For the proof of achievability, consider a truncated DMS $[X]$ such that $P\{X^n \ne [X]^n\}$ tends to zero as $n \to \infty$.)

3.17. *Rate–distortion function.* Consider the lossy source coding for a DMS X with distortion measure $d(x, \hat{x})$.

(a) Using the operational definition, show that the rate–distortion function $R(D)$ is nonincreasing and convex for $D \ge 0$.

(b) Show that the information rate–distortion function $R(D)$ is nonincreasing, convex, and continuous for $D \ge 0$.

3.18. *Bounds on the quadratic rate–distortion function.* Let X be an arbitrary memoryless (stationary) source with variance P, and let $d(x, \hat{x}) = (x - \hat{x})^2$ be the quadratic distortion measure.

(a) Show that the rate–distortion function is bounded as

$$h(X) - \frac{1}{2}\log(2\pi e D) \le R(D) \le \frac{1}{2}\log\left(\frac{P}{D}\right)$$

with equality iff X is a WGN(P) source. (Hint: For the upper bound, consider $\hat{X} = (P - D)X/P + Z$, where $Z \sim N(0, D(P - D)/P)$ is independent of X.)

Remark: The lower bound is referred to as the *Shannon lower bound*.

(b) Is the Gaussian source harder or easier to describe than other sources with the same variance?

3.19. *Lossy source coding from a noisy observation.* Let $X \sim p(x)$ be a DMS and Y be another DMS obtained by passing X through a DMC $p(y|x)$. Let $d(x, \hat{x})$ be a distortion measure and consider a lossy source coding problem in which Y (instead of X) is encoded and sent to the decoder who wishes to reconstruct X with a prescribed distortion D.

Unlike the regular lossy source coding setup, the encoder maps each y^n sequence to an index $m \in [1 : 2^{nR})$. Otherwise, the definitions of $(2^{nR}, n)$ codes, achievability, and rate–distortion function are the same as before.

Let $D_{\min} = \min_{\hat{x}(y)} E[d(X, \hat{x}(Y))]$. Show that the rate–distortion function for this setting is

$$R(D) = \min_{p(\hat{x}|y):E(d(X,\hat{X}))\le D} I(Y; \hat{X}) \quad \text{for } D \ge D_{\min}.$$

(Hint: Define a new distortion measure $d'(y, \hat{x}) = E(d(X, \hat{x}) \mid Y = y)$, and show that

$$E[d(X^n, \hat{x}^n(m(Y^n)))] = E[d'(Y^n, \hat{x}^n(m(Y^n)))].)$$

3.20. *To code or not to code.* Consider a WGN(P) source U and a Gaussian channel with output $Y = gX + Z$, where $Z \sim N(0, 1)$. We wish to communicate the source over the channel at rate $r = 1$ symbol/transmission with the smallest possible squared error distortion. Assume an *expected* average power constraint

$$\frac{1}{n} \sum_{i=1}^{n} E(x_i^2(U^n)) \leq nP.$$

(a) Find the minimum distortion achieved by separate source and channel coding.

(b) Find the distortion achieved when the sender transmits $X_i = U_i$, $i \in [1 : n]$, i.e., performs no coding, and the receiver uses the (linear) MMSE estimate $\hat{U}_i$ of U_i given Y_i. Compare this to the distortion in part (a) and comment on the results.

3.21. *Two reconstructions.* Let X be a DMS, and $d_1(x, \hat{x}_1)$, $\hat{x}_1 \in \hat{\mathcal{X}}_1$, and $d_2(x, \hat{x}_2)$, $\hat{x}_2 \in \hat{\mathcal{X}}_2$, be two distortion measures. We wish to reconstruct X under both distortion measures from the same description as depicted in Figure 3.16.

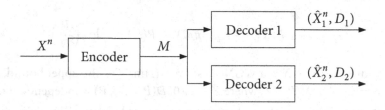

Figure 3.16. Lossy source coding with two reconstructions.

Define a $(2^{nR}, n)$ code, achievability of the rate–distortion triple (R, D_1, D_2), and the rate–distortion function $R(D_1, D_2)$ in the standard way. Show that

$$R(D_1, D_2) = \min_{p(\hat{x}_1, \hat{x}_2 \mid x): E(d_j(x, \hat{x}_j)) \leq D_j, \, j=1,2} I(X; \hat{X}_1, \hat{X}_2).$$

3.22. *Lossy source coding with reconstruction cost.* Let X be a DMS and $d(x, \hat{x})$ be a distortion measure. Further let $b(\hat{x}) \geq 0$ be a cost function on $\hat{\mathcal{X}}$. Suppose that there is an average cost constraint on each reconstruction sequence $\hat{x}^n(m)$,

$$b(\hat{x}^n(m)) \leq \sum_{i=1}^{n} b(\hat{x}_i(m)) \leq nB \quad \text{for every } m \in [1 : 2^{nR}),$$

in addition to the distortion constraint $E(d(X^n, \hat{X}^n)) \leq D$. Define a $(2^{nR}, n)$ code, achievability of the triple (R, D, B), and rate–distortion–cost function $R(D, B)$ in the standard way. Show that

$$R(D, B) = \min_{p(\hat{x}|x):\mathsf{E}(d(X,\hat{X}))\leq D,\, \mathsf{E}(b(\hat{X}))\leq B} I(X; \hat{X}).$$

Note that this problem is not a special case of the above two-reconstruction problem.

APPENDIX 3A   PROOF OF LEMMA 3.2

We first note that $I([X]_j; [Y_j]_k) \to I([X]_j; Y_j) = h(Y_j) - h(Z)$ as $k \to \infty$. This follows since $([Y_j]_k - Y_j)$ tends to zero as $k \to \infty$; recall Section 2.3. Hence it suffices to show that

$$\liminf_{j\to\infty} h(Y_j) \geq h(Y).$$

First note that the pdf of Y_j converges pointwise to that of $Y \sim \mathrm{N}(0, S + 1)$. To prove this, consider

$$f_{Y_j}(y) = \int f_Z(y - x)\, dF_{[X]_j}(x) = \mathsf{E}(f_Z(y - [X]_j)).$$

Since the Gaussian pdf $f_Z(z)$ is continuous and bounded, $f_{Y_j}(y)$ converges to $f_Y(y)$ for every y by the weak convergence of $[X]_j$ to X. Furthermore, we have

$$f_{Y_j}(y) = \mathsf{E}(f_Z(y - [X]_j)) \leq \max_z f_Z(z) = \frac{1}{\sqrt{2\pi}}.$$

Hence, for each $a > 0$, by the dominated convergence theorem (Appendix B),

$$h(Y_j) = \int_{-\infty}^{\infty} -f_{Y_j}(y) \log f_{Y_j}(y)\, dy$$

$$\geq \int_{-a}^{a} -f_{Y_j}(y) \log f_{Y_j}(y)\, dy + \mathsf{P}\{|Y_j| \geq a\} \cdot \min_y (-\log f_{Y_j}(y)),$$

which converges to

$$\int_{-a}^{a} -f(y) \log f(y)\, dy + \mathsf{P}\{|Y| \geq a\} \cdot \min_y (-\log f(y))$$

as $j \to \infty$. Taking $a \to \infty$, we obtain the desired result.

PART II

SINGLE-HOP NETWORKS

CHAPTER 4

Multiple Access Channels

We introduce the multiple access channel as a simple model for noisy many-to-one communication, such as the uplink of a cellular system, medium access in a local area network (LAN), or multiple reporters asking questions in a press conference. We then establish the main result of this chapter, which is a computable characterization of the capacity region of the two-sender multiple access channel with independent messages. The proof involves the new techniques of successive cancellation decoding, time sharing, simultaneous decoding, and coded time sharing. This result is extended to establish the capacity region of the Gaussian multiple access channel. We show that successive cancellation decoding outperforms the point-to-point based coding schemes of time-division multiple access and treating the other sender's signal as noise. Finally, we extend these results to multiple access channels with more than two senders.

4.1 DISCRETE MEMORYLESS MULTIPLE ACCESS CHANNEL

Consider the multiple access communication system model depicted in Figure 4.1. Each sender wishes to communicate an *independent* message reliably to a common receiver. As such, sender $j = 1, 2$ encodes its message M_j into a codeword X_j^n and transmits it over the shared channel. Upon receiving the sequence Y^n, the decoder finds estimates $\hat{M}_j$, $j = 1, 2$, of the messages. Since the senders transmit over a common noisy channel, a tradeoff arises between the rates of reliable communication for the two messages—when one sender transmits at a high rate, the other sender may need to back off its rate to ensure reliable communication of both messages. As in the point-to-point communication case, we study the limit on this tradeoff when there is no constraint on the code block length n.

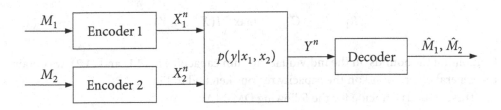

Figure 4.1. Multiple access communication system with independent messages.

We first consider a 2-sender *discrete memoryless multiple access channel* (DM-MAC) model $(\mathcal{X}_1 \times \mathcal{X}_2, p(y|x_1, x_2), \mathcal{Y})$ that consists of three finite sets $\mathcal{X}_1, \mathcal{X}_2, \mathcal{Y}$, and a collection of conditional pmfs $p(y|x_1, x_2)$ on $\mathcal{Y}$ (one for each input symbol pair (x_1, x_2)).

A $(2^{nR_1}, 2^{nR_2}, n)$ code for the DM-MAC consists of

- two message sets $[1 : 2^{nR_1}]$ and $[1 : 2^{nR_2}]$,

- two encoders, where encoder 1 assigns a codeword $x_1^n(m_1)$ to each message $m_1 \in [1 : 2^{nR_1}]$ and encoder 2 assigns a codeword $x_2^n(m_2)$ to each message $m_2 \in [1 : 2^{nR_2}]$, and

- a decoder that assigns an estimate $(\hat{m}_1, \hat{m}_2) \in [1 : 2^{nR_1}] \times [1 : 2^{nR_2}]$ or an error message e to each received sequence y^n.

We assume that the message pair (M_1, M_2) is uniformly distributed over $[1 : 2^{nR_1}] \times [1 : 2^{nR_2}]$. Consequently, $x_1^n(M_1)$ and $x_2^n(M_2)$ are independent. The average probability of error is defined as

$$P_e^{(n)} = \mathsf{P}\{(\hat{M}_1, \hat{M}_2) \neq (M_1, M_2)\}.$$

A rate pair (R_1, R_2) is said to be *achievable* for the DM-MAC if there exists a sequence of $(2^{nR_1}, 2^{nR_2}, n)$ codes such that $\lim_{n \to \infty} P_e^{(n)} = 0$. The *capacity region* $\mathscr{C}$ of the DM-MAC is the closure of the set of achievable rate pairs (R_1, R_2). Sometimes we are interested in the sum-capacity C_{sum} of the DM-MAC defined as $C_{\text{sum}} = \max\{R_1 + R_2 : (R_1, R_2) \in \mathscr{C}\}$.

4.2 SIMPLE BOUNDS ON THE CAPACITY REGION

We begin with simple bounds on the capacity region of the DM-MAC. For this discussion, note that the maximum achievable individual rates are

$$
\begin{aligned}
C_1 &= \max_{x_2, p(x_1)} I(X_1; Y | X_2 = x_2), \\
C_2 &= \max_{x_1, p(x_2)} I(X_2; Y | X_1 = x_1).
\end{aligned}
\tag{4.1}
$$

Using these rates, we can readily achieve any rate pairs below the line segment between $(C_1, 0)$ and $(0, C_2)$ using *time division* (or *frequency division*), which yields the inner bound sketched in Figure 4.2. Following similar steps to the converse proof of the channel coding theorem in Section 3.1.4, we can show that the sum-rate is upper bounded as

$$R_1 + R_2 \leq C_{12} = \max_{p(x_1)p(x_2)} I(X_1, X_2; Y).
\tag{4.2}$$

Combining the bounds on the individual and sum capacities in (4.1) and (4.2), we obtain the general outer bound on the capacity region sketched in Figure 4.2.

These bounds coincide for the following DM-MAC.

Example 4.1 (Binary multiplier MAC). Suppose that the inputs X_1 and X_2 are binary and the output $Y = X_1 \cdot X_2$. Then it can be easily checked that $C_1 = C_2 = C_{12} = 1$. Thus

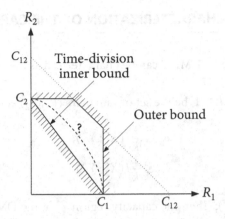

Figure 4.2. Time-division inner bound and an outer bound on the capacity region.

the inner and outer bounds in Figure 4.2 coincide and the capacity region is the set of rate pairs (R_1, R_2) such that

$$R_1 \le 1,$$
$$R_2 \le 1,$$
$$R_1 + R_2 \le 1.$$

This is plotted in Figure 4.3a.

The bounds in Figure 4.2 do not coincide in general, however.

Example 4.2 (Binary erasure MAC). Suppose that the inputs X_1 and X_2 are binary and the output $Y = X_1 + X_2$ is ternary. Again it can be easily checked that $C_1 = C_2 = 1$ and $C_{12} = 3/2$. Hence, the inner and outer bounds do not coincide for this channel. We will shortly see that the outer bound is the capacity region. This is plotted in Figure 4.3b.

In general, neither the inner bound nor the outer bound in Figure 4.2 is tight.

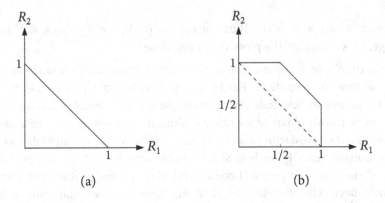

Figure 4.3. Capacity regions of (a) the binary multiplier MAC, and (b) the binary erasure MAC.

4.3*   MULTILETTER CHARACTERIZATION OF THE CAPACITY REGION

The capacity region of the DM-MAC can be characterized as follows.

Theorem 4.1. Let $\mathscr{C}^{(k)}$, $k \geq 1$, be the set of rate pairs (R_1, R_2) such that

$$R_1 \leq \frac{1}{k} I(X_1^k; Y^k),$$

$$R_2 \leq \frac{1}{k} I(X_2^k; Y^k)$$

for some pmf $p(x_1^k)p(x_2^k)$. Then the capacity region $\mathscr{C}$ of the DM-MAC $p(y|x_1, x_2)$ is the closure of the set

$$\bigcup_k \mathscr{C}^{(k)}.$$

Proof of achievability. We code over k symbols of $\mathcal{X}_1$ and $\mathcal{X}_2$ (super-symbols) together. Fix a product pmf $p(x_1^k)p(x_2^k)$. Randomly and independently generate 2^{nkR_1} sequences $x_1^{nk}(m_1)$, $m_1 \in [1 : 2^{nkR_1}]$, each according to $\prod_{i=1}^n p_{X_1^k}(x_{1,(i-1)k+1}^{ik})$. Similarly generate 2^{nkR_2} sequences $x_2^{nk}(m_2)$, $m_2 \in [1 : 2^{nkR_2}]$, each according to $\prod_{i=1}^n p_{X_2^k}(x_{2,(i-1)k+1}^{ik})$. Upon receiving the sequence y^{nk}, the decoder uses joint typicality decoding to find an estimate for each message separately. It can be readily shown that the average probability of error tends to zero as $n \to \infty$ if $kR_1 < I(X_1^k; Y^k) - \delta(\epsilon)$ and $kR_2 < I(X_2^k; Y^k) - \delta(\epsilon)$.

Proof of the converse. Using Fano's inequality, we can show that

$$R_1 \leq \frac{1}{n} I(X_1^n; Y^n) + \epsilon_n,$$

$$R_2 \leq \frac{1}{n} I(X_2^n; Y^n) + \epsilon_n,$$

where ϵ_n tends to zero as $n \to \infty$. Thus, for any $\epsilon > 0$, $(R_1 - \epsilon, R_2 - \epsilon) \in \mathscr{C}^{(n)}$ for n sufficiently large. This completes the proof of the converse.

Although the above *multiletter* characterization of the capacity region is well-defined, it is not clear how to compute it. Furthermore, this characterization does not provide any insight into how to best code for the multiple access channel. Such multiletter expressions can be readily obtained for other multiuser channels and sources, leading to a fairly complete but unsatisfactory theory. Consequently, we seek computable *single-letter* characterizations of capacity, such as Shannon's capacity formula for the point-to-point channel, that shed light on practical coding techniques. Single-letter characterizations, however, have been difficult to find for most multiuser channels and sources. The DM-MAC is one of rare channels for which a complete single-letter characterization of the capacity region is known. We develop this characterization over the next several sections.

4.4 TIME SHARING

In time/frequency division, only one sender transmits in each time slot/frequency band. Time sharing generalizes time division by allowing the senders to transmit simultaneously at different nonzero rates in each slot.

> **Proposition 4.1.** If the rate pairs (R_{11}, R_{21}) and (R_{12}, R_{22}) are achievable for the DM-MAC $p(y|x_1, x_2)$, then the rate pair $(R_1, R_2) = (\alpha R_{11} + \bar{\alpha} R_{12}, \alpha R_{21} + \bar{\alpha} R_{22})$ is achievable for every $\alpha \in [0, 1]$.

Proof. Consider two sequences of codes, one achieving (R_{11}, R_{21}) and the other achieving (R_{12}, R_{22}). For each block length n, assume without loss of generality that αn is an integer. Consider the $(2^{\alpha n R_{11}}, 2^{\alpha n R_{21}}, \alpha n)$ and $(2^{\bar{\alpha} n R_{12}}, 2^{\bar{\alpha} n R_{22}}, \bar{\alpha} n)$ codes from the given first and second sequences of codes, respectively.

To send the message pair (M_1, M_2), we perform *rate splitting*. We represent the message M_1 by independent messages M_{11} at rate αR_{11} and M_{12} at rate $\bar{\alpha} R_{12}$. Similarly, we represent the message M_2 by M_{21} at rate αR_{21} and M_{22} at rate $\bar{\alpha} R_{22}$. Thus, $R_1 = \alpha R_{11} + \bar{\alpha} R_{12}$ and $R_2 = \alpha R_{21} + \bar{\alpha} R_{22}$.

For the first αn transmissions, sender $j = 1, 2$ transmits its codeword for M_{j1} from the $(2^{\alpha n R_{11}}, 2^{\alpha n R_{21}}, \alpha n)$ code and for the rest of the transmissions, it transmits its codeword for M_{j2} from the $(2^{\bar{\alpha} n R_{12}}, 2^{\bar{\alpha} n R_{22}}, \bar{\alpha} n)$ code. Upon receiving y^n, the receiver decodes $y^{\alpha n}$ using the decoder of the first code and $y^n_{\alpha n+1}$ using the decoder of the second code.

By assumption, the probability of error for each decoder tends to zero as $n \to \infty$. Hence, by the union of events bound, the probability of decoding error tends to zero as $n \to \infty$ and the rate pair $(R_1, R_2) = (\alpha R_{11} + \bar{\alpha} R_{12}, \alpha R_{21} + \bar{\alpha} R_{22})$ is achievable. This completes the proof of the proposition.

Remark 4.1. Time division and frequency division are special cases of time sharing, in which the senders transmit at rate pairs $(R_1, 0)$ and $(0, R_2)$.

Remark 4.2. The above *time-sharing argument* shows that the capacity region of the DM-MAC is convex. Note that this proof uses the *operational* definition of the capacity region (as opposed to the information definition in terms of mutual information).

Remark 4.3. Similar time-sharing arguments can be used to show the convexity of the capacity region of *any* (synchronous) communication channel for which capacity is defined as the optimal rate of block codes, e.g., capacity with cost constraint, as well as optimal rate regions for source coding problems, e.g., rate–distortion function in Chapter 3. As we will see in Chapter 24, when the sender transmissions in the DM-MAC are not synchronized, time sharing becomes infeasible and consequently the capacity region is not necessarily convex.

Remark 4.4. The rate-splitting technique will be used in other coding schemes later; for example, see Section 6.5.

4.5   SINGLE-LETTER CHARACTERIZATION OF THE CAPACITY REGION

We are now ready to present a single-letter characterization of the DM-MAC capacity region. Let $(X_1, X_2) \sim p(x_1)p(x_2)$. Let $\mathscr{R}(X_1, X_2)$ be the set of rate pairs (R_1, R_2) such that

$$R_1 \le I(X_1; Y | X_2),$$
$$R_2 \le I(X_2; Y | X_1),$$
$$R_1 + R_2 \le I(X_1, X_2; Y).$$

As shown in Figure 4.4, this set is in general a pentagonal region with a 45° side because

$$\max\{I(X_1; Y | X_2), I(X_2; Y | X_1)\} \le I(X_1, X_2; Y) \le I(X_1; Y | X_2) + I(X_2; Y | X_1).$$

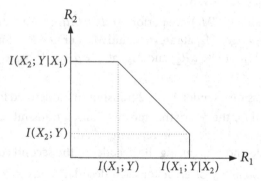

Figure 4.4. The region $\mathscr{R}(X_1, X_2)$ for a typical DM-MAC.

For example, consider the binary erasure MAC in Example 4.2. Setting $X_1, X_2 \sim$ Bern$(1/2)$ to be independent, the corresponding $\mathscr{R}(X_1, X_2)$ is the set of rate pairs (R_1, R_2) such that

$$R_1 \le H(Y | X_2) = 1,$$
$$R_2 \le H(Y | X_1) = 1,$$
$$R_1 + R_2 \le H(Y) = \frac{3}{2}.$$

This region coincides with the outer bound in Figure 4.2; hence it is the capacity region.

In general, the capacity region of the DM-MAC includes multiple $\mathscr{R}(X_1, X_2)$ regions as depicted in Figure 4.5.

Theorem 4.2. The capacity region $\mathscr{C}$ of the DM-MAC $p(y|x_1, x_2)$ is the convex closure of $\bigcup_{p(x_1)p(x_2)} \mathscr{R}(X_1, X_2)$.

We prove Theorem 4.2 in the following two subsections.

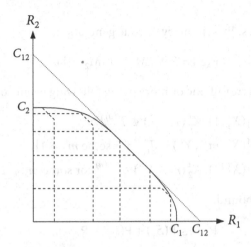

Figure 4.5. The capacity region of a typical DM-MAC. The individual capacities C_1, C_2, and the sum-capacity C_{12} are defined in (4.1) and (4.2), respectively.

4.5.1 Proof of Achievability

Let $(X_1, X_2) \sim p(x_1)p(x_2)$. We show that every rate pair (R_1, R_2) in the interior of the region $\mathcal{R}(X_1, X_2)$ is achievable. The rest of the capacity region is achieved using time sharing between points in different $\mathcal{R}(X_1, X_2)$ regions. Assume nR_1 and nR_2 to be integers.

Codebook generation. Randomly and independently generate 2^{nR_1} sequences $x_1^n(m_1)$, $m_1 \in [1 : 2^{nR_1}]$, each according to $\prod_{i=1}^{n} p_{X_1}(x_{1i})$. Similarly generate 2^{nR_2} sequences $x_2^n(m_2)$, $m_2 \in [1 : 2^{nR_2}]$, each according to $\prod_{i=1}^{n} p_{X_2}(x_{2i})$. These codewords constitute the codebook, which is revealed to the encoders and the decoder.

Encoding. To send message m_1, encoder 1 transmits $x_1^n(m_1)$. Similarly, to send m_2, encoder 2 transmits $x_2^n(m_2)$.

In the following, we consider two decoding rules.

Successive cancellation decoding. This decoding rule aims to achieve one of the two *corner points* of the pentagonal region $\mathcal{R}(X_1, X_2)$, for example,

$$R_1 < I(X_1; Y),$$
$$R_2 < I(X_2; Y|X_1).$$

Decoding is performed in two steps:

1. The decoder declares that $\hat{m}_1$ is sent if it is the unique message such that $(x_1^n(\hat{m}_1), y^n) \in \mathcal{T}_\epsilon^{(n)}$; otherwise it declares an error.

2. If such $\hat{m}_1$ is found, the decoder finds the unique $\hat{m}_2$ such that $(x_1^n(\hat{m}_1), x_2^n(\hat{m}_2), y^n) \in \mathcal{T}_\epsilon^{(n)}$; otherwise it declares an error.

Analysis of the probability of error. We bound the probability of error averaged over codebooks and messages. By symmetry of code generation,

$$P(\mathcal{E}) = P(\mathcal{E} \mid M_1 = 1, M_2 = 1).$$

The decoder makes an error iff one or more of the following events occur:

$$\mathcal{E}_1 = \{(X_1^n(1), X_2^n(1), Y^n) \notin \mathcal{T}_\epsilon^{(n)}\},$$
$$\mathcal{E}_2 = \{(X_1^n(m_1), Y^n) \in \mathcal{T}_\epsilon^{(n)} \text{ for some } m_1 \neq 1\},$$
$$\mathcal{E}_3 = \{(X_1^n(1), X_2^n(m_2), Y^n) \in \mathcal{T}_\epsilon^{(n)} \text{ for some } m_2 \neq 1\}.$$

By the union of events bound,

$$P(\mathcal{E}) \leq P(\mathcal{E}_1) + P(\mathcal{E}_2) + P(\mathcal{E}_3).$$

We bound each term. By the LLN, the first term $P(\mathcal{E}_1)$ tends to zero as $n \to \infty$. For the second term, note that for $m_1 \neq 1$, $(X_1^n(m_1), Y^n) \sim \prod_{i=1}^n p_{X_1}(x_{1i}) p_Y(y_i)$. Hence by the packing lemma in Section 3.2 with $\mathcal{A} = [2 : 2^{nR_1}]$, $X \leftarrow X_1$, and $U = \emptyset$, $P(\mathcal{E}_2)$ tends to zero as $n \to \infty$ if $R_1 < I(X_1; Y) - \delta(\epsilon)$.

For the third term, note that for $m_2 \neq 1$, $X_2^n(m_2) \sim \prod_{i=1}^n p_{X_2}(x_{2i})$ is independent of $(X_1^n(1), Y^n) \sim \prod_{i=1}^n p_{X_1,Y}(x_{1i}, y_i)$. Hence, again by the packing lemma with $\mathcal{A} = [2 : 2^{nR_2}]$, $X \leftarrow X_2$, $Y \leftarrow (X_1, Y)$, and $U = \emptyset$, $P(\mathcal{E}_3)$ tends to zero as $n \to \infty$ if $R_2 < I(X_2; Y, X_1) - \delta(\epsilon)$, or equivalently—since X_1 and X_2 are independent—if $R_2 < I(X_2; Y|X_1) - \delta(\epsilon)$. Thus, the total average probability of decoding error $P(\mathcal{E})$ tends to zero as $n \to \infty$ if $R_1 < I(X_1; Y) - \delta(\epsilon)$ and $R_2 < I(X_2; Y|X_1) - \delta(\epsilon)$. Since the probability of error averaged over codebooks, $P(\mathcal{E})$, tends to zero as $n \to \infty$, there must exist a sequence of $(2^{nR_1}, 2^{nR_2}, n)$ codes such that $\lim_{n \to \infty} P_e^{(n)} = 0$.

Achievability of the other corner point of $\mathcal{R}(X_1, X_2)$ follows by changing the decoding order. To show achievability of other points in $\mathcal{R}(X_1, X_2)$, we use time sharing between corner points and points on the axes. Finally, to show achievability of points in $\mathscr{C}$ that are not in any single $\mathcal{R}(X_1, X_2)$ region, we use time sharing between points in these regions.

Simultaneous decoding. We can prove achievability of *every* rate pair in the interior of $\mathcal{R}(X_1, X_2)$ *without* time sharing. The decoder declares that $(\hat{m}_1, \hat{m}_2)$ is sent if it is the unique message pair such that $(x_1^n(\hat{m}_1), x_2^n(\hat{m}_2), y^n) \in \mathcal{T}_\epsilon^{(n)}$; otherwise it declares an error.

Analysis of the probability of error. As before, we bound the probability of error averaged over codebooks and messages. To analyze this probability, consider all possible pmfs induced on the triple $(X_1^n(m_1), X_2^n(m_2), Y^n)$ as listed in Table 4.1.

Then the error event $\mathcal{E}$ occurs iff one or more of the following events occur:

$$\mathcal{E}_1 = \{(X_1^n(1), X_2^n(1), Y^n) \notin \mathcal{T}_\epsilon^{(n)}\},$$
$$\mathcal{E}_2 = \{(X_1^n(m_1), X_2^n(1), Y^n) \in \mathcal{T}_\epsilon^{(n)} \text{ for some } m_1 \neq 1\},$$
$$\mathcal{E}_3 = \{(X_1^n(1), X_2^n(m_2), Y^n) \in \mathcal{T}_\epsilon^{(n)} \text{ for some } m_2 \neq 1\},$$
$$\mathcal{E}_4 = \{(X_1^n(m_1), X_2^n(m_2), Y^n) \in \mathcal{T}_\epsilon^{(n)} \text{ for some } m_1 \neq 1, m_2 \neq 1\}.$$

m_1	m_2	Joint pmf
1	1	$p(x_1^n)p(x_2^n)p(y^n\|x_1^n, x_2^n)$
$*$	1	$p(x_1^n)p(x_2^n)p(y^n\|x_2^n)$
1	$*$	$p(x_1^n)p(x_2^n)p(y^n\|x_1^n)$
$*$	$*$	$p(x_1^n)p(x_2^n)p(y^n)$

Table 4.1. The joint pmfs induced by different (m_1, m_2) pairs. The $*$ symbol corresponds to message m_1 or $m_2 \neq 1$.

Thus by the union of events bound,

$$P(\mathcal{E}) \leq P(\mathcal{E}_1) + P(\mathcal{E}_2) + P(\mathcal{E}_3) + P(\mathcal{E}_4).$$

We now bound each term. By the LLN, $P(\mathcal{E}_1)$ tends to zero as $n \to \infty$. By the packing lemma, $P(\mathcal{E}_2)$ tends to zero as $n \to \infty$ if $R_1 < I(X_1; Y|X_2) - \delta(\epsilon)$. Similarly, $P(\mathcal{E}_3)$ tends to zero as $n \to \infty$ if $R_2 < I(X_2; Y|X_1) - \delta(\epsilon)$. Finally, since for $m_1 \neq 1, m_2 \neq 1$, $(X_1^n(m_1), X_2^n(m_2))$ is independent of $(X_1^n(1), X_2^n(1), Y^n)$, again by the packing lemma with $\mathcal{A} = [2 : 2^{nR_1}] \times [2 : 2^{nR_2}]$, $X \leftarrow (X_1, X_2)$, and $U = \emptyset$, $P(\mathcal{E}_4)$ tends to zero as $n \to \infty$ if $R_1 + R_2 < I(X_1, X_2; Y) - \delta(\epsilon)$.

As in successive cancellation decoding, to show achievability of points in $\mathscr{C}$ that are not in any single $\mathscr{R}(X_1, X_2)$ region, we use time sharing between points in different such regions.

Remark 4.5. For the DM-MAC, simultaneous decoding does not achieve higher rates than successive cancellation decoding with time sharing. We will encounter several scenarios throughout the book, where simultaneous decoding achieves *strictly* higher rates than sequential decoding schemes.

Remark 4.6. Unlike the capacity of the DMC, the capacity region of the DM-MAC with the maximal probability of error can be strictly *smaller* than that with the average probability of error. However, by allowing randomization at the encoders, the capacity region with the maximal probability of error can be shown to be equal to that with the average probability of error.

4.5.2   Proof of the Converse

We need to show that given any sequence of $(2^{nR_1}, 2^{nR_2}, n)$ codes with $\lim_{n \to \infty} P_e^{(n)} = 0$, then we must have $(R_1, R_2) \in \mathscr{C}$ as defined in Theorem 4.2. First note that each code induces the joint pmf

$$(M_1, M_2, X_1^n, X_2^n, Y^n) \sim 2^{-n(R_1 + R_2)} p(x_1^n|m_1)p(x_2^n|m_2) \prod_{i=1}^{n} p_{Y|X_1, X_2}(y_i|x_{1i}, x_{2i}).$$

By Fano's inequality,

$$H(M_1, M_2|Y^n) \leq n(R_1 + R_2)P_e^{(n)} + 1 = n\epsilon_n, \tag{4.3}$$

where ϵ_n tends to zero as $n \to \infty$. Now, using similar steps to the converse proof of the channel coding theorem in Section 3.1.1, it is easy to show that

$$n(R_1 + R_2) \leq \sum_{i=1}^{n} I(X_{1i}, X_{2i}; Y_i) + n\epsilon_n.$$

Next, note from (4.3) that $H(M_1|Y^n, M_2) \leq H(M_1, M_2|Y^n) \leq n\epsilon_n$. Hence

$$
\begin{aligned}
nR_1 &= H(M_1) \\
&= H(M_1|M_2) \\
&= I(M_1; Y^n|M_2) + H(M_1|Y^n, M_2) \\
&\leq I(M_1; Y^n|M_2) + n\epsilon_n \\
&= \sum_{i=1}^{n} I(M_1; Y_i|Y^{i-1}, M_2) + n\epsilon_n \\
&\overset{(a)}{=} \sum_{i=1}^{n} I(M_1; Y_i|Y^{i-1}, M_2, X_{2i}) + n\epsilon_n \\
&\leq \sum_{i=1}^{n} I(M_1, M_2, Y^{i-1}; Y_i|X_{2i}) + n\epsilon_n \\
&\overset{(b)}{=} \sum_{i=1}^{n} I(X_{1i}, M_1, M_2, Y^{i-1}; Y_i|X_{2i}) + n\epsilon_n \\
&= \sum_{i=1}^{n} I(X_{1i}; Y_i|X_{2i}) + \sum_{i=1}^{n} I(M_1, M_2, Y^{i-1}; Y_i|X_{1i}, X_{2i}) + n\epsilon_n \\
&\overset{(c)}{=} \sum_{i=1}^{n} I(X_{1i}; Y_i|X_{2i}) + n\epsilon_n,
\end{aligned}
$$

where (a) and (b) follow since X_{ji} is a function of M_j for $j = 1, 2$, respectively, and (c) follows by the memoryless property of the channel, which implies that $(M_1, M_2, Y^{i-1}) \to (X_{1i}, X_{2i}) \to Y_i$ form a Markov chain. Similarly bounding R_2, we have shown that

$$R_1 \leq \frac{1}{n} \sum_{i=1}^{n} I(X_{1i}; Y_i|X_{2i}) + \epsilon_n,$$

$$R_2 \leq \frac{1}{n} \sum_{i=1}^{n} I(X_{2i}; Y_i|X_{1i}) + \epsilon_n,$$

$$R_1 + R_2 \leq \frac{1}{n} \sum_{i=1}^{n} I(X_{1i}, X_{2i}; Y_i) + \epsilon_n.$$

Since M_1 and M_2 are independent, so are $X_{1i}(M_1)$ and $X_{2i}(M_2)$ for all i. Note that bounding each of the above terms by its corresponding capacity, i.e., C_1, C_2, and C_{12} in (4.1) and (4.2), respectively, yields the simple outer bound in Figure 4.2, which is in general larger than the inner bound we established.

Continuing the proof, let the random variable $Q \sim \text{Unif}[1 : n]$ be independent of (X_1^n, X_2^n, Y^n). Then, we can write

$$R_1 \le \frac{1}{n} \sum_{i=1}^{n} I(X_{1i}; Y_i \mid X_{2i}) + \epsilon_n$$

$$= \frac{1}{n} \sum_{i=1}^{n} I(X_{1i}; Y_i \mid X_{2i}, Q = i) + \epsilon_n$$

$$= I(X_{1Q}; Y_Q \mid X_{2Q}, Q) + \epsilon_n.$$

We now observe that $Y_Q \mid \{X_{1Q} = x_1, X_{2Q} = x_2\} \sim p(y \mid x_1, x_2)$, that is, it is distributed according to the channel conditional pmf. Hence, we identify $X_1 = X_{1Q}$, $X_2 = X_{2Q}$, and $Y = Y_Q$ to obtain

$$R_1 \le I(X_1; Y \mid X_2, Q) + \epsilon_n.$$

We can similarly bound R_2 and $R_1 + R_2$. Note that by the independence of X_1^n and X_2^n, X_1 and X_2 are conditionally independent given Q. Thus, the rate pair (R_1, R_2) must satisfy the inequalities

$$R_1 \le I(X_1; Y \mid X_2, Q) + \epsilon_n,$$
$$R_2 \le I(X_2; Y \mid X_1, Q) + \epsilon_n,$$
$$R_1 + R_2 \le I(X_1, X_2; Y \mid Q) + \epsilon_n$$

for $Q \sim \text{Unif}[1 : n]$ and some pmf $p(x_1 \mid q)p(x_2 \mid q)$ and hence for some pmf $p(q)p(x_1 \mid q)$ $p(x_2 \mid q)$ with Q taking values in some finite set (independent of n). Since ϵ_n tends to zero as $n \to \infty$ by the assumption that $\lim_{n \to \infty} P_e^{(n)} = 0$, we have shown that the rate pair (R_1, R_2) must be in $\mathscr{C}'$, which is the closure of the set of rate pairs (R_1, R_2) such that

$$R_1 \le I(X_1; Y \mid X_2, Q),$$
$$R_2 \le I(X_2; Y \mid X_1, Q), \tag{4.4}$$
$$R_1 + R_2 \le I(X_1, X_2; Y \mid Q)$$

for some pmf $p(q)p(x_1 \mid q)p(x_2 \mid q)$ with Q in some finite set.

The random variable Q is referred to as a *time-sharing* random variable. It is an *auxiliary* random variable, that is, a random variable that is not part of the channel variables. Note that $Q \to (X_1, X_2) \to Y$ form a Markov chain.

To complete the proof of the converse, we need to show that the above region $\mathscr{C}'$ is identical to the region $\mathscr{C}$ in Theorem 4.2. We already know that $\mathscr{C} \subseteq \mathscr{C}'$ (because we proved that every achievable rate pair must be in $\mathscr{C}'$). We can see this directly also by noting that $\mathscr{C}'$ contains all $\mathscr{R}(X_1, X_2)$ regions by definition as well as all points in the convex closure of the union of these regions, since any such point can be represented as a convex combination of points in these regions using the time-sharing random variable Q.

It can be also shown that $\mathscr{C}' \subseteq \mathscr{C}$. Indeed, any joint pmf $p(q)p(x_1 \mid q)p(x_2 \mid q)$ defines a pentagonal region with a 45° side. Thus it suffices to check if the corner points of this pentagonal region are in $\mathscr{C}$. First consider the corner point $(I(X_1; Y|Q), I(X_2; Y|X_1, Q))$. It is easy to see that this corner point belongs to $\mathscr{C}$, since it is a finite convex combination of the points $(I(X_1; Y|Q = q), I(X_2; Y|X_1, Q = q))$, $q \in \mathcal{Q}$, each of which in turn belongs to $\mathscr{R}(X_{1q}, X_{2q})$ with $(X_{1q}, X_{2q}) \sim p(x_1|q)p(x_2|q)$. We can similarly show that the other corner points of the pentagonal region also belong to $\mathscr{C}$. This shows that $\mathscr{C}' = \mathscr{C}$, which completes the proof of Theorem 4.2.

However, neither the characterization $\mathscr{C}$ nor $\mathscr{C}'$ seems to be "computable." How many $\mathscr{R}(X_1, X_2)$ sets do we need to consider in computing each point on the boundary of $\mathscr{C}$? How large must the cardinality of $\mathcal{Q}$ be to compute each point on the boundary of $\mathscr{C}'$? By the convex cover method in Appendix C, we can show that it suffices to take the cardinality of the time-sharing random variable $|\mathcal{Q}| \le 3$ and hence at most three $\mathscr{R}(X_1, X_2)$ regions need to be considered for each point on the boundary of $\mathscr{C}$. By exploiting the special structure of the $\mathscr{R}(X_1, X_2)$ regions, we show in Appendix 4A that $|\mathcal{Q}| \le 2$ is sufficient. This yields the following computable characterization of the capacity region.

Theorem 4.3. The capacity region of the DM-MAC $p(y|x_1, x_2)$ is the set of rate pairs (R_1, R_2) such that

$$R_1 \le I(X_1; Y|X_2, Q),$$
$$R_2 \le I(X_2; Y|X_1, Q),$$
$$R_1 + R_2 \le I(X_1, X_2; Y|Q)$$

for some pmf $p(q)p(x_1|q)p(x_2|q)$ with the cardinality of Q bounded as $|\mathcal{Q}| \le 2$.

Remark 4.7. In the above characterization of the capacity region, it should be understood that $p(q, x_1, x_2, y) = p(q)p(x_1|q)p(x_2|q)p(y|x_1, x_2)$, i.e., $Q \to (X_1, X_2) \to Y$. Throughout the book we will encounter many such characterizations. We will not explicitly include the given source or channel pmf in the characterization. It should be understood that the joint pmf is the product of the given source or channel pmf and the pmf over which we are optimizing.

Remark 4.8. The region in the theorem is closed and convex.

Remark 4.9. It can be shown that time sharing is required for some DM-MACs, that is, setting $|\mathcal{Q}| = 1$ is not sufficient in general; see the push-to-talk MAC in Problem 4.2.

Remark 4.10. The sum-capacity of the DM-MAC can be expressed as

$$C_{\text{sum}} = \max_{p(q)p(x_1|q)p(x_2|q)} I(X_1, X_2; Y|Q)$$
$$\overset{(a)}{=} \max_{p(x_1)p(x_2)} I(X_1, X_2; Y),$$

where (a) follows since the average is upper bounded by the maximum. Thus the sum-capacity is "computable" even without any cardinality bound on Q. However, the second maximization problem is not convex in general, so there does not exist an efficient algorithm to compute C_{sum} from it.

4.5.3   Coded Time Sharing

It turns out that we can achieve the capacity region characterization in Theorem 4.3 directly without explicit time sharing. The proof involves the new technique of *coded time sharing* as described in the following alternative proof of achievability.

Codebook generation. Fix a pmf $p(q)p(x_1|q)p(x_2|q)$. Randomly generate a *time-sharing* sequence q^n according to $\prod_{i=1}^{n} p_Q(q_i)$. Randomly and conditionally independently generate 2^{nR_1} sequences $x_1^n(m_1)$, $m_1 \in [1 : 2^{nR_1}]$, each according to $\prod_{i=1}^{n} p_{X_1|Q}(x_{1i}|q_i)$. Similarly generate 2^{nR_2} sequences $x_2^n(m_2)$, $m_2 \in [1 : 2^{nR_2}]$, each according to $\prod_{i=1}^{n} p_{X_2|Q}(x_{2i}|q_i)$. The chosen codebook, including q^n, is revealed to the encoders and the decoder.

Encoding. To send (m_1, m_2), transmit $x_1^n(m_1)$ and $x_2^n(m_2)$.

Decoding. The decoder declares that $(\hat{m}_1, \hat{m}_2)$ is sent if it is the unique message pair such that $(q^n, x_1^n(\hat{m}_1), x_2^n(\hat{m}_2), y^n) \in \mathcal{T}_\epsilon^{(n)}$; otherwise it declares an error.

Analysis of the probability of error. The decoder makes an error iff one or more of the following events occur:

$$\mathcal{E}_1 = \{(Q^n, X_1^n(1), X_2^n(1), Y^n) \notin \mathcal{T}_\epsilon^{(n)}\},$$
$$\mathcal{E}_2 = \{(Q^n, X_1^n(m_1), X_2^n(1), Y^n) \in \mathcal{T}_\epsilon^{(n)} \text{ for some } m_1 \neq 1\},$$
$$\mathcal{E}_3 = \{(Q^n, X_1^n(1), X_2^n(m_2), Y^n) \in \mathcal{T}_\epsilon^{(n)} \text{ for some } m_2 \neq 1\},$$
$$\mathcal{E}_4 = \{(Q^n, X_1^n(m_1), X_2^n(m_2), Y^n) \in \mathcal{T}_\epsilon^{(n)} \text{ for some } m_1 \neq 1, m_2 \neq 1\}.$$

Hence $\mathsf{P}(\mathcal{E}) \leq \sum_{j=1}^{4} \mathsf{P}(\mathcal{E}_j)$. By the LLN, $\mathsf{P}(\mathcal{E}_1)$ tends to zero as $n \to \infty$. Since for $m_1 \neq 1$, $X_1^n(m_1)$ is conditionally independent of $(X_2^n(1), Y^n)$ given Q^n, by the packing lemma with $\mathcal{A} = [2 : 2^{nR_1}]$, $Y \leftarrow (X_2, Y)$, and $U \leftarrow Q$, $\mathsf{P}(\mathcal{E}_2)$ tends to zero as $n \to \infty$ if $R_1 < I(X_1; X_2, Y|Q) - \delta(\epsilon) = I(X_1; Y|X_2, Q) - \delta(\epsilon)$. Similarly, $\mathsf{P}(\mathcal{E}_3)$ tends to zero as $n \to \infty$ if $R_2 < I(X_2; Y|X_1, Q) - \delta(\epsilon)$. Finally, since for $m_1 \neq 1, m_2 \neq 1$, $(X_1^n(m_1), X_2^n(m_2))$ is conditionally independent of Y^n given Q^n, again by the packing lemma, $\mathsf{P}(\mathcal{E}_4)$ tends to zero as $n \to \infty$ if $R_1 + R_2 < I(X_1, X_2; Y|Q) - \delta(\epsilon)$. The rest of the proof follows as before.

Remark 4.11. As we will see, for example in Chapter 6, coded time sharing is necessary when an achievable rate region cannot be equivalently expressed as the convex closure of the union of rate regions that are achievable without time sharing; see Problem 4.4.

4.6   GAUSSIAN MULTIPLE ACCESS CHANNEL

Consider the 2-sender discrete-time additive white Gaussian noise MAC (Gaussian MAC in short) depicted in Figure 4.6, which is a simple model, for example, for uplink (handset-to-base station) channels in cellular systems. The channel output corresponding to the inputs X_1 and X_2 is

$$Y = g_1 X_1 + g_2 X_2 + Z,$$

where g_1 and g_2 are the channel gains and $Z \sim N(0, N_0/2)$ is the noise. Thus in transmission time $i \in [1 : n]$, the channel output is

$$Y_i = g_1 X_{1i} + g_2 X_{2i} + Z_i,$$

where $\{Z_i\}$ is a WGN$(N_0/2)$ process, independent of the channel inputs X_1^n and X_2^n (when no feedback is present). We assume average transmission power constraints

$$\sum_{i=1}^{n} x_{ji}^2(m_j) \le nP, \quad m_j \in [1 : 2^{nR_j}], \ j = 1, 2.$$

Assume without loss of generality that $N_0/2 = 1$ and define the received powers (SNRs) as $S_j = g_j^2 P$, $j = 1, 2$. As we will see, the capacity region depends only on S_1 and S_2.

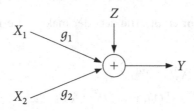

Figure 4.6. Gaussian multiple access channel.

4.6.1 Capacity Region of the Gaussian MAC

The capacity region of the Gaussian MAC is a pentagonal region with corner points characterized by point-to-point Gaussian channel capacities.

Theorem 4.4. The capacity region of the Gaussian MAC is the set of rate pairs (R_1, R_2) such that

$$R_1 \le C(S_1),$$
$$R_2 \le C(S_2),$$
$$R_1 + R_2 \le C(S_1 + S_2),$$

where $C(x)$ is the Gaussian capacity function.

This region is sketched in Figure 4.7. Note that the capacity region coincides with the simple outer bound in Section 4.2.

Proof of achievability. Consider the $\mathcal{R}(X_1, X_2)$ region where X_1 and X_2 are $N(0, P)$,

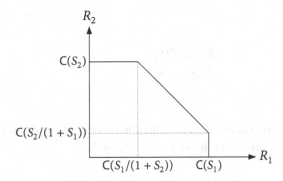

Figure 4.7. Capacity region of the Gaussian MAC.

independent of each other. Then

$$
\begin{aligned}
I(X_1; Y \mid X_2) &= h(Y \mid X_2) - h(Y \mid X_1, X_2) \\
&= h(g_1 X_1 + g_2 X_2 + Z \mid X_2) - h(g_1 X_1 + g_2 X_2 + Z \mid X_1, X_2) \\
&= h(g_1 X_1 + Z) - h(Z) \\
&= \frac{1}{2} \log(2\pi e(S_1 + 1)) - \frac{1}{2} \log(2\pi e) \\
&= C(S_1).
\end{aligned}
$$

The other two mutual information terms follow similarly. Hence the capacity region coincides with a single $\mathcal{R}(X_1, X_2)$ region and there is no need to use time sharing. The rest of the achievability proof follows by first establishing a coding theorem for the DM-MAC with input costs (see Problem 4.8), and then applying the discretization procedure used in the achievability proof for the point-to-point Gaussian channel in Section 3.4.

The converse proof is similar to that for the point-to-point Gaussian channel.

4.6.2 Comparison with Point-to-Point Coding Schemes

We compare the capacity region of the Gaussian MAC with the suboptimal rate regions achieved by practical schemes that use point-to-point Gaussian channel codes. We further show that such codes, when used with successive cancellation decoding and time sharing, can achieve the entire capacity region.

Treating other codeword as noise. In this scheme, Gaussian random codes are used and each message is decoded while treating the other codeword as noise. This scheme achieves the set of rate pairs (R_1, R_2) such that

$$
R_1 < C\left(\frac{S_1}{1 + S_2}\right),
$$

$$
R_2 < C\left(\frac{S_2}{1 + S_1}\right).
$$

Time-division multiple access. A naive time-division scheme achieves the set of rate pairs (R_1, R_2) such that

$$R_1 < \alpha\, C(S_1),$$
$$R_2 < \bar{\alpha}\, C(S_2)$$

for some $\alpha \in [0, 1]$.

Note that when the channel SNRs are sufficiently low, treating the other codeword as noise can outperform time division as illustrated in Figure 4.8.

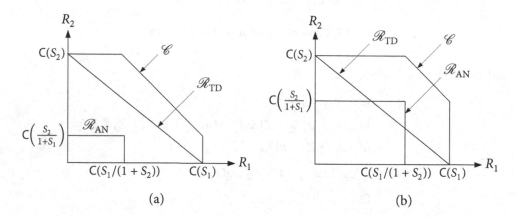

Figure 4.8. Comparison between time division (region $\mathscr{R}_{TD}$) and treating the other codeword as noise (region $\mathscr{R}_{AN}$): (a) high SNR, (b) low SNR.

Time division with power control. The average power used by the senders in time division is strictly lower than the average power constraint P for $\alpha \in (0, 1)$. If the senders are allowed to use higher powers during their transmission periods (without violating the power constraint over the entire transmission block), strictly higher rates can be achieved. We divide the transmission block into two subblocks, one of length αn and the other of length $\bar{\alpha} n$ (assuming αn is an integer). During the first subblock, sender 1 transmits using Gaussian random codes at average power P/α (rather than P) and sender 2 does not transmit. During the second subblock, sender 2 transmits at average power $P/\bar{\alpha}$ and sender 1 does not transmit. Note that the average power constraints are satisfied. This scheme achieves the set of rate pairs (R_1, R_2) such that

$$R_1 < \alpha\, C(S_1/\alpha),$$
$$R_2 < \bar{\alpha}\, C(S_2/\bar{\alpha}) \tag{4.5}$$

for some $\alpha \in [0, 1]$.

Now set $\alpha = S_1/(S_1 + S_2)$. Substituting in (4.5) and adding yields a point (R_1, R_2) on the boundary of the inner bound that lies on the sum-capacity line $C_{12} = C(S_1 + S_2)$ as shown in Figure 4.9!

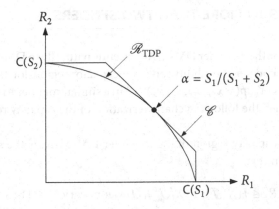

Figure 4.9. Time division with power control (region $\mathscr{R}_{\mathrm{TDP}}$).

Remark 4.12. It can be shown that time division with power control *always* outperforms treating the other codeword as noise.

Successive cancellation decoding. As in the DM-MAC case, the corner points of the Gaussian MAC capacity region can be achieved using successive cancellation decoding as depicted in Figure 4.10.

- Upon receiving $y^n = g_2 x_2^n(m_2) + g_1 x_1^n(m_1) + z^n$, the receiver recovers m_2 while treating the received signal $g_1 x_1^n(m_1)$ from sender 1 as part of the noise. The probability of error for this step tends to zero as $n \to \infty$ if $R_2 < C(S_2/(S_1 + 1))$.

- The receiver then subtracts $g_2 x_2^n(m_2)$ from y^n and decodes $g_1 x_1^n(m_1) + z^n$ to recover the message m_1. The probability of error for this step tends to zero as $n \to \infty$ if the first decoding step is successful and $R_1 < C(S_1)$.

The other corner point can be achieved by changing the decoding order and any point on the $R_1 + R_2 = C(S_1 + S_2)$ line can be achieved by time sharing between the two corner points. Thus, any point inside the capacity region can be achieved using good point-to-point Gaussian channel codes.

As before, the above argument can be made rigorous via the discretization procedure used in the proof of achievability for the point-to-point Gaussian channel.

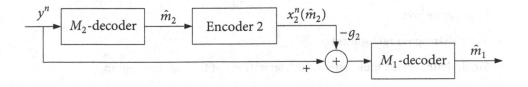

Figure 4.10. Successive cancellation decoding assuming $\hat{m}_2$ is unique.

4.7   EXTENSIONS TO MORE THAN TWO SENDERS

The capacity region for the 2-sender DM-MAC extends naturally to DM-MACs with more senders. Defining a code, achievability, and the capacity region for the k-sender DM-MAC $(\mathcal{X}_1 \times \mathcal{X}_2 \times \cdots \times \mathcal{X}_k, p(y|x_1, x_2, \ldots, x_k), \mathcal{Y})$ in a similar manner to the 2-sender case, we can readily establish the following characterization of the capacity region.

Theorem 4.5. The capacity region of the k-sender DM-MAC is the set of rate tuples $(R_1, R_2, \ldots, R_k)$ such that

$$\sum_{j \in \mathcal{J}} R_j \leq I(X(\mathcal{J}); Y | X(\mathcal{J}^c), Q) \quad \text{for every } \mathcal{J} \subseteq [1:k]$$

for some pmf $p(q) \prod_{j=1}^{k} p(x_j | q)$ with $|\mathcal{Q}| \leq k$.

For the k-sender Gaussian MAC with received SNRs S_j for $j \in [1:k]$, the capacity region is the set of rate tuples such that

$$\sum_{j \in \mathcal{J}} R_j \leq C\left(\sum_{j \in \mathcal{J}} S_j \right) \quad \text{for every } \mathcal{J} \subseteq [1:k].$$

This region is a polymatroid in the k-dimensional space and each of $k!$ corner points can be achieved by using point-to-point Gaussian codes and successive cancellation decoding in some message decoding order.

Remark 4.13. The multiple access channel is one of the few examples in network information theory in which a straightforward extension of the optimal results for a small number of users is optimal for an arbitrarily large number of users. Other examples include the degraded broadcast channel discussed in Chapter 5 and the distributed lossless source coding problem discussed in Chapter 10. In other cases, e.g., the broadcast channel with degraded message sets in Chapter 8, such straightforward extensions will be shown to be strictly suboptimal.

SUMMARY

- Discrete memoryless multiple access channel (DM-MAC)
- Capacity region
- Time sharing via rate splitting
- Single-letter versus multiletter characterizations of the capacity region
- Successive cancellation decoding
- Simultaneous decoding is more powerful than successive cancellation decoding

- Time-sharing random variable

- Bounding the cardinality of the time-sharing random variable

- Coded time sharing is more powerful than time sharing

- Gaussian multiple access channel:

  - Time division with power control achieves the sum-capacity

  - Capacity region with average power constraints achieved via optimal point-to-point codes, successive cancellation decoding, and time sharing

BIBLIOGRAPHIC NOTES

The multiple access channel was first alluded to in Shannon (1961). The multiletter characterization of the capacity region in Theorem 4.1 was established by van der Meulen (1971b). The single-letter characterization in Theorem 4.2 was established by Ahlswede (1971, 1974) and Liao (1972). The characterization in Theorem 4.3 is due to Slepian and Wolf (1973b) and the cardinality bound $|Q| \leq 2$ appears in Csiszár and Körner (1981b). The original proof of achievability uses successive cancellation decoding. Simultaneous decoding first appeared in El Gamal and Cover (1980). Coded time sharing is due to Han and Kobayashi (1981). Dueck (1981b) established the strong converse. The capacity region of the Gaussian MAC in Theorem 4.4 is due to Cover (1975c) and Wyner (1974). Han (1979) and Tse and Hanly (1998) studied the polymatroidal structure of the capacity region of the k-sender MAC. Surveys of the early literature on the multiple access channel can be found in van der Meulen (1977, 1985).

PROBLEMS

4.1. Show that the multiletter characterization of the DM-MAC capacity region $\mathcal{C}$ in Theorem 4.1 is convex.

4.2. *Capacity region of multiple access channels.*

(a) Consider the binary multiplier MAC in Example 4.1. We established the capacity region using time division between two individual capacities. Show that the capacity region can be also expressed as the union of $\mathcal{R}(X_1, X_2)$ sets (with no time sharing) and specify the set of pmfs $p(x_1)p(x_2)$ on (X_1, X_2) that achieve the boundary of the capacity region.

(b) Find the capacity region of the modulo-2 sum MAC, where X_1 and X_2 are binary and $Y = X_1 \oplus X_2$. Again show that the capacity region can be expressed as the union of $\mathcal{R}(X_1, X_2)$ sets and therefore time sharing is not necessary.

(c) The capacity regions of the above two MACs and the Gaussian MAC can each be expressed as the union of $\mathcal{R}(X_1, X_2)$ sets and no time sharing is necessary. Is time sharing ever necessary? Find the capacity of the *push-to-talk* MAC with binary inputs and output, given by $p(0|0,0) = p(1|0,1) = p(1|1,0) = 1$ and $p(0|1,1) = 1/2$. Why is this channel called "push-to-talk"? Show that the capacity region *cannot* be completely expressed as the union of $\mathscr{R}(X_1, X_2)$ sets and that time sharing is necessary.

Remark: This example is due to Csiszár and Körner (1981b, Problem 3.2.6). Another simple example for which time sharing is needed can be found in Bierbaum and Wallmeier (1979).

4.3. *Average vs. maximal probability of error.* We proved the channel coding theorem for a discrete memoryless channel under the average probability of error criterion. It is straightforward to show that the capacity is the same if we instead deal with the maximal probability of error. If we have a $(2^{nR}, n)$ code with $P_e^{(n)} \le \epsilon$, then by the Markov inequality, at least half its codewords have $\lambda_m \le 2\epsilon$ and this half has a rate $R - 1/n \to R$. Such an argument cannot be used to show that the capacity region of an arbitrary MAC under the maximal probability of error criterion is the same as that under the average probability of error criterion.

(a) Argue that simply discarding half of the codeword pairs with the highest probability of error does not work in general.

(b) How about throwing out the worst half of each sender's codewords? Show that this does not work either. (Hint: Provide a simple example of a set of probabilities $\lambda_{m_1 m_2} \in [0,1]$, $(m_1, m_2) \in [1:2^{nR}] \times [1:2^{nR}]$, such that the average probability of error $2^{-2nR} \sum_{m_1, m_2} \lambda_{m_1 \dot{m}_2} \le \epsilon$ for some $\epsilon \in (0, 1/4)$, yet there are no subsets $\mathcal{J}, \mathcal{K} \subseteq [1:2^{nR}]$ with cardinalities $|\mathcal{J}|, |\mathcal{K}| \ge 2^{nR}/2$ that satisfy $\lambda_{m_1 m_2} \le 4\epsilon$ for all $(m_1, m_2) \in \mathcal{J} \times \mathcal{K}$.)

Remark: A much stronger statement holds. Dueck (1978) provided an example of a MAC for which the capacity region with maximal probability of error is strictly smaller than that with average probability of error. This gives yet another example in which a result from point-to-point information theory does not necessarily carry over to the multiuser case.

4.4. *Convex closure of the union of sets.* Let

$$\mathscr{R}_1 = \{(r_1, r_2): r_1 \le I_1\ r_2 \le I_2,\ r_1 + r_2 \le I_{12}\}$$

for some $I_1, I_2, I_{12} \ge 0$ and

$$\mathscr{R}_1' = \{(r_1, r_2): r_1 \le I_1',\ r_2 \le I_2',\ r_1 + r_2 \le I_{12}'\}$$

for some $I_1', I_2', I_{12}' \ge 0$. Let $\mathscr{R}_2$ be the convex closure of the union of $\mathscr{R}_1$ and $\mathscr{R}_1'$, and $\mathscr{R}_3 = \{(r_1, r_2): r_1 \le \alpha I_1 + \bar{\alpha} I_1',\ r_2 \le \alpha I_2 + \bar{\alpha} I_2',\ r_1 + r_2 \le \alpha I_{12} + \bar{\alpha} I_{12}'$ for some $\alpha \in [0,1]\}$.

(a) Show that $\mathscr{R}_2 \subseteq \mathscr{R}_3$.

(b) Provide a counterexample showing that the inclusion can be strict.

(c) Under what condition on I_1, I_2, I_{12}, I_1', I_2', and I_{12}', is $\mathscr{R}_2 = \mathscr{R}_3$?

(Hint: Consider the case $I_1 = 2$, $I_2 = 5$, $I_{12} = 6$, $I_1' = 1$, $I_2' = 4$, and $I_{12}' = 7$.)

4.5. *Another multiletter characterization.* Consider a DM-MAC $p(y|x_1, x_2)$. Let $\mathscr{R}^{(k)}$, $k \geq 1$, be the set of rate pairs (R_1, R_2) such that

$$R_1 \leq \frac{1}{k} I(X_1^k; Y^k | X_2^k),$$

$$R_2 \leq \frac{1}{k} I(X_2^k; Y^k | X_1^k),$$

$$R_1 + R_2 \leq \frac{1}{k} I(X_1^k, X_2^k; Y^k)$$

for some pmf $p(x_1^k)p(x_2^k)$. Show that the capacity region $\mathscr{C}$ can be characterized as the closure of $\bigcup_k \mathscr{R}^{(k)}$.

4.6. *From multiletter to single-letter.* Show that the multiletter characterization in Theorem 4.1 can be directly reduced to Theorem 4.3 (without a cardinality bound).

4.7. *Converse for the Gaussian MAC.* Prove the weak converse for the Gaussian MAC with average power constraint P, starting from the inequalities

$$nR_1 \leq \sum_{i=1}^{n} I(X_{1i}; Y_i | X_{2i}) + n\epsilon_n,$$

$$nR_2 \leq \sum_{i=1}^{n} I(X_{2i}; Y_i | X_{1i}) + n\epsilon_n,$$

$$n(R_1 + R_2) \leq \sum_{i=1}^{n} I(X_{1i}, X_{2i}; Y_i) + n\epsilon_n.$$

4.8. *DM-MAC with input costs.* Consider the DM-MAC $p(y|x_1, x_2)$ and let $b_1(x_1)$ and $b_2(x_2)$ be input cost functions with $b_j(x_{j0}) = 0$ for some $x_{j0} \in \mathcal{X}_j$, $j = 1, 2$. Assume that there are cost constraints $\sum_{i=1}^{n} b_j(x_{ji}(m_j)) \leq nB_j$ for $m_j \in [1 : 2^{nR}]$, $j = 1, 2$. Show that the capacity region is the set of rate pairs (R_1, R_2) such that

$$R_1 \leq I(X_1; Y | X_2, Q),$$
$$R_2 \leq I(X_2; Y | X_1, Q),$$
$$R_1 + R_2 \leq I(X_1, X_2; Y | Q)$$

for some pmf $p(q)p(x_1|q)p(x_2|q)$ that satisfies $\mathsf{E}(b_j(X_j)) \leq B_j$, $j = 1, 2$.

4.9. *Cooperative capacity of a MAC.* Consider a DM-MAC $p(y|x_1, x_2)$. Assume that both senders have access to both messages $m_1 \in [1 : 2^{nR_1}]$ and $m_2 \in [1 : 2^{nR_2}]$; thus each of the codewords $x_1^n(m_1, m_2)$ and $x_2^n(m_1, m_2)$ can depend on both messages.

(a) Find the capacity region.

(b) Evaluate the region for the Gaussian MAC with noise power 1, channel gains g_1 and g_2, and average power constraint P on each of X_1 and X_2.

4.10. *Achievable SNR region.* Consider a Gaussian multiple access channel $Y = gX_1 + X_2 + Z$ with $Z \sim N(0, 1)$, $g \geq 1$, and average power constraints P_1 on X_1 and P_2 on X_2.

(a) Specify the capacity region of this channel with SNRs $S_1 = g^2 P_1$ and $S_2 = P_2$.

(b) Suppose we wish to communicate reliably at a fixed rate pair (R_1, R_2). Specify the *achievable SNR region* $\mathscr{S}(R_1, R_2)$ consisting of all SNR pairs (s_1, s_2) such that (R_1, R_2) is achievable.

(c) Find the SNR pair $(s_1^*, s_2^*) \in \mathscr{S}(R_1, R_2)$ that minimizes the total average transmission power $P_{\text{sum}} = P_1 + P_2$. (Hint: You can use a simple geometric argument.)

(d) Can (R_1, R_2) be achieved with minimum total average power P_{sum} using only successive cancellation decoding (i.e., without time sharing)? If so, what is the order of message decoding? Can (R_1, R_2) be achieved by time division with power control? (Hint: Treat the cases $g = 1$ and $g > 1$ separately.)

(e) Find the minimum-energy-per-bit region of the Gaussian MAC, that is, the set of all energy pairs $(E_1, E_2) = (P_1/R_1, P_2/R_2)$ such that the rate pair (R_1, R_2) is achievable with average code power pair (P_1, P_2).

4.11. *Two-sender two-receiver channel.* Consider the DM 2-sender 2-receiver channel depicted in Figure 4.11 with message pair (M_1, M_2) uniformly distributed over $[1 : 2^{nR_1}] \times [1 : 2^{nR_2}]$. Sender $j = 1, 2$ encodes the message M_j. Each receiver wishes to recover both messages (how is this channel related to the multiple access channel?). The average probability of error is defined as $P_e^{(n)} = P\{(\hat{M}_{11}, \hat{M}_{21}) \neq (M_1, M_2) \text{ or } (\hat{M}_{12}, \hat{M}_{22}) \neq (M_1, M_2)\}$. Achievability and the capacity region are defined as for the DM-MAC. Find a single-letter characterization of the capacity region of this 2-sender 2-receiver channel.

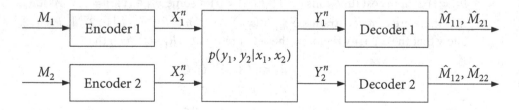

Figure 4.11. Two-sender two-receiver channel.

4.12. *MAC with list codes.* A $(2^{nR_1}, 2^{nR_2}, 2^{nL_2}, n)$ list code for a DM-MAC $p(y|x_1, x_2)$ consists of two message sets $[1 : 2^{nR_1}]$ and $[1 : 2^{nR_2}]$, two encoders, and a *list* decoder. The message pair (M_1, M_2) is uniformly distributed over $[1 : 2^{nR_1}] \times [1 : 2^{nR_2}]$. Upon receiving y^n, the decoder finds $\hat{m}_1$ and a list of messages $\mathcal{L}_2 \subseteq [1 : 2^{nR_2}]$ of size $|\mathcal{L}_2| \leq 2^{nL_2}$. An error occurs if $\hat{M}_1 \neq M_1$ or if the list $\mathcal{L}_2$ does not contain the transmitted message M_2, i.e., $P_e^{(n)} = P\{\hat{M}_1 \neq M_1 \text{ or } M_2 \notin \mathcal{L}_2\}$. A triple (R_1, R_2, L_2)

is said to be achievable if there exists a sequence of $(2^{nR_1}, 2^{nR_2}, 2^{nL_2}, n)$ list codes such that $\lim_{n\to\infty} P_e^{(n)} = 0$. The list capacity region of the DM-MAC is the closure of the set of achievable triples (R_1, R_2, L_2).

(a) Find a single-letter characterization of the list capacity region of the DM-MAC.

(b) Prove achievability and the converse.

4.13. *MAC with dependent messages.* Consider a DM-MAC $p(y|x_1, x_2)$ with message pair (M_1', M_2') distributed according to an arbitrary pmf on $[1 : 2^{nR_1}] \times [1 : 2^{nR_2}]$. In particular, M_1' and M_2' are not independent in general. Show that if the rate pair (R_1, R_2) satisfies the inequalities in Theorem 4.2, then there exists a sequence of $(2^{nR_1}, 2^{nR_2}, n)$ codes with $\lim_{n\to\infty} P\{(\hat{M}_1', \hat{M}_2') \neq (M_1, M_2)\} = 0$.

APPENDIX 4A   CARDINALITY BOUND ON Q

We already know that the region $\mathscr{C}'$ of Theorem 4.3 without any bound on the cardinality of Q is the capacity region. To establish the bound on $|Q|$, we first consider some properties of the region $\mathscr{C}$ in Theorem 4.2.

- Any pair (R_1, R_2) in $\mathscr{C}$ is in the convex closure of the union of no more than two $\mathscr{R}(X_1, X_2)$ regions. This follows by the Fenchel–Eggleston–Carathéodory theorem in Appendix A. Indeed, since the union of the $\mathscr{R}(X_1, X_2)$ sets is a connected compact set in $\mathbb{R}^2$, each point in its convex closure can be represented as a convex combination of at most 2 points in the union, and thus each point is in the convex closure of the union of no more than two $\mathscr{R}(X_1, X_2)$ sets.

- Any point on the boundary of $\mathscr{C}$ is either a boundary point of some $\mathscr{R}(X_1, X_2)$ set or a convex combination of the "upper-diagonal" or "lower-diagonal" corner points of two $\mathscr{R}(X_1, X_2)$ sets as shown in Figure 4.12.

These two properties imply that $|Q| \leq 2$ suffices, which completes the proof.

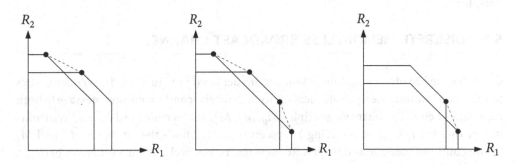

Figure 4.12. Boundary points of $\mathscr{C}$ as convex combinations of the corner points of $\mathscr{R}(X_1, X_2)$.

CHAPTER 5

Degraded Broadcast Channels

We introduce the broadcast channel as a model for noisy one-to-many communication, such as the downlink of a cellular system, digital TV broadcasting, or a lecture to a group of people with diverse backgrounds. Unlike the multiple access channel, no single-letter characterization is known for the capacity region of the broadcast channel. However, there are several interesting coding schemes and corresponding inner bounds on the capacity region that are tight in some special cases.

In this chapter, we introduce the technique of superposition coding and use it to establish an inner bound on the capacity region of the broadcast channel. We show that this inner bound is tight for the class of degraded broadcast channels, which includes the Gaussian broadcast channel. The converse proofs involve the new technique of auxiliary random variable identification and the use of Mrs. Gerber's lemma, a symmetrization argument, and the entropy power inequality. We then show that superposition coding is optimal for the more general classes of less noisy and more capable broadcast channels. The converse proof uses the Csiszár sum identity.

We will resume the discussion of the broadcast channel in Chapters 8 and 9. In these chapters, we will present more general inner and outer bounds on the capacity region and show that they coincide for other classes of channels. The reason for postponing the presentation of these results is pedagogical—we will need coding techniques that can be introduced at a more elementary level through the discussion of channels with state in Chapter 7.

5.1 DISCRETE MEMORYLESS BROADCAST CHANNEL

Consider the broadcast communication system depicted in Figure 5.1. The sender wishes to reliably communicate a private message to each receiver and a common message to both receivers. It encodes the message triple (M_0, M_1, M_2) into a codeword X^n and transmits it over the channel. Upon receiving Y_j^n, receiver $j = 1, 2$ finds the estimates $\hat{M}_{0j}$ and $\hat{M}_j$ of the common message and its private message, respectively. A tradeoff arises between the rates of the three messages—when one of the rates is high, the other rates may need to be reduced to ensure reliable communication of all three messages. As before, we study the asymptotic limit on this tradeoff.

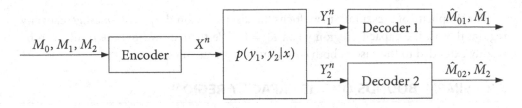

Figure 5.1. Two-receiver broadcast communication system.

We first consider a 2-receiver *discrete memoryless broadcast channel* (DM-BC) model $(\mathcal{X}, p(y_1, y_2|x), \mathcal{Y}_1 \times \mathcal{Y}_2)$ that consists of three finite sets $\mathcal{X}, \mathcal{Y}_1, \mathcal{Y}_2$, and a collection of conditional pmfs $p(y_1, y_2|x)$ on $\mathcal{Y}_1 \times \mathcal{Y}_2$ (one for each input symbol x).

A $(2^{nR_0}, 2^{nR_1}, 2^{nR_2}, n)$ code for a DM-BC consists of

- three message sets $[1 : 2^{nR_0}]$, $[1 : 2^{nR_1}]$, and $[1 : 2^{nR_2}]$,

- an encoder that assigns a codeword $x^n(m_0, m_1, m_2)$ to each message triple $(m_0, m_1, m_2) \in [1 : 2^{nR_0}] \times [1 : 2^{nR_1}] \times [1 : 2^{nR_2}]$, and

- two decoders, where decoder 1 assigns an estimate $(\hat{m}_{01}, \hat{m}_1) \in [1 : 2^{nR_0}] \times [1 : 2^{nR_1}]$ or an error message e to each received sequence y_1^n, and decoder 2 assigns an estimate $(\hat{m}_{02}, \hat{m}_2) \in [1 : 2^{nR_0}] \times [1 : 2^{nR_2}]$ or an error message e to each received sequence y_2^n.

We assume that the message triple (M_0, M_1, M_2) is uniformly distributed over $[1 : 2^{nR_0}] \times [1 : 2^{nR_1}] \times [1 : 2^{nR_2}]$. The average probability of error is defined as

$$P_e^{(n)} = \mathsf{P}\{(\hat{M}_{01}, \hat{M}_1) \ne (M_0, M_1) \text{ or } (\hat{M}_{02}, \hat{M}_2) \ne (M_0, M_2)\}.$$

A rate triple (R_0, R_1, R_2) is said to be achievable for the DM-BC if there exists a sequence of $(2^{nR_0}, 2^{nR_1}, 2^{nR_2}, n)$ codes such that $\lim_{n \to \infty} P_e^{(n)} = 0$. The capacity region $\mathscr{C}$ of the DM-BC is the closure of the set of achievable rate triples (R_0, R_1, R_2).

The following simple observation, which results from the lack of cooperation between the two receivers, will prove useful.

Lemma 5.1. The capacity region of the DM-BC depends on the channel conditional pmf $p(y_1, y_2|x)$ only through the conditional marginal pmfs $p(y_1|x)$ and $p(y_2|x)$.

Proof. Consider the individual probabilities of error

$$P_{ej}^{(n)} = \mathsf{P}\{(\hat{M}_{0j}, \hat{M}_j) \ne (M_0, M_j)\}, \quad j = 1, 2.$$

Note that each term depends only on its corresponding conditional marginal pmf $p(y_j|x)$, $j = 1, 2$. By the union of events bound, $P_e^{(n)} \le P_{e1}^{(n)} + P_{e2}^{(n)}$. Also $P_e^{(n)} \ge \max\{P_{e1}^{(n)}, P_{e2}^{(n)}\}$. Hence $\lim_{n \to \infty} P_e^{(n)} = 0$ iff $\lim_{n \to \infty} P_{e1}^{(n)} = 0$ and $\lim_{n \to \infty} P_{e2}^{(n)} = 0$, which implies that the capacity region of a DM-BC depends only on the conditional marginal pmfs. This completes the proof of the lemma.

For simplicity of presentation, we focus the discussion on the *private-message* capacity region, that is, the capacity region when $R_0 = 0$. We then show how the results can be readily extended to the case of both common and private messages.

5.2 SIMPLE BOUNDS ON THE CAPACITY REGION

Consider a DM-BC $p(y_1, y_2|x)$. Let $C_j = \max_{p(x)} I(X; Y_j)$, $j = 1, 2$, be the capacities of the DMCs $p(y_1|x)$ and $p(y_2|x)$. These capacities define the time-division inner bound in Figure 5.2. By allowing full cooperation between the receivers, we obtain the bound on the sum-rate

$$R_1 + R_2 \le C_{12} = \max_{p(x)} I(X; Y_1, Y_2).$$

Combining this bound with the individual capacity bounds gives the outer bound on the capacity region in Figure 5.2.

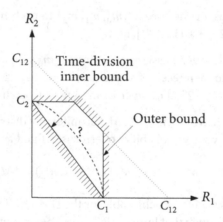

Figure 5.2. Time-division inner bound and an outer bound on the capacity region.

The time-division bound can be tight in some cases.

Example 5.1. Consider a symmetric DM-BC, where $\mathcal{Y}_1 = \mathcal{Y}_2 = \mathcal{Y}$ and $p_{Y_1|X}(y|x) = p_{Y_2|X}(y|x) = p(y|x)$. In this example, $C_1 = C_2 = \max_{p(x)} I(X; Y)$. Since the capacity region depends only on the marginals of $p(y_1, y_2|x)$, we can assume that $Y_1 = Y_2 = Y$. Thus

$$R_1 + R_2 \le C_{12} = \max_{p(x)} I(X; Y) = C_1 = C_2.$$

This shows that the time-division inner bound and the outer bound sometimes coincide.

The following example shows that the bounds do not always coincide.

Example 5.2. Consider a DM-BC with orthogonal components, where $\mathcal{X} = \mathcal{X}_1 \times \mathcal{X}_2$ and $p(y_1, y_2|x_1, x_2) = p(y_1|x_1)p(y_2|x_2)$. For this example, the capacity region is the set of rate pairs (R_1, R_2) such that $R_1 \le C_1$ and $R_2 \le C_2$; thus the outer bound is tight, but the time-division inner bound is not.

As we will see, neither bound is tight in general.

5.3   SUPERPOSITION CODING INNER BOUND

The superposition coding technique is motivated by broadcast channels where one receiver is "stronger" than the other, for example, because it is closer to the sender, such that it can always recover the "weaker" receiver's message. This suggests a *layered* coding approach in which the weaker receiver's message is treated as a "public" (common) message and the stronger receiver's message is treated as a "private" message. The weaker receiver recovers only the public message, while the stronger receiver recovers both messages using successive cancellation decoding as for the MAC; see Section 4.5.1. We illustrate this coding scheme in the following.

Example 5.3 (Binary symmetric broadcast channel). The binary symmetric broadcast channel (BS-BC) consists of a $\text{BSC}(p_1)$ and a $\text{BSC}(p_2)$ as depicted in Figure 5.3. Assume that $p_1 < p_2 < 1/2$.

Using time division, we can achieve the straight line between the capacities $1 - H(p_1)$ and $1 - H(p_2)$ of the individual BSCs. Can we achieve higher rates than time division? To answer this question, consider the following *superposition coding* technique illustrated in Figure 5.4.

For $\alpha \in [0, 1/2]$, let $U \sim \text{Bern}(1/2)$ and $V \sim \text{Bern}(\alpha)$ be independent, and $X = U \oplus V$. Randomly and independently generate 2^{nR_2} sequences $u^n(m_2)$, each i.i.d. $\text{Bern}(1/2)$ (cloud centers). Randomly and independently generate 2^{nR_1} sequences $v^n(m_1)$, each i.i.d. $\text{Bern}(\alpha)$. The sender transmits $x^n(m_1, m_2) = u^n(m_2) \oplus v^n(m_1)$ (satellite codeword).

To recover m_2, receiver 2 decodes $y_2^n = u^n(m_2) \oplus (v^n(m_1) \oplus z_2^n)$ while treating $v^n(m_1)$ as noise. The probability of decoding error tends to zero as $n \to \infty$ if $R_2 < 1 - H(\alpha * p_2)$, where $\alpha * p_2 = \alpha \bar{p}_2 + \bar{\alpha} p_2$.

Receiver 1 uses successive cancellation decoding—it first decodes $y_1^n = u^n(m_2) \oplus (v^n(m_1) \oplus z_1^n)$ to recover m_2 while treating $v^n(m_1)$ as part of the noise, subtracts off $u^n(m_2)$,

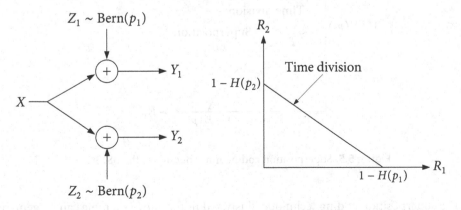

Figure 5.3. Binary symmetric broadcast channel and its time-division inner bound.

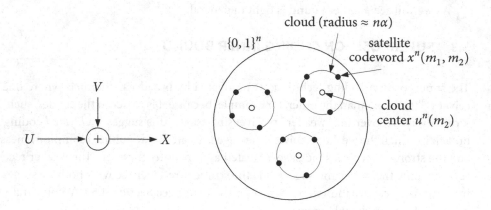

Figure 5.4. Superposition coding for the BS-BC.

and then decodes $v^n(m_1) \oplus z_1^n$ to recover m_1. The probability of decoding error tends to zero as $n \to \infty$ if $R_1 < I(V; V \oplus Z_1) = H(\alpha * p_1) - H(p_1)$ and $R_2 < 1 - H(\alpha * p_1)$. The latter condition is already satisfied by the rate constraint for receiver 2 since $p_1 < p_2$.

Thus, superposition coding leads to an inner bound consisting of the set of rate pairs (R_1, R_2) such that

$$R_1 \leq H(\alpha * p_1) - H(p_1),$$
$$R_2 \leq 1 - H(\alpha * p_2)$$

for some $\alpha \in [0, 1/2]$. This inner bound is larger than the time-division inner bound as sketched in Figure 5.5. We will show later that this bound is the capacity region of the BS-BC.

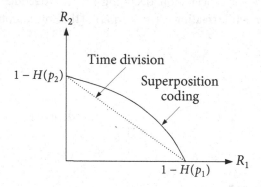

Figure 5.5. Superposition coding inner bound for the BS-BC.

The superposition coding technique illustrated in the above example can be generalized to obtain the following inner bound on the capacity region of the general DM-BC.

Theorem 5.1 (Superposition Coding Inner Bound). A rate pair (R_1, R_2) is achievable for the DM-BC $p(y_1, y_2|x)$ if

$$R_1 < I(X; Y_1|U),$$
$$R_2 < I(U; Y_2),$$
$$R_1 + R_2 < I(X; Y_1)$$

for some pmf $p(u, x)$.

Remark 5.1. The random variable U in the theorem is an *auxiliary* random variable that does not correspond to any of the channel variables. While the time-sharing random variable encountered in Chapter 4 is also an auxiliary random variable, U actually carries message information and as such plays a more essential role in the characterization of the capacity region.

Remark 5.2. The superposition coding inner bound is evaluated for the joint pmf $p(u, x)$ $p(y_1, y_2|x)$, i.e., $U \rightarrow X \rightarrow (Y_1, Y_2)$; see Remark 4.7.

Remark 5.3. It can be shown that the inner bound is convex and therefore there is no need for further convexification via a time-sharing random variable.

5.3.1 Proof of the Superposition Coding Inner Bound

We begin with a sketch of achievability; see Figure 5.6. Fix a pmf $p(u, x)$ and randomly generate 2^{nR_2} "cloud centers" $u^n(m_2)$. For each cloud center $u^n(m_2)$, generate 2^{nR_1} "satellite" codewords $x^n(m_1, m_2)$. Receiver 2 decodes for the cloud center $u^n(m_2)$ and receiver 1 decodes for the satellite codeword.

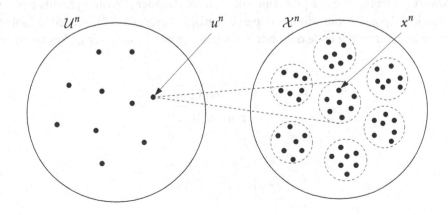

Figure 5.6. Superposition coding.

We now provide details of the proof.

Codebook generation. Fix a pmf $p(u)p(x|u)$. Randomly and independently generate 2^{nR_2} sequences $u^n(m_2)$, $m_2 \in [1:2^{nR_2}]$, each according to $\prod_{i=1}^n p_U(u_i)$. For each $m_2 \in [1:2^{nR_2}]$, randomly and conditionally independently generate 2^{nR_1} sequences $x^n(m_1, m_2)$, $m_1 \in [1:2^{nR_1}]$, each according to $\prod_{i=1}^n p_{X|U}(x_i|u_i(m_2))$.

Encoding. To send (m_1, m_2), transmit $x^n(m_1, m_2)$.

Decoding. Decoder 2 declares that $\hat{m}_2$ is sent if it is the unique message such that $(u^n(\hat{m}_2), y_2^n) \in \mathcal{T}_\epsilon^{(n)}$; otherwise it declares an error. Since decoder 1 is interested only in recovering m_1, it uses simultaneous decoding without requiring the recovery of m_2, henceforth referred to as *simultaneous nonunique decoding*. Decoder 1 declares that $\hat{m}_1$ is sent if it is the unique message such that $(u^n(m_2), x^n(\hat{m}_1, m_2), y_1^n) \in \mathcal{T}_\epsilon^{(n)}$ for some m_2; otherwise it declares an error.

Analysis of the probability of error. Assume without loss of generality that $(M_1, M_2) = (1, 1)$ is sent. First consider the average probability of error for decoder 2. This decoder makes an error iff one or both of the following events occur:

$$\mathcal{E}_{21} = \{(U^n(1), Y_2^n) \notin \mathcal{T}_\epsilon^{(n)}\},$$
$$\mathcal{E}_{22} = \{(U^n(m_2), Y_2^n) \in \mathcal{T}_\epsilon^{(n)} \text{ for some } m_2 \neq 1\}.$$

Thus the probability of error for decoder 2 is upper bounded as

$$\mathsf{P}(\mathcal{E}_2) \leq \mathsf{P}(\mathcal{E}_{21}) + \mathsf{P}(\mathcal{E}_{22}).$$

By the LLN, the first term $\mathsf{P}(\mathcal{E}_{21})$ tends to zero as $n \to \infty$. For the second term, since $U^n(m_2)$ is independent of $(U^n(1), Y_2^n)$ for $m_2 \neq 1$, by the packing lemma, $\mathsf{P}(\mathcal{E}_{22})$ tends to zero as $n \to \infty$ if $R_2 < I(U; Y_2) - \delta(\epsilon)$.

Next consider the average probability of error for decoder 1. To analyze this probability, consider all possible pmfs for the triple $(U^n(m_2), X^n(m_1, m_2), Y_1^n)$ as listed in Table 5.1. Since the last case in the table does not result in an error, the decoder makes an error iff

m_1	m_2	Joint pmf
1	1	$p(u^n, x^n)p(y_1^n|x^n)$
*	1	$p(u^n, x^n)p(y_1^n|u^n)$
*	*	$p(u^n, x^n)p(y_1^n)$
1	*	$p(u^n, x^n)p(y_1^n)$

Table 5.1. The joint pmfs induced by different (m_1, m_2) pairs. The $*$ symbol corresponds to message $m_1 \neq 1$ or $m_2 \neq 1$.

one or more of the following three events occur:

$$\mathcal{E}_{11} = \{(U^n(1), X^n(1,1), Y_1^n) \notin \mathcal{T}_{\epsilon}^{(n)}\},$$

$$\mathcal{E}_{12} = \{(U^n(1), X^n(m_1,1), Y_1^n) \in \mathcal{T}_{\epsilon}^{(n)} \text{ for some } m_1 \neq 1\},$$

$$\mathcal{E}_{13} = \{(U^n(m_2), X^n(m_1,m_2), Y_1^n) \in \mathcal{T}_{\epsilon}^{(n)} \text{ for some } m_1 \neq 1, m_2 \neq 1\}.$$

Thus the probability of error for decoder 1 is upper bounded as

$$\mathsf{P}(\mathcal{E}_1) \leq \mathsf{P}(\mathcal{E}_{11}) + \mathsf{P}(\mathcal{E}_{12}) + \mathsf{P}(\mathcal{E}_{13}).$$

We now bound each term. By the LLN, $\mathsf{P}(\mathcal{E}_{11})$ tends to zero as $n \to \infty$. For the second term, note that if $m_1 \neq 1$, then $X^n(m_1, 1)$ is conditionally independent of $(X^n(1,1), Y_1^n)$ given $U^n(1)$ and is distributed according to $\prod_{i=1}^{n} p_{X|U}(x_i|u_i(1))$. Hence, by the packing lemma, $\mathsf{P}(\mathcal{E}_{12})$ tends to zero as $n \to \infty$ if $R_1 < I(X; Y_1|U) - \delta(\epsilon)$.

Finally, for the third term, note that for $m_2 \neq 1$ (and any m_1), $(U^n(m_2), X^n(m_1, m_2))$ is independent of $(U^n(1), X^n(1, 1), Y_1^n)$. Hence, by the packing lemma, $\mathsf{P}(\mathcal{E}_{13})$ tends to zero as $n \to \infty$ if $R_1 + R_2 < I(U, X; Y_1) - \delta(\epsilon) = I(X; Y_1) - \delta(\epsilon)$ (recall that $U \to X \to Y_1$ form a Markov chain). This completes the proof of achievability.

Remark 5.4. Consider the error event

$$\mathcal{E}_{14} = \{(U^n(m_2), X^n(1, m_2), Y_1^n) \in \mathcal{T}_{\epsilon}^{(n)} \text{ for some } m_2 \neq 1\}.$$

By the packing lemma, $\mathsf{P}(\mathcal{E}_{14})$ tends to zero as $n \to \infty$ if $R_2 < I(U, X; Y_1) - \delta(\epsilon) = I(X; Y_1) - \delta(\epsilon)$, which is already satisfied. Therefore, the inner bound does not change if we use simultaneous (unique) decoding and require decoder 1 to also recover M_2.

Remark 5.5. We can obtain a second superposition coding inner bound by having receiver 1 decode for the cloud center (which would now represent M_1) and receiver 2 decode for the satellite codeword. This yields the set of rate pairs (R_1, R_2) such that

$$R_1 < I(U; Y_1),$$
$$R_2 < I(X; Y_2|U),$$
$$R_1 + R_2 < I(X; Y_2)$$

for some pmf $p(u, x)$. The convex closure of the union of the two superposition coding inner bounds constitutes a generally tighter inner bound on the capacity region.

Remark 5.6. Superposition coding is optimal for several classes of broadcast channels for which one receiver is stronger than the other, as we detail in the following sections. It is not, however, optimal in general since requiring one of the receivers to recover both messages (even nonuniquely for the other receiver's message) can unduly constrain the set of achievable rates. In Chapter 8, we present Marton's coding scheme, which can outperform superposition coding by not requiring either receiver to recover both messages.

5.4 DEGRADED DM-BC

In a degraded broadcast channel, one of the receivers is *statistically* stronger than the other. Formally, a DM-BC is said to be *physically degraded* if

$$p(y_1, y_2 | x) = p(y_1 | x) p(y_2 | y_1),$$

i.e., $X \to Y_1 \to Y_2$ form a Markov chain.

More generally, a DM-BC $p(y_1, y_2 | x)$ is said to be *stochastically degraded* (or simply degraded) if there exists a random variable $\tilde{Y}_1$ such that $\tilde{Y}_1 | \{X = x\} \sim p_{Y_1 | X}(\tilde{y}_1 | x)$, i.e., $\tilde{Y}_1$ has the same conditional pmf as Y_1 (given X), and $X \to \tilde{Y}_1 \to Y_2$ form a Markov chain.

Since the capacity region of a DM-BC depends only on the conditional marginal pmfs, the capacity region of a stochastically degraded DM-BC is the same as that of its corresponding physically degraded channel. This observation will be used later in the proof of the converse.

Note that the BS-BC in Example 5.3 is a degraded DM-BC. Again assume that $p_1 < p_2 < 1/2$. The channel is degraded since we can write $Y_2 = X \oplus \tilde{Z}_1 \oplus \tilde{Z}_2$, where $\tilde{Z}_1 \sim$ Bern(p_1) and $\tilde{Z}_2 \sim$ Bern$(\tilde{p}_2)$ are independent, and

$$\tilde{p}_2 = \frac{p_2 - p_1}{1 - 2p_1}.$$

Figure 5.7 shows the physically degraded BS-BC with the same conditional marginal pmfs. The capacity region of the original BS-BC is the same as that of the physically degraded version.

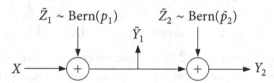

Figure 5.7. Corresponding physically degraded BS-BC.

The superposition coding inner bound is tight for the class of degraded broadcast channels.

Theorem 5.2. The capacity region of the degraded DM-BC $p(y_1, y_2 | x)$ is the set of rate pairs (R_1, R_2) such that

$$R_1 \leq I(X; Y_1 | U),$$
$$R_2 \leq I(U; Y_2)$$

for some pmf $p(u, x)$, where the cardinality of the auxiliary random variable U satisfies $|\mathcal{U}| \leq \min\{|\mathcal{X}|, |\mathcal{Y}_1|, |\mathcal{Y}_2|\} + 1$.

To prove achievability, note that the sum of the first two inequalities in the superposition coding inner bound gives $R_1 + R_2 < I(U; Y_2) + I(X; Y_1|U)$. Since the channel is degraded, $I(U; Y_1) \geq I(U; Y_2)$ for all $p(u, x)$. Hence, $I(U; Y_2) + I(X; Y_1|U) \leq I(U; Y_1) + I(X; Y_1|U) = I(X; Y_1)$ and the third inequality is automatically satisfied.

5.4.1 Proof of the Converse

We need to show that given any sequence of $(2^{nR_1}, 2^{nR_2}, n)$ codes with $\lim_{n\to\infty} P_e^{(n)} = 0$, we must have $R_1 \leq I(X; Y_1|U)$ and $R_2 \leq I(U; Y_2)$ for some pmf $p(u, x)$ such that $U \to X \to (Y_1, Y_2)$ form a Markov chain. The key new idea in the converse proof is the identification of the auxiliary random variable U. Every $(2^{nR_1}, 2^{nR_2}, n)$ code (by definition) induces a joint pmf on $(M_1, M_2, X^n, Y_1^n, Y_2^n)$ of the form

$$p(m_1, m_2, x^n, y_1^n, y_2^n) = 2^{-n(R_1+R_2)} p(x^n|m_1, m_2) \prod_{i=1}^{n} p_{Y_1,Y_2|X}(y_{1i}, y_{2i}|x_i).$$

By Fano's inequality,

$$H(M_1|Y_1^n) \leq nR_1 P_e^{(n)} + 1 \leq n\epsilon_n,$$
$$H(M_2|Y_2^n) \leq nR_2 P_e^{(n)} + 1 \leq n\epsilon_n$$

for some ϵ_n that tends to zero as $n \to \infty$. Hence

$$\begin{aligned} nR_1 &\leq I(M_1; Y_1^n) + n\epsilon_n, \\ nR_2 &\leq I(M_2; Y_2^n) + n\epsilon_n. \end{aligned} \tag{5.1}$$

First consider the following intuitive identification of the auxiliary random variable. Since U represents M_2 in the superposition coding scheme, it is natural to choose $U = M_2$, which also satisfies the desired Markov chain condition $U \to X_i \to (Y_{1i}, Y_{2i})$. Using this identification, consider the first mutual information term in (5.1),

$$\begin{aligned} I(M_1; Y_1^n) &\leq I(M_1; Y_1^n|M_2) \\ &= I(M_1; Y_1^n|U) \\ &= \sum_{i=1}^{n} I(M_1; Y_{1i}|U, Y_1^{i-1}) \tag{5.2} \\ &\leq \sum_{i=1}^{n} I(M_1, Y_1^{i-1}; Y_{1i}|U) \\ &\leq \sum_{i=1}^{n} I(X_i, M_1, Y_1^{i-1}; Y_{1i}|U) \\ &= \sum_{i=1}^{n} I(X_i; Y_{1i}|U). \tag{5.3} \end{aligned}$$

Note that this bound has the same structure as the inequality on R_1 in Theorem 5.2. Next, consider the second mutual information term in (5.1),

$$I(M_2; Y_2^n) = \sum_{i=1}^{n} I(M_2; Y_{2i}|Y_2^{i-1})$$

$$= \sum_{i=1}^{n} I(U; Y_{2i}|Y_2^{i-1}).$$

However, $I(U; Y_{2i}|Y_2^{i-1})$ is not necessarily less than or equal to $I(U; Y_{2i})$. Thus setting $U = M_2$ does not yield an inequality that has the same form as the one on R_2 in Theorem 5.2.

Step (5.2) in the above series of inequalities suggests the identification of the auxiliary random variable as $U_i = (M_2, Y_1^{i-1})$, which also satisfies $U_i \to X_i \to (Y_{1i}, Y_{2i})$. This gives

$$I(M_1; Y_1^n) \le I(M_1; Y_1^n|M_2)$$

$$= \sum_{i=1}^{n} I(M_1; Y_{1i}|Y_1^{i-1}, M_2)$$

$$= \sum_{i=1}^{n} I(X_i, M_1; Y_{1i}|U_i)$$

$$= \sum_{i=1}^{n} I(X_i; Y_{1i}|U_i). \tag{5.4}$$

Now consider the second term in (5.1)

$$I(M_2; Y_2^n) = \sum_{i=1}^{n} I(M_2; Y_{2i}|Y_2^{i-1})$$

$$\le \sum_{i=1}^{n} I(M_2, Y_2^{i-1}; Y_{2i})$$

$$\le \sum_{i=1}^{n} I(M_2, Y_2^{i-1}, Y_1^{i-1}; Y_{2i}). \tag{5.5}$$

However, $I(M_2, Y_2^{i-1}, Y_1^{i-1}; Y_{2i})$ is not necessarily equal to $I(M_2, Y_1^{i-1}; Y_{2i})$.

Using the observation that the capacity region of the general degraded DM-BC is the same as that of the corresponding physically degraded BC, we can assume without loss of generality that $X \to Y_1 \to Y_2$ form a Markov chain. This implies that $Y_2^{i-1} \to (M_2, Y_1^{i-1}) \to Y_{2i}$ also form a Markov chain and hence (5.5) implies that

$$I(M_2; Y_2^n) \le \sum_{i=1}^{n} I(U_i; Y_{2i}). \tag{5.6}$$

To complete the proof, define the time-sharing random variable Q to be uniformly distributed over $[1:n]$ and independent of $(M_1, M_2, X^n, Y_1^n, Y_2^n)$, and identify $U = (Q, U_Q)$,

$X = X_Q$, $Y_1 = Y_{1Q}$, and $Y_2 = Y_{2Q}$. Clearly, $U \to X \to (Y_1, Y_2)$ form a Markov chain. Using these definitions and substituting from (5.4) and (5.6) into (5.1), we have

$$nR_1 \leq \sum_{i=1}^{n} I(X_i; Y_{1i}|U_i) + n\epsilon_n$$
$$= nI(X_Q; Y_{1Q}|U_Q, Q) + n\epsilon_n$$
$$= nI(X; Y_1|U) + n\epsilon_n,$$

$$nR_2 \leq \sum_{i=1}^{n} I(U_i; Y_{2i}) + n\epsilon_n$$
$$= nI(U_Q; Y_{2Q}|Q) + n\epsilon_n$$
$$\leq nI(U; Y_2) + n\epsilon_n.$$

The bound on the cardinality of U can be established using the convex cover method in Appendix C. This completes the proof of Theorem 5.2.

Remark 5.7. The proof works also for $U_i = (M_2, Y_2^{i-1})$ or $U_i = (M_2, Y_1^{i-1}, Y_2^{i-1})$; both identifications satisfy the Markov condition $U_i \to X_i \to (Y_{1i}, Y_{2i})$.

5.4.2 Capacity Region of the BS-BC

We show that the superposition coding inner bound for the BS-BC presented in Example 5.3 is tight. Since the BS-BC is degraded, Theorem 5.2 characterizes its capacity region. We show that this region can be simplified to the set of rate pairs (R_1, R_2) such that

$$R_1 \leq H(\alpha * p_1) - H(p_1),$$
$$R_2 \leq 1 - H(\alpha * p_2) \tag{5.7}$$

for some $\alpha \in [0, 1/2]$.

In Example 5.3, we argued that the interior of this rate region is achievable via superposition coding by taking $U \sim \text{Bern}(1/2)$ and $X = U \oplus V$, where $V \sim \text{Bern}(\alpha)$ is independent of U. We now show that the characterization of the capacity region in Theorem 5.2 reduces to (5.7) using two alternative methods.

Proof via Mrs. Gerber's lemma. First recall that the capacity region is the same as that of the physically degraded DM-BC $X \to \tilde{Y}_1 \to Y_2$. Thus we assume that it is physically degraded, i.e., $Y_2 = Y_1 \oplus \tilde{Z}_2$, where $Z_1 \sim \text{Bern}(p_1)$ and $\tilde{Z}_2 \sim \text{Bern}(\tilde{p}_2)$ are independent and $p_2 = p_1 * \tilde{p}_2$.

Consider the second inequality in the capacity region characterization in Theorem 5.2

$$I(U; Y_2) = H(Y_2) - H(Y_2|U) \leq 1 - H(Y_2|U).$$

Since $1 \geq H(Y_2|U) \geq H(Y_2|X) = H(p_2)$, there exists an $\alpha \in [0, 1/2]$ such that $H(Y_2|U) = H(\alpha * p_2)$. Next consider

$$I(X; Y_1|U) = H(Y_1|U) - H(Y_1|X) = H(Y_1|U) - H(p_1).$$

Now let $0 \leq H^{-1}(v) \leq 1/2$ be the inverse of the binary entropy function. By physical degradedness and the scalar MGL in Section 2.1,

$$H(Y_2|U) = H(Y_1 \oplus \tilde{Z}_2|U) \geq H(H^{-1}(H(Y_1|U)) * \tilde{p}_2).$$

But $H(Y_2|U) = H(\alpha * p_2) = H(\alpha * p_1 * \tilde{p}_2)$, and thus

$$H(Y_1|U) \leq H(\alpha * p_1).$$

Proof via a symmetrization argument. Since for the BS-BC $|\mathcal{U}| \leq |\mathcal{X}| + 1 = 3$, assume without loss of generality that $\mathcal{U} = \{1, 2, 3\}$. Given any $(U, X) \sim p(u, x)$, define $(\tilde{U}, \tilde{X}) \sim p(\tilde{u}, \tilde{x})$ as

$$p_{\tilde{U}}(u) = p_{\tilde{U}}(-u) = \frac{1}{2} p_U(u), \quad u \in \{1, 2, 3\},$$
$$p_{\tilde{X}|\tilde{U}}(x|u) = p_{\tilde{X}|\tilde{U}}(1 - x| - u) = p_{X|U}(x|u), \quad (u, x) \in \{1, 2, 3\} \times \{0, 1\}.$$

Further define $\tilde{Y}_1$ to be the output of the BS-BC when the input is $\tilde{X}$. Given $\{|\tilde{U}| = u\}$, $u = 1, 2, 3$, the channel from $\tilde{U}$ to $\tilde{X}$ is a BSC with parameter $p(x|u)$ (with input alphabet $\{-u, u\}$ instead of $\{0, 1\}$). Thus, $H(Y_1|U = u) = H(\tilde{Y}_1|\tilde{U} = u) = H(\tilde{Y}_1|\tilde{U} = -u)$ for $u = 1, 2, 3$, which implies that $H(Y_1|U) = H(\tilde{Y}_1|\tilde{U})$ and the first bound in the capacity region is preserved. Similarly, $H(Y_2|U) = H(\tilde{Y}_2|\tilde{U})$. Also note that $\tilde{X} \sim \text{Bern}(1/2)$ and so are the corresponding outputs $\tilde{Y}_1$ and $\tilde{Y}_2$. Hence,

$$H(Y_2) - H(Y_2|U) \leq H(\tilde{Y}_2) - H(\tilde{Y}_2|\tilde{U}).$$

Therefore, it suffices to evaluate the capacity region characterization in Theorem 5.2 with the above symmetric input pmfs $p(\tilde{u}, \tilde{x})$.

Next consider the weighted sum

$$\lambda I(\tilde{X}; \tilde{Y}_1|\tilde{U}) + (1 - \lambda)I(\tilde{U}; \tilde{Y}_2)$$
$$= \lambda H(\tilde{Y}_1|\tilde{U}) - (1 - \lambda)H(\tilde{Y}_2|\tilde{U}) - \lambda H(p_1) + (1 - \lambda)$$
$$= \sum_{u=1}^{3} (\lambda H(\tilde{Y}_1|\tilde{U}, |\tilde{U}| = u) - (1 - \lambda)H(\tilde{Y}_2|\tilde{U}, |\tilde{U}| = u))p(u) - \lambda H(p_1) + (1 - \lambda)$$
$$\leq \max_{u \in \{1,2,3\}} (\lambda H(\tilde{Y}_1|\tilde{U}, |\tilde{U}| = u) - (1 - \lambda)H(\tilde{Y}_2|\tilde{U}, |\tilde{U}| = u)) - \lambda H(p_1) + (1 - \lambda)$$

for some $\lambda \in [0, 1]$. Since the maximum is attained by a single u (i.e., $p_{\tilde{U}}(u) = p_{\tilde{U}}(-u) = 1/2$) and given $|\tilde{U}| = u$, the channel from $\tilde{U}$ to $\tilde{X}$ is a BSC, the set of rate pairs (R_1, R_2) such that

$$\lambda R_1 + (1 - \lambda)R_2 \leq \max_{\alpha} [\lambda(H(\alpha * p_1) - H(p_1)) + (1 - \lambda)(1 - H(\alpha * p_2))]$$

constitutes an outer bound on the capacity region given by the supporting line corresponding to each $\lambda \in [0, 1]$. Finally, since the rate region in (5.7) is convex (see Problem 5.3), by Lemma A.1 it coincides with the capacity region.

Remark 5.8. A brute-force optimization of the weighted sum $R_{sum}(\lambda) = \lambda I(\tilde{X}; \tilde{Y}_1 | \tilde{U}) + (1 - \lambda) I(\tilde{U}; \tilde{Y}_2)$ can lead to the same conclusion. In this approach, it suffices to optimize over all $p(u, x)$ with $|\mathcal{U}| \le 2$. The above symmetrization argument provides a more elegant approach to performing this three-dimensional optimization.

Remark 5.9. The second converse proof does not require physical degradedness. In fact, this symmetrization argument can be used to evaluate the superposition coding inner bound in Theorem 5.1 for *any* nondegraded binary-input DM-BC with $p_{Y_1,Y_2|X}(y_1, y_2|0) = p_{Y_1,Y_2|X}(-y_1, -y_2|1)$ for all y_1, y_2 (after proper relabeling of the symbols) such as the BSC–BEC BC in Example 5.4 of Section 5.6.

5.5 GAUSSIAN BROADCAST CHANNEL

Consider the 2-receiver discrete-time additive white Gaussian noise (Gaussian in short) BC model depicted in Figure 5.8. The channel outputs corresponding to the input X are

$$Y_1 = g_1 X + Z_1,$$
$$Y_2 = g_2 X + Z_2,$$

where g_1 and g_2 are channel gains, and $Z_1 \sim N(0, N_0/2)$ and $Z_2 \sim N(0, N_0/2)$ are noise components. Thus in transmission time i, the channel outputs are

$$Y_{1i} = g_1 X_i + Z_{1i},$$
$$Y_{2i} = g_2 X_i + Z_{2i},$$

where $\{Z_{1i}\}$ and $\{Z_{2i}\}$ are WGN($N_0/2$) processes, independent of the channel input X^n. Note that only the marginal distributions of $\{(Z_{1i}, Z_{2i})\}$ are relevant to the capacity region and hence we do not need to specify their joint distribution. Assume without loss of generality that $|g_1| \ge |g_2|$ and $N_0/2 = 1$. Further assume an average transmission power constraint

$$\sum_{i=1}^{n} x_i^2(m_1, m_2) \le nP, \quad (m_1, m_2) \in [1 : 2^{nR_1}] \times [1 : 2^{nR_2}].$$

For notational convenience, we consider the equivalent Gaussian BC channel

$$Y_1 = X + Z_1,$$
$$Y_2 = X + Z_2.$$

In this model, the channel gains are normalized to 1 and the *transmitter-referred* noise components Z_1 and Z_2 are zero-mean Gaussian with powers $N_1 = 1/g_1^2$ and $N_2 = 1/g_2^2$, respectively. This equivalence can be seen by first multiplying both sides of the equations of the original channel model by $1/g_1$ and $1/g_2$, respectively, and then scaling the resulting channel outputs by g_1 and g_2. Note that the equivalent broadcast channel is

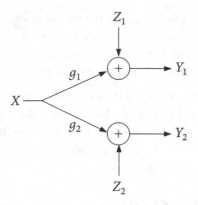

Figure 5.8. Gaussian broadcast channel.

stochastically degraded. Hence, its capacity region is the same as that of the physically degraded Gaussian BC (see Figure 5.9)

$$Y_1 = X + Z_1,$$
$$Y_2 = Y_1 + \tilde{Z}_2,$$

where Z_1 and $\tilde{Z}_2$ are independent Gaussian random variables with average powers N_1 and $(N_2 - N_1)$, respectively. Define the channel received signal-to-noise ratios as $S_1 = P/N_1$ and $S_2 = P/N_2$.

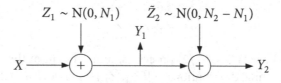

Figure 5.9. Corresponding physically degraded Gaussian BC.

5.5.1 Capacity Region of the Gaussian BC

The capacity region of the Gaussian BC is a function only of the channel SNRs and a power allocation parameter.

Theorem 5.3. The capacity region of the Gaussian BC is the set of rate pairs (R_1, R_2) such that

$$R_1 \leq C(\alpha S_1),$$
$$R_2 \leq C\left(\frac{\bar{\alpha} S_2}{\alpha S_2 + 1}\right)$$

for some $\alpha \in [0, 1]$, where $C(x)$ is the Gaussian capacity function.

Achievability follows by setting $U \sim N(0, \bar{\alpha}P)$ and $V \sim N(0, \alpha P)$, independent of each other, $X = U + V \sim N(0, P)$ in the superposition coding inner bound. With this choice of (U, X), it can be readily to shown that

$$I(X; Y_1|U) = C\left(\frac{\alpha P}{N_1}\right) = C(\alpha S_1),$$

$$I(U; Y_2) = C\left(\frac{\bar{\alpha}P}{\alpha P + N_2}\right) = C\left(\frac{\bar{\alpha}S_2}{\alpha S_2 + 1}\right),$$

which are the expressions in the theorem.

As for the BS-BC in Example 5.3, the superposition coding scheme for the Gaussian BC can be described more explicitly. Randomly and independently generate 2^{nR_2} sequences $u^n(m_2)$, $m_2 \in [1 : 2^{nR_2}]$, each i.i.d. $N(0, \bar{\alpha}P)$, and 2^{nR_1} sequences $v^n(m_1)$, $m_1 \in [1 : 2^{nR_1}]$, each i.i.d. $N(0, \alpha P)$. To send the message pair (m_1, m_2), the encoder transmits $x^n(m_1, m_2) = u^n(m_2) + v^n(m_1)$.

Receiver 2 recovers m_2 from $y_2^n = u^n(m_2) + (v^n(m_1) + z_2^n)$ while treating $v^n(m_1)$ as noise. The probability of decoding error tends to zero as $n \to \infty$ if $R_2 < C(\bar{\alpha}P/(\alpha P + N_2))$.

Receiver 1 uses successive cancellation—it first decodes $y_1^n = u^n(m_2) + (v^n(m_1) + z_1^n)$ to recover m_2 while treating $v^n(m_1)$ as part of the noise, subtracts off $u^n(m_2)$, and then decodes $v^n(m_1) + z_1^n$ to recover m_1. The probability of decoding error tends to zero as $n \to \infty$ if $R_1 < C(\alpha P/N_1)$ and $R_2 < C(\bar{\alpha}P/(\alpha P + N_1))$. Since $N_1 < N_2$, the latter condition is already satisfied by the rate constraint for receiver 2.

Remark 5.10. To make the above arguments rigorous, we can use the discretization procedure detailed in the achievability proof for the point-to-point Gaussian channel in Section 3.4.

Remark 5.11. As for the Gaussian MAC in Section 4.6, every point in the capacity region can be achieved simply using good point-to-point Gaussian channel codes and successive cancellation decoding.

5.5.2 Proof of the Converse

Since the capacity region of the Gaussian BC is the same as that of its corresponding physically degraded Gaussian BC, we prove the converse for the physically degraded Gaussian BC shown in Figure 5.9. By Fano's inequality, we have

$$nR_1 \le I(M_1; Y_1^n|M_2) + n\epsilon_n,$$

$$nR_2 \le I(M_2; Y_2^n) + n\epsilon_n.$$

We need to show that there exists an $\alpha \in [0, 1]$ such that

$$I(M_1; Y_1^n|M_2) \le n\,C(\alpha S_1) = n\,C\left(\frac{\alpha P}{N_1}\right)$$

and

$$I(M_2; Y_2^n) \le n\,C\left(\frac{\bar{\alpha}S_2}{\alpha S_2 + 1}\right) = n\,C\left(\frac{\bar{\alpha}P}{\alpha P + N_2}\right).$$

Consider

$$I(M_2; Y_2^n) = h(Y_2^n) - h(Y_2^n|M_2) \leq \frac{n}{2}\log(2\pi e(P + N_2)) - h(Y_2^n|M_2).$$

Since

$$\frac{n}{2}\log(2\pi e N_2) = h(Z_2^n) = h(Y_2^n|M_2, X^n) \leq h(Y_2^n|M_2) \leq h(Y_2^n) \leq \frac{n}{2}\log(2\pi e(P + N_2)),$$

there must exist an $\alpha \in [0, 1]$ such that

$$h(Y_2^n|M_2) = \frac{n}{2}\log(2\pi e(\alpha P + N_2)). \tag{5.8}$$

Next consider

$$\begin{aligned}
I(M_1; Y_1^n|M_2) &= h(Y_1^n|M_2) - h(Y_1^n|M_1, M_2) \\
&= h(Y_1^n|M_2) - h(Y_1^n|M_1, M_2, X^n) \\
&= h(Y_1^n|M_2) - h(Y_1^n|X^n) \\
&= h(Y_1^n|M_2) - \frac{n}{2}\log(2\pi e N_1).
\end{aligned}$$

Now using the conditional EPI in Section 2.1, we obtain

$$\begin{aligned}
h(Y_2^n|M_2) &= h(Y_1^n + \tilde{Z}_2^n|M_2) \\
&\geq \frac{n}{2}\log(2^{2h(Y_1^n|M_2)/n} + 2^{2h(\tilde{Z}_2^n|M_2)/n}) \\
&= \frac{n}{2}\log(2^{2h(Y_1^n|M_2)/n} + 2\pi e(N_2 - N_1)).
\end{aligned}$$

Combining this inequality with (5.8) implies that

$$2\pi e(\alpha P + N_2) \geq 2^{2h(Y_1^n|M_2)/n} + 2\pi e(N_2 - N_1).$$

Thus, $h(Y_1^n|M_2) \leq (n/2)\log(2\pi e(\alpha P + N_1))$ and hence

$$I(M_1; Y_1^n|M_2) \leq \frac{n}{2}\log(2\pi e(\alpha P + N_1)) - \frac{n}{2}\log(2\pi e N_1) = n\,\mathsf{C}\left(\frac{\alpha P}{N_1}\right).$$

This completes the proof of Theorem 5.3.

Remark 5.12. The converse can be proved directly starting with the single-letter characterization $R_1 \leq I(X; Y_1|U)$, $R_2 \leq I(U; Y_2)$ with the power constraint $\mathsf{E}(X^2) \leq P$. The proof then follows similar steps to the above converse proof and uses the scalar version of the conditional EPI.

Remark 5.13. The similarity between the above converse proof and the converse proof for the BS-BC hints at some form of duality between Mrs. Gerber's lemma for binary random variables and the entropy power inequality for continuous random variables.

5.6   LESS NOISY AND MORE CAPABLE BROADCAST CHANNELS

Superposition coding is optimal for the following two classes of broadcast channels, which are more general than the class of degraded broadcast channels.

Less noisy DM-BC. A DM-BC $p(y_1, y_2|x)$ is said to be *less noisy* if $I(U; Y_1) \geq I(U; Y_2)$ for all $p(u, x)$. In this case we say that receiver 1 is less noisy than receiver 2. The private-message capacity region of the less noisy DM-BC is the set of rate pairs (R_1, R_2) such that

$$R_1 \leq I(X; Y_1|U),$$
$$R_2 \leq I(U; Y_2)$$

for some pmf $p(u, x)$, where $|\mathcal{U}| \leq \min\{|\mathcal{X}|, |\mathcal{Y}_2|\} + 1$. Note that the less noisy condition guarantees that in the superposition coding scheme, receiver 1 can recover the message intended for receiver 2.

More capable DM-BC. A DM-BC $p(y_1, y_2|x)$ is said to be *more capable* if $I(X; Y_1) \geq I(X; Y_2)$ for all $p(x)$. In this case we say that receiver 1 is more capable than receiver 2. The private-message capacity region of the more capable DM-BC is the set of rate pairs (R_1, R_2) such that

$$R_1 \leq I(X; Y_1|U),$$
$$R_2 \leq I(U; Y_2), \tag{5.9}$$
$$R_1 + R_2 \leq I(X; Y_1)$$

for some pmf $p(u, x)$, where $|\mathcal{U}| \leq \min\{|\mathcal{X}|, |\mathcal{Y}_1| + |\mathcal{Y}_2|\} + 1$.

It can be easily shown that if a DM-BC is degraded then it is less noisy, and that if a DM-BC is less noisy then it is more capable. The converse to each of these statements does not hold in general as illustrated in the following example.

Example 5.4 (A BSC and a BEC). Consider a DM-BC with $X \in \{0, 1\}$, $Y_1 \in \{0, 1\}$, and $Y_2 \in \{0, 1, e\}$, where the channel from X to Y_1 is a BSC(p), $p \in (0, 1/2)$, and the channel from X to Y_2 is a BEC(ϵ), $\epsilon \in (0, 1)$. Then it can be shown that the following hold:

1. For $0 < \epsilon \leq 2p$, Y_1 is a *degraded* version of Y_2.

2. For $2p < \epsilon \leq 4p(1 - p)$, Y_2 is *less noisy* than Y_1, but Y_1 is not a degraded version of Y_2.

3. For $4p(1 - p) < \epsilon \leq H(p)$, Y_2 is *more capable* than Y_1, but not less noisy.

4. For $H(p) < \epsilon < 1$, the channel does not belong to *any* of the three classes.

The capacity region for each of these cases is achieved using superposition coding. The converse proofs for the first three cases follow by evaluating the capacity expressions for the degraded BC, less noisy BC, and more capable BC, respectively. The converse proof for the last case will be given in Chapter 8.

5.6.1 Proof of the Converse for the More Capable DM-BC

It is difficult to find an identification of the auxiliary random variable that satisfies the desired properties for the capacity region characterization in (5.9). Instead, we prove the converse for the equivalent region consisting of all rate pairs (R_1, R_2) such that

$$R_2 \le I(U; Y_2),$$
$$R_1 + R_2 \le I(X; Y_1|U) + I(U; Y_2),$$
$$R_1 + R_2 \le I(X; Y_1)$$

for some pmf $p(u, x)$.

The converse proof for this alternative region involves a tricky identification of the auxiliary random variable and the application of the Csiszár sum identity in Section 2.3.

By Fano's inequality, it is straightforward to show that

$$nR_2 \le I(M_2; Y_2^n) + n\epsilon_n,$$
$$n(R_1 + R_2) \le I(M_1; Y_1^n|M_2) + I(M_2; Y_2^n) + n\epsilon_n,$$
$$n(R_1 + R_2) \le I(M_1; Y_1^n) + I(M_2; Y_2^n|M_1) + n\epsilon_n.$$

Consider the mutual information terms in the second inequality.

$$
\begin{aligned}
I(M_1; Y_1^n|M_2) + I(M_2; Y_2^n) &= \sum_{i=1}^{n} I(M_1; Y_{1i}|M_2, Y_1^{i-1}) + \sum_{i=1}^{n} I(M_2; Y_{2i}|Y_{2,i+1}^n) \\
&\le \sum_{i=1}^{n} I(M_1, Y_{2,i+1}^n; Y_{1i}|M_2, Y_1^{i-1}) + \sum_{i=1}^{n} I(M_2, Y_{2,i+1}^n; Y_{2i}) \\
&= \sum_{i=1}^{n} I(M_1, Y_{2,i+1}^n; Y_{1i}|M_2, Y_1^{i-1}) + \sum_{i=1}^{n} I(M_2, Y_{2,i+1}^n, Y_1^{i-1}; Y_{2i}) \\
&\quad - \sum_{i=1}^{n} I(Y_1^{i-1}; Y_{2i}|M_2, Y_{2,i+1}^n) \\
&= \sum_{i=1}^{n} I(M_1; Y_{1i}|M_2, Y_1^{i-1}, Y_{2,i+1}^n) + \sum_{i=1}^{n} I(M_2, Y_{2,i+1}^n, Y_1^{i-1}; Y_{2i}) \\
&\quad - \sum_{i=1}^{n} I(Y_1^{i-1}; Y_{2i}|M_2, Y_{2,i+1}^n) + \sum_{i=1}^{n} I(Y_{2,i+1}^n; Y_{1i}|M_2, Y_1^{i-1}) \\
&\overset{(a)}{=} \sum_{i=1}^{n} \big(I(M_1; Y_{1i}|U_i) + I(U_i; Y_{2i}) \big) \\
&\le \sum_{i=1}^{n} \big(I(X_i; Y_{1i}|U_i) + I(U_i; Y_{2i}) \big),
\end{aligned}
$$

where $Y_1^0, Y_{2,n+1}^n = \emptyset$ and (a) follows by the Csiszár sum identity and the auxiliary random variable identification $U_i = (M_2, Y_1^{i-1}, Y_{2,i+1}^n)$.

Next consider the mutual information term in the first inequality

$$I(M_2; Y_2^n) = \sum_{i=1}^{n} I(M_2; Y_{2i} | Y_{2,i+1}^n)$$

$$\leq \sum_{i=1}^{n} I(M_2, Y_1^{i-1}, Y_{2,i+1}^n; Y_{2i}) = \sum_{i=1}^{n} I(U_i; Y_{2i}).$$

For the third inequality, define $V_i = (M_1, Y_1^{i-1}, Y_{2,i+1}^n)$. Following similar steps to the bound for the second inequality, we have

$$I(M_1; Y_1^n) + I(M_2; Y_2^n | M_1) \leq \sum_{i=1}^{n} \big(I(V_i; Y_{1i}) + I(X_i; Y_{2i} | V_i) \big)$$

$$\overset{(a)}{\leq} \sum_{i=1}^{n} \big(I(V_i; Y_{1i}) + I(X_i; Y_{1i} | V_i) \big) = \sum_{i=1}^{n} I(X_i; Y_{1i}),$$

where (a) follows by the more capable condition, which implies $I(X; Y_2 | V) \leq I(X; Y_1 | V)$ whenever $V \to X \to (Y_1, Y_2)$ form a Markov chain.

The rest of the proof follows by introducing a time-sharing random variable $Q \sim$ Unif$[1 : n]$ independent of $(M_1, M_2, X^n, Y_1^n, Y_2^n)$ and defining $U = (Q, U_Q)$, $X = X_Q$, $Y_1 = Y_{1Q}$, and $Y_2 = Y_{2Q}$. The bound on the cardinality of U can be proved using the convex cover method in Appendix C. This completes the proof of the converse.

Remark 5.14. The converse proof for the more capable DM-BC also establishes the converse for the less noisy and degraded DM-BCs.

5.7   EXTENSIONS

We extend the results in the previous sections to the setup with common message and to channels with more than two receivers.

Capacity region with common and private messages. Our discussion so far has focused on the private-message capacity region. As mentioned in Remark 5.4, the superposition coding inner bound is still achievable if we require the "stronger" receiver to also recover the message intended for the "weaker" receiver. Therefore, if the rate pair (R_1, R_2) is achievable for private messages, then the rate triple $(R_0, R_1, R_2 - R_0)$ is achievable for private and common messages.

Using this observation, we can readily show that the capacity region with common message of the more capable DM-BC is the set of rate triples (R_0, R_1, R_2) such that

$$R_1 \leq I(X; Y_1 | U),$$
$$R_0 + R_2 \leq I(U; Y_2),$$
$$R_0 + R_1 + R_2 \leq I(X; Y_1)$$

for some pmf $p(u, x)$. Unlike achievability, the converse is not implied automatically by the converse for the private-message capacity region. The proof, however, follows similar steps to that for the private-message capacity region.

k-Receiver degraded DM-BC. Consider a k-receiver degraded DM-BC $p(y_1, \ldots, y_k|x)$, where $X \to Y_1 \to Y_2 \to \cdots \to Y_k$ form a Markov chain. The private-message capacity region is the set of rate tuples $(R_1, \ldots, R_k)$ such that

$$R_1 \le I(X; Y_1|U_2),$$
$$R_j \le I(U_j; Y_j|U_{j+1}), \quad j \in [2:k],$$

for some pmf $p(u_k, u_{k-1})p(u_{k-2}|u_{k-1}) \cdots p(x|u_2)$ and $U_{k+1} = \emptyset$. The capacity region for the 2-receiver Gaussian BC also extends to more than two receivers.

Remark 5.15. The capacity region is not known in general for the less noisy DM-BC with $k > 3$ receivers and for the more capable DM-BC with $k > 2$ receivers.

SUMMARY

- Discrete memoryless broadcast channel (DM-BC)
- Capacity region depends only on the channel marginal pmfs
- Superposition coding
- Simultaneous nonunique decoding
- Physically and stochastically degraded BCs
- Capacity region of degraded BCs is achieved by superposition coding
- Identification of the auxiliary random variable in the proof of the converse
- Bounding the cardinality of the auxiliary random variable
- Proof of the converse for the BS-BC:
  - Mrs. Gerber's lemma
  - Symmetrization argument
- Gaussian BC is always degraded
- Use of EPI in the proof of the converse for the Gaussian BC
- Less noisy and more capable BCs:
  - Degraded $\Rightarrow$ less noisy $\Rightarrow$ more capable
  - Superposition coding is optimal
- Use of Csiszár sum identity in the proof of the converse for more capable BCs

- **Open problems:**

  5.1. What is the capacity region of less noisy BCs with four or more receivers?

  5.2. What is the capacity region of more capable BCs with three or more receivers?

BIBLIOGRAPHIC NOTES

The broadcast channel was first introduced by Cover (1972), who demonstrated super-position coding through the BS-BC and Gaussian BC examples, and conjectured the characterization of the capacity region of the stochastically degraded broadcast channel using the auxiliary random variable U. Bergmans (1973) proved achievability of the capacity region of the degraded DM-BC. Subsequently, Gallager (1974) proved the converse by providing the nonintuitive identification of the auxiliary random variable discussed in Section 5.4. He also provided a bound on the cardinality of U. Wyner (1973) proved the converse for the capacity region of the BS-BC using Mrs. Gerber's lemma. The symmetrization argument in the alternative converse proof for the BS-BC is due to Nair (2010), who also applied it to binary-input symmetric-output BCs. Bergmans (1974) established the converse for the capacity region of the Gaussian BC using the entropy power inequality. A strong converse for the capacity region of the DM-BC was proved by Ahlswede and Körner (1975) for the maximal probability of error. A technique by Willems (1990) can be used to extend this strong converse to the average probability of error; see also Csiszár and Körner (1981b) for an indirect proof based on the correlation elimination technique by Ahlswede (1978).

The classes of less noisy and more capable DM-BCs were introduced by Körner and Marton (1977a), who provided operational definitions of these classes and established the capacity region for the less noisy case. The capacity region for the more capable case was established by El Gamal (1979). Nair (2010) established the classification the BSC–BEC broadcast channel and showed that superposition coding is optimal for all classes. The capacity region of the 3-receiver less noisy DM-BC is due to Wang and Nair (2010), who showed that superposition coding is optimal for this case as well. Surveys of the literature on the broadcast channel can be found in van der Meulen (1977, 1981) and Cover (1998).

PROBLEMS

5.1. Show that the converse for the degraded DM-BC can be proved with the auxiliary random variable identification $U_i = (M_2, Y_2^{i-1})$ or $U_i = (M_2, Y_1^{i-1}, Y_2^{i-1})$.

5.2. Prove the converse for the capacity region of the Gaussian BC by starting from the single-letter characterization

$$R_1 \le I(X; Y_1 | U),$$
$$R_2 \le I(U; Y_2)$$

for some cdf $F(u, x)$ such that $\mathsf{E}(X^2) \le P$.

5.3.   Verify that the characterizations of the capacity region for the BS-BC and the Gaussian BC in (5.7) and in Theorem 5.3, respectively, are convex.

5.4.   Show that if a DM-BC is degraded, then it is also less noisy.

5.5.   Show that if a DM-BC is less noisy, then it is also more capable.

5.6.   Given a DM-BC $p(y_1, y_2|x)$, let $D(p(x)) = I(X; Y_1) - I(X; Y_2)$. Show that $D(p(x))$ is concave in $p(x)$ iff Y_1 is less noisy than Y_2.

5.7.   Prove the classification of the BSC–BEC broadcast channel in Example 5.4.

5.8.   Show that the two characterizations of the capacity region for the more capable DM-BC in Section 5.6 are equivalent. (Hint: One direction is trivial. For the other direction, show that the corner points of these regions are the same.)

5.9.   *Another simple outer bound.* Consider the DM-BC with three messages.

   (a) Show that if a rate triple (R_0, R_1, R_2) is achievable, then it must satisfy the inequalities
$$R_0 + R_1 \le I(X; Y_1),$$
$$R_0 + R_2 \le I(X; Y_2)$$
   for some pmf $p(x)$.

   (b) Show that this outer bound on the capacity region can be strictly tighter than the simple outer bound in Figure 5.2. (Hint: Consider two antisymmetric Z channels.)

   (c) Set $R_1 = R_2 = 0$ in the outer bound in part (a) to show that the *common-message capacity* is
$$C_0 = \max_{p(x)} \min\{I(X; Y_1),\ I(X; Y_2)\}.$$
   Argue that C_0 is in general strictly smaller than $\min\{C_1, C_2\}$.

5.10. *Binary erasure broadcast channel.* Consider a DM-BC $p(y_1, y_2|x)$ where the channel from X to Y_1 is a BEC(p_1) and the channel from X to Y_2 is a BEC(p_2) with $p_1 \le p_2$. Find the capacity region in terms of p_1 and p_2.

5.11. *Product of two degraded broadcast channels.* Consider two degraded DM-BCs $p(y_{11}|x_1)p(y_{21}|y_{11})$ and $p(y_{12}|x_2)p(y_{22}|y_{12})$. The product of these two degraded DM-BCs depicted in Figure 5.10 is a DM-BC with $X = (X_1, X_2)$, $Y_1 = (Y_{11}, Y_{12})$, $Y_2 = (Y_{21}, Y_{22})$, and $p(y_1, y_2|x) = p(y_{11}|x_1)p(y_{21}|y_{11})p(y_{12}|x_2)p(y_{22}|y_{12})$. Show that the private-message capacity region of the product DM-BC is the set of rate pairs (R_1, R_2) such that
$$R_1 \le I(X_1; Y_{11}|U_1) + I(X_2; Y_{12}|U_2),$$
$$R_2 \le I(U_1; Y_{21}) + I(U_2; Y_{22})$$

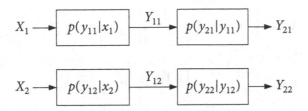

Figure 5.10. Product of two degraded broadcast channels.

for some pmf $p(u_1, x_1)p(u_2, x_2)$. Thus, the capacity region is the Minkowski sum of the capacity regions of the two component DM-BCs.

Remark: This result is due to Poltyrev (1977).

5.12. *Product of two Gaussian broadcast channels.* Consider two Gaussian BCs

$$Y_{1j} = X_j + Z_{1j},$$
$$Y_{2j} = Y_{1j} + \tilde{Z}_{2j}, \quad j = 1, 2,$$

where $Z_{11} \sim N(0, N_{11})$, $Z_{12} \sim N(0, N_{12})$, $\tilde{Z}_{21} \sim N(0, \tilde{N}_{21})$, and $\tilde{Z}_{22} \sim N(0, \tilde{N}_{22})$ are independent noise components. Assume the average transmission power constraint

$$\sum_{i=1}^{n} (x_{1i}^2(m_1, m_2) + x_{2i}^2(m_1, m_2)) \le nP.$$

Find the private-message capacity region of the product Gaussian BC with $X = (X_1, X_2)$, $Y_1 = (Y_{11}, Y_{12})$, and $Y_2 = (Y_{21}, Y_{22})$, in terms of the noise powers, P, and power allocation parameters $\alpha_1, \alpha_2 \in [0, 1]$ for each subchannel and $\beta \in [0, 1]$ between two channels.

Remark: This result is due to Hughes-Hartogs (1975).

5.13. *Minimum-energy-per-bit region.* Consider the Gaussian BC with noise powers N_1 and N_2. Find the minimum-energy-per-bit region, that is, the set of all energy pairs $(E_1, E_2) = (P/R_1, P/R_2)$ such that the rate pair (R_1, R_2) is achievable with average code power P.

5.14. *Reversely degraded broadcast channels with common message.* Consider two reversely degraded DM-BCs $p(y_{11}|x_1)p(y_{21}|y_{11})$ and $p(y_{22}|x_2)p(y_{12}|y_{22})$. The product of these two degraded DM-BCs depicted in Figure 5.11 is a DM-BC with $X = (X_1, X_2)$, $Y_1 = (Y_{11}, Y_{12})$, $Y_2 = (Y_{21}, Y_{22})$, and $p(y_1, y_2|x) = p(y_{11}|x_1)p(y_{21}|y_{11}) \cdot p(y_{22}|x_2)p(y_{12}|y_{22})$. A common message $M_0 \in [1 : 2^{nR_0}]$ is to be communicated to both receivers. Show that the common-message capacity is

$$C_0 = \max_{p(x_1)p(x_2)} \min\{I(X_1; Y_{11}) + I(X_2; Y_{12}), I(X_1; Y_{21}) + I(X_2; Y_{22})\}.$$

Remark: This channel is studied in more detail in Section 9.4.

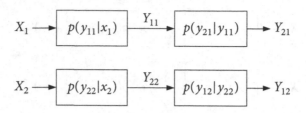

Figure 5.11. Product of reversely degraded broadcast channels.

5.15. *Duality between Gaussian broadcast and multiple access channels.* Consider the following Gaussian BC and Gaussian MAC:

- Gaussian BC: $Y_1 = g_1 X + Z_1$ and $Y_2 = g_2 X + Z_2$, where $Z_1 \sim N(0, 1)$ and $Z_2 \sim N(0, 1)$. Assume average power constraint P on X.

- Gaussian MAC: $Y = g_1 X_1 + g_2 X_2 + Z$, where $Z \sim N(0, 1)$. Assume the average sum-power constraint

$$\sum_{i=1}^{n} (x_{1i}^2(m_1) + x_{2i}^2(m_2)) \le nP, \quad (m_1, m_2) \in [1 : 2^{nR_1}] \times [1 : 2^{nR_2}].$$

(a) Characterize the (private-message) capacity regions of these two channels in terms of P, g_1, g_2, and power allocation parameter $\alpha \in [0, 1]$.

(b) Show that the two capacity regions are equal.

(c) Show that every point (R_1, R_2) on the boundary of the capacity region of the above Gaussian MAC is achievable using random coding and successive cancellation decoding. That is, time sharing is not needed in this case.

(d) Argue that the sequence of codes that achieves the rate pairs (R_1, R_2) on the boundary of the Gaussian MAC capacity region can be used to achieve the same point on the capacity region of the above Gaussian BC.

Remark: This result is a special case of a general duality result between the Gaussian vector BC and MAC presented in Chapter 9.

5.16. *k-receiver Gaussian BC.* Consider the k-receiver Gaussian BC

$$Y_1 = X + Z_1,$$
$$Y_j = Y_{j-1} + \tilde{Z}_j, \quad j \in [2 : k],$$

where $Z_1, \tilde{Z}_2, \ldots, \tilde{Z}_k$ are independent Gaussian noise components with powers $N_1, \tilde{N}_2, \ldots, \tilde{N}_k$, respectively. Assume average power constraint P on X. Provide a characterization of the private-message capacity region in terms of the noise powers, P, and power allocation parameters $\alpha_1, \ldots, \alpha_k \ge 0$ with $\sum_{j=1}^{k} \alpha_j = 1$.

5.17. *Three-receiver less noisy BC.* Consider the 3-receiver DM-BC $p(y_1, y_2, y_3|x)$ such that $I(U; Y_1) \ge I(U; Y_2) \ge I(U; Y_3)$ for all $p(u, x)$.

(a) Suppose that $W \to X^n \to (Y_1^n, Y_2^n)$ form a Markov chain. Show that

$$I(Y_2^{i-1}; Y_{1i}|W) \le I(Y_1^{i-1}; Y_{1i}|W), \quad i \in [2:n].$$

(b) Suppose that $W \to X^n \to (Y_2^n, Y_3^n)$ form a Markov chain. Show that

$$I(Y_3^{i-1}; Y_{3i}|W) \le I(Y_2^{i-1}; Y_{3i}|W), \quad i \in [2:n].$$

(c) Using superposition coding and parts (a) and (b), show that the capacity region is the set of rate triples (R_1, R_2, R_3) such that

$$R_1 \le I(X; Y_1|U, V),$$
$$R_2 \le I(V; Y_2|U),$$
$$R_3 \le I(U; Y_3)$$

for some pmf $p(u, v, x)$. (Hint: Identify $U_i = (M_3, Y_2^{i-1})$ and $V_i = M_2$.)

Remark: This result is due to Wang and Nair (2010).

5.18. *MAC with degraded message sets.* Consider a DM-MAC $p(y|x_1, x_2)$ with message pair (M_0, M_1) uniformly distributed over $[1 : 2^{nR_0}] \times [1 : 2^{nR_1}]$. Sender 1 encodes (M_0, M_1), while sender 2 encodes only M_0. The receiver wishes to recover both messages. The probability of error, achievability, and the capacity region are defined as for the DM-MAC with private messages.

(a) Show that the capacity region is the set of rate pairs (R_0, R_1) such that

$$R_1 \le I(X_1; Y|X_2),$$
$$R_0 + R_1 \le I(X_1, X_2; Y)$$

for some pmf $p(x_1, x_2)$.

(b) Characterize the capacity region of the Gaussian MAC with noise power 1, channel gains g_1 and g_2, and average power constraint P on each of X_1 and X_2.

5.19. *MAC with common message.* Consider a DM-MAC $p(y|x_1, x_2)$ with message triple (M_0, M_1, M_2) uniformly distributed over $[1 : 2^{nR_0}] \times [1 : 2^{nR_1}] \times [1 : 2^{nR_2}]$. Sender 1 encodes (M_0, M_1) and sender 2 encodes (M_0, M_2). Thus, the common message M_0 is available to both senders, while the private messages M_1 and M_2 are available only to the respective senders. The receiver wishes to recover all three messages. The probability of error, achievability, and the capacity region are defined as for the DM-MAC with private messages.

(a) Show that the capacity region is the set of rate triples (R_0, R_1, R_2) such that

$$R_1 \le I(X_1; Y|U, X_2),$$
$$R_2 \le I(X_2; Y|U, X_1),$$
$$R_1 + R_2 \le I(X_1, X_2; Y|U),$$
$$R_0 + R_1 + R_2 \le I(X_1, X_2; Y)$$

for some pmf $p(u)p(x_1|u)p(x_2|u)$.

(b) Show that the capacity region of the Gaussian MAC with average power constraint P on each of X_1 and X_2 is the set of rate triples (R_0, R_1, R_2) such that

$$R_1 \le C(\alpha_1 S_1),$$
$$R_2 \le C(\alpha_2 S_2),$$
$$R_1 + R_2 \le C(\alpha_1 S_1 + \alpha_2 S_2),$$
$$R_0 + R_1 + R_2 \le C\left(S_1 + S_2 + 2\sqrt{\bar{\alpha}_1 \bar{\alpha}_2 S_1 S_2}\right)$$

for some $\alpha_1, \alpha_2 \in [0, 1]$.

Remark: The capacity region of the DM-MAC with common message is due to Slepian and Wolf (1973b). The converse for the Gaussian case is due to Bross, Lapidoth, and Wigger (2008).

5.20. *MAC with a helper.* Consider the 3-sender DM-MAC $p(y|x_1, x_2, x_3)$ with message pair (M_1, M_2) uniformly distributed over $[1 : 2^{nR_1}] \times [1 : 2^{nR_2}]$ as depicted in Figure 5.12. Sender 1 encodes M_1, sender 2 encodes M_2, and sender 3 encodes (M_1, M_2). The receiver wishes to recover both messages. Show that the capacity region is the set of rate pairs (R_1, R_2) such that

$$R_1 \le I(X_1, X_3; Y | X_2, Q),$$
$$R_2 \le I(X_2, X_3; Y | X_1, Q),$$
$$R_1 + R_2 \le I(X_1, X_2, X_3; Y | Q)$$

for some pmf $p(q)p(x_1|q)p(x_2|q)p(x_3|x_1, x_2, q)$ with $|\mathcal{Q}| \le 2$.
Remark: This is a special case of the capacity region of the general l-message k-sender DM-MAC established by Han (1979).

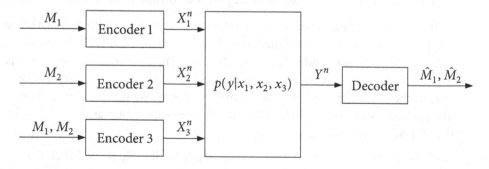

Figure 5.12. Three-sender DM-MAC with two messages.

CHAPTER 6

Interference Channels

We introduce the interference channel as a model for single-hop multiple one-to-one communications, such as pairs of base stations–handsets communicating over a frequency band that suffers from intercell interference, pairs of DSL modems communicating over a bundle of telephone lines that suffers from crosstalk, or pairs of people talking to each other in a cocktail party. The capacity region of the interference channel is not known in general. In this chapter, we focus on coding schemes for the two sender–receiver pair interference channel that are optimal or close to optimal in some special cases.

We first study simple coding schemes that use point-to-point channel codes, namely time division, treating interference as noise, and simultaneous decoding. We show that simultaneous decoding is optimal under strong interference, that is, when the interfering signal at each receiver is stronger than the signal from its respective sender. These inner bounds are compared for the Gaussian interference channel. We extend the strong interference result to the Gaussian case and show that treating interference as noise is sum-rate optimal when the interference is sufficiently weak. The converse proof of the latter result uses the new idea of a genie that provides side information to each receiver about its intended codeword.

We then present the Han–Kobayashi coding scheme, which generalizes the aforementioned simple schemes by also using rate splitting (see Section 4.4) and superposition coding (see Section 5.3). We show that the Han–Kobayashi scheme is optimal for the class of injective deterministic interference channels. The converse proof of this result is extended to establish an outer bound on the capacity region of the class of injective semideterministic interference channels, which includes the Gaussian interference channel. The outer bound for the Gaussian case, and hence the capacity region, is shown to be within half a bit per dimension of the Han–Kobayashi inner bound. This gap vanishes in the limit of high signal and interference to noise ratios for the normalized symmetric capacity (degrees of freedom). We discuss an interesting correspondence to q-ary expansion deterministic (QED) interference channels in this limit.

Finally, we introduce the new idea of interference alignment through a QED interference channel with many sender–receiver pairs. Interference alignment for wireless fading channels will be illustrated in Section 23.7.

6.1 DISCRETE MEMORYLESS INTERFERENCE CHANNEL

Consider the two sender–receiver pair communication system depicted in Figure 6.1, where each sender wishes to communicate a message to its respective receiver over a shared *interference channel*. Each message M_j, $j = 1, 2$, is separately encoded into a codeword X_j^n and transmitted over the channel. Upon receiving the sequence Y_j^n, receiver $j = 1, 2$ finds an estimate $\hat{M}_j$ of the message M_j. Because communication takes place over a shared channel, the signal at each receiver can suffer not only from the noise in the channel, but also from interference by the other transmitted codeword. This leads to a tradeoff between the rates at which both messages can be reliably communicated. We seek to determine the limits on this tradeoff.

We first consider a two sender–receiver (2-user) pair *discrete memoryless interference channel* (DM-IC) model $(\mathcal{X}_1 \times \mathcal{X}_2, p(y_1, y_2 | x_1, x_2), \mathcal{Y}_1 \times \mathcal{Y}_2)$ that consists of four finite sets $\mathcal{X}_1, \mathcal{X}_2, \mathcal{Y}_1, \mathcal{Y}_2$, and a collection of conditional pmfs $p(y_1, y_2 | x_1, x_2)$ on $\mathcal{Y}_1 \times \mathcal{Y}_2$. A $(2^{nR_1}, 2^{nR_2}, n)$ code for the interference channel consists of

- two message sets $[1 : 2^{nR_1}]$ and $[1 : 2^{nR_2}]$,

- two encoders, where encoder 1 assigns a codeword $x_1^n(m_1)$ to each message $m_1 \in [1 : 2^{nR_1}]$ and encoder 2 assigns a codeword $x_2^n(m_2)$ to each message $m_2 \in [1 : 2^{nR_2}]$, and

- two decoders, where decoder 1 assigns an estimate $\hat{m}_1$ or an error message e to each received sequence y_1^n and decoder 2 assigns an estimate $\hat{m}_2$ or an error message e to each received sequence y_2^n.

We assume that the message pair (M_1, M_2) is uniformly distributed over $[1 : 2^{nR_1}] \times [1 : 2^{nR_2}]$. The average probability of error is defined as

$$P_e^{(n)} = \mathsf{P}\{(\hat{M}_1, \hat{M}_2) \neq (M_1, M_2)\}.$$

A rate pair (R_1, R_2) is said to be achievable for the DM-IC if there exists a sequence of $(2^{nR_1}, 2^{nR_2}, n)$ codes such that $\lim_{n\to\infty} P_e^{(n)} = 0$. The capacity region $\mathscr{C}$ of the DM-IC is the closure of the set of achievable rate pairs (R_1, R_2) and the sum-capacity C_{sum} of the DM-IC is defined as $C_{\text{sum}} = \max\{R_1 + R_2 : (R_1, R_2) \in \mathscr{C}\}$.

As for the broadcast channel, the capacity region of the DM-IC depends on the channel conditional pmf $p(y_1, y_2 | x_1, x_2)$ only through the conditional marginals $p(y_1 | x_1, x_2)$ and $p(y_2 | x_1, x_2)$. The capacity region of the DM-IC is not known in general.

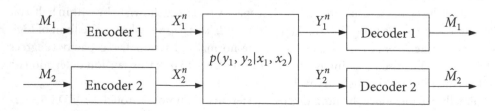

Figure 6.1. Two sender–receiver pair communication system.

6.2   SIMPLE CODING SCHEMES

We first consider several simple coding schemes for the interference channel.

Time division. The maximum achievable individual rates for the two sender–receiver pairs are

$$C_1 = \max_{p(x_1),\, x_2} I(X_1; Y_1 | X_2 = x_2),$$

$$C_2 = \max_{p(x_2),\, x_1} I(X_2; Y_2 | X_1 = x_1).$$

These capacities define the time-division inner bound consisting of all rate pairs (R_1, R_2) such that

$$\begin{aligned} R_1 &< \alpha C_1, \\ R_2 &< \bar{\alpha} C_2 \end{aligned} \tag{6.1}$$

for some $\alpha \in [0, 1]$. This bound is tight in some special cases.

Example 6.1 (Modulo-2 sum IC). Consider a DM-IC where the channel inputs X_1, X_2 and outputs Y_1, Y_2 are binary, and $Y_1 = Y_2 = X_1 \oplus X_2$. The time-division inner bound reduces to the set of rate pairs (R_1, R_2) such that $R_1 + R_2 < 1$. In the other direction, by allowing cooperation between the receivers, we obtain the upper bound on the sum-rate

$$R_1 + R_2 \le C_{12} = \max_{p(x_1)p(x_2)} I(X_1, X_2; Y_1, Y_2).$$

Since in our example, $Y_1 = Y_2$, this bound reduces to the set of rate pairs (R_1, R_2) such that $R_1 + R_2 \le 1$. Hence the time-division inner bound is tight.

The time-division inner bound is not tight in general, however.

Example 6.2 (No interference). Consider an interference channel with orthogonal components $p(y_1, y_2 | x_1, x_2) = p(y_1 | x_1)p(y_2 | x_2)$. In this case, the channel can be viewed simply as two separate DMCs and the capacity region is the set of rate pairs (R_1, R_2) such that $R_1 \le C_1$ and $R_2 \le C_2$. This is clearly larger than the time-division inner bound.

Treating interference as noise. Another inner bound on the capacity region of the interference channel can be achieved using point-to-point codes, time sharing, and treating interference as noise. This yields the *interference-as-noise* inner bound consisting of all rate pairs (R_1, R_2) such that

$$\begin{aligned} R_1 &< I(X_1; Y_1 | Q), \\ R_2 &< I(X_2; Y_2 | Q) \end{aligned} \tag{6.2}$$

for some pmf $p(q)p(x_1|q)p(x_2|q)$.

Simultaneous decoding. At the opposite extreme of treating interference as noise, we can have each receiver recover both messages. Following the achievability proof for the

DM-MAC using simultaneous decoding and coded time sharing in Section 4.5.3 (also see Problem 4.4), we can easily show that this scheme yields the *simultaneous-decoding* inner bound on the capacity region of the DM-IC consisting of all rate pairs (R_1, R_2) such that

$$R_1 < \min\{I(X_1; Y_1|X_2, Q), I(X_1; Y_2|X_2, Q)\},$$
$$R_2 < \min\{I(X_2; Y_1|X_1, Q), I(X_2; Y_2|X_1, Q)\}, \qquad (6.3)$$
$$R_1 + R_2 < \min\{I(X_1, X_2; Y_1|Q), I(X_1, X_2; Y_2|Q)\}$$

for some pmf $p(q)p(x_1|q)p(x_2|q)$.

Remark 6.1. Let $\mathscr{R}(X_1, X_2)$ be the set of rate pairs (R_1, R_2) such that

$$R_1 < \min\{I(X_1; Y_1|X_2), I(X_1; Y_2|X_2)\},$$
$$R_2 < \min\{I(X_2; Y_2|X_1), I(X_2; Y_1|X_1)\},$$
$$R_1 + R_2 < \min\{I(X_1, X_2; Y_1), I(X_1, X_2; Y_2)\}$$

for some pmf $p(x_1)p(x_2)$. Unlike the DM-MAC, the inner bound in (6.3) can be strictly larger than the convex closure of the union of $\mathscr{R}(X_1, X_2)$ over all $p(x_1)p(x_2)$. Hence, coded time sharing can achieve higher rates than (uncoded) time sharing, and is needed to achieve the inner bound in (6.3).

The simultaneous-decoding inner bound is sometimes tight.

Example 6.3. Consider a DM-IC with output alphabets $\mathcal{Y}_1 = \mathcal{Y}_2$ and $p_{Y_1|X_1,X_2}(y|x_1, x_2) = p_{Y_2|X_1,X_2}(y|x_1, x_2)$. The simultaneous-decoding inner bound reduces to the set of rate pairs (R_1, R_2) such that

$$R_1 < I(X_1; Y_1|X_2, Q),$$
$$R_2 < I(X_2; Y_2|X_1, Q),$$
$$R_1 + R_2 < I(X_1, X_2; Y_1|Q)$$

for some pmf $p(q)p(x_1|q)p(x_2|q)$. Now, using standard converse proof techniques, we can establish the outer bound on the capacity region of the general DM-IC consisting of all rate pairs (R_1, R_2) such that

$$R_1 \le I(X_1; Y_1|X_2, Q),$$
$$R_2 \le I(X_2; Y_2|X_1, Q),$$
$$R_1 + R_2 \le I(X_1, X_2; Y_1, Y_2|Q)$$

for some pmf $p(q)p(x_1|q)p(x_2|q)$. This bound can be further improved by using the fact that the capacity region depends only on the marginals of $p(y_1, y_2|x_1, x_2)$. If a rate pair (R_1, R_2) is achievable, then it must satisfy the inequalities

$$R_1 \le I(X_1; Y_1|X_2, Q),$$
$$R_2 \le I(X_2; Y_2|X_1, Q), \qquad (6.4)$$
$$R_1 + R_2 \le \min_{\tilde{p}(y_1, y_2|x_1, x_2)} I(X_1, X_2; Y_1, Y_2|Q)$$

for some pmf $p(q)p(x_1|q)p(x_2|q)$, where the minimum in the third inequality is over all conditional pmfs $\tilde{p}(y_1, y_2|x_1, x_2)$ with the same marginals $p(y_1|x_1, x_2)$ and $p(y_2|x_1, x_2)$ as the given channel conditional pmf $p(y_1, y_2|x_1, x_2)$.

Now, since the marginals of the channel in our example are identical, the minimum in the third inequality of the outer bound in (6.4) is attained for $Y_1 = Y_2$, and the bound reduces to the set of all rate pairs (R_1, R_2) such that

$$R_1 \leq I(X_1; Y_1|X_2, Q),$$
$$R_2 \leq I(X_2; Y_2|X_1, Q),$$
$$R_1 + R_2 \leq I(X_1, X_2; Y_1|Q)$$

for some pmf $p(q)p(x_1|q)p(x_2|q)$. Hence, simultaneous decoding is optimal for the DM-IC in this example.

Simultaneous nonunique decoding. We can improve upon the simultaneous-decoding inner bound via nonunique decoding, that is, by not requiring each receiver to recover the message intended for the other receiver. This yields the *simultaneous-nonunique-decoding* inner bound consisting of all rate pairs (R_1, R_2) such that

$$R_1 < I(X_1; Y_1|X_2, Q),$$
$$R_2 < I(X_2; Y_2|X_1, Q), \tag{6.5}$$
$$R_1 + R_2 < \min\{I(X_1, X_2; Y_1|Q), I(X_1, X_2; Y_2|Q)\}$$

for some pmf $p(q)p(x_1|q)p(x_2|q)$.

The achievability proof of this inner bound uses techniques we have already encountered in Sections 4.5 and 5.3. Fix a pmf $p(q)p(x_1|q)p(x_2|q)$. Randomly generate a sequence $q^n \sim \prod_{i=1}^n p_Q(q_i)$. Randomly and conditionally independently generate 2^{nR_1} sequences $x_1^n(m_1)$, $m_1 \in [1:2^{nR_1}]$, each according to $\prod_{i=1}^n p_{X_1|Q}(x_{1i}|q_i)$, and 2^{nR_2} sequences $x_2^n(m_2)$, $m_2 \in [1:2^{nR_2}]$, each according to $\prod_{i=1}^n p_{X_2|Q}(x_{2i}|q_i)$. To send (m_1, m_2), encoder $j = 1, 2$ transmits $x_j^n(m_j)$.

Decoder 1 finds the unique message $\hat{m}_1$ such that $(q^n, x_1^n(\hat{m}_1), x_2^n(m_2), y_1^n) \in \mathcal{T}_\epsilon^{(n)}$ for some m_2. By the LLN and the packing lemma, the probability of error for decoder 1 tends to zero as $n \to \infty$ if $R_1 < I(X_1; Y_1, X_2|Q) - \delta(\epsilon) = I(X_1; Y_1|X_2, Q) - \delta(\epsilon)$ and $R_1 + R_2 < I(X_1, X_2; Y_1|Q) - \delta(\epsilon)$. Similarly, decoder 2 finds the unique message $\hat{m}_2$ such that $(q^n, x_1^n(m_1), x_2^n(\hat{m}_2), y_2^n) \in \mathcal{T}_\epsilon^{(n)}$ for some m_1. Again by the LLN and the packing lemma, the probability of error for decoder 2 tends to zero as $n \to \infty$ if $R_2 < I(X_2; Y_2|X_1, Q) - \delta(\epsilon)$ and $R_1 + R_2 < I(X_1, X_2; Y_2|Q) - \delta(\epsilon)$. This completes the achievability proof of the simultaneous-nonunique-decoding inner bound.

6.3 STRONG INTERFERENCE

Suppose that each receiver in an interference channel is physically closer to the interfering transmitter than to its own transmitter and hence the received signal from the interfering transmitter is stronger than that from its transmitter. Under such strong interference condition, each receiver can essentially recover the message of the interfering transmitter without imposing an additional constraint on its rate. We define two notions of strong interference for the DM-IC and show that simultaneous decoding is optimal under both notions.

Very strong interference. A DM-IC is said to have *very strong interference* if

$$I(X_1; Y_1 | X_2) \leq I(X_1; Y_2),$$
$$I(X_2; Y_2 | X_1) \leq I(X_2; Y_1) \tag{6.6}$$

for all $p(x_1)p(x_2)$. The capacity region of the DM-IC with very strong interference is the set of rate pairs (R_1, R_2) such that

$$R_1 \leq I(X_1; Y_1 | X_2, Q),$$
$$R_2 \leq I(X_2; Y_2 | X_1, Q)$$

for some pmf $p(q)p(x_1|q)p(x_2|q)$. The converse proof is quite straightforward, since this region constitutes an outer bound on the capacity region of the general DM-IC. The proof of achievability follows by noting that under the very strong interference condition, the sum-rate inequality in the simultaneous (unique or nonunique) decoding inner bound is inactive. Note that the capacity region can be achieved also via successive cancellation decoding and time sharing. Each decoder successively recovers the other message and then its own message. Because of the very strong interference condition, only the requirements on the achievable rates for the second decoding step matter.

Strong interference. A DM-IC is said to have *strong interference* if

$$I(X_1; Y_1 | X_2) \leq I(X_1; Y_2 | X_2),$$
$$I(X_2; Y_2 | X_1) \leq I(X_2; Y_1 | X_1) \tag{6.7}$$

for all $p(x_1)p(x_2)$. Note that this is an extension of the more capable condition for the DM-BC. In particular, Y_2 is more capable than Y_1 given X_2, and Y_1 is more capable than Y_2 given X_1. Clearly, if the channel has very strong interference, then it also has strong interference. The converse is not necessarily true as illustrated by the following.

Example 6.4. Consider the DM-IC with binary inputs X_1, X_2 and ternary outputs $Y_1 = Y_2 = X_1 + X_2$. Then

$$I(X_1; Y_1 | X_2) = I(X_1; Y_2 | X_2) = H(X_1),$$
$$I(X_2; Y_2 | X_1) = I(X_2; Y_1 | X_1) = H(X_2).$$

Therefore, this DM-IC has strong interference. However,

$$I(X_1; Y_1 | X_2) = H(X_1) \geq H(X_1) - H(X_1 | Y_2) = I(X_1; Y_2),$$
$$I(X_2; Y_2 | X_1) = H(X_2) \geq H(X_2) - H(X_2 | Y_1) = I(X_2; Y_1)$$

with strict inequality for some pmf $p(x_1)p(x_2)$. Therefore, this channel does not satisfy the very strong interference condition.

We now show that the simultaneous-nonunique-decoding inner bound is tight under the strong interference condition.

Theorem 6.1. The capacity region of the DM-IC $p(y_1, y_2|x_1, x_2)$ with strong interference is the set of rate pairs (R_1, R_2) such that

$$R_1 \leq I(X_1; Y_1|X_2, Q),$$
$$R_2 \leq I(X_2; Y_2|X_1, Q),$$
$$R_1 + R_2 \leq \min\{I(X_1, X_2; Y_1|Q), I(X_1, X_2; Y_2|Q)\}$$

for some pmf $p(q)p(x_1|q)p(x_2|q)$ with $|\mathcal{Q}| \leq 4$.

Proof of the converse. The first two inequalities can be easily established. By symmetry it suffices to show that $R_1 + R_2 \leq I(X_1, X_2; Y_2|Q)$. Consider

$$
\begin{aligned}
n(R_1 + R_2) &= H(M_1) + H(M_2) \\
&\stackrel{(a)}{\leq} I(M_1; Y_1^n) + I(M_2; Y_2^n) + n\epsilon_n \\
&\stackrel{(b)}{=} I(X_1^n; Y_1^n) + I(X_2^n; Y_2^n) + n\epsilon_n \\
&\leq I(X_1^n; Y_1^n|X_2^n) + I(X_2^n; Y_2^n) + n\epsilon_n \\
&\stackrel{(c)}{\leq} I(X_1^n; Y_2^n|X_2^n) + I(X_2^n; Y_2^n) + n\epsilon_n \\
&= I(X_1^n, X_2^n; Y_2^n) + n\epsilon_n \\
&\leq \sum_{i=1}^{n} I(X_{1i}, X_{2i}; Y_{2i}) + n\epsilon_n \\
&= nI(X_1, X_2; Y_2|Q) + n\epsilon_n,
\end{aligned}
$$

where (a) follows by Fano's inequality and (b) follows since $M_j \to X_j^n \to Y_j^n$ for $j = 1, 2$ (by the independence of M_1 and M_2). Step (c) is established using the following.

Lemma 6.1. For a DM-IC $p(y_1, y_2|x_1, x_2)$ with strong interference, $I(X_1^n; Y_1^n|X_2^n) \leq I(X_1^n; Y_2^n|X_2^n)$ for all $(X_1^n, X_2^n) \sim p(x_1^n)p(x_2^n)$ and all $n \geq 1$.

This lemma can be proved by noting that the strong interference condition implies that $I(X_1; Y_1|X_2, U) \leq I(X_1; Y_2|X_2, U)$ for all $p(u)p(x_1|u)p(x_2|u)$ and using induction on n.

The other bound $R_1 + R_2 \leq I(X_1, X_2; Y_1|Q)$ follows similarly, which completes the proof of the theorem.

6.4 GAUSSIAN INTERFERENCE CHANNEL

Consider the 2-user-pair Gaussian interference channel depicted in Figure 6.2, which is a simple model for a wireless interference channel or a DSL cable bundle. The channel outputs corresponding to the inputs X_1 and X_2 are

$$Y_1 = g_{11}X_1 + g_{12}X_2 + Z_1,$$
$$Y_2 = g_{21}X_1 + g_{22}X_2 + Z_2,$$

where g_{jk}, $j, k = 1, 2$, is the channel gain from sender k to receiver j, and $Z_1 \sim N(0, N_0/2)$ and $Z_2 \sim N(0, N_0/2)$ are noise components. Assume average power constraint P on each of X_1 and X_2. We assume without loss of generality that $N_0/2 = 1$ and define the received SNRs as $S_1 = g_{11}^2 P$ and $S_2 = g_{22}^2 P$ and the received *interference-to-noise ratios* (INRs) as $I_1 = g_{12}^2 P$ and $I_2 = g_{21}^2 P$.

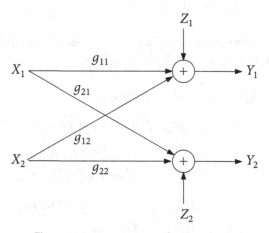

Figure 6.2. Gaussian interference channel.

The capacity region of the Gaussian IC is not known in general.

6.4.1 Inner Bounds

We specialize the inner bounds in Section 6.2 to the Gaussian case.

Time division with power control. Using time division and power control, we obtain the time-division inner bound on the capacity region of the Gaussian IC that consists of all rate pairs (R_1, R_2) such that

$$R_1 < \alpha\, C(S_1/\alpha),$$
$$R_2 < \bar{\alpha}\, C(S_2/\bar{\alpha})$$

for some $\alpha \in [0, 1]$.

Treating interference as noise. Consider the inner bound in (6.2) subject to the power constraints. By setting $X_1 \sim N(0, P)$, $X_2 \sim N(0, P)$, and $Q = \emptyset$, we obtain the inner bound on the capacity region of the Gaussian IC consisting of all rate pairs (R_1, R_2) such that

$$R_1 < C(S_1/(1 + I_1)),$$
$$R_2 < C(S_2/(1 + I_2)).$$

Note, however, that Gaussian input signals are not necessarily optimal when evaluating the mutual information characterization in (6.2) under the power constraints. Also note that the above inner bound can be further improved via time sharing and power control.

Simultaneous nonunique decoding. The inner bound in (6.5) subject to the power constraints is optimized by setting $X_1 \sim N(0, P)$, $X_2 \sim N(0, P)$, and $Q = \emptyset$. This gives the inner bound on the capacity region of the Gaussian IC that consists of all rate pairs (R_1, R_2) such that

$$R_1 < C(S_1),$$
$$R_2 < C(S_2),$$
$$R_1 + R_2 < \min\{C(S_1 + I_1), C(S_2 + I_2)\}.$$

Although this bound is again achieved using optimal point-to-point Gaussian codes, it cannot be achieved in general via successive cancellation decoding.

   The above inner bounds are compared in Figure 6.3 for symmetric Gaussian ICs with SNRs $S_1 = S_2 = S = 1$ and increasing INRs $I_1 = I_2 = I$. When interference is weak (Figure 6.3a), treating interference as noise can outperform time division and simultaneous nonunique decoding, and is in fact sum-rate optimal as we show in Section 6.4.3. As interference becomes stronger (Figure 6.3b), simultaneous nonunique decoding and time division begin to outperform treating interference as noise. As interference becomes even stronger, simultaneous nonunique decoding outperforms the other two coding schemes (Figures 6.3c,d), ultimately achieving the interference-free rate region consisting of all rate pairs (R_1, R_2) such that $R_1 < C_1$ and $R_2 < C_2$ (Figure 6.3d).

6.4.2   Capacity Region of the Gaussian IC with Strong Interference

A Gaussian IC is said to have *strong interference* if $|g_{21}| \geq |g_{11}|$ and $|g_{12}| \geq |g_{22}|$, or equivalently, $I_2 \geq S_1$ and $I_1 \geq S_2$.

Theorem 6.2. The capacity region of the Gaussian IC with strong interference is the set of rate pairs (R_1, R_2) such that

$$R_1 \leq C(S_1),$$
$$R_2 \leq C(S_2),$$
$$R_1 + R_2 \leq \min\{C(S_1 + I_1), C(S_2 + I_2)\}.$$

The proof of achievability follows by using simultaneous nonunique decoding. The proof of the converse follows by noting that the above condition is equivalent to the strong interference condition for the DM-IC in (6.7) and showing that $X_1 \sim N(0, P)$ and $X_2 \sim N(0, P)$ optimize the mutual information terms.

   The nontrivial step is to show that the condition $I_2 \geq S_1$ and $I_1 \geq S_2$ is equivalent to the condition $I(X_1; Y_1|X_2) \leq I(X_1; Y_2|X_2)$ and $I(X_2; Y_2|X_1) \leq I(X_2; Y_1|X_1)$ for every $F(x_1)F(x_2)$. If $I_2 \geq S_1$ and $I_1 \geq S_2$, then it can be easily shown that the Gaussian BC from

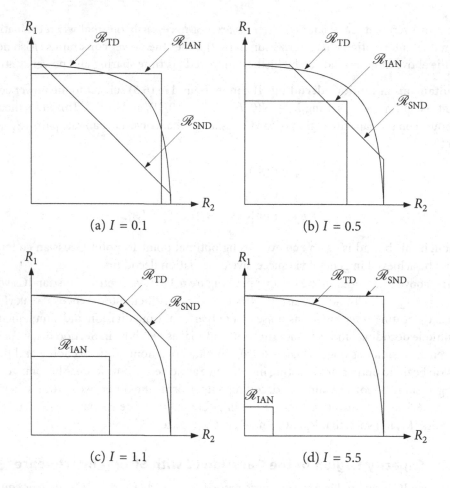

Figure 6.3. Comparison of time division (region $\mathcal{R}_{TD}$), treating interference as noise (region $\mathcal{R}_{IAN}$), and simultaneous nonunique decoding (region $\mathcal{R}_{SND}$) for $S = 1$ and different values of I. Treating interference as noise achieves the sum-capacity for case (a), while $\mathcal{R}_{SND}$ is the capacity region for cases (c) and (d).

X_1 to $(Y_2 - g_{22}X_2, Y_1 - g_{12}X_2)$ given X_2 is degraded and the Gaussian BC from X_2 to $(Y_1 - g_{11}X_1, Y_2 - g_{21}X_1)$ given X_1 is degraded, and hence each is more capable. This proves one direction of the equivalence. To prove the other direction, assume that $h(g_{11}X_1 + Z_1) \le h(g_{21}X_1 + Z_2)$ and $h(g_{22}X_2 + Z_2) \le h(g_{12}X_2 + Z_1)$. Substituting $X_1 \sim N(0, P)$ and $X_2 \sim N(0, P)$ shows that $I_2 \ge S_1$ and $I_1 \ge S_2$, respectively.

Remark 6.2. A Gaussian IC is said to have *very strong interference* if $S_2 \le I_1/(1 + S_1)$ and $S_1 \le I_2/(1 + S_2)$. It can be shown that this condition is the same as the very strong interference condition for the DM-IC in (6.6) when restricted to Gaussian inputs. Under this condition, the capacity region is the set of rate pairs (R_1, R_2) such that $R_1 \le C(S_1)$ and $R_2 \le C(S_2)$ and hence interference does not impair communication.

6.4.3   Sum-Capacity of the Gaussian IC with Weak Interference

A Gaussian IC is said to have *weak interference* if for some $\rho_1, \rho_2 \in [0, 1]$,

$$\sqrt{I_1/S_2}\,(1 + I_2) \le \rho_2 \sqrt{1 - \rho_1^2},$$
$$\sqrt{I_2/S_1}\,(1 + I_1) \le \rho_1 \sqrt{1 - \rho_2^2}. \tag{6.8}$$

Under this weak interference condition, treating interference as noise is optimal for the sum-rate.

> **Theorem 6.3.** The sum-capacity of the Gaussian IC with weak interference is
>
> $$C_{\text{sum}} = C\left(\frac{S_1}{1 + I_1}\right) + C\left(\frac{S_2}{1 + I_2}\right).$$

The interesting part of the proof is the converse. It involves the use of a *genie* to establish an upper bound on the sum-capacity. For simplicity of presentation, we consider the symmetric case with $I_1 = I_2 = I$ and $S_1 = S_2 = S$. In this case, the weak interference condition in (6.8) reduces to

$$\sqrt{I/S}\,(1 + I) \le \frac{1}{2} \tag{6.9}$$

and the sum-capacity is $C_{\text{sum}} = 2\,C(S/(1 + I))$.

Proof of the converse. Consider the *genie-aided* Gaussian IC depicted in Figure 6.4 with side information

$$T_1 = \sqrt{I/P}\,(X_1 + \eta W_1),$$
$$T_2 = \sqrt{I/P}\,(X_2 + \eta W_2),$$

where $W_1 \sim N(0, 1)$ and $W_2 \sim N(0, 1)$ are independent noise components with $E(Z_1 W_1) = E(Z_2 W_2) = \rho$ and $\eta \ge 0$. Suppose that a genie reveals T_1 to decoder 1 and T_2 to decoder 2. Clearly, the sum-capacity of this channel $\tilde{C}_{\text{sum}} \ge C_{\text{sum}}$.

We first show that if $\eta^2 I \le (1 - \rho^2)P$ (useful genie), then the sum-capacity of the genie-aided channel is achieved by using Gaussian inputs and treating interference as noise. We then show that if in addition, $\eta \rho \sqrt{S/P} = 1 + I$ (smart genie), then the sum-capacity of the genie-aided channel is the same as that of the original channel. Since $\tilde{C}_{\text{sum}} \ge C_{\text{sum}}$, this also shows that C_{sum} is achieved by using Gaussian inputs and treating interference as noise. Using the second condition to eliminate η from the first condition gives $\sqrt{I/S}(1 + I) \le \rho \sqrt{1 - \rho^2}$. Taking $\rho = \sqrt{1/2}$, which maximizes the range of I, gives the weak interference condition in the theorem. The proof steps involve properties of differential entropy, including the maximum differential entropy lemma; the fact that Gaussian is the worst noise with a given average power in an additive noise channel with Gaussian input (see Problem 2.12); and properties of jointly Gaussian random variables (see Appendix B).

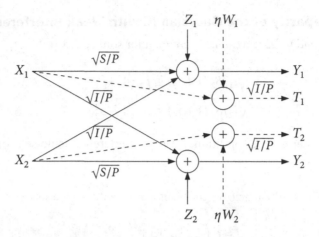

Figure 6.4. Genie-aided Gaussian interference channel.

Let X_1^* and X_2^* be independent $N(0, P)$ random variables, and Y_1^*, Y_2^* and T_1^*, T_2^* be the corresponding channel outputs and side information. Then we can establish the following condition under which $\tilde{C}_{\text{sum}}$ is achieved by treating interference as Gaussian noise.

Lemma 6.2 (Useful Genie). If $\eta^2 I \leq (1 - \rho^2)P$, then the sum-capacity of the above genie-aided channel is

$$\tilde{C}_{\text{sum}} = I(X_1^*; Y_1^*, T_1^*) + I(X_2^*; Y_2^*, T_2^*).$$

The proof of this lemma is in Appendix 6A.

Remark 6.3. If $\rho = 0$, $\eta = 1$, and $I \leq P$, then the genie is always useful.

Continuing the proof of the converse, suppose that the following *smart genie* condition

$$\eta \rho \sqrt{S/P} = 1 + I$$

holds. Note that combined with the (useful genie) condition for the lemma, the smart genie gives the weak interference condition in (6.9). Now by the smart genie condition,

$$
\begin{aligned}
\mathsf{E}(T_1^* \mid X_1^*, Y_1^*) &= \mathsf{E}(T_1^* \mid X_1^*, \sqrt{I/P}\, X_2^* + Z_1) \\
&= \sqrt{I/P}\, X_1^* + \eta \sqrt{I/P}\, \mathsf{E}(W_1 \mid \sqrt{I/P}\, X_2^* + Z_1) \\
&= \sqrt{I/P}\, X_1^* + \frac{\eta \rho \sqrt{I/P}}{1 + I}(\sqrt{I/P}\, X_2^* + Z_1) \\
&= \sqrt{I/S}\, Y_1^* \\
&= \mathsf{E}(T_1^* \mid Y_1^*).
\end{aligned}
$$

Since all random variables involved are jointly Gaussian, this implies that $X_1^* \to Y_1^* \to T_1^*$

form a Markov chain, or equivalently, $I(X_1^*; T_1^*|Y_1^*) = 0$. Similarly $I(X_2^*; T_2^*|Y_2^*) = 0$. Finally, by the useful genie lemma,

$$C_{\text{sum}} \le \tilde{C}_{\text{sum}} = I(X_1^*; Y_1^*, T_1^*) + I(X_2^*; Y_2^*, T_2^*) = I(X_1^*; Y_1^*) + I(X_2^*; Y_2^*).$$

This completes the proof of the converse.

Remark 6.4. The idea of a genie providing each receiver with side information about its intended codeword can be used to obtain outer bounds on the capacity region of the general Gaussian IC; see Problem 6.17. This same idea will be used also in the converse proof for the injective deterministic IC in Section 6.6.

6.5 HAN–KOBAYASHI INNER BOUND

The Han–Kobayashi inner bound is the best-known bound on the capacity region of the DM-IC. It includes all the inner bounds we discussed so far, and is tight for all interference channels with known capacity regions. We consider the following characterization of this inner bound.

Theorem 6.4 (Han–Kobayashi Inner Bound). A rate pair (R_1, R_2) is achievable for the DM-IC $p(y_1, y_2|x_1, x_2)$ if

$$R_1 < I(X_1; Y_1|U_2, Q),$$
$$R_2 < I(X_2; Y_2|U_1, Q),$$
$$R_1 + R_2 < I(X_1, U_2; Y_1|Q) + I(X_2; Y_2|U_1, U_2, Q),$$
$$R_1 + R_2 < I(X_2, U_1; Y_2|Q) + I(X_1; Y_1|U_1, U_2, Q),$$
$$R_1 + R_2 < I(X_1, U_2; Y_1|U_1, Q) + I(X_2, U_1; Y_2|U_2, Q),$$
$$2R_1 + R_2 < I(X_1, U_2; Y_1|Q) + I(X_1; Y_1|U_1, U_2, Q) + I(X_2, U_1; Y_2|U_2, Q),$$
$$R_1 + 2R_2 < I(X_2, U_1; Y_2|Q) + I(X_2; Y_2|U_1, U_2, Q) + I(X_1, U_2; Y_1|U_1, Q)$$

for some pmf $p(q)p(u_1, x_1|q)p(u_2, x_2|q)$, where $|\mathcal{U}_1| \le |\mathcal{X}_1| + 4$, $|\mathcal{U}_2| \le |\mathcal{X}_2| + 4$, and $|\mathcal{Q}| \le 7$.

Remark 6.5. The Han–Kobayashi inner bound reduces to the interference-as-noise inner bound in (6.2) by setting $U_1 = U_2 = \emptyset$. At the other extreme, the Han–Kobayashi inner bound reduces to the simultaneous-nonunique-decoding inner bound in (6.5) by setting $U_1 = X_1$ and $U_2 = X_2$. Thus, the bound is tight for the class of DM-ICs with strong interference.

Remark 6.6. The Han–Kobayashi inner bound can be readily extended to the Gaussian IC with average power constraints and evaluated using Gaussian (U_j, X_j), $j = 1, 2$; see Problem 6.16. It is not known, however, if the restriction to the Gaussian distribution is sufficient.

6.5.1 Proof of the Han–Kobayashi Inner Bound

The proof uses rate splitting. We represent each message M_j, $j = 1, 2$, by independent "public" message M_{j0} at rate R_{j0} and "private" message M_{jj} at rate R_{jj}. Thus, $R_j = R_{j0} + R_{jj}$. These messages are sent via superposition coding, whereby the cloud center U_j represents the public message M_{j0} and the satellite codeword X_j represents the message pair (M_{j0}, M_{jj}). The public messages are to be recovered by both receivers, while each private message is to be recovered only by its intended receiver. We first show that $(R_{10}, R_{20}, R_{11}, R_{22})$ is achievable if

$$R_{11} < I(X_1; Y_1 | U_1, U_2, Q),$$
$$R_{11} + R_{10} < I(X_1; Y_1 | U_2, Q),$$
$$R_{11} + R_{20} < I(X_1, U_2; Y_1 | U_1, Q),$$
$$R_{11} + R_{10} + R_{20} < I(X_1, U_2; Y_1 | Q),$$
$$R_{22} < I(X_2; Y_2 | U_1, U_2, Q),$$
$$R_{22} + R_{20} < I(X_2; Y_2 | U_1, Q),$$
$$R_{22} + R_{10} < I(X_2, U_1; Y_2 | U_2, Q),$$
$$R_{22} + R_{20} + R_{10} < I(X_2, U_1; Y_2 | Q)$$

for some pmf $p(q)p(u_1, x_1 | q)p(u_2, x_2 | q)$.

Codebook generation. Fix a pmf $p(q)p(u_1, x_1 | q)p(u_2, x_2 | q)$. Generate a sequence $q^n \sim \prod_{i=1}^{n} p_Q(q_i)$. For $j = 1, 2$, randomly and conditionally independently generate $2^{nR_{j0}}$ sequences $u_j^n(m_{j0})$, $m_{j0} \in [1 : 2^{nR_{j0}}]$, each according to $\prod_{i=1}^{n} p_{U_j | Q}(u_{ji} | q_i)$. For each m_{j0}, randomly and conditionally independently generate $2^{nR_{jj}}$ sequences $x_j^n(m_{j0}, m_{jj})$, $m_{jj} \in [1 : 2^{nR_{jj}}]$, each according to $\prod_{i=1}^{n} p_{X_j | U_j, Q}(x_{ji} | u_{ji}(m_{j0}), q_i)$.

Encoding. To send $m_j = (m_{j0}, m_{jj})$, encoder $j = 1, 2$ transmits $x_j^n(m_{j0}, m_{jj})$.

Decoding. We use simultaneous nonunique decoding. Upon receiving y_1^n, decoder 1 finds the unique message pair $(\hat{m}_{10}, \hat{m}_{11})$ such that $(q^n, u_1^n(\hat{m}_{10}), u_2^n(m_{20}), x_1^n(\hat{m}_{10}, \hat{m}_{11}), y_1^n) \in \mathcal{T}_\epsilon^{(n)}$ for some $m_{20} \in [1 : 2^{nR_{20}}]$; otherwise it declares an error. Decoder 2 finds the message pair $(\hat{m}_{20}, \hat{m}_{22})$ similarly.

Analysis of the probability of error. Assume message pair $((1, 1), (1, 1))$ is sent. We bound the average probability of error for each decoder. First consider decoder 1. As shown in Table 6.1, we have eight cases to consider (here conditioning on q^n is suppressed). Cases 3 and 4, and 6 and 7, respectively, share the same pmf, and case 8 does not cause an error. Thus, we are left with only five error events and decoder 1 makes an error only if one or more of the following events occur:

$$\mathcal{E}_{10} = \{(Q^n, U_1^n(1), U_2^n(1), X_1^n(1, 1), Y_1^n) \notin \mathcal{T}_\epsilon^{(n)}\},$$
$$\mathcal{E}_{11} = \{(Q^n, U_1^n(1), U_2^n(1), X_1^n(1, m_{11}), Y_1^n) \in \mathcal{T}_\epsilon^{(n)} \text{ for some } m_{11} \neq 1\},$$
$$\mathcal{E}_{12} = \{(Q^n, U_1^n(m_{10}), U_2^n(1), X_1^n(m_{10}, m_{11}), Y_1^n) \in \mathcal{T}_\epsilon^{(n)} \text{ for some } m_{10} \neq 1, m_{11}\},$$

	m_{10}	m_{20}	m_{11}	Joint pmf
1	1	1	1	$p(u_1^n, x_1^n)p(u_2^n)p(y_1^n \mid x_1^n, u_2^n)$
2	1	1	*	$p(u_1^n, x_1^n)p(u_2^n)p(y_1^n \mid u_1^n, u_2^n)$
3	*	1	*	$p(u_1^n, x_1^n)p(u_2^n)p(y_1^n \mid u_2^n)$
4	*	1	1	$p(u_1^n, x_1^n)p(u_2^n)p(y_1^n \mid u_2^n)$
5	1	*	*	$p(u_1^n, x_1^n)p(u_2^n)p(y_1^n \mid u_1^n)$
6	*	*	1	$p(u_1^n, x_1^n)p(u_2^n)p(y_1^n)$
7	*	*	*	$p(u_1^n, x_1^n)p(u_2^n)p(y_1^n)$
8	1	*	1	$p(u_1^n, x_1^n)p(u_2^n)p(y_1^n \mid x_1^n)$

Table 6.1. The joint pmfs induced by different (m_{10}, m_{20}, m_{11}) triples.

$$\mathcal{E}_{13} = \{(Q^n, U_1^n(1), U_2^n(m_{20}), X_1^n(1, m_{11}), Y_1^n) \in \mathcal{T}_\epsilon^{(n)} \text{ for some } m_{20} \neq 1, m_{11} \neq 1\},$$
$$\mathcal{E}_{14} = \{(Q^n, U_1^n(m_{10}), U_2^n(m_{20}), X_1^n(m_{10}, m_{11}), Y_1^n) \in \mathcal{T}_\epsilon^{(n)}$$
$$\text{for some } m_{10} \neq 1, m_{20} \neq 1, m_{11}\}.$$

Hence, the average probability of error for decoder 1 is upper bounded as

$$\mathsf{P}(\mathcal{E}_1) \leq \mathsf{P}(\mathcal{E}_{10}) + \mathsf{P}(\mathcal{E}_{11}) + \mathsf{P}(\mathcal{E}_{12}) + \mathsf{P}(\mathcal{E}_{13}) + \mathsf{P}(\mathcal{E}_{14}).$$

We bound each term. By the LLN, $\mathsf{P}(\mathcal{E}_{10})$ tends to zero as $n \to \infty$. By the packing lemma, $\mathsf{P}(\mathcal{E}_{11})$ tends to zero as $n \to \infty$ if $R_{11} < I(X_1; Y_1 \mid U_1, U_2, Q) - \delta(\epsilon)$. Similarly, by the packing lemma, $\mathsf{P}(\mathcal{E}_{12})$, $\mathsf{P}(\mathcal{E}_{13})$, and $\mathsf{P}(\mathcal{E}_{14})$ tend to zero as $n \to \infty$ if the conditions $R_{11} + R_{10} < I(X_1; Y_1 \mid U_2, Q) - \delta(\epsilon)$, $R_{11} + R_{20} < I(X_1, U_2; Y_1 \mid U_1, Q) - \delta(\epsilon)$, and $R_{11} + R_{10} + R_{20} < I(X_1, U_2; Y_1 \mid Q) - \delta(\epsilon)$ are satisfied, respectively. The average probability of error for decoder 2 can be bounded similarly. Finally, substituting $R_{11} = R_1 - R_{10}$ and $R_{22} = R_2 - R_{20}$, and using the Fourier–Motzkin procedure with the constraints $0 \leq R_{j0} \leq R_j$, $j = 1, 2$, to eliminate R_{10} and R_{20} (see Appendix D for the details), we obtain the seven inequalities in Theorem 6.4 and two additional inequalities $R_1 < I(X_1; Y_1 \mid U_1, U_2, Q) + I(X_2, U_1; Y_2 \mid U_2, Q)$ and $R_2 < I(X_1, U_2; Y_1 \mid U_1, Q) + I(X_2; Y_2 \mid U_1, U_2, Q)$. The corresponding inner bound can be shown to be equivalent to the inner bound in Theorem 6.4; see Problem 6.12. The cardinality bound on $\mathcal{Q}$ can be proved using the convex cover method in Appendix C. This completes the proof of the Han–Kobayashi inner bound.

6.6 INJECTIVE DETERMINISTIC IC

Consider the deterministic interference channel depicted in Figure 6.5. The channel outputs are given by the functions

$$Y_1 = y_1(X_1, T_2),$$
$$Y_2 = y_2(X_2, T_1),$$

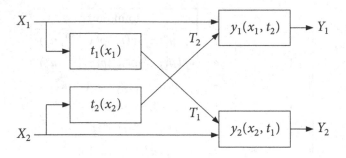

Figure 6.5. Injective deterministic interference channel.

where $T_1 = t_1(X_1)$ and $T_2 = t_2(X_2)$ are functions of X_1 and X_2, respectively. We assume that the functions y_1 and y_2 are injective in t_2 and t_1, respectively, that is, for every $x_1 \in \mathcal{X}_1$, $y_1(x_1, t_2)$ is a one-to-one function of t_2 and for every $x_2 \in \mathcal{X}_2$, $y_2(x_2, t_1)$ is a one-to-one function of t_1. Note that these conditions imply that $H(Y_1|X_1) = H(T_2)$ and $H(Y_2|X_2) = H(T_1)$.

This class of interference channels is motivated by the Gaussian IC, where the functions y_1 and y_2 are additions. Unlike the Gaussian IC, however, the channel is noiseless and its capacity region can be fully characterized.

Theorem 6.5. The capacity region of the injective deterministic interference channel is the set of rate pairs (R_1, R_2) such that

$$R_1 \le H(Y_1|T_2, Q),$$
$$R_2 \le H(Y_2|T_1, Q),$$
$$R_1 + R_2 \le H(Y_1|Q) + H(Y_2|T_1, T_2, Q),$$
$$R_1 + R_2 \le H(Y_1|T_1, T_2, Q) + H(Y_2|Q),$$
$$R_1 + R_2 \le H(Y_1|T_1, Q) + H(Y_2|T_2, Q),$$
$$2R_1 + R_2 \le H(Y_1|Q) + H(Y_1|T_1, T_2, Q) + H(Y_2|T_2, Q),$$
$$R_1 + 2R_2 \le H(Y_1|T_1, Q) + H(Y_2|Q) + H(Y_2|T_1, T_2, Q)$$

for some pmf $p(q)p(x_1|q)p(x_2|q)$.

The proof of achievability follows by noting that the above region coincides with the Han–Kobayashi inner bound (take $U_1 = T_1$, $U_2 = T_2$).

Remark 6.7. By the one-to-one conditions on the functions y_1 and y_2, decoder 1 knows T_2^n after decoding for X_1^n and decoder 2 knows T_1^n after decoding for X_2^n. As such, the interference random variables T_1 and T_2 can be naturally considered as the auxiliary random variables that represent the public messages in the Han–Kobayashi scheme.

Proof of the converse. Consider the first two inequalities in the characterization of the capacity region. By specializing the outer bound in (6.4), we obtain

$$nR_1 \le nI(X_1; Y_1 | X_2, Q) + n\epsilon_n = nH(Y_1 | T_2, Q) + n\epsilon_n,$$
$$nR_2 \le nH(Y_2 | T_1, Q) + n\epsilon_n,$$

where Q is the usual time-sharing random variable.

Now consider the third inequality. By Fano's inequality,

$$n(R_1 + R_2) \le I(M_1; Y_1^n) + I(M_2; Y_2^n) + n\epsilon_n$$

$$\overset{(a)}{\le} I(M_1; Y_1^n) + I(M_2; Y_2^n, T_2^n) + n\epsilon_n$$

$$\le I(X_1^n; Y_1^n) + I(X_2^n; Y_2^n, T_2^n) + n\epsilon_n$$

$$\overset{(b)}{\le} I(X_1^n; Y_1^n) + I(X_2^n; T_2^n, Y_2^n | T_1^n) + n\epsilon_n$$

$$= H(Y_1^n) - H(Y_1^n | X_1^n) + I(X_2^n; T_2^n | T_1^n) + I(X_2^n; Y_2^n | T_1^n, T_2^n) + n\epsilon_n$$

$$\overset{(c)}{=} H(Y_1^n) + H(Y_2^n | T_1^n, T_2^n) + n\epsilon_n$$

$$\le \sum_{i=1}^{n} (H(Y_{1i}) + H(Y_{2i} | T_{1i}, T_{2i})) + n\epsilon_n$$

$$= n(H(Y_1 | Q) + H(Y_2 | T_1, T_2, Q)) + n\epsilon_n.$$

Here step (a) is the key step in the proof. Even if a "genie" gives receiver 2 its common message T_2 as side information to help it find X_2, the capacity region does not change! Step (b) follows by the fact that X_2^n and T_1^n are independent, and (c) follows by the equalities $H(Y_1^n | X_1^n) = H(T_2^n)$ and $I(X_2^n; T_2^n | T_1^n) = H(T_2^n)$. Similarly, for the fourth inequality,

$$n(R_1 + R_2) \le n(H(Y_2 | Q) + H(Y_1 | T_1, T_2, Q)) + n\epsilon_n.$$

Consider the fifth inequality

$$n(R_1 + R_2) \le I(X_1^n; Y_1^n) + I(X_2^n; Y_2^n) + n\epsilon_n$$

$$= H(Y_1^n) - H(Y_1^n | X_1^n) + H(Y_2^n) - H(Y_2^n | X_1^n) + n\epsilon_n$$

$$\overset{(a)}{=} H(Y_1^n) - H(T_2^n) + H(Y_2^n) - H(T_1^n) + n\epsilon_n$$

$$\le H(Y_1^n | T_1^n) + H(Y_2^n | T_2^n) + n\epsilon_n$$

$$\le n(H(Y_1 | T_1, Q) + H(Y_2 | T_2, Q)) + n\epsilon_n,$$

where (a) follows by the one-to-one conditions of the injective deterministic IC. Following similar steps, consider the sixth inequality

$$n(2R_1 + R_2) \le 2I(M_1; Y_1^n) + I(M_2; Y_2^n) + n\epsilon_n$$

$$\le I(X_1^n; Y_1^n) + I(X_1^n; Y_1^n, T_1^n | T_2^n) + I(X_2^n; Y_2^n) + n\epsilon_n$$

$$= H(Y_1^n) - H(T_2^n) + H(T_1^n) + H(Y_1^n | T_1^n, T_2^n) + H(Y_2^n) - H(T_1^n) + n\epsilon_n$$

$$= H(Y_1^n) - H(T_2^n) + H(Y_1^n | T_1^n, T_2^n) + H(Y_2^n) + n\epsilon_n$$

$$\le H(Y_1^n) + H(Y_1^n | T_1^n, T_2^n) + H(Y_2^n | T_2^n) + n\epsilon_n$$

$$\le n(H(Y_1 | Q) + H(Y_1 | T_1, T_2, Q) + H(Y_2 | T_2, Q)) + n\epsilon_n.$$

Similarly, for the last inequality, we have

$$n(R_1 + 2R_2) \le n\big(H(Y_1|T_1, Q) + H(Y_2|Q) + H(Y_2|T_1, T_2, Q)\big) + n\epsilon_n.$$

This completes the proof of the converse.

6.7 CAPACITY REGION OF THE GAUSSIAN IC WITHIN HALF A BIT

As we have seen, the capacity (region) of the Gaussian IC is known only under certain strong and weak interference conditions and is achieved by extreme special cases of the Han–Kobayashi scheme where no rate splitting is used. How close is the Han–Kobayashi inner bound in its full generality to the capacity region?

We show that even a suboptimal evaluation of the Han–Kobayashi inner bound differs by no more than half a bit per rate component from the capacity region, independent of the channel parameters! We prove this result by first establishing bounds on the capacity region of a class of semideterministic ICs that include both the Gaussian IC and the injective deterministic IC in Section 6.6 as special cases.

6.7.1 Injective Semideterministic IC

Consider the semideterministic interference channel depicted in Figure 6.6. Here again the functions y_1, y_2 satisfy the condition that for every $x_1 \in \mathcal{X}_1$, $y_1(x_1, t_2)$ is a one-to-one function of t_2 and for every $x_2 \in \mathcal{X}_2$, $y_2(x_2, t_1)$ is a one-to-one function of t_1. The generalization comes from making the mappings from X_1 to T_1 and from X_2 to T_2 random.

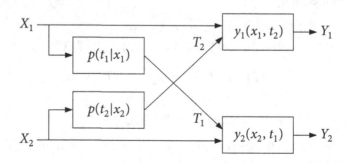

Figure 6.6. Injective semideterministic interference channel.

Note that if we assume the channel variables to be real-valued instead of finite, the Gaussian IC becomes a special case of this semideterministic IC with $T_1 = g_{21}X_1 + Z_2$ and $T_2 = g_{12}X_2 + Z_1$.

Outer bound on the capacity region. Consider the following outer bound on the capacity region of the injective semideterministic IC.

Proposition 6.1. Any achievable rate pair (R_1, R_2) for the injective semideterministic IC must satisfy the inequalities

$$R_1 \leq H(Y_1|X_2, Q) - H(T_2|X_2),$$
$$R_2 \leq H(Y_2|X_1, Q) - H(T_1|X_1),$$
$$R_1 + R_2 \leq H(Y_1|Q) + H(Y_2|U_2, X_1, Q) - H(T_1|X_1) - H(T_2|X_2),$$
$$R_1 + R_2 \leq H(Y_1|U_1, X_2, Q) + H(Y_2|Q) - H(T_1|X_1) - H(T_2|X_2),$$
$$R_1 + R_2 \leq H(Y_1|U_1, Q) + H(Y_2|U_2, Q) - H(T_1|X_1) - H(T_2|X_2),$$
$$2R_1 + R_2 \leq H(Y_1|Q) + H(Y_1|U_1, X_2, Q) + H(Y_2|U_2, Q) - H(T_1|X_1) - 2H(T_2|X_2),$$
$$R_1 + 2R_2 \leq H(Y_2|Q) + H(Y_2|U_2, X_1, Q) + H(Y_1|U_1, Q) - 2H(T_1|X_1) - H(T_2|X_2)$$

for some pmf $p(q)p(x_1|q)p(x_2|q)p_{T_1|X_1}(u_1|x_1)p_{T_2|X_2}(u_2|x_2)$.

This outer bound is established by extending the proof of the converse for the injective deterministic IC. We again use a genie argument with U_j conditionally independent of T_j given X_j, $j = 1, 2$. The details are given in Appendix 6B.

Remark 6.8. If we replace each channel $p(t_j|x_j)$, $j = 1, 2$, with a deterministic function $t_j(X_j)$, the above outer bound reduces to the capacity region of the injective deterministic IC in Theorem 6.5 by setting $U_j = T_j$, $j = 1, 2$.

Remark 6.9. The above outer bound is *not* tight under the strong interference condition in (6.7), and tighter outer bounds can be established.

Remark 6.10. We can obtain a corresponding outer bound for the Gaussian IC with differential entropies in place of entropies in the above outer bound.

Inner bound on the capacity region. The Han–Kobayashi inner bound with the restriction that $p(u_1, u_2|q, x_1, x_2) = p_{T_1|X_1}(u_1|x_1) \, p_{T_2|X_2}(u_2|x_2)$ reduces to the following.

Proposition 6.2. A rate pair (R_1, R_2) is achievable for the injective semideterministic IC if

$$R_1 < H(Y_1|U_2, Q) - H(T_2|U_2, Q),$$
$$R_2 < H(Y_2|U_1, Q) - H(T_1|U_1, Q),$$
$$R_1 + R_2 < H(Y_1|Q) + H(Y_2|U_1, U_2, Q) - H(T_1|U_1, Q) - H(T_2|U_2, Q),$$
$$R_1 + R_2 < H(Y_1|U_1, U_2, Q) + H(Y_2|Q) - H(T_1|U_1, Q) - H(T_2|U_2, Q),$$
$$R_1 + R_2 < H(Y_1|U_1, Q) + H(Y_2|U_2, Q) - H(T_1|U_1, Q) - H(T_2|U_2, Q),$$
$$2R_1 + R_2 < H(Y_1|Q) + H(Y_1|U_1, U_2, Q) + H(Y_2|U_2, Q)$$
$$\qquad\qquad - H(T_1|U_1, Q) - 2H(T_2|U_2, Q),$$
$$R_1 + 2R_2 < H(Y_2|Q) + H(Y_2|U_1, U_2, Q) + H(Y_1|U_1, Q)$$
$$\qquad\qquad - 2H(T_1|U_1, Q) - H(T_2|U_2, Q)$$

for some pmf $p(q)p(x_1|q)p(x_2|q)p_{T_1|X_1}(u_1|x_1)p_{T_2|X_2}(u_2|x_2)$.

Considering the Gaussian IC, we obtain a corresponding inner bound with differential entropies in place of entropies. This inner bound coincides with the outer bound for the injective deterministic interference channel discussed in Section 6.6, where T_1 is a deterministic function of X_1 and T_2 is a deterministic function of X_2 (thus $U_1 = T_1$ and $U_2 = T_2$).

Gap between the inner and outer bounds. For a fixed $(Q, X_1, X_2) \sim p(q)p(x_1|q)p(x_2|q)$, let $\mathscr{R}_o(Q, X_1, X_2)$ be the region defined by the set of inequalities in Proposition 6.1, and let $\mathscr{R}_i(Q, X_1, X_2)$ denote the closure of the region defined by the set of inequalities in Proposition 6.2.

Lemma 6.3. If $(R_1, R_2) \in \mathscr{R}_o(Q, X_1, X_2)$, then

$$(R_1 - I(X_2; T_2|U_2, Q), R_2 - I(X_1; T_1|U_1, Q)) \in \mathscr{R}_i(Q, X_1, X_2).$$

To prove this lemma, we first construct the rate region $\overline{\mathscr{R}}_o(Q, X_1, X_2)$ from the outer bound $\mathscr{R}_o(Q, X_1, X_2)$ by replacing X_j in every positive conditional entropy term in $\mathscr{R}_o(Q, X_1, X_2)$ with U_j for $j = 1, 2$. Clearly $\overline{\mathscr{R}}_o(Q, X_1, X_2) \supseteq \mathscr{R}_o(Q, X_1, X_2)$. Observing that

$$I(X_j; T_j|U_j) = H(T_j|U_j) - H(T_j|X_j), \quad j = 1, 2,$$

and comparing the rate region $\overline{\mathscr{R}}_o(Q, X_1, X_2)$ to the inner bound $\mathscr{R}_i(Q, X_1, X_2)$, we see that $\overline{\mathscr{R}}_o(Q, X_1, X_2)$ can be equivalently characterized as the set of rate pairs (R_1, R_2) that satisfy the statement in Lemma 6.3.

6.7.2   Half-Bit Theorem for the Gaussian IC

We show that the outer bound in Proposition 6.1, when specialized to the Gaussian IC, is achievable within half a bit per dimension. For the Gaussian IC, the auxiliary random variables in the outer bound can be expressed as

$$\begin{aligned}
U_1 &= g_{21}X_1 + Z_2', \\
U_2 &= g_{12}X_2 + Z_1',
\end{aligned} \tag{6.10}$$

where Z_1' and Z_2' are N(0, 1), independent of each other and of (X_1, X_2, Z_1, Z_2). Substituting in the outer bound in Proposition 6.1, we obtain an outer bound $\mathscr{R}_o$ on the capacity region of the Gaussian IC that consists of all rate pairs (R_1, R_2) such that

$$
\begin{aligned}
R_1 &\leq C(S_1), \\
R_2 &\leq C(S_2), \\
R_1 + R_2 &\leq C\left(\frac{S_1}{1 + I_2}\right) + C(I_2 + S_2), \\
R_1 + R_2 &\leq C\left(\frac{S_2}{1 + I_1}\right) + C(I_1 + S_1), \\
R_1 + R_2 &\leq C\left(\frac{S_1 + I_1 + I_1 I_2}{1 + I_2}\right) + C\left(\frac{S_2 + I_2 + I_1 I_2}{1 + I_1}\right), \\
2R_1 + R_2 &\leq C\left(\frac{S_1}{1 + I_2}\right) + C(S_1 + I_1) + C\left(\frac{S_2 + I_2 + I_1 I_2}{1 + I_1}\right), \\
R_1 + 2R_2 &\leq C\left(\frac{S_2}{1 + I_1}\right) + C(S_2 + I_2) + C\left(\frac{S_1 + I_1 + I_1 I_2}{1 + I_2}\right).
\end{aligned}
\tag{6.11}
$$

Now we show that $\mathscr{R}_o$ is achievable within half a bit.

Theorem 6.6 (Half-Bit Theorem). For the Gaussian IC, if $(R_1, R_2) \in \mathscr{R}_o$, then $(R_1 - 1/2, R_2 - 1/2)$ is achievable.

To prove this theorem, consider Lemma 6.3 for the Gaussian IC with the auxiliary random variables in (6.10). Then, for $j = 1, 2$,

$$
\begin{aligned}
I(X_j; T_j | U_j, Q) &= h(T_j | U_j, Q) - h(T_j | U_j, X_j, Q) \\
&\leq h(T_j - U_j) - h(Z_j) \\
&= \frac{1}{2}.
\end{aligned}
$$

6.7.3 Symmetric Degrees of Freedom

Consider the symmetric Gaussian IC with $S_1 = S_2 = S$ and $I_1 = I_2 = I$. Note that S and I fully characterize the channel. Define the *symmetric capacity* of the channel as $C_{\text{sym}} = \max\{R : (R, R) \in \mathscr{C}\}$ and the *normalized* symmetric capacity as

$$
d_{\text{sym}} = \frac{C_{\text{sym}}}{C(S)}.
$$

We find the *symmetric degrees of freedom* (DoF) d_{sym}^*, which is the limit of d_{sym} as the SNR and INR approach infinity. Note that in taking the limit, we are considering a sequence of channels rather than any particular channel. This limit, however, sheds light on the optimal coding strategies under different regimes of high SNR/INR.

Specializing the outer bound $\mathscr{R}_o$ in (6.11) to the symmetric case yields

$$C_{sym} \leq \overline{C}_{sym} = \min\left\{C(S), \frac{1}{2}C\left(\frac{S}{1+I}\right) + \frac{1}{2}C(S+I), C\left(\frac{S+I+I^2}{1+I}\right),\right.$$
$$\left.\frac{2}{3}C\left(\frac{S}{1+I}\right) + \frac{1}{3}C(S+2I+I^2)\right\}.$$

By the half-bit theorem,

$$\frac{\overline{C}_{sym} - 1/2}{C(S)} \leq d_{sym} \leq \frac{\overline{C}_{sym}}{C(S)}.$$

Thus, the difference between the upper and lower bounds converges to zero as $S \to \infty$, and the normalized symmetric capacity converges to the degrees of freedom d_{sym}^*. This limit, however, depends on how I scales as $S \to \infty$. Since it is customary to measure SNR and INR in decibels (dBs), we consider the limit for a constant ratio between the logarithms of the INR and SNR

$$\alpha = \frac{\log I}{\log S},$$

or equivalently, $I = S^\alpha$. Then, as $S \to \infty$, the normalized symmetric capacity d_{sym} converges to

$$d_{sym}^*(\alpha) = \lim_{S \to \infty} \frac{\overline{C}_{sym}|_{I=S^\alpha}}{C(S)}$$
$$= \min\{1, \max\{\alpha/2, 1 - \alpha/2\}, \max\{\alpha, 1 - \alpha\},$$
$$\max\{2/3, 2\alpha/3\} + \max\{1/3, 2\alpha/3\} - 2\alpha/3\}.$$

Since the fourth bound inside the minimum is redundant, we have

$$d_{sym}^*(\alpha) = \min\{1, \max\{\alpha/2, 1 - \alpha/2\}, \max\{\alpha, 1 - \alpha\}\}. \tag{6.12}$$

The symmetric DoF as a function of α is plotted in Figure 6.7. Note the unexpected W (instead of V) shape of the DoF curve. When interference is negligible ($\alpha \leq 1/2$), the DoF is $1 - \alpha$ and corresponds to the limit of the normalized rates achieved by treating interference as noise. For strong interference ($\alpha \geq 1$), the DoF is $\min\{1, \alpha/2\}$ and corresponds to simultaneous decoding. In particular, when interference is very strong ($\alpha \geq 2$), it does not impair the DoF. For moderate interference ($1/2 \leq \alpha \leq 1$), the DoF corresponds to the Han–Kobayashi rate splitting; see Problem 6.16. However, the DoF first increases until $\alpha = 2/3$ and then decreases to $1/2$ as α is increased to 1. Note that for $\alpha = 1/2$ and $\alpha = 1$, time division is also optimal.

Remark 6.11. In the above analysis, we scaled the channel gains under a fixed power constraint. Alternatively, we can fix the channel gains and scale the power P to infinity. It is not difficult to see that under this high power regime, $\lim_{P \to \infty} d^* = 1/2$, regardless of the values of the channel gains. Thus time division is asymptotically optimal.

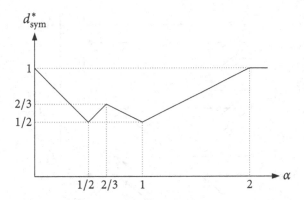

Figure 6.7. Degrees of freedom for symmetric Gaussian IC versus $\alpha = \log I / \log S$.

6.8 DETERMINISTIC APPROXIMATION OF THE GAUSSIAN IC

We introduce the q-ary expansion deterministic (QED) interference channel and show that it closely approximates the Gaussian IC in the limit of high SNR. The inputs to the QED-IC are q-ary L-vectors X_1 and X_2 for some "q-ary digit pipe" number L. We express X_1 as $[X_{1,L-1}, X_{1,L-2}, X_{1,L-3}, \ldots, X_{10}]^T$, where $X_{1l} \in [0 : q-1]$ for $l \in [0 : L-1]$, and similarly for X_2. Consider the symmetric case where the interference is specified by the parameter $\alpha \in [0, 2]$ such that αL is an integer. Define the "shift" parameter $s = (\alpha - 1)L$. The output of the channel depends on whether the shift is negative or positive.

Downshift. Here $s < 0$, i.e., $0 \le \alpha < 1$, and Y_1 is a q-ary L-vector with

$$Y_{1l} = \begin{cases} X_{1l} & \text{if } L + s \le l \le L - 1, \\ X_{1l} + X_{2,l-s} \ (\text{mod } q) & \text{if } 0 \le l \le L + s - 1. \end{cases}$$

This case is depicted in Figure 6.8. The outputs of the channel can be represented as

$$\begin{aligned} Y_1 &= X_1 + G_s X_2, \\ Y_2 &= G_s X_1 + X_2, \end{aligned} \tag{6.13}$$

where G_s is an $L \times L$ (down)shift matrix with $G_s(j, k) = 1$ if $k = j - s$ and $G_s(j, k) = 0$, otherwise.

Upshift. Here $s \ge 0$, i.e., $1 \le \alpha \le 2$, and Y_1 is a q-ary αL-vector with

$$Y_{1l} = \begin{cases} X_{2,l-s} & \text{if } L \le l \le L + s - 1, \\ X_{1l} + X_{2,l-s} \ (\text{mod } q) & \text{if } s \le l \le L - 1, \\ X_{1l} & \text{if } 0 \le l \le s - 1. \end{cases}$$

Again the outputs of the channel can be represented as in (6.13), where G_s is now an $(L + s) \times L$ (up)shift matrix with $G_s(j, k) = 1$ if $j = k$ and $G_s(j, k) = 0$, otherwise.

The capacity region of the symmetric QED-IC can be obtained by a straightforward

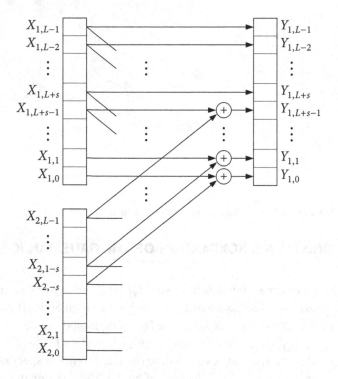

Figure 6.8. The q-ary expansion deterministic interference channel with downshift.

evaluation of the capacity region of the injective deterministic IC in Theorem 6.5. Let $R'_j = R_j/(L \log q)$, $j = 1, 2$. The normalized capacity region $\mathscr{C}'$ is the set of rate pairs (R'_1, R'_2) such that

$$
\begin{aligned}
R'_1 &\leq 1, \\
R'_2 &\leq 1, \\
R'_1 + R'_2 &\leq \max\{2\alpha,\ 2 - 2\alpha\}, \\
R'_1 + R'_2 &\leq \max\{\alpha,\ 2 - \alpha\}, \\
2R'_1 + R'_2 &\leq 2, \\
R'_1 + 2R'_2 &\leq 2
\end{aligned}
\tag{6.14}
$$

for $\alpha \in [1/2, 1]$, and

$$
\begin{aligned}
R'_1 &\leq 1, \\
R'_2 &\leq 1, \\
R'_1 + R'_2 &\leq \max\{2\alpha,\ 2 - 2\alpha\}, \\
R'_1 + R'_2 &\leq \max\{\alpha,\ 2 - \alpha\}
\end{aligned}
\tag{6.15}
$$

for $\alpha \in [0, 1/2) \cup (1, 2]$.

Surprisingly, the capacity region of the symmetric QED-IC can be achieved error-free using a simple single-letter linear coding scheme. We illustrate this scheme for the normalized symmetric capacity $C'_{sym} = \max\{R : (R,R) \in \mathscr{C}'\}$. Encoder $j = 1, 2$ represents its "single-letter" message by a q-ary LC'_{sym}-vector U_j and transmits $X_j = AU_j$, where A is an $L \times LC'_{sym}$ q-ary matrix. Decoder j multiplies its received symbol Y_j by a corresponding $LC'_{sym} \times L$ matrix B to recover U_j perfectly! For example, consider a binary expansion deterministic IC with $q = 2$, $L = 12$, and $\alpha = 5/6$. The symmetric capacity for this case is $C_{sym} = 7$ bits/transmission. For encoding, we use the matrix

$$A = \begin{bmatrix} 1 & 0 & 0 & 0 & 0 & 0 & 0 \\ 0 & 1 & 0 & 0 & 0 & 0 & 0 \\ 0 & 0 & 1 & 0 & 0 & 0 & 0 \\ 0 & 0 & 0 & 1 & 0 & 0 & 0 \\ 0 & 0 & 0 & 0 & 1 & 0 & 0 \\ 0 & 0 & 1 & 0 & 0 & 0 & 0 \\ 0 & 1 & 0 & 0 & 0 & 0 & 0 \\ 1 & 0 & 0 & 0 & 0 & 0 & 0 \\ 0 & 0 & 0 & 0 & 0 & 0 & 0 \\ 0 & 0 & 0 & 0 & 0 & 0 & 0 \\ 0 & 0 & 0 & 0 & 0 & 1 & 0 \\ 0 & 0 & 0 & 0 & 0 & 0 & 1 \end{bmatrix}.$$

Note that the first 3 bits of U_j are sent twice, while $X_{1,2} = X_{1,3} = 0$. The transmitted symbol X_j and the two signal components of the received vector Y_j are illustrated in Figure 6.9. Decoding for U_1 can also be performed sequentially as follows (see Figure 6.9):

1. $U_{1,6} = Y_{1,11}$, $U_{1,5} = Y_{1,10}$, $U_{1,1} = Y_{1,1}$, and $U_{1,0} = Y_{1,0}$. Also $U_{2,5} = Y_{1,3}$ and $U_{2,6} = Y_{1,2}$.

2. $U_{1,3} = Y_{1,8} \oplus U_{2,5}$ and $U_{1,4} = Y_{1,9} \oplus U_{2,6}$. Also $U_{2,4} = Y_{1,4} \oplus U_{1,6}$.

3. $U_{1,2} = Y_{1,7} \oplus U_{2,4}$.

This decoding procedure corresponds to multiplying the output by the matrix

$$B = \begin{bmatrix} 1 & 0 & 0 & 0 & 0 & 0 & 0 & 0 & 0 & 0 & 0 & 0 \\ 0 & 1 & 0 & 0 & 0 & 0 & 0 & 0 & 0 & 0 & 0 & 0 \\ 0 & 0 & 1 & 0 & 0 & 0 & 0 & 0 & 0 & 1 & 0 & 0 \\ 0 & 0 & 0 & 1 & 0 & 0 & 0 & 0 & 1 & 0 & 0 & 0 \\ 1 & 0 & 0 & 0 & 1 & 0 & 0 & 1 & 0 & 0 & 0 & 0 \\ 0 & 0 & 0 & 0 & 0 & 0 & 0 & 0 & 0 & 0 & 1 & 0 \\ 0 & 0 & 0 & 0 & 0 & 0 & 0 & 0 & 0 & 0 & 0 & 1 \end{bmatrix}.$$

Note that $BA = I$ and $BG_sA = 0$, and hence interference is canceled out while the intended signal is recovered perfectly.

Under the choice of the input $X_j = AU_j$, $j = 1, 2$, where U_j is uniformly distributed over the set of binary vectors of length LC'_{sym}, the symmetric capacity can be expressed as

$$C_{sym} = H(U_j) = I(U_j; Y_j) = I(X_j; Y_j), \quad j = 1, 2.$$

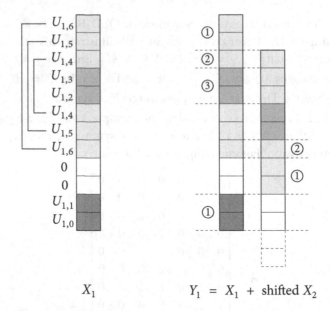

$$X_1 \qquad\qquad Y_1 = X_1 + \text{shifted } X_2$$

Figure 6.9. Transmitted symbol X_j and the received vector Y_j. The circled numbers denote the order of decoding.

Hence, the symmetric capacity is achieved error-free simply by treating interference as noise! In fact, the same linear coding technique can achieve the entire capacity region, which is generally characterized as the set of rate pairs (R_1, R_2) such that

$$R_1 < I(X_1; Y_1),$$
$$R_2 < I(X_2; Y_2) \tag{6.16}$$

for some pmf $p(x_1)p(x_2)$. A similar linear coding technique can be readily developed for any q-ary alphabet, dimension L, and $\alpha \in [0, 2]$ that achieves the entire capacity region by treating interference as noise.

6.8.1*   QED-IC Approximation of the Gaussian IC

Considering the normalized capacity region characterization in (6.14) and (6.15), we can show that the normalized symmetric capacity of the symmetric QED-IC is

$$C'_{\text{sym}} = \min\{1, \max\{\alpha/2, 1 - \alpha/2\}, \max\{\alpha, 1 - \alpha\}\} \tag{6.17}$$

for $\alpha \in [0, 2]$. This matches the DoF $d^*_{\text{sym}}(\alpha)$ of the symmetric Gaussian IC in (6.12) exactly. It can be shown that the Gaussian IC can be closely approximated by a QED-IC. Therefore, if a normalized rate pair is achievable for the QED-IC, then it is achievable for the corresponding Gaussian IC in the high SNR/INR limit, and vice versa.

We only sketch the proof that achievability carries over from the QED-IC to the Gaussian IC. Consider the q-ary expansions of the inputs and outputs of the Gaussian IC, e.g., $X_1 = X_{1,L-1}X_{1,L-2} \cdots X_{1,1}X_{1,0} . X_{1,-1}X_{1,-2} \cdots$, where $X_{1l} \in [0 : q-1]$ are q-ary digits.

Assuming $P = 1$, we express the channel outputs as $Y_1 = \sqrt{S}X_1 + \sqrt{I}X_2 + Z_1$ and $Y_2 = \sqrt{I}X_1 + \sqrt{S}X_2 + Z_2$. Suppose that $\sqrt{S}$ and $\sqrt{I}$ are powers of q. Then the digits of X_1, X_2, Y_1, and Y_2 align with each other. We further assume that the noise Z_1 is peak-power-constrained. Then, only the least-significant digits of Y_1 are affected by the noise. These digits are considered unusable for transmission. Now, we restrict each input digit to values from $[0 : \lfloor(q - 1)/2\rfloor]$. Thus, the signal additions at the q-ary digit-level are independent of each other, that is, there are no carry-overs, and the additions are effectively modulo-q. Note that this assumption does not affect the rate significantly because $\log\left(\lfloor(q - 1)/2\rfloor\right) / \log q$ can be made arbitrarily close to one by choosing q sufficiently large. Under the above assumptions, we arrive at a QED-IC, whereby the (random coding) achievability proof for rate pairs in (6.16) carries over to the Gaussian IC.

Remark 6.12. Recall that the capacity region of the QED-IC can be achieved by a simple single-letter linear coding technique (treating interference as noise) without using the full Han–Kobayashi coding scheme. Hence, the approximate capacity region and the DoF of the Gaussian IC can be both achieved simply by treating interference as noise. The resulting approximation gap, however, is significantly larger than half a bit.

6.9   EXTENSIONS TO MORE THAN TWO USER PAIRS

Interference channels with more than two user pairs are far less understood. For example, the notion of strong interference does not seem to naturally extend to more than two user pairs. These channels also exhibit the interesting property that decoding at each receiver is impaired by the *joint* effect of interference from the other senders rather by each sender's signal separately. Consequently, coding schemes that deal directly with the effect of the *combined interference signal* are expected to achieve higher rates. One such coding scheme is *interference alignment*, whereby the code is designed so that the combined interfering signal at each receiver is confined (*aligned*) to a subset of the receiver signal space. The subspace that contains the combined interference is discarded, while the desired signal is reconstructed from the orthogonal subspace. We illustrate this scheme in the following example.

Example 6.5 (*k*-User-pair symmetric QED-IC). Consider the k-user-pair QED-IC

$$Y_j = X_j + G_s \sum_{j' \neq j} X_{j'}, \quad j \in [1:k],$$

where $X_1, \ldots, X_k$ are q-ary L vectors, $Y_1, \ldots, Y_k$ are q-ary L_s vectors, $L_s = \max\{L, L + s\}$, and G_s is the $L_s \times L$ s-shift matrix for some $s \in [-L, L]$. As before, let $\alpha = (L + s)/L$. If $\alpha = 1$, then the received signals are identical and the normalized symmetric capacity is $C'_{\text{sym}} = 1/k$, which is achieved via time division. However, if $\alpha \neq 1$, then the normalized symmetric capacity is

$$C'_{\text{sym}} = \min\{1, \max\{\alpha, 1 - \alpha\}, \max\{\alpha/2, 1 - \alpha/2\}\},$$

which is equal to the normalized symmetric capacity for the 2-user-pair case, regardless of k! To show this, consider the single-letter linear coding technique described earlier for the 2-user-pair case. Then it is easy to check that the symmetric capacity is achievable (error-free), since the interfering signals from other senders are aligned in the same subspace and can be filtered out simultaneously.

Using the same approximation procedure detailed for the 2-user-pair case, this deterministic IC example shows that the DoF of the symmetric k-user-pair Gaussian IC is

$$d_{sym}^*(\alpha) = \begin{cases} 1/k & \text{if } \alpha = 1, \\ \min\{1, \max\{\alpha, 1 - \alpha\}, \max\{\alpha/2, 1 - \alpha/2\}\} & \text{otherwise.} \end{cases}$$

The DoF is achieved simply by treating interference as noise with a carefully chosen input pmf.

SUMMARY

- Discrete memoryless interference channel (DM-IC)
- Simultaneous nonunique decoding is optimal under strong interference
- Coded time sharing can strictly outperform time sharing
- Han–Kobayashi coding scheme:
  - Rate splitting and superposition coding
  - Fourier–Motzkin elimination
  - Optimal for injective deterministic ICs
- Gaussian interference channel:
  - Capacity region under strong interference achieved via simultaneous decoding
  - Sum-capacity under weak interference achieved by treating interference as noise
  - Genie-based converse proof
  - Han–Kobayashi coding scheme achieves within half a bit of the capacity region
  - Symmetric degrees of freedom
  - Approximation by the q-ary expansion deterministic IC in high SNR
  - Interference alignment
- **Open problems:**
  6.1. What is the capacity region of the Gaussian IC with weak interference?
  6.2. What is the generalization of strong interference to three or more user pairs?

6.3. What is the capacity region of the 3-user-pair injective deterministic IC?

6.4. Is the Han–Kobayashi inner bound tight in general?

BIBLIOGRAPHIC NOTES

The interference channel was first studied by Ahlswede (1974), who established basic inner and outer bounds including the simultaneous decoding inner bound in (6.3). The outer bound in (6.4) is based on a simple observation by L. Coviello that improves upon the outer bound in Sato (1977) by reversing the order of the union (over the input pmfs) and the intersection (over the channel pmfs). Carleial (1975) introduced the notion of very strong interference for the Gaussian IC and showed that the capacity region is the intersection of the capacity regions for the two component Gaussian MACs. The capacity region of the Gaussian IC with strong interference was established by Sato (1978b) and Han and Kobayashi (1981). Costa and El Gamal (1987) extended these results to the DM-IC.

Carleial (1978) introduced the idea of rate splitting and established an inner bound using successive cancellation decoding and (uncoded) time sharing. His inner bound was improved through simultaneous decoding and coded time sharing by Han and Kobayashi (1981). The inner bound in the Han–Kobayashi paper used four auxiliary random variables representing public and private messages and involved more inequalities than in Theorem 6.4. The equivalent characterization with only two auxiliary random variables and a reduced set of inequalities in Theorem 6.4 is due to Chong, Motani, Garg, and El Gamal (2008). The injective deterministic IC in Section 6.6 was introduced by El Gamal and Costa (1982), who used the genie argument to show that the Han–Kobayashi inner bound is tight.

Kramer (2004) developed a genie-based outer bound for the Gaussian IC. Shang, Kramer, and Chen (2009), Annapureddy and Veeravalli (2009), and Motahari and Khandani (2009) independently established the sum-capacity of the Gaussian IC with weak interference in Theorem 6.3. Our proof using the genie method follows the one by Annapureddy and Veeravalli (2009). The half-bit theorem was first established by Etkin, Tse, and Wang (2008) using the Han–Kobayashi inner bound and a variant of the genie-based outer bound by Kramer (2004). The proof in Section 6.7 using the injective semideterministic IC is due to Telatar and Tse (2007).

The approximation of the Gaussian IC by the q-ary expansion deterministic channel was first proposed by Avestimehr, Diggavi, and Tse (2011). Bresler, Parekh, and Tse (2010) applied this approach to approximate the many-to-one Gaussian IC. This approximation method was further refined by Jafar and Vishwanath (2010) and Bresler and Tse (2008). The symmetric capacity achieving linear coding scheme for the QED-IC is due to Jafar and Vishwanath (2010). Bandemer (2009) showed that the entire capacity region can be achieved by this linear coding scheme.

Interference alignment has been investigated for several classes of Gaussian channels by Maddah-Ali, Motahari, and Khandani (2008), Cadambe and Jafar (2008), Ghasemi,

Motahari, and Khandani (2010), Motahari, Gharan, Maddah-Ali, and Khandani (2009), Gou and Jafar (2010), and Nazer, Gastpar, Jafar, and Vishwanath (2009), and for QED-ICs by Jafar and Vishwanath (2010), Cadambe, Jafar, and Shamai (2009) and Bandemer, Vazquez-Vilar, and El Gamal (2009). Depending on the specific channel, this alignment is achieved via linear subspaces (Maddah-Ali, Motahari, and Khandani 2008), signal scale levels (Cadambe, Jafar, and Shamai 2009), time delay slots (Cadambe and Jafar 2008), or number-theoretic irrational bases (Motahari, Gharan, Maddah-Ali, and Khandani 2009). In each case, the subspace that contains the combined interference is disregarded, while the desired signal is reconstructed from the orthogonal subspace.

There are very few results on the IC with more than two user pairs beyond interference alignment. A straightforward extension of the Han–Kobayashi coding scheme is shown to be optimal for the deterministic IC (Gou and Jafar 2009), where the received signal is one-to-one to *all* interference signals given the intended signal. More interestingly, each receiver can decode for the combined (not individual) interference, which is achieved using structured codes for the many-to-one Gaussian IC (Bresler, Parekh, and Tse 2010). Decoding for the combined interference has been also applied to deterministic ICs with more than two user pairs (Bandemer and El Gamal 2011).

PROBLEMS

6.1. Establish the interference-as-noise inner bound in (6.2).

6.2. Prove the outer bound in (6.4).

6.3. Prove Lemma 6.1.

6.4. Verify the outer bound on the capacity region of the Gaussian IC in (6.11).

6.5. Show that the normalized capacity region of the QED-IC reduces to the regions in (6.14) and (6.15) and that the normalized symmetric capacity is given by (6.17).

6.6. *Successive cancellation decoding vs. simultaneous decoding.* Consider a DM-IC $p(y_1, y_2 | x_1, x_2)$. As in the simple coding schemes discussed in Section 6.2, suppose that point-to-point codes are used. Consider the successive cancellation decoding scheme, where receiver 1 first decodes for M_2 and then decodes for its own message M_1. Likewise receiver 2 first decodes for M_1 and then for M_2.

(a) Find the rate region achieved by successive cancellation decoding.

(b) Show that this region is always contained in the simultaneous-decoding inner bound in (6.3).

6.7. *Successive cancellation decoding for the Gaussian IC.* In Chapter 4 we found that for the DM-MAC, successive cancellation decoding with time sharing achieves the same inner bound as simultaneous decoding. In this problem, we show that this is not the case for the interference channel.

Consider the Gaussian IC with SNRs S_1 and S_2 and INRs I_1 and I_2.

(a) Write down the rate region achieved by successive cancellation decoding with Gaussian codes and no power control.

(b) Under what conditions is this region equal to the simultaneous-nonunique-decoding inner bound in Section 6.4?

(c) How much worse can successive cancellation decoding be than simultaneous nonunique decoding?

6.8. *Handoff.* Consider two symmetric Gaussian ICs, one with SNR S and INR $I > S$, and the other with SNR I and INR S. Thus, the second Gaussian IC is equivalent to the setting where the messages are sent to the other receivers in the first Gaussian IC. Which channel has a larger capacity region?

6.9. *Power control.* Consider the symmetric Gaussian IC with SNR S and INR I.

(a) Write down the rate region achieved by treating interference as noise with time sharing between two transmission subblocks and power control. Express the region in terms of three parameters: time-sharing fraction $\alpha \in [0, 1]$ and two power allocation parameters $\beta_1, \beta_2 \in [0, 1]$.

(b) Similarly, write down the rate region achieved by simultaneous nonunique decoding with time sharing between two transmission subblocks and power control in terms of α, β_1, β_2.

6.10. *Gaussian Z interference channel.* Consider the Gaussian IC depicted in Figure 6.10 with SNRs S_1, S_2, and INR I_1. (Here the INR $I_2 = 0$.)

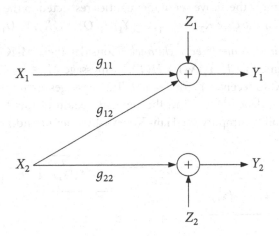

Figure 6.10. Gaussian interference channel with $I_2 = 0$.

(a) Find the capacity region when $S_2 \leq I_1$.

(b) Find the sum-capacity when $I_1 \leq S_2$.

(c) Find the capacity region when $S_2 \leq I_1/(1 + S_1)$.

6.11. *Minimum-energy-per-bit region.* Consider the Gaussian IC with channel gains g_{11}, g_{12}, g_{21}, and g_{22}. Find the minimum-energy-per-bit region, that is, the set of all energy pairs $(E_1, E_2) = (P_1/R_1, P_2/R_2)$ such that the rate pair (R_1, R_2) is achievable with average code power pair (P_1, P_2).

6.12. *An equivalent characterization of the Han–Kobayashi inner bound.* Consider the inner bound on the capacity region of the DM-IC that consists of all rate pairs (R_1, R_2) such that

$$R_1 < I(X_1; Y_1 | U_2, Q),$$
$$R_1 < I(X_1; Y_1 | U_1, U_2, Q) + I(X_2, U_1; Y_2 | U_2, Q),$$
$$R_2 < I(X_2; Y_2 | U_1, Q),$$
$$R_2 < I(X_1, U_2; Y_1 | U_1, Q) + I(X_2; Y_2 | U_1, U_2, Q),$$
$$R_1 + R_2 < I(X_1, U_2; Y_1 | Q) + I(X_2; Y_2 | U_1, U_2, Q),$$
$$R_1 + R_2 < I(X_2, U_1; Y_2 | Q) + I(X_1; Y_1 | U_1, U_2, Q),$$
$$R_1 + R_2 < I(X_1, U_2; Y_1 | U_1, Q) + I(X_2, U_1; Y_2 | U_2, Q),$$
$$2R_1 + R_2 < I(X_1, U_2; Y_1 | Q) + I(X_1; Y_1 | U_1, U_2, Q) + I(X_2, U_1; Y_2 | U_2, Q),$$
$$R_1 + 2R_2 < I(X_2, U_1; Y_2 | Q) + I(X_2; Y_2 | U_1, U_2, Q) + I(X_1, U_2; Y_1 | U_1, Q)$$

for some pmf $p(q)p(u_1, x_1 | q)p(u_2, x_2 | q)$. Show that this inner bound is equivalent to the characterization of the Han–Kobayashi inner bound in Theorem 6.4. (Hint: Show that if $R_1 \geq I(X_1; Y_1 | U_1, U_2, Q) + I(X_2, U_1; Y_2 | U_2, Q)$ then the inequalities in Theorem 6.4 imply the above set of inequalities restricted to the choice of $U_1 = \emptyset$, and similarly for the case $R_2 \geq I(X_1, U_2; Y_1 | U_1, Q) + I(X_2; Y_2 | U_1, U_2, Q)$.)

6.13. *A semideterministic interference channel.* Consider the DM-IC depicted in Figure 6.11. Assume that $H(Y_2 | X_2) = H(T)$. A message $M_j \in [1 : 2^{nR_j}]$ is to be sent from sender X_j to receiver Y_j for $j = 1, 2$. The messages are uniformly distributed and mutually independent. Find the capacity region of this DM-IC. (Hint: To prove achievability, simplify the Han–Kobayashi inner bound.)

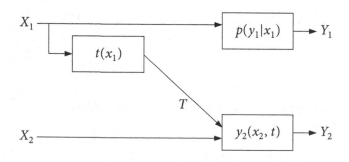

Figure 6.11. Semideterministic DM-IC.

6.14. *Binary injective deterministic interference channel.* Consider an injective deterministic IC with binary inputs X_1 and X_2 and ternary outputs $Y_1 = X_1 + X_2$ and $Y_2 = X_1 - X_2 + 1$. Find the capacity region of the channel.

6.15. *Deterministic interference channel with strong interference.* Find the conditions on the functions of the injective deterministic IC in Section 6.6 under which the channel has strong interference.

6.16. *Han–Kobayashi inner bound for the Gaussian IC.* Consider the Gaussian IC with SNRs S_1, S_2 and INRs I_1, I_2.

(a) Show that the Han–Kobayashi inner bound, when evaluated with Gaussian inputs (without power control), reduces to the set of rate pairs (R_1, R_2) such that

$$R_1 < E_Q\left[C\left(\frac{S_1}{1 + \lambda_{2Q}I_1}\right)\right],$$

$$R_2 < E_Q\left[C\left(\frac{S_2}{1 + \lambda_{1Q}I_2}\right)\right],$$

$$R_1 + R_2 < E_Q\left[C\left(\frac{S_1 + \bar{\lambda}_{2Q}I_1}{1 + \lambda_{2Q}I_1}\right) + C\left(\frac{\lambda_{2Q}S_2}{1 + \lambda_{1Q}I_2}\right)\right],$$

$$R_1 + R_2 < E_Q\left[C\left(\frac{S_2 + \bar{\lambda}_{1Q}I_2}{1 + \lambda_{1Q}I_2}\right) + C\left(\frac{\lambda_{1Q}S_1}{1 + \lambda_{2Q}I_1}\right)\right],$$

$$R_1 + R_2 < E_Q\left[C\left(\frac{\lambda_{1Q}S_1 + \bar{\lambda}_{2Q}I_1}{1 + \lambda_{2Q}I_1}\right) + C\left(\frac{\lambda_{2Q}S_2 + \bar{\lambda}_{1Q}I_2}{1 + \lambda_{1Q}I_2}\right)\right],$$

$$2R_1 + R_2 \le E_Q\left[C\left(\frac{S_1 + \bar{\lambda}_{2Q}I_1}{1 + \lambda_{2Q}I_1}\right) + C\left(\frac{\lambda_{1Q}S_1}{1 + \lambda_{2Q}I_1}\right) + C\left(\frac{\lambda_{2Q}S_2 + \bar{\lambda}_{1Q}I_2}{1 + \lambda_{1Q}I_2}\right)\right],$$

$$R_1 + 2R_2 \le E_Q\left[C\left(\frac{S_2 + \bar{\lambda}_{1Q}I_2}{1 + \lambda_{1Q}I_2}\right) + C\left(\frac{\lambda_{2Q}S_2}{1 + \lambda_{1Q}I_2}\right) + C\left(\frac{\lambda_{1Q}S_1 + \bar{\lambda}_{2Q}I_1}{1 + \lambda_{2Q}I_1}\right)\right]$$

for some $\lambda_{1Q}, \lambda_{2Q} \in [0, 1]$ and pmf $p(q)$ with $|\mathcal{Q}| \le 7$.

(b) Suppose that $S_1 = S_2 = S$ and $I_1 = I_2 = I$. By further specializing the inner bound in part (a), show that the symmetric capacity is lower bounded as

$$C_{\text{sym}} \ge \max_{\lambda \in [0,1]} \min\left\{C\left(\frac{S}{1 + \lambda I}\right), C\left(\frac{\lambda S + \bar{\lambda}I}{1 + \lambda I}\right),\right.$$
$$\left.\frac{1}{2}\left(C\left(\frac{S + \bar{\lambda}I}{1 + \lambda I}\right) + C\left(\frac{\lambda S}{1 + \lambda I}\right)\right)\right\}.$$

(c) Use part (b) to show that the symmetric DoF is lower bounded as

$$d^*_{\text{sym}}(\alpha) \ge \max\{1 - \alpha, \min\{1 - \alpha/2, \alpha\}, \min\{1, \alpha/2\}\},$$

which coincides with (6.12). (Hint: Consider $\lambda = 0, 1,$ and $1/(1 + S^\alpha)$.)

6.17. *Genie-aided outer bound for the Gaussian IC.* Consider the symmetric Gaussian IC with SNR S and INR I. Establish the outer bound on the capacity region that consists of the set of rate pairs (R_1, R_2) such that

$$R_1 \le C(S),$$
$$R_2 \le C(S),$$
$$R_1 + R_2 \le C(S) + C(S/(1+I)),$$
$$R_1 + R_2 \le 2\,C(I + S/(1+I)),$$
$$2R_1 + R_2 \le C(S+I) + C(I + S/(1+I)) + C(S) - C(I),$$
$$R_1 + 2R_2 \le C(S+I) + C(I + S/(1+I)) + C(S) - C(I).$$

(Hint: For the last two inequalities, suppose that receiver 1 has side information $T_1 = \sqrt{I/P}X_1 + W_1$ and receiver 2 has side information $T_2 = \sqrt{I/P}X_2 + W_2$, where W_1 and W_2 are i.i.d. N(0, 1), independent of (Z_1, Z_2).)
Remark: This bound, which is tighter than the outer bound in (6.11), is due to Etkin, Tse, and Wang (2008).

6.18. *Rate splitting for the more capable DM-BC.* Consider the alternative characterization of the capacity region in Section 5.6. Prove achievability of this region using rate splitting and Fourier–Motzkin elimination. (Hint: Divide M_1 into two independent messages M_{10} at rate R_{10} and M_{11} at rate R_{11}. Represent (M_{10}, M_2) by U and (M_{10}, M_{11}, M_2) by X.)

APPENDIX 6A   PROOF OF LEMMA 6.2

The sum-capacity $\tilde{C}_{sum}$ is achieved by treating interference as Gaussian noise. Thus we only need to prove the converse. Let $Q \sim \text{Unif}[1:n]$ be a time-sharing random variable independent of all other random variables and define $(T_1, T_2, Y_1, Y_2) = (T_{1Q}, T_{2Q}, Y_{1Q}, Y_{2Q})$. Thus, $(T_1, T_2, Y_1, Y_2) = (T_{1i}, T_{2i}, Y_{1i}, Y_{2i})$ with probability $1/n$ for $i \in [1:n]$. Suppose that a rate pair $(\tilde{R}_1, \tilde{R}_2)$ is achievable for the genie-aided channel. Then by Fano's inequality,

$$n\tilde{R}_1 \le I(X_1^n; Y_1^n, T_1^n) + n\epsilon_n$$
$$= I(X_1^n; T_1^n) + I(X_1^n; Y_1^n | T_1^n) + n\epsilon_n$$
$$= h(T_1^n) - h(T_1^n | X_1^n) + h(Y_1^n | T_1^n) - h(Y_1^n | T_1^n, X_1^n) + n\epsilon_n$$
$$\le h(T_1^n) - h(T_1^n | X_1^n) + \sum_{i=1}^n h(Y_{1i} | T_1^n) - h(Y_1^n | T_1^n, X_1^n) + n\epsilon_n$$
$$\overset{(a)}{\le} h(T_1^n) - h(T_1^n | X_1^n) + \sum_{i=1}^n h(Y_{1i} | T_1) - h(Y_1^n | T_1^n, X_1^n) + n\epsilon_n$$
$$\overset{(b)}{\le} h(T_1^n) - h(T_1^n | X_1^n) + nh(Y_1^* | T_1^*) - h(Y_1^n | T_1^n, X_1^n) + n\epsilon_n$$
$$\overset{(c)}{=} h(T_1^n) - nh(T_1^* | X_1^*) + nh(Y_1^* | T_1^*) - h(Y_1^n | T_1^n, X_1^n) + n\epsilon_n,$$

where (a) follows since $h(Y_{1i}|T_1^n) = h(Y_{1i}|T_1^n, Q) \leq h(Y_{1i}|T_{1Q}, Q) \leq h(Y_{1i}|T_{1Q})$, (b) follows by the maximum differential entropy lemma and concavity, and (c) follows since $h(T_1^n|X_1^n) = h(\eta\sqrt{I/P}\,W_1^n) = nh(\eta\sqrt{I/P}\,W_1) = nh(T_1^*|X_1^*)$. Similarly,

$$n\tilde{R}_2 \leq h(T_2^n) - nh(T_2^*|X_2^*) + nh(Y_2^*|T_2^*) - h(Y_2^n|T_2^n, X_2^n) + n\epsilon_n.$$

Thus, we can upper bound the sum-rate as

$$n(\tilde{R}_1 + \tilde{R}_2) \leq h(T_1^n) - h(Y_2^n|T_2^n, X_2^n) - nh(T_1^*|X_1^*) + nh(Y_1^*|T_1^*)$$
$$+ h(T_2^n) - h(Y_1^n|T_1^n, X_1^n) - nh(T_2^*|X_2^*) + nh(Y_2^*|T_2^*) + n\epsilon_n.$$

Evaluating the first two terms, we obtain

$$h(T_1^n) - h(Y_2^n|T_2^n, X_2^n) = h(\sqrt{I/P}\,X_1^n + \eta\sqrt{I/P}\,W_1^n) - h(\sqrt{I/P}\,X_1^n + Z_2^n \mid W_2^n)$$
$$= h(\sqrt{I/P}\,X_1^n + V_1^n) - h(\sqrt{I/P}\,X_1^n + V_2^n),$$

where $V_1^n = \eta\sqrt{I/P}\,W_1^n$ is i.i.d. $N(0, \eta^2 I/P)$ and $V_2^n = Z_2^n - E(Z_2^n|W_2^n)$ is i.i.d. $N(0, 1 - \rho^2)$. Given the useful genie condition $\eta^2 I/P \leq 1 - \rho^2$, express $V_2^n = V_1^n + V^n$, where V^n is i.i.d. $N(0, 1 - \rho^2 - \eta^2 I/P)$, independent of V_1^n. Now let $(V, V_1, V_2, X_1) = (V_Q, V_{1Q}, V_{2Q}, X_{1Q})$ and consider

$$h(T_1^n) - h(Y_2^n|T_2^n, X_2^n) = h(\sqrt{I/P}\,X_1^n + V_1^n) - h(\sqrt{I/P}\,X_1^n + V_1^n + V^n)$$
$$= -I\left(V^n;\ \sqrt{I/P}\,X_1^n + V_1^n + V^n\right)$$
$$= -nh(V) + h(V^n \mid \sqrt{I/P}\,X_1^n + V_1^n + V^n)$$
$$\leq -nh(V) + \sum_{i=1}^{n} h(V_i \mid \sqrt{I/P}\,X_1^n + V_1^n + V^n)$$
$$\leq -nh(V) + \sum_{i=1}^{n} h(V_i \mid \sqrt{I/P}\,X_{1i} + V_{1i} + V_i)$$
$$\leq -nh(V) + nh(V \mid \sqrt{I/P}\,X_1 + V_1 + V)$$
$$\overset{(a)}{\leq} -nI(V;\ \sqrt{I/P}\,X_1^* + V_1 + V)$$
$$= nh(\sqrt{I/P}\,X_1^* + V_1) - nh(\sqrt{I/P}\,X_1^* + V_1 + V)$$
$$= nh(T_1^*) - nh(Y_2^*|T_2^*, X_2^*),$$

where (a) follows since Gaussian is the worst noise with a given average power in an additive noise channel with Gaussian input; see Problem 2.12. The other terms $h(T_2^n) - h(Y_1^n|T_1^n, X_1^n)$ can be bounded in the same manner. This completes the proof of the lemma.

APPENDIX 6B   PROOF OF PROPOSITION 6.1

Consider a sequence of $(2^{nR_1}, 2^{nR_2})$ codes with $\lim_{n\to\infty} P_e^{(n)} = 0$. Furthermore, let $X_1^n, X_2^n, T_1^n, T_2^n, Y_1^n, Y_2^n$ denote the random variables resulting from encoding and transmitting the independent messages M_1 and M_2. Define random variables U_1^n, U_2^n such that U_{ji} is jointly distributed with X_{ji} according to $p_{T_j|X_j}(u_{ji}|x_{ji})$, conditionally independent of T_{ji} given X_{ji} for $j = 1, 2$ and $i \in [1 : n]$. By Fano's inequality,

$$nR_j = H(M_j)$$
$$\leq I(M_j; Y_j^n) + n\epsilon_n$$
$$\leq I(X_j^n; Y_j^n) + n\epsilon_n.$$

This directly yields a multiletter outer bound of the capacity region. We are looking for a nontrivial single-letter upper bound.

Observe that

$$I(X_1^n; Y_1^n) = H(Y_1^n) - H(Y_1^n|X_1^n)$$
$$= H(Y_1^n) - H(T_2^n|X_1^n)$$
$$= H(Y_1^n) - H(T_2^n)$$
$$\leq \sum_{i=1}^{n} H(Y_{1i}) - \boxed{H(T_2^n)},$$

since Y_1^n and T_2^n are one-to-one given X_1^n, and T_2^n is independent of X_1^n. The second term $H(T_2^n)$, however, is not easily upper-bounded in a single-letter form. Now consider the following augmentation

$$I(X_1^n; Y_1^n) \leq I(X_1^n; Y_1^n, U_1^n, X_2^n)$$
$$= I(X_1^n; U_1^n) + I(X_1^n; X_2^n|U_1^n) + I(X_1^n; Y_1^n|U_1^n, X_2^n)$$
$$= H(U_1^n) - H(U_1^n|X_1^n) + H(Y_1^n|U_1^n, X_2^n) - H(Y_1^n|X_1^n, U_1^n, X_2^n)$$
$$\overset{(a)}{=} H(T_1^n) - H(U_1^n|X_1^n) + H(Y_1^n|U_1^n, X_2^n) - H(T_2^n|X_2^n)$$
$$\leq \boxed{H(T_1^n)} - \sum_{i=1}^{n} H(U_{1i}|X_{1i}) + \sum_{i=1}^{n} H(Y_{1i}|U_{1i}, X_{2i}) - \sum_{i=1}^{n} H(T_{2i}|X_{2i}).$$

The second and fourth terms in (a) represent the output of a memoryless channel given its input. Thus they readily single-letterize with equality. The third term can be upper-bounded in a single-letter form. The first term $H(T_1^n)$ will be used to cancel boxed terms such as $H(T_2^n)$ above. Similarly, we can write

$$I(X_1^n; Y_1^n) \leq I(X_1^n; Y_1^n, U_1^n)$$
$$= I(X_1^n; U_1^n) + I(X_1^n; Y_1^n|U_1^n)$$
$$= H(U_1^n) - H(U_1^n|X_1^n) + H(Y_1^n|U_1^n) - H(Y_1^n|X_1^n, U_1^n)$$
$$= H(T_1^n) - H(U_1^n|X_1^n) + H(Y_1^n|U_1^n) - H(T_2^n)$$
$$\leq \boxed{H(T_1^n)} - \boxed{H(T_2^n)} - \sum_{i=1}^{n} H(U_{1i}|X_{1i}) + \sum_{i=1}^{n} H(Y_{1i}|U_{1i}),$$

and

$$
\begin{aligned}
I(X_1^n; Y_1^n) &\leq I(X_1^n; Y_1^n, X_2^n) \\
&= I(X_1^n; X_2^n) + I(X_1^n; Y_1^n | X_2^n) \\
&= H(Y_1^n | X_2^n) - H(Y_1^n | X_1^n, X_2^n) \\
&= H(Y_1^n | X_2^n) - H(T_2^n | X_2^n) \\
&\leq \sum_{i=1}^n H(Y_{1i} | X_{2i}) - \sum_{i=1}^n H(T_{2i} | X_{2i}).
\end{aligned}
$$

By symmetry, similar bounds can be established for $I(X_2^n; Y_2^n)$, namely,

$$
I(X_2^n; Y_2^n) \leq \sum_{i=1}^n H(Y_{2i}) - \boxed{H(T_1^n)},
$$

$$
I(X_2^n; Y_2^n) \leq \boxed{H(T_2^n)} - \sum_{i=1}^n H(U_{2i} | X_{2i}) + \sum_{i=1}^n H(Y_{2i} | U_{2i}, X_{1i}) - \sum_{i=1}^n H(T_{1i} | X_{1i}),
$$

$$
I(X_2^n; Y_2^n) \leq \boxed{H(T_2^n)} - \boxed{H(T_1^n)} - \sum_{i=1}^n H(U_{2i} | X_{2i}) + \sum_{i=1}^n H(Y_{2i} | U_{2i}),
$$

$$
I(X_2^n; Y_2^n) \leq \sum_{i=1}^n H(Y_{2i} | X_{1i}) - \sum_{i=1}^n H(T_{1i} | X_{1i}).
$$

Now consider linear combinations of the above inequalities where all boxed terms are canceled. Combining them with the bounds using Fano's inequality and using a time-sharing variable $Q \sim \mathrm{Unif}[1:n]$ completes the proof of the outer bound.

CHAPTER 7

Channels with State

In previous chapters, we assumed that the channel statistics do not change over transmissions and are completely known to the senders and the receivers. In this chapter, we study channels with state, which model communication settings where the channel statistics are not fully known or vary over transmissions, such as a wireless channel with fading, a write-once memory with programmed cells, a memory with stuck-at faults, or a communication channel with an adversary (jammer). The uncertainty about the channel statistics in such settings is captured by a state that may be fixed throughout the transmission block or vary randomly (or arbitrarily) over transmissions. The information about the channel state may be fully or partially available at the sender, the receiver, or both. In each setting, the channel capacity can be defined as before, but taking into consideration the state model and the availability of state information at the sender and/or the receiver.

We first discuss the compound channel model, where the state is selected from a given set and fixed throughout transmission. We then briefly discuss the arbitrarily varying channel, where the state varies over transmissions in an unknown manner. The rest of the chapter is dedicated to studying channels for which the state varies randomly over transmissions according to an i.i.d. process. We establish the capacity under various assumptions on state information availability at the encoder and/or the decoder. The most interesting case is when the state information is available causally or noncausally only at the encoder. For the causal case, the capacity is achieved by the Shannon strategy where each transmitted symbol is a function of a codeword symbol and the current state. For the noncausal case, the capacity is achieved by the Gelfand–Pinsker coding scheme, which involves joint typicality encoding and the new technique of multicoding (subcodebook generation). When specialized to the Gaussian channel with additive Gaussian state, the Gelfand–Pinsker scheme leads to the writing on dirty paper result, which shows that the effect of the state can be completely canceled. We also discuss several extensions of these results to multiuser channels with random state.

In Chapter 23, we will discuss the application of some of the above results to Gaussian fading channels, which are popular models for wireless communication. The multicoding technique will be used to establish an inner bound on the capacity region of the general broadcast channel in Chapter 8. Writing on dirty paper will be used to establish an alternative achievability proof for the Gaussian BC in Chapter 8, which will be extended to establish the capacity region of the vector Gaussian BC in Chapter 9.

7.1   DISCRETE MEMORYLESS CHANNEL WITH STATE

Consider the point-to-point communication system with state depicted in Figure 7.1. The sender wishes to communicate a message $M \in [1 : 2^{nR}]$ over a channel with state to the receiver with possible *side information* about the state sequence s^n available at the encoder and/or the decoder. We consider a *discrete-memoryless channel with state* model ($\mathcal{X} \times \mathcal{S}, p(y|x, s), \mathcal{Y}$) that consists of a finite input alphabet $\mathcal{X}$, a finite output alphabet $\mathcal{Y}$, a finite state alphabet $\mathcal{S}$, and a collection of conditional pmfs $p(y|x, s)$ on $\mathcal{Y}$. The channel is memoryless in the sense that, without feedback, $p(y^n|x^n, s^n, m) = \prod_{i=1}^{n} p_{Y|X,S}(y_i|x_i, s_i)$.

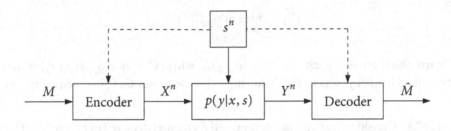

Figure 7.1. Point-to-point communication system with state.

In the following sections, we study special cases of this general setup.

7.2   COMPOUND CHANNEL

The compound channel is a DMC with state, where the channel is selected arbitrarily from a set of possible DMCs and fixed throughout the transmission block as depicted in Figure 7.2. It models communication in the presence of uncertainty about channel statistics. For clarity of notation, we use the equivalent definition of a compound channel as consisting of a set of DMCs ($\mathcal{X}, p(y_s|x), \mathcal{Y}$), where $y_s \in \mathcal{Y}$ for every state s in the finite set $\mathcal{S}$. The state s remains the same throughout the transmission block, i.e., $p(y^n|x^n, s^n) = \prod_{i=1}^{n} p_{Y_s|X}(y_i|x_i) = \prod_{i=1}^{n} p_{Y|X,S}(y_i|x_i, s)$.

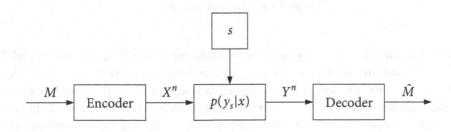

Figure 7.2. The compound channel.

Consider the case where the state is not available at either the encoder or the decoder. The definition of a $(2^{nR}, n)$ code under this assumption is the same as for the DMC. The average probability of error, however, is defined as

$$P_e^{(n)} = \max_{s \in \mathcal{S}} P\{M \neq \hat{M} \mid s \text{ is the selected channel state}\}.$$

Achievability and capacity are also defined as for the DMC.

Theorem 7.1. The capacity of the compound channel $(\mathcal{X}, \{p(y_s|x): s \in \mathcal{S}\}, \mathcal{Y})$ with no state information available at either the encoder or the decoder is

$$C_{CC} = \max_{p(x)} \min_{s \in \mathcal{S}} I(X; Y_s).$$

Clearly, the channel capacity $C_{CC} \leq \min_{s \in \mathcal{S}} C_s$, where $C_s = \max_{p(x)} I(X; Y_s)$ is the capacity of the channel $p(y_s|x)$. The following example shows that this inequality can be strict.

Example 7.1. Consider the two-state compound Z channel depicted in Figure 7.3. If $s = 1$, $p_{Y_s|X}(0|1) = 1/2$, and if $s = 2$, $p_{Y_s|X}(1|0) = 1/2$. The capacity of this compound channel is

$$C_{CC} = H\left(\frac{1}{4}\right) - \frac{1}{2} = 0.3113,$$

and is attained by $X \sim \text{Bern}(1/2)$. Note that the capacity is strictly less than $C_1 = C_2 = H(1/5) - 2/5 = \log(5/4) = 0.3219$.

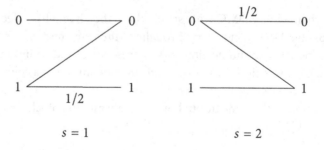

Figure 7.3. Z channel with state.

Remark 7.1. Note the similarity between the compound channel setup and the DM-BC with input X and outputs Y_s, $s \in \mathcal{S}$, when only a common message is to be sent to all the receivers. The main difference between these two setups is that in the broadcast channel case each receiver knows the statistics of its channel from the sender, while in the compound channel case with no state information available at the encoder or the decoder, the receiver knows only that the channel is one of several possible DMCs. Theorem 7.1 shows that the capacity is the same for these two setups, which in turn shows that the capacity of the compound channel does not increase when the state s is available at the decoder. This is not surprising since the receiver can *learn* the state from a relatively short training sequence.

Remark 7.2. It can be easily shown that $\min_{s \in \mathcal{S}} C_s$ is the capacity of the compound channel when the state s is available at the encoder.

Converse proof of Theorem 7.1. For every sequence of codes with $\lim_{n \to \infty} P_e^{(n)} = 0$, by Fano's inequality, we must have $H(M|Y_s^n) \le n\epsilon_n$ for some ϵ_n that tends to zero as $n \to \infty$ for every $s \in \mathcal{S}$. As in the converse proof for the DMC,

$$nR \le I(M; Y_s^n) + n\epsilon_n$$

$$\le \sum_{i=1}^{n} I(X_i; Y_{si}) + n\epsilon_n.$$

Now we introduce the time-sharing random variable $Q \sim \text{Unif}[1:n]$ independent of (M, X^n, Y_s^n, s) and define $X = X_Q$ and $Y_s = Y_{sQ}$. Then, $Q \to X \to Y_s$ form a Markov chain and

$$nR \le nI(X_Q; Y_{sQ}|Q) + n\epsilon_n$$
$$\le nI(X, Q; Y_s) + n\epsilon_n$$
$$= nI(X; Y_s) + n\epsilon_n$$

for every $s \in \mathcal{S}$. By taking $n \to \infty$, we have

$$R \le \min_{s \in \mathcal{S}} I(X; Y_s)$$

for some pmf $p(x)$. This completes the proof of the converse.

Achievability proof of Theorem 7.1. The proof uses random codebook generation and joint typicality decoding. We fix $p(x)$ and randomly and independently generate 2^{nR} sequences $x^n(m)$, $m \in [1:2^{nR}]$, each according to $\prod_{i=1}^{n} p_X(x_i)$. To send m, the encoder transmits $x^n(m)$. Upon receiving y^n, the decoder finds a unique message $\hat{m}$ such that $(x^n(\hat{m}), y^n) \in T_\epsilon^{(n)}(X, Y_s)$ for some $s \in \mathcal{S}$.

We now analyze the probability of error. Assume without loss of generality that $M = 1$ is sent. The decoder makes an error only if one or both of the following events occur:

$$\mathcal{E}_1 = \{(X^n(1), Y^n) \notin T_\epsilon^{(n)}(X, Y_{s'}) \text{ for all } s' \in \mathcal{S}\},$$
$$\mathcal{E}_2 = \{(X^n(m), Y^n) \in T_\epsilon^{(n)}(X, Y_{s'}) \text{ for some } m \neq 1, s' \in \mathcal{S}\}.$$

Then, the average probability of error is upper bounded as $P(\mathcal{E}) \le P(\mathcal{E}_1) + P(\mathcal{E}_2)$. By the LLN, $P\{(X^n(1), Y^n) \notin T_\epsilon^{(n)}(X, Y_s)\}$ tends to zero as $n \to \infty$. Thus $P(\mathcal{E}_1)$ tends to zero as $n \to \infty$. By the packing lemma, for each $s' \in \mathcal{S}$,

$$\lim_{n \to \infty} P\{(X^n(m), Y^n) \in T_\epsilon^{(n)}(X, Y_{s'}) \text{ for some } m \neq 1\} = 0,$$

if $R < I(X; Y_{s'}) - \delta(\epsilon)$. (Recall that the packing lemma applies to an arbitrary output pmf

$p(y^n)$.) Hence, by the union of events bound,

$$P(\mathcal{E}_2) = P\{(X^n(m), Y^n) \in \mathcal{T}_\epsilon^{(n)}(X, Y_{s'}) \text{ for some } m \neq 1, s' \in \mathcal{S}\}$$
$$\leq |\mathcal{S}| \cdot \max_{s' \in \mathcal{S}} P\{(X^n(m), Y^n) \in \mathcal{T}_\epsilon^{(n)}(X, Y_{s'}) \text{ for some } m \neq 1\},$$

which tends to zero as $n \to \infty$ if $R < I(X; Y_{s'}) - \delta(\epsilon)$ for all $s' \in \mathcal{S}$, or equivalently, $R < \min_{s' \in \mathcal{S}} I(X; Y_{s'}) - \delta(\epsilon)$. This completes the achievability proof.

Theorem 7.1 can be generalized to the case where $\mathcal{S}$ is arbitrary (not necessarily finite), and the capacity is

$$C_{\text{CC}} = \max_{p(x)} \inf_{s \in \mathcal{S}} I(X; Y_s).$$

This can be proved, for example, by noting that the probability of error for the packing lemma decays exponentially fast in n and the *effective* number of states is polynomial in n (there are only polynomially many empirical pmfs $p(y_s^n|x^n)$); see Problem 7.5.

Example 7.2. Consider the compound BSC(s) with $s \in [0, p]$, $p < 1/2$. Then, $C_{\text{CC}} = 1 - H(p)$, which is attained by $X \sim \text{Bern}(1/2)$. In particular, if $p = 1/2$, then $C_{\text{CC}} = 0$. This example demonstrates that the compound channel model is quite pessimistic, being robust against the worst-case channel.

Remark 7.3. The compound channel model can be readily extended to channels with input cost and to Gaussian channels. The extension to Gaussian channels with state is particularly interesting because these channels are used to model wireless communication channels with fading; see Chapter 23.

7.3*   ARBITRARILY VARYING CHANNEL

In an arbitrarily varying channel (AVC), the state sequence s^n changes over transmissions in an unknown and possibly adversarial manner. The capacity of the AVC depends on the availability of *common randomness* between the encoder and the decoder (deterministic code versus randomized code as defined in Problem 3.6), the performance criterion (average versus maximal probability of error), and knowledge of the adversary (codebook and/or the actual codeword transmitted). For example, suppose that X and S are binary and $Y = X + S$ is ternary. If the adversary knows the codebook, then it can always choose S^n to be one of the codewords. Given the sum of two codewords, the decoder has no way of distinguishing the true codeword from the interference. Hence, the probability of error is close to $1/2$ and the capacity is equal to zero. By contrast, if the encoder and the decoder can use common randomness, they can use a randomized code to combat the adversary. In this case, the capacity is $1/2$ bit/transmission, which is the capacity of a BEC($1/2$) that corresponds to $S \sim \text{Bern}(1/2)$, for both the average and maximal error probability criteria. If the adversary, however, knows the actual codeword transmitted, the shared randomness again becomes useless since the adversary can make the output equal to the all ones sequence, and the capacity is again equal to zero.

    Suppose that the encoder and the decoder can use shared common randomness to randomize the encoding and decoding operations, and the adversary has no knowledge of the actual codeword transmitted. In this case, the performance criterion of average or maximal probability of error does not affect the capacity, which is

$$C_{\text{AVC}} = \max_{p(x)} \min_{p(s)} I(X; Y_S) = \min_{p(s)} \max_{p(x)} I(X; Y_S). \tag{7.1}$$

Hence, the capacity is the saddle point of the game played by the encoder and the state selector with randomized strategies $p(x)$ and $p(s)$.

7.4 CHANNELS WITH RANDOM STATE

We now turn our attention to the less adversarial setup where the state is randomly chosen by nature. We consider the *DMC with DM state* model $(\mathcal{X} \times \mathcal{S}, p(y|x, s)p(s), \mathcal{Y})$, where the state sequence $(S_1, S_2, \ldots)$ is i.i.d. with $S_i \sim p_S(s_i)$. We are interested in finding the capacity of this channel under various scenarios of state information availability at the encoder and/or the decoder. The fact that the state changes over transmissions provides a temporal dimension to state information availability. The state may be available *causally* (that is, only S^i is known before transmission i takes place), *noncausally* (that is, S^n is known before communication commences), or with some delay or lookahead. For state availability at the decoder, the capacity under these different temporal constraints is the same. This is not the case, however, for state availability at the encoder as we will see.

We first consider two simple special cases. More involved cases are discussed in Sections 7.4.1, 7.5, and 7.6.

State information not available at either the encoder or the decoder. Let $p(y|x) = \sum_s p(s)p(y|x, s)$ be the DMC obtained by averaging the DMCs $p(y|x, s)$ over the state. Then it is easy to see that the capacity when the state information is not available at the encoder or the decoder is

$$C = \max_{p(x)} I(X; Y).$$

State information available only at the decoder. When the state sequence is available only at the decoder, the capacity is

$$C_{\text{SI-D}} = \max_{p(x)} I(X; Y, S) = \max_{p(x)} I(X; Y|S). \tag{7.2}$$

Achievability follows by treating (Y, S) as the output of the DMC $p(y, s|x) = p(s)p(y|x, s)$. The converse proof is straightforward.

7.4.1 State Information Available at Both the Encoder and the Decoder

Suppose that the state information is available causally and/or noncausally at both the encoder and the decoder as depicted in Figure 7.4. Then the capacity is the same for all four combinations and is given by

$$C_{\text{SI-ED}} = \max_{p(x|s)} I(X; Y|S). \tag{7.3}$$

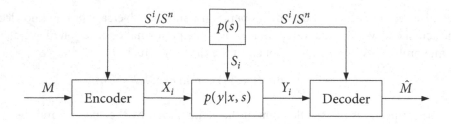

Figure 7.4. State information available at both the encoder and the decoder.

Achievability of $C_{\text{SI-ED}}$ can be proved by rate splitting and treating S^n as a time-sharing sequence.

Rate splitting. Divide the message M into independent messages $M_s \in [1 : 2^{nR_s}]$, $s \in S$. Thus $R = \sum_s R_s$.

Codebook generation. Fix the conditional pmf $p(x|s)$ that achieves the capacity and let $0 < \epsilon < 1$. For every $s \in S$, randomly and independently generate 2^{nR_s} sequences $x^n(m_s, s)$, $m_s \in [1 : 2^{nR_s}]$, each according to $\prod_{i=1}^{n} p_{X|S}(x_i|s)$.

Encoding. To send message $m = (m_s : s \in S)$, consider the corresponding codeword tuple $(x^n(m_s, s) : s \in S)$. Store each of these codewords in a first-in first-out (FIFO) buffer of length n. In time $i \in [1 : n]$, the encoder transmits the first untransmitted symbol from the FIFO buffer corresponding to the state s_i.

Decoding and the analysis of the probability of error. The decoder demultiplexes the received sequence into subsequences $(y^{n_s}(s), s \in S)$, where $\sum_s n_s = n$. Assuming $s^n \in T_\epsilon^{(n)}$, and hence $n_s \geq n(1 - \epsilon)p(s)$ for all $s \in S$, it finds for each s a unique $\hat{m}_s$ such that the codeword subsequence $x^{n(1-\epsilon)p(s)}(\hat{m}_s, s)$ is jointly typical with $y^{np(s)(1-\epsilon)}(s)$. By the LLN and the packing lemma, the probability of error for each decoding step tends to zero as $n \to \infty$ if $R_s < (1 - \epsilon)p(s)I(X; Y|S = s) - \delta(\epsilon)$. Thus, the total probability of error tends to zero as $n \to \infty$ if $R < (1 - \epsilon)I(X; Y|S) - \delta(\epsilon)$. This completes the proof of achievability.

The converse for (7.3) when the state is noncausally available is quite straightforward. Note that this converse also establishes the capacity for the causal case.

Remark 7.4. The capacity expressions in (7.2) for the case with state information available only at the decoder and in (7.3) for the case with state information available at both the encoder and the decoder continue to hold when $\{S_i\}$ is a stationary ergodic process. This key observation will be used in the discussion of fading channels in Chapter 23.

7.4.2 Extensions to the DM-MAC with State

The above coding theorems can be extended to some multiuser channels with DM state. An interesting example is the DM-MAC with 2-DM state components $(\mathcal{X}_1 \times \mathcal{X}_2 \times S, p(y|x_1, x_2, s)p(s), \mathcal{Y})$, where $S = (S_1, S_2)$. The motivation for assuming two state components is that in certain practical settings, such as the MAC with fading studied in Chapter 23 and the random access channel studied in Chapter 24, the effect of the channel uncertainty or variation with time can be modeled by a separate state for the path from each sender to the receiver. The DM-MAC with a single DM state is a special case of this model with $S = S_1 = S_2$.

- When no state information is available at the encoders or the decoder, the capacity region is that for the average DM-MAC $p(y|x_1, x_2) = \sum_s p(s)p(y|x_1, x_2, s)$.

- When the state sequence is available only at the decoder, the capacity region $\mathscr{C}_{\text{SI-D}}$ is the set of rate pairs (R_1, R_2) such that

$$R_1 \leq I(X_1; Y|X_2, Q, S),$$
$$R_2 \leq I(X_2; Y|X_1, Q, S),$$
$$R_1 + R_2 \leq I(X_1, X_2; Y|Q, S)$$

for some pmf $p(q)p(x_1|q)p(x_2|q)$.

- When the state sequence is available at both encoders and the decoder, the capacity region $\mathscr{C}_{\text{SI-ED}}$ is the set of rate pairs (R_1, R_2) such that

$$R_1 \leq I(X_1; Y|X_2, Q, S),$$
$$R_2 \leq I(X_2; Y|X_1, Q, S),$$
$$R_1 + R_2 \leq I(X_1, X_2; Y|Q, S)$$

for some pmf $p(q)p(x_1|s, q)p(x_2|s, q)$ and the encoders can adapt their codebooks according to the state sequence.

- More interestingly, when the state sequence is available at the decoder and each state component is available causally or noncausally at its respective encoder, that is, the encoders are specified by $x_j^n(m_j, s_j^n)$ for $j = 1, 2$, the capacity region with such *distributed* state information, $\mathscr{C}_{\text{DSI-ED}}$, is the set of rate pairs (R_1, R_2) such that

$$R_1 \leq I(X_1; Y|X_2, Q, S),$$
$$R_2 \leq I(X_2; Y|X_1, Q, S), \tag{7.4}$$
$$R_1 + R_2 \leq I(X_1, X_2; Y|Q, S)$$

for some pmf $p(q)p(x_1|s_1, q)p(x_2|s_2, q)$ and each encoder can *adapt* its codebook to its state component sequence.

The proofs of achievability and the converse for the above results follow the proofs for the DM-MAC and the DMC with DM state.

7.5   CAUSAL STATE INFORMATION AVAILABLE AT THE ENCODER

We consider yet another special case of state availability for the DMC with DM state $p(y|x, s)\, p(s)$. Suppose that the state sequence is available only at the encoder. In this case, the capacity depends on whether the state information is available causally or noncausally. We first study the causal case, that is, when the encoder knows S^i before transmission i, as depicted in Figure 7.5. A $(2^{nR}, n)$ code for this setup is specified by a message set $[1 : 2^{nR}]$, an encoder $x_i(m, s^i)$, $i \in [1 : n]$, and a decoder $\hat{m}(y^n)$. The definitions of the average probability of error, achievability, and capacity for this case are otherwise the same as for the DMC without state.

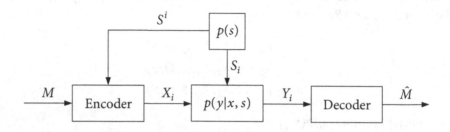

Figure 7.5. State information causally available only at the encoder.

The capacity for this case is given in the following.

Theorem 7.2. The capacity of the DMC with DM state $p(y|x, s)p(s)$ when the state information is available causally only at the encoder is

$$C_{\text{CSI-E}} = \max_{p(u),\, x(u,s)} I(U; Y),$$

where U is an auxiliary random variable independent of S with $|\mathcal{U}| \le \min\{(|\mathcal{X}| - 1)|\mathcal{S}| + 1, |\mathcal{Y}|\}$.

Proof of achievability. Fix the pmf $p(u)$ and function $x(u, s)$ that attain $C_{\text{CSI-E}}$. Randomly and independently generate 2^{nR} sequences $u^n(m)$, $m \in [1 : 2^{nR}]$, each according to $\prod_{i=1}^n p_U(u_i)$. To send m given the state s_i, the encoder transmits $x(u_i(m), s_i)$ at time $i \in [1 : n]$. The decoder finds the unique message $\hat{m}$ such that $(u^n(\hat{m}), y^n) \in \mathcal{T}_\epsilon^{(n)}$. By the LLN and the packing lemma, the probability of error tends to zero as $n \to \infty$ if $R < I(U; Y) - \delta(\epsilon)$, which completes the proof of achievability.

Remark 7.5. The above coding scheme corresponds to attaching a deterministic "physical device" $x(u, s)$ with two inputs U and S and one output X in front of the actual channel input as depicted in Figure 7.6. This induces a new DMC $p(y|u) = \sum_s p(y|x(u, s), s)p(s)$ with input U and output Y with capacity $C_{\text{CSI-E}}$.

Remark 7.6 (Shannon strategy). We can view the encoding as being performed over the set of all functions $\{x_u(s) : S \to \mathcal{X}\}$ indexed by u as the input alphabet. This technique of coding over functions onto $\mathcal{X}$ instead of actual symbols in $\mathcal{X}$ is referred to as

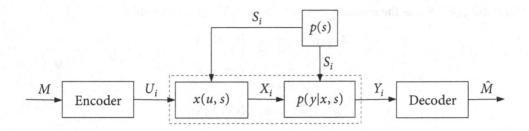

Figure 7.6. Coding with causal state information at the encoder.

the *Shannon strategy*. Note that the above cardinality bound shows that we need only $\min\{(|\mathcal{X}|-1)|\mathcal{S}|+1, |\mathcal{Y}|\}$ (instead of $|\mathcal{X}|^{|\mathcal{S}|}$) functions.

Proof of the converse. The key to the proof is the identification of the auxiliary random variable U. We would like X_i to be a function of (U_i, S_i). In general, X_i is a function of (M, S^{i-1}, S_i). So we define $U_i = (M, S^{i-1})$. This identification also satisfies the requirements that U_i is independent of S_i and $U_i \to (X_i, S_i) \to Y_i$ form a Markov chain for every $i \in [1:n]$. Now, by Fano's inequality,

$$nR \le I(M; Y^n) + n\epsilon_n$$
$$= \sum_{i=1}^{n} I(M; Y_i | Y^{i-1}) + n\epsilon_n$$
$$\le \sum_{i=1}^{n} I(M, Y^{i-1}; Y_i) + n\epsilon_n$$
$$\le \sum_{i=1}^{n} I(M, S^{i-1}, Y^{i-1}; Y_i) + n\epsilon_n$$
$$\overset{(a)}{=} \sum_{i=1}^{n} I(M, S^{i-1}, X^{i-1}, Y^{i-1}; Y_i) + n\epsilon_n$$
$$\overset{(b)}{=} \sum_{i=1}^{n} I(M, S^{i-1}, X^{i-1}; Y_i) + n\epsilon_n$$
$$\overset{(c)}{=} \sum_{i=1}^{n} I(U_i; Y_i) + n\epsilon_n$$
$$\le n \max_{p(u), x(u,s)} I(U; Y) + n\epsilon_n,$$

where (a) and (c) follow since X^{i-1} is a function of (M, S^{i-1}) and (b) follows since $Y^{i-1} \to (X^{i-1}, S^{i-1}) \to Y_i$. This completes the proof of the converse.

Remark 7.7. Consider the case where the state sequence is available causally at both the encoder and the decoder in Section 7.4.1. Note that we can use the Shannon strategy to provide an alternative proof of achievability to the time-sharing proof discussed earlier.

Treating (Y^n, S^n) as the equivalent channel output, Theorem 7.2 reduces to

$$C_{\text{SI-ED}} = \max_{p(u),\, x(u,s)} I(U; Y, S)$$

$$\stackrel{(a)}{=} \max_{p(u),\, x(u,s)} I(U; Y|S)$$

$$\stackrel{(b)}{=} \max_{p(u),\, x(u,s)} I(X; Y|S)$$

$$\stackrel{(c)}{=} \max_{p(x|s)} I(X; Y|S),$$

where (a) follows by the independence of U and S, and (b) follows since X is a function of (U, S) and $U \rightarrow (S, X) \rightarrow Y$ form a Markov chain. Step (c) follows by the *functional representation lemma* in Appendix B, which states that every conditional pmf $p(x|s)$ can be represented as a deterministic function $x(u, s)$ of S and a random variable U that is independent of S.

7.6 NONCAUSAL STATE INFORMATION AVAILABLE AT THE ENCODER

We now consider the case where the state sequence is available noncausally only at the encoder. In other words, a $(2^{nR}, n)$ code is specified by a message set $[1 : 2^{nR}]$, an encoder $x^n(m, s^n)$, and a decoder $\hat{m}(y^n)$. The definitions of the average probability of error, achievability, and capacity are otherwise the same as for the DMC without state.

Assuming noncausal state information availability at the encoder, however, may appear unrealistic. How can the encoder know the state sequence before it is generated? The following example demonstrates a real-world scenario in which the state can be available noncausally at the encoder (other scenarios will be discussed later).

Example 7.3 (Memory with stuck-at faults). Consider the DMC with DM state depicted in Figure 7.7, which is a model for a digital memory with "stuck-at" faults or a write-once memory (WOM) such as a CD-ROM. As shown in the figure, the state $S = 0$ corresponds to a faulty memory cell that outputs a 0 independent of its stored input value, the state $S = 1$ corresponds to a faulty memory cell that always outputs a 1, and the state $S = 2$ corresponds to a nonfaulty cell that outputs the same value as its stored input. The probabilities of these states are $p/2$, $p/2$, and $1 - p$, respectively.

The writer (encoder) who knows the locations of the faults (by testing the memory) wishes to reliably store information in a way that does not require the reader (decoder) to know the locations of the faults. How many bits per memory cell can be reliably stored?

* If neither the writer nor the reader knows the fault locations, we can store up to

$$C = \max_{p(x)} I(X; Y) = 1 - H\left(\frac{p}{2}\right) \quad \text{bits/cell.}$$

* If both the writer and the reader know the fault locations, we can store up to

$$C_{\text{SI-ED}} = \max_{p(x|s)} I(X; Y|S) = 1 - p \quad \text{bits/cell.}$$

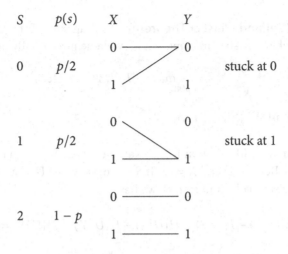

S	$p(s)$	X		Y	

Figure 7.7. Memory with stuck-at faults.

- If the reader knows the fault locations (erasure channel), we can also store

$$C_{\text{SI-D}} = \max_{p(x)} I(X; Y|S) = 1 - p \quad \text{bits/cell.}$$

Surprisingly, even if only the writer knows the fault locations we can still reliably store up to $1 - p$ bits/cell! Consider the following *multicoding* scheme. Assume that the memory has n cells. Randomly and independently assign each binary n-sequence to one of 2^{nR} *subcodebooks* $C(m)$, $m \in [1 : 2^{nR}]$; see Figure 7.8. To store message m, the writer searches in subcodebook $C(m)$ for a sequence that matches the pattern of the faulty cells and stores it; otherwise it declares an error. The reader declares as the message estimate the index of the subcodebook that the stored sequence belongs to. Since there are roughly np faulty cells for n sufficiently large, there are roughly $2^{n(1-p)}$ sequences that match any given fault pattern. Hence for n sufficiently large, any given subcodebook has at least one matching sequence with high probability if $R < C_{\text{SI-E}} = (1 - p)$, and asymptotically up to $n(1 - p)$ bits can be reliably stored in an n-bit memory with np faulty cells.

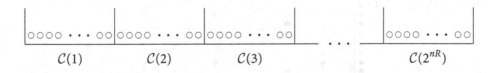

Figure 7.8. Multicoding scheme for the memory with stuck-at faults. Each binary n-sequence is randomly assigned to a subcodebook $C(m)$, $m \in [1 : 2^{nR}]$.

The following theorem generalizes the above example to DMCs with DM state using the same basic coding idea.

Theorem 7.3 (Gelfand–Pinsker Theorem). The capacity of the DMC with DM state $p(y|x, s)p(s)$ when the state information is available noncausally only at the encoder is

$$C_{\text{SI-E}} = \max_{p(u|s),\, x(u,s)} (I(U;Y) - I(U;S)),$$

where $|\mathcal{U}| \leq \min\{|\mathcal{X}|\cdot|\mathcal{S}|, |\mathcal{Y}| + |\mathcal{S}| - 1\}$.

For the memory with stuck-at faults example, given $S = 2$, that is, when there is no fault, we set $U \sim \text{Bern}(1/2)$ and $X = U$. If $S = 1$ or 0, we set $U = X = S$. Since $Y = X = U$ under this choice of $p(u|s)$ and $x(u, s)$, we have

$$I(U;Y) - I(U;S) = H(U|S) - H(U|Y) = H(U|S) = 1 - p.$$

We prove the Gelfand–Pinsker theorem in the next two subsections.

7.6.1 Proof of Achievability

The Gelfand–Pinsker coding scheme is illustrated in Figure 7.9. It uses multicoding and joint typicality encoding and decoding. For each message m, we generate a *subcodebook* $\mathcal{C}(m)$ of $2^{n(\tilde{R}-R)}$ sequences $u^n(l)$. To send m given the state sequence s^n, we find a sequence $u^n(l) \in \mathcal{C}(m)$ that is jointly typical with s^n and transmit $x^n(u^n(l), s^n)$. The receiver recovers $u^n(l)$ and then declares its subcodebook index as the message estimate.

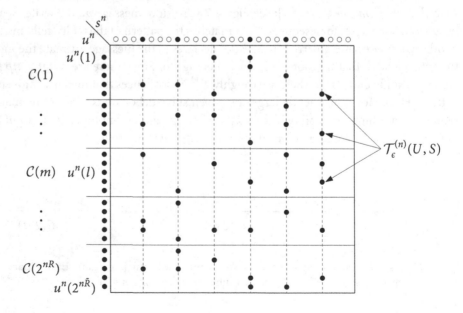

Figure 7.9. Gelfand–Pinsker coding scheme. Each subcodebook $\mathcal{C}(m), m \in [1:2^{nR}]$, consists of $2^{n(\tilde{R}-R)}$ sequences $u^n(l)$.

We now provide the details of the achievability proof.

Codebook generation. Fix the conditional pmf $p(u|s)$ and function $x(u, s)$ that attain the capacity and let $\tilde{R} > R$. For each message $m \in [1 : 2^{nR}]$ generate a subcodebook $C(m)$ consisting of $2^{n(\tilde{R}-R)}$ randomly and independently generated sequences $u^n(l)$, $l \in [(m - 1)2^{n(\tilde{R}-R)} + 1 : m2^{n(\tilde{R}-R)}]$, each according to $\prod_{i=1}^{n} p_U(u_i)$.

Encoding. To send message $m \in [1 : 2^{nR}]$ with the state sequence s^n observed, the encoder chooses a sequence $u^n(l) \in C(m)$ such that $(u^n(l), s^n) \in T_{\epsilon'}^{(n)}$. If no such sequence exists, it picks $l = 1$. The encoder then transmits $x_i = x(u_i(l), s_i)$ at time $i \in [1 : n]$.

Decoding. Let $\epsilon > \epsilon'$. Upon receiving y^n, the decoder declares that $\hat{m} \in [1 : 2^{nR}]$ is sent if it is the unique message such that $(u^n(l), y^n) \in T_{\epsilon}^{(n)}$ for some $u^n(l) \in C(\hat{m})$; otherwise it declares an error.

Analysis of the probability of error. Assume without loss of generality that $M = 1$ and let L denote the index of the chosen U^n sequence for $M = 1$ and S^n. The decoder makes an error only if one or more of the following events occur:

$$\mathcal{E}_1 = \{(U^n(l), S^n) \notin T_{\epsilon'}^{(n)} \text{ for all } U^n(l) \in C(1)\},$$
$$\mathcal{E}_2 = \{(U^n(L), Y^n) \notin T_{\epsilon}^{(n)}\},$$
$$\mathcal{E}_3 = \{(U^n(l), Y^n) \in T_{\epsilon}^{(n)} \text{ for some } l \notin [1 : 2^{n(\tilde{R}-R)}]\}.$$

Thus, the probability of error is upper bounded as

$$P(\mathcal{E}) \le P(\mathcal{E}_1) + P(\mathcal{E}_1^c \cap \mathcal{E}_2) + P(\mathcal{E}_3).$$

By the covering lemma in Section 3.7 with $U = \emptyset$, $X \leftarrow S$, and $\hat{X} \leftarrow U$, $P(\mathcal{E}_1)$ tends to zero as $n \to \infty$ if $\tilde{R} - R > I(U; S) + \delta(\epsilon')$. Next, note that

$$\mathcal{E}_1^c = \{(U^n(L), S^n) \in T_{\epsilon'}^{(n)}\} = \{(U^n(L), X^n, S^n) \in T_{\epsilon'}^{(n)}\}$$

and

$$Y^n \mid \{U^n(L) = u^n, X^n = x^n, S^n = s^n\} \sim \prod_{i=1}^{n} p_{Y|U,X,S}(y_i | u_i, x_i, s_i)$$

$$= \prod_{i=1}^{n} p_{Y|X,S}(y_i | x_i, s_i).$$

Hence, by the conditional typicality lemma in Section 2.5 (see also Problem 2.13), $P(\mathcal{E}_1^c \cap \mathcal{E}_2)$ tends to zero as $n \to \infty$ (recall the standing assumption that $\epsilon' < \epsilon$).

Finally, since every $U^n(l) \notin C(1)$ is distributed according to $\prod_{i=1}^{n} p_U(u_i)$ and is independent of Y^n, by the packing lemma in Section 3.2, $P(\mathcal{E}_3)$ tends to zero as $n \to \infty$ if $\tilde{R} < I(U; Y) - \delta(\epsilon)$. Combining these results, we have shown $P(\mathcal{E})$ tends to zero as $n \to \infty$ if $R < I(U; Y) - I(U; S) - \delta(\epsilon) - \delta(\epsilon')$. This completes the proof of achievability.

Remark 7.8. In the bound on $P(\mathcal{E}_3)$, the sequence Y^n is not generated i.i.d. However, we can still use the packing lemma.

7.6.2 Proof of the Converse

Again the key is to identify U_i such that $U_i \to (X_i, S_i) \to Y_i$ form a Markov chain. By Fano's inequality, for any code with $\lim_{n \to \infty} P_e^{(n)} = 0$, $H(M|Y^n) \le n\epsilon_n$, where ϵ_n tends to zero as $n \to \infty$. Consider

$$nR \le I(M; Y^n) + n\epsilon_n$$

$$= \sum_{i=1}^{n} I(M; Y_i | Y^{i-1}) + n\epsilon_n$$

$$\le \sum_{i=1}^{n} I(M, Y^{i-1}; Y_i) + n\epsilon_n$$

$$= \sum_{i=1}^{n} I(M, Y^{i-1}, S_{i+1}^n; Y_i) - \sum_{i=1}^{n} I(Y_i; S_{i+1}^n | M, Y^{i-1}) + n\epsilon_n$$

$$\overset{(a)}{=} \sum_{i=1}^{n} I(M, Y^{i-1}, S_{i+1}^n; Y_i) - \sum_{i=1}^{n} I(Y^{i-1}; S_i | M, S_{i+1}^n) + n\epsilon_n$$

$$\overset{(b)}{=} \sum_{i=1}^{n} I(M, Y^{i-1}, S_{i+1}^n; Y_i) - \sum_{i=1}^{n} I(M, Y^{i-1}, S_{i+1}^n; S_i) + n\epsilon_n,$$

where (a) follows by the Csiszár sum identity and (b) follows since (M, S_{i+1}^n) is independent of S_i. Now identifying $U_i = (M, Y^{i-1}, S_{i+1}^n)$, we have $U_i \to (X_i, S_i) \to Y_i$ for $i \in [1:n]$ as desired. Hence

$$nR \le \sum_{i=1}^{n} \left(I(U_i; Y_i) - I(U_i; S_i) \right) + n\epsilon_n$$

$$\le n \max_{p(u,x|s)} \left(I(U; Y) - I(U; S) \right) + n\epsilon_n. \tag{7.5}$$

The cardinality bound on $\mathcal{U}$ can be proved using the convex cover method in Appendix C.

Finally, we show that it suffices to maximize over $p(u|s)$ and functions $x(u, s)$. For a fixed $p(u|s)$, note that

$$p(y|u) = \sum_{x,s} p(s|u)p(x|u, s)p(y|x, s)$$

is linear in $p(x|u, s)$. Since $p(u|s)$ is fixed, the maximum in (7.5) is only over $I(U; Y)$, which is convex in $p(y|u)$ (since $p(u)$ is fixed) and hence in $p(x|u, s)$. This implies that the maximum is attained at an extreme point of the set of pmfs $p(x|u, s)$, that is, using one of the deterministic mappings $x(u, s)$. This completes the proof of the Gelfand–Pinsker theorem.

7.6.3 Comparison with the Causal Case

Recall that when the state information is available noncausally only at the encoder, the capacity is

$$C_{\text{SI-E}} = \max_{p(u|s),\, x(u,s)} \left(I(U; Y) - I(U; S) \right).$$

By comparison, when the state information is available *causally* only at the encoder, the capacity can be expressed as

$$C_{\text{CSI-E}} = \max_{p(u),\, x(u,s)} \left(I(U; Y) - I(U; S) \right),$$

since $I(U; S) = 0$. Note that these two expressions have the same form, except that in the causal case the maximum is over $p(u)$ instead of $p(u|s)$.

In addition, the coding schemes for both scenarios are the same except that in the noncausal case the encoder knows the state sequence S^n in advance and hence can choose a sequence U^n jointly typical with S^n; see Figure 7.10. Thus the cost of causality (that is, the gap between the causal and the noncausal capacities) is captured entirely by the more restrictive independence condition between U and S.

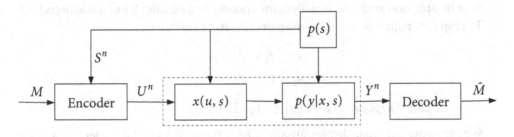

Figure 7.10. Alternative view of Gelfand–Pinsker coding.

7.6.4 Extensions to the Degraded DM-BC with State

The coding schemes for channels with state information available at the encoder can be extended to multiuser channels. For example, Shannon strategy can be used whenever the state information is available causally at some of the encoders. Similarly, Gelfand–Pinsker coding can be used whenever the state sequence is available noncausally. The optimality of these extensions, however, is not known in most cases.

In this subsection, we consider the degraded DM-BC with DM state $p(y_1|x, s)p(y_2|y_1)$ $p(s)$ depicted in Figure 7.11. The capacity region of this channel is known for the following state information availability scenarios:

- State information available causally at the encoder: In this case, the encoder at transmission i assigns a symbol x_i to each message pair (m_1, m_2) and state sequence s^i and decoder $j = 1, 2$ assigns an estimate $\hat{m}_j$ to each output sequence y_j^n. The capacity region is the set of rate pairs (R_1, R_2) such that

$$R_1 \le I(U_1; Y_1 | U_2),$$
$$R_2 \le I(U_2; Y_2)$$

for some pmf $p(u_1, u_2)$ and function $x(u_1, u_2, s)$, where (U_1, U_2) is independent of S with $|\mathcal{U}_1| \le |\mathcal{X}| \cdot |\mathcal{S}|(|\mathcal{X}| \cdot |\mathcal{S}| + 1)$ and $|\mathcal{U}_2| \le |\mathcal{X}| \cdot |\mathcal{S}| + 1$.

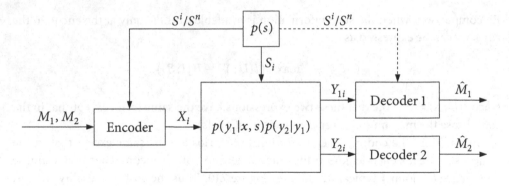

Figure 7.11. Degraded DM-BC with state.

- State information available causally at the encoder and decoder 1, but not at decoder 2: The capacity region is the set of rate pairs (R_1, R_2) such that

$$R_1 \le I(X; Y_1 | U, S),$$
$$R_2 \le I(U; Y_2)$$

for some pmf $p(u)p(x|u, s)$ with $|\mathcal{U}| \le |\mathcal{X}| \cdot |\mathcal{S}| + 1$.

- State information available noncausally at both the encoder and decoder 1, but not at decoder 2: The capacity region is the set of rate pairs (R_1, R_2) such that

$$R_1 \le I(X; Y_1 | U, S),$$
$$R_2 \le I(U; Y_2) - I(U; S)$$

for some pmf $p(u, x|s)$ with $|\mathcal{U}| \le |\mathcal{X}| \cdot |\mathcal{S}| + 1$.

However, the capacity region is not known for other scenarios, for example, when the state information is available noncausally only at the encoder.

7.7   WRITING ON DIRTY PAPER

Consider the Gaussian channel with additive Gaussian state depicted in Figure 7.12. The output of the channel is

$$Y = X + S + Z,$$

where the state $S \sim N(0, Q)$ and the noise $Z \sim N(0, 1)$ are independent. Assume an expected average transmission power constraint

$$\sum_{i=1}^{n} E(x_i^2(m, S^n)) \le nP \quad \text{for every } m \in [1 : 2^{nR}],$$

where the expectation is over the random state sequence S^n.

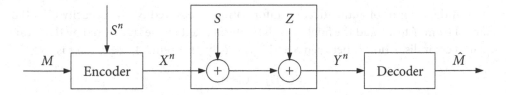

Figure 7.12. Gaussian channel with additive Gaussian state available noncausally at the encoder.

If the state is not available at either the encoder or the decoder, the state becomes part of the additive noise and the capacity is

$$C = C\left(\frac{P}{1 + Q}\right).$$

If the state is available at the decoder, it can be simply subtracted off, which in effect reduces the channel to a Gaussian channel with no state. Hence the capacity for this case is

$$C_{\text{SI-D}} = C_{\text{SI-ED}} = C(P). \tag{7.6}$$

Now suppose that the state information is available noncausally only at the encoder. Can the *effect* of the state still be canceled in this case? Note that the encoder cannot simply presubtract the state because the resulting codeword may violate the power constraint. Nevertheless, it turns out that the effect of the state can still be completely canceled!

Theorem 7.4. The capacity of the Gaussian channel with additive Gaussian state when the state information is available noncausally only at the encoder is

$$C_{\text{SI-E}} = C(P).$$

This result is known as *writing on dirty paper*. Imagine a sheet of paper with independent dirt spots of normal intensity. The writer knows the spot locations and intensities, but the reader cannot distinguish between the message and the dirt. The writing on dirty paper result shows that using an "optimal writing scheme," the reader can still recover the message as if the paper had no dirt at all.

Writing on dirty paper will prove useful in coding for Gaussian BCs as will be discussed in Chapters 8 and 9. In addition, it is used in information hiding and digital watermarking.

Example 7.4 (Digital watermarking). Suppose that a publisher embeds a signal (watermark) X in a host image or text S. A viewer receives a noisy version of the watermarked image $Y = X + S + Z$, where $Z \sim N(0, 1)$ is an additive Gaussian noise. Given a host image S^n, the authentication message $M \in [1 : 2^{nR}]$ is encoded into a watermark sequence $X^n(M, S^n)$, which is added to the image to produce the watermarked image $S^n + X^n$. An authenticator wishes to retrieve the message M from Y^n. What is the optimal tradeoff between the amount of authentication information (measured by the capacity C of the channel from X to Y) and the fidelity of the watermarked image (measured by the mean squared error distortion D between S and $S + X$)? By Theorem 7.4, the answer is

$$C(D) = \mathsf{C}(D),$$

where D corresponds to the average power of the watermark.

Proof of Theorem 7.4. The proof of the converse follows immediately from (7.6). To prove achievability, first consider the extension of the Gelfand–Pinsker theorem to the DMC with DM state and nonnegative cost function $b(x)$, $x \in \mathcal{X}$. We assume an input cost constraint

$$\sum_{i=1}^{n} \mathsf{E}[b(x_i(m, S^n))] \leq nB \quad \text{for every } m \in [1:2^{nR}].$$

In this case, the capacity is

$$C_{\text{SI-E}} = \max_{p(u|s),\, x(u,s):\mathsf{E}(b(X))\leq B} (I(U;Y) - I(U;S)). \tag{7.7}$$

Now we return to the Gaussian channel with additive Gaussian state. Since the channel is Gaussian and linear, it is natural to expect the optimal distribution on (U, X) given S that attains the capacity to be also Gaussian. This turns out to be the case, but with the nonintuitive choice of $X \sim \mathsf{N}(0, P)$ as independent of S and $U = X + \alpha S$. Evaluating the mutual information terms in (7.7) with this choice, we obtain

$$
\begin{aligned}
I(U;Y) &= h(X + S + Z) - h(X + S + Z \mid X + \alpha S) \\
&= h(X + S + Z) + h(X + \alpha S) - h(X + S + Z, X + \alpha S) \\
&= \frac{1}{2} \log \left(\frac{(P + Q + 1)(P + \alpha^2 Q)}{PQ(1 - \alpha)^2 + (P + \alpha^2 Q)} \right)
\end{aligned}
$$

and

$$I(U;S) = \frac{1}{2} \log \left(\frac{P + \alpha^2 Q}{P} \right).$$

Thus

$$
\begin{aligned}
R(\alpha) &= I(U;Y) - I(U;S) \\
&= \frac{1}{2} \log \left(\frac{P(P + Q + 1)}{PQ(1 - \alpha)^2 + (P + \alpha^2 Q)} \right).
\end{aligned}
$$

Maximizing over α gives $\max_{\alpha} R(\alpha) = \mathsf{C}(P)$, which is attained by $\alpha^* = P/(P + 1)$. Finally, applying the discretization procedure in Section 3.4 completes the proof of the theorem.

Alternative proof of Theorem 7.4. In the proof of the writing on dirty paper result, we found that the optimal coefficient for S is $\alpha^* = P/(P + 1)$, which is the weight of the linear minimum mean squared error (MMSE) estimate of X given $X + Z$. The following alternative proof uses the connection to MMSE estimation to generalize the result to non-Gaussian state S.

As before, let $U = X + \alpha S$, where $X \sim N(0, P)$ is independent of S and $\alpha = P/(P + 1)$. From the Gelfand–Pinsker theorem, we can achieve

$$I(U; Y) - I(U; S) = h(U|S) - h(U|Y).$$

We wish to show that this is equal to

$$I(X; X + Z) = h(X) - h(X|X + Z)$$
$$= h(X) - h(X - \alpha(X + Z)|X + Z)$$
$$= h(X) - h(X - \alpha(X + Z)),$$

where the last step follows by the fact that $(X - \alpha(X + Z))$, which is the error of the MMSE estimate of X given $(X + Z)$, and $(X + Z)$ are orthogonal and thus independent because X and $(X + Z)$ are jointly Gaussian.

First consider

$$h(U|S) = h(X + \alpha S|S) = h(X|S) = h(X).$$

Since $X - \alpha(X + Z)$ is independent of $(X + Z, S)$ and thus of $Y = X + S + Z$,

$$h(U|Y) = h(X + \alpha S|Y)$$
$$= h(X + \alpha S - \alpha Y|Y)$$
$$= h(X - \alpha(X + Z)|Y)$$
$$= h(X - \alpha(X + Z)).$$

This completes the proof.

Remark 7.9. This derivation does not require S to be Gaussian. Hence, $C_{\text{SI-E}} = C(P)$ for *any* independent state S with finite power!

7.7.1 Writing on Dirty Paper for the Gaussian MAC

The writing on dirty paper result can be extended to several multiuser Gaussian channels. Consider the Gaussian MAC with additive Gaussian state $Y = X_1 + X_2 + S + Z$, where the state $S \sim N(0, Q)$ and the noise $Z \sim N(0, 1)$ are independent. Assume average power constraints P_1 on X_1 and P_2 on X_2. Further assume that the state sequence S^n is available noncausally at both encoders. The capacity region of this channel is the set of rate pairs (R_1, R_2) such that

$$R_1 \le C(P_1),$$
$$R_2 \le C(P_2),$$
$$R_1 + R_2 \le C(P_1 + P_2)$$

This is the capacity region when the state sequence is available also at the decoder (so the interference S^n can be canceled out). Thus the proof of the converse is trivial.

For the proof of achievability, consider the "writing on dirty paper" channel $Y = X_1 + S + (X_2 + Z)$ with input X_1, known state S, and unknown noise $(X_2 + Z)$ that will be shortly shown to be independent of S. We first prove achievability of the corner point $(R_1, R_2) = (C(P_1/(P_2 + 1)), C(P_2))$ of the capacity region. By taking $U_1 = X_1 + \alpha_1 S$, where $X_1 \sim N(0, P_1)$ independent of S and $\alpha_1 = P_1/(P_1 + P_2 + 1)$, we can achieve $R_1 = I(U_1; Y) - I(U_1; S) = C(P_1/(P_2 + 1))$. Now as in successive cancellation decoding for the Gaussian MAC, once u_1^n is decoded correctly, the decoder can subtract it from y^n to obtain the effective channel $\tilde{y}^n = y^n - u_1^n = x_2^n + (1 - \alpha_1)s^n + z^n$. Thus, for sender 2, the channel is another "writing on dirty paper" channel with input X_2, known state $(1 - \alpha_1)S$, and unknown noise Z. By taking $U_2 = X_2 + \alpha_2 S$, where $X_2 \sim N(0, P_2)$ is independent of S and $\alpha_2 = (1 - \alpha_1)P_2/(P_2 + 1) = P_2/(P_1 + P_2 + 1)$, we can achieve $R_2 = I(U_2; \tilde{Y}) - I(U_2; S) = C(P_2)$. The other corner point can be achieved by reversing the role of two encoders. The rest of the capacity region can then be achieved using time sharing.

Achievability can be proved alternatively by considering the DM-MAC with DM state $p(y|x_1, x_2, s)p(s)$ when the state information is available noncausally at both encoders and evaluating the inner bound on the capacity region consisting of all rate pairs (R_1, R_2) such that

$$R_1 < I(U_1; Y|U_2) - I(U_1; S|U_2),$$
$$R_2 < I(U_2; Y|U_1) - I(U_2; S|U_1),$$
$$R_1 + R_2 < I(U_1, U_2; Y) - I(U_1, U_2; S)$$

for some pmf $p(u_1|s)p(u_2|s)$ and functions $x_1(u_1, s)$ and $x_2(u_2, s)$. By choosing (U_1, U_2, X_1, X_2) as before, it can be easily checked that this inner bound simplifies to the capacity region. It is not known, however, if this inner bound is tight for the general DM-MAC with DM state.

7.7.2 Writing on Dirty Paper for the Gaussian BC

Now consider the Gaussian BC with additive Gaussian state $Y_j = X + S_j + Z_j$, $j = 1, 2$, where the state components $S_1 \sim N(0, Q_1)$ and $S_2 \sim N(0, Q_2)$ and the noise components $Z_1 \sim N(0, N_1)$ are $Z_2 \sim N(0, N_2)$ are independent, and $N_2 \geq N_1$. Assume average power constraint P on X. We assume that the state sequences S_1^n and S_2^n are available noncausally at the encoder. The capacity region of this channel is the set of rate pairs (R_1, R_2) such that

$$R_1 \leq C\left(\frac{\alpha P}{N_1}\right),$$
$$R_2 \leq C\left(\frac{\bar{\alpha} P}{\alpha P + N_2}\right)$$

for some $\alpha \in [0, 1]$. This is the same as the capacity region when the state sequences are available also at their respective decoders, which establishes the converse.

The proof of achievability follows closely that of the MAC case. We split the input into two independent parts $X_1 \sim N(0, \alpha P)$ and $X_2 \sim N(0, \bar{\alpha} P)$ such that $X = X_1 + X_2$. For the weaker receiver, consider the "writing on dirty paper" channel $Y_2 = X_2 + S_2 + (X_1 + Z_2)$ with input X_2, known state S_2, and noise $(X_1 + Z_2)$. Then by taking $U_2 = X_2 + \beta_2 S_2$, where $X_2 \sim N(0, \bar{\alpha} P)$ independent of S_2 and $\beta_2 = \bar{\alpha} P / (P + N_2)$, we can achieve any rate $R_2 < I(U_2; Y_2) - I(U_2; S_2) = C(\bar{\alpha} P / (\alpha P + N_2))$. For the stronger receiver, consider another "writing on dirty paper" channel $Y_1 = X_1 + (X_2 + S_1) + Z_1$ with input X_1, known state $(X_2 + S_1)$, and noise Z_{1i}. Using the writing on dirty paper result with $U_1 = X_1 + \beta_1 (X_2 + S_1)$ and $\beta_1 = \alpha P / (\alpha P + N_1)$, we can achieve any rate $R_1 < C(\alpha P / N_1)$.

As in the MAC case, achievability can be proved alternatively by considering the DM-BC with state $p(y_1, y_2 | x, s) p(s)$ when the state information is available noncausally at the encoder and evaluating the inner bound on the capacity region consisting of all rate pairs (R_1, R_2) such that

$$R_1 < I(U_1; Y_1) - I(U_1; S),$$
$$R_2 < I(U_2; Y_2) - I(U_2; S),$$
$$R_1 + R_2 < I(U_1; Y_1) + I(U_2; Y_1) - I(U_1; U_2) - I(U_1, U_2; S)$$

for some pmf $p(u_1, u_2 | s)$ and function $x(u_1, u_2, s)$. Taking $S = (S_1, S_2)$ and choosing (U_1, U_2, X) as before, this inner bound simplifies to the capacity region. Again it is not known if this inner bound is tight for the general DM-BC with DM state.

7.8   CODED STATE INFORMATION

So far we have considered settings where the state sequence is available perfectly (causally or noncausally) at the encoder and/or the decoder. Providing such perfect information about the state sequence is often not feasible, however, because the channel over which such information is to be sent has limited capacity. This motivates the coded side information setup in which a rate-limited description of the state is provided to the encoder and/or to the decoder. What is the optimal tradeoff between the communication rate and the rate at which the state is described?

The answer to this question is known only in a few special cases, for example, when the state sequence is available at the decoder, but only a coded version of it is available at the encoder as depicted in Figure 7.13, Formally, consider a DMC with DM state and define a $(2^{nR}, 2^{nR_s}, n)$ code to consist of a state encoder $m_s(s^n) \in [1 : 2^{nR_s}]$, a message encoder $x^n(m, m_s)$, and a decoder $\hat{m}(y^n, s^n)$. The probability of error is defined as $P_e^{(n)} = P\{\hat{M} \neq M\}$. A rate pair (R, R_s) is said to be achievable if there exists a sequence of $(2^{nR}, 2^{nR_s}, n)$ codes such that $\lim_{n \to \infty} P_e^{(n)} = 0$. We define the capacity–rate function $C(R_s)$ as the supremum over rates R such that (R, R_s) is achievable.

Theorem 7.5. The capacity–rate function for the DMC with DM state $p(y|x, s) p(s)$ when the state information is available at the decoder and coded state information of rate R_s is available at the encoder is

$$C(R_s) = \max_{p(u|s) p(x|u): R_s \geq I(U;S)} I(X; Y | S, U).$$

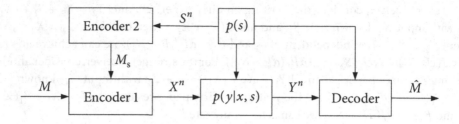

Figure 7.13. Coded state information available at the encoder.

Achievability follows by using joint typicality encoding to describe the state sequence S^n by U^n at rate R_s and using time sharing on the "compressed state" sequence U^n as for the case where the state sequence is available at both the encoder and the decoder; see Section 7.4.1.

To prove the converse, we identify the auxiliary random variable as $U_i = (M_s, S^{i-1})$ and consider

$$nR_s \geq H(M_s)$$
$$= I(M_s; S^n)$$
$$= \sum_{i=1}^{n} I(M_s; S_i | S^{i-1})$$
$$= \sum_{i=1}^{n} I(M_s, S^{i-1}; S_i)$$
$$= \sum_{i=1}^{n} I(U_i; S_i).$$

In addition, by Fano's inequality,

$$nR \leq I(M; Y^n, S^n) + n\epsilon_n$$
$$= I(M; Y^n | S^n) + n\epsilon_n$$
$$= \sum_{i=1}^{n} I(M; Y_i | Y^{i-1}, S^n) + n\epsilon_n$$
$$= \sum_{i=1}^{n} I(M; Y_i | Y^{i-1}, S^n, M_s) + n\epsilon_n$$
$$\leq \sum_{i=1}^{n} I(X_i, M, Y^{i-1}, S_{i+1}^n; Y_i | S_i, S^{i-1}, M_s) + n\epsilon_n$$
$$\leq \sum_{i=1}^{n} I(X_i; Y_i | S_i, U_i) + n\epsilon_n.$$

Since $S_i \to U_i \to X_i$ and $U_i \to (X_i, S_i) \to Y_i$ for $i \in [1:n]$, we can complete the proof by introducing the usual time-sharing random variable.

SUMMARY

- State-dependent channel models:

  - Compound channel

  - Arbitrarily varying channel

  - DMC with DM state

- Channel coding with side information

- Shannon strategy

- Gelfand–Pinsker coding:

  - Multicoding (subcodebook generation)

  - Use of joint typicality encoding in channel coding

- Writing on dirty paper

- **Open problems:**

  7.1. What is the capacity region of the DM-MAC with DM state when the state information is available causally or noncausally at the encoders?

  7.2. What is the common-message capacity of the DM-BC with DM state when the state information is available noncausally at the encoder?

BIBLIOGRAPHIC NOTES

The compound channel was first studied by Blackwell, Breiman, and Thomasian (1959), Dobrushin (1959b), and Wolfowitz (1960), who established Theorem 7.1 for arbitrary state spaces using maximum likelihood decoding or joint typicality decoding. An alternative proof was given by Csiszár and Körner (1981b) using maximum mutual information decoding. Generalizations of the compound channel to the multiple access channel can be found in Ahlswede (1974), Han (1998), and Lapidoth and Narayan (1998). The arbitrarily varying channel was first studied by Blackwell, Breiman, and Thomasian (1960) who established the capacity characterization in (7.1). Coding theorems for other AVC settings can be found in Ahlswede and Wolfowitz (1969, 1970), Ahlswede (1978), Wolfowitz (1978), Csiszár and Narayan (1988), Csiszár and Körner (1981b), and Csiszár and Körner (1981a).

Shannon (1958) introduced the DMC with DM state and established Theorem 7.2 using the idea of coding over the function space (Shannon strategy). Goldsmith and Varaiya (1997) used time sharing on the state sequence to establish the capacity when the state information is available at both the encoder and the decoder. The capacity region in (7.4) of the DM-MAC with two state components when the state information is available in a distributed manner is due to Jafar (2006).

The memory with stuck-at faults was studied by Kuznetsov and Tsybakov (1974), who established the capacity using multicoding. This result was generalized to DMCs with DM state by Gelfand and Pinsker (1980b) and Heegard and El Gamal (1983). Our proofs follow Heegard and El Gamal (1983). The capacity results on the DM-BC with DM state are due to Steinberg (2005).

The writing on dirty paper result is due to Costa (1983). The alternative proof follows Cohen and Lapidoth (2002). A similar result was also obtained for the case of nonstationary, nonergodic Gaussian noise and state (Yu, Sutivong, Julian, Cover, and Chiang 2001). Erez, Shamai, and Zamir (2005) studied lattice coding strategies for writing on dirty paper. Writing on dirty paper for the Gaussian MAC and the Gaussian BC was studied by Gelfand and Pinsker (1984). Applications to information hiding and digital watermarking can be found in Moulin and O'Sullivan (2003).

The DMC with coded state information was first studied by Heegard and El Gamal (1983), who considered a more general setup in which encoded versions of the state information are separately available at the encoder and the decoder. Rosenzweig, Steinberg, and Shamai (2005) studied a similar setup with coded state information as well as a quantized version of the state sequence available at the encoder. Extensions to the DM-MAC with coded state information were studied in Cemal and Steinberg (2005). Surveys of the literature on channels with state can be found in Lapidoth and Narayan (1998) and Keshet, Steinberg, and Merhav (2008).

PROBLEMS

7.1.   Consider the DMC with DM state $p(y|x, s)p(s)$ and let $b(x) \geq 0$ be an input cost function with $b(x_0) = 0$ for some $x_0 \in \mathcal{X}$. Assume that the state information is noncausally available at the encoder. Establish the capacity–cost function in (7.7).

7.2.   Establish the capacity region of the DM-MAC with 2-DM state components $p(y|x_1, x_2, s)p(s)$, where $S = (S_1, S_2)$, (a) when the state information is available only at the decoder and (b) when the state information is available at the decoder and each state component S_j, $j = 1, 2$, is available at encoder j.

7.3.   Establish the capacity region of the degraded DM-BC with DM state $p(y_1|x, s)$ $p(y_2|y_1)p(s)$ (a) when the state information is causally available only at the encoder, (b) when the state information is causally available at the encoder and decoder 1, and (c) when the state information is *noncausally* available at the encoder and decoder 1.

7.4.   *Compound channel with state information available at the encoder.* Suppose that the encoder knows the state s before communication commences over a compound channel $p(y_s|x)$. Find the capacity.

7.5.   *Compound channel with arbitrary state space.* Consider the compound channel $p(y_s|x)$ with state space $\mathcal{S}$ of infinite cardinality.

(a) Show that there is a finite subset $\tilde{S}_n \subset S$ with cardinality at most $(n + 1)^{2|\mathcal{X}| \cdot |\mathcal{Y}|}$ such that for every $s \in S$, there exists $\tilde{s} \in \tilde{S}_n$ with

$$T_{\epsilon}^{(n)}(X, Y_s) = T_{\epsilon}^{(n)}(X, Y_{\tilde{s}}).$$

(Hint: Recall that a typical set is defined by the upper and lower bounds on the number of occurrences of each pair $(x, y) \in \mathcal{X} \times \mathcal{Y}$ in an n sequence.)

(b) Use part (a) to prove achievability for Theorem 7.1 with infinite state space.

7.6.   *No state information.* Show that the capacity of the DMC with DM state $p(y|x, s)$ $p(s)$ when no state information is available at either the encoder or decoder is

$$C = \max_{p(x)} I(X; Y),$$

where $p(y|x) = \sum_s p(s)p(y|x, s)$. Further show that any $(2^{nR}, n)$ code for the DMC $p(y|x)$ achieves the same average probability of error when used over the DMC with DM state $p(y|x, s)p(s)$, and vice versa.

7.7.   *Stationary ergodic state process.* Assume that $\{S_i\}$ is a discrete stationary ergodic process.

(a) Show that the capacity of the DMC with state $p(y|x, s)$ when the state information is available only at the decoder is

$$C_{\text{SI-D}} = \max_{p(x)} I(X; Y|S).$$

(Hint: If $X^n \sim \prod_{i=1}^{n} p_X(x_i)$, then (X^n, S^n, Y^n) is jointly stationary ergodic.)

(b) Show that the capacity of the DMC with state $p(y|x, s)$ when the state information is available at both the encoder and the decoder is

$$C_{\text{SI-ED}} = \max_{p(x|s)} I(X; Y|S).$$

7.8.   *Strictly causal state information.* Consider the DMC with DM state $p(y|x, s)p(s)$. Suppose that the state information is available strictly causally at the encoder, that is, the encoder is specified by $x_i(m, s^{i-1})$, $i \in [1 : n]$. Establish the capacity (a) when the state information is not available at the decoder and (b) when the state information is also available at the decoder.

7.9.   *DM-MAC with strictly causal state information.* Consider the DM-MAC with DM state $Y = (X_1 \oplus S, X_2)$, where the inputs X_1 and X_2 are binary and the state $S \sim$ Bern$(1/2)$. Establish the capacity region (a) when the state information is not available at either the encoders or the decoder, (b) when the state information is available strictly causally only at encoder 2, and (c) when the state information is available strictly causally at both encoders.

Remark: The above two problems demonstrate that while strictly causal state information does not increase the capacity of point-to-point channels, it can increase the capacity of some multiuser channels. We will see similar examples for channels with feedback in Chapter 17.

7.10. *Value of state information.* Consider the DMC with DM state $p(y|x, s)p(s)$. Quantify how much state information can help by proving the following statements:

(a) $C_{\text{SI-D}} - C \le \max_{p(x)} H(S|Y)$.

(b) $C_{\text{SI-ED}} - C_{\text{SI-E}} \le C_{\text{SI-ED}} - C_{\text{CSI-E}} \le \max_{p(x|s)} H(S|Y)$.

Thus, the state information at the decoder is worth at most $H(S)$ bits. Show that the state information at the encoder can be much more valuable by providing an example for which $C_{\text{SI-E}} - C > H(S)$.

7.11. *Ternary-input memory with state.* Consider the DMC with DM state $p(y|x, s)p(s)$ depicted in Figure 7.14.

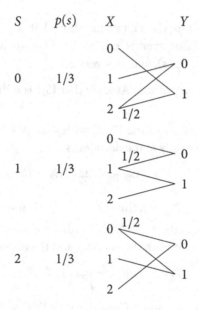

Figure 7.14. Ternary-input memory with state.

(a) Show that the capacity when the state information is not available at either the encoder or the decoder is $C = 0$.

(b) Show that the capacity when the state information is available only at the decoder is $C_{\text{SI-D}} = 2/3$.

(c) Show that the capacity when the state information is (causally or noncausally) available at the encoder (regardless of whether it is available at the decoder) is $C_{\text{CSI-E}} = C_{\text{SI-E}} = C_{\text{SI-ED}} = 1$.

7.12. *BSC with state.* Consider the DMC with DM state $p(y|x, s)p(s)$, where $S \sim \text{Bern}(q)$ and $p(y|x, s)$ is a BSC(p_s), $s = 0, 1$.

(a) Show that the capacity when the state information is not available at either the encoder or the decoder is

$$C = 1 - H(p_0\bar{q} + p_1 q).$$

(b) Show that the capacity when the state information is available at the decoder (regardless of whether it is available at the encoder) is

$$C_{\text{SI-D}} = C_{\text{SI-ED}} = 1 - \bar{q}H(p_0) - qH(p_1).$$

(c) Show that the capacity when the state information is (causally or noncausally) available only at the encoder is

$$C_{\text{CSI-E}} = C_{\text{SI-E}} = \begin{cases} 1 - H(p_0\bar{q} + p_1 q) & \text{if } p_0, p_1 \in [0, 1/2] \text{ or } p_0, p_1 \in [1/2, 1], \\ 1 - H(\tilde{p}_0\bar{q} + p_1 q) & \text{otherwise.} \end{cases}$$

(Hint: Use a symmetrization argument similar to that in Section 5.4.2 with $x(u, s) = 1 - x(-u, s)$ and show that it suffices to consider $U \sim \text{Bern}(1/2)$ independent of S.)

Remark: The above two problems appeared as examples in Heegard and El Gamal (1983). These problems demonstrate that there is no universal ordering between $C_{\text{SI-E}}$ and $C_{\text{SI-D}}$, each of which is strictly less than $C_{\text{SI-ED}}$ in general.

7.13.   *Common-message broadcasting with state information.* Consider the DM-BC with DM state $p(y_1, y_2 | x, s)p(s)$. Establish the common-message capacity C_0 (that is, the maximum achievable common rate R_0 when $R_1 = R_2 = 0$) for the following settings:

(a) The state information is available only at decoder 1.

(b) The state information is *causally* available at both the encoder and decoder 1.

(c) The state information is *noncausally* available at the encoder and decoder 1.

(d) The state information is *causally* available only at the encoder.

7.14.   *Memory with stuck-at faults and noise.* Recall the memory with stuck-at faults in Example 7.3. Assume that the memory now has temporal noise in addition to stuck-at faults such that for state $s = 2$, the memory is modeled by the BSC(p) for $p \in [0, 1/2]$. Find the capacity of this channel (a) when the state information is *causally* available only at the encoder and (b) when the state information is *noncausally* available only at the encoder.

7.15.   *MMSE estimation via writing on dirty paper.* Consider the additive noise channel with output (observation)

$$Y = X + S + Z,$$

where X is the transmitted signal and has mean μ and variance P, S is the state and has zero mean and variance Q, and Z is the noise and has zero mean and variance N. Assume that X, S, and Z are uncorrelated. The sender knows S and wishes to transmit a signal U, but instead transmits X such that $U = X + \alpha S$ for some constant α.

(a) Find the mean squared error (MSE) of the linear MMSE estimate of U given Y in terms only of μ, α, P, Q, and N.

(b) Find the value of α that minimizes the MSE in part (a).

(c) How does the minimum MSE obtained in part (b) compare to the MSE of the linear MMSE when there is no state at all, i.e., $S = 0$? Interpret the result.

7.16. *Cognitive radio.* Consider the Gaussian IC

$$Y_1 = g_{11}X_1 + g_{12}X_2 + Z_1,$$
$$Y_2 = g_{22}X_2 + Z_2,$$

where g_{11}, g_{12}, and g_{22} are channel gains, and $Z_1 \sim N(0, 1)$ and $Z_2 \sim N(0, 1)$ are noise components. Sender 1 encodes two independent messages M_1 and M_2, while sender 2 encodes only M_2. Receiver 1 wishes to recover M_1 and receiver 2 wishes to recover M_2. Assume average power constraint P on each of X_1 and X_2. Find the capacity region.

Remark: This is a simple example of cognitive radio channel models studied, for example, in Devroye, Mitran, and Tarokh (2006), Wu, Vishwanath, and Arapostathis (2007), and Jovičić and Viswanath (2009).

7.17. *Noisy state information.* Consider the DMC with DM state $p(y|x, s)p(s)$, where the state has three components $S = (T_0, T_1, T_2) \sim p(t_0, t_1, t_2)$. Suppose that T_1 is available at the encoder, T_2 is available at the decoder, and T_0 is hidden from both.

(a) Suppose that T_1 is a function of T_2. Show that the capacity (for both causal and noncausal cases) is

$$C_{\text{NSI}} = \max_{p(x|t_1)} I(X; Y|T_2).$$

(b) Show that any $(2^{nR}, n)$ code (in either the causal or noncausal case) for this channel achieves the same probability of error when used over the DMC with state $p(y'|x, t_1)p(t_1)$, where $Y' = (Y, T_2)$ and

$$p(y, t_2|x, t_1) = \sum_{t_0} p(y|x, t_0, t_1, t_2)p(t_0, t_2|t_1),$$

and vice versa.

Remark: This result is due to Caire and Shamai (1999).

CHAPTER 8

General Broadcast Channels

We resume the discussion of broadcast channels started in Chapter 5. Again consider the 2-receiver DM-BC $p(y_1, y_2|x)$ with private and common messages depicted in Figure 8.1. The definitions of a code, achievability, and capacity regions are the same as in Chapter 5. As mentioned before, the capacity region of the DM-BC is not known in general. In Chapter 5, we presented the superposition coding scheme and showed that it is optimal for several classes of channels in which one receiver is stronger than the other. In this chapter, we study coding schemes that can outperform superposition coding and present the tightest known inner and outer bounds on the capacity region of the general broadcast channel.

We first show that superposition coding is optimal for the 2-receiver DM-BC with degraded message sets, that is, when either $R_1 = 0$ or $R_2 = 0$. We then show that super-position coding is not optimal for BCs with more than two receivers. In particular, we establish the capacity region of the 3-receiver multilevel BC. The achievability proof in-volves the new idea of indirect decoding, whereby a receiver who wishes to recover only the common message still uses satellite codewords in decoding for the cloud center.

We then present Marton's inner bound on the private-message capacity region of the 2-receiver DM-BC and show that it is optimal for the class of semideterministic BCs. The coding scheme involves the multicoding technique introduced in Chapter 7 and the new idea of joint typicality codebook generation to construct dependent codewords for inde-pendent messages without the use of a superposition structure. The proof of the inner bound uses the mutual covering lemma, which is a generalization of the covering lemma in Section 3.7. Marton's coding scheme is then combined with superposition coding to establish an inner bound on the capacity region of the DM-BC that is tight for all classes of DM-BCs with known capacity regions. Next, we establish the Nair–El Gamal outer bound on the capacity region of the DM-BC. We show through an example that there is

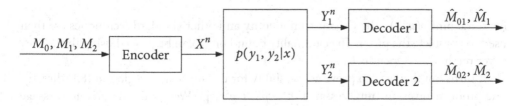

Figure 8.1. Two-receiver broadcast communication system.

a gap between these inner and outer bounds. Finally, we discuss extensions of the afore-mentioned coding techniques to broadcast channels with more than two receivers and with arbitrary messaging requirements.

8.1   DM-BC WITH DEGRADED MESSAGE SETS

A broadcast channel with *degraded message sets* is a model for a *layered* communication setup in which a sender wishes to communicate a common message to *all* receivers, a first private message to a first subset of the receivers, a second private message to a second subset of the first subset, and so on. Such a setup arises, for example, in video or music broadcasting over a wireless network at varying levels of quality. The common message represents the lowest quality description of the data to be sent to all receivers, and each private message represents a refinement over the common message and previous private messages.

We first consider the 2-receiver DM-BC with degraded message sets, i.e., with $R_2 = 0$. The capacity region for this case is known.

Theorem 8.1. The capacity region of the 2-receiver DM-BC $p(y_1, y_2|x)$ with degraded message sets is the set of rate pairs (R_0, R_1) such that

$$R_0 \le I(U; Y_2),$$
$$R_1 \le I(X; Y_1|U),$$
$$R_0 + R_1 \le I(X; Y_1)$$

for some pmf $p(u, x)$ with $|\mathcal{U}| \le \min\{|\mathcal{X}|, |\mathcal{Y}_1| + |\mathcal{Y}_2|\} + 1$.

This capacity region is achieved using superposition coding (see Section 5.3) by noting that receiver 1 can recover both messages M_0 and M_1. The converse proof follows similar lines to that for the more capable BC in Section 5.6. It is proved by first considering the alternative characterization of the capacity region consisting of all rate pairs (R_0, R_1) such that

$$R_0 \le I(U; Y_2),$$
$$R_0 + R_1 \le I(X; Y_1), \tag{8.1}$$
$$R_0 + R_1 \le I(X; Y_1|U) + I(U; Y_2)$$

for some pmf $p(u, x)$. The Csiszár sum identity and other standard techniques are then used to complete the proof. The cardinality bound on $\mathcal{U}$ can be proved using the convex cover method in Appendix C.

It turns out that we can prove achievability for the alternative region in (8.1) directly. The proof involves an (unnecessary) *rate splitting* step. We divide the private message M_1 into two independent messages M_{10} at rate R_{10} and M_{11} at rate R_{11}. We use super-position coding whereby the cloud center U represents the message pair (M_0, M_{10}) and the satellite codeword X represents the message triple (M_0, M_{10}, M_{11}). Following similar steps to the achievability proof of the superposition coding inner bound, we can show that (R_0, R_{10}, R_{11}) is achievable if

$$R_0 + R_{10} < I(U; Y_2),$$
$$R_{11} < I(X; Y_1|U),$$
$$R_0 + R_1 < I(X; Y_1)$$

for some pmf $p(u, x)$. Substituting $R_{11} = R_1 - R_{10}$, we have the conditions

$$R_{10} \geq 0,$$
$$R_{10} < I(U; Y_2) - R_0,$$
$$R_{10} > R_1 - I(X; Y_1|U),$$
$$R_0 + R_1 < I(X; Y_1),$$
$$R_{10} \leq R_1.$$

Eliminating R_{10} by the Fourier–Motzkin procedure in Appendix D yields the desired characterization.

Rate splitting turns out to be crucial when the DM-BC has more than 2 receivers.

8.2   THREE-RECEIVER MULTILEVEL DM-BC

The degraded message set capacity region of the DM-BC with more than two receivers is not known in general. We show that the straightforward extension of the superposition coding inner bound in Section 5.3 to more than two receivers is not optimal in general.

Consider the 3-receiver *multilevel* DM-BC $(\mathcal{X}, p(y_1, y_3|x)p(y_2|y_1), \mathcal{Y}_1 \times \mathcal{Y}_2 \times \mathcal{Y}_3)$ depicted in Figure 8.2, which is a 3-receiver DM-BC where Y_2 is a degraded version of Y_1. We consider the case of two degraded message sets, where a common message $M_0 \in [1 : 2^{nR_0}]$ is to be communicated to all receivers, and a private message $M_1 \in [1 : 2^{nR_1}]$ is to be communicated only to receiver 1. The definitions of a code, probability of error, achievability, and capacity region for this setting are as before.

A straightforward extension of the superposition coding inner bound to the 3-receiver multilevel DM-BC, where receivers 2 and 3 decode for the cloud center and receiver 1

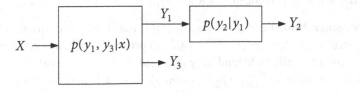

Figure 8.2. Three-receiver multilevel broadcast channel.

decodes for the satellite codeword, gives the set of rate pairs (R_0, R_1) such that

$$R_0 < \min\{I(U; Y_2), I(U; Y_3)\},$$
$$R_1 < I(X; Y_1 | U) \tag{8.2}$$

for some pmf $p(u, x)$. Note that the last inequality $R_0 + R_1 < I(X; Y_1)$ in the superposition coding inner bound in Theorem 5.1 drops out by the assumption that Y_2 is a degraded version of Y_1.

This region turns out not to be optimal in general.

Theorem 8.2. The capacity region of the 3-receiver multilevel DM-BC $p(y_1, y_3 | x)$ $p(y_2 | y_1)$ is the set of rate pairs (R_0, R_1) such that

$$R_0 \leq \min\{I(U; Y_2), I(V; Y_3)\},$$
$$R_1 \leq I(X; Y_1 | U),$$
$$R_0 + R_1 \leq I(V; Y_3) + I(X; Y_1 | V)$$

for some pmf $p(u, v)p(x|v)$ with $|\mathcal{U}| \leq |\mathcal{X}| + 4$ and $|\mathcal{V}| \leq (|\mathcal{X}| + 1)(|\mathcal{X}| + 4)$.

The proof of the converse uses steps from the converse proofs for the degraded BC and the 2-receiver BC with degraded message sets. The cardinality bounds on $\mathcal{U}$ and $\mathcal{V}$ can be proved using the extension of the convex cover method to multiple random variables in Appendix C.

8.2.1   Proof of Achievability

The achievability proof of Theorem 8.2 uses the new idea of *indirect decoding*. We divide the private message M_1 into two messages M_{10} and M_{11}. We generate a codebook of u^n sequences for the common message M_0, and use superposition coding to generate a codebook of v^n sequences for the message pair (M_0, M_{10}). We then use superposition coding again to generate a codebook of x^n sequences for the message triple $(M_0, M_{10}, M_{11}) = (M_0, M_1)$. Receiver 1 finds (M_0, M_1) by decoding for x^n, receiver 2 finds M_0 by decoding for u^n, and receiver 3 finds M_0 indirectly by simultaneous decoding for (u^n, v^n). We now provide details of the proof.

Rate splitting. Divide the private message M_1 into two independent messages M_{10} at rate R_{10} and M_{11} at rate R_{11}. Hence $R_1 = R_{10} + R_{11}$.

Codebook generation. Fix a pmf $p(u, v)p(x|v)$. Randomly and independently generate 2^{nR_0} sequences $u^n(m_0)$, $m_0 \in [1 : 2^{nR_0}]$, each according to $\prod_{i=1}^n p_U(u_i)$. For each m_0, randomly and conditionally independently generate $2^{nR_{10}}$ sequences $v^n(m_0, m_{10})$, $m_{10} \in [1 : 2^{nR_{10}}]$, each according to $\prod_{i=1}^n p_{V|U}(v_i | u_i(m_0))$. For each pair (m_0, m_{10}), randomly and conditionally independently generate $2^{nR_{11}}$ sequences $x^n(m_0, m_{10}, m_{11})$, $m_{11} \in [1 : 2^{nR_{11}}]$, each according to $\prod_{i=1}^n p_{X|V}(x_i | v_i(m_0, m_{10}))$.

Encoding. To send the message pair $(m_0, m_1) = (m_0, m_{10}, m_{11})$, the encoder transmits $x^n(m_0, m_{10}, m_{11})$.

Decoding and analysis of the probability of error for decoders 1 and 2. Decoder 2 declares that $\hat{m}_{02} \in [1 : 2^{nR_0}]$ is sent if it is the unique message such that $(u^n(\hat{m}_{02}), y_2^n) \in T_\epsilon^{(n)}$. By the LLN and the packing lemma, the probability of error tends to zero as $n \to \infty$ if $R_0 < I(U; Y_2) - \delta(\epsilon)$. Decoder 1 declares that $(\hat{m}_{01}, \hat{m}_{10}, \hat{m}_{11})$ is sent if it is the unique message triple such that $(u^n(\hat{m}_{01}), v^n(\hat{m}_{01}, \hat{m}_{10}), x^n(\hat{m}_{01}, \hat{m}_{10}, \hat{m}_{11}), y_1^n) \in T_\epsilon^{(n)}$. By the LLN and the packing lemma, the probability of error tends to zero as $n \to \infty$ if

$$R_{11} < I(X; Y_1|V) - \delta(\epsilon),$$
$$R_{10} + R_{11} < I(X; Y_1|U) - \delta(\epsilon),$$
$$R_0 + R_{10} + R_{11} < I(X; Y_1) - \delta(\epsilon).$$

Decoding and analysis of the probability of error for decoder 3. If receiver 3 decodes for m_0 directly by finding the unique message $\hat{m}_{03}$ such that $(u^n(\hat{m}_{03}), y_3^n) \in T_\epsilon^{(n)}$, we obtain the condition $R_0 < I(U; Y_3) - \delta(\epsilon)$, which together with the previous conditions gives the extended superposition coding inner bound in (8.2).

To achieve the capacity region, receiver 3 decodes for m_0 *indirectly*. It declares that $\hat{m}_{03}$ is sent if it is the unique message such that $(u^n(\hat{m}_{03}), v^n(\hat{m}_{03}, m_{10}), y_3^n) \in T_\epsilon^{(n)}$ for some $m_{10} \in [1 : 2^{nR_{10}}]$. Assume that $(M_0, M_{10}) = (1, 1)$ is sent and consider all possible pmfs for the triple $(U^n(m_0), V^n(m_0, m_{10}), Y_3^n)$ as listed in Table 8.1.

m_0	m_{10}	Joint pmf
1	1	$p(u^n, v^n)p(y_3^n\|v^n)$
1	*	$p(u^n, v^n)p(y_3^n\|u^n)$
*	*	$p(u^n, v^n)p(y_3^n)$
*	1	$p(u^n, v^n)p(y_3^n)$

Table 8.1. The joint distribution induced by different (m_0, m_{10}) pairs.

The second case does not result in an error and the last two cases have the same pmf. Thus, the decoding error occurs iff one or both of the following events occur:

$$\mathcal{E}_{31} = \{(U^n(1), V^n(1, 1), Y_3^n) \notin T_\epsilon^{(n)}\},$$
$$\mathcal{E}_{32} = \{(U^n(m_0), V^n(m_0, m_{10}), Y_3^n) \in T_\epsilon^{(n)} \text{ for some } m_0 \neq 1, m_{10}\}.$$

Then, the probability of error for decoder 3 averaged over codebooks $P(\mathcal{E}_3) \le P(\mathcal{E}_{31}) + P(\mathcal{E}_{32})$. By the LLN, $P(\mathcal{E}_{31})$ tends to zero as $n \to \infty$. By the packing lemma (with $\mathcal{A} = [2 : 2^{n(R_0 + R_{10})}]$, $X \leftarrow (U, V)$, and $U = \emptyset$), $P(\mathcal{E}_{32})$ tends to zero as $n \to \infty$ if $R_0 + R_{10} < I(U, V; Y_3) - \delta(\epsilon) = I(V; Y_3) - \delta(\epsilon)$. Combining the bounds, substituting $R_{10} + R_{11} = R_1$, and eliminating R_{10} and R_{11} by the Fourier–Motzkin procedure completes the proof of achievability.

Indirect decoding is illustrated in Figures 8.3 and 8.4. Suppose that $R_0 > I(U; Y_3)$ as shown in Figure 8.3. Then receiver 3 cannot decode for the cloud center $u^n(1)$ directly. Now suppose that $R_0 + R_{10} < I(V; Y_3)$ in addition. Then receiver 3 can decode for the cloud center *indirectly* by finding the unique message $\hat{m}_0$ such that $(u^n(\hat{m}_0), v^n(\hat{m}_0, m_{10}), y_3^n) \in \mathcal{T}_\epsilon^{(n)}$ for some m_{10} as shown in Figure 8.4. Note that the condition $R_0 + R_{10} < I(V; Y_3)$ suffices in general (even when $R_0 \leq I(U; Y_3)$).

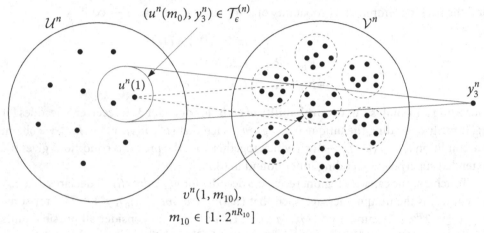

Figure 8.3. Direct decoding for the cloud center $u^n(1)$.

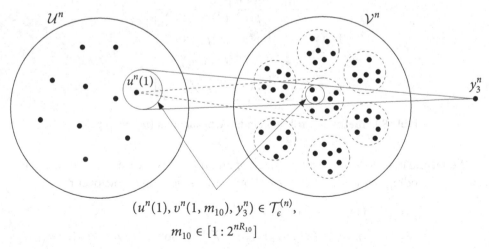

Figure 8.4. Indirect decoding for $u^n(1)$ via $v^n(1, m_{10})$.

Remark 8.1. Although it seems surprising that higher rates can be achieved by having receiver 3 recover more than it needs to, the reason we can do better than superposition coding can be explained by the observation that for a 2-receiver BC $p(y_2, y_3|x)$, the conditions $I(U; Y_2) < I(U; Y_3)$ and $I(V; Y_2) > I(V; Y_3)$ can hold simultaneously for some $U \to V \to X$; see discussions on less noisy and more capable BCs in Section 5.6. Now, considering our 3-receiver BC scenario, suppose we have a choice of U such that $I(U; Y_3) < I(U; Y_2)$. In this case, requiring receivers 2 and 3 to directly decode for u^n necessitates that $R_0 < I(U; Y_3)$. From the above observation, a V may exist such that $U \to V \to X$ and $I(V; Y_3) > I(V; Y_2)$, in which case the rate of the common message can be increased to $I(U; Y_2)$ and receiver 3 can still find u^n indirectly by decoding for (u^n, v^n). Thus, the additional "degree-of-freedom" introduced by V helps increase the rates in spite of the fact that receiver 3 may need to recover more than just the common message.

Remark 8.2. If we require receiver 3 to recover M_{10}, we obtain the region consisting of all rate pairs (R_0, R_1) such that

$$
\begin{aligned}
R_0 &< \min\{I(U; Y_2), I(V; Y_3)\}, \\
R_1 &< I(X; Y_1|U), \\
R_0 + R_1 &< I(V; Y_3) + I(X; Y_1|V), \\
R_1 &< I(V; Y_3|U) + I(X; Y_1|V)
\end{aligned}
\tag{8.3}
$$

for some pmf $p(u)p(v|u)p(x|v)$. While this region involves one more inequality than the capacity region in Theorem 8.2, it can be shown by optimizing the choice of V for each given U that it coincides with the capacity region. However, requiring decoder 3 to recover M_{10} is unnecessary and leads to a region with more inequalities for which the converse is difficult to establish.

8.2.2 Multilevel Product DM-BC

We show via an example that the extended superposition coding inner bound in (8.2) can be strictly smaller than the capacity region in Theorem 8.2 for the 3-receiver multilevel DM-BC. Consider the product of two 3-receiver BCs specified by the Markov chains

$$
\begin{aligned}
X_1 &\to Y_{31} \to Y_{11} \to Y_{21}, \\
X_2 &\to Y_{12} \to Y_{22}.
\end{aligned}
$$

For this channel, the extended superposition coding inner bound in (8.2) reduces to the set of rate pairs (R_0, R_1) such that

$$
\begin{aligned}
R_0 &\le I(U_1; Y_{21}) + I(U_2; Y_{22}), \\
R_0 &\le I(U_1; Y_{31}), \\
R_1 &\le I(X_1; Y_{11}|U_1) + I(X_2; Y_{12}|U_2)
\end{aligned}
\tag{8.4}
$$

for some pmf $p(u_1, x_1)p(u_2, x_2)$. Similarly, it can be shown that the capacity region in Theorem 8.2 reduces to the set of rate pairs (R_0, R_1) such that

$$
\begin{aligned}
R_0 &\le I(U_1; Y_{21}) + I(U_2; Y_{22}), \\
R_0 &\le I(V_1; Y_{31}), \\
R_1 &\le I(X_1; Y_{11}|U_1) + I(X_2; Y_{12}|U_2), \\
R_0 + R_1 &\le I(V_1; Y_{31}) + I(X_1; Y_{11}|V_1) + I(X_2; Y_{12}|U_2)
\end{aligned}
\tag{8.5}
$$

for some pmf $p(u_1, v_1)p(x_1|v_1)p(u_2, x_2)$.

In the following, we compare the extended superposition coding inner bound in (8.4) to the capacity region in (8.5).

Example 8.1. Consider the multilevel product DM-BC depicted in Figure 8.5, where X_1, X_2, Y_{12}, and Y_{31} are binary, and Y_{11}, Y_{21}, $Y_{22} \in \{0, 1, e\}$.

The extended superposition coding inner bound in (8.4) can be simplified to the set of rate pairs (R_0, R_1) such that

$$R_0 \le \min \left\{ \frac{\alpha}{6} + \frac{\beta}{2}, \alpha \right\},$$

$$R_1 \le \frac{\bar{\alpha}}{2} + \bar{\beta}$$

(8.6)

for some $\alpha, \beta \in [0, 1]$. It is straightforward to show that $(R_0, R_1) = (1/2, 5/12)$ lies on the boundary of this region. By contrast, the capacity region in (8.5) can be simplified to the set of rate pairs (R_0, R_1) such that

$$R_0 \le \min \left\{ \frac{r}{6} + \frac{s}{2}, t \right\},$$

$$R_1 \le \frac{1-r}{2} + 1 - s,$$

(8.7)

$$R_0 + R_1 \le t + \frac{1-t}{2} + 1 - s$$

for some $0 \le r \le t \le 1, 0 \le s \le 1$. Note that substituting $r = t$ in (8.7) yields the extended superposition coding inner bound in (8.6). By setting $r = 0$, $s = 1$, $t = 1$, it can be readily checked that $(R_0, R_1) = (1/2, 1/2)$ lies on the boundary of the capacity region. For $R_0 = 1/2$, however, the maximum achievable R_1 in the extended superposition coding inner bound in (8.6) is 5/12. Thus the capacity region is strictly larger than the extended superposition coding inner bound.

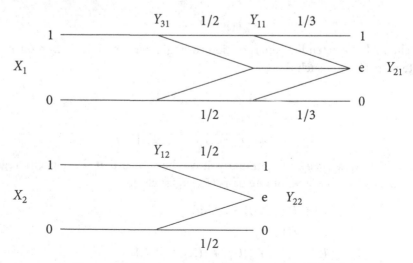

Figure 8.5. Binary erasure multilevel product DM-BC.

8.3 MARTON'S INNER BOUND

We now turn our attention to the 2-receiver DM-BC with only private messages, i.e., when $R_0 = 0$. First we show that a rate pair (R_1, R_2) is achievable for the DM-BC $p(y_1, y_2|x)$ if

$$R_1 < I(U_1; Y_1),$$
$$R_2 < I(U_2; Y_2) \tag{8.8}$$

for some pmf $p(u_1)p(u_2)$ and function $x(u_1, u_2)$. The proof of achievability for this inner bound is straightforward.

Codebook generation. Fix a pmf $p(u_1)p(u_2)$ and $x(u_1, u_2)$. Randomly and independently generate 2^{nR_1} sequences $u_1^n(m_1)$, $m_1 \in [1 : 2^{nR_1}]$, each according to $\prod_{i=1}^n p_{U_1}(u_{1i})$, and 2^{nR_2} sequences $u_2^n(m_2)$, $m_2 \in [1 : 2^{nR_2}]$, each according to $\prod_{i=1}^n p_{U_2}(u_{2i})$.

Encoding. To send (m_1, m_2), transmit $x_i(u_{1i}(m_1), u_{2i}(m_2))$ at time $i \in [1 : n]$.

Decoding and analysis of the probability of error. Decoder $j = 1, 2$ declares that $\hat{m}_j$ is sent if it is the unique message such that $(U_j^n(\hat{m}_j), Y_j^n) \in \mathcal{T}_\epsilon^{(n)}$. By the LLN and the packing lemma, the probability of decoding error tends to zero as $n \to \infty$ if $R_j < I(U_j; Y_j) - \delta(\epsilon)$ for $j = 1, 2$. This completes the proof of achievability.

Marton's inner bound allows U_1 and U_2 in (8.8) to be arbitrarily dependent (even though the messages themselves are independent). This comes at an apparent penalty term in the sum-rate.

Theorem 8.3 (Marton's Inner Bound). A rate pair (R_1, R_2) is achievable for the DM-BC $p(y_1, y_2|x)$ if

$$R_1 < I(U_1; Y_1),$$
$$R_2 < I(U_2; Y_2),$$
$$R_1 + R_2 < I(U_1; Y_1) + I(U_2; Y_2) - I(U_1; U_2)$$

for some pmf $p(u_1, u_2)$ and function $x(u_1, u_2)$.

Before proving the theorem, we make a few remarks and show that Marton's inner bound is tight for the semideterministic BC.

Remark 8.3. Note that Marton's inner bound reduces to the inner bound in (8.8) if U_1 and U_2 are restricted to be independent.

Remark 8.4. As for the Gelfand–Pinsker theorem in Section 7.6, Marton's inner bound does not become larger if we evaluate it using general conditional pmfs $p(x|u_1, u_2)$. To show this, fix a pmf $p(u_1, u_2, x)$. By the functional representation lemma in Appendix B, there exists a random variable V independent of (U_1, U_2) such that X is a function of

(U_1, U_2, V). Now defining $U_1' = (U_1, V)$, we have

$$I(U_1; Y_1) \le I(U_1'; Y_1),$$
$$I(U_1; Y_1) + I(U_2; Y_2) - I(U_1; U_2) \le I(U_1'; Y_1) + I(U_2; Y_2) - I(U_1'; U_2),$$

and X is a function of (U_1', U_2). Thus there is no loss of generality in restricting X to be a deterministic function of (U_1, U_2).

Remark 8.5. Marton's inner bound is not convex in general, but can be readily convexified via a time-sharing random variable Q.

8.3.1 Semideterministic DM-BC

Marton's inner bound is tight for the class of *semideterministic* BCs for which Y_1 is a function of X, i.e., $Y_1 = y_1(X)$. Hence, we can set the auxiliary random variable $U_1 = Y_1$, which simplifies Marton's inner bound to the set of rate pairs (R_1, R_2) such that

$$R_1 \le H(Y_1),$$
$$R_2 \le I(U; Y_2),$$
$$R_1 + R_2 \le H(Y_1|U) + I(U; Y_2)$$

for some pmf $p(u, x)$. This region turns out to be the capacity region. The converse follows by the general outer bound on the capacity region of the broadcast channel presented in Section 8.5.

For the special case of fully deterministic DM-BCs, where $Y_1 = y_1(X)$ and $Y_2 = y_2(X)$, the capacity region further simplifies to the set of rate pairs (R_1, R_2) such that

$$R_1 \le H(Y_1),$$
$$R_2 \le H(Y_2),$$
$$R_1 + R_2 \le H(Y_1, Y_2)$$

for some pmf $p(x)$. This region is evaluated for the following simple channel.

Example 8.2 (Blackwell channel). Consider the deterministic BC depicted in Figure 8.6. The capacity region, which is plotted in Figure 8.7, is the union of the two regions

$$\mathcal{R}_1 = \{(R_1, R_2): R_1 \le H(\alpha), R_2 \le H(\alpha/2), R_1 + R_2 \le H(\alpha) + \alpha \text{ for } \alpha \in [1/2, 2/3]\},$$
$$\mathcal{R}_2 = \{(R_1, R_2): R_1 \le H(\alpha/2), R_2 \le H(\alpha), R_1 + R_2 \le H(\alpha) + \alpha \text{ for } \alpha \in [1/2, 2/3]\}.$$

The first region is attained by setting $p_X(0) = p_X(2) = \alpha/2$ and $p_X(1) = \bar{\alpha}$, while the second region is attained by setting $p_X(0) = \bar{\alpha}$ and $p_X(1) = p_X(2) = \alpha/2$. It can be shown that this capacity region is strictly larger than both the superposition coding inner bound and the inner bound in (8.8).

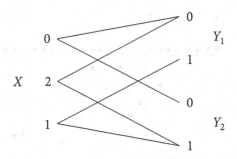

Figure 8.6. Blackwell channel. When $X = 0$, $Y_1 = Y_2 = 0$. When $X = 1$, $Y_1 = Y_2 = 1$. However, when $X = 2$, $Y_1 = 0$ while $Y_2 = 1$.

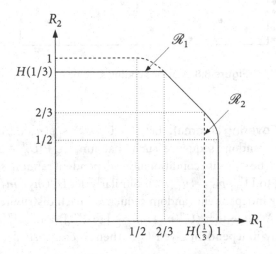

Figure 8.7. The capacity region of the Blackwell channel: $\mathscr{C} = \mathscr{R}_1 \cup \mathscr{R}_2$. Note that the sum-rate $H(1/3) + 2/3 = \log 3$ is achievable.

8.3.2 Achievability of Marton's Inner Bound

The proof of achievability uses multicoding and joint typicality codebook generation. The idea is illustrated in Figure 8.8. For each message m_j, $j = 1, 2$, we generate a subcodebook $C_j(m_j)$ consisting of independently generated u_j^n sequences. For each message pair (m_1, m_2), we find a jointly typical sequence pair (u_1^n, u_2^n) in the product subcodebook $C_1(m_1) \times C_2(m_2)$. Note that although these codeword pairs are dependent, they represent independent messages! To send (m_1, m_2), we transmit a symbol-by-symbol function $x(u_{1i}, u_{2i})$, $i \in [1 : n]$, of the selected sequence pair (u_1^n, u_2^n). Each receiver decodes for its intended codeword using joint typicality decoding.

A crucial requirement for Marton's coding scheme to succeed is the existence of at least one jointly typical sequence pair (u_1^n, u_2^n) in the chosen product subcodebook $C_1(m_1) \times C_2(m_2)$. A sufficient condition on the subcodebook sizes to guarantee the existence of such a sequence pair is provided in the following.

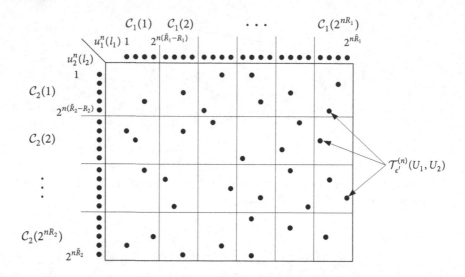

Figure 8.8. Marton's coding scheme.

Lemma 8.1 (Mutual Covering Lemma). Let $(U_0, U_1, U_2) \sim p(u_0, u_1, u_2)$ and $\epsilon' < \epsilon$. Let $U_0^n \sim p(u_0^n)$ be a random sequence such that $\lim_{n \to \infty} \mathsf{P}\{U_0^n \in \mathcal{T}_{\epsilon'}^{(n)}\} = 1$. Let $U_1^n(m_1)$, $m_1 \in [1 : 2^{nr_1}]$, be pairwise conditionally independent random sequences, each distributed according to $\prod_{i=1}^n p_{U_1 | U_0}(u_{1i} | u_{0i})$. Similarly, let $U_2^n(m_2)$, $m_2 \in [1 : 2^{nr_2}]$, be pairwise conditionally independent random sequences, each distributed according to $\prod_{i=1}^n p_{U_2 | U_0}(u_{2i} | u_{0i})$. Assume that $(U_1^n(m_1) : m_1 \in [1 : 2^{nr_1}])$ and $(U_2^n(m_2) : m_2 \in [1 : 2^{nr_2}])$ are conditionally independent given U_0^n. Then, there exists $\delta(\epsilon)$ that tends to zero as $\epsilon \to 0$ such that

$$\lim_{n \to \infty} \mathsf{P}\{(U_0^n, U_1^n(m_1), U_2^n(m_2)) \notin \mathcal{T}_\epsilon^{(n)} \text{ for all } (m_1, m_2) \in [1 : 2^{nr_1}] \times [1 : 2^{nr_2}]\} = 0,$$

if $r_1 + r_2 > I(U_1; U_2 | U_0) + \delta(\epsilon)$.

The proof of this lemma is given in Appendix 8A. Note that the mutual covering lemma extends the covering lemma in two ways:

- By considering a single U_1^n sequence ($r_1 = 0$), we obtain the same rate requirement $r_2 > I(U_1; U_2 | U_0) + \delta(\epsilon)$ as in the covering lemma.

- The lemma requires only pairwise conditional independence among the sequences $U_1^n(m_1)$, $m_1 \in [1 : 2^{nr_1}]$ (and among the sequences $U_2^n(m_2)$, $m_2 \in [1 : 2^{nr_2}]$). This implies, for example, that it suffices to use linear codes for lossy source coding of a binary symmetric source with Hamming distortion.

We are now ready to present the details of the proof of Marton's inner bound in Theorem 8.3.

Codebook generation. Fix a pmf $p(u_1, u_2)$ and function $x(u_1, u_2)$ and let $\tilde{R}_1 \geq R_1, \tilde{R}_2 \geq R_2$. For each message $m_1 \in [1 : 2^{nR_1}]$ generate a *subcodebook* $\mathcal{C}_1(m_1)$ consisting of $2^{n(\tilde{R}_1 - R_1)}$ randomly and independently generated sequences $u_1^n(l_1)$, $l_1 \in [(m_1 - 1)2^{n(\tilde{R}_1 - R_1)} + 1 : m_1 2^{n(\tilde{R}_1 - R_1)}]$, each according to $\prod_{i=1}^{n} p_{U_1}(u_{1i})$. Similarly, for each message $m_2 \in [1 : 2^{nR_2}]$ generate a subcodebook $\mathcal{C}_2(m_2)$ consisting of $2^{n(\tilde{R}_2 - R_2)}$ randomly and independently generated sequences $u_2^n(l_2)$, $l_2 \in [(m_2 - 1)2^{n(\tilde{R}_2 - R_2)} + 1 : m_2 2^{n(\tilde{R}_2 - R_2)}]$, each according to $\prod_{i=1}^{n} p_{U_2}(u_{2i})$.

For each $(m_1, m_2) \in [1 : 2^{nR_1}] \times [1 : 2^{nR_2}]$, find an index pair (l_1, l_2) such that $u_1^n(l_1) \in \mathcal{C}_1(m_1)$, $u_2^n(l_2) \in \mathcal{C}_2(m_2)$, and $(u_1^n(l_1), u_2^n(l_2)) \in \mathcal{T}_{\epsilon'}^{(n)}$. If there is more than one such pair, choose an arbitrary one among those. If no such pair exists, choose $(l_1, l_2) = (1, 1)$. Then generate $x^n(m_1, m_2)$ as $x_i(m_1, m_2) = x(u_{1i}(l_1), u_{2i}(l_2))$, $i \in [1 : n]$.

Encoding. To send the message pair (m_1, m_2), transmit $x^n(m_1, m_2)$.

Decoding. Let $\epsilon > \epsilon'$. Decoder 1 declares that $\hat{m}_1$ is sent if it is the unique message such that $(u_1^n(l_1), y_1^n) \in \mathcal{T}_{\epsilon}^{(n)}$ for some $u_1^n(l_1) \in \mathcal{C}_1(\hat{m}_1)$; otherwise it declares an error. Similarly, decoder 2 finds the unique message $\hat{m}_2$ such that $(u_2^n(l_2), y_2^n) \in \mathcal{T}_{\epsilon}^{(n)}$ for some $u_2^n(l_2) \in \mathcal{C}_2(\hat{m}_2)$.

Analysis of the probability of error. Assume without loss of generality that $(M_1, M_2) = (1, 1)$ and let (L_1, L_2) be the index pair of the chosen sequences $(U_1^n(L_1), U_2^n(L_2)) \in \mathcal{C}_1^n(1) \times \mathcal{C}_2^n(1)$. Then decoder 1 makes an error only if one or more of the following events occur:

$$\mathcal{E}_0 = \{(U_1^n(l_1), U_2^n(l_2)) \notin \mathcal{T}_{\epsilon'}^{(n)} \text{ for all } (U_1^n(l_1), U_2^n(l_2)) \in \mathcal{C}_1(1) \times \mathcal{C}_2(1)\},$$

$$\mathcal{E}_{11} = \{(U_1^n(L_1), Y_1^n) \notin \mathcal{T}_{\epsilon}^{(n)}\},$$

$$\mathcal{E}_{12} = \{(U_1^n(l_1), Y_1^n) \in \mathcal{T}_{\epsilon}^{(n)}(U_1, Y_1) \text{ for some } l_1 \notin [1 : 2^{n(\tilde{R}_1 - R_1)}]\}.$$

Thus the probability of error for decoder 1 is upper bounded as

$$P(\mathcal{E}_1) \leq P(\mathcal{E}_0) + P(\mathcal{E}_0^c \cap \mathcal{E}_{11}) + P(\mathcal{E}_{12}).$$

To bound $P(\mathcal{E}_0)$, we note that the subcodebook $\mathcal{C}_1(1)$ consists of $2^{n(\tilde{R}_1 - R_1)}$ i.i.d. $U_1^n(l_1)$ sequences and the subcodebook $\mathcal{C}_2(1)$ consists of $2^{n(\tilde{R}_2 - R_2)}$ i.i.d. $U_2^n(l_2)$ sequences. Hence, by the mutual covering lemma (with $U_0 = \emptyset$, $r_1 = \tilde{R}_1 - R_1$, and $r_2 = \tilde{R}_2 - R_2$), $P(\mathcal{E}_0)$ tends to zero as $n \to \infty$ if $(\tilde{R}_1 - R_1) + (\tilde{R}_2 - R_2) > I(U_1; U_2) + \delta(\epsilon')$.

To bound $P(\mathcal{E}_0^c \cap \mathcal{E}_{11})$, note that since $(U_1^n(L_1), U_2^n(L_2)) \in \mathcal{T}_{\epsilon'}^{(n)}$ and $\epsilon' < \epsilon$, then by the conditional typicality lemma in Section 2.5, $P\{(U_1^n(L_1), U_2^n(L_2), X^n, Y_1^n) \notin \mathcal{T}_{\epsilon}^{(n)}\}$ tends to zero as $n \to \infty$. Hence, $P(\mathcal{E}_0^c \cap \mathcal{E}_{11})$ tends to zero as $n \to \infty$.

To bound $P(\mathcal{E}_{12})$, note that since $U_1^n(l_1) \sim \prod_{i=1}^{n} p_{U_1}(u_{1i})$ and Y_1^n is independent of every $U_1^n(l_1) \notin \mathcal{C}(1)$, then by the packing lemma, $P(\mathcal{E}_{12})$ tends to zero as $n \to \infty$ if $\tilde{R}_1 < I(U_1; Y_1) - \delta(\epsilon)$. Similarly, the average probability of error $P(\mathcal{E}_2)$ for decoder 2 tends to zero as $n \to \infty$ if $\tilde{R}_2 < I(U_2; Y_2) + \delta(\epsilon)$ and $(\tilde{R}_1 - R_1) + (\tilde{R}_2 - R_2) > I(U_1; U_2) + \delta(\epsilon')$.

Thus, the average probability of error $P(\mathcal{E})$ tends to zero as $n \to \infty$ if the rate pair (R_1, R_2) satisfies the inequalities

$$
\begin{aligned}
R_1 &\le \tilde{R}_1, \\
R_2 &\le \tilde{R}_2, \\
\tilde{R}_1 &< I(U_1; Y_1) - \delta(\epsilon), \\
\tilde{R}_2 &< I(U_2; Y_2) - \delta(\epsilon), \\
R_1 + R_2 &< \tilde{R}_1 + \tilde{R}_2 - I(U_1; U_2) - \delta(\epsilon')
\end{aligned}
$$

for some $(\tilde{R}_1, \tilde{R}_2)$, or equivalently, if

$$
\begin{aligned}
R_1 &< I(U_1; Y_1) - \delta(\epsilon), \\
R_2 &< I(U_2; Y_2) - \delta(\epsilon), \\
R_1 + R_2 &< I(U_1; Y_1) + I(U_2; Y_2) - I(U_1; U_2) - \delta(\epsilon').
\end{aligned}
$$

This completes the proof of achievability.

8.3.3 Relationship to Gelfand–Pinsker Coding

Marton's coding scheme is closely related to the Gelfand–Pinsker coding scheme for the DMC with DM state when the state information is available noncausally at the encoder; see Section 7.6. Fix a pmf $p(u_1, u_2)$ and function $x(u_1, u_2)$ in Marton's inner bound and consider the achievable rate pair $R_1 < I(U_1; Y_1) - I(U_1; U_2)$ and $R_2 < I(U_2; Y_2)$. As shown in Figure 8.9, the coding scheme for communicating M_1 to receiver 1 is identical to that for communicating M_1 over the channel $p(y_1 | u_1, u_2) = p(y_1 | x(u_1, u_2))$ with state U_2 available noncausally at the encoder.

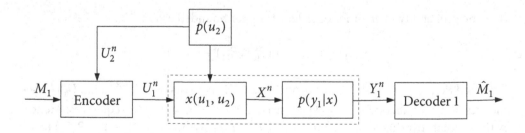

Figure 8.9. An interpretation of Marton's coding scheme at a corner point.

Application to the Gaussian Broadcast Channel. We revisit the Gaussian BC studied in Section 5.5, with outputs

$$
\begin{aligned}
Y_1 &= X + Z_1, \\
Y_2 &= X + Z_2,
\end{aligned}
$$

where $Z_1 \sim N(0, N_1)$ and $Z_2 \sim N(0, N_2)$. As before, assume average power constraint P on X and define the received SNRs as $S_j = P/N_j$, $j = 1, 2$. We show that a rate pair (R_1, R_2) is achievable if

$$R_1 < C(\alpha S_1),$$

$$R_2 < C\left(\frac{\bar{\alpha} S_2}{\alpha S_2 + 1}\right)$$

for some $\alpha \in [0, 1]$ without using superposition coding and successive cancellation decoding, and even when $N_1 > N_2$.

Decompose the channel input X into the sum of two independent parts $X_1 \sim N(0, \alpha P)$ and $X_2 \sim N(0, \bar{\alpha} P)$ as shown in Figure 8.10. To send M_2 to Y_2, consider the channel $Y_2 = X_2 + X_1 + Z_2$ with input X_2, Gaussian interference signal X_1, and Gaussian noise Z_2. Treating the interference signal X_1 as noise, M_2 can be communicated reliably to receiver 2 if $R_2 < C(\bar{\alpha} S_2/(\alpha S_2 + 1))$. To send M_1 to receiver 1, consider the channel $Y_1 = X_1 + X_2 + Z_1$ with input X_1, additive Gaussian state X_2, and Gaussian noise Z_1, where the state $X_2^n(M_2)$ is known noncausally at the encoder. By the writing on dirty paper result (which is a special case of Gelfand–Pinsker coding and hence Marton coding), M_1 can be communicated reliably to receiver 1 if $R_1 < C(\alpha S_1)$.

The above heuristic argument can be made rigorous by considering Marton's inner bound with input cost, setting $U_2 = X_2$, $U_1 = \beta U_2 + X_1$, and $X = X_1 + X_2 = (U_1 - \beta U_2) + U_2$, where $X_1 \sim N(0, \alpha P)$ and $X_2 \sim N(0, \bar{\alpha} P)$ are independent, and $\beta = \alpha P/(\alpha P + N_1)$, and using the discretization procedure in Section 3.4.

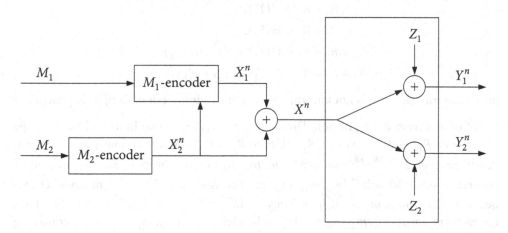

Figure 8.10. Writing on dirty paper for the Gaussian BC.

In Section 9.3, we extend this scheme to the multiple-input multiple-output (MIMO) Gaussian BC, which is not degraded in general, and show that a vector version of writing on dirty paper achieves the capacity region.

8.4 MARTON'S INNER BOUND WITH COMMON MESSAGE

Marton's inner bound can be extended to the case of common and private messages by using superposition coding and rate splitting in addition to multicoding and joint typicality codebook generation. This yields the following inner bound on the capacity of the DM-BC.

> **Theorem 8.4 (Marton's Inner Bound with Common Message).** A rate triple (R_0, R_1, R_2) is achievable for the DM-BC $p(y_1, y_2|x)$ if
>
> $$R_0 + R_1 < I(U_0, U_1; Y_1),$$
> $$R_0 + R_2 < I(U_0, U_2; Y_2),$$
> $$R_0 + R_1 + R_2 < I(U_0, U_1; Y_1) + I(U_2; Y_2|U_0) - I(U_1; U_2|U_0),$$
> $$R_0 + R_1 + R_2 < I(U_1; Y_1|U_0) + I(U_0, U_2; Y_2) - I(U_1; U_2|U_0),$$
> $$2R_0 + R_1 + R_2 < I(U_0, U_1; Y_1) + I(U_0, U_2; Y_2) - I(U_1; U_2|U_0)$$
>
> for some pmf $p(u_0, u_1, u_2)$ and function $x(u_0, u_1, u_2)$, where $|\mathcal{U}_0| \le |\mathcal{X}| + 4$, $|\mathcal{U}_1| \le |\mathcal{X}|$, and $|\mathcal{U}_2| \le |\mathcal{X}|$.

This inner bound is tight for all classes of DM-BCs with known capacity regions. In particular, the capacity region of the deterministic BC $(Y_1 = y_1(X), Y_2 = y_2(X))$ with common message is the set of rate triples (R_0, R_1, R_2) such that

$$R_0 \le \min\{I(U; Y_1), I(U; Y_2)\},$$
$$R_0 + R_1 < H(Y_1),$$
$$R_0 + R_2 < H(Y_2),$$
$$R_0 + R_1 + R_2 < H(Y_1) + H(Y_2|U, Y_1),$$
$$R_0 + R_1 + R_2 < H(Y_1|U, Y_2) + H(Y_2)$$

for some pmf $p(u, x)$. It is not known, however, if this inner bound is tight in general.

Proof of Theorem 8.4 (outline). Divide M_j, $j = 1, 2$, into two independent messages M_{j0} at rate R_{j0} and M_{jj} at rate R_{jj}. Hence $R_j = R_{j0} + R_{jj}$. Randomly and independently generate $2^{n(R_0 + R_{10} + R_{20})}$ sequences $u_0^n(m_0, m_{10}, m_{20})$. For each $(m_0, m_{10}, m_{20}, m_{11})$, generate a subcodebook $\mathcal{C}_1(m_0, m_{10}, m_{20}, m_{11})$ consisting of $2^{n(\tilde{R}_{11} - R_{11})}$ independent sequences $u_1^n(m_0, m_{10}, m_{20}, l_{11})$, $l_{11} \in [(m_{11} - 1)2^{n(\tilde{R}_{11} - R_{11})} + 1 : m_{11}2^{n(\tilde{R}_{11} - R_{11})}]$. Similarly, for each $(m_0, m_{10}, m_{20}, m_{22})$, generate a subcodebook $\mathcal{C}_2(m_0, m_{10}, m_{20}, m_{22})$ consisting of $2^{n(\tilde{R}_{22} - R_{22})}$ independent sequences $u_2^n(m_0, m_{10}, m_{20}, l_{22})$, $l_{22} \in [(m_{22} - 1)2^{n(\tilde{R}_{22} - R_{22})} + 1 : m_{22}2^{n(\tilde{R}_{22} - R_{22})}]$. As in the codebook generation step in Marton's coding scheme for the private-message case, for each $(m_0, m_{10}, m_{11}, m_{20}, m_{22})$, find an index pair (l_{11}, l_{22}) such that $u_j^n(m_0, m_{10}, m_{20}, l_{jj}) \in \mathcal{C}_j(m_0, m_{10}, m_{20}, m_{jj})$, $j = 1, 2$, and

$$\left(u_1^n(m_0, m_{10}, m_{20}, l_{11}), u_2^n(m_0, m_{10}, m_{20}, l_{22})\right) \in \mathcal{T}_{\epsilon'}^{(n)},$$

and generate $x^n(m_0, m_{10}, m_{11}, m_{20}, m_{22})$ as

$$x_i = x\big(u_{0i}(m_0, m_{10}, m_{20}), u_{1i}(m_0, m_{10}, m_{20}, l_{11}), u_{2i}(m_0, m_{10}, m_{20}, l_{22})\big)$$

for $i \in [1:n]$. To send the message triple $(m_0, m_1, m_2) = (m_0, m_{10}, m_{11}, m_{20}, m_{22})$, transmit $x^n(m_0, m_{10}, m_{20}, m_{11}, m_{22})$.

Receiver $j = 1, 2$ uses joint typicality decoding to find the unique message triple $(\hat{m}_{0j}, \hat{m}_{j0}, \hat{m}_{jj})$. Following similar steps to the proof for the private-message case and using the Fourier–Motzkin elimination procedure, it can be shown that the probability of decoding error tends to zero as $n \to \infty$ if the inequalities in Theorem 8.4 are satisfied. The cardinality bounds of the auxiliary random variables can be proved using the *perturbation method* in Appendix C.

Remark 8.6. The inner bound in Theorem 8.4 can be equivalently characterized as the set of rate triples (R_0, R_1, R_2) such that

$$
\begin{aligned}
R_0 &< \min\{I(U_0; Y_1), I(U_0; Y_2)\}, \\
R_0 + R_1 &< I(U_0, U_1; Y_1), \\
R_0 + R_2 &< I(U_0, U_2; Y_2), \\
R_0 + R_1 + R_2 &< I(U_0, U_1; Y_1) + I(U_2; Y_2|U_0) - I(U_1; U_2|U_0), \\
R_0 + R_1 + R_2 &< I(U_1; Y_1|U_0) + I(U_0, U_2; Y_2) - I(U_1; U_2|U_0)
\end{aligned}
\tag{8.9}
$$

for some pmf $p(u_0, u_1, u_2)$ and function $x(u_0, u_1, u_2)$. This region can be achieved *without* rate splitting.

Marton's private-message inner bound with U_0. By setting $R_0 = 0$ in Theorem 8.4, we obtain the following inner bound.

Proposition 8.1. A rate pair (R_1, R_2) is achievable for the DM-BC $p(y_1, y_2|x)$ if

$$
\begin{aligned}
R_1 &< I(U_0, U_1; Y_1), \\
R_2 &< I(U_0, U_2; Y_2), \\
R_1 + R_2 &< I(U_0, U_1; Y_1) + I(U_2; Y_2|U_0) - I(U_1; U_2|U_0), \\
R_1 + R_2 &< I(U_1; Y_1|U_0) + I(U_0, U_2; Y_2) - I(U_1; U_2|U_0)
\end{aligned}
$$

for some pmf $p(u_0, u_1, u_2)$ and function $x(u_0, u_1, u_2)$.

This inner bound is tight for all classes of broadcast channels with known private-message capacity regions. Furthermore, it can be shown that the special case of $U_0 = \emptyset$ (that is, Marton's inner bound in Theorem 8.3) is not tight in general for the degraded BC.

8.5 OUTER BOUNDS

First consider the following outer bound on the private-message capacity region of the DM-BC.

Theorem 8.5. If a rate pair (R_1, R_2) is achievable for the DM-BC $p(y_1, y_2|x)$, then it must satisfy the inequalities

$$R_1 \leq I(U_1; Y_1),$$
$$R_2 \leq I(U_2; Y_2),$$
$$R_1 + R_2 \leq \min\{I(U_1; Y_1) + I(X; Y_2|U_1), \ I(U_2; Y_2) + I(X; Y_1|U_2)\}$$

for some pmf $p(u_1, u_2, x)$.

The proof of this theorem follows similar steps to the converse proof for the more capable DM-BC in Section 5.6. The outer bound in the theorem is tight for all classes of broadcast channels with known capacity regions presented in this chapter and in Chapter 5, but not tight in general. In the following, we discuss two applications of this outer bound.

8.5.1 A BSC and A BEC

We revisit Example 5.4 in Section 5.6, where the channel from X to Y_1 is a BSC(p) and the channel from X to Y_2 is a BEC(ϵ). As mentioned in the example, when $H(p) < \epsilon < 1$, this DM-BC does not belong to any class with known capacity region. Nevertheless, the private-message capacity region for this range of parameter values is still achievable using superposition coding and is given by the set of rate pairs (R_1, R_2) such that

$$R_2 \leq I(U; Y_2),$$
$$R_1 + R_2 \leq I(U; Y_2) + I(X; Y_1|U) \tag{8.10}$$

for some pmf $p(u, x)$ with $X \sim \text{Bern}(1/2)$.

Proof of achievability. Recall the superposition coding inner bound that consists of all rate pairs (R_1, R_2) such that

$$R_2 < I(U; Y_2),$$
$$R_1 + R_2 < I(U; Y_2) + I(X; Y_1|U),$$
$$R_1 + R_2 < I(X; Y_1)$$

for some pmf $p(u, x)$. Now it can be easily shown that if $H(p) < \epsilon < 1$ and $X \sim \text{Bern}(1/2)$, then $I(U; Y_2) \leq I(U; Y_1)$ for all $p(u|x)$. Hence, the third inequality in the superposition coding inner bound is inactive and any rate pair in the capacity region is achievable.

Proof of the converse. We relax the inequalities in Theorem 8.5 by replacing the first inequality with $R_1 \le I(X; Y_1)$ and dropping the first term inside the minimum in the third inequality. Now setting $U_2 = U$, we obtain an outer bound on the capacity region that consists of all rate pairs (R_1, R_2) such that

$$R_2 \le I(U; Y_2),$$
$$R_1 + R_2 \le I(U; Y_2) + I(X; Y_1 | U), \qquad (8.11)$$
$$R_1 \le I(X; Y_1)$$

for some pmf $p(u, x)$. We first show that it suffices to consider only joint pmfs $p(u, x)$ with $X \sim \text{Bern}(1/2)$. Given a pair $(U, X) \sim p(u, x)$, let $W' \sim \text{Bern}(1/2)$ and $(U', \tilde{X})$ be defined as

$$P\{U' = u, \tilde{X} = x \mid W' = w\} = p_{U,X}(u, x \oplus w).$$

Let $(\tilde{Y}_1, \tilde{Y}_2)$ be the channel output pair corresponding to the input $\tilde{X}$. Then, by the symmetries in the input construction and the channels, it is easy to check that $\tilde{X} \sim \text{Bern}(1/2)$ and

$$I(U; Y_2) = I(U'; \tilde{Y}_2 | W') \le I(U', W'; \tilde{Y}_2) = I(\tilde{U}; \tilde{Y}_2),$$
$$I(X; Y_1 | U) = I(\tilde{X}; \tilde{Y}_1 | U', W') = I(\tilde{X}; \tilde{Y}_1 | \tilde{U}),$$
$$I(X; Y_1) = I(\tilde{X}; \tilde{Y}_1 | W') \le I(\tilde{X}; \tilde{Y}_1),$$

where $\tilde{U} = (U', W')$. This proves the sufficiency of $X \sim \text{Bern}(1/2)$. As mentioned in the proof of achievability, this implies that $I(U; Y_2) \le I(U; Y_1)$ for all $p(u|x)$ and hence the outer bound in (8.11) reduces to the characterization of the capacity region in (8.10).

8.5.2 Binary Skew-Symmetric Broadcast Channel

The binary skew-symmetric broadcast channel consists of two symmetric Z channels as depicted in Figure 8.11.

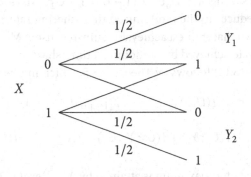

Figure 8.11. Binary skew-symmetric broadcast channel.

We show that the private-message sum-capacity $C_{\text{sum}} = \max\{R_1 + R_2 : (R_1, R_2) \in \mathscr{C}\}$ for this channel is bounded as

$$0.3616 \le C_{\text{sum}} \le 0.3726. \qquad (8.12)$$

The lower bound is achieved by the following *randomized time-sharing* technique. We divide message M_j, $j = 1, 2$, into two independent messages M_{j0} and M_{jj}.

Codebook generation. Randomly and independently generate $2^{n(R_{10}+R_{20})}$ sequences $u^n(m_{10}, m_{20})$, each i.i.d. Bern(1/2). For each (m_{10}, m_{20}), let $k(m_{10}, m_{20})$ be the number of locations where $u_i(m_{10}, m_{20}) = 1$. Randomly and conditionally independently generate $2^{nR_{11}}$ sequences $x^{k(m_{10},m_{20})}(m_{10}, m_{20}, m_{11})$, each i.i.d. Bern($\alpha$). Similarly, randomly and conditionally independently generate $2^{nR_{22}}$ sequences $x^{n-k(m_{10},m_{20})}(m_{10}, m_{20}, m_{22})$, each i.i.d. Bern($\bar{\alpha}$).

Encoding. To send the message pair (m_1, m_2), represent it by the message quadruple $(m_{10}, m_{20}, m_{11}, m_{22})$. Transmit $x^{k(m_{10},m_{20})}(m_{10}, m_{20}, m_{11})$ in the locations where $u_i(m_{10}, m_{20}) = 1$ and $x^{n-k(m_{10},m_{20})}(m_{10}, m_{20}, m_{22})$ in the locations where $u_i(m_{10}, m_{20}) = 0$. Thus the messages m_{11} and m_{22} are transmitted using time sharing with respect to $u^n(m_{10}, m_{20})$.

Decoding. Each decoder $j = 1, 2$ first decodes for $u^n(m_{10}, m_{20})$ and then proceeds to recover m_{jj} from the output subsequence that corresponds to the respective symbol locations in $u^n(m_{10}, m_{20})$. It is straightforward to show that a rate pair (R_1, R_2) is achievable using this scheme if

$$R_1 < \min\{I(U; Y_1), I(U; Y_2)\} + \frac{1}{2}I(X; Y_1|U = 1),$$

$$R_2 < \min\{I(U; Y_1), I(U; Y_2)\} + \frac{1}{2}I(X; Y_2|U = 0), \qquad (8.13)$$

$$R_1 + R_2 < \min\{I(U; Y_1), I(U; Y_2)\} + \frac{1}{2}\big(I(X; Y_1|U = 1) + I(X; Y_2|U = 0)\big)$$

for some $\alpha \in [0, 1]$. Taking the maximum of the sum-rate bound over α establishes the sum-capacity lower bound $C_{\text{sum}} \geq 0.3616$.

Note that by considering $U_0 \sim$ Bern(1/2), $U_1 \sim$ Bern(α), and $U_2 \sim$ Bern(α), independent of each other, and setting $X = U_0 U_1 + (1 - U_0)(1 - U_2)$, Marton's inner bound with U_0 in Proposition 8.1 reduces to the randomized time-sharing rate region in (8.13). Furthermore, it can be shown that the best achievable sum-rate using Marton's coding scheme is 0.3616, which is the rate achieved by randomized time-sharing.

The upper bound in (8.12) follows by Theorem 8.5, which implies that

$$C_{\text{sum}} \leq \max_{p(u_1, u_2, x)} \min\{I(U_1; Y_1) + I(U_2; Y_2|U_1), I(U_2; Y_2) + I(U_1; Y_1|U_2)\}$$

$$\leq \max_{p(u_1, u_2, x)} \frac{1}{2}(I(U_1; Y_1) + I(U_2; Y_2|U_1) + I(U_2; Y_2) + I(U_1; Y_1|U_2)).$$

It can be shown that the latter maximum is attained by $X \sim$ Bern(1/2) and binary U_1 and U_2, which gives the upper bound on the sum-capacity $C_{\text{sum}} \leq 0.3726$.

8.5.3 Nair–El Gamal Outer Bound

We now consider the following outer bound on the capacity region of the DM-BC with common and private messages.

Theorem 8.6 (Nair–El Gamal Outer Bound). If a rate triple (R_0, R_1, R_2) is achievable for the DM-BC $p(y_1, y_2|x)$, then it must satisfy the inequalities

$$R_0 \leq \min\{I(U_0; Y_1),\ I(U_0; Y_2)\},$$
$$R_0 + R_1 \leq I(U_0, U_1; Y_1),$$
$$R_0 + R_2 \leq I(U_0, U_2; Y_2),$$
$$R_0 + R_1 + R_2 \leq I(U_0, U_1; Y_1) + I(U_2; Y_2|U_0, U_1),$$
$$R_0 + R_1 + R_2 \leq I(U_1; Y_1|U_0, U_2) + I(U_0, U_2; Y_2)$$

for some pmf $p(u_1)p(u_2)p(u_0|u_1, u_2)$ and function $x(u_0, u_1, u_2)$.

The proof of this outer bound is quite similar to the converse proof for the more capable DM-BC in Section 5.6 and is given in Appendix 8B.

The Nair–El Gamal bound is tight for all broadcast channels with known capacity regions that we discussed so far. Note that the outer bound coincides with Marton's inner bound with common message in Theorem 8.4 (or more directly, the alternative characterization in (8.9)) if $I(U_1; U_2|U_0, Y_1) = I(U_1; U_2|U_0, Y_2) = 0$ for every pmf $p(u_0, u_1, u_2, x)$ that attains a boundary point of the outer bound. This bound is not tight in general, however.

Remark 8.7. It can be shown that the Nair–El Gamal outer bound with no common message, i.e., with $R_0 = 0$, simplifies to the outer bound in Theorem 8.5, that is, no U_0 is needed when evaluating the outer bound in Theorem 8.6 with $R_0 = 0$. This is in contrast to Marton's inner bound, where the private-message region without U_0 in Theorem 8.3 can be strictly smaller than that with U_0 in Proposition 8.1.

8.6 INNER BOUNDS FOR MORE THAN TWO RECEIVERS

Marton's inner bound can be easily extended to more than two receivers. For example, consider a 3-receiver DM-BC $p(y_1, y_2, y_3|x)$ with three private messages $M_j \in [1 : 2^{nR_j}]$, $j = 1, 2, 3$. A rate triple (R_1, R_2, R_3) is achievable for the 3-receiver DM-BC if

$$R_1 < I(U_1; Y_1),$$
$$R_2 < I(U_2; Y_2),$$
$$R_3 < I(U_3; Y_3),$$
$$R_1 + R_2 < I(U_1; Y_1) + I(U_2; Y_2) - I(U_1; U_2),$$
$$R_1 + R_3 < I(U_1; Y_1) + I(U_3; Y_3) - I(U_1; U_3),$$
$$R_2 + R_3 < I(U_2; Y_2) + I(U_3; Y_3) - I(U_2; U_3),$$
$$R_1 + R_2 + R_3 < I(U_1; Y_1) + I(U_2; Y_2) + I(U_3; Y_3) - I(U_1; U_2) - I(U_1, U_2; U_3)$$

for some pmf $p(u_1, u_2, u_3)$ and function $x(u_1, u_2, u_3)$. This region can be readily extended to any number of receivers k.

It can be readily shown that the extended Marton inner bound is tight for deterministic DM-BCs with an arbitrary number of receivers, where we substitute $U_j = Y_j$ for $j \in [1 : k]$. The proof of achievability follows similar steps to the case of two receivers using the following extension of the mutual covering lemma.

Lemma 8.2 (Multivariate Covering Lemma). Let $(U_0, U_1, \ldots, U_k) \sim p(u_0, u_1, \ldots, u_k)$ and $\epsilon' < \epsilon$. Let $U_0^n \sim p(u_0^n)$ be a random sequence with $\lim_{n\to\infty} \mathsf{P}\{U_0^n \in \mathcal{T}_{\epsilon'}^{(n)}\} = 1$. For each $j \in [1 : k]$, let $U_j^n(m_j), m_j \in [1 : 2^{nr_j}]$, be pairwise conditionally independent random sequences, each distributed according to $\prod_{i=1}^{n} p_{U_j|U_0}(u_{ji}|u_{0i})$. Assume that $(U_1^n(m_1): m_1 \in [1 : 2^{nr_1}]), (U_2^n(m_2): m_2 \in [1 : 2^{nr_2}]), \ldots, (U_k^n(m_k): m_k \in [1 : 2^{nr_k}])$ are mutually conditionally independent given U_0^n. Then, there exists $\delta(\epsilon)$ that tends to zero as $\epsilon \to 0$ such that

$$\lim_{n\to\infty} \mathsf{P}\{(U_0^n, U_1^n(m_1), U_2^n(m_2), \ldots, U_k^n(m_k)) \notin \mathcal{T}_{\epsilon}^{(n)} \text{ for all } (m_1, m_2, \ldots, m_k)\} = 0,$$

if $\sum_{j\in\mathcal{J}} r_j > \sum_{j\in\mathcal{J}} H(U_j|U_0) - H(U(\mathcal{J})|U_0) + \delta(\epsilon)$ for all $\mathcal{J} \subseteq [1 : k]$ with $|\mathcal{J}| \geq 2$.

The proof of this lemma is a straightforward extension of that for the $k = 2$ case presented in Appendix 8A. When $k = 3$ and $U_0 = \emptyset$, the conditions for joint typicality become

$$r_1 + r_2 > I(U_1; U_2) + \delta(\epsilon),$$
$$r_1 + r_3 > I(U_1; U_3) + \delta(\epsilon),$$
$$r_2 + r_3 > I(U_2; U_3) + \delta(\epsilon),$$
$$r_1 + r_2 + r_3 > I(U_1; U_2) + I(U_1, U_2; U_3) + \delta(\epsilon).$$

In general, combining Marton's coding with superposition coding, rate splitting, and indirect decoding, we can construct an inner bound for DM-BCs with more than two receivers for any given messaging requirement. We illustrate this construction via the 3-receiver DM-BC $p(y_1, y_2, y_3|x)$ with two degraded message sets, where a common message $M_0 \in [1 : 2^{nR_0}]$ is to be communicated to receivers 1, 2, and 3, and a private message $M_1 \in [1 : 2^{nR_1}]$ is to be communicated only to receiver 1.

Proposition 8.2. A rate pair (R_0, R_1) is achievable for the 3-receiver DM-BC $p(y_1, y_2, y_3|x)$ with 2 degraded message sets if

$$R_0 < \min\{I(V_2; Y_2), I(V_3; Y_3)\},$$
$$2R_0 < I(V_2; Y_2) + I(V_3; Y_3) - I(V_2; V_3|U),$$
$$R_0 + R_1 < \min\{I(X; Y_1), I(V_2; Y_2) + I(X; Y_1|V_2), I(V_3; Y_3) + I(X; Y_1|V_3)\},$$
$$2R_0 + R_1 < I(V_2; Y_2) + I(V_3; Y_3) + I(X; Y_1|V_2, V_3) - I(V_2; V_3|U),$$
$$2R_0 + 2R_1 < I(V_2; Y_2) + I(V_3; Y_3) + I(X; Y_1|V_2) + I(X; Y_1|V_3) - I(V_2; V_3|U),$$
$$2R_0 + 2R_1 < I(V_2; Y_2) + I(V_3; Y_3) + I(X; Y_1|U) + I(X; Y_1|V_2, V_3) - I(V_2; V_3|U)$$

for some pmf $p(u, v_2, v_3, x) = p(u)p(v_2|u)p(x, v_3|v_2) = p(u)p(v_3|u)p(x, v_2|v_3)$.

It can be shown that this inner bound is tight when Y_1 is less noisy than Y_2, which is a generalization of the class of multilevel DM-BCs discussed earlier in Section 8.2.

Proof of Proposition 8.2 (outline). Divide M_1 into four independent messages M_{10} at rate R_{10}, M_{11} at rate R_{11}, M_{12} at rate R_{12}, and M_{13} at rate R_{13}. Let $\tilde{R}_{12} \geq R_{12}$ and $\tilde{R}_{13} \geq R_{13}$. Randomly and independently generate $2^{n(R_0+R_{10})}$ sequences $u^n(m_0, m_{10})$, $(m_0, m_{10}) \in [1 : 2^{nR_0}] \times [1 : 2^{nR_{10}}]$. As in the achievability proof of Theorem 8.4, for each (m_0, m_{10}), we use the codebook generation step in the proof of Marton's inner bound in Theorem 8.3 to randomly and conditionally independently generate $2^{n\tilde{R}_{12}}$ sequences $(v_2^n(m_0, m_{10}, l_2))$, $l_2 \in [1 : 2^{n\tilde{R}_{12}}]$, and $2^{n\tilde{R}_{13}}$ sequences $v_3^n(m_0, m_{10}, l_3))$, $l_3 \in [1 : 2^{n\tilde{R}_{13}}]$. For each $(m_0, m_{10}, m_{12}, m_{13})$, find a jointly typical sequence pair $(v_2^n(m_0, m_{10}, l_2), v_3^n(m_0, m_{10}, l_3))$, where $l_2 \in [(m_{12}-1)2^{n(\tilde{R}_{12}-R_{12})} + 1 : m_{12}2^{n(\tilde{R}_{12}-R_{12})}]$ and $l_3 \in [(m_{13}-1)2^{n(\tilde{R}_{13}-R_{13})} + 1 : m_{13}2^{n(\tilde{R}_{13}-R_{13})}]$, and randomly and conditionally independently generate $2^{nR_{11}}$ sequences $x^n(m_0, m_{10}, m_{11}, m_{12}, m_{13})$. To send the message pair (m_0, m_1), transmit $x^n(m_0, m_{10}, m_{11}, m_{12}, m_{13})$.

Receiver 1 uses joint typicality decoding to find the unique tuple $(\hat{m}_{01}, \hat{m}_{10}, \hat{m}_{11}, \hat{m}_{12}, \hat{m}_{13})$. Receiver $j = 2, 3$ uses indirect decoding to find the unique $\hat{m}_{0j}$ through (u^n, v_j^n). Following a standard analysis of the probability of error and using the Fourier–Motzkin elimination procedure, it can be shown that the probability of error tends to zero as $n \to \infty$ if the inequalities in Proposition 8.2 are satisfied.

SUMMARY

- Indirect decoding
- Marton's inner bound:
  - Multidimensional subcodebook generation
  - Generating correlated codewords for independent messages
  - Tight for all BCs with known capacity regions
  - A common auxiliary random variable is needed even when coding only for private messages
- Mutual covering lemma and its multivariate extension
- Connection between Marton coding and Gelfand–Pinsker coding
- Writing on dirty paper achieves the capacity region of the Gaussian BC
- Randomized time sharing for the binary skew-symmetric BC
- Nair–El Gamal outer bound:
  - Does not always coincide with Marton's inner bound
  - Not tight in general

- **Open problems:**

   8.1. What is the capacity region of the general 3-receiver DM-BC with one common message to all three receivers and one private message to one receiver?

   8.2. Is superposition coding optimal for the general 3-receiver DM-BC with one message to all three receivers and another message to two receivers?

   8.3. What is the sum-capacity of the binary skew-symmetric broadcast channel?

   8.4. Is Marton's inner bound tight in general?

BIBLIOGRAPHIC NOTES

The capacity region of the 2-receiver DM-BC with degraded message sets was established by Körner and Marton (1977b), who proved the strong converse using the technique of images of sets (Körner and Marton 1977c). The multilevel BC was introduced by Borade, Zheng, and Trott (2007), who conjectured that the straightforward generalization of Theorem 8.1 to three receivers in (8.2) is optimal. The capacity region of the multilevel BC in Theorem 8.2 was established using indirect decoding by Nair and El Gamal (2009). They also established the general inner bound for the 3-receiver DM-BC with degraded message sets in Section 8.6.

The inner bound on the private-message capacity region in (8.8) is a special case of a more general inner bound by Cover (1975b) and van der Meulen (1975), which consists of all rate pairs (R_1, R_2) that satisfy the inequalities in Proposition 8.1 for some pmf $p(u_0)p(u_1)p(u_2)$ and $x(u_0, u_1, u_2)$. The deterministic broadcast channel in Example 8.2 is attributed to D. Blackwell. The capacity region of the Blackwell channel was established by Gelfand (1977). The capacity region of the deterministic broadcast channel was established by Pinsker (1978). The capacity region of the semideterministic broadcast channel was established independently by Marton (1979) and Gelfand and Pinsker (1980a). The inner bounds in Theorem 8.3 and in (8.9) are due to Marton (1979). The mutual covering lemma and its application in the proof of achievability are due to El Gamal and van der Meulen (1981). The inner bound on the capacity region of the DM-BC with common message in Theorem 8.4 is due to Liang (2005). The equivalence to Marton's characterization in (8.9) was shown by Liang, Kramer, and Poor (2011). Gohari, El Gamal, and Anantharam (2010) showed that Marton's inner bound without U_0 is not optimal even for a degraded DM-BC with no common message.

The outer bound in Theorem 8.5 is a direct consequence of the converse proof for the more capable BC by El Gamal (1979). Nair and El Gamal (2007) showed through the binary skew-symmetric BC example that it is strictly tighter than the earlier outer bounds by Sato (1978a) and Körner and Marton (Marton 1979). The outer bound in Theorem 8.6 is due to Nair and El Gamal (2007). Other outer bounds can be found in Liang, Kramer, and Shamai (2008), Gohari and Anantharam (2008), and Nair (2008). These bounds coincide with the bound in Theorem 8.5 when there is no common message. It is not known, however, if they are strictly tighter than the Nair–El Gamal outer bound. The application of the outer bound in Theorem 8.5 to the BSC–BEC broadcast channel is due to Nair (2010). The binary skew-symmetric channel was first studied by Hajek and Pursley (1979), who showed through randomized time sharing that U_0 is needed for the Cover–van der Meulen inner bound even when there is no common message. Gohari and Anantharam (2009), Nair and Wang (2008), and Jog and Nair (2010) showed that the maximum sum-rate of Marton's inner bound (with U_0) coincides with the randomized time-sharing lower bound, establishing a strict gap between Marton's inner bound with U_0 and the outer bound in Theorem 8.5. Geng, Gohari, Nair, and Yu (2011) established the capacity region of the product of reversely semideterministic or more capable DM-BCs and provided an ingenious counterexample that shows that the Nair–El Gamal outer bound is not tight.

PROBLEMS

8.1. Prove achievability of the extended superposition coding inner bound in (8.2) for the 3-receiver multilevel DM-BC.

8.2. Consider the binary erasure multilevel product DM-BC in Example 8.1. Show that the extended superposition coding inner bound in (8.2) can be simplified to (8.6).

8.3. Prove the converse for the capacity region of the 3-receiver multilevel DM-BC in Theorem 8.2. (Hint: Use the converse proof for the degraded DM-BC for Y_1 and Y_2 and the converse proof for the 2-receiver DM-BC with degraded message sets for Y_1 and Y_3.)

8.4. Consider the BSC–BEC broadcast channel in Section 8.5.1. Show that if $H(p) < \epsilon < 1$ and $X \sim \text{Bern}(1/2)$, then $I(U; Y_2) \le I(U; Y_1)$ for all $p(u|x)$.

8.5. Complete the proof of the mutual covering lemma for the case $U_0 \ne \emptyset$.

8.6. Show that the inner bound on the capacity region of the 3-receiver DM-BC with two degraded message sets in Proposition 8.2 is tight when Y_1 is less noisy that Y_2.

8.7. *Complete decoding for the 3-receiver multilevel DM-BC.* Show that the fourth inequality in (8.3) is redundant and thus that the region simplifies to the capacity region in Theorem 8.2. (Hint: Given the choice of U, consider two cases $I(U; Y_2) \le I(U; Y_3)$ and $I(U; Y_2) > I(U; Y_3)$ separately, and optimize V.)

8.8. *Sato's outer bound.* Show that if a rate pair (R_1, R_2) is achievable for the DM-BC $p(y_1, y_2|x)$, then it must satisfy

$$R_1 \le I(X; Y_1),$$
$$R_2 \le I(X; Y_2),$$
$$R_1 + R_2 \le \min_{\tilde{p}(y_1, y_2|x)} I(X; Y_1, Y_2)$$

for some pmf $p(x)$, where the minimum is over all conditional pmfs $\tilde{p}(y_1, y_2|x)$

with the same conditional marginal pmfs $p(y_1|x)$ and $p(y_2|x)$ as the given channel conditional pmf $p(y_1, y_2|x)$.

8.9. *Write-once memory.* Consider a memory that can store bits without any noise. But unlike a regular memory, each cell of the memory is permanently programmed to zero once a zero is stored. Thus, the memory can be modeled as a DMC with state $Y = XS$, where S is the previous state of the memory, X is the input to the memory, and Y is the output of the memory.

Suppose that the initial state of the memory is $S_i = 1$, $i \in [1:n]$. We wish to use the memory twice at rates R_1 and R_2, respectively. For the first write, we store data with input X_1 and output $Y_1 = X_1$. For the second write, we store data with input X_2 and output $Y_2 = X_1 X_2$.

(a) Show that the capacity region for two writes over the memory is the set of rate pairs (R_1, R_2) such that

$$R_1 \le H(\alpha),$$
$$R_2 \le 1 - \alpha$$

for some $\alpha \in [0, 1/2]$.

(b) Find the sum-capacity.

(c) Now suppose we use the memory k times at rates $R_1, \ldots, R_k$. The memory at stage k has $Y_k = X_1 X_2 \cdots X_k$. Find the capacity region and sum-capacity.

Remark: This result is due to Wolf, Wyner, Ziv, and Körner (1984).

8.10. *Marton inner bound with common message.* Consider the Marton inner bound with common message in Theorem 8.4. Provide the details of the coding scheme outlined below.

(a) Using the packing lemma and the mutual covering lemma, show that the probability of error tends to zero as $n \to \infty$ if

$$R_0 + R_{01} + R_{02} + \tilde{R}_{11} < I(U_0, U_1; Y_1),$$
$$R_0 + R_{01} + R_{02} + \tilde{R}_{22} < I(U_0, U_2; Y_2),$$
$$\tilde{R}_{11} < I(U_1; Y_1|U_0),$$
$$\tilde{R}_{22} < I(U_2; Y_2|U_0),$$
$$\tilde{R}_{11} + \tilde{R}_{22} - R_{11} - R_{22} > I(U_1; U_2|U_0).$$

(b) Establish the inner bound by using the Fourier–Motzkin elimination procedure to reduce the set of inequalities in part (a).

(c) Show that the inner bound is convex.

8.11. *Maximal probability of error.* Consider a $(2^{nR_1}, 2^{nR_2}, n)$ code $\mathcal{C}$ for the DM-BC $p(y_1, y_2|x)$ with average probability of error $P_e^{(n)}$. In this problem, we show that there exists a $(2^{nR_1}/n^2, 2^{nR_2}/n^2, n)$ code with maximal probability of error less than

$(P_e^{(n)})^{1/2}$. Consequently, the capacity region with maximal probability of error is the same as that with average probability of error. This is in contrast to the DM-MAC case, where the capacity region for maximal probability of error can be strictly smaller than that for average probability of error; see Problem 4.3.

(a) A codeword $x^n(m_1, m_2)$ is said to be "bad" if its probability of error $\lambda_{m_1 m_2} = P\{(\hat{M}_1, \hat{M}_2) \neq (M_1, M_2) \,|\, (M_1, M_2) = (m_1, m_2)\} > (P_e^{(n)})^{1/2}$. Show that there are at most $2^{n(R_1+R_2)}(P_e^{(n)})^{1/2}$ "bad" codewords $x^n(m_1, m_2)$.

(b) Randomly and independently permute the message indices m_1 and m_2 to generate a new codebook $\bar{C}$ consisting of codewords $x^n(\sigma_1(m_1), \sigma(m_2))$, where σ_1 and σ_2 denote the random permutations. Then partition the codebook $\bar{C}$ into subcodebooks $\bar{C}(m_1', m_2')$, $(m_1', m_2') \in [1 : 2^{nR_1}/n^2] \times [1 : 2^{nR_2}/n^2]$, each consisting of $n^2 \times n^2$ sequences. Show that the probability that all n^4 codewords in the subcodebook $\bar{C}(m_1', m_2')$ are "bad" is upper bounded by $(2P_e^{(n)})^{n^2/2}$ for n sufficiently large. (Hint: This probability is upper bounded by the probability that all n^2 "diagonal" codewords are "bad." Upper bound the latter probability using part (a) and the independence of the permutations.)

(c) Suppose that a subcodebook $\bar{C}(m_1', m_2')$ is said to be "bad" if all sequences in the subcodebook are "bad." Show that the expected number of "bad" subcodebooks is upper bounded by

$$\frac{2^{n(R_1+R_2)}}{n^4}(2P_e^{(n)})^{n^2/2}$$

for n sufficiently large. Further show that this upper bound tends to zero as $n \to \infty$ if $P_e^{(n)}$ tends to zero as $n \to \infty$.

(d) Argue that there exists at least one permutation pair (σ_1, σ_2) such that there is no "bad" subcodebook. Conclude that there exists a $(2^{nR_1}/n^2, 2^{nR_2}/n^2, n)$ code with maximal probability of error less than or equal to $(P_e^{(n)})^{1/2}$ for n sufficiently large, if $P_e^{(n)}$ tends to zero as $n \to \infty$.

Remark: This argument is due to İ. E. Telatar, which is a simplified version of the original proof by Willems (1990).

APPENDIX 8A   PROOF OF THE MUTUAL COVERING LEMMA

We prove the case $U_0 = \emptyset$ only. The proof for the general case follows similarly.

Let $\mathcal{A} = \{(m_1, m_2) \in [1 : 2^{nr_1}] \times [1 : 2^{nr_2}] : (U_1^n(m_1), U_2^n(m_2)) \in \mathcal{T}_\epsilon^{(n)}(U_1, U_2)\}$. Then by the Chebyshev lemma in Appendix B, the probability of the event of interest can be bounded as

$$P\{|\mathcal{A}| = 0\} \le P\{(|\mathcal{A}| - E|\mathcal{A}|)^2 \ge (E|\mathcal{A}|)^2\} \le \frac{\mathrm{Var}(|\mathcal{A}|)}{(E|\mathcal{A}|)^2}.$$

We now show that $\mathrm{Var}(|\mathcal{A}|)/(\mathrm{E}\,|\mathcal{A}|)^2$ tends to zero as $n \to \infty$ if

$$r_1 > 3\delta(\epsilon),$$
$$r_2 > 3\delta(\epsilon),$$
$$r_1 + r_2 > I(U_1; U_2) + \delta(\epsilon).$$

Using indicator random variables, we can express $|\mathcal{A}|$ as

$$|\mathcal{A}| = \sum_{m_1=1}^{2^{nr_1}} \sum_{m_2=1}^{2^{nr_2}} E(m_1, m_2),$$

where

$$E(m_1, m_2) = \begin{cases} 1 & \text{if } (U_1^n(m_1), U_2^n(m_2)) \in \mathcal{T}_\epsilon^{(n)}, \\ 0 & \text{otherwise} \end{cases}$$

for each $(m_1, m_2) \in [1:2^{nr_1}] \times [1:2^{nr_2}]$. Let

$$p_1 = \mathsf{P}\{(U_1^n(1), U_2^n(1)) \in \mathcal{T}_\epsilon^{(n)}\},$$
$$p_2 = \mathsf{P}\{(U_1^n(1), U_2^n(1)) \in \mathcal{T}_\epsilon^{(n)}, (U_1^n(1), U_2^n(2)) \in \mathcal{T}_\epsilon^{(n)}\},$$
$$p_3 = \mathsf{P}\{(U_1^n(1), U_2^n(1)) \in \mathcal{T}_\epsilon^{(n)}, (U_1^n(2), U_2^n(1)) \in \mathcal{T}_\epsilon^{(n)}\},$$
$$p_4 = \mathsf{P}\{(U_1^n(1), U_2^n(1)) \in \mathcal{T}_\epsilon^{(n)}, (U_1^n(2), U_2^n(2)) \in \mathcal{T}_\epsilon^{(n)}\} = p_1^2.$$

Then

$$\mathsf{E}(|\mathcal{A}|) = \sum_{m_1,m_2} \mathsf{P}\{(U_1^n(m_1), U_2^n(m_2)) \in \mathcal{T}_\epsilon^{(n)}\} = 2^{n(r_1+r_2)} p_1,$$

$$\mathsf{E}(|\mathcal{A}|^2) = \sum_{m_1,m_2} \mathsf{P}\{(U_1^n(m_1), U_2^n(m_2)) \in \mathcal{T}_\epsilon^{(n)}\}$$

$$+ \sum_{m_1,m_2} \sum_{m_2' \neq m_2} \mathsf{P}\{(U_1^n(m_1), U_2^n(m_2)) \in \mathcal{T}_\epsilon^{(n)}, (U_1^n(m_1), U_2^n(m_2')) \in \mathcal{T}_\epsilon^{(n)}\}$$

$$+ \sum_{m_1,m_2} \sum_{m_1' \neq m_1} \mathsf{P}\{(U_1^n(m_1), U_2^n(m_2)) \in \mathcal{T}_\epsilon^{(n)}, (U_1^n(m_1'), U_2^n(m_2)) \in \mathcal{T}_\epsilon^{(n)}\}$$

$$+ \sum_{m_1,m_2} \sum_{m_1' \neq m_1} \sum_{m_2' \neq m_2} \mathsf{P}\{(U_1^n(m_1), U_2^n(m_2)) \in \mathcal{T}_\epsilon^{(n)}, (U_1^n(m_1'), U_2^n(m_2')) \in \mathcal{T}_\epsilon^{(n)}\}$$

$$\leq 2^{n(r_1+r_2)} p_1 + 2^{n(r_1+2r_2)} p_2 + 2^{n(2r_1+r_2)} p_3 + 2^{2n(r_1+r_2)} p_4.$$

Hence

$$\mathrm{Var}(|\mathcal{A}|) \leq 2^{n(r_1+r_2)} p_1 + 2^{n(r_1+2r_2)} p_2 + 2^{n(2r_1+r_2)} p_3.$$

Now by the joint typicality lemma, for sufficiently large n, we have

$$p_1 \geq 2^{-n(I(U_1;U_2)+\delta(\epsilon))},$$
$$p_2 \leq 2^{-n(2I(U_1;U_2)-\delta(\epsilon))},$$
$$p_3 \leq 2^{-n(2I(U_1;U_2)-\delta(\epsilon))},$$

and hence

$$p_2/p_1^2 \le 2^{3n\delta(\epsilon)},$$
$$p_3/p_1^2 \le 2^{3n\delta(\epsilon)}.$$

Therefore

$$\frac{\text{Var}(|\mathcal{A}|)}{(\text{E}\,|\mathcal{A}|)^2} \le 2^{-n(r_1+r_2-I(U_1;U_2)-\delta(\epsilon))} + 2^{-n(r_1-3\delta(\epsilon))} + 2^{-n(r_2-3\delta(\epsilon))},$$

which tends to zero as $n \to \infty$ if

$$r_1 > 3\delta(\epsilon),$$
$$r_2 > 3\delta(\epsilon),$$
$$r_1 + r_2 > I(U_1;U_2) + \delta(\epsilon).$$

It can be similarly shown that $P\{|\mathcal{A}| = 0\}$ tends to zero as $n \to \infty$ if $r_1 = 0$ and $r_2 > I(U_1;U_2) + \delta(\epsilon)$, or if $r_1 > I(U_1;U_2) + \delta(\epsilon)$ and $r_2 = 0$. Combining these three sets of inequalities, we have shown that $P\{|\mathcal{A}| = 0\}$ tends to zero as $n \to \infty$ if $r_1 + r_2 > I(U_1;U_2) + 4\delta(\epsilon)$.

APPENDIX 8B   PROOF OF THE NAIR–EL GAMAL OUTER BOUND

By Fano's inequality and following standard steps, we have

$$nR_0 \le \min\{I(M_0;Y_1^n), I(M_0;Y_2^n)\} + n\epsilon_n,$$
$$n(R_0 + R_1) \le I(M_0, M_1;Y_1^n) + n\epsilon_n,$$
$$n(R_0 + R_2) \le I(M_0, M_2;Y_2^n) + n\epsilon_n,$$
$$n(R_0 + R_1 + R_2) \le I(M_0, M_1;Y_1^n) + I(M_2;Y_2^n|M_0, M_1) + n\epsilon_n,$$
$$n(R_0 + R_1 + R_2) \le I(M_1;Y_1^n|M_0, M_2) + I(M_0, M_2;Y_2^n) + n\epsilon_n.$$

We bound the mutual information terms in the above bounds. First consider the terms in the fourth inequality

$$I(M_0, M_1;Y_1^n) + I(M_2;Y_2^n|M_0, M_1)$$

$$= \sum_{i=1}^n I(M_0, M_1;Y_{1i}|Y_1^{i-1}) + \sum_{i=1}^n I(M_2;Y_{2i}|M_0, M_1, Y_{2,i+1}^n).$$

Now consider

$$\sum_{i=1}^n I(M_0, M_1;Y_{1i}|Y_1^{i-1}) \le \sum_{i=1}^n I(M_0, M_1, Y_1^{i-1};Y_{1i})$$

$$= \sum_{i=1}^n I(M_0, M_1, Y_1^{i-1}, Y_{2,i+1}^n;Y_{1i}) - \sum_{i=1}^n I(Y_{2,i+1}^n;Y_{1i}|M_0, M_1, Y_1^{i-1}).$$

Also

$$\sum_{i=1}^{n} I(M_2; Y_{2i} | M_0, M_1, Y_{2,i+1}^n) \le \sum_{i=1}^{n} I(M_2, Y_1^{i-1}; Y_{2i} | M_0, M_1, Y_{2,i+1}^n)$$

$$= \sum_{i=1}^{n} I(Y_1^{i-1}; Y_{2i} | M_0, M_1, Y_{2,i+1}^n)$$

$$+ \sum_{i=1}^{n} I(M_2; Y_{2i} | M_0, M_1, Y_{2,i+1}^n, Y_1^{i-1}).$$

Combining the above results and identifying auxiliary random variables as $U_{0i} = (M_0, Y_1^{i-1}, Y_{2,i+1}^n)$, $U_{1i} = M_1$, and $U_{2i} = M_2$, we obtain

$$I(M_0, M_1; Y_1^n) + I(M_2; Y_2^n | M_0, M_1)$$

$$\le \sum_{i=1}^{n} I(M_0, M_1, Y_1^{i-1}, Y_{2,i+1}^n; Y_{1i}) - \sum_{i=1}^{n} I(Y_{2,i+1}^n; Y_{1i} | M_0, M_1, Y_1^{i-1})$$

$$+ \sum_{i=1}^{n} I(Y_1^{i-1}; Y_{2i} | M_0, M_1, Y_{2,i+1}^n) + \sum_{i=1}^{n} I(M_2; Y_{2i} | M_0, M_1, Y_{2,i+1}^n, Y_1^{i-1})$$

$$\overset{(a)}{=} \sum_{i=1}^{n} I(M_0, M_1, Y_1^{i-1}, Y_{2,i+1}^n; Y_{1i}) + \sum_{i=1}^{n} I(M_2; Y_{2i} | M_0, M_1, Y_{2,i+1}^n, Y_1^{i-1})$$

$$= \sum_{i=1}^{n} I(U_{0i}, U_{1i}; Y_{1i}) + \sum_{i=1}^{n} I(U_{2i}; Y_{2i} | U_{0i}, U_{1i}),$$

where (a) follows by the Csiszár sum identity. Similarly

$$I(M_1; Y_1^n | M_0, M_2) + I(M_0, M_2; Y_2^n) \le \sum_{i=1}^{n} I(U_{1i}; Y_{1i} | U_{0i}, U_{2i}) + \sum_{i=1}^{n} I(U_{0i}, U_{2i}; Y_{2i}).$$

It can be also easily shown that

$$I(M_0; Y_j^n) \le \sum_{i=1}^{n} I(U_{0i}; Y_{ji}),$$

$$I(M_0, M_j; Y_j^n) \le \sum_{i=1}^{n} I(U_{0i}, U_{ji}; Y_{ji}), \quad j = 1, 2.$$

The rest of the proof follows by introducing a time-sharing random variable Q independent of $(M_0, M_1, M_2, X^n, Y_1^n, Y_2^n)$, and uniformly distributed over $[1 : n]$, and defining $U_0 = (Q, U_{0Q})$, $U_1 = U_{1Q}$, $U_2 = U_{2Q}$, $X = X_Q$, $Y_1 = Y_{1Q}$, and $Y_2 = Y_{2Q}$. Using arguments similar to Remark 8.4, it can be easily verified that taking a function $x(u_0, u_1, u_2)$ suffices. Note that the independence of the messages M_1 and M_2 implies the independence of the auxiliary random variables U_1 and U_2 as specified.

CHAPTER 9

Gaussian Vector Channels

Gaussian vector channels are models for multiple-input multiple-output (MIMO) wireless communication systems in which each transmitter and each receiver can have more than a single antenna. The use of multiple antennas provides several benefits in a wireless multipath medium, including higher received power via beamforming, higher channel capacity via spatial multiplexing without increasing bandwidth or transmission power, and improved transmission robustness via diversity coding. In this chapter, we focus on the limits on spatial multiplexing of MIMO communication.

We first establish the capacity of the Gaussian vector point-to-point channel. We show that this channel is equivalent to the Gaussian product channel discussed in Section 3.4 and thus the capacity is achieved via water-filling. We then establish the capacity region of the Gaussian vector multiple access channel as a straightforward extension of the scalar case and show that the sum-capacity is achieved via iterative water-filling.

The rest of the chapter is dedicated to studying Gaussian vector broadcast channels. We first establish the capacity region with common message for the special case of the Gaussian product broadcast channel, which is not in general degraded. Although the capacity region of this channel is achieved via superposition coding with a product input pmf on the channel components, the codebook generation for the common message and decoding must each be performed jointly across the channel components. For the special case of private messages only, the capacity region can be expressed as the intersection of the capacity regions of two enhanced degraded broadcast channels. Next, we turn our attention to the general Gaussian vector broadcast channel and establish the private-message capacity region. We first describe a vector extension of writing on dirty paper for the Gaussian BC discussed in Section 8.3. We show that this scheme is optimal by constructing an enhanced degraded Gaussian vector broadcast channel for every corner point on the boundary of the capacity region. The proof uses Gaussian vector BC–MAC duality, convex optimization techniques (Lagrange duality and the KKT condition), and the entropy power inequality.

In Chapter 23, we show that in the presence of fading, the capacity of the vector Gaussian channel grows linearly with the number of antennas.

9.1 GAUSSIAN VECTOR POINT-TO-POINT CHANNEL

Consider the point-to-point communication system depicted in Figure 9.1. The sender

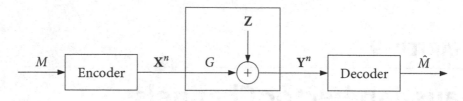

Figure 9.1. MIMO point-to-point communication system.

wishes to reliably communicate a message M to the receiver over a MIMO communication channel.

We model the MIMO communication channel as a Gaussian vector channel, where the output of the channel $\mathbf{Y}$ corresponding to the input $\mathbf{X}$ is

$$\mathbf{Y} = G\mathbf{X} + \mathbf{Z}.$$

Here $\mathbf{Y}$ is an r-dimensional vector, $\mathbf{X}$ is a t-dimensional vector, $\mathbf{Z} \sim N(0, K_{\mathbf{Z}})$, $K_{\mathbf{Z}} > 0$, is an r-dimensional noise vector, and G is an $r \times t$ constant channel gain matrix with its element G_{jk} representing the gain of the channel from transmitter antenna k to receiver antenna j. The channel is discrete-time and the noise vector process $\{\mathbf{Z}(i)\}$ is i.i.d. with $\mathbf{Z}(i) \sim N(0, K_{\mathbf{Z}})$ for every transmission $i \in [1 : n]$. We assume average transmission power constraint P on every codeword $\mathbf{x}^n(m) = (\mathbf{x}(m, 1), \ldots, \mathbf{x}(m, n))$, i.e.,

$$\sum_{i=1}^{n} \mathbf{x}^T(m, i)\mathbf{x}(m, i) \le nP, \quad m \in [1 : 2^{nR}].$$

Remark 9.1. We can assume without loss of generality that $K_{\mathbf{Z}} = I_r$, since the channel $\mathbf{Y} = G\mathbf{X} + \mathbf{Z}$ with a general $K_{\mathbf{Z}} > 0$ can be transformed into the channel

$$\tilde{\mathbf{Y}} = K_{\mathbf{Z}}^{-1/2}\mathbf{Y}$$
$$= K_{\mathbf{Z}}^{-1/2}G\mathbf{X} + \tilde{\mathbf{Z}},$$

where $\tilde{\mathbf{Z}} = K_{\mathbf{Z}}^{-1/2}\mathbf{Z} \sim N(0, I_r)$, and vice versa.

Remark 9.2. The Gaussian vector channel reduces to the Gaussian product channel in Section 3.4 when $r = t = d$, $G = \text{diag}(g_1, g_2, \ldots, g_d)$, and $K_{\mathbf{Z}} = I_d$.

The capacity of the Gaussian vector channel is obtained by a straightforward evaluation of the formula for channel capacity with input cost in Theorem 3.2.

Theorem 9.1. The capacity of the Gaussian vector channel is

$$C = \max_{K_{\mathbf{X}} \ge 0 : \text{tr}(K_{\mathbf{X}}) \le P} \frac{1}{2} \log |GK_{\mathbf{X}}G^T + I_r|.$$

Proof. First note that the capacity with power constraint is upper bounded as

$$
\begin{aligned}
C &\le \sup_{F(\mathbf{x}):\mathrm{E}(\mathbf{X}^T\mathbf{X})\le P} I(\mathbf{X};\mathbf{Y}) \\
&= \sup_{F(\mathbf{x}):\mathrm{E}(\mathbf{X}^T\mathbf{X})\le P} h(\mathbf{Y}) - h(\mathbf{Z}) \\
&= \max_{K_\mathbf{X}\succeq 0:\mathrm{tr}(K_\mathbf{X})\le P} \frac{1}{2}\log|GK_\mathbf{X}G^T + I_r|,
\end{aligned}
$$

where the last step follows by the maximum differential entropy lemma in Section 2.2. In particular, the supremum is attained by a Gaussian $\mathbf{X}$ with zero mean and covariance matrix $K_\mathbf{X}$. With this choice of $\mathbf{X}$, the output $\mathbf{Y}$ is also Gaussian with covariance matrix $GK_\mathbf{X}G^T + I_r$. To prove achievability, we resort to the DMC with input cost in Section 3.3 and use the discretization procedure in Section 3.4. This completes the proof of Theorem 9.1.

The optimal covariance matrix $K_\mathbf{X}^*$ can be characterized more explicitly. Suppose that G has rank d and singular value decomposition $G = \Phi\Gamma\Psi^T$ with $\Gamma = \mathrm{diag}(\gamma_1, \gamma_2, \dots, \gamma_d)$. Then

$$
\begin{aligned}
C &= \max_{K_\mathbf{X}\succeq 0:\mathrm{tr}(K_\mathbf{X})\le P} \frac{1}{2}\log|GK_\mathbf{X}G^T + I_r| \\
&= \max_{K_\mathbf{X}\succeq 0:\mathrm{tr}(K_\mathbf{X})\le P} \frac{1}{2}\log|\Phi\Gamma\Psi^T K_\mathbf{X}\Psi\Gamma\Phi^T + I_r| \\
&\overset{(a)}{=} \max_{K_\mathbf{X}\succeq 0:\mathrm{tr}(K_\mathbf{X})\le P} \frac{1}{2}\log|\Phi^T\Phi\Gamma\Psi^T K_\mathbf{X}\Psi\Gamma + I_d| \\
&\overset{(b)}{=} \max_{K_\mathbf{X}\succeq 0:\mathrm{tr}(K_\mathbf{X})\le P} \frac{1}{2}\log|\Gamma\Psi^T K_\mathbf{X}\Psi\Gamma + I_d| \\
&\overset{(c)}{=} \max_{\tilde{K}_\mathbf{X}\succeq 0:\mathrm{tr}(\tilde{K}_\mathbf{X})\le P} \frac{1}{2}\log|\Gamma\tilde{K}_\mathbf{X}\Gamma + I_d|,
\end{aligned}
$$

where (a) follows since $|AB + I| = |BA + I|$ with $A = \Phi\Gamma\Psi^T K_\mathbf{X}\Psi\Gamma$ and $B = \Phi^T$, and (b) follows since $\Phi^T\Phi = I_d$ (recall the definition of the singular value decomposition in Notation), and (c) follows since the maximization problem is equivalent to that in (b) via the transformations $\tilde{K}_\mathbf{X} = \Psi^T K_\mathbf{X}\Psi$ and $K_\mathbf{X} = \Psi\tilde{K}_\mathbf{X}\Psi^T$. By Hadamard's inequality, the optimal $\tilde{K}_\mathbf{X}^*$ is a diagonal matrix $\mathrm{diag}(P_1, P_2, \dots, P_d)$ such that the water-filling condition is satisfied, i.e.,

$$
P_j = \left[\lambda - \frac{1}{\gamma_i^2}\right]^+,
$$

where λ is chosen so that $\sum_{j=1}^{d} P_j = P$. Finally, from the transformation between $K_\mathbf{X}$ and $\tilde{K}_\mathbf{X}$, the optimal $K_\mathbf{X}^*$ is $K_\mathbf{X}^* = \Psi\tilde{K}_\mathbf{X}^*\Psi^T$. Thus the transmitter should align its signal *direction* with the singular vectors of the effective channel and allocate an appropriate amount of power in each direction to *water-fill* over the singular values.

9.1.1 Equivalent Gaussian Product Channel

The role of the singular values of the Gaussian vector channel gain matrix can be seen more directly. Let $\mathbf{Y} = G\mathbf{X} + \mathbf{Z}$, where the channel gain matrix G has a singular value decomposition $G = \Phi\Gamma\Psi^T$ and rank d. We show that this channel is *equivalent* to the Gaussian product channel

$$\tilde{Y}_j = \gamma_j \tilde{X}_j + \tilde{Z}_j, \quad j \in [1:d],$$

where $\tilde{Z}_j$, $j \in [1:d]$, are independent Gaussian additive noise with common average power of 1.

First consider the channel $\mathbf{Y} = G\mathbf{X} + \mathbf{Z}$ with the input transformation $\mathbf{X} = \Psi\tilde{\mathbf{X}}$ and the output transformation $\tilde{\mathbf{Y}} = \Phi^T\mathbf{Y}$. This gives a Gaussian product channel

$$
\begin{aligned}
\tilde{\mathbf{Y}} &= \Phi^T G\mathbf{X} + \Phi^T\mathbf{Z} \\
&= \Phi^T(\Phi\Gamma\Psi^T)\Psi\tilde{\mathbf{X}} + \Phi^T\mathbf{Z} \\
&= \Gamma\tilde{\mathbf{X}} + \tilde{\mathbf{Z}},
\end{aligned}
$$

where the zero-mean Gaussian noise vector $\tilde{\mathbf{Z}} = \Phi^T\mathbf{Z}$ has covariance matrix $\Phi^T\Phi = I_d$. Let $\tilde{K}_{\mathbf{X}} = \mathrm{E}(\tilde{\mathbf{X}}\tilde{\mathbf{X}}^T)$ and $K_{\mathbf{X}} = \mathrm{E}(\mathbf{X}\mathbf{X}^T)$. Then

$$\mathrm{tr}(K_{\mathbf{X}}) = \mathrm{tr}(\Psi\tilde{K}_{\mathbf{X}}\Psi^T) = \mathrm{tr}(\Psi^T\Psi\tilde{K}_{\mathbf{X}}) = \mathrm{tr}(\tilde{K}_{\mathbf{X}}).$$

Hence, every code for the Gaussian product channel $\tilde{\mathbf{Y}} = \Gamma\tilde{\mathbf{X}} + \tilde{\mathbf{Z}}$ can be transformed into a code for the Gaussian vector channel $\mathbf{Y} = G\mathbf{X} + \mathbf{Z}$ with the same probability of error.

Conversely, given the channel $\tilde{\mathbf{Y}} = \Gamma\tilde{\mathbf{X}} + \tilde{\mathbf{Z}}$, we perform the input transformation $\tilde{\mathbf{X}} = \Psi^T\mathbf{X}$ and the output transformation $\mathbf{Y}' = \Phi\tilde{\mathbf{Y}}$ to obtain the channel

$$\mathbf{Y}' = \Phi\Gamma\Psi^T\mathbf{X} + \Phi\tilde{\mathbf{Z}} = G\mathbf{X} + \Phi\tilde{\mathbf{Z}}.$$

Noting that $\Phi\Phi^T \preceq I_r$, we then add an independent Gaussian noise $\mathbf{Z}' \sim \mathrm{N}(0, I_r - \Phi\Phi^T)$ to $\mathbf{Y}'$, which yields

$$\mathbf{Y} = \mathbf{Y}' + \mathbf{Z}' = G\mathbf{X} + \Phi\tilde{\mathbf{Z}} + \mathbf{Z}' = G\mathbf{X} + \mathbf{Z},$$

where $\mathbf{Z} = \Phi\tilde{\mathbf{Z}} + \mathbf{Z}'$ has the covariance matrix $\Phi\Phi^T + (I_r - \Phi\Phi^T) = I_r$. Also since $\Psi\Psi^T \preceq I_t$, $\mathrm{tr}(\tilde{K}_{\mathbf{X}}) = \mathrm{tr}(\Psi^T K_{\mathbf{X}}\Psi) = \mathrm{tr}(\Psi\Psi^T K_{\mathbf{X}}) \leq \mathrm{tr}(K_{\mathbf{X}})$. Hence every code for the Gaussian vector channel $\mathbf{Y} = G\mathbf{X} + \mathbf{Z}$ can be transformed into a code for the Gaussian product channel $\tilde{\mathbf{Y}} = \Gamma\tilde{\mathbf{X}} + \tilde{\mathbf{Z}}$ with the same probability of error. Consequently, the two channels have the same capacity.

9.1.2 Reciprocity

Since the channel gain matrices G and G^T have the same set of (nonzero) singular values, the channels with gain matrices G and G^T have the same capacity. In fact, both channels are equivalent to the same Gaussian product channel, and hence are equivalent to each other. The following result is an immediate consequence of this equivalence, and will be useful later in proving the Gaussian vector BC–MAC duality.

Lemma 9.1 (Reciprocity Lemma). For every $r \times t$ channel gain matrix G and $t \times t$ matrix $K \succeq 0$, there exists an $r \times r$ matrix $\overline{K} \succeq 0$ such that

$$\text{tr}(\overline{K}) \le \text{tr}(K),$$
$$|G^T \overline{K} G + I_t| = |G K G^T + I_r|.$$

To prove this lemma, consider the singular value decomposition $G = \Phi \Gamma \Psi^T$ and let $\overline{K} = \Phi \Psi^T K \Psi \Phi^T$. Now, we check that $\overline{K}$ satisfies both properties. Indeed, since $\Psi \Psi^T \preceq I_t$,

$$\text{tr}(\overline{K}) = \text{tr}(\Psi \Phi^T \Phi \Psi^T K) = \text{tr}(\Psi \Psi^T K) \le \text{tr}(K).$$

To check the second property, let $d = \text{rank}(G)$ and consider

$$
\begin{aligned}
|G^T \overline{K} G + I_t| &= |\Psi \Gamma \Phi^T (\Phi \Psi^T K \Psi \Phi^T) \Phi \Gamma \Psi^T + I_t| \\
&\overset{(a)}{=} |(\Psi^T \Psi) \Gamma \Psi^T K \Psi \Gamma + I_d| \\
&= |\Gamma \Psi^T K \Psi \Gamma + I_d| \\
&= |(\Phi^T \Phi) \Gamma \Psi^T K \Psi \Gamma + I_d| \\
&\overset{(b)}{=} |\Phi \Gamma \Psi^T K \Psi \Gamma \Phi^T + I_r| \\
&= |G K G^T + I_r|,
\end{aligned}
$$

where (a) follows since if $A = \Psi \Gamma \Psi^T K \Psi \Gamma$ and $B = \Psi^T$, then $|I + AB| = |I + BA|$, and (b) follows similarly. This completes the proof of the reciprocity lemma.

9.1.3 Alternative Characterization of $K_{\mathbf{X}}^*$

Consider the Gaussian vector channel $\mathbf{Y} = G\mathbf{X} + \mathbf{Z}$, where $\mathbf{Z}$ has an arbitrary covariance matrix $K_{\mathbf{Z}} \succ 0$. We have already characterized the optimal input covariance matrix $K_{\mathbf{X}}^*$ for the effective channel gain matrix $K_{\mathbf{Z}}^{-1/2} G$ via singular value decomposition and water-filling. Here we give an alternative characterization of $K_{\mathbf{X}}^*$ via Lagrange duality.

First note that the optimal input covariance matrix $K_{\mathbf{X}}^*$ is the solution to the convex optimization problem

$$
\begin{aligned}
\text{maximize} \quad & \frac{1}{2} \log |G K_{\mathbf{X}} G^T + K_{\mathbf{Z}}| \\
\text{subject to} \quad & K_{\mathbf{X}} \succeq 0, \\
& \text{tr}(K_{\mathbf{X}}) \le P.
\end{aligned}
\tag{9.1}
$$

With dual variables

$$
\begin{aligned}
\text{tr}(K_{\mathbf{X}}) \le P \quad &\Leftrightarrow \quad \lambda \ge 0, \\
K_{\mathbf{X}} \succeq 0 \quad &\Leftrightarrow \quad Y \succeq 0,
\end{aligned}
$$

we can form the Lagrangian

$$L(K_{\mathbf{X}}, Y, \lambda) = \frac{1}{2} \log |G K_{\mathbf{X}} G^T + K_{\mathbf{Z}}| + \text{tr}(Y K_{\mathbf{X}}) - \lambda(\text{tr}(K_{\mathbf{X}}) - P).$$

Recall that if a convex optimization problem satisfies *Slater's condition* (that is, the feasible region has an interior point), then the *Karush–Kuhn–Tucker (KKT) condition* provides necessary and sufficient condition for the optimal solution; see Appendix E. For the problem in (9.1), Slater's condition is satisfied for any $P > 0$. Thus, a solution K_X^* is primal optimal iff there exists a dual optimal solution (λ^*, Υ^*) that satisfies the KKT condition:

- The Lagrangian is stationary (zero differential with respect to K_X), i.e.,

$$\frac{1}{2}G^T(GK_X^*G^T + K_Z)^{-1}G + \Upsilon^* - \lambda^* I_r = 0.$$

- The complementary slackness conditions

$$\lambda^*\left(\operatorname{tr}(K_X^*) - P\right) = 0,$$
$$\operatorname{tr}(\Upsilon^* K_X^*) = 0$$

are satisfied.

In particular, fixing $\Upsilon^* = 0$, any solution K_X^* with $\operatorname{tr}(K_X^*) = P$ is optimal if

$$\frac{1}{2}G^T(GK_X^*G^T + K_Z)^{-1}G = \lambda^* I_r$$

for some $\lambda^* > 0$. Such a covariance matrix K_X^* corresponds to water-filling with all subchannels under water (which occurs when the SNR is sufficiently high).

9.2   GAUSSIAN VECTOR MULTIPLE ACCESS CHANNEL

Consider the MIMO multiple access communication system depicted in Figure 9.2, where each sender wishes to communicate an independent message to the receiver. Assume a Gaussian vector multiple access channel (GV-MAC) model

$$Y = G_1 X_1 + G_2 X_2 + Z,$$

where Y is an r-dimensional output vector, X_1 and X_2 are t-dimensional input vectors, G_1 and G_2 are $r \times t$ channel gain matrices, and $Z \sim N(0, K_Z)$ is an r-dimensional noise vector. As before, we assume without loss of generality that $K_Z = I_r$. We further assume average power constraint P on each of X_1 and X_2, i.e.,

$$\sum_{i=1}^{n} x_j^T(m_j, i)x_j(m_j, i) \le nP, \quad m_j \in [1:2^{nR_j}], \; j = 1, 2.$$

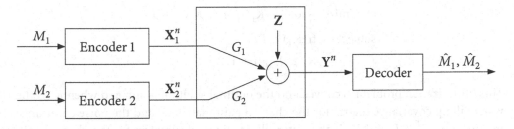

Figure 9.2. MIMO multiple access communication system.

The capacity region of the GV-MAC is given in the following.

Theorem 9.2. The capacity region of the GV-MAC is the set of rate pairs (R_1, R_2) such that

$$R_1 \le \frac{1}{2} \log |G_1 K_1 G_1^T + I_r|,$$

$$R_2 \le \frac{1}{2} \log |G_2 K_2 G_2^T + I_r|,$$

$$R_1 + R_2 \le \frac{1}{2} \log |G_1 K_1 G_1^T + G_2 K_2 G_2^T + I_r|$$

for some $K_1, K_2 \succeq 0$ with $\mathrm{tr}(K_j) \le P$, $j = 1, 2$.

To prove the theorem, note that the capacity region of the GV-MAC can be simply characterized by the capacity region of the MAC with input costs in Problem 4.8 as the set of rate pairs (R_1, R_2) such that

$$R_1 \le I(\mathbf{X}_1; \mathbf{Y}|\mathbf{X}_2, Q),$$
$$R_2 \le I(\mathbf{X}_2; \mathbf{Y}|\mathbf{X}_1, Q),$$
$$R_1 + R_2 \le I(\mathbf{X}_1, \mathbf{X}_2; \mathbf{Y}|Q)$$

for some conditionally independent $\mathbf{X}_1$ and $\mathbf{X}_2$ given Q that satisfy the constraints $E(\mathbf{X}_j^T \mathbf{X}_j) \le P$, $j = 1, 2$. Furthermore, it is easy to show that $|\mathcal{Q}| = 1$ suffices and that among all input distributions with given correlation matrices $K_1 = E(\mathbf{X}_1 \mathbf{X}_1^T)$ and $K_2 = E(\mathbf{X}_2 \mathbf{X}_2^T)$, Gaussian input vectors $\mathbf{X}_1 \sim \mathrm{N}(0, K_1)$ and $\mathbf{X}_2 \sim \mathrm{N}(0, K_2)$ simultaneously maximize all three mutual information bounds. The rest of the achievability proof follows by the discretization procedure.

Remark 9.3. Unlike the point-to-point Gaussian vector channel, which is always equivalent to a product of Gaussian channels, the GV-MAC cannot be factorized into a product of Gaussian MACs in general.

Remark 9.4. The sum-capacity of the GV-MAC can be found by solving the maximization problem

$$\text{maximize} \quad \frac{1}{2} \log |G_1 K_1 G_1^T + G_2 K_2 G_2^T + I_r|$$

$$\text{subject to} \quad \text{tr}(K_j) \le P,$$

$$K_j \succeq 0, \quad j = 1, 2.$$

This optimization problem is convex and the optimal solution is attained when K_1 is the water-filling covariance matrix for the channel gain matrix G_1 and the noise covariance matrix $G_2 K_2 G_2^T + I_r$ and K_2 is the water-filling covariance matrix for the channel gain matrix G_2 and the noise covariance matrix $G_1 K_1 G_1^T + I_r$. The following *iterative water-filling* algorithm finds the optimal (K_1, K_2):

repeat

$$\Sigma_1 \leftarrow G_2 K_2 G_2^T + I_r$$

$$K_1 \leftarrow \arg \max_{K:\text{tr}(K) \le P} \log |G_1 K G_1^T + \Sigma_1|$$

$$\Sigma_2 \leftarrow G_1 K_1 G_1^T + I_r$$

$$K_2 \leftarrow \arg \max_{K:\text{tr}(K) \le P} \log |G_2 K G_2^T + \Sigma_2|$$

until the desired accuracy is reached.

It can be shown that this algorithm converges to the optimal solution (sum-capacity) from any initial assignment of K_1 and K_2.

9.2.1 GV-MAC with More than Two Senders

The capacity region of the GV-MAC can be extended to any number of senders. Consider the k-sender GV-MAC model

$$\mathbf{Y} = \sum_{j=1}^{k} G_j \mathbf{X}_j + \mathbf{Z},$$

where $\mathbf{Z} \sim N(0, K_{\mathbf{Z}})$ is the noise vector. Assume an average power constraint P on each $\mathbf{X}_j$. It can be shown that the capacity region is the set of rate tuples $(R_1, \dots, R_k)$ such that

$$\sum_{j \in \mathcal{J}} R_j \le \frac{1}{2} \log \left| \sum_{j \in \mathcal{J}} G_j K_j G_j^T + I_r \right|, \quad \mathcal{J} \subseteq [1:k],$$

for some $K_1, \dots, K_k \succeq 0$ with $\text{tr}(K_j) \le P$, $j \in [1:k]$. The iterative water-filling algorithm can be easily extended to find the sum-capacity achieving covariance matrices $K_1, \dots, K_k$.

9.3 GAUSSIAN VECTOR BROADCAST CHANNEL

Consider the MIMO broadcast communication system depicted in Figure 9.3. The sender wishes to communicate a common message M_0 to the two receivers and a private message

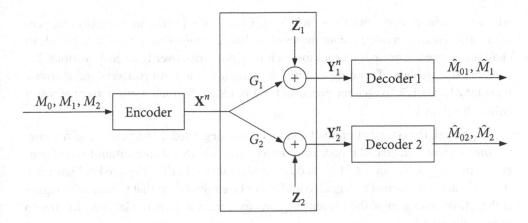

Figure 9.3. MIMO broadcast communication system.

M_j to receiver $j = 1, 2$. The channel is modeled by a Gaussian vector broadcast channel (GV-BC)

$$Y_1 = G_1 X + Z_1,$$
$$Y_2 = G_2 X + Z_2,$$

where G_1, G_2 are $r \times t$ channel gain matrices and $Z_1 \sim N(0, I_r)$ and $Z_2 \sim N(0, I_r)$. Assume the average transmission power constraint $\sum_{i=1}^{n} x^T(m_0, m_1, m_2, i)x(m_0, m_1, m_2, i) \le nP$ for $(m_0, m_1, m_2) \in [1 : 2^{nR_0}] \times [1 : 2^{nR_1}] \times [1 : 2^{nR_2}]$.

Note that unlike the scalar Gaussian BC, the Gaussian vector BC is not in general degraded, and the capacity region is known only in several special cases.

- If $t = r$ and G_1 and G_2 are diagonal, then the channel is a product of Gaussian BCs and the capacity region is known. This case is discussed in the following section.

- If $M_0 = \emptyset$, then the (private-message) capacity region is known. The achievability proof uses a generalization of the writing on dirty paper scheme to the vector case, which is discussed in Section 9.5. The converse proof is quite involved. In fact, even if the channel is degraded, the converse proof does not follow simply by using the EPI as for the scalar case in Section 5.5. The converse proof is given in Section 9.6.

- If $M_1 = \emptyset$ (or $M_2 = \emptyset$), then the (degraded message sets) capacity region is known.

9.4 GAUSSIAN PRODUCT BROADCAST CHANNEL

The Gaussian product broadcast channel is a special case of the Gaussian vector BC depicted in Figure 9.3. The channel consists of a set of d parallel Gaussian BCs

$$Y_{1k} = X_k + Z_{1k},$$
$$Y_{2k} = X_k + Z_{2k}, \quad k \in [1 : d],$$

where the noise components $Z_{jk} \sim N(0, N_{jk})$, $j = 1, 2$, $k \in [1 : d]$, are mutually independent. This channel models a continuous-time (waveform/spectral) Gaussian broadcast channel, where the noise spectrum for each receiver varies over frequency or time. We consider the problem of sending a common message M_0 to both receivers and a private message M_j, $j = 1, 2$, to each receiver over this broadcast channel. Assume average power constraint P on X^d.

In general, the Gaussian product channel is not degraded, but a product of *reversely* (or *inconsistently*) degraded broadcast channels. As such, we assume without loss of generality that $N_{1k} \le N_{2k}$ for $k \in [1 : r]$ and $N_{1k} > N_{2k}$ for $[r + 1 : d]$ as depicted in Figure 9.4. If $r = d$, then the channel is degraded and it can be easily shown that the capacity region is the Minkowski sum of the capacity regions for each component Gaussian BC (up to power allocation).

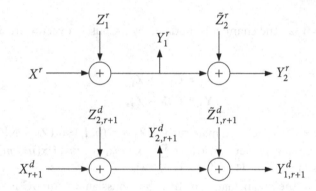

Figure 9.4. Gaussian product broadcast channel. Here $\tilde{Z}_2^r = Z_1^r + \check{Z}_2^r$ and $Z_{1,r+1}^d = \tilde{Z}_{1,r+1}^d + Z_{2,r+1}^d$.

Although the channel under consideration is not in general degraded, the capacity region is established using superposition coding. Coding for the common message, however, requires special attention.

Theorem 9.3. The capacity region of the Gaussian product broadcast channel is the set of rate triples (R_0, R_1, R_2) such that

$$R_0 + R_1 \le \sum_{k=1}^{r} C\left(\frac{\beta_k P}{N_{1k}}\right) + \sum_{k=r+1}^{d} C\left(\frac{\alpha_k \beta_k P}{\bar{\alpha}_k \beta_k P + N_{1k}}\right),$$

$$R_0 + R_2 \le \sum_{k=1}^{r} C\left(\frac{\alpha_k \beta_k P}{\bar{\alpha}_k \beta_k P + N_{2k}}\right) + \sum_{k=r+1}^{d} C\left(\frac{\beta_k P}{N_{2k}}\right),$$

$$R_0 + R_1 + R_2 \le \sum_{k=1}^{r} C\left(\frac{\beta_k P}{N_{1k}}\right) + \sum_{k=r+1}^{d} \left(C\left(\frac{\alpha_k \beta_k P}{\bar{\alpha}_k \beta_k P + N_{1k}}\right) + C\left(\frac{\bar{\alpha}_k \beta_k P}{N_{2k}}\right)\right),$$

$$R_0 + R_1 + R_2 \leq \sum_{k=1}^{r} \left(C\left(\frac{\alpha_k \beta_k P}{\bar{\alpha}_k \beta_k P + N_{2k}}\right) + C\left(\frac{\bar{\alpha}_k \beta_k P}{N_{1k}}\right) \right) + \sum_{k=r+1}^{d} C\left(\frac{\beta_k P}{N_{2k}}\right)$$

for some $\alpha_k, \beta_k \in [0, 1]$, $k \in [1 : d]$, with $\sum_{k=1}^{d} \beta_k = 1$.

To prove Theorem 9.3, we first consider a product DM-BC consisting of two reversely degraded DM-BCs $(\mathcal{X}_1, p(y_{11}|x_1)p(y_{21}|y_{11}), \mathcal{Y}_{11} \times \mathcal{Y}_{21})$ and $(\mathcal{X}_2, p(y_{22}|x_2)p(y_{12}|y_{22}), \mathcal{Y}_{12} \times \mathcal{Y}_{22})$ depicted in Figure 9.5. The capacity of this channel is as follows.

Proposition 9.1. The capacity region of the product of two reversely degraded DM-BCs is the set of rate triples (R_0, R_1, R_2) such that

$$R_0 + R_1 \leq I(X_1; Y_{11}) + I(U_2; Y_{12}),$$
$$R_0 + R_2 \leq I(X_2; Y_{22}) + I(U_1; Y_{21}),$$
$$R_0 + R_1 + R_2 \leq I(X_1; Y_{11}) + I(U_2; Y_{12}) + I(X_2; Y_{22}|U_2),$$
$$R_0 + R_1 + R_2 \leq I(X_2; Y_{22}) + I(U_1; Y_{21}) + I(X_1; Y_{11}|U_1)$$

for some pmf $p(u_1, x_1)p(u_2, x_2)$.

The converse proof of this proposition follows from the Nair–El Gamal outer bound in Chapter 8. The proof of achievability uses rate splitting and superposition coding.

Rate splitting. Divide M_j, $j = 1, 2$, into two independent messages: M_{j0} at rate R_{j0} and M_{jj} at rate R_{jj}.

Codebook generation. Fix a pmf $p(u_1, x_1)p(u_2, x_2)$. Randomly and independently generate $2^{n(R_0 + R_{10} + R_{20})}$ sequence pairs $(u_1^n, u_2^n)(m_0, m_{10}, m_{20})$, $(m_0, m_{10}, m_{20}) \in [1 : 2^{nR_0}] \times [1 : 2^{nR_{10}}] \times [1 : 2^{nR_{20}}]$, each according to $\prod_{i=1}^{n} p_{U_1}(u_{1i})p_{U_2}(u_{2i})$. For each (m_0, m_{10}, m_{20}), randomly and conditionally independently generate $2^{nR_{jj}}$ sequences $x_j^n(m_0, m_{10}, m_{20}, m_{jj})$, $m_{jj} \in [1 : 2^{nR_{jj}}]$, $j = 1, 2$, each according to $\prod_{i=1}^{n} p_{X_j|U_j}(x_{ji}|u_{ji}(m_0, m_{10}, m_{20}))$.

Encoding. To send the message triple $(m_0, m_1, m_2) = (m_0, (m_{10}, m_{11}), (m_{20}, m_{22}))$, the encoder transmits $(x_1^n(m_0, m_{10}, m_{20}, m_{11}), x_2^n(m_0, m_{10}, m_{20}, m_{22}))$.

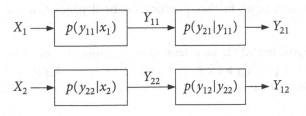

Figure 9.5. Product of two reversely degraded DM-BCs.

Decoding and analysis of the probability of error. Decoder 1 finds the unique triple $(\hat{m}_{01}, \hat{m}_{10}, \hat{m}_{11})$ such that $((u_1^n, u_2^n)(\hat{m}_{01}, \hat{m}_{10}, m_{20}), x_1^n(\hat{m}_{01}, \hat{m}_{10}, m_{20}, \hat{m}_{11}), y_{11}^n, y_{12}^n) \in T_\epsilon^{(n)}$ for some m_{20}. Similarly, decoder 2 finds the unique triple $(\hat{m}_{02}, \hat{m}_{20}, \hat{m}_{22})$ such that $((u_1^n, u_2^n)(\hat{m}_{02}, m_{10}, \hat{m}_{20}), x_2^n(\hat{m}_{02}, m_{10}, \hat{m}_{20}, \hat{m}_{22}), y_{21}^n, y_{22}^n) \in T_\epsilon^{(n)}$ for some m_{10}.

Using standard arguments, it can be shown that the probability of error for decoder 1 tends to zero as $n \to \infty$ if

$$R_0 + R_1 + R_{20} < I(U_1, U_2, X_1; Y_{11}, Y_{12}) - \delta(\epsilon)$$
$$= I(X_1; Y_{11}) + I(U_2; Y_{12}) - \delta(\epsilon),$$
$$R_{11} < I(X_1; Y_{11} | U_1) - \delta(\epsilon).$$

Similarly, the probability of error for decoder 2 tends to zero as $n \to \infty$ if

$$R_0 + R_{10} + R_2 < I(X_2; Y_{22}) + I(U_1; Y_{21}) - \delta(\epsilon),$$
$$R_{22} < I(X_2; Y_{22} | U_2) - \delta(\epsilon).$$

Substituting $R_{jj} = R_j - R_{j0}$, $j = 1, 2$, combining with the constraints $R_{10}, R_{20} \geq 0$, and eliminating R_{10} and R_{20} by the Fourier–Motzkin procedure in Appendix D completes the achievability proof of Proposition 9.1.

Remark 9.5. It is interesting to note that even though U_1 and U_2 are statistically independent, the codebook for the common message is generated simultaneously, and not as the product of two independent codebooks. Decoding for the common message at each receiver is also performed simultaneously.

Remark 9.6. Recall that the capacity region $\mathscr{C}$ of the product of two degraded DM-BCs $p(y_{11}|x_1)p(y_{21}|y_{11})$ and $p(y_{12}|x_2)p(y_{22}|y_{12})$ is the Minkowski sum of the individual capacity regions $\mathscr{C}_1$ and $\mathscr{C}_2$, i.e., $\mathscr{C} = \{(R_{01} + R_{02}, R_{11} + R_{12}, R_{21} + R_{22}) : (R_{01}, R_{11}, R_{21}) \in \mathscr{C}_1, (R_{02}, R_{12}, R_{22}) \in \mathscr{C}_2\}$; see Problem 5.11. However, the capacity region in Proposition 9.1 is in general larger than the Minkowski sum of the individual capacity regions. For example, consider the special case of reversely degraded BC with $Y_{21} = Y_{12} = \emptyset$. The common-message capacities of the component BCs are $C_{01} = C_{02} = 0$, while the common-message capacity of the product BC is

$$C_0 = \min\left\{\max_{p(x_1)} I(X_1; Y_{11}),\ \max_{p(x_2)} I(X_2; Y_{22})\right\}.$$

This shows that simultaneous codebook generation and decoding for the common message can be much more powerful than product codebook generation and separate decoding.

Remark 9.7. Proposition 9.1 can be extended to the product of any number of channels. Suppose that $X_k \to Y_{1k} \to Y_{2k}$ for $k \in [1:r]$ and $X_k \to Y_{2k} \to Y_{1k}$ for $k \in [r+1:d]$. Then the capacity region of the product of these d reversely degraded DM-BCs is

$$R_0 + R_1 \leq \sum_{k=1}^{r} I(X_k; Y_{1k}) + \sum_{k=r+1}^{d} I(U_k; Y_{1k}),$$

$$R_0 + R_2 \le \sum_{k=1}^{r} I(U_k; Y_{2k}) + \sum_{k=r+1}^{d} I(X_k; Y_{2k}),$$

$$R_0 + R_1 + R_2 \le \sum_{k=1}^{r} I(X_k; Y_{1k}) + \sum_{k=r+1}^{d} (I(U_k; Y_{1k}) + I(X_k; Y_{2k}|U_k)),$$

$$R_0 + R_1 + R_2 \le \sum_{k=1}^{r} (I(U_k; Y_{2k}) + I(X_k; Y_{1k}|U_k)) + \sum_{k=r+1}^{d} I(X_k; Y_{2k})$$

for some pmf $p(u_1, x_1)p(u_2, x_2)$.

Proof of Theorem 9.3. Achievability follows immediately by extending Proposition 9.1 to the case with input cost, taking $U_k \sim N(0, \alpha_k \beta_k P)$, $V_k \sim N(0, \bar{\alpha}_k \beta_k P)$, $k \in [1:d]$, independent of each other, and $X_k = U_k + V_k$, $k \in [1:d]$, and using the discretization procedure in Section 3.4. The converse follows by considering the rate region in Proposition 9.1 with average power constraint and appropriately applying the EPI.

Remark 9.8. The capacity region of the product of reversely degraded DM-BCs for more than two receivers is not known in general.

Specialization to the private-message capacity region. By setting $R_0 = 0$, Proposition 9.1 yields the private-message capacity region $\mathscr{C}$ of the product of two reversely degraded DM-BC that consists of all rate pairs (R_1, R_2) such that

$$R_1 \le I(X_1; Y_{11}) + I(U_2; Y_{12}),$$
$$R_2 \le I(X_2; Y_{22}) + I(U_1; Y_{21}),$$
$$R_1 + R_2 \le I(X_1; Y_{11}) + I(U_2; Y_{12}) + I(X_2; Y_{22}|U_2),$$
$$R_1 + R_2 \le I(X_2; Y_{22}) + I(U_1; Y_{21}) + I(X_1; Y_{11}|U_1)$$

for some pmf $p(u_1, x_1)p(u_2, x_2)$. This can be further simplified as follows.

Let $C_{11} = \max_{p(x_1)} I(X_1; Y_{11})$ and $C_{22} = \max_{p(x_2)} I(X_2; Y_{22})$. Then the private-message capacity region $\mathscr{C}$ is the set of rate pairs (R_1, R_2) such that

$$\begin{aligned} R_1 &\le C_{11} + I(U_2; Y_{12}), \\ R_2 &\le C_{22} + I(U_1; Y_{21}), \\ R_1 + R_2 &\le C_{11} + I(U_2; Y_{12}) + I(X_2; Y_{22}|U_2), \\ R_1 + R_2 &\le C_{22} + I(U_1; Y_{21}) + I(X_1; Y_{11}|U_1) \end{aligned} \qquad (9.2)$$

for some pmf $p(u_1, x_1)p(u_2, x_2)$. This region can be expressed alternatively as the intersection of two regions (see Figure 9.6):

- $\mathscr{C}_1$ that consists of all rate pairs (R_1, R_2) such that

$$R_1 \le C_{11} + I(U_2; Y_{12}),$$
$$R_1 + R_2 \le C_{11} + I(U_2; Y_{12}) + I(X_2; Y_{22}|U_2)$$

for some pmf $p(u_2, x_2)$, and

- $\mathscr{C}_2$ that consists of all rate pairs (R_1, R_2) such that

$$R_2 \leq C_{22} + I(U_1; Y_{21}),$$
$$R_1 + R_2 \leq C_{22} + I(U_1; Y_{21}) + I(X_1; Y_{11}|U_1)$$

for some pmf $p(u_1, x_1)$.

Note that $\mathscr{C}_1$ is the capacity region of the *enhanced* degraded product BC with $Y_{21} = Y_{11}$, which is in general larger than the capacity region of the original product BC. Similarly $\mathscr{C}_2$ is the capacity region of the enhanced degraded product BC with $Y_{12} = Y_{22}$. Thus $\mathscr{C} \subseteq \mathscr{C}_1 \cap \mathscr{C}_2$.

To establish the other direction, i.e., $\mathscr{C} \supseteq \mathscr{C}_1 \cap \mathscr{C}_2$, we show that each boundary point of $\mathscr{C}_1 \cap \mathscr{C}_2$ lies on the boundary of the capacity region $\mathscr{C}$. To show this, first note that $(C_{11}, C_{22}) \in \mathscr{C}$ is on the boundary of $\mathscr{C}_1 \cap \mathscr{C}_2$; see Figure 9.6. Moreover, each boundary point (R_1^*, R_2^*) such that $R_1^* \geq C_{11}$ is on the boundary of $\mathscr{C}_1$ and satisfies the conditions $R_1^* = C_{11} + I(U_2; Y_{12})$, $R_2^* = I(X_2; Y_{22}|U_2)$ for some pmf $p(u_2, x_2)$. By evaluating $\mathscr{C}$ with the same $p(u_2, x_2)$, $U_1 = \emptyset$, and $p(x_1)$ that attains C_{11}, it follows that (R_1^*, R_2^*) lies on the boundary of $\mathscr{C}$. We can similarly show every boundary point (R_1^*, R_2^*) with $R_1^* \leq C_{11}$ also lies on the boundary of $\mathscr{C}$. This shows that $\mathscr{C} = \mathscr{C}_1 \cap \mathscr{C}_2$.

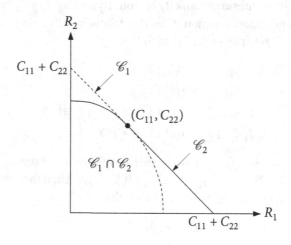

Figure 9.6. The private-message capacity region of a reversely degraded DM-BC: $\mathscr{C} = \mathscr{C}_1 \cap \mathscr{C}_2$.

Remark 9.9. The above argument establishes the converse for the private-message capacity region immediately, since the capacity region of the original BC is contained in that of each enhanced degraded BC. As we will see in Section 9.6.3, this *channel enhancement* approach turns out to be crucial for proving the converse for the general Gaussian vector BC.

Remark 9.10. Unlike the capacity region for the general setting with common message in Proposition 9.1, the private-message capacity region in (9.2) can be generalized to more than two receivers.

9.5   VECTOR WRITING ON DIRTY PAPER

Consider the problem of communicating a message over a Gaussian vector channel with additive Gaussian vector state when the state sequence is available noncausally at the encoder as depicted in Figure 9.7. The channel output corresponding to the input $\mathbf{X}$ is

$$\mathbf{Y} = G\mathbf{X} + \mathbf{S} + \mathbf{Z},$$

where the state $\mathbf{S} \sim N(0, K_\mathbf{S})$ and the noise $\mathbf{Z} \sim N(0, I_r)$ are independent. Assume an average power constraint $\sum_{i=1}^{n} E(\mathbf{x}^T(m, \mathbf{S}, i)\mathbf{x}(m, \mathbf{S}, i)) \le nP$, $m \in [1 : 2^{nR}]$.

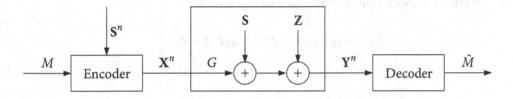

Figure 9.7. Gaussian vector channel with additive Gaussian state available non-causally at the decoder.

This setup is a generalization of the scalar case discussed in Section 7.7, for which the capacity is the same as when the state is not present and is achieved via writing on dirty paper (that is, Gelfand–Pinsker coding specialized to this case).

It turns out that the same result holds for the vector case, that is, the capacity is the same as if $\mathbf{S}$ were not present, and is given by

$$C = \max_{K_\mathbf{X} \succeq 0:\, \mathrm{tr}(K_\mathbf{X}) \le P} \frac{1}{2} \log |GK_\mathbf{X}G^T + I_r|.$$

As for the scalar case, this result follows by appropriately evaluating the Gelfand–Pinsker capacity expression

$$C = \sup_{F(\mathbf{u}|\mathbf{s}),\, \mathbf{x}(\mathbf{u},\mathbf{s}):\, E(\mathbf{X}^T\mathbf{X}) \le P} \big(I(\mathbf{U}; \mathbf{Y}) - I(\mathbf{U}; \mathbf{S}) \big).$$

Let $\mathbf{U} = \mathbf{X} + A\mathbf{S}$, where $\mathbf{X} \sim N(\mathbf{0}, K_\mathbf{X})$ is independent of $\mathbf{S}$ and

$$A = K_\mathbf{X}G^T(GK_\mathbf{X}G^T + I_r)^{-1}.$$

We can easily check that the matrix A is chosen such that $A(G\mathbf{X} + \mathbf{Z})$ is the MMSE estimate of $\mathbf{X}$. Thus, $\mathbf{X} - A(G\mathbf{X} + \mathbf{Z})$ is independent of $G\mathbf{X} + \mathbf{Z}$, $\mathbf{S}$, and hence $\mathbf{Y} = G\mathbf{X} + \mathbf{Z} + \mathbf{S}$. Finally consider

$$h(\mathbf{U}|\mathbf{S}) = h(\mathbf{X} + A\mathbf{S}|\mathbf{S})$$
$$= h(\mathbf{X})$$

and

$$h(\mathbf{U}|\mathbf{Y}) = h(\mathbf{X} + A\mathbf{S}|\mathbf{Y})$$
$$= h(\mathbf{X} + A\mathbf{S} - A\mathbf{Y}|\mathbf{Y})$$
$$= h(\mathbf{X} - A(G\mathbf{X} + \mathbf{Z})|\mathbf{Y})$$
$$= h(\mathbf{X} - A(G\mathbf{X} + \mathbf{Z}))$$
$$= h(\mathbf{X} - A(G\mathbf{X} + \mathbf{Z})|G\mathbf{X} + \mathbf{Z})$$
$$= h(\mathbf{X}|G\mathbf{X} + \mathbf{Z}).$$

Combining the above results implies that we can achieve

$$h(\mathbf{X}) - h(\mathbf{X}|G\mathbf{X} + \mathbf{Z}) = I(\mathbf{X}; G\mathbf{X} + \mathbf{Z})$$
$$= \frac{1}{2} \log |GK_{\mathbf{X}}G^T + I_r|$$

for every $K_{\mathbf{X}}$ with $\operatorname{tr}(K_{\mathbf{X}}) \le P$. Note that this vector writing on dirty paper result holds for any additive (non-Gaussian) state $\mathbf{S}$ independent of the Gaussian noise $\mathbf{Z}$.

9.6   GAUSSIAN VECTOR BC WITH PRIVATE MESSAGES

Again consider the MIMO broadcast communication system depicted in Figure 9.3. Suppose that $M_0 = \emptyset$, that is, the sender only wishes to communicate a private message M_j to receiver $j = 1, 2$. Define the following two rate regions:

- $\mathcal{R}_1$ that consists of all rate pairs (R_1, R_2) such that

$$R_1 < \frac{1}{2} \log \frac{|G_1 K_1 G_1^T + G_1 K_2 G_1^T + I_r|}{|G_1 K_2 G_1^T + I_r|},$$
$$R_2 < \frac{1}{2} \log |G_2 K_2 G_2^T + I_r|$$

  for some $K_1, K_2 \succeq 0$ with $\operatorname{tr}(K_1 + K_2) \le P$, and

- $\mathcal{R}_2$ that consists of all rate pair (R_1, R_2) such that

$$R_1 < \frac{1}{2} \log |G_1 K_1 G_1^T + I_r|,$$
$$R_2 < \frac{1}{2} \log \frac{|G_2 K_2 G_2^T + G_2 K_1 G_2^T + I_r|}{|G_2 K_1 G_2^T + I_r|}$$

  for some $K_1, K_2 \succeq 0$ with $\operatorname{tr}(K_1 + K_2) \le P$.

Let $\mathscr{R}_{\text{WDP}}$ be the convex hull of the union of $\mathscr{R}_1$ and $\mathscr{R}_2$, as illustrated in Figure 9.8. This characterizes the capacity region.

> **Theorem 9.4.** The private-message capacity region of the Gaussian vector BC is
>
> $$\mathscr{C} = \mathscr{R}_{\text{WDP}}.$$

This capacity region coincides with Marton's inner bound without U_0 (see Section 8.3) with properly chosen Gaussian random vectors $(U_1, U_2, \mathbf{X})$. Equivalently and more intuitively, achievability can be established via the vector writing on dirty paper scheme discussed in Section 8.3. The converse proof uses the channel enhancement idea introduced in Section 9.4, convex optimization techniques (see Appendix E for a brief description of these techniques), Gaussian BC–MAC duality, and the conditional vector EPI.

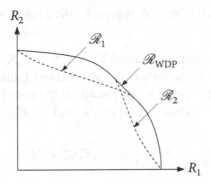

Figure 9.8. The capacity region of a GV-BC: $\mathscr{C} = \mathscr{R}_{\text{WDP}} = \text{co}(\mathscr{R}_1 \cup \mathscr{R}_2)$.

9.6.1   Proof of Achievability

We illustrate the achievability of every rate pair in the interior of $\mathscr{R}_1$ (or $\mathscr{R}_2$). The rest of the capacity region is achieved using time sharing between points in $\mathscr{R}_1$ and $\mathscr{R}_2$. First consider a rate pair (R_1, R_2) in the interior of $\mathscr{R}_2$. The writing on dirty paper scheme is illustrated in Figure 9.9. Fix the covariance matrices K_1 and K_2 such that $\text{tr}(K_1 + K_2) \le P$, and let $\mathbf{X} = \mathbf{X}_1 + \mathbf{X}_2$, where $\mathbf{X}_1 \sim N(0, K_1)$ and $\mathbf{X}_2 \sim N(0, K_2)$ are independent. Encoding is performed successively. In the figure M_2 is encoded first.

- To send M_2 to $\mathbf{Y}_2$, we consider the channel $\mathbf{Y}_2 = G_2\mathbf{X}_2 + G_2\mathbf{X}_1 + \mathbf{Z}_2$ with input $\mathbf{X}_2$, additive independent Gaussian interference signal $G_2\mathbf{X}_1$, and additive Gaussian noise $\mathbf{Z}_2$. Treating the interference signal $G_2\mathbf{X}_1$ as noise, by Theorem 9.1 for the Gaussian vector channel, M_2 can be sent reliably to $\mathbf{Y}_2$ if

$$R_2 < \frac{1}{2} \log \frac{|G_2 K_2 G_2^T + G_2 K_1 G_2^T + I_r|}{|G_2 K_1 G_2^T + I_r|}.$$

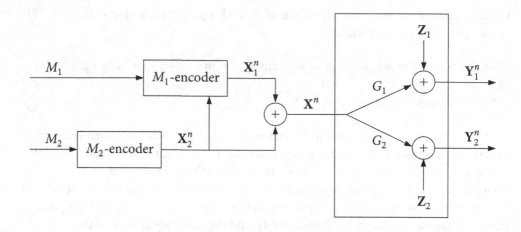

Figure 9.9. Writing on dirty paper for the Gaussian vector BC.

- To send M_1 to $\mathbf{Y}_1$, consider the channel $\mathbf{Y}_1 = G_1\mathbf{X}_1 + G_1\mathbf{X}_2 + \mathbf{Z}_1$, with input $\mathbf{X}_1$, independent additive Gaussian state $G_1\mathbf{X}_2$, and additive Gaussian noise $\mathbf{Z}_1$, where the state vector $G_1\mathbf{X}_2^n(M_2)$ is available noncausally at the encoder. By the vector writing on dirty paper result in Section 9.5, M_1 can be sent reliably to $\mathbf{Y}_1$ if

$$R_1 < \frac{1}{2}\log|G_1 K_1 G_1^T + I_r|.$$

Now, by considering the other message encoding order, we can similarly obtain the constraints on the rates

$$R_1 < \frac{1}{2}\log\frac{|G_1 K_1 G_1^T + G_1 K_2 G_1^T + I_r|}{|G_1 K_2 G_1^T + I_r|},$$
$$R_2 < \frac{1}{2}\log|G_2 K_2 G_2^T + I_r|.$$

The rest of the achievability proof follows as for the scalar case.

Computing the capacity region is difficult because the rate terms for R_j, $j = 1, 2$, are not concave functions of K_1 and K_2. This difficulty can be overcome by the following duality result between the Gaussian vector BC and the Gaussian vector MAC. This result will prove useful in the proof of the converse.

9.6.2 Gaussian Vector BC–MAC Duality

Given the GV-BC (referred to as the *original BC*) with channel gain matrices G_1 and G_2 and power constraint P, consider a GV-MAC with channel gain matrices G_1^T and G_2^T (referred to as the *dual MAC*) as depicted in Figure 9.10.

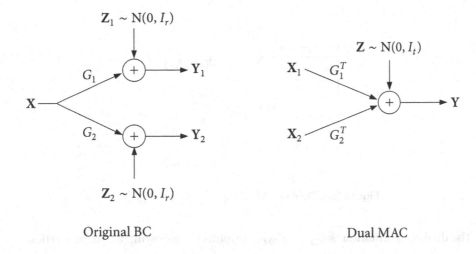

Original BC   Dual MAC

Figure 9.10. Original BC and its dual MAC.

The following states the dual relationship between these two channels.

Lemma 9.2 (BC–MAC Duality Lemma). Let $\mathscr{C}_{\text{DMAC}}$ denote the capacity of the dual MAC under the *sum-power* constraint

$$\sum_{i=1}^{n}(\mathbf{x}_1^T(m_1, i)\mathbf{x}_1(m_1, i) + \mathbf{x}_2^T(m_2, i)\mathbf{x}_2(m_2, i)) \le nP \quad \text{for every } (m_1, m_2).$$

Then

$$\mathscr{R}_{\text{WDP}} = \mathscr{C}_{\text{DMAC}}.$$

Note that this lemma generalizes the scalar Gaussian BC–MAC duality result in Problem 5.15. The proof is given in Appendix 9A.

It is easy to characterize $\mathscr{C}_{\text{DMAC}}$ (see Figure 9.11). Let K_1 and K_2 be the covariance matrices for each sender and $\mathscr{R}(K_1, K_2)$ be the set of rate pairs (R_1, R_2) such that

$$R_1 \le \frac{1}{2} \log |G_1^T K_1 G_1 + I_t|,$$

$$R_2 \le \frac{1}{2} \log |G_2^T K_2 G_2 + I_t|,$$

$$R_1 + R_2 \le \frac{1}{2} \log |G_1^T K_1 G_1 + G_2^T K_2 G_2 + I_t|.$$

Then any rate pair in the interior of $\mathscr{R}(K_1, K_2)$ is achievable and thus the capacity region of the dual MAC under the sum-power constraint is

$$\mathscr{C}_{\text{DMAC}} = \bigcup_{K_1, K_2 \succeq 0: \text{tr}(K_1) + \text{tr}(K_2) \le P} \mathscr{R}(K_1, K_2).$$

The converse proof of this statement follows the same steps as that for the GV-MAC with individual power constraints.

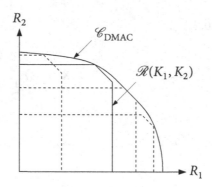

Figure 9.11. Dual GV-MAC capacity region $\mathscr{C}_{\text{DMAC}}$.

The dual representation $\mathscr{R}_{\text{WDP}} = \mathscr{C}_{\text{DMAC}}$ exhibits the following useful properties:

- The rate constraint terms for $\mathscr{R}(K_1, K_2)$ are concave functions of (K_1, K_2), so we can use tools from convex optimization.

- The region $\bigcup_{K_1, K_2} \mathscr{R}(K_1, K_2)$ is closed and convex.

Consequently, each boundary point (R_1^*, R_2^*) of $\mathscr{C}_{\text{DMAC}}$ lies on the boundary of $\mathscr{R}(K_1, K_2)$ for some K_1 and K_2, and is a solution to the convex optimization problem

$$\text{maximize}\quad \alpha R_1 + \bar{\alpha} R_2$$
$$\text{subject to}\quad (R_1, R_2) \in \mathscr{C}_{\text{DMAC}}$$

for some $\alpha \in [0, 1]$. That is, (R_1^*, R_2^*) lies on the supporting line with slope $-\alpha/\bar{\alpha}$ as shown in Figure 9.12. Now it can be easily seen that each boundary point (R_1^*, R_2^*) of $\mathscr{C}_{\text{DMAC}}$ is a corner point of $\mathscr{R}(K_1, K_2)$ for some K_1 and K_2 (we refer to such a corner point (R_1^*, R_2^*) as a *boundary corner point*) or a convex combination of boundary corner points (or both). Furthermore, if a boundary corner point (R_1^*, R_2^*) is inside the positive quadrant, i.e., $R_1^*, R_2^* > 0$, then it has a *unique* supporting line; see Appendix 9B for the proof. In other words, $\mathscr{C}_{\text{DMAC}}$ does not have a *kink* inside the positive quadrant.

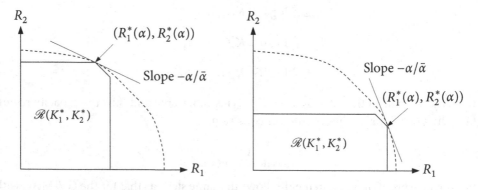

Figure 9.12. Boundary corner points of the dual MAC capacity region.

9.6.3   Proof of the Converse

To prove the converse for Theorem 9.4, by the BC–MAC duality, it suffices to show that every achievable rate pair (R_1, R_2) for the original GV-BC is in $\mathscr{C}_{\text{DMAC}}$. First note that the end points of $\mathscr{C}_{\text{DMAC}}$ are $(C_1, 0)$ and $(0, C_2)$, where

$$C_j = \max_{K_j : \text{tr}(K_j) \leq P} \frac{1}{2} \log |G_j^T K_j G_j + I_t|, \quad j = 1, 2,$$

is the capacity of the channel to each receiver. By the reciprocity lemma (Lemma 9.1), these corner points correspond to the individual capacity bounds for the GV-BC and hence are on the boundary of its capacity region $\mathscr{C}$.

We now focus on proving the optimality of $\mathscr{C}_{\text{DMAC}}$ inside the positive quadrant. The proof consists of the following three steps:

- First we characterize the boundary corner point $(R_1^*(\alpha), R_2^*(\alpha))$ associated with the unique supporting line of slope $-\alpha/\bar{\alpha}$ via Lagrange duality.

- Next we construct an *enhanced* degraded Gaussian vector BC, referred to as DBC(α), for each boundary corner point $(R_1^*(\alpha), R_2^*(\alpha))$ such that the capacity region $\mathscr{C}$ (of the original BC) is contained in the enhanced BC capacity region $\mathscr{C}_{\text{DBC}(\alpha)}$ (Lemma 9.3) as illustrated in Figure 9.13.

- We then show that the boundary corner point $(R_1^*(\alpha), R_2^*(\alpha))$ is on the boundary of $\mathscr{C}_{\text{DBC}(\alpha)}$ (Lemma 9.4). Since $\mathscr{C}_{\text{DMAC}} = \mathscr{R}_{\text{WDP}} \subseteq \mathscr{C} \subseteq \mathscr{C}_{\text{DBC}(\alpha)}$, we conclude that each boundary corner point $(R_1^*(\alpha), R_2^*(\alpha))$ is on the boundary of $\mathscr{C}$ as illustrated in Figure 9.14.

Finally, since every boundary corner point $(R_1^*(\alpha), R_2^*(\alpha))$ of $\mathscr{C}_{\text{DMAC}}$ inside the positive quadrant has a unique supporting line, the boundary of $\mathscr{C}_{\text{DMAC}}$ must coincide with that of $\mathscr{C}$ by Lemma A.2, which completes the proof of the converse. We now give the details of the above three steps.

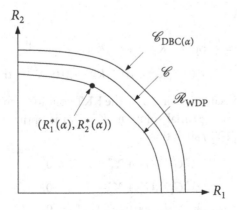

Figure 9.13. Illustration of step 2: $\mathscr{R}_{\text{WDP}} \subseteq \mathscr{C} \subseteq \mathscr{C}_{\text{DBC}(\alpha)}$.

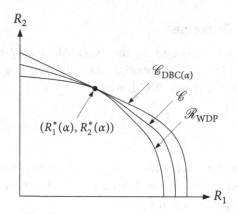

Figure 9.14. Illustration of step 3: Optimality of $(R_1^*(\alpha), R_2^*(\alpha))$.

Step 1 (Boundary corner points of $\mathcal{C}_{\mathrm{DMAC}}$ via Lagrange duality). Recall that every boundary corner point (R_1^*, R_2^*) inside the positive quadrant maximizes $\alpha R_1 + \bar{\alpha} R_2$ for some $\alpha \in [0, 1]$. Assume without loss of generality that $\alpha \le 1/2 \le \bar{\alpha}$. Then the rate pair

$$R_1^* = \frac{1}{2} \log \frac{|G_1^T K_1^* G_1 + G_2^T K_2^* G_2 + I_t|}{|G_2^T K_2^* G_2 + I_t|},$$

$$R_2^* = \frac{1}{2} \log |G_2^T K_2^* G_2 + I_t|$$

uniquely corresponds to an optimal solution to the convex optimization problem

$$\text{maximize } \frac{\alpha}{2} \log |G_1^T K_1 G_1 + G_2^T K_2 G_2 + I_t| + \frac{\bar{\alpha} - \alpha}{2} \log |G_2^T K_2 G_2 + I_t|$$

$$\text{subject to } \mathrm{tr}(K_1) + \mathrm{tr}(K_2) \le P, \quad K_1, K_2 \succeq 0.$$

Introducing the dual variables

$$\mathrm{tr}(K_1) + \mathrm{tr}(K_2) \le P \quad \Leftrightarrow \quad \lambda \ge 0,$$

$$K_1, K_2 \succeq 0 \quad \Leftrightarrow \quad Y_1, Y_2 \succeq 0,$$

we form the Lagrangian

$$L(K_1, K_2, Y_1, Y_2, \lambda) = \frac{\alpha}{2} \log |G_1^T K_1 G_1 + G_2^T K_2 G_2 + I_t| + \frac{\bar{\alpha} - \alpha}{2} \log |G_2^T K_2 G_2 + I_t|$$

$$+ \mathrm{tr}(Y_1 K_1) + \mathrm{tr}(Y_2 K_2) - \lambda(\mathrm{tr}(K_1) + \mathrm{tr}(K_2) - P).$$

Since Slater's condition is satisfied for $P > 0$, the KKT condition characterizes the optimal solution (K_1^*, K_2^*), that is, a primal solution (K_1^*, K_2^*) is optimal iff there exists a dual optimal solution $(\lambda^*, Y_1^*, Y_2^*)$ such that

$$\lambda^* G_1 \Sigma_1 G_1^T + Y_1^* - \lambda^* I_r = 0,$$
$$\lambda^* G_2 \Sigma_2 G_2^T + Y_2^* - \lambda^* I_r = 0,$$
$$\lambda^* (\mathrm{tr}(K_1^*) + \mathrm{tr}(K_2^*) - P) = 0,$$
$$\mathrm{tr}(Y_1^* K_1^*) = \mathrm{tr}(Y_2^* K_2^*) = 0,$$

(9.3)

where

$$\Sigma_1 = \frac{\alpha}{2\lambda^*}(G_1^T K_1^* G_1 + G_2^T K_2^* G_2 + I_t)^{-1},$$

$$\Sigma_2 = \frac{\alpha}{2\lambda^*}(G_1^T K_1^* G_1 + G_2^T K_2^* G_2 + I_t)^{-1} + \frac{\bar{\alpha} - \alpha}{2\lambda^*}(G_2^T K_2^* G_2 + I_t)^{-1}.$$

Note that $\lambda^* > 0$ by the first two equality conditions and the positive definite property of Σ_1 and Σ_2. Further define

$$K_1^{**} = \frac{\alpha}{2\lambda^*}(G_2^T K_2^* G_2 + I_t)^{-1} - \Sigma_1,$$

$$K_2^{**} = \frac{\bar{\alpha}}{2\lambda^*}I_t - K_1^{**} - \Sigma_2.$$
(9.4)

It can be easily verified that

1. $K_1^{**}, K_2^{**} \succeq 0$,

2. $\mathrm{tr}(K_1^{**}) + \mathrm{tr}(K_2^{**}) = P$ (since for a matrix A, $I - (I + A)^{-1} = (I + A)^{-1}A$), and

3. the boundary corner point (R_1^*, R_2^*) can be written as

$$R_1^* = \frac{1}{2}\log\frac{|K_1^{**} + \Sigma_1|}{|\Sigma_1|},$$

$$R_2^* = \frac{1}{2}\log\frac{|K_1^{**} + K_2^{**} + \Sigma_2|}{|K_1^{**} + \Sigma_2|}.$$
(9.5)

Step 2 (Construction of the enhanced degraded GV-BC). For each boundary corner point $(R_1^*(\alpha), R_2^*(\alpha))$ corresponding to the supporting line $(\alpha, \bar{\alpha})$, we define the GV-BC DBC(α)

$$\tilde{Y}_1 = X + W_1,$$

$$\tilde{Y}_2 = X + W_2,$$

where $W_1 \sim N(0, \Sigma_1)$ and $W_2 \sim N(0, \Sigma_2)$ are the noise vector components. As in the original BC, we assume average power constraint P on X. Since $\Sigma_2 \succeq \Sigma_1 \succ 0$, the enhanced channel DBC(α) is a degraded BC. Assume without loss of generality that it is *physically degraded*, as shown in Figure 9.15.

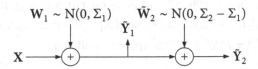

Figure 9.15. DBC(α).

We now show the following.

Lemma 9.3. The capacity region $\mathscr{C}_{\mathrm{DBC}(\alpha)}$ of DBC(α) is an outer bound on the capacity region $\mathscr{C}$ of the original BC, i.e.,

$$\mathscr{C} \subseteq \mathscr{C}_{\mathrm{DBC}(\alpha)}.$$

To prove this lemma, we show that the original BC is a degraded version of the enhanced DBC(α). This clearly implies that any code for the original BC achieves the same or smaller probability of error when used for the DBC(α); hence every rate pair in $\mathscr{C}$ is also in $\mathscr{C}_{\mathrm{DBC}(\alpha)}$. By the KKT optimality conditions for (K_1^*, K_2^*) in (9.3),

$$I_r - G_j \Sigma_j G_j^T \geq 0, \quad j = 1, 2.$$

Now each receiver $j = 1, 2$ of DBC(α) can multiply the received signal $\tilde{\mathbf{Y}}_j$ by G_j and add an independent noise vector $\mathbf{W}_j' \sim \mathrm{N}(0, I_r - G_j \Sigma_j G_j^T)$ to form the new output

$$\mathbf{Y}_j' = G_j \tilde{\mathbf{Y}}_j + \mathbf{W}_j'$$
$$= G_j \mathbf{X} + G_j \mathbf{W}_j + \mathbf{W}_j',$$

where the zero-mean Gaussian noise $G_j \mathbf{W}_j + \mathbf{W}_j'$ has covariance matrix $G_j \Sigma_j G_j^T + (I_r - G_j \Sigma_j G_j^T) = I_r$. Thus the transformed received signal $\mathbf{Y}_j'$ has the same distribution as the received signal $\mathbf{Y}_j$ for the original BC with channel gain matrix G_j and noise covariance matrix I_r.

Step 3 (Optimality of $(R_1^*(\alpha), R_2^*(\alpha))$). We are now ready to establish the final step in the proof of the converse.

Lemma 9.4. Every boundary corner point $(R_1^*(\alpha), R_2^*(\alpha))$ in $\mathscr{C}_{\mathrm{DMAC}}$ in the positive quadrant corresponding to the supporting line of slope $-\alpha/\tilde{\alpha}$ is on the boundary of the capacity region of DBC(α).

To prove this lemma, we follow similar steps to the proof of the converse for the scalar Gaussian BC in Section 5.5.1. First recall the representation of the boundary corner point (R_1^*, R_2^*) in (9.5). Consider a $(2^{nR_1}, 2^{nR_2}, n)$ code for DBC(α) with $\lim_{n \to \infty} P_e^{(n)} = 0$. To prove the optimality of (R_1^*, R_2^*), we show that if $R_1 > R_1^*$, then $R_2 \leq R_2^*$.

First note that $h(\tilde{\mathbf{Y}}_2^n) = h(\mathbf{X}^n + \mathbf{W}_2^n)$ is upper bounded by

$$\max_{K_\mathbf{X}: \mathrm{tr}(K_\mathbf{X}) \leq P} \frac{n}{2} \log((2\pi e)^t |K_\mathbf{X} + \Sigma_2|).$$

Recalling the properties

$$\text{tr}(K_1^{**} + K_2^{**}) = P,$$

$$K_1^{**} + K_2^{**} + \Sigma_2 = \frac{\bar{\alpha}}{2\lambda^*} I_t,$$

we see that the covariance matrix $(K_1^{**} + K_2^{**})$ satisfies the KKT condition for the point-to-point channel $\mathbf{Y}_2 = \mathbf{X} + \mathbf{Z}_2$. Therefore

$$h(\tilde{\mathbf{Y}}_2^n) \le \frac{n}{2} \log((2\pi e)^t |K_1^{**} + K_2^{**} + \Sigma_2|).$$

As in the converse proof for the scalar Gaussian BC, we use the conditional vector EPI to lower bound $h(\tilde{\mathbf{Y}}_2^n | M_2)$. By Fano's inequality and the assumption $R_1^* < R_1$,

$$\frac{n}{2} \log \frac{|K_1^{**} + \Sigma_1|}{|\Sigma_1|} = nR_1^*$$

$$\le I(M_1; \mathbf{Y}_1^n | M_2) + n\epsilon_n$$

$$= h(\tilde{\mathbf{Y}}_1^n | M_2) - h(\tilde{\mathbf{Y}}_1^n | M_1, M_2) + n\epsilon_n$$

$$\le h(\tilde{\mathbf{Y}}_1^n | M_2) - \frac{n}{2} \log((2\pi e)^t |\Sigma_1|) + n\epsilon_n,$$

or equivalently,

$$h(\tilde{\mathbf{Y}}_1^n | M_2) \ge \frac{n}{2} \log((2\pi e)^t |K_1^{**} + \Sigma_1|) - n\epsilon_n.$$

Since $\tilde{\mathbf{Y}}_2^n = \tilde{\mathbf{Y}}_1^n + \tilde{\mathbf{W}}_2^n$, and $\tilde{\mathbf{Y}}_1^n$ and $\tilde{\mathbf{W}}_2^n$ are independent and have densities, by the conditional vector EPI,

$$h(\tilde{\mathbf{Y}}_2^n | M_2) \ge \frac{nt}{2} \log(2^{2h(\tilde{\mathbf{Y}}_1^n | M_2)/(nt)} + 2^{2h(\tilde{\mathbf{W}}_2^n | M_2)/(nt)})$$

$$\ge \frac{nt}{2} \log((2\pi e)|K_1^{**} + \Sigma_1|^{1/t} + (2\pi e)|\Sigma_2 - \Sigma_1|^{1/t}) - n\epsilon_n'$$

for some ϵ_n' that tends to zero as $n \to \infty$. But from the definitions of $\Sigma_1, \Sigma_2, K_1^{**}, K_2^{**}$, the matrices

$$K_1^{**} + \Sigma_1 = \frac{\alpha}{2\lambda^*} (G_2^T K_2^* G_2 + I_t)^{-1},$$

$$\Sigma_2 - \Sigma_1 = \frac{\bar{\alpha} - \alpha}{2\lambda^*} (G_2^T K_2^* G_2 + I_t)^{-1}$$

are scaled versions of each other. Hence

$$|K_1^{**} + \Sigma_1|^{1/t} + |\Sigma_2 - \Sigma_1|^{1/t} = |(K_1^{**} + \Sigma_1) + (\Sigma_2 - \Sigma_1)|^{1/t}$$

$$= |K_1^{**} + \Sigma_2|^{1/t}.$$

Therefore

$$h(\tilde{\mathbf{Y}}_2^n | M_2) \ge \frac{n}{2} \log((2\pi e)^t |K_1^{**} + \Sigma_2|) - n\epsilon_n'.$$

Combining the bounds and taking $n \to \infty$, we finally obtain

$$R_2 \le \frac{1}{2} \log \frac{|K_1^{**} + K_2^{**} + \Sigma_2|}{|K_1^{**} + \Sigma_2|} = R_2^*.$$

This completes the proof of Lemma 9.4. Combining the results from the previous steps completes the proof of the converse and Theorem 9.4.

9.6.4   GV-BC with More than Two Receivers

The capacity region of the 2-receiver GV-BC can be extended to an arbitrary number of receivers. Consider the k-receiver Gaussian vector broadcast channel

$$\mathbf{Y}_j = G_j \mathbf{X} + \mathbf{Z}_j, \quad j \in [1:k],$$

where $\mathbf{Z}_j \sim N(0, I_r)$, $j \in [1:k]$. Assume average power constraint P on $\mathbf{X}$. The capacity region of this channel is as follows.

Theorem 9.5. The capacity region of the k-receiver GV-BC is the convex hull of the set of rate tuples $(R_1, R_2, \ldots, R_k)$ such that

$$R_{\sigma(j)} \le \frac{1}{2} \log \frac{\left|\sum_{j' \ge j} G_{\sigma(j')} K_{\sigma(j')} G_{\sigma(j')}^T + I_r\right|}{\left|\sum_{j' > j} G_{\sigma(j')} K_{\sigma(j')} G_{\sigma(j')}^T + I_r\right|}$$

for some permutation σ on $[1:k]$ and positive semidefinite matrices $K_1, \ldots, K_k$ with $\sum_j \mathrm{tr}(K_j) \le P$.

For a given σ, the corresponding rate region is achievable by writing on dirty paper with decoding order $\sigma(1) \to \cdots \to \sigma(k)$. As in the 2-receiver case, the converse proof hinges on the BC–MAC duality, which can be easily extended to more than two receivers. First, it can be shown that the corresponding dual MAC capacity region $\mathscr{C}_{\mathrm{DMAC}}$ consists of all rate tuples $(R_1, R_2, \ldots, R_k)$ such that

$$\sum_{j \in \mathcal{J}} R_j \le \frac{1}{2} \log \left| \sum_{j \in \mathcal{J}} G_j^T K_j G_j + I_t \right|, \quad \mathcal{J} \subseteq [1:k],$$

for some positive semidefinite matrices $K_1, \ldots, K_k$ with $\sum_j \mathrm{tr}(K_j) \le P$. The optimality of $\mathscr{C}_{\mathrm{DMAC}}$ can be proved by induction on k. For the case $R_j = 0$ for some $j \in [1:k]$, the problem reduces to proving the optimality of $\mathscr{C}_{\mathrm{DMAC}}$ for $k - 1$ receivers. Therefore, we can consider $(R_1, \ldots, R_k)$ in the positive orthant only. Now as in the 2-receiver case, each boundary corner point can be shown to be optimal by constructing a corresponding degraded GV-BC, and showing that the boundary corner point is on the boundary of the capacity region of the degraded GV-BC and hence on the boundary of the capacity region of the original GV-BC. It can be also shown that $\mathscr{C}_{\mathrm{DMAC}}$ does not have a kink, that is, every boundary corner point has a unique supporting hyperplane. Therefore, the boundary of $\mathscr{C}_{\mathrm{DMAC}}$ must coincide with the capacity region, which proves the optimality of writing on dirty paper.

SUMMARY

- Gaussian vector channel models

- Reciprocity via singular value decomposition

- Water-filling and iterative water-filling

- Vector writing on dirty paper:

  - Marton's coding scheme

  - Optimal for the Gaussian vector BC with private messages

- BC–MAC duality

- Proof of the converse for the Gaussian vector BC:

  - Converse for each boundary corner point

  - Characterization of the corner point via Lagrange duality and KKT condition

  - Construction of an enhanced degraded BC

  - Use of the vector EPI for the enhanced degraded BC

- **Open problems:**

  9.1. What is the capacity region of the Gaussian product BC with more than two receivers?

  9.2. Can the converse for the Gaussian vector BC be proved directly by optimizing the Nair–El Gamal outer bound?

  9.3. What is the capacity region of the 2-receiver Gaussian vector BC with common message?

BIBLIOGRAPHIC NOTES

The Gaussian vector channel was first considered by Telatar (1999) and Foschini (1996). The capacity region of the Gaussian vector MAC in Theorem 9.2 was established by Cheng and Verdú (1993). The iterative water-filling algorithm in Section 9.2 is due to Yu, Rhee, Boyd, and Cioffi (2004). The capacity region of the degraded Gaussian product broadcast channel was established by Hughes-Hartogs (1975). The more general Theorem 9.3 and Proposition 9.1 appeared in El Gamal (1980). The private-message capacity region in (9.2) is due to Poltyrev (1977).

Caire and Shamai (2003) devised a writing on dirty paper scheme for the Gaussian vector broadcast channel and showed that it achieves the sum-capacity for $r = 2$ and $t = 1$ antennas. The sum-capacity for an arbitrary number of antennas was established by Vishwanath, Jindal, and Goldsmith (2003) and Viswanath and Tse (2003) using the BC–MAC duality and by Yu and Cioffi (2004) using a minimax argument. The capacity region in Theorem 9.4 was established by Weingarten, Steinberg, and Shamai (2006), using the technique of channel enhancement. Our proof is a simplified version of the proof by Mohseni (2006).

A recent survey of the literature on Gaussian vector channels can be found in Biglieri, Calderbank, Constantinides, Goldsmith, Paulraj, and Poor (2007).

PROBLEMS

9.1. Show that the capacity region of the reversely degraded DM-BC in Proposition 9.1 coincides with both the Marton inner bound and the Nair–El Gamal outer bound in Chapter 8.

9.2. Establish the converse for Theorem 9.3 starting from the rate region in Proposition 9.1 with average power constraint.

9.3. *Time division for the Gaussian vector MAC.* Consider a 2-sender Gaussian vector MAC with channel output $\mathbf{Y} = (Y_1, Y_2)$ given by

$$Y_1 = X_1 + Z_1$$
$$Y_2 = X_1 + X_2 + Z_2,$$

where $Z_1 \sim N(0, 1)$ and $Z_2 \sim N(0, 1)$ are independent. Assume average power constraint P on each of X_1 and X_2.

(a) Find the capacity region.

(b) Find the time-division inner bound (with power control). Is it possible to achieve any point on the boundary of the capacity region (except for the end points)?

9.4. *Sato's outer bound for the Gaussian IC.* Consider the Gaussian IC in Section 6.4 with channel gain matrix

$$G = \begin{bmatrix} g_{11} & g_{12} \\ g_{21} & g_{22} \end{bmatrix}$$

and SNRs S_1 and S_2. Show that if a rate pair (R_1, R_2) is achievable, then it must satisfy the inequalities

$$R_1 \le C(S_1),$$
$$R_2 \le C(S_2),$$
$$R_1 + R_2 \le \min_K \frac{1}{2} \log |PGG^T + K|,$$

where the minimum is over all covariance matrices K of the form

$$K = \begin{bmatrix} 1 & \rho \\ \rho & 1 \end{bmatrix}.$$

APPENDIX 9A   PROOF OF THE BC–MAC DUALITY LEMMA

We show that $\mathscr{R}_{\text{WDP}} \subseteq \mathscr{C}_{\text{DMAC}}$. The other direction of inclusion can be proved similarly.

Consider the writing on dirty paper scheme in which the message M_1 is encoded before M_2. Denote by K_1 and K_2 the covariance matrices for $\mathbf{X}_{11}^n(M_1)$ and $\mathbf{X}_{21}^n(\mathbf{X}_{11}^n, M_1)$, respectively. We know that the following rates are achievable:

$$R_1 = \frac{1}{2} \log \frac{|G_1(K_1 + K_2)G_1^T + I_r|}{|G_1 K_2 G_1^T + I_r|},$$

$$R_2 = \frac{1}{2} \log |G_2 K_2 G_2^T + I_r|.$$

We show that (R_1, R_2) is achievable in the dual MAC with the sum power constraint $P = \text{tr}(K_1) + \text{tr}(K_2)$ using successive cancellation decoding for M_2 before M_1.

Let

$$K_1' = \Sigma_1^{-1/2} \overline{K}_1 \Sigma_1^{-1/2},$$

where $\Sigma_1^{-1/2}$ is the symmetric square root inverse of $\Sigma_1 = G_1 K_2 G_1^T + I_r$ and $\overline{K}_1$ is obtained by the reciprocity lemma in Section 9.1.2 for the covariance matrix K_1 and the channel matrix $\Sigma_1^{-1/2} G_1$ such that $\text{tr}(\overline{K}_1) \leq \text{tr}(K_1)$ and

$$|\Sigma_1^{-1/2} G_1 K_1 G_1^T \Sigma_1^{-1/2} + I_r| = |G_1^T \Sigma_1^{-1/2} \overline{K}_1 \Sigma_1^{-1/2} G_1 + I_t|.$$

Further let

$$K_2' = \Sigma_2^{1/2} K_2 \Sigma_2^{1/2},$$

where $\Sigma_2^{1/2}$ is the symmetric square root of $\Sigma_2 = G_1^T K_1' G_1 + I_t$ and the bar over $\Sigma_2^{1/2} K_2 \Sigma_2^{1/2}$ means K_2' is obtained by the reciprocity lemma for the covariance matrix $\Sigma_2^{1/2} K_2 \Sigma_2^{1/2}$ and the channel matrix $G_2 \Sigma_2^{-1/2}$ such that

$$\text{tr}(K_2') \leq \text{tr}(\Sigma_2^{1/2} K_2 \Sigma_2^{1/2})$$

and

$$|G_2 \Sigma_2^{-1/2}(\Sigma_2^{1/2} K_2 \Sigma_2^{1/2})\Sigma_2^{-1/2} G_2^T + I_r| = |\Sigma_2^{-1/2} G_2^T K_2' G_2 \Sigma_2^{-1/2} + I_t|.$$

Using the covariance matrices K_1' and K_2' for senders 1 and 2, respectively, the following rates are achievable for the dual MAC when the receiver decodes for M_2 before M_1 (the reverse of the encoding order for the BC):

$$R_1' = \frac{1}{2} \log |G_1^T K_1' G_1 + I_t|,$$

$$R_2' = \frac{1}{2} \log \frac{|G_1^T K_1' G_1 + G_2^T K_2' G_2 + I_t|}{|G_1^T K_1' G_1 + I_t|}.$$

From the definitions of K_1', K_2', Σ_1, and Σ_2, we have

$$R_1 = \frac{1}{2} \log \frac{|G_1(K_1 + K_2)G_1^T + I_r|}{|G_1 K_2 G_1^T + I_r|}$$

$$= \frac{1}{2} \log \frac{|G_1 K_1 G_1^T + \Sigma_1|}{|\Sigma_1|}$$

$$= \frac{1}{2} \log |\Sigma_1^{-1/2} G_1 K_1 G_1^T \Sigma_1^{-1/2} + I_r|$$

$$= \frac{1}{2} \log |G_1^T \Sigma_1^{-1/2} \overline{K}_1 \Sigma_1^{-1/2} G_1 + I_t|$$

$$= R_1',$$

and similarly we can show that $R_2 = R_2'$. Furthermore

$$\mathrm{tr}(K_2') = \mathrm{tr}\left(\overline{\Sigma_2^{1/2} K_2 \Sigma_2^{1/2}}\right)$$

$$\leq \mathrm{tr}(\Sigma_2^{1/2} K_2 \Sigma_2^{1/2})$$

$$= \mathrm{tr}(\Sigma_2 K_2)$$

$$= \mathrm{tr}\left((G_1^T K_1' G_1 + I_t)K_2\right)$$

$$= \mathrm{tr}(K_2) + \mathrm{tr}(K_1' G_1 K_2 G_1^T)$$

$$= \mathrm{tr}(K_2) + \mathrm{tr}\left(K_1'(\Sigma_1 - I_r)\right)$$

$$= \mathrm{tr}(K_2) + \mathrm{tr}(\Sigma_1^{1/2} K_1' \Sigma_1^{1/2}) - \mathrm{tr}(K_1')$$

$$= \mathrm{tr}(K_2) + \mathrm{tr}(\overline{K}_1) - \mathrm{tr}(K_1')$$

$$\leq \mathrm{tr}(K_2) + \mathrm{tr}(K_1) - \mathrm{tr}(K_1').$$

Therefore, any point in $\mathscr{R}_{\mathrm{WDP}}$ is also achievable for the dual MAC under the sum-power constraint. The proof for the case where M_2 is encoded before M_1 follows similarly.

APPENDIX 9B   UNIQUENESS OF THE SUPPORTING LINE

We show that every boundary corner point (R_1^*, R_2^*) with $R_1^*, R_2^* > 0$ has a unique supporting line $(\alpha, \bar{\alpha})$. Consider a boundary corner point (R_1^*, R_2^*) in the positive quadrant. Assume without loss of generality that $0 \leq \alpha \leq 1/2$ and

$$R_1^* = \frac{1}{2} \log \frac{|G_1^T K_1^* G_1 + G_2^T K_2^* G_2 + I_t|}{|G_2^T K_2^* G_2 + I_t|},$$

$$R_2^* = \frac{1}{2} \log |G_2^T K_2^* G_2 + I_t|,$$

where (K_1^*, K_2^*) is an optimal solution to the convex optimization problem

$$\text{maximize} \quad \frac{\alpha}{2} \log |G_1^T K_1 G_1 + G_2^T K_2 G_2 + I_t| + \frac{\bar{\alpha} - \alpha}{2} \log |G_2^T K_2 G_2 + I_t|$$

$$\text{subject to} \quad \mathrm{tr}(K_1) + \mathrm{tr}(K_2) \leq P$$

$$K_1, K_2 \succeq 0.$$

We prove the uniqueness by contradiction. Suppose that (R_1^*, R_2^*) has another supporting line $(\beta, \bar{\beta}) \neq (\alpha, \bar{\alpha})$. First note that α and β must be nonzero. Otherwise, $K_1^* = 0$, which contradicts the assumption that $R_1^* > 0$. Also by the assumption that $R_1^* > 0$, we must have $G_1^T K_1^* G_1 \neq 0$ as well as $K_1^* \neq 0$. Now consider a feasible solution of the optimization problem at $((1 - \epsilon)K_1^*, \epsilon K_1^* + K_2^*)$, given by

$$\frac{\alpha}{2} \log \left| (1 - \epsilon)G_1^T K_1^* G_1 + G_2^T (\epsilon K_1^* + K_2^*)G_2 + I_t \right| + \frac{\bar{\alpha} - \alpha}{2} \log \left| G_2^T (\epsilon K_1^* + K_2^*)G_2 + I_t \right|.$$

Taking the derivative (Boyd and Vandenberghe 2004) at $\epsilon = 0$ and using the optimality of (K_1^*, K_2^*), we obtain

$$\alpha \operatorname{tr}((G_1^T K_1^* G_1 + G_2^T K_2^* G_2 + I_t)^{-1}(G_2^T K_1^* G_2 - G_1^T K_1^* G_1))$$
$$+ (\bar{\alpha} - \alpha) \operatorname{tr}((G_2^T K_2^* G_2 + I_t)^{-1}(G_2^T K_1^* G_2)) = 0,$$

and similarly

$$\beta \operatorname{tr}((G_1^T K_1^* G_1 + G_2^T K_2^* G_2 + I_t)^{-1}(G_2^T K_1^* G_2 - G_1^T K_1^* G_1))$$
$$+ (\bar{\beta} - \beta) \operatorname{tr}((G_2^T K_2^* G_2 + I_t)^{-1}(G_2^T K_1^* G_2)) = 0.$$

But since $\alpha \neq \beta$, this implies that

$$\operatorname{tr}((G_1^T K_1^* G_1 + G_2^T K_2^* G_2 + I_t)^{-1}(G_2^T K_1^* G_2 - G_1^T K_1^* G_1))$$
$$= \operatorname{tr}((G_2^T K_2^* G_2 + I_t)^{-1}(G_2^T K_1^* G_2)) = 0,$$

which in turn implies that $G_1^T K_1^* G_1 = 0$ and that $R_1^* = 0$. But this contradicts the hypothesis that $R_1^* > 0$, which completes the proof.

CHAPTER 10

Distributed Lossless Compression

In this chapter, we begin the discussion on communication of uncompressed sources over multiple noiseless links. We consider the limits on lossless compression of separately encoded sources, which is motivated by distributed sensing problems. For example, consider a sensor network for measuring the temperature at different locations across a city. Suppose that each sensor node compresses its measurement and transmits it to a common base station via a noiseless link. What is the minimum total transmission rate needed so that the base station can losslessly recover the measurements from all the sensors? If the sensor measurements are independent of each other, then the answer to this question is straightforward; each sensor compresses its measurement to the entropy of its respective temperature process, and the limit on the total rate is the sum of the individual entropies. The temperature processes at the sensors, however, can be highly correlated. Can such correlation be exploited to achieve a lower rate than the sum of the individual entropies? Slepian and Wolf showed that the total rate can be reduced to the joint entropy of the processes, that is, the limit on distributed lossless compression is the same as that on centralized compression, where the sources are jointly encoded. The achievability proof of this surprising result uses the new idea of random binning.

We then consider lossless source coding with helpers. Suppose that the base station in our sensor network example wishes to recover the temperature measurements from only a subset of the sensors while using the information sent by the rest of the sensor nodes to help achieve this goal. What is the optimal tradeoff between the rates from the different sensors? We establish the optimal rate region for the case of a single helper node.

In Chapter 20, we continue the discussion of distributed lossless source coding by considering more general networks modeled by graphs.

10.1 DISTRIBUTED LOSSLESS SOURCE CODING FOR A 2-DMS

Consider the distributed compression system depicted in Figure 10.1, where two sources X_1 and X_2 are separately encoded (described) at rates R_1 and R_2, respectively, and the descriptions are communicated over noiseless links to a decoder who wishes to recover both sources losslessly. What is the set of simultaneously achievable description rate pairs (R_1, R_2)?

We assume a 2-component DMS (2-DMS) $(\mathcal{X}_1 \times \mathcal{X}_2, p(x_1, x_2))$, informally referred to as *correlated sources* or 2-DMS (X_1, X_2), that consists of two finite alphabets $\mathcal{X}_1$, $\mathcal{X}_2$ and

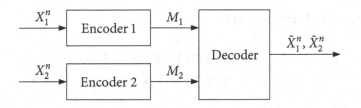

Figure 10.1. Distributed lossless compression system.

a joint pmf $p(x_1, x_2)$ over $\mathcal{X}_1 \times \mathcal{X}_2$. The 2-DMS (X_1, X_2) generates a jointly i.i.d. random process $\{(X_{1i}, X_{2i})\}$ with $(X_{1i}, X_{2i}) \sim p_{X_1, X_2}(x_{1i}, x_{2i})$.

A $(2^{nR_1}, 2^{nR_2}, n)$ *distributed lossless source code* for the 2-DMS (X_1, X_2) consists of

- two encoders, where encoder 1 assigns an index $m_1(x_1^n) \in [1 : 2^{nR_1})$ to each sequence $x_1^n \in \mathcal{X}_1^n$ and encoder 2 assigns an index $m_2(x_2^n) \in [1 : 2^{nR_2})$ to each sequence $x_2^n \in \mathcal{X}_2^n$, and

- a decoder that assigns an estimate $(\hat{x}_1^n, \hat{x}_2^n) \in \mathcal{X}_1^n \times \mathcal{X}_2^n$ or an error message e to each index pair $(m_1, m_2) \in [1 : 2^{nR_1}) \times [1 : 2^{nR_2})$.

The probability of error for a distributed lossless source code is defined as

$$P_e^{(n)} = \mathsf{P}\{(\hat{X}_1^n, \hat{X}_2^n) \neq (X_1^n, X_2^n)\}.$$

A rate pair (R_1, R_2) is said to be *achievable* for distributed lossless source coding if there exists a sequence of $(2^{nR_1}, 2^{nR_2}, n)$ codes such that $\lim_{n \to \infty} P_e^{(n)} = 0$. The *optimal rate region* $\mathcal{R}^*$ is the closure of the set of achievable rate pairs.

Remark 10.1. As for the capacity region of a multiuser channel, we can readily show that $\mathcal{R}^*$ is convex by using time sharing.

10.2   INNER AND OUTER BOUNDS ON THE OPTIMAL RATE REGION

By the lossless source coding theorem in Section 3.5, a rate pair (R_1, R_2) is achievable for distributed lossless source coding if

$$R_1 > H(X_1),$$
$$R_2 > H(X_2).$$

This gives the inner bound in Figure 10.2.

Also, by the lossless source coding theorem, a rate $R \geq H(X_1, X_2)$ is necessary and sufficient to send a pair of sources (X_1, X_2) together to a receiver. This yields the sum-rate bound

$$R_1 + R_2 \geq H(X_1, X_2). \tag{10.1}$$

Furthermore, using similar steps to the converse proof of the lossless source coding theorem, we can show that any achievable rate pair for distributed lossless source coding must satisfy the conditional entropy bounds

$$R_1 \geq H(X_1|X_2),$$
$$R_2 \geq H(X_2|X_1).$$

(10.2)

Combining the bounds in (10.1) and (10.2), we obtain the outer bound on the optimal rate region in Figure 10.2. In the following, we show that the outer bound is tight.

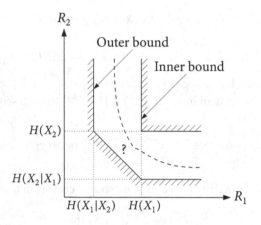

Figure 10.2. Inner and outer bounds on the optimal rate region $\mathcal{R}^*$.

10.3 SLEPIAN–WOLF THEOREM

First suppose that sender 1 observes both X_1 and X_2. Then, by a conditional version of the lossless source coding theorem (see Problem 10.1), it can be readily shown that the corner point $(R_1, R_2) = (H(X_1|X_2), H(X_2))$ on the boundary of the outer bound in Figure 10.2 is achievable. Slepian and Wolf showed that this corner point is achievable even when sender 1 does *not* know X_2. Thus the sum-rate $R_1 + R_2 > H(X_1, X_2)$ is achievable even when the sources are separately encoded and the outer bound in Figure 10.2 is tight!

Theorem 10.1 (Slepian–Wolf Theorem). The optimal rate region $\mathcal{R}^*$ for distributed lossless source coding of a 2-DMS (X_1, X_2) is the set of rate pairs (R_1, R_2) such that

$$R_1 \geq H(X_1|X_2),$$
$$R_2 \geq H(X_2|X_1),$$
$$R_1 + R_2 \geq H(X_1, X_2).$$

The following example illustrates the saving in rate using Slepian–Wolf coding.

Example 10.1. Consider a *doubly symmetric binary source* (DSBS(p)) (X_1, X_2), where X_1 and X_2 are binary random variables with $p_{X_1,X_2}(0,0) = p_{X_1,X_2}(1,1) = (1-p)/2$ and $p_{X_1,X_2}(0,1) = p_{X_1,X_2}(1,0) = p/2$, $p \in [0, 1/2]$. Thus, $X_1 \sim$ Bern($1/2$), $X_2 \sim$ Bern($1/2$), and their modulo-2 sum $Z = X_1 \oplus X_2 \sim$ Bern(p) is independent of each of X_1 and X_2. Equivalently, $X_2 = X_1 \oplus Z$ is the output of a BSC(p) with input X_1 and vice versa.

Suppose that $p = 0.01$. Then the sources X_1 and X_2 are highly dependent. If we compress the sources independently, then we need to send 2 bits/symbol-pair. However, if we use Slepian–Wolf coding instead, we need to send only $H(X_1) + H(X_2|X_1) = H(1/2) + H(0.01) = 1 + 0.0808 = 1.0808$ bits/symbol-pair.

In general, using Slepian–Wolf coding, the optimal rate region for the DSBS(p) is the set of all rate pairs (R_1, R_2) such that

$$R_1 \geq H(p),$$
$$R_2 \geq H(p),$$
$$R_1 + R_2 \geq 1 + H(p).$$

The converse of the Slepian–Wolf theorem is already established by the fact the rate region in the theorem coincides with the outer bound given by (10.1) and (10.2) in Figure 10.2. Achievability involves the new idea of *random binning*. We first illustrate this idea through an alternative achievability proof of the lossless source coding theorem.

10.3.1   Lossless Source Coding via Random Binning

Consider the following random binning achievability proof of the lossless source coding theorem in Section 3.5.

Codebook generation. Randomly and independently assign an index $m(x^n) \in [1:2^{nR}]$ to each sequence $x^n \in \mathcal{X}^n$ according to a uniform pmf over $[1:2^{nR}]$. We refer to each subset of sequences with the same index m as a *bin* $\mathcal{B}(m)$, $m \in [1:2^{nR}]$. This binning scheme is illustrated in Figure 10.3. The chosen bin assignments are revealed to the encoder and the decoder.

Encoding. Upon observing $x^n \in \mathcal{B}(m)$, the encoder sends the bin index m.

Decoding. Upon receiving m, the decoder declares $\hat{x}^n$ to be the estimate of the source sequence if it is the unique typical sequence in $\mathcal{B}(m)$; otherwise it declares an error.

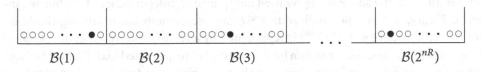

$\quad\quad \mathcal{B}(1) \quad\quad\quad\quad \mathcal{B}(2) \quad\quad\quad\quad \mathcal{B}(3) \quad\quad\quad\quad\quad\quad\quad\quad \mathcal{B}(2^{nR})$

Figure 10.3. Random binning for a single source. The black circles denote the typical x^n sequences and the white circles denote the rest.

Analysis of the probability of error. We show that the probability of error averaged over bin assignments tends to zero as $n \to \infty$ if $R > H(X)$.

Let M denote the random bin index of X^n, i.e., $X^n \in \mathcal{B}(M)$. Note that $M \sim \text{Unif}[1 : 2^{nR}]$ is independent of X^n. The decoder makes an error iff one or both of the following events occur:

$$\mathcal{E}_1 = \{X^n \notin \mathcal{T}_\epsilon^{(n)}\},$$
$$\mathcal{E}_2 = \{\tilde{x}^n \in \mathcal{B}(M) \text{ for some } \tilde{x}^n \neq X^n, \tilde{x}^n \in \mathcal{T}_\epsilon^{(n)}\}.$$

Then, by the symmetry of codebook construction and the union of events bound, the average probability of error is upper bounded as

$$P(\mathcal{E}) \leq P(\mathcal{E}_1) + P(\mathcal{E}_2)$$
$$= P(\mathcal{E}_1) + P(\mathcal{E}_2 \mid X^n \in \mathcal{B}(1)).$$

We now bound each probability of error term. By the LLN, $P(\mathcal{E}_1)$ tends to zero as $n \to \infty$. For the second term, consider

$$P(\mathcal{E}_2 \mid X^n \in \mathcal{B}(1)) = \sum_{x^n} P\{X^n = x^n \mid X^n \in \mathcal{B}(1)\}$$
$$\cdot P\{\tilde{x}^n \in \mathcal{B}(1) \text{ for some } \tilde{x}^n \neq x^n, \tilde{x}^n \in \mathcal{T}_\epsilon^{(n)} \mid x^n \in \mathcal{B}(1), X^n = x^n\}$$
$$\overset{(a)}{\leq} \sum_{x^n} p(x^n) \sum_{\substack{\tilde{x}^n \in \mathcal{T}_\epsilon^{(n)} \\ \tilde{x}^n \neq x^n}} P\{\tilde{x}^n \in \mathcal{B}(1) \mid x^n \in \mathcal{B}(1), X^n = x^n\}$$
$$\overset{(b)}{=} \sum_{x^n} p(x^n) \sum_{\substack{\tilde{x}^n \in \mathcal{T}_\epsilon^{(n)} \\ \tilde{x}^n \neq x^n}} P\{\tilde{x}^n \in \mathcal{B}(1)\}$$
$$\leq |\mathcal{T}_\epsilon^{(n)}| \cdot 2^{-nR}$$
$$\leq 2^{n(H(X)+\delta(\epsilon))} 2^{-nR},$$

where (a) and (b) follow since for every $\tilde{x}^n \neq x^n$, the events $\{x^n \in \mathcal{B}(1)\}$, $\{\tilde{x}^n \in \mathcal{B}(1)\}$, and $\{X^n = x^n\}$ are mutually independent. Thus, the probability of error averaged over bin assignments tends to zero as $n \to \infty$ if $R > H(X) + \delta(\epsilon)$. Hence, there must exist a sequence of bin assignments with $\lim_{n\to\infty} P_e^{(n)} = 0$. This completes the achievability proof of the lossless source coding theorem via random binning.

Remark 10.2. In the above proof, we used only *pairwise* independence of the bin assignments. Hence, if X is a Bernoulli source, we can use random *linear* binning (hashing) $m(x^n) = H x^n$, where $m \in [1 : 2^{nR}]$ is represented by a vector of nR bits and the elements of the $nR \times n$ "parity-check" random binary matrix H are generated i.i.d. Bern(1/2). Note that this is the dual of the linear channel coding in which encoding is performed using a random generator matrix; see Section 3.1.3. Such linear binning can be extended to the more general case where $\mathcal{X}$ is the set of elements of a finite field.

10.3.2   Achievability Proof of the Slepian–Wolf Theorem

We now use random binning to prove achievability of the Slepian–Wolf theorem.

Codebook generation. Randomly and independently assign an index $m_1(x_1^n)$ to each sequence $x_1^n \in \mathcal{X}_1^n$ according to a uniform pmf over $[1 : 2^{nR_1}]$. The sequences with the same index m_1 form a bin $\mathcal{B}_1(m_1)$. Similarly assign an index $m_2(x_2^n) \in [1 : 2^{nR_2}]$ to each sequence $x_2^n \in \mathcal{X}_2^n$. The sequences with the same index m_2 form a bin $\mathcal{B}_2(m_2)$. This binning scheme is illustrated in Figure 10.4. The bin assignments are revealed to the encoders and the decoder.

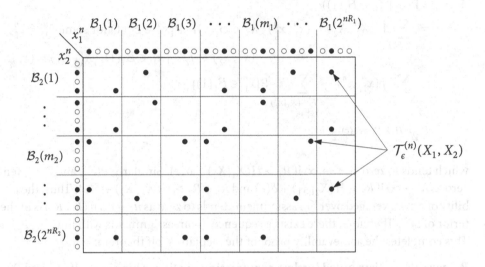

Figure 10.4. Random binning for two correlated sources. The black circles correspond to typical sequences.

Encoding. Upon observing $x_1^n \in \mathcal{B}_1(m_1)$, encoder 1 sends m_1. Similarly, upon observing $x_2^n \in \mathcal{B}_2(m_2)$, encoder 2 sends m_2.

Decoding. Given the received index pair (m_1, m_2), the decoder declares $(\hat{x}_1^n, \hat{x}_2^n)$ to be the estimate of the source pair if it is the unique jointly typical pair in the product bin $\mathcal{B}_1(m_1) \times \mathcal{B}_2(m_2)$; otherwise it declares an error.

Analysis of the probability of error. We bound the probability of error averaged over bin assignments. Let M_1 and M_2 denote the random bin indices for X_1^n and X_2^n, respectively. The decoder makes an error iff one or more of the following events occur:

$$\mathcal{E}_1 = \{(X_1^n, X_2^n) \notin \mathcal{T}_\epsilon^{(n)}\},$$

$$\mathcal{E}_2 = \{\tilde{x}_1^n \in \mathcal{B}_1(M_1) \text{ for some } \tilde{x}_1^n \neq X_1^n, (\tilde{x}_1^n, X_2^n) \in \mathcal{T}_\epsilon^{(n)}\},$$

$$\mathcal{E}_3 = \{\tilde{x}_2^n \in \mathcal{B}_2(M_2) \text{ for some } \tilde{x}_2^n \neq X_2^n, (X_1^n, \tilde{x}_2^n) \in \mathcal{T}_\epsilon^{(n)}\},$$

$$\mathcal{E}_4 = \{\tilde{x}_1^n \in \mathcal{B}_1(M_1), \tilde{x}_2^n \in \mathcal{B}_2(M_2) \text{ for some } \tilde{x}_1^n \neq X_1^n, \tilde{x}_2^n \neq X_2^n, (\tilde{x}_1^n, \tilde{x}_2^n) \in \mathcal{T}_\epsilon^{(n)}\}.$$

Then, the average probability of error is upper bounded as

$$P(\mathcal{E}) \le P(\mathcal{E}_1) + P(\mathcal{E}_2) + P(\mathcal{E}_3) + P(\mathcal{E}_4).$$

Now we bound each probability of error term. By the LLN, $P(\mathcal{E}_1)$ tends to zero as $n \to \infty$. Now consider the second term. Using the symmetry of the codebook construction and following similar steps to the achievability proof of the lossless source coding theorem in Section 10.3.1, we have

$$
\begin{aligned}
P(\mathcal{E}_2) &= P(\mathcal{E}_2 | X_1^n \in \mathcal{B}_1(1)) \\
&= \sum_{(x_1^n, x_2^n)} P\{(X_1^n, X_2^n) = (x_1^n, x_2^n) \mid X_1^n \in \mathcal{B}_1(1)\} \cdot P\{\tilde{x}_1^n \in \mathcal{B}_1(1) \text{ for some } \tilde{x}_1^n \ne x_1^n, \\
&\qquad\qquad (\tilde{x}_1^n, x_2^n) \in \mathcal{T}_\epsilon^{(n)} \mid x_1^n \in \mathcal{B}_1(1), (X_1^n, X_2^n) = (x_1^n, x_2^n)\} \\
&\le \sum_{(x_1^n, x_2^n)} p(x_1^n, x_2^n) \sum_{\substack{\tilde{x}_1^n \in \mathcal{T}_\epsilon^{(n)}(X_1 | x_2^n) \\ \tilde{x}_1^n \ne x_1^n}} P\{\tilde{x}_1^n \in \mathcal{B}_1(1)\} \\
&\le 2^{n(H(X_1|X_2)+\delta(\epsilon))} 2^{-nR_1},
\end{aligned}
$$

which tends to zero as $n \to \infty$ if $R_1 > H(X_1|X_2) + \delta(\epsilon)$. Similarly, $P(\mathcal{E}_3)$ and $P(\mathcal{E}_4)$ tend to zero as $n \to \infty$ if $R_2 > H(X_2|X_1) + \delta(\epsilon)$ and $R_1 + R_2 > H(X_1, X_2) + \delta(\epsilon)$. Thus, the probability of error averaged over bin assignments tends to zero as $n \to \infty$ if (R_1, R_2) is in the interior of $\mathscr{R}^*$. Therefore, there exists a sequence of bin assignments with $\lim_{n\to\infty} P_e^{(n)} = 0$. This completes the achievability proof of the Slepian–Wolf theorem.

Remark 10.3 (Distributed lossless compression via linear binning). If X_1 and X_2 are Bernoulli sources, we can use random linear binnings $m_1(x_1^n) = H_1 x_1^n$ and $m_2(x_2^n) = H_2 x_2^n$, where the entries of H_1 and H_2 are generated i.i.d. Bern(1/2). If in addition, (X_1, X_2) is DSBS(p) (i.e., $Z = X_1 \oplus X_2 \sim$ Bern(p)), then we can use the following coding scheme based on single-source linear binning. Consider the corner point $(R_1, R_2) = (1, H(p))$ of the optimal rate region. Suppose that X_1^n is sent uncoded while X_2^n is encoded as HX_2^n with a randomly generated $n(H(p) + \delta(\epsilon)) \times n$ parity-check matrix H. The decoder can calculate $HX_1^n \oplus HX_2^n = HZ^n$, from which Z^n can be recovered with high probability as in the single-source case. Hence, $X_2^n = X_1^n \oplus Z^n$ can also be recovered with high probability.

10.4   LOSSLESS SOURCE CODING WITH A HELPER

Consider the distributed compression system depicted in Figure 10.5, where only one of the two sources is to be recovered losslessly and the encoder for the other source (helper) provides *coded side information* to the decoder to help reduce the first encoder's rate.

The definitions of a code, achievability, and optimal rate region $\mathscr{R}^*$ are the same as for the distributed lossless source coding setup in Section 10.1, except that the probability of error is defined as $P_e^{(n)} = P\{\hat{X}^n \ne X^n\}$.

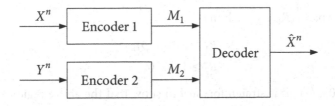

Figure 10.5. Distributed lossless compression system with a helper.

If there is no helper, i.e., $R_2 = 0$, then $R_1 \geq H(X)$ is necessary and sufficient for loss-lessly recovering X at the decoder. At the other extreme, if the helper sends Y losslessly to the decoder, i.e., $R_2 \geq H(Y)$, then $R_1 \geq H(X|Y)$ is necessary and sufficient by the Slepian–Wolf theorem. These two extreme points define the time-sharing inner bound and the trivial outer bound in Figure 10.6. Neither bound is tight in general, however.

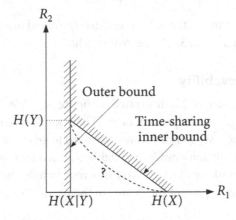

Figure 10.6. Inner and outer bounds on the optimal rate region for lossless source coding with a helper.

The optimal rate region for the one-helper problem can be characterized as follows.

Theorem 10.2. Let (X, Y) be a 2-DMS. The optimal rate region $\mathscr{R}^*$ for lossless source coding of X with a helper observing Y is the set of rate pairs (R_1, R_2) such that

$$R_1 \geq H(X|U),$$
$$R_2 \geq I(Y; U)$$

for some conditional pmf $p(u|y)$, where $|\mathcal{U}| \leq |\mathcal{Y}| + 1$.

We illustrate this result in the following.

Example 10.2. Let (X, Y) be a DSBS(p), $p \in [0, 1/2]$. The optimal rate region simplifies to the set of rate pairs (R_1, R_2) such that

$$R_1 \geq H(\alpha * p),$$
$$R_2 \geq 1 - H(\alpha) \tag{10.3}$$

for some $\alpha \in [0, 1/2]$. It is straightforward to show that the above region is attained by setting the backward test channel from Y to U as a BSC(α). The proof of optimality uses Mrs. Gerber's lemma as in the converse for the binary symmetric BC in Section 5.4.2. First, note that $H(p) \leq H(X|U) \leq 1$. Thus there exists an $\alpha \in [0, 1/2]$ such that $H(X|U) = H(\alpha * p)$. By the scalar MGL, $H(X|U) = H(Y \oplus Z|U) \geq H(H^{-1}(H(Y|U)) * p)$, since given U, Z and Y remain independent. Thus, $H^{-1}(H(Y|U)) \leq \alpha$, which implies that

$$I(Y; U) = H(Y) - H(Y|U)$$
$$= 1 - H(Y|U)$$
$$\geq 1 - H(\alpha).$$

Optimality of the rate region in (10.3) can be alternatively proved using a symmetrization argument similar to that in Section 5.4.2; see Problem 10.7.

10.4.1 Proof of Achievability

The coding scheme for Theorem 10.2 is illustrated in Figure 10.7. We use random binning and joint typicality encoding. The helper (encoder 2) uses joint typicality encoding (as in the achievability proof of the lossy source coding theorem) to generate a description of the source Y represented by the auxiliary random variable U. Encoder 1 uses random binning as in the achievability proof of the Slepian–Wolf theorem to help the decoder recover X given that it knows U. We now present the details of the proof.

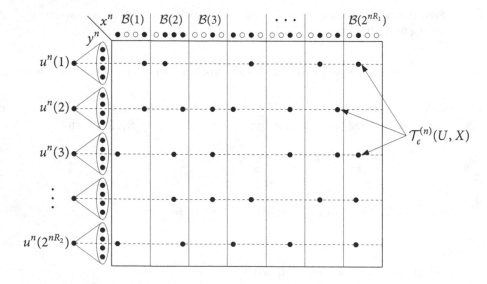

Figure 10.7. Coding scheme for lossless source coding with a helper.

Codebook generation. Fix a conditional pmf $p(u|y)$ and let $p(u) = \sum_y p(y)p(u|y)$. Randomly and independently assign an index $m_1(x^n) \in [1:2^{nR_1}]$ to each sequence $x^n \in \mathcal{X}^n$. The set of sequences with the same index m_1 form a bin $\mathcal{B}(m_1)$. Randomly and independently generate 2^{nR_2} sequences $u^n(m_2)$, $m_2 \in [1:2^{nR_2}]$, each according to $\prod_{i=1}^n p_U(u_i)$.

Encoding. If $x^n \in \mathcal{B}(m_1)$, encoder 1 sends m_1. Encoder 2 finds an index m_2 such that $(u^n(m_2), y^n) \in \mathcal{T}_{\epsilon'}^{(n)}$. If there is more than one such index, it sends the smallest one among them. If there is no such index, set $m_2 = 1$. Encoder 2 then sends the index m_2 to the decoder.

Decoding. The receiver finds the unique $\hat{x}^n \in \mathcal{B}(m_1)$ such that $(\hat{x}^n, u^n(m_2)) \in \mathcal{T}_{\epsilon}^{(n)}$. If there is none or more than one, it declares an error.

Analysis of the probability of error. Assume that M_1 and M_2 are the chosen indices for encoding X^n and Y^n, respectively. The decoder makes an error iff one or more of the following events occur:

$$\mathcal{E}_1 = \{(U^n(m_2), Y^n) \notin \mathcal{T}_{\epsilon'}^{(n)} \text{ for all } m_2 \in [1:2^{nR_2})\},$$
$$\mathcal{E}_2 = \{(X^n, U^n(M_2)) \notin \mathcal{T}_{\epsilon}^{(n)}\},$$
$$\mathcal{E}_3 = \{\tilde{x}^n \in \mathcal{B}(M_1), (\tilde{x}^n, U^n(M_2)) \in \mathcal{T}_{\epsilon}^{(n)} \text{ for some } \tilde{x}^n \neq X^n\}.$$

Thus, the probability of error is upper bounded as

$$\mathsf{P}(\mathcal{E}) \leq \mathsf{P}(\mathcal{E}_1) + \mathsf{P}(\mathcal{E}_1^c \cap \mathcal{E}_2) + \mathsf{P}(\mathcal{E}_3 \,|\, X^n \in \mathcal{B}(1)).$$

We now bound each term. By the covering lemma in Section 3.7 with $X \leftarrow Y$, $\hat{X} \leftarrow U$, and $U \leftarrow \emptyset$, $\mathsf{P}(\mathcal{E}_1)$ tends to zero as $n \to \infty$ if $R_2 > I(Y;U) + \delta(\epsilon')$. Next, note that $\mathcal{E}_1^c = \{(U^n(M_2), Y^n) \in \mathcal{T}_{\epsilon'}^{(n)}\}$ and $X^n \,|\, \{U^n(M_2) = u^n, Y^n = y^n\} \sim \prod_{i=1}^n p_{X|Y}(x_i|y_i)$. Hence, by the conditional typicality lemma in Section 2.5 (see also Problem 2.13), $\mathsf{P}(\mathcal{E}_1^c \cap \mathcal{E}_2)$ tends to zero as $n \to \infty$. Finally for the third term, consider

$$\mathsf{P}(\mathcal{E}_3 | X^n \in \mathcal{B}(1)) = \sum_{(x^n, u^n)} \mathsf{P}\{(X^n, U^n) = (x^n, u^n) \,|\, X^n \in \mathcal{B}(1)\} \mathsf{P}\{\tilde{x}^n \in \mathcal{B}(1) \text{ for some}$$
$$\tilde{x}^n \neq x^n, (\tilde{x}^n, u^n) \in \mathcal{T}_{\epsilon}^{(n)} \,|\, X^n \in \mathcal{B}(1), (X^n, U^n) = (x^n, u^n)\}$$
$$\leq \sum_{(x^n, u^n)} p(x^n, u^n) \sum_{\substack{\tilde{x}^n \in \mathcal{T}_{\epsilon}^{(n)}(X|u^n) \\ \tilde{x}^n \neq x^n}} \mathsf{P}\{\tilde{x}^n \in \mathcal{B}(1)\}$$
$$\leq 2^{n(H(X|U)+\delta(\epsilon))} 2^{-nR_1},$$

which tends to zero as $n \to \infty$ if $R_1 > H(X|U) + \delta(\epsilon)$. This completes the proof of achievability.

10.4.2   Proof of the Converse

Let M_1 and M_2 denote the indices from encoders 1 and 2, respectively. By Fano's inequality, $H(X^n|M_1, M_2) \leq n\epsilon_n$ for some ϵ_n that tends to zero as $n \to \infty$.

First consider

$$nR_2 \geq H(M_2)$$
$$\geq I(M_2; Y^n)$$
$$= \sum_{i=1}^{n} I(M_2; Y_i | Y^{i-1})$$
$$= \sum_{i=1}^{n} I(M_2, Y^{i-1}; Y_i)$$
$$\stackrel{(a)}{=} \sum_{i=1}^{n} I(M_2, Y^{i-1}, X^{i-1}; Y_i),$$

where (a) follows since $X^{i-1} \to (M_2, Y^{i-1}) \to Y_i$ form a Markov chain. Identifying $U_i = (M_2, Y^{i-1}, X^{i-1})$ and noting that $U_i \to Y_i \to X_i$ form a Markov chain, we have shown that

$$nR_2 \geq \sum_{i=1}^{n} I(U_i; Y_i).$$

Next consider

$$nR_1 \geq H(M_1)$$
$$\geq H(M_1 | M_2)$$
$$= H(M_1 | M_2) + H(X^n | M_1, M_2) - H(X^n | M_1, M_2)$$
$$\geq H(X^n, M_1 | M_2) - n\epsilon_n$$
$$= H(X^n | M_2) - n\epsilon_n$$
$$= \sum_{i=1}^{n} H(X_i | M_2, X^{i-1}) - n\epsilon_n$$
$$\geq \sum_{i=1}^{n} H(X_i | M_2, X^{i-1}, Y^{i-1}) - n\epsilon_n$$
$$= \sum_{i=1}^{n} H(X_i | U_i) - n\epsilon_n.$$

Using a time-sharing random variable $Q \sim \mathrm{Unif}[1:n]$, independent of (X^n, Y^n, U^n), we obtain

$$\frac{1}{n} \sum_{i=1}^{n} H(X_i | U_i, Q = i) = H(X_Q | U_Q, Q),$$

$$\frac{1}{n} \sum_{i=1}^{n} I(Y_i; U_i | Q = i) = I(Y_Q; U_Q | Q).$$

Since Q is independent of Y_Q, $I(Y_Q; U_Q | Q) = I(Y_Q; U_Q, Q)$. Thus, defining $X = X_Q$, $Y = Y_Q$, and $U = (U_Q, Q)$ and letting $n \to \infty$, we have shown that $R_1 \geq H(X|U)$ and $R_2 \geq I(Y; U)$ for some conditional pmf $p(u|y)$. The cardinality bound on $\mathcal{U}$ can be proved using the convex cover method in Appendix C. This completes the proof of Theorem 10.2.

10.5   EXTENSIONS TO MORE THAN TWO SOURCES

The Slepian–Wolf theorem can be extended to distributed lossless source coding for an arbitrary number of sources.

Theorem 10.3. The optimal rate region $\mathcal{R}^*(X_1, X_2, \ldots, X_k)$ for distributed lossless source coding of a k-DMS $(X_1, \ldots, X_k)$ is the set of rate tuples $(R_1, R_2, \ldots, R_k)$ such that

$$\sum_{j \in \mathcal{S}} R_j \geq H(X(\mathcal{S}) | X(\mathcal{S}^c)) \quad \text{for all } \mathcal{S} \subseteq [1:k].$$

For example, $\mathcal{R}^*(X_1, X_2, X_3)$ is the set of rate triples (R_1, R_2, R_3) such that

$$R_1 \geq H(X_1 | X_2, X_3),$$
$$R_2 \geq H(X_2 | X_1, X_3),$$
$$R_3 \geq H(X_3 | X_1, X_2),$$
$$R_1 + R_2 \geq H(X_1, X_2 | X_3),$$
$$R_1 + R_3 \geq H(X_1, X_3 | X_2),$$
$$R_2 + R_3 \geq H(X_2, X_3 | X_1),$$
$$R_1 + R_2 + R_3 \geq H(X_1, X_2, X_3).$$

Remark 10.4. It is interesting to note that the optimal rate region for distributed lossless source coding and the capacity region of the multiple access channel, both of which are for many-to-one communication, have single-letter characterizations for an arbitrary number of senders. This is not the case for dual one-to-many communication settings such as the broadcast channel (even with only two receivers), as we have seen in Chapter 5.

Theorem 10.2 can be similarly generalized to k sources $(X_1, X_2, \ldots, X_k)$ and a single helper Y.

Theorem 10.4. Let $(X_1, X_2, \ldots, X_k, Y)$ be a $(k+1)$-DMS. The optimal rate region for the lossless source coding of $(X_1, X_2, \ldots, X_k)$ with helper Y is the set of rate tuples $(R_1, R_2, \ldots, R_k, R_{k+1})$ such that

$$\sum_{j \in \mathcal{S}} R_j \geq H(X(\mathcal{S}) | U, X(\mathcal{S}^c)) \quad \text{for all } \mathcal{S} \in [1:k],$$

$$R_{k+1} \geq I(Y; U)$$

for some conditional pmf $p(u|y)$ with $|\mathcal{U}| \leq |\mathcal{Y}| + 2^k - 1$.

The optimal rate region for the lossless source coding with more than one helper is not known in general even when there is only one source to be recovered.

SUMMARY

- k-Component discrete memoryless source (k-DMS)
- Distributed lossless source coding for a k-DMS:
  - Slepian–Wolf optimal rate region
  - Random binning
  - Source coding via random linear codes
- Lossless source coding with a helper:
  - Joint typicality encoding for lossless source coding
  - Use of Mrs. Gerber's lemma in the proof of the converse for the doubly symmetric binary source (DSBS)

BIBLIOGRAPHIC NOTES

The Slepian–Wolf theorem was first proved in Slepian and Wolf (1973a). The random binning proof is due to Cover (1975a), who also showed that the proof can be extended to any pair of stationary ergodic sources X_1 and X_2 with joint entropy rates $\overline{H}(X_1, X_2)$. Since the converse can also be easily extended, the Slepian–Wolf theorem for this larger class of correlated sources is the set of rate pairs (R_1, R_2) such that

$$R_1 \geq \overline{H}(X_1 | X_2) = \overline{H}(X_1, X_2) - \overline{H}(X_2),$$
$$R_2 \geq \overline{H}(X_2 | X_1),$$
$$R_1 + R_2 \geq \overline{H}(X_1, X_2).$$

The random linear binning technique for the DSBS in Remark 10.2 appeared in Wyner (1974).

Error-free distributed lossless source coding has been studied, for example, in Witsenhausen (1976b), Ahlswede (1979), and El Gamal and Orlitsky (1984). As discussed in the bibliographic notes of Chapter 3, the rate of lossless source coding is the same as that of error-free compression using variable-length codes. This result does not hold in general for distributed source coding, that is, the Slepian–Wolf theorem *does not* hold in general for error-free distributed compression (using variable-length codes). In fact, for many 2-DMS (X_1, X_2), the optimal error-free rate region is

$$R_1 \geq H(X_1),$$
$$R_2 \geq H(X_2).$$

For example, error-free distributed compression for a DSBS(0.01) requires that $R_1 = R_2 = 1$ bit per symbol; in other words, no compression is possible.

The problem of lossless source coding with a helper was first studied by Wyner (1973), who established the optimal rate region for the DSBS in Example 10.2. The optimal rate region for a general DMS in Theorem 10.2 was established independently by Ahlswede and Körner (1975) and Wyner (1975b). The strong converse was proved via the blowing-up lemma by Ahlswede, Gács, and Körner (1976).

PROBLEMS

10.1. *Conditional lossless source coding.* Consider the lossless source coding setup depicted in Figure 10.8. Let (X, Y) be a 2-DMS. The source sequence X^n is to be sent losslessly to a decoder with side information Y^n available at both the encoder and the decoder. Thus, a $(2^{nR}, n)$ code is defined by an encoder $m(x^n, y^n)$ and a decoder $\hat{x}^n(m, y^n)$, and the probability of error is defined as $P_e^{(n)} = \mathsf{P}\{\hat{X}^n \neq X^n\}$.

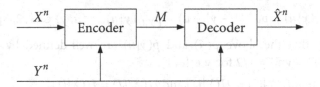

Figure 10.8. Source coding with side information.

(a) Find the optimal rate R^*.

(b) Prove achievability using $|\mathcal{T}_\epsilon^{(n)}(X|y^n)| \leq 2^{n(H(X|Y)+\delta(\epsilon))}$ for $y^n \in \mathcal{T}_\epsilon^{(n)}(Y)$.

(c) Prove the converse using Fano's inequality.

(d) Using part (c), establish the inequalities on the rates in (10.2) for distributed lossless source coding.

10.2. Prove the converse for the Slepian–Wolf theorem by establishing the outer bound given by (10.1) and (10.2).

10.3. Provide the details of the achievability proof for lossless source coding of a Bern(p) source using random linear binning described in Remark 10.2.

10.4. Show that the rate region in Theorem 10.2 is convex.

10.5. Prove Theorem 10.2 for a 3-DMS (X_1, X_2, Y).

10.6. *Lossless source coding with degraded source sets.* Consider the distributed lossless source coding setup depicted in Figure 10.9. Let (X, Y) be a 2-DMS. Sender 1 encodes (X^n, Y^n) into an index M_1, while sender 2 encodes only Y^n into an index M_2. Find the optimal rate region (a) when the decoder wishes to recover both X and Y losslessly, and (b) when the decoder wishes to recover only Y.

10.7. *Converse for lossless source coding of a DSBS with a helper via symmetrization.* Consider distributed lossless coding of the DSBS with a helper in Example 10.2.

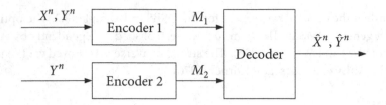

Figure 10.9. Distributed lossless source coding with degraded source sets.

Here we evaluate the optimal rate region in Theorem 10.2 via a symmetrization argument similar to that in Section 5.4.2. Given $p(u|y)$ (and corresponding $p(u)$ and $p(y|u)$), define $p(\tilde{u}|y)$ by

$$p_{\tilde{U}}(u) = p_{\tilde{U}}(-u) = \frac{1}{2}p_U(u), \quad u = 1, 2, 3,$$

$$p_{Y|\tilde{U}}(y|u) = p_{Y|\tilde{U}}(1 - y| - u) = p_{Y|U}(y|u), \quad (u, y) \in \{1, 2, 3\} \times \{0, 1\}.$$

(a) Verify that the above $p(\tilde{u})$ and $p(y|\tilde{u})$ are well-defined by checking that $\sum_{\tilde{u}} p(\tilde{u})p(y|\tilde{u}) = 1/2$ for $y \in \{0, 1\}$.

(b) Show that $H(Y|U) = H(Y|\tilde{U})$ and $H(X|U) = H(X|\tilde{U})$.

(c) Show that for any $\lambda \in [0, 1]$, the weighted sum $\lambda H(X|\tilde{U}) + (1 - \lambda)I(\tilde{U}; Y)$ is minimized by some BSC $p(\tilde{u}|y)$.

(d) Show that the rate region in (10.3) is convex.

(e) Combining parts (b), (c), and (d), conclude that the optimal rate region is characterized by (10.3).

10.8. *Lossless source coding with two decoders and side information.* Consider the lossless source coding setup for a 3-DMS (X, Y_1, Y_2) depicted in Figure 10.10. Source X is encoded into an index M, which is broadcast to two decoders. Decoder $j = 1, 2$ wishes to recover X losslessly from the index M and side information Y_j. The probability of error is defined as $\mathrm{P}\{\hat{X}_1^n \neq X^n \text{ or } \hat{X}_2^n \neq X^n\}$. Find the optimal lossless source coding rate.

10.9. *Correlated sources with side information.* Consider the distributed lossless source coding setup for a 3-DMS (X_1, X_2, Y) depicted in Figure 10.11. Sources X_1 and X_2 are separately encoded into M_1 and M_2, respectively, and the decoder wishes to recover both sources from M_1, M_2, and side information Y is available. Find the optimal rate region $\mathscr{R}^*$.

10.10. *Helper to both the encoder and the decoder.* Consider the variant on the one-helper problem for a 2-DMS (X, Y) in Section 10.4, as depicted in Figure 10.12. Suppose that coded side information of Y is available at both the encoder and the decoder, and the decoder wishes to recover X losslessly. Find the optimal rate region $\mathscr{R}^*$.

10.11. *Cascade lossless source coding.* Consider the distributed lossless source coding setup depicted in Figure 10.13. Let (X_1, X_2) be a 2-DMS. Source X_1 is encoded into an index M_1. Then source X_2 and the index M_1 are encoded into an index M_2. The decoder wishes to recover both sources losslessly only from M_2. Find the optimal rate region $\mathscr{R}^*$.

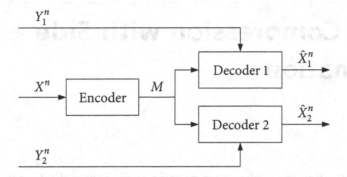

Figure 10.10. Lossless source coding with multiple side information.

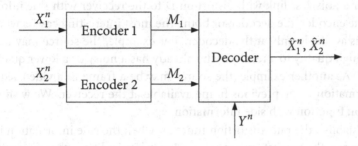

Figure 10.11. Distributed lossless source coding with side information.

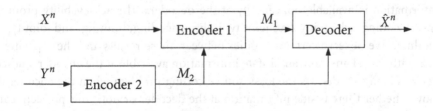

Figure 10.12. Lossless source coding with a helper.

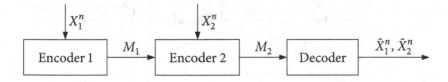

Figure 10.13. Cascade lossless source coding.

CHAPTER 11

Lossy Compression with Side Information

We turn our attention to distributed lossy compression. In this chapter, we consider the special case of lossy source coding with side information depicted in Figure 11.1. Let (X, Y) be a 2-DMS and $d(x, \hat{x})$ be a distortion measure. The sender wishes to communicate the source X over a noiseless link with distortion D to the receiver with side information Y available at the encoder, the decoder, or both. The most interesting case is when the side information is available only at the decoder. For example, the source may be an image to be sent at high quality to a receiver who already has a noisy or a lower quality version of the image. As another example, the source may be a frame in a video sequence and the side information is the previous frame available at the receiver. We wish to find the rate–distortion function with side information.

We first establish the rate–distortion function when the side information is available at the decoder and the reconstruction sequence can depend only causally on the side information. We show that in the lossless limit, causal side information may not reduce the optimal compression rate. We then establish the rate–distortion function when the side information is available noncausally at the decoder. The achievability proof uses Wyner–Ziv coding, which involves joint typicality encoding, binning, and joint typicality decoding. We observe certain dualities between these results and the capacities for channels with causal and noncausal state information available at the encoder studied in Chapter 7. Finally we discuss the lossy source coding problem when the encoder does not know whether there is side information at the decoder or not. This problem can be viewed as a source coding dual to the broadcast channel approach to coding for fading channels discussed in Chapter 23.

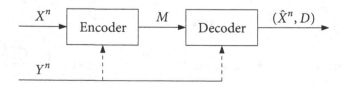

Figure 11.1. Lossy compression system with side information.

11.1   SIMPLE SPECIAL CASES

By the lossy source coding theorem in Section 3.6, the rate–distortion function with no side information at either the encoder or the decoder is

$$R(D) = \min_{p(\hat{x}|x):E(d(X,\hat{X}))\le D} I(X; \hat{X}).$$

In the lossless case, the optimal compression rate is $R^* = H(X)$. This corresponds to $R(0)$ under the Hamming distortion measure as discussed in Section 3.6.4.

It can be easily shown that when side information is available only at the encoder, the rate–distortion function remains the same, i.e.,

$$R_{\text{SI-E}}(D) = \min_{p(\hat{x}|x):E(d(X,\hat{X}))\le D} I(X; \hat{X}) = R(D). \tag{11.1}$$

Thus for the lossless case, $R^*_{\text{SI-E}} = H(X) = R^*$.

When side information is available at both the encoder and the decoder (causally or noncausally), the rate–distortion function is a conditional version of that with no side information, i.e.,

$$R_{\text{SI-ED}}(D) = \min_{p(\hat{x}|x,y):E(d(X,\hat{X}))\le D} I(X; \hat{X}|Y). \tag{11.2}$$

The proof of this result is a simple extension of the proof of the lossy source coding theorem. For the lossless case, the optimal rate is $R^*_{\text{SI-ED}} = H(X|Y)$, which follows also by the conditional version of the lossless source coding theorem; see Problem 10.1.

11.2   CAUSAL SIDE INFORMATION AVAILABLE AT THE DECODER

We consider yet another special case of side information availability. Suppose that the side information sequence is available only at the decoder. Here the rate–distortion function depends on whether the side information is available causally or noncausally.

We first consider the case where the reconstruction of each symbol $\hat{X}_i(M, Y^i)$ can depend causally on the side information sequence Y^i as depicted in Figure 11.2. This setting is motivated, for example, by denoising with side information. Suppose that X^n is the trajectory of a target including the terrain in which it is moving. The tracker (decoder) has a description M of the terrain available prior to tracking. At time i, the tracker obtains a noisy observation Y_i of the target's location and wishes to output an estimate $\hat{X}_i(M, Y^i)$ of the location based on the terrain description M and the noisy observations Y^i. The rate–distortion function characterizes the performance limit of the optimal sequential target tracker (filter). The causal side information setup is also motivated by an attempt to better understand the duality between channel coding and source coding as discussed in Section 11.3.3.

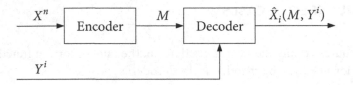

Figure 11.2. Lossy source coding with causal side information at the decoder.

A $(2^{nR}, n)$ lossy source code with side information available causally at the decoder consists of

1. an encoder that assigns an index $m(x^n) \in [1 : 2^{nR})$ to each $x^n \in \mathcal{X}^n$ and

2. a decoder that assigns an estimate $\hat{x}_i(m, y^i)$ to each received index m and side information sequence y^i for $i \in [1 : n]$.

The rate–distortion function with causal side information available at the decoder $R_{\text{CSI-D}}(D)$ is the infimum of rates R such that there exists a sequence of $(2^{nR}, n)$ codes with $\limsup_{n \to \infty} \mathsf{E}(d(X^n, \hat{X}^n)) \le D$.

The rate–distortion function has a simple characterization.

Theorem 11.1. Let (X, Y) be a 2-DMS and $d(x, \hat{x})$ be a distortion measure. The rate–distortion function for X with side information Y causally available at the decoder is

$$R_{\text{CSI-D}}(D) = \min I(X; U) \quad \text{for } D \ge D_{\min},$$

where the minimum is over all conditional pmfs $p(u|x)$ with $|\mathcal{U}| \le |\mathcal{X}| + 1$ and functions $\hat{x}(u, y)$ such that $\mathsf{E}[d(X, \hat{X})] \le D$, and $D_{\min} = \min_{\hat{x}(y)} \mathsf{E}[d(X, \hat{x}(Y))]$.

Note that $R_{\text{CSI-D}}(D)$ is nonincreasing, convex, and thus continuous in D. In the following example, we compare the rate–distortion function in the theorem with the rate–distortion functions with no side information and with side information available at both the encoder and the decoder.

Example 11.1. Let (X, Y) be a DSBS(p), $p \in [0, 1/2]$, and d be a Hamming distortion measure. When there is no side information (see Example 3.4), the rate–distortion function is

$$R(D) = \begin{cases} 1 - H(D) & \text{for } 0 \le D \le 1/2, \\ 0 & \text{for } D > 1/2. \end{cases}$$

By comparison, when side information is available at both the encoder and the decoder, the rate–distortion function can be found by evaluating (11.2), which yields

$$R_{\text{SI-ED}}(D) = \begin{cases} H(p) - H(D) & \text{for } 0 \le D \le p, \\ 0 & \text{for } D > p. \end{cases} \tag{11.3}$$

Now suppose that the side information is causally available at the decoder. Then it can be shown by evaluating the rate–distortion function in Theorem 11.1 that

$$R_{\text{CSI-D}}(D) = \begin{cases} 1 - H(D) & \text{for } 0 \le D \le D_c, \\ (p - D)H'(D_c) & \text{for } D_c < D \le p, \\ 0 & \text{for } D > p, \end{cases}$$

where H' is the derivative of the binary entropy function, and D_c is the solution to the equation $(1 - H(D_c))/(p - D_c) = H'(D_c)$. Thus $R_{\text{CSI-D}}(D)$ coincides with $R(D)$ for $0 \le D \le D_c$, and is otherwise given by the tangent to the curve of $R(D)$ that passes through the point $(p, 0)$ for $D_c \le D \le p$ as shown in Figure 11.3. In other words, the optimum performance is achieved by time sharing between rate–distortion coding with no side information and zero-rate decoding that uses only the side information. For sufficiently small distortion, $R_{\text{CSI-D}}(D)$ is achieved simply by ignoring the side information.

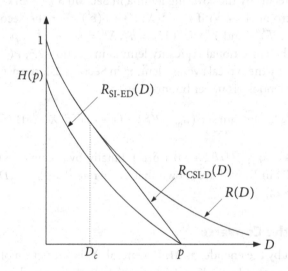

Figure 11.3. Comparison of $R(D)$, $R_{\text{CSI-D}}(D)$, and $R_{\text{SI-ED}}(D)$ for a DSBS(p).

11.2.1   Proof of Achievability

To prove achievability for Theorem 11.1, we use joint typicality encoding to describe X by U as in the achievability proof of the lossy source coding theorem. The reconstruction $\hat{X}$ is a function of U and the side information Y. The following provides the details.

Codebook generation. Fix the conditional pmf $p(u|x)$ and function $\hat{x}(u, y)$ that attain $R_{\text{CSI-D}}(D/(1 + \epsilon))$, where D is the desired distortion. Randomly and independently generate 2^{nR} sequences $u^n(m)$, $m \in [1 : 2^{nR}]$, each according to $\prod_{i=1}^{n} p_U(u_i)$.

Encoding. Given a source sequence x^n, find an index m such that $(u^n(m), x^n) \in \mathcal{T}_{\epsilon'}^{(n)}$. If there is more than one such index, choose the smallest one among them. If there is no such index, set $m = 1$. The encoder sends the index m to the decoder.

Decoding. The decoder finds the reconstruction sequence $\hat{x}^n(m, y^n)$ by setting $\hat{x}_i = \hat{x}(u_i(m), y_i)$ for $i \in [1:n]$.

Analysis of expected distortion. Denote the chosen index by M and let $\epsilon > \epsilon'$. Define the "error" event

$$\mathcal{E} = \{(U^n(M), X^n, Y^n) \notin \mathcal{T}_\epsilon^{(n)}\}.$$

Then by the union of events bound,

$$\mathsf{P}(\mathcal{E}) \leq \mathsf{P}(\mathcal{E}_0) + \mathsf{P}(\mathcal{E}_0^c \cap \mathcal{E}),$$

where

$$\mathcal{E}_0 = \{(U^n(m), X^n) \notin \mathcal{T}_{\epsilon'}^{(n)} \text{ for all } m \in [1:2^{nR}]\} = \{(U^n(M), X^n) \notin \mathcal{T}_{\epsilon'}^{(n)}\}.$$

We now bound each term. By the covering lemma in Section 3.7 (with $U \leftarrow \emptyset$ and $\hat{X} \leftarrow U$), $\mathsf{P}(\mathcal{E}_0)$ tends to zero as $n \to \infty$ if $R > I(X; U) + \delta(\epsilon')$. For the second term, since $\epsilon > \epsilon'$, $(U^n(M), X^n) \in \mathcal{T}_{\epsilon'}^{(n)}$, and $Y^n | \{U^n(M) = u^n, X^n = x^n\} \sim \prod_{i=1}^n p_{Y|U,X}(y_i|u_i, x_i) = \prod_{i=1}^n p_{Y|X}(y_i|x_i)$, by the conditional typicality lemma in Section 2.5, $\mathsf{P}(\mathcal{E}_0^c \cap \mathcal{E})$ tends to zero as $n \to \infty$. Thus, by the typical average lemma in Section 2.4, the asymptotic distortion averaged over codebooks is upper bounded as

$$\limsup_{n\to\infty} \mathsf{E}(d(X^n; \hat{X}^n)) \leq \limsup_{n\to\infty} \left(d_{\max} \mathsf{P}(\mathcal{E}) + (1 + \epsilon) \mathsf{E}(d(X, \hat{X})) \mathsf{P}(\mathcal{E}^c) \right) \leq D,$$

if $R > I(X; U) + \delta(\epsilon') = R_{\text{CSI-D}}(D/(1 + \epsilon)) + \delta(\epsilon')$. Finally, by taking $\epsilon \to 0$ and using the continuity of $R_{\text{CSI-D}}(D)$ in D, we have shown that every rate $R > R_{\text{CSI-D}}(D)$ is achievable. This completes the proof.

11.2.2 Proof of the Converse

Denote the index sent by the encoder as M. In general, $\hat{X}_i$ is a function of (M, Y^i), so we set $U_i = (M, Y^{i-1})$. Note that $U_i \to X_i \to Y_i$ form a Markov chain and $\hat{X}_i$ is a function of (U_i, Y_i) as desired. Consider

$$nR \geq H(M)$$
$$= I(X^n; M)$$
$$= \sum_{i=1}^n I(X_i; M|X^{i-1})$$
$$= \sum_{i=1}^n I(X_i; M, X^{i-1})$$
$$\overset{(a)}{=} \sum_{i=1}^n I(X_i; M, X^{i-1}, Y^{i-1})$$
$$\geq \sum_{i=1}^n I(X_i; U_i)$$

$$\geq \sum_{i=1}^{n} R_{\text{CSI-D}}(\mathsf{E}(d(X_i, \hat{X}_i)))$$

$$\overset{(b)}{\geq} n R_{\text{CSI-D}}\left(\frac{1}{n} \sum_{i=1}^{n} \mathsf{E}(d(X_i, \hat{X}_i))\right),$$

where (a) follows since $X_i \to (M, X^{i-1}) \to Y^{i-1}$ form a Markov chain and (b) follows by the convexity of $R_{\text{CSI-D}}(D)$. Since $\limsup_{n \to \infty}(1/n) \sum_{i=1}^{n} \mathsf{E}(d(X_i, \hat{X}_i)) \leq D$ (by assumption) and $R_{\text{CSI-D}}(D)$ is nonincreasing, $R \geq R_{\text{CSI-D}}(D)$. The cardinality bound on $\mathcal{U}$ can be proved using the convex cover method in Appendix C. This completes the proof of Theorem 11.1.

11.2.3 Lossless Source Coding with Causal Side Information

Consider the source coding with causal side information setting in which the decoder wishes to recover X losslessly, that is, with probability of error $\mathsf{P}\{X^n \neq \hat{X}^n\}$ that tends to zero as $n \to \infty$. Denote the optimal compression rate by $R_{\text{CSI-D}}^*$. Clearly $H(X|Y) \leq R_{\text{CSI-D}}^* \leq H(X)$. First consider some special cases:

- If $X = Y$, then $R_{\text{CSI-D}}^* = H(X|Y)$ $(= 0)$.

- If X and Y are independent, then $R_{\text{CSI-D}}^* = H(X|Y)$ $(= H(X))$.

- If $X = (Y, Z)$, where Y and Z are independent, then $R_{\text{CSI-D}}^* = H(X|Y)$ $(= H(Z))$.

The following theorem establishes the optimal lossless rate in general.

Theorem 11.2. Let (X, Y) be a 2-DMS. The optimal lossless compression rate for the source X with causal side information Y available at the decoder is

$$R_{\text{CSI-D}}^* = \min I(X; U),$$

where the minimum is over all conditional pmfs $p(u|x)$ with $|\mathcal{U}| \leq |\mathcal{X}| + 1$ such that $H(X|U, Y) = 0$.

To prove the theorem, we show that $R_{\text{CSI-D}}^* = R_{\text{CSI-D}}(0)$ under Hamming distortion measure as in the random coding proof of the lossless source coding theorem (without side information) in Section 3.6.4. The details are as follows.

Proof of the converse. Consider the lossy source coding problem with $\mathcal{X} = \hat{\mathcal{X}}$ and Hamming distortion measure d, where the minimum in Theorem 11.1 is evaluated at $D = 0$. Note that requiring that the block error probability $\mathsf{P}\{X^n \neq \hat{X}^n\}$ tend to zero as $n \to \infty$ is stronger than requiring that the average bit-error probability $(1/n) \sum_{i=1}^{n} \mathsf{P}\{X_i \neq \hat{X}_i\} = \mathsf{E}(d(X^n, \hat{X}^n))$ tend to zero. Hence, by the converse proof for the lossy coding case, it follows that $R_{\text{CSI-D}}^* \geq R_{\text{CSI-D}}(0) = \min I(X; U)$, where the minimum is over all conditional pmfs $p(u|x)$ such that $\mathsf{E}(d(X, \hat{x}(U, Y))) = 0$, or equivalently, $\hat{x}(u, y) = x$ (hence $p(x|u, y) = 1$) for all (x, y, u) with $p(x, y, u) > 0$, or equivalently, $H(X|U, Y) = 0$.

Proof of achievability. Consider the proof of achievability for the lossy case. If $(x^n, y^n, u^n(m)) \in T_\epsilon^{(n)}$ for some $m \in [1 : 2^{nR})$, then $p_{X,Y,U}(x_i, y_i, u_i(m)) > 0$ for all $i \in [1 : n]$ and hence $\hat{x}(u_i(m), y_i) = x_i$ for all $i \in [1 : n]$. Thus

$$P\{\hat{X}^n = X^n \,|\, (U^n(m), X^n, Y^n) \in T_\epsilon^{(n)} \text{ for some } m \in [1 : 2^{nR})\} = 1.$$

But, since $P\{(U^n(m), X^n, Y^n) \notin T_\epsilon^{(n)} \text{ for all } m \in [1 : 2^{nR})\}$ tends to zero as $n \to \infty$, so does $P\{\hat{X}^n \neq X^n\}$ as $n \to \infty$. This completes the proof of achievability.

Compared to the Slepian–Wolf theorem, Theorem 11.2 shows that causality of side information can severely limit how encoders can leverage correlation between the sources. The following proposition shows that under a fairly general condition causal side information does not help at all.

> **Proposition 11.1.** Let (X, Y) be a 2-DMS such that $p(x, y) > 0$ for all $(x, y) \in \mathcal{X} \times \mathcal{Y}$. Then, $R_{\text{CSI-D}}^* = H(X)$.

To prove the proposition, we show by contradiction that if $p(x, y) > 0$ for all (x, y), then the condition $H(X|U, Y) = 0$ in Theorem 11.2 implies that $H(X|U) = 0$. Suppose $H(X|U) > 0$. Then $p(u) > 0$ and $0 < p(x|u) < 1$ for some (x, u), which also implies that $0 < p(x'|u) < 1$ for some $x' \neq x$. Since $d(x, \hat{x}(u, y)) = 0$ and $d(x', \hat{x}(u, y)) = 0$ cannot both hold at the same time, and $p_{U,X,Y}(u, x, y) > 0$ and $p_{U,X,Y}(u, x', y) > 0$ (because $p(x, y) > 0$), we must have $E[d(X, \hat{x}(U, Y))] > 0$. But this contradicts the assumption of $E[d(X, \hat{x}(U, Y))] = 0$. Thus, $H(X|U) = 0$ and $R_{\text{CSI-D}}^* = H(X)$. This completes the proof of the proposition.

11.3   NONCAUSAL SIDE INFORMATION AVAILABLE AT THE DECODER

We now consider the source coding setting with side information available *noncausally* at the decoder as depicted in Figure 11.4. In other words, a $(2^{nR}, n)$ code is defined by an encoder $m(x^n)$ and a decoder $\hat{x}^n(m, y^n)$.

The rate–distortion function is also known for this case.

> **Theorem 11.3 (Wyner–Ziv Theorem).** Let (X, Y) be a 2-DMS and $d(x, \hat{x})$ be a distortion measure. The rate–distortion function for X with side information Y available noncausally at the decoder is
>
> $$R_{\text{SI-D}}(D) = \min\left(I(X; U) - I(Y; U)\right) = \min I(X; U|Y) \quad \text{for } D \geq D_{\min},$$
>
> where the minimum is over all conditional pmfs $p(u|x)$ with $|\mathcal{U}| \leq |\mathcal{X}| + 1$ and functions $\hat{x}(u, y)$ such that $E[d(X, \hat{X})] \leq D$, and $D_{\min} = \min_{\hat{x}(y)} E[d(X, \hat{x}(Y))]$.

Note that $R_{\text{SI-D}}(D)$ is nonincreasing, convex, and thus continuous in D.

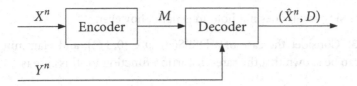

Figure 11.4. Lossy source coding with noncausal side information at the decoder.

Clearly, $R_{\text{SI-ED}}(D) \le R_{\text{SI-D}}(D) \le R_{\text{CSI-D}}(D) \le R(D)$. The difference between $R_{\text{SI-D}}(D)$ and $R_{\text{CSI-D}}(D)$ is in the subtracted term $I(Y;U)$. Also recall that with side information at both the encoder and the decoder,

$$R_{\text{SI-ED}}(D) = \min_{p(u|x,y),\, \hat{x}(u,y):\mathsf{E}(d(X,\hat{X}))\le D} I(X;U|Y).$$

Hence the difference between $R_{\text{SI-D}}(D)$ and $R_{\text{SI-ED}}(D)$ is in taking the minimum over $p(u|x)$ versus $p(u|x, y)$.

We know from the Slepian–Wolf theorem that the optimal lossless compression rate with side information Y only at the decoder, $R_{\text{SI-D}}^* = H(X|Y)$, is the same as the rate when the side information Y is available at both the encoder and the decoder. Does this equality also hold in general for the lossy case? The following example shows that in some cases the rate–distortion function with side information available only at the decoder in Theorem 11.3 is the same as when it is available at both the encoder and the decoder.

Example 11.2 (Quadratic Gaussian source coding with noncausal side information).
We show that $R_{\text{SI-D}}(D) = R_{\text{SI-ED}}(D)$ for a 2-WGN source (X, Y) and squared error distortion. Assume without loss of generality that $X \sim \mathrm{N}(0, P)$ and the side information $Y = X + Z$, where $Z \sim \mathrm{N}(0, N)$ is independent of X. It is easy to show that the rate–distortion function with noncausal side information Y available at both the encoder and the decoder is

$$R_{\text{SI-ED}}(D) = \mathsf{R}\left(\frac{P'}{D}\right), \tag{11.4}$$

where $P' = \mathrm{Var}(X|Y) = PN/(P + N)$ and $\mathsf{R}(x)$ is the quadratic Gaussian rate function.

Now consider the case in which the side information is available only at the decoder. Clearly, $R_{\text{SI-D}}(D) = 0$ for $D \ge P'$. For $0 < D \le P'$, we take the auxiliary random variable to be $U = X + V$, where $V \sim \mathrm{N}(0, Q)$ is independent of X and Y, that is, we use a Gaussian test channel from X to U. Setting $Q = P'D/(P' - D)$, we can easily see that

$$I(X;U|Y) = \mathsf{R}\left(\frac{P'}{D}\right).$$

Now let the reconstruction $\hat{X}$ be the MMSE estimate of X given U and Y. Then, $\mathsf{E}((X - \hat{X})^2) = \mathrm{Var}(X|U, Y)$, which can be easily shown to be equal to D. Thus the rate–distortion function for this case is again

$$R_{\text{SI-D}}(D) = \mathsf{R}\left(\frac{P'}{D}\right) = R_{\text{SI-ED}}(D). \tag{11.5}$$

This surprising result does not hold in general, however.

Example 11.3. Consider the case of a DSBS(p), $p \in [0, 1/2]$, and Hamming distortion measure. It can be shown that the rate–distortion function for this case is

$$R_{\text{SI-D}}(D) = \begin{cases} g(D) & \text{for } 0 \le D \le D_c', \\ (p - D)g'(D_c') & \text{for } D_c' < D \le p, \\ 0 & \text{for } D > p, \end{cases}$$

where $g(D) = H(p * D) - H(D)$, g' is the derivative of g, and D_c' is the solution to the equation $g(D_c')/(p - D_c') = g'(D_c')$. It can be easily shown that $R_{\text{SI-D}}(D) > R_{\text{SI-ED}}(D) = H(p) - H(D)$ for all $D \in (0, p)$. Thus there is a nonzero cost for the lack of side information at the encoder. Figure 11.5 compares the rate–distortion functions with and without side information.

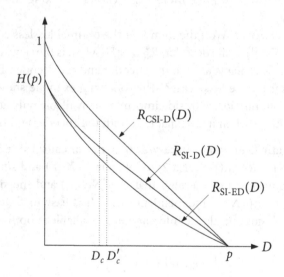

Figure 11.5. Comparison of $R_{\text{CSI-D}}(D)$, $R_{\text{SI-D}}(D)$, and $R_{\text{SI-ED}}(D)$ for a DSBS(p).

11.3.1 Proof of Achievability

The Wyner–Ziv coding scheme uses the *compress–bin* idea illustrated in Figure 11.6. As in the causal case, we use joint typicality encoding to describe X by U. Since U is correlated with Y, however, binning can be used to reduce its description rate. The bin index of U is sent to the receiver. The decoder uses joint typicality decoding with Y to recover U and then reconstructs $\hat{X}$ from U and Y. We now provide the details.

Codebook generation. Fix the conditional pmf $p(u|x)$ and function $\hat{x}(u, y)$ that attain the rate–distortion function $R_{\text{SI-D}}(D/(1 + \epsilon))$, where D is the desired distortion. Randomly and independently generate $2^{n\tilde{R}}$ sequences $u^n(l)$, $l \in [1 : 2^{n\tilde{R}}]$, each according to $\prod_{i=1}^{n} p_U(u_i)$. Partition the set of indices $l \in [1 : 2^{n\tilde{R}}]$ into equal-size subsets referred to as

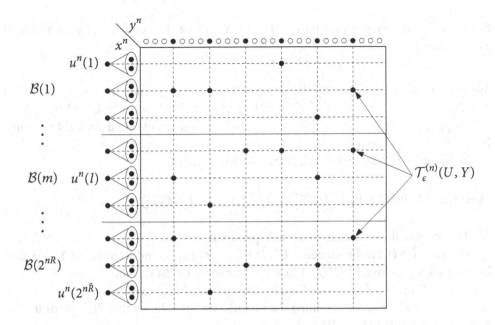

Figure 11.6. Wyner–Ziv coding scheme. Each bin $\mathcal{B}(m)$, $m \in [1 : 2^{nR}]$, consists of $2^{n(\tilde{R}-R)}$ indices.

bins $\mathcal{B}(m) = [(m-1)2^{n(\tilde{R}-R)} + 1 : m2^{n(\tilde{R}-R)}]$, $m \in [1 : 2^{nR}]$. The codebook is revealed to the encoder and the decoder.

Encoding. Given x^n, the encoder finds an index l such that $(x^n, u^n(l)) \in \mathcal{T}_{\epsilon'}^{(n)}$. If there is more than one such index, it selects one of them uniformly at random. If there is no such index, it selects an index from $[1 : 2^{n\tilde{R}}]$ uniformly at random. The encoder sends the bin index m such that $l \in \mathcal{B}(m)$. Note that randomized encoding here is used only to simplify the analysis and should be viewed as a part of random codebook generation; see Problem 11.4.

Decoding. Let $\epsilon > \epsilon'$. Upon receiving m, the decoder finds the unique index $\hat{l} \in \mathcal{B}(m)$ such that $(u^n(\hat{l}), y^n) \in \mathcal{T}_{\epsilon}^{(n)}$; otherwise it sets $\hat{l} = 1$. It then computes the reconstruction sequence as $\hat{x}_i = \hat{x}(u_i(\hat{l}), y_i)$ for $i \in [1 : n]$.

Analysis of expected distortion. Let (L, M) denote the chosen indices at the encoder and $\hat{L}$ be the index estimate at the decoder. Define the "error" event

$$\mathcal{E} = \{(U^n(\hat{L}), X^n, Y^n) \notin \mathcal{T}_{\epsilon}^{(n)}\}$$

and consider the events

$$\mathcal{E}_1 = \{(U^n(l), X^n) \notin \mathcal{T}_{\epsilon'}^{(n)} \text{ for all } l \in [1 : 2^{n\tilde{R}}]\},$$
$$\mathcal{E}_2 = \{(U^n(L), X^n, Y^n) \notin \mathcal{T}_{\epsilon}^{(n)}\},$$
$$\mathcal{E}_3 = \{(U^n(\tilde{l}), Y^n) \in \mathcal{T}_{\epsilon}^{(n)} \text{ for some } \tilde{l} \in \mathcal{B}(M), \tilde{l} \neq L\}.$$

Since the "error" event occurs only if $(U^n(L), X^n, Y^n) \notin T_\epsilon^{(n)}$ or $\hat{L} \neq L$, by the union of events bound,

$$P(\mathcal{E}) \leq P(\mathcal{E}_1) + P(\mathcal{E}_1^c \cap \mathcal{E}_2) + P(\mathcal{E}_3).$$

We now bound each term. By the covering lemma, $P(\mathcal{E}_1)$ tends to zero as $n \to \infty$ if $\tilde{R} > I(X; U) + \delta(\epsilon')$. Since $\epsilon > \epsilon'$, $\mathcal{E}_1^c = \{(U^n(L), X^n) \in T_{\epsilon'}^{(n)}\}$, and $Y^n | \{U^n(L) = u^n, X^n = x^n\} \sim \prod_{i=1}^n p_{Y|U,X}(y_i|u_i, x_i) = \prod_{i=1}^n p_{Y|X}(y_i|x_i)$, by the conditional typicality lemma, $P(\mathcal{E}_1^c \cap \mathcal{E}_2)$ tends to zero as $n \to \infty$.

To bound $P(\mathcal{E}_3)$, we first establish the following bound.

Lemma 11.1. $P(\mathcal{E}_3) \leq P\{(U^n(\tilde{l}), Y^n) \in T_\epsilon^{(n)} \text{ for some } \tilde{l} \in \mathcal{B}(1)\}.$

The proof of this lemma is given in Appendix 11A.

For each $\tilde{l} \in \mathcal{B}(1)$, the sequence $U^n(\tilde{l}) \sim \prod_{i=1}^n p_U(u_i)$ is independent of Y^n. Hence, by the packing lemma, $P\{(U^n(\tilde{l}), Y^n) \in T_\epsilon^{(n)} \text{ for some } \tilde{l} \in \mathcal{B}(1)\}$ tends to zero as $n \to \infty$ if $\tilde{R} - R < I(Y; U) - \delta(\epsilon)$. Therefore, by Lemma 11.1, $P(\mathcal{E}_3)$ tends to zero as $n \to \infty$ if $\tilde{R} - R < I(Y; U) - \delta(\epsilon)$. Combining the bounds, we have shown that $P(\mathcal{E})$ tends to zero as $n \to \infty$ if $R > I(X; U) - I(Y; U) + \delta(\epsilon) + \delta(\epsilon')$.

When there is no "error," $(U^n(L), X^n, Y^n) \in T_\epsilon^{(n)}$. Thus, by the law of total expectation and the typical average lemma, the asymptotic distortion averaged over the random codebook and encoding is upper bounded as

$$\limsup_{n\to\infty} E(d(X^n, \hat{X}^n)) \leq \limsup_{n\to\infty} \left(d_{\max} P(\mathcal{E}) + (1 + \epsilon) E(d(X, \hat{X})) P(\mathcal{E}^c)\right) \leq D,$$

if $R > I(X; U) - I(Y; U) + \delta(\epsilon) + \delta(\epsilon') = R_{\text{SI-D}}(D/(1 + \epsilon)) + \delta(\epsilon) + \delta(\epsilon')$. Finally from the continuity of $R_{\text{SI-D}}(D)$ in D, taking $\epsilon \to 0$ shows that any rate–distortion pair (R, D) with $R > R_{\text{SI-D}}(D)$ is achievable, which completes the proof of achievability.

Remark 11.1 (Deterministic versus random binning). Note that the above proof uses *deterministic* instead of random binning. This is because we are given a set of *randomly* generated sequences instead of a set of deterministic sequences; hence, there is no need to also randomize the binning. Following similar arguments to the proof of the lossless source coding with causal side information, we can readily show that the corner points of the Slepian–Wolf region are special cases of the Wyner–Ziv theorem. As such, random binning is not required for either.

Remark 11.2 (Binning versus multicoding). There is an interesting duality between multicoding and binning. On the one hand, the multicoding technique used in the achievability proofs of the Gelfand–Pinsker theorem and Marton's inner bound is a *channel coding* technique—we are given a set of messages and we generate a set of codewords (subcodebook) for each message, which *increases* the rate. To send a message, we transmit a codeword in its subcodebook that satisfies a desired property. On the other hand, binning is a *source coding* technique—we are given a set of indices/sequences and we compress them into a smaller number of bin indices, which *reduces* the rate. To send an index/sequence, we send its bin index. Thus, although in both techniques we partition a set into equal-size subsets, multicoding and binning are in some general sense dual to each other. We discuss other dualities between channel and source coding shortly.

11.3.2   Proof of the Converse

Let M denote the encoded index of X^n. The key to the proof is to identify U_i. In general $\hat{X}_i$ is a function of (M, Y^n). We would like $\hat{X}_i$ to be a function of (U_i, Y_i), so we identify the auxiliary random variable $U_i = (M, Y^{i-1}, Y_{i+1}^n)$. Note that this is a valid choice since $U_i \to X_i \to Y_i$ form a Markov chain. Consider

$$
\begin{aligned}
nR &\geq H(M) \\
&\geq H(M|Y^n) \\
&= I(X^n; M|Y^n) \\
&= \sum_{i=1}^{n} I(X_i; M|Y^n, X^{i-1}) \\
&\overset{(a)}{=} \sum_{i=1}^{n} I(X_i; M, Y^{i-1}, Y_{i+1}^n, X^{i-1}|Y_i) \\
&\geq \sum_{i=1}^{n} I(X_i; U_i|Y_i) \\
&\geq \sum_{i=1}^{n} R_{\text{SI-D}}(\mathbb{E}[d(X_i, \hat{X}_i)]) \\
&\overset{(b)}{\geq} nR_{\text{SI-D}} \left(\frac{1}{n} \sum_{i=1}^{n} \mathbb{E}(d(X_i, \hat{X}_i)) \right),
\end{aligned}
$$

where (a) follows since (X_i, Y_i) is independent of $(Y^{i-1}, Y_{i+1}^n, X^{i-1})$ and (b) follows by the convexity of $R_{\text{SI-D}}(D)$. Since $\limsup_{n\to\infty}(1/n)\sum_{i=1}^{n} \mathbb{E}(d(X_i, \hat{X}_i)) \leq D$ (by assumption) and $R_{\text{SI-D}}(D)$ is nonincreasing, $R \geq R_{\text{SI-D}}(D)$. The cardinality bound on $\mathcal{U}$ can be proved using the convex cover method in Appendix C. This completes the proof of the Wyner–Ziv theorem.

11.3.3   Source–Channel Coding Dualities

First recall the two fundamental limits on point-to-point communication—the rate–distortion function for a DMS X, $R(D) = \min_{p(\hat{x}|x)} I(\hat{X}; X)$, and the capacity for a DMC $p(y|x)$, $C = \max_{p(x)} I(X; Y)$. By comparing these two expressions, we observe interesting dualities between the given source $X \sim p(x)$ and channel $p(y|x)$, the test channel $p(\hat{x}|x)$ and channel input $X \sim p(x)$, and the minimum and maximum. Thus, roughly speaking, these two solutions (and underlying problems) are dual to each other.

Now, let us compare the rate–distortion functions with side information at the decoder

$$
R_{\text{CSI-D}}(D) = \min I(X; U),
$$
$$
R_{\text{SI-D}}(D) = \min \left(I(X; U) - I(Y; U) \right)
$$

with the capacity expressions for the DMC with DM state available at the encoder presented in Section 7.6

$$C_{\text{CSI-E}} = \max I(U; Y),$$
$$C_{\text{SI-E}} = \max (I(U; Y) - I(U; S)).$$

There are obvious dualities between the maximum and minimum, and multicoding and binning. But perhaps the most intriguing duality is that $R_{\text{SI-D}}(D)$ is the difference between the "covering rate" $I(U; X)$ and the "packing rate" $I(U; Y)$, while $C_{\text{SI-E}}$ is the difference between the "packing rate" $I(U; Y)$ and the "covering rate" $I(U; S)$.

These types of dualities are abundant in network information theory. For example, we can observe a similar duality between distributed lossless source coding and the deterministic broadcast channel. Although not as mathematically precise as Lagrange duality in convex analysis or the BC–MAC duality in Chapter 9, this covering–packing duality (or the source coding–channel coding duality in general) can lead to a better understanding of the underlying coding techniques.

11.4   SOURCE CODING WHEN SIDE INFORMATION MAY BE ABSENT

Consider the lossy source coding setup depicted in Figure 11.7, where the side information Y may be not be available at the decoder; thus the encoder needs to send a robust description of the source X under this uncertainty. Let (X, Y) be a 2-DMS and $d_j(x, \hat{x}_j)$, $j = 1, 2$, be two distortion measures. The encoder generates a description of X so that decoder 1 who does not have any side information can reconstruct X with distortion D_1 and decoder 2 who has side information Y can reconstruct X with distortion D_2.

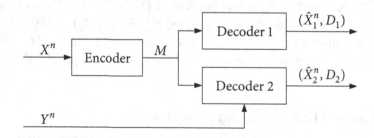

Figure 11.7. Source coding when side information may be absent.

We investigate this problem for the following side information availability scenarios. The definitions of a $(2^{nR}, n)$ code, achievability for distortion pair (D_1, D_2), and the rate–distortion function for each scenario can be defined as before.

Causal side information at decoder 2. If side information is available *causally* only at decoder 2, then

$$R_{\text{CSI-D2}}(D_1, D_2) = \min I(X; U),$$

where the minimum is over all conditional pmfs $p(u|x)$ with $|\mathcal{U}| \leq |\mathcal{X}| + 2$ and functions $\hat{x}_1(u)$ and $\hat{x}_2(u, y)$ such that $E(d_j(X, \hat{X}_j)) \leq D_j$, $j = 1, 2$.

Noncausal side information at decoder 2. If the side information is available noncausally at decoder 2, then

$$R_{\text{SI-D2}}(D_1, D_2) = \min \left(I(X; \hat{X}_1) + I(X; U|\hat{X}_1, Y) \right),$$

where the minimum is over all conditional pmfs $p(u, \hat{x}_1|x)$ with $|\mathcal{U}| \leq |\mathcal{X}| \cdot |\hat{\mathcal{X}}_1| + 2$ and functions $\hat{x}_2(u, \hat{x}_1, y)$ such that $E(d_j(X, \hat{X}_j)) \leq D_j$, $j = 1, 2$. Achievability follows by using lossy source coding for the source X with reconstruction $\hat{X}_1$ at rate $I(X; \hat{X}_1)$ and then using Wyner–Ziv coding of the pair $(X, \hat{X}_1)$ with side information $(\hat{X}_1, Y)$ available at decoder 2. To prove the converse, let M be the encoded index of X^n, identify the auxiliary random variable $U_i = (M, Y^{i-1}, Y^n_{i+1}, X^{i-1})$, and consider

$$nR \geq H(M)$$
$$= I(M; X^n, Y^n)$$
$$= I(M; X^n|Y^n) + I(M; Y^n)$$
$$= \sum_{i=1}^{n} (I(X_i; M|Y^n, X^{i-1}) + I(Y_i; M|Y^{i-1}))$$
$$= \sum_{i=1}^{n} (I(X_i; M, X^{i-1}, Y^{i-1}, Y^n_{i+1}|Y_i) + I(Y_i; M, Y^{i-1}))$$
$$\geq \sum_{i=1}^{n} (I(X_i; U_i, \hat{X}_{1i}|Y_i) + I(Y_i; \hat{X}_{1i}))$$
$$\geq \sum_{i=1}^{n} (I(X_i; U_i|\hat{X}_{1i}, Y_i) + I(X_i; \hat{X}_{1i})).$$

The rest of the proof follows similar steps to the proof of the Wyner–Ziv theorem.

Side information at the encoder and decoder 2. If side information is available (causally or noncausally) at both the encoder and decoder 2, then

$$R_{\text{SI-ED2}}(D_1, D_2) = \min \left(I(X, Y; \hat{X}_1) + I(X; \hat{X}_2|\hat{X}_1, Y) \right),$$

where the minimum is over all conditional pmfs $p(\hat{x}_1, \hat{x}_2|x, y)$ such that $E(d_j(X, \hat{X}_j)) \leq D_j$, $j = 1, 2$. Achievability follows by using lossy source coding for the source pair (X, Y) with reconstruction $\hat{X}_1$ at rate $I(X, Y; \hat{X}_1)$ and then using conditional lossy source coding for the source X and reconstruction $\hat{X}_2$ with side information $(\hat{X}_1, Y)$ available at both the encoder and decoder 2 (see Section 11.1). To prove the converse, first observe that $I(X, Y; \hat{X}_1) + I(X; \hat{X}_2|\hat{X}_1, Y) = I(\hat{X}_1; Y) + I(X; \hat{X}_1, \hat{X}_2|Y)$. Using the same first steps as in the proof of the case of side information only at decoder 2, we have

$$nR \geq \sum_{i=1}^{n}(I(X_i; M, X^{i-1}, Y^{i-1}, Y_{i+1}^n | Y_i) + I(Y_i; M, Y^{i-1}))$$

$$\geq \sum_{i=1}^{n}(I(X_i; \hat{X}_{1i}, \hat{X}_{2i} | Y_i) + I(Y_i; \hat{X}_{1i})).$$

The rest of the proof follows as before.

SUMMARY

- Lossy source coding with causal side information at the decoder:

  - Lossless source coding with causal side information does not reduce lossless encoding rate when $p(x, y) > 0$ for all (x, y)

- Lossy source coding with noncausal side information at the decoder:

  - Wyner–Ziv coding (compress–bin)

  - Deterministic binning

  - Use of the packing lemma in source coding

- Corner points of the Slepian–Wolf region are special cases of the Wyner–Ziv rate–distortion function

- Dualities between source coding with side information at the decoder and channel coding with state information at the encoder

BIBLIOGRAPHIC NOTES

Weissman and El Gamal (2006) established Theorem 11.1 and its lossless counterpart in Theorem 11.2. Theorem 11.3 is due to Wyner and Ziv (1976).

The duality between source and channel coding was first observed by Shannon (1959). This duality has been made more precise in, for example, Gupta and Verdú (2011). Duality between source coding with side information and channel coding with state information was further explored by Cover and Chiang (2002) and Pradhan, Chou, and Ramchandran (2003).

The problem of lossy source coding when side information may be absent was studied by Heegard and Berger (1985) and Kaspi (1994). Heegard and Berger (1985) also considered the case when different side information is available at several decoders. The rate–distortion function with side information available at both the encoder and decoder 2 was established by Kaspi (1994).

PROBLEMS

11.1. Let (X, Y) be a 2-DMS. Consider lossy source coding for X (a) when side information Y is available only at the encoder and (b) when Y is available at both the encoder and the decoder. Provide the details of the proof of the rate–distortion functions $R_{\text{SI-E}}(D)$ in (11.1) and $R_{\text{SI-ED}}(D)$ in (11.2).

11.2. Consider lossy source coding with side information for a DSBS(p) in Example 11.1. Derive the rate–distortion function $R_{\text{SI-ED}}(D)$ in (11.3).

11.3. Consider quadratic Gaussian source coding with side information in Example 11.2.

(a) Provide the details of the proof of the rate–distortion function $R_{\text{SI-ED}}$ in (11.4).

(b) Provide the details of the proof of the rate–distortion function $R_{\text{SI-D}}$ in (11.5).

11.4. *Randomized source code.* Suppose that in the definition of the $(2^{nR}, n)$ lossy source code with side information available noncausally at the decoder, we allow the encoder and the decoder to use random mappings. Specifically, let W be an arbitrary random variable independent of the source X and the side information Y. The encoder generates an index $m(x^n, W)$ and the decoder generates an estimate $\hat{x}^n(m, y^n, W)$. Show that this randomization does not decrease the rate–distortion function.

11.5. *Quadratic Gaussian source coding with causal side information.* Consider the causal version of quadratic Gaussian source coding with side information in Example 11.2. Although the rate–distortion function $R_{\text{CSI-D}}(D)$ is characterized in Theorem 11.1, the conditional distribution and the reconstruction function that attain the minimum are not known. Suppose we use the Gaussian test channel $U = X + Z$, where $Z \sim N(0, N)$ independent of (X, Y), and the MMSE estimate $\hat{X} = \mathsf{E}(X|U, Y)$. Find the corresponding upper bound on the rate–distortion function.
Remark: This problem was studied in Weissman and El Gamal (2006), who also showed that Gaussian test channels (even with memory) are suboptimal.

11.6. *Side information with occasional erasures.* Let X be a Bern(1/2) source, Y be the output of a BEC(p) with input X, and d be a Hamming distortion measure. Find a simple expression for the rate–distortion function $R_{\text{SI-D}}(D)$ for X with side information Y.

11.7. *Side information dependent distortion measure.* Consider lossy source coding with noncausal side information in Section 11.3. Suppose that the distortion measure $d(x, \hat{x}, y)$ depends on the side information Y as well. Show that the rate–distortion function is characterized by the Wyner–Ziv theorem, except that the minimum is over all $p(u|x)$ and $\hat{x}(u, y)$ such that $\mathsf{E}(d(X, \hat{X}, Y)) \leq D$.
Remark: This problem was studied by Linder, Zamir, and Zeger (2000).

11.8. *Lossy source coding from a noisy observation with side information.* Let $(X, Y, Z) \sim p(x, y, z)$ be a 3-DMS and $d(x, \hat{x})$ be a distortion measure. Consider the lossy source coding problem, where the encoder has access to a noisy version Y of the source X instead of X itself and the decoder has side information Z. Thus, unlike the standard lossy source coding setup with side information, the encoder maps each y^n sequence to an index $m \in [1 : 2^{nR})$. Otherwise, the definitions of a $(2^{nR}, n)$ code, achievability, and rate–distortion function are the same as before.

(a) Suppose that the side information Z is causally available at the decoder. This setup corresponds to the tracking scenario mentioned in Section 11.2, where X is the location of the target, Y is a description of the terrain information, and Z is the information from the tracking sensor. Show that the rate–distortion function is

$$R_{\text{CSI-D}}(D) = \min I(Y; U) \quad \text{for } D \geq D_{\min},$$

where the minimum is over all conditional pmfs $p(u|y)$ and functions $\hat{x}(u, z)$ such that $\mathsf{E}(d(X, \hat{X})) \leq D$, and $D_{\min} = \min_{\hat{x}(y,z)} \mathsf{E}[d(X, \hat{x}(Y, Z))]$.

(b) Now suppose that the side information Z is noncausally available at the decoder. Show that the rate–distortion function is

$$R_{\text{SI-D}}(D) = \min I(Y; U|Z) \quad \text{for } D \geq D_{\min},$$

where the minimum is over all conditional pmfs $p(u|y)$ and functions $\hat{x}(u, z)$ such that $\mathsf{E}(d(X, \hat{X})) \leq D$.

11.9. *Quadratic Gaussian source coding when side information may be absent.* Consider the source coding setup when side information may be absent. Let (X, Y) be a 2-WGN source with $X \sim N(0, P)$ and $Y = X + Z$, where $Z \sim N(0, N)$ is independent of X, and let d_1 and d_2 be squared error distortion measures. Show that the rate–distortion function is

$$R_{\text{SI-D2}}(D_1, D_2) = \begin{cases} \mathsf{R}\left(\frac{PN}{D_2(D_1+N)}\right) & \text{if } D_1 < P \text{ and } D_2 < D_1 N/(D_1 + N), \\ \mathsf{R}\left(\frac{P}{D_1}\right) & \text{if } D_1 < P \text{ and } D_2 \geq D_1 N/(D_1 + N), \\ \mathsf{R}\left(\frac{PN}{D_2(P+N)}\right) & \text{if } D_1 \geq P, D_2 < D_1 N/(D_1 + N), \\ 0 & \text{otherwise.} \end{cases}$$

11.10. *Lossy source coding with coded side information at the encoder and the decoder.* Let (X, Y) be a 2-DMS and $d(x, \hat{x})$ be a distortion measure. Consider the lossy source coding setup depicted in Figure 11.8, where coded side information of Y is available at both the encoder and the decoder. Define $\mathcal{R}_1(D)$ as the set of rate pairs (R_1, R_2) such that

$$R_1 \geq I(X; \hat{X}|U),$$
$$R_2 \geq I(Y; U)$$

for some conditional pmf $p(u|y)p(\hat{x}|u, x)$ that satisfies $\mathsf{E}(d(X, \hat{X})) \leq D$. Further define $\mathcal{R}_2(D)$ to be the set of rate pairs (R_1, R_2) that satisfy the above inequalities for some conditional pmf $p(u|y)p(\hat{x}|u, x, y)$ such that $\mathsf{E}(d(X, \hat{X})) \leq D$.

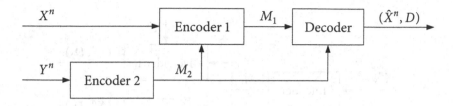

Figure 11.8. Lossy source coding with coded side information.

(a) Show that $\mathscr{R}_1(D)$ is an inner bound on the rate–distortion region.

(b) Show that $\mathscr{R}_1(D) = \mathscr{R}_2(D)$.

(c) Prove that $\mathscr{R}_2(D)$ is an outer bound on the rate–distortion region. Conclude that $\mathscr{R}_1(D)$ is the rate–distortion region.

11.11. *Source coding with degraded side information.* Consider the lossy source coding setup depicted in Figure 11.9. Let (X, Y_1, Y_2) be a 3-DMS such that $X \to Y_1 \to Y_2$ form a Markov chain. The encoder sends a description M of the source X so that decoder $j = 1, 2$ can reconstruct X with distortion D_j using side information Y_j. Find the rate–distortion function.

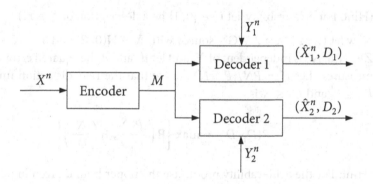

Figure 11.9. Lossy source coding with degraded side information.

Remark: This result is due to Heegard and Berger (1985).

11.12. *Complementary delivery.* Consider the lossy source coding with side information setup in Figure 11.10. The encoder sends a description $M \in [1 : 2^{nR}]$ of the 2-DMS (X, Y) to both decoders so that decoder 1 can reconstruct Y with distortion D_2 and decoder 2 can reconstruct X with distortion D_1. Define the rate–distortion function $R(D_1, D_2)$ to be the infimum of rates R such that the rate–distortion triple (R, D_1, D_2) is achievable.

(a) Establish the upper bound

$$R(D_1, D_2) \le \min \max\{I(X; U|Y), I(Y; U|X)\},$$

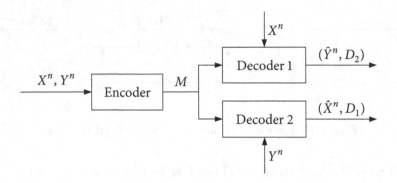

Figure 11.10. Complementary delivery.

where the minimum is over all conditional pmfs $p(u|x, y)$ and functions $\hat{x}(u, y)$ and $\hat{y}(u, x)$ such that $\mathsf{E}(d_1(X, \hat{X})) \le D_1$ and $\mathsf{E}(d_2(Y, \hat{Y})) \le D_2$.

(b) Let (X, Y) be a DSBS(p), and d_1 and d_2 be Hamming distortion measures. Show that the rate–distortion function for $D_1, D_2 \le p$ is

$$R(D_1, D_2) = H(p) - H(\min\{D_1, D_2\}).$$

(Hint: For achievability, let $U \in \{0, 1\}$ be a description of $X \oplus Y$.)

(c) Now let (X, Y) be a 2-WGN source with $X \sim \mathrm{N}(0, P)$ and $Y = X + Z$, where $Z \sim \mathrm{N}(0, N)$ is independent of X, and let d_1 and d_2 be squared error distortion measures. Let $P' = PN/(P + N)$. Show that the rate–distortion function for $D_1 \le P'$ and $D_2 \le N$ is

$$R(D_1, D_2) = \max\left\{ \mathsf{R}\left(\frac{P'}{D_1}\right), \mathsf{R}\left(\frac{N}{D_2}\right)\right\}.$$

(Hint: For the achievability proof, use the upper bound given in part (a). Let $U = (V_1, V_2)$ such that $V_1 \to X \to Y \to V_2$ and choose Gaussian test channels from X to V_1 and from Y to V_2, and the functions $\hat{x}$ and $\hat{y}$ to be the MMSE estimate of X given (V_1, Y) and Y given (V_2, X), respectively.)

Remark: This problem was first considered in Kimura and Uyematsu (2006).

APPENDIX 11A   PROOF OF LEMMA 11.1

We first show that

$$\mathsf{P}\{(U^n(\tilde{l}), Y^n) \in \mathcal{T}_{\epsilon}^{(n)} \text{ for some } \tilde{l} \in \mathcal{B}(m), \tilde{l} \ne L \mid M = m\}$$
$$\le \mathsf{P}\{(U^n(\tilde{l}), Y^n) \in \mathcal{T}_{\epsilon}^{(n)} \text{ for some } \tilde{l} \in \mathcal{B}(1) \mid M = m\}.$$

This holds trivially when $m = 1$. For $m \neq 1$, consider

$$P\{(U^n(\tilde{l}), Y^n) \in \mathcal{T}_\epsilon^{(n)} \text{ for some } \tilde{l} \in \mathcal{B}(m), \tilde{l} \neq L \,|\, M = m\}$$

$$= \sum_{l \in \mathcal{B}(m)} p(l|m) \, P\{(U^n(\tilde{l}), Y^n) \in \mathcal{T}_\epsilon^{(n)} \text{ for some } \tilde{l} \in \mathcal{B}(m), \tilde{l} \neq l \,|\, L = l, M = m\}$$

$$\overset{(a)}{=} \sum_{l \in \mathcal{B}(m)} p(l|m) \, P\{(U^n(\tilde{l}), Y^n) \in \mathcal{T}_\epsilon^{(n)} \text{ for some } \tilde{l} \in \mathcal{B}(m), \tilde{l} \neq l \,|\, L = l\}$$

$$\overset{(b)}{=} \sum_{l \in \mathcal{B}(m)} p(l|m) \, P\{(U^n(\tilde{l}), Y^n) \in \mathcal{T}_\epsilon^{(n)} \text{ for some } \tilde{l} \in [1 : 2^{n(\tilde{R}-R)} - 1] \,|\, L = l\}$$

$$\leq \sum_{l \in \mathcal{B}(m)} p(l|m) \, P\{(U^n(\tilde{l}), Y^n) \in \mathcal{T}_\epsilon^{(n)} \text{ for some } \tilde{l} \in \mathcal{B}(1) \,|\, L = l\}$$

$$\overset{(c)}{=} \sum_{l \in \mathcal{B}(m)} p(l|m) \, P\{(U^n(\tilde{l}), Y^n) \in \mathcal{T}_\epsilon^{(n)} \text{ for some } \tilde{l} \in \mathcal{B}(1) \,|\, L = l, M = m\}$$

$$= P\{(U^n(\tilde{l}), Y^n) \in \mathcal{T}_\epsilon^{(n)} \text{ for some } \tilde{l} \in \mathcal{B}(1) \,|\, M = m\},$$

where (a) and (c) follow since M is a function of L and (b) follows since given $L = l$, any collection of $2^{n(\tilde{R}-R)} - 1$ codewords $U^n(\tilde{l})$ with $\tilde{l} \neq l$ has the same distribution. Hence, we have

$$P(\mathcal{E}_3) = \sum_m p(m) \, P\{(U^n(\tilde{l}), Y^n) \in \mathcal{T}_\epsilon^{(n)} \text{ for some } \tilde{l} \in \mathcal{B}(m), \tilde{l} \neq L \,|\, M = m\}$$

$$\leq \sum_m p(m) \, P\{(U^n(\tilde{l}), Y^n) \in \mathcal{T}_\epsilon^{(n)} \text{ for some } \tilde{l} \in \mathcal{B}(1) \,|\, M = m\}$$

$$= P\{(U^n(\tilde{l}), Y^n) \in \mathcal{T}_\epsilon^{(n)} \text{ for some } \tilde{l} \in \mathcal{B}(1)\}.$$

CHAPTER 12

Distributed Lossy Compression

In this chapter, we consider the general distributed lossy compression system depicted in Figure 12.1. Let (X_1, X_2) be a 2-DMS and $d_j(x_j, \hat{x}_j)$, $j = 1, 2$, be two distortion measures. Sender $j = 1, 2$ wishes to communicate its source X_j over a noiseless link to the receiver with a prescribed distortion D_j. We wish to find the rate–distortion region, which is the set of description rates needed to achieve the desired distortions. It is not difficult to see that this problem includes, as special cases, distributed lossless source coding in Chapter 10 and lossy source coding with noncausal side information available at the decoder in Chapter 11. Unlike these special cases, however, the rate–distortion region of the distributed lossy source coding problem is not known in general.

We first extend the Wyner–Ziv coding scheme to establish the Berger–Tung inner bound on the rate–distortion region. The proof involves two new lemmas—the Markov lemma, which is, in a sense, a stronger version of the conditional typicality lemma, and the mutual packing lemma, which is a dual to the mutual covering lemma in Section 8.3. We also establish an outer bound on the rate–distortion region, which is tight in some cases. The rest of the chapter is largely dedicated to establishing the optimality of the Berger–Tung coding scheme for the quadratic Gaussian case. The proof of the converse involves the entropy power inequality, identification of a common-information random variable, and results from MMSE estimation. We then study the quadratic Gaussian CEO problem in which each sender observes a noisy version of the same Gaussian source and the receiver wishes to reconstruct the source with a prescribed distortion. We show that the Berger–Tung coding scheme is again optimal for this problem. We show via an example, however, that the Berger–Tung coding scheme is not optimal in general.

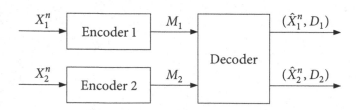

Figure 12.1. Distributed lossy compression system.

12.1   BERGER–TUNG INNER BOUND

A $(2^{nR_1}, 2^{nR_2}, n)$ *distributed lossy source code* consists of

- two encoders, where encoder 1 assigns an index $m_1(x_1^n) \in [1 : 2^{nR_1})$ to each sequence $x_1^n \in \mathcal{X}_1^n$ and encoder 2 assigns an index $m_2(x_2^n) \in [1 : 2^{nR_2})$ to each sequence $x_2^n \in \mathcal{X}_2^n$, and

- a decoder that assigns a pair of estimates $(\hat{x}_1^n, \hat{x}_2^n)$ to each index pair $(m_1, m_2) \in [1 : 2^{nR_1}) \times [1 : 2^{nR_2})$.

A rate–distortion quadruple (R_1, R_2, D_1, D_2) is said to be achievable (and a rate pair (R_1, R_2) is achievable for distortion pair (D_1, D_2)) if there exists a sequence of $(2^{nR_1}, 2^{nR_2}, n)$ codes with

$$\limsup_{n \to \infty} \mathsf{E}(d_j(X_j^n, \hat{X}_j^n)) \le D_j, \quad j = 1, 2.$$

The rate–distortion region $\mathscr{R}(D_1, D_2)$ for distributed lossy source coding is the closure of the set of all rate pairs (R_1, R_2) such that (R_1, R_2, D_1, D_2) is achievable.

The rate–distortion region for distributed lossy source coding is not known in general. The following is an inner bound on the rate–distortion region.

> **Theorem 12.1 (Berger–Tung Inner Bound).** Let (X_1, X_2) be a 2-DMS and $d_1(x_1, \hat{x}_1)$ and $d_2(x_2, \hat{x}_2)$ be two distortion measures. A rate pair (R_1, R_2) is achievable with distortion pair (D_1, D_2) for distributed lossy source coding if
>
> $$R_1 > I(X_1; U_1 | U_2, Q),$$
> $$R_2 > I(X_2; U_2 | U_1, Q),$$
> $$R_1 + R_2 > I(X_1, X_2; U_1, U_2 | Q)$$
>
> for some conditional pmf $p(q)p(u_1|x_1, q)p(u_2|x_2, q)$ with $|\mathcal{Q}| \le 4$, $|\mathcal{U}_j| \le |\mathcal{X}_j| + 4$, $j = 1, 2$, and functions $\hat{x}_1(u_1, u_2, q)$ and $\hat{x}_2(u_1, u_2, q)$ such that $\mathsf{E}(d_j(X_j, \hat{X}_j)) \le D_j, j = 1, 2$.

The Berger–Tung inner bound is tight in several special cases.

- It reduces to the Slepian–Wolf region when d_1 and d_2 are Hamming distortion measures and $D_1 = D_2 = 0$ (set $U_1 = X_1$ and $U_2 = X_2$).

- It reduces to the Wyner–Ziv rate–distortion function when there is no rate limit on describing X_2, i.e., $R_2 \ge H(X_2)$. In this case, the only active constraint $I(X_1; U_1 | U_2, Q)$ is minimized by $U_2 = X_2$ and $Q = \emptyset$.

- More generally, suppose that d_2 is a Hamming distortion measure and $D_2 = 0$. Then the Berger–Tung inner bound is tight and reduces to the set of rate pairs (R_1, R_2) such that

$$R_1 \geq I(X_1; U_1 | X_2),$$
$$R_2 \geq H(X_2 | U_1), \tag{12.1}$$
$$R_1 + R_2 \geq I(X_1; U_1 | X_2) + H(X_2) = I(X_1; U_1) + H(X_2 | U_1)$$

for some conditional pmf $p(u_1|x_1)$ and function $\hat{x}_1(u_1, x_2)$ that satisfy the constraint $\mathsf{E}(d_1(X_1, \hat{X}_1)) \leq D_1$.

- The Berger–Tung inner bound is tight for the quadratic Gaussian case, that is, for a 2-WGN source and squared error distortion measures. We will discuss this result in detail in Section 12.3.

However, the Berger–Tung inner bound is not tight in general as we show via an example in Section 12.5.

12.1.1   Markov and Mutual Packing Lemmas

We will need the following two lemmas in the proof of achievability.

Lemma 12.1 (Markov Lemma). Suppose that $X \to Y \to Z$ form a Markov chain. Let $(x^n, y^n) \in \mathcal{T}_{\epsilon'}^{(n)}(X, Y)$ and $Z^n \sim p(z^n|y^n)$, where the conditional pmf $p(z^n|y^n)$ satisfies the following conditions:

1. $\lim_{n\to\infty} \mathsf{P}\{(y^n, Z^n) \in \mathcal{T}_{\epsilon'}^{(n)}(Y, Z)\} = 1.$

2. For every $z^n \in \mathcal{T}_{\epsilon'}^{(n)}(Z|y^n)$ and n sufficiently large

$$2^{-n(H(Z|Y)+\delta(\epsilon'))} \leq p(z^n|y^n) \leq 2^{-n(H(Z|Y)-\delta(\epsilon'))}$$

for some $\delta(\epsilon')$ that tends to zero as $\epsilon' \to 0$.

Then, for some sufficiently small $\epsilon' < \epsilon$,

$$\lim_{n\to\infty} \mathsf{P}\{(x^n, y^n, Z^n) \in \mathcal{T}_{\epsilon}^{(n)}(X, Y, Z)\} = 1.$$

The proof of this lemma involves a counting argument and is given in Appendix 12A. Note that if $p(z^n|y^n) = \prod_{i=1}^{n} p_{Z|Y}(z_i|y_i)$, then the conclusion in the lemma holds by the conditional typicality lemma in Section 2.5, which was sufficient for the achievability proof of the Wyner–Ziv theorem in Section 11.3. In the proof of the Berger–Tung inner bound, however, joint typicality encoding is used twice, necessitating the above lemma, which is applicable to more general conditional pmfs $p(z^n|y^n)$.

The second lemma is a straightforward generalization of the packing lemma in Section 3.2. It can be viewed also as a dual to the mutual covering lemma in Section 8.3.

Lemma 12.2 (Mutual Packing Lemma). Let $(U_1, U_2) \sim p(u_1, u_2)$. Let $U_1^n(l_1)$, $l_1 \in [1 : 2^{nr_1}]$, be random sequences, each distributed according to $\prod_{i=1}^n p_{U_1}(u_{1i})$ with arbitrary dependence on the rest of the $U_1^n(l_1)$ sequences. Similarly, let $U_2^n(l_2)$, $l_2 \in [1 : 2^{nr_2}]$, be random sequences, each distributed according to $\prod_{i=1}^n p_{U_2}(u_{2i})$ with arbitrary dependence on the rest of the $U_2^n(l_2)$ sequences. Assume that $(U_1^n(l_1) \colon l_1 \in [1 : 2^{nr_1}])$ and $(U_2^n(l_2) \colon l_2 \in [1 : 2^{nr_2}])$ are independent. Then there exists $\delta(\epsilon)$ that tends to zero as $\epsilon \to 0$ such that

$$\lim_{n \to \infty} \mathsf{P}\{(U_1^n(l_1), U_2^n(l_2)) \in \mathcal{T}_\epsilon^{(n)} \text{ for some } (l_1, l_2) \in [1 : 2^{nr_1}] \times [1 : 2^{nr_2}]\} = 0,$$

if $r_1 + r_2 < I(U_1; U_2) - \delta(\epsilon)$.

12.1.2   Proof of the Berger–Tung Inner Bound

Achievability of the Berger–Tung inner bound uses the *distributed* compress–bin scheme illustrated in Figure 12.2. As in Wyner–Ziv coding, the scheme uses joint typicality encoding and binning at each encoder and joint typicality decoding and symbol-by-symbol reconstructions at the decoder.

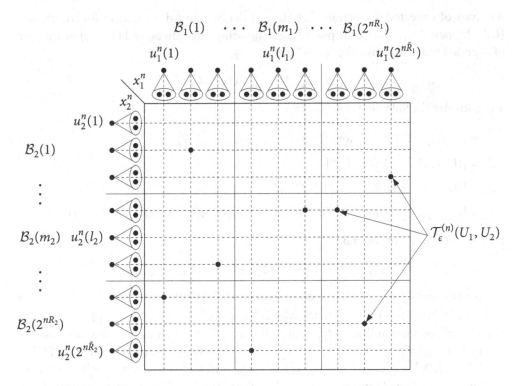

Figure 12.2. Berger–Tung coding scheme. Each bin $\mathcal{B}_j(m_j)$, $m_j \in [1 : 2^{nR_j}]$, $j = 1, 2$, consists of $2^{n(\tilde{R}_j - R_j)}$ indices.

We prove achievability for $|Q| = 1$; the rest of the proof follows using time sharing. In the following we assume that $\epsilon > \epsilon' > \epsilon''$.

Codebook generation. Fix a conditional pmf $p(u_1|x_1)p(u_2|x_2)$ and functions $\hat{x}_1(u_1, u_2)$ and $\hat{x}_2(u_1, u_2)$ such that $\mathsf{E}(d_j(X_j, \hat{X}_j)) \le D_j/(1 + \epsilon)$, $j = 1, 2$. Let $\tilde{R}_1 \ge R_1$, $\tilde{R}_2 \ge R_2$. For $j = 1, 2$, randomly and independently generate $2^{n\tilde{R}_j}$ sequences $u_j^n(l_j)$, $l_j \in [1 : 2^{n\tilde{R}_j}]$, each according to $\prod_{i=1}^{n} p_{U_j}(u_{ji})$. Partition the set of indices $l_j \in [1 : 2^{n\tilde{R}_j}]$ into equal-size bins $\mathcal{B}_j(m_j) = [(m_j - 1)2^{n(\tilde{R}_j - R_j)} + 1 : m_j 2^{n(\tilde{R}_j - R_j)}]$, $m_j \in [1 : 2^{nR_j}]$. The codebook is revealed to the encoders and the decoder.

Encoding. Upon observing x_j^n, encoder $j = 1, 2$ finds an index $l_j \in [1 : 2^{n\tilde{R}_j}]$ such that $(x_j^n, u_j^n(l_j)) \in \mathcal{T}_{\epsilon''}^{(n)}$. If there is more than one such index l_j, encoder j selects one of them uniformly at random. If there is no such index l_j, encoder j selects an index from $[1 : 2^{n\tilde{R}_j}]$ uniformly at random. Encoder $j = 1, 2$ sends the index m_j such that $l_j \in \mathcal{B}_j(m_j)$.

Decoding. The decoder finds the unique index pair $(\hat{l}_1, \hat{l}_2) \in \mathcal{B}_1(m_1) \times \mathcal{B}_2(m_2)$ such that $(u_1^n(\hat{l}_1), u_2^n(\hat{l}_2)) \in \mathcal{T}_{\epsilon}^{(n)}$. If there is such a unique index pair $(\hat{l}_1, \hat{l}_2)$, the reconstructions are computed as $\hat{x}_{1i}(u_{1i}(\hat{l}_1), u_{2i}(\hat{l}_2))$ and $\hat{x}_{2i}(u_{1i}(\hat{l}_1), u_{2i}(\hat{l}_2))$ for $i \in [1 : n]$; otherwise $\hat{x}_1^n$ and $\hat{x}_2^n$ are set to arbitrary sequences in $\hat{\mathcal{X}}_1^n$ and $\hat{\mathcal{X}}_2^n$, respectively.

Analysis of expected distortion. Let (L_1, L_2) denote the pair of indices for the chosen (U_1^n, U_2^n) pair, (M_1, M_2) be the pair of corresponding bin indices, and $(\hat{L}_1, \hat{L}_2)$ be the pair of decoded indices. Define the "error" event

$$\mathcal{E} = \{(U_1^n(\hat{L}_1), U_2^n(\hat{L}_2), X_1^n, X_2^n) \notin \mathcal{T}_{\epsilon}^{(n)}\}$$

and consider the following events

$$\mathcal{E}_1 = \{(U_j^n(l_j), X_j^n) \notin \mathcal{T}_{\epsilon''}^{(n)} \text{ for all } l_j \in [1 : 2^{n\tilde{R}_j}] \text{ for } j = 1, 2\},$$
$$\mathcal{E}_2 = \{(U_1^n(L_1), X_1^n, X_2^n) \notin \mathcal{T}_{\epsilon'}^{(n)}\},$$
$$\mathcal{E}_3 = \{(U_1^n(L_1), U_2^n(L_2), X_1^n, X_2^n) \notin \mathcal{T}_{\epsilon}^{(n)}\},$$
$$\mathcal{E}_4 = \{(U_1^n(\tilde{l}_1), U_2^n(\tilde{l}_2)) \in \mathcal{T}_{\epsilon}^{(n)} \text{ for some } (\tilde{l}_1, \tilde{l}_2) \in \mathcal{B}_1(M_1) \times \mathcal{B}_2(M_2), (\tilde{l}_1, \tilde{l}_2) \ne (L_1, L_2)\}.$$

By the union of events bound,

$$\mathsf{P}(\mathcal{E}) \le \mathsf{P}(\mathcal{E}_1) + \mathsf{P}(\mathcal{E}_1^c \cap \mathcal{E}_2) + \mathsf{P}(\mathcal{E}_2^c \cap \mathcal{E}_3) + \mathsf{P}(\mathcal{E}_4).$$

We bound each term. By the covering lemma, $\mathsf{P}(\mathcal{E}_1)$ tends to zero as $n \to \infty$ if $\tilde{R}_j > I(U_j; X_j) + \delta(\epsilon'')$ for $j = 1, 2$. Since $X_2^n | \{U_1^n(L_1) = u_1^n, X_1^n = x_1^n\} \sim \prod_{i=1}^{n} p_{X_2|X_1}(x_{2i}|x_{1i})$ and $\epsilon' > \epsilon''$, by the conditional typicality lemma, $\mathsf{P}(\mathcal{E}_1^c \cap \mathcal{E}_2)$ tends to zero as $n \to \infty$.

To bound $\mathsf{P}(\mathcal{E}_2^c \cap \mathcal{E}_3)$, let $(u_1^n, x_1^n, x_2^n) \in \mathcal{T}_{\epsilon'}^{(n)}(U_1, X_1, X_2)$ and consider $\mathsf{P}\{U_2^n(L_2) = u_2^n | X_2^n = x_2^n, X_1^n = x_1^n, U_1^n(L_1) = u_1^n\} = \mathsf{P}\{U_2^n(L_2) = u_2^n | X_2^n = x_2^n\} = p(u_2^n|x_2^n)$. First note that by the covering lemma, $\mathsf{P}\{U_2^n(L_2) \in \mathcal{T}_{\epsilon'}^{(n)}(U_2|x_2^n) | X_2^n = x_2^n\}$ converges to 1 as $n \to \infty$, that is, $p(u_2^n|x_2^n)$ satisfies the first condition in the Markov lemma. In Appendix 12B we show that it also satisfies the second condition.

Lemma 12.3. For every $u_2^n \in \mathcal{T}_{\epsilon'}^{(n)}(U_2|x_2^n)$ and n sufficiently large,

$$p(u_2^n|x_2^n) \doteq 2^{-nH(U_2|X_2)}.$$

Hence, by the Markov lemma with $Z \leftarrow U_2$, $Y \leftarrow X_2$, and $X \leftarrow (U_1, X_1)$, we have

$$\lim_{n\to\infty} \mathsf{P}\{(u_1^n, x_1^n, x_2^n, U_2^n(L_2)) \in \mathcal{T}_{\epsilon}^{(n)} \mid U_1^n(L_1) = u_1^n, X_1^n = x_1^n, X_2^n = x_2^n\} = 1,$$

if $(u_1^n, x_1^n, x_2^n) \in \mathcal{T}_{\epsilon'}^{(n)}(U_1, X_1, X_2)$ and $\epsilon' < \epsilon$ is sufficiently small. Therefore, $\mathsf{P}(\mathcal{E}_2^c \cap \mathcal{E}_3)$ tends to zero as $n \to \infty$.

Following a similar argument to Lemma 11.1 in the proof of the Wyner–Ziv theorem, we have

$$\mathsf{P}(\mathcal{E}_4) \le \mathsf{P}\{(U_1^n(\tilde{l}_1), U_2^n(\tilde{l}_2)) \in \mathcal{T}_{\epsilon}^{(n)} \text{ for some } (\tilde{l}_1, \tilde{l}_2) \in \mathcal{B}_1(1) \times \mathcal{B}_2(1)\}.$$

Hence, by the mutual packing lemma, $\mathsf{P}(\mathcal{E}_4)$ tends to zero as $n \to \infty$ if $(\tilde{R}_1 - R_1) + (\tilde{R}_2 - R_2) < I(U_1; U_2) - \delta(\epsilon)$. Combining the bounds and eliminating $\tilde{R}_1$ and $\tilde{R}_2$, we have shown that $\mathsf{P}(\mathcal{E})$ tends to zero as $n \to \infty$ if

$$R_1 > I(U_1; X_1) - I(U_1; U_2) + \delta(\epsilon'') + \delta(\epsilon)$$
$$= I(U_1; X_1|U_2) + \delta'(\epsilon),$$
$$R_2 > I(U_2; X_2|U_1) + \delta'(\epsilon), \qquad (12.2)$$
$$R_1 + R_2 > I(U_1; X_1) + I(U_2; X_2) - I(U_1; U_2) + 2\delta'(\epsilon)$$
$$= I(U_1, U_2; X_1, X_2) + 2\delta'(\epsilon).$$

As in previous lossy source coding achievability proofs, by the typical average lemma , it can be shown that the asymptotic distortions averaged over the random codebook and encoding, and over (X_1^n, X_2^n) are bounded as

$$\limsup_{n\to\infty} \mathsf{E}(d_j(X_j^n, \hat{X}_j^n)) \le D_j, \quad j = 1, 2,$$

if the inequalities in (12.2) are satisfied. Now using time sharing establishes the achievability of every rate pair (R_1, R_2) that satisfies the inequalities in the theorem for some conditional pmf $p(q)p(u_1|x_1, q)p(u_2|x_2, q)$ and functions $\hat{x}_1(u_1, u_2, q)$, $\hat{x}_2(u_1, u_2, q)$ such that $\mathsf{E}(d_j(X_j, \hat{X}_j)) < D_j$, $j = 1, 2$. Finally, using the continuity of mutual information completes the proof with nonstrict distortion inequalities.

12.2 BERGER–TUNG OUTER BOUND

The following is an outer bound on the rate–distortion region for distributed lossy source coding.

Theorem 12.2 (Berger–Tung Outer Bound). Let (X_1, X_2) be a 2-DMS and $d_1(x_1, \hat{x}_1)$ and $d_2(x_2, \hat{x}_2)$ be two distortion measures. If a rate pair (R_1, R_2) is achievable with distortion pair (D_1, D_2) for distributed lossy source coding, then it must satisfy the inequalities

$$R_1 \geq I(X_1, X_2; U_1 | U_2),$$
$$R_2 \geq I(X_1, X_2; U_2 | U_1),$$
$$R_1 + R_2 \geq I(X_1, X_2; U_1, U_2)$$

for some conditional pmf $p(u_1, u_2 | x_1, x_2)$ and functions $\hat{x}_1(u_1, u_2)$ and $\hat{x}_2(u_1, u_2)$ such that $U_1 \to X_1 \to X_2$ and $X_1 \to X_2 \to U_2$ form Markov chains and $\mathsf{E}(d_j(X_j, \hat{X}_j)) \leq D_j$, $j = 1, 2$.

This outer bound is similar to the Berger–Tung inner bound except that the region is convex without the use of a time-sharing random variable and the Markov conditions are weaker than the condition $U_1 \to X_1 \to X_2 \to U_2$ in the inner bound. The proof of the outer bound uses standard arguments with the auxiliary random variable identifications $U_{1i} = (M_1, X_1^{i-1}, X_2^{i-1})$ and $U_{2i} = (M_2, X_1^{i-1}, X_2^{i-1})$.

The Berger–Tung outer bound is again tight when d_2 is a Hamming distortion measure with $D_2 = 0$, which includes as special cases distributed lossless source coding (Slepian–Wolf) and lossy source coding with noncausal side information available at the decoder (Wyner–Ziv). The outer bound is also tight when X_2 is a function of X_1. In this case, the Berger–Tung inner bound is not tight as discussed in Section 12.5. The outer bound, however, is not tight in general. For example, it is not tight for the quadratic Gaussian case discussed next.

12.3   QUADRATIC GAUSSIAN DISTRIBUTED SOURCE CODING

Consider the distributed lossy source coding problem for a 2-WGN(1, ρ) source (X_1, X_2) and squared error distortion measures $d_j(x_j, \hat{x}_j) = (x_j - \hat{x}_j)^2$, $j = 1, 2$. Assume without loss of generality that $\rho = \mathsf{E}(X_1 X_2) \geq 0$.

The Berger–Tung inner bound is tight for this case and reduces to the following.

Theorem 12.3. The rate–distortion region $\mathscr{R}(D_1, D_2)$ for distributed lossy source coding a 2-WGN(1, ρ) source (X_1, X_2) and squared error distortion is the intersection of the following three regions:

$$\mathscr{R}_1(D_1) = \{(R_1, R_2): R_1 \geq g(R_2, D_1)\},$$
$$\mathscr{R}_2(D_2) = \{(R_1, R_2): R_2 \geq g(R_1, D_2)\},$$
$$\mathscr{R}_{12}(D_1, D_2) = \left\{(R_1, R_2): R_1 + R_2 \geq \mathsf{R}\left(\frac{(1 - \rho^2)\phi(D_1, D_2)}{2D_1 D_2}\right)\right\},$$

where

$$g(R,D) = \mathsf{R}\left(\frac{1-\rho^2+\rho^2 2^{-2R}}{D}\right),$$

$$\phi(D_1, D_2) = 1 + \sqrt{1 + 4\rho^2 D_1 D_2/(1-\rho^2)^2},$$

and $\mathsf{R}(x)$, $x \geq 1$, is the quadratic Gaussian rate function.

The rate–distortion region $\mathscr{R}(D_1, D_2)$ is sketched in Figure 12.3.

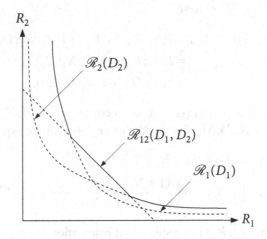

Figure 12.3. The rate–distortion region for quadratic Gaussian distributed source coding; $\mathscr{R}(D_1, D_2) = \mathscr{R}_1(D_1) \cap \mathscr{R}_2(D_2) \cap \mathscr{R}_{12}(D_1, D_2)$.

We prove Theorem 12.3 in the next two subsections.

12.3.1   Proof of Achievability

Assume without loss of generality that $D_1 \leq D_2$. We set the auxiliary random variables in the Berger–Tung inner bound to $U_1 = X_1 + V_1$ and $U_2 = X_2 + V_2$, where $V_1 \sim \mathrm{N}(0, N_1)$ and $V_2 \sim \mathrm{N}(0, N_2)$, $N_2 \geq N_1$, are independent of each other and of (X_1, X_2). Let the reconstructions $\hat{X}_1$ and $\hat{X}_2$ be the MMSE estimates $\mathsf{E}(X_1|U_1, U_2)$ and $\mathsf{E}(X_2|U_1, U_2)$ of X_1 and X_2, respectively. We refer to such choice of (U_1, U_2) as a *distributed Gaussian test channel* characterized by (N_1, N_2).

Each distributed Gaussian test channel corresponds to the distortion pair

$$D_1 = \mathsf{E}\left((X_1 - \hat{X}_1)^2\right) = \frac{N_1(1 + N_2 - \rho^2)}{(1 + N_1)(1 + N_2) - \rho^2},$$

$$D_2 = \mathsf{E}\left((X_2 - \hat{X}_2)^2\right) = \frac{N_2(1 + N_1 - \rho^2)}{(1 + N_1)(1 + N_2) - \rho^2} \geq D_1.$$

Evaluating the Berger–Tung inner bound with the above choices of test channels and reconstruction functions, we obtain

$$R_1 > I(X_1; U_1 | U_2) = \mathsf{R}\left(\frac{(1+N_1)(1+N_2)-\rho^2}{N_1(1+N_2)}\right),$$

$$R_2 > I(X_2; U_2 | U_1) = \mathsf{R}\left(\frac{(1+N_1)(1+N_2)-\rho^2}{N_2(1+N_1)}\right), \tag{12.3}$$

$$R_1 + R_2 > I(X_1, X_2; U_1, U_2) = \mathsf{R}\left(\frac{(1+N_1)(1+N_2)-\rho^2}{N_1 N_2}\right).$$

Since $U_1 \to X_1 \to X_2 \to U_2$, we have

$$I(X_1, X_2; U_1, U_2) = I(X_1, X_2; U_1 | U_2) + I(X_1, X_2; U_2)$$
$$= I(X_1; U_1 | U_2) + I(X_2; U_2)$$
$$= I(X_1; U_1) + I(X_2; U_2 | U_1).$$

Thus in general, the rate region in (12.3) has two corner points $(I(X_1; U_1 | U_2), I(X_2; U_2))$ and $(I(X_1; U_1), I(X_2; U_2 | U_1))$. The first (left) corner point can be expressed as

$$R_2' = I(X_2; U_2) = \mathsf{R}\big((1+N_2)/N_2\big),$$

$$R_1' = I(X_1; U_1 | U_2) = \mathsf{R}\left(\frac{(1+N_1)(1+N_2)-\rho^2}{N_1(1+N_2)}\right) = g(R_2', D_1).$$

The other corner point (R_1'', R_2'') has a similar representation.

We now consider the following two cases.

Case 1: $(1 - D_2) \le \rho^2(1 - D_1)$. For this case, the characterization for $\mathscr{R}(D_1, D_2)$ can be simplified as follows.

Lemma 12.4. If $(1 - D_2) \le \rho^2(1 - D_1)$, then $\mathscr{R}_1(D_1) \subseteq \mathscr{R}_2(D_2) \cap \mathscr{R}_{12}(D_1, D_2)$.

The proof of this lemma is in Appendix 12C.

Consider a distributed Gaussian test channel with $N_1 = D_1/(1 - D_1)$ and $N_2 = \infty$ (i.e., $U_2 = \emptyset$). Then, the left corner point of the rate region in (12.3) becomes

$$R_2' = 0,$$

$$R_1' = \mathsf{R}\left(\frac{1+N_1}{N_1}\right) = \mathsf{R}\left(\frac{1}{D_1}\right) = g(R_2', D_1).$$

Also it can be easily verified that the distortion constraints are satisfied as

$$\mathsf{E}\big((X_1 - \hat{X}_1)^2\big) = 1 - \frac{1}{1+N_1} = D_1,$$

$$\mathsf{E}\big((X_2 - \hat{X}_2)^2\big) = 1 - \frac{\rho^2}{1+N_1} = 1 - \rho^2 + \rho^2 D_1 \le D_2.$$

Now consider a test channel with $(\tilde{N}_1, \tilde{N}_2)$, where $\tilde{N}_2 < \infty$ and $\tilde{N}_1 \geq N_1$ such that

$$\frac{\tilde{N}_1(1 + \tilde{N}_2 - \rho^2)}{(1 + \tilde{N}_1)(1 + \tilde{N}_2) - \rho^2} = \frac{N_1}{1 + N_1} = D_1.$$

Then, the corresponding distortion pair $(\tilde{D}_1, \tilde{D}_2)$ satisfies $\tilde{D}_1 = D_1$ and

$$\frac{\tilde{D}_2}{\tilde{D}_1} = \frac{1/\tilde{N}_1 + 1/(1 - \rho^2)}{1/\tilde{N}_2 + 1/(1 - \rho^2)} \leq \frac{1/N_1 + 1/(1 - \rho^2)}{1/(1 - \rho^2)} = \frac{D_2}{D_1},$$

i.e., $\tilde{D}_2 \leq D_2$. Furthermore, as shown before, the left corner point of the corresponding rate region is

$$\tilde{R}_2 = I(X_2; U_2) = \mathsf{R}\left((1 + \tilde{N}_2)/\tilde{N}_2\right),$$
$$\tilde{R}_1 = g(\tilde{R}_2, \tilde{D}_1).$$

Hence, by varying $0 < \tilde{N}_2 < \infty$, we can achieve the entire region $\mathscr{R}_1(D_1)$.

Case 2: $(1 - D_2) > \rho^2(1 - D_1)$. In this case, there exists a distributed Gaussian test channel with some (N_1, N_2) such that both distortion constraints are tight. To show this, we use the following result, which is a straightforward consequence of the matrix inversion lemma in Appendix B.

Lemma 12.5. Let $U_j = X_j + V_j$, $j = 1, 2$, where $\mathbf{X} = (X_1, X_2)$ and $\mathbf{V} = (V_1, V_2)$ are independent zero-mean Gaussian random vectors with respective covariance matrices $K_{\mathbf{X}}, K_{\mathbf{V}} \succ 0$. Let $K_{\mathbf{X}|\mathbf{U}} = K_{\mathbf{X}} - K_{\mathbf{XU}}K_{\mathbf{U}}^{-1}K_{\mathbf{XU}}^T$ be the error covariance matrix of the (linear) MMSE estimate of $\mathbf{X}$ given $\mathbf{U}$. Then

$$K_{\mathbf{X}|\mathbf{U}}^{-1} = K_{\mathbf{X}}^{-1} + K_{\mathbf{X}}^{-1}K_{\mathbf{XU}}^T\left(K_{\mathbf{U}} - K_{\mathbf{XU}}K_{\mathbf{X}}^{-1}K_{\mathbf{XU}}^T\right)^{-1}K_{\mathbf{XU}}K_{\mathbf{X}}^{-1} = K_{\mathbf{X}}^{-1} + K_{\mathbf{V}}^{-1}.$$

Hence, V_1 and V_2 are independent iff $K_{\mathbf{V}}^{-1} = K_{\mathbf{X}|\mathbf{U}}^{-1} - K_{\mathbf{X}}^{-1}$ is diagonal.

Now let

$$K = \begin{bmatrix} D_1 & \theta\sqrt{D_1 D_2} \\ \theta\sqrt{D_1 D_2} & D_2 \end{bmatrix},$$

where $\theta \in [0, 1]$ is chosen so that $K^{-1} - K_{\mathbf{X}}^{-1} = \mathrm{diag}(N_1, N_2)$ for some $N_1, N_2 > 0$. By simple algebra, if $(1 - D_2) > \rho^2(1 - D_1)$, then such a choice of θ exists and is given by

$$\theta = \frac{\sqrt{(1 - \rho^2)^2 + 4\rho^2 D_1 D_2} - (1 - \rho^2)}{2\rho\sqrt{D_1 D_2}}. \tag{12.4}$$

Hence, by Lemma 12.5, we have the covariance matrix $K = K_{\mathbf{X}-\hat{\mathbf{X}}} = K_{\mathbf{X}|\mathbf{U}}$ with $U_j = X_j + V_j$, $j = 1, 2$, where $V_1 \sim \mathsf{N}(0, N_1)$ and $V_2 \sim \mathsf{N}(0, N_2)$ are independent of each other and of (X_1, X_2). In other words, there exists a distributed Gaussian test channel characterized by (N_1, N_2) with corresponding distortion pair (D_1, D_2).

With such choice of the distributed Gaussian test channel, it can be readily verified that the left corner point (R'_1, R'_2) of the rate region is

$$R'_1 + R'_2 = \mathsf{R}\left(\frac{(1-\rho^2)\phi(D_1, D_2)}{2D_1 D_2}\right),$$

$$R'_2 = \mathsf{R}\left(\frac{1+N_2}{N_2}\right),$$

$$R'_1 = g(R'_2, D_1).$$

Therefore, (R'_1, R'_2) is on the boundary of both $\mathscr{R}_{12}(D_1, D_2)$ and $\mathscr{R}_1(D_1)$. Following similar steps to case 1, we can show that any rate pair $(\tilde{R}_1, \tilde{R}_2)$ such that $\tilde{R}_1 = g(\tilde{R}_2, D_1)$ and $\tilde{R}_2 \geq R'_2$ is achievable; see Figure 12.4. Similarly, the right corner point (R''_1, R''_2) is on the boundary of $\mathscr{R}_{12}(D_1, D_2)$ and $\mathscr{R}_2(D_2)$, and any rate pair $(\tilde{R}_1, \tilde{R}_2)$ such that $\tilde{R}_2 = g(\tilde{R}_1, D_2)$ and $\tilde{R}_1 \geq R''_1$ is achievable. Using time sharing between the two corner points completes the proof for case 2.

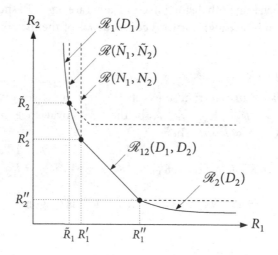

Figure 12.4. Rate region for case 2. Here $\mathscr{R}(N_1, N_2)$ denotes the rate region for (N_1, N_2) in (12.3) with left corner point (R'_1, R'_2) and right corner point (R''_1, R''_2), and $\mathscr{R}(\tilde{N}_1, \tilde{N}_2)$ denotes the rate region for $(\tilde{N}_1, \tilde{N}_2)$ with left corner point $(\tilde{R}_1, \tilde{R}_2)$.

The rest of the achievability proof follows by the discretization procedure described in the proof of the quadratic Gaussian source coding theorem in Section 3.8.

12.3.2   Proof of the Converse

We first show that the rate–distortion region is contained in both $\mathscr{R}_1(D_1)$ and $\mathscr{R}_2(D_2)$, i.e., $\mathscr{R}(D_1, D_2) \subseteq \mathscr{R}_1(D_1) \cap \mathscr{R}_2(D_2)$. For a given sequence of $(2^{nR_1}, 2^{nR_2}, n)$ codes that achieves distortions D_1 and D_2, consider

$$nR_1 \geq H(M_1)$$
$$\geq H(M_1|M_2)$$
$$= I(M_1; X_1^n|M_2)$$
$$= h(X_1^n|M_2) - h(X_1^n|M_1, M_2)$$
$$\geq h(X_1^n|M_2) - h(X_1^n|\hat{X}_1^n)$$
$$\geq h(X_1^n|M_2) - \sum_{i=1}^{n} h(X_{1i}|\hat{X}_{1i})$$
$$\overset{(a)}{\geq} \boxed{h(X_1^n|M_2)} - \frac{n}{2} \log(2\pi e D_1),$$

where (a) follows by Jensen's inequality and the distortion constraint. Next consider

$$nR_2 \geq I(M_2; X_2^n)$$
$$= h(X_2^n) - h(X_2^n|M_2)$$
$$= \frac{n}{2} \log(2\pi e) - \boxed{h(X_2^n|M_2)}.$$

Since the sources X_1 and X_2 are jointly Gaussian, we can express X_1^n as $X_1^n = \rho X_2^n + W^n$, where $\{W_i\}$ is a WGN$(1 - \rho^2)$ process independent of $\{X_{2i}\}$ and hence of M_2. Now by the conditional EPI,

$$2^{2h(X_1^n|M_2)/n} \geq 2^{2h(\rho X_2^n|M_2)/n} + 2^{2h(W^n|M_2)/n}$$
$$= \rho^2 2^{2h(X_2^n|M_2)/n} + 2\pi e(1 - \rho^2)$$
$$\geq 2\pi e(\rho^2 2^{-2R_2} + (1 - \rho^2)).$$

Therefore

$$nR_1 \geq \frac{n}{2} \log(2\pi e(1 - \rho^2 + \rho^2 2^{-2R_2})) - \frac{n}{2} \log(2\pi e D_1)$$
$$= ng(R_2, D_1)$$

and $(R_1, R_2) \in \mathcal{R}_1(D_1)$. Similarly, $\mathcal{R}(D_1, D_2) \subseteq \mathcal{R}_2(D_2)$.

Remark 12.1. The above argument shows that the rate–distortion region when only X_1 is to be reconstructed with distortion $D_1 < 1$ is $\mathcal{R}(D_1, 1) = \mathcal{R}_1(D_1)$. Similarly, when only X_2 is to be reconstructed, $\mathcal{R}(1, D_2) = \mathcal{R}_2(D_2)$.

Proof of $\mathcal{R}(D_1, D_2) \subseteq \mathcal{R}_{12}(D_1, D_2)$. We now proceed to show that the rate–distortion region is contained in $\mathcal{R}_{12}(D_1, D_2)$. By assumption and in light of Lemma 12.4, we assume without loss of generality that $(1 - D_1) \geq (1 - D_2) \geq \rho^2(1 - D_1)$ and the sum-rate bound is *active*. We establish the following two lower bounds $R'(\theta)$ and $R''(\theta)$ on the sum-rate parametrized by $\theta \in [-1, 1]$ and show that

$$\min_{\theta} \max\{R'(\theta), R''(\theta)\} = \mathsf{R}\left(\frac{(1-\rho^2)\phi(D_1, D_2)}{2D_1 D_2}\right),$$

which implies that $\mathscr{R}(D_1, D_2) \subseteq \mathscr{R}_{12}(D_1, D_2)$.

Cooperative lower bound. The sum-rate for the distributed source coding problem is lower bounded by the sum-rate for the cooperative (centralized) lossy source coding setting depicted in Figure 12.5.

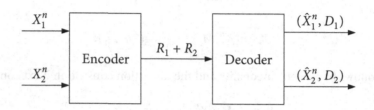

Figure 12.5. Cooperative lossy source coding.

For any sequence of codes that achieves the desired distortions, consider

$$n(R_1 + R_2) \geq H(M_1, M_2)$$
$$= I(X_1^n, X_2^n; M_1, M_2)$$
$$= h(X_1^n, X_2^n) - h(X_1^n, X_2^n | M_1, M_2)$$
$$\geq \frac{n}{2} \log\left((2\pi e)^2 |K_{\mathbf{X}}|\right) - \sum_{i=1}^{n} h(X_{1i}, X_{2i} | M_1, M_2)$$
$$\geq \frac{n}{2} \log\left((2\pi e)^2 |K_{\mathbf{X}}|\right) - \sum_{i=1}^{n} h(X_{1i} - \hat{X}_{1i}, X_{2i} - \hat{X}_{2i})$$
$$\geq \frac{n}{2} \log\left((2\pi e)^2 |K_{\mathbf{X}}|\right) - \sum_{i=1}^{n} \frac{1}{2} \log\left((2\pi e)^2 |\tilde{K}_i|\right)$$
$$\geq \frac{n}{2} \log\left((2\pi e)^2 |K_{\mathbf{X}}|\right) - \frac{n}{2} \log\left((2\pi e)^2 |\tilde{K}|\right)$$
$$= \frac{n}{2} \log \frac{|K_{\mathbf{X}}|}{|\tilde{K}|},$$

where $\tilde{K}_i = K_{\mathbf{X}(i)-\hat{\mathbf{X}}(i)}$ and $\tilde{K} = (1/n) \sum_{i=1}^{n} \tilde{K}_i$. Since $\tilde{K}(j, j) \leq D_j, j = 1, 2$,

$$\tilde{K} \preceq K(\theta) = \begin{bmatrix} D_1 & \theta\sqrt{D_1 D_2} \\ \theta\sqrt{D_1 D_2} & D_2 \end{bmatrix} \tag{12.5}$$

for some $\theta \in [-1, 1]$. Hence,

$$R_1 + R_2 \geq R'(\theta) = \mathsf{R}\left(\frac{|K_{\mathbf{X}}|}{|K(\theta)|}\right) = \mathsf{R}\left(\frac{1-\rho^2}{D_1 D_2 (1-\theta^2)}\right). \tag{12.6}$$

μ-Sum lower bound. Let $Y_i = \mu_1 X_{1i} + \mu_2 X_{2i} + Z_i = \mu^T \mathbf{X}(i) + Z_i$, where $\{Z_i\}$ is a WGN(N) process independent of $\{(X_{1i}, X_{2i})\}$. Then for every (μ, N),

$$
\begin{aligned}
n(R_1 + R_2) &\geq H(M_1, M_2) \\
&= I(\mathbf{X}^n, Y^n; M_1, M_2) \\
&\geq h(\mathbf{X}^n, Y^n) - h(Y^n | M_1, M_2) - h(\mathbf{X}^n | M_1, M_2, Y^n) \\
&\geq \sum_{i=1}^n \left(h(\mathbf{X}(i), Y_i) - h(Y_i | M_1, M_2) - h(\mathbf{X}(i) | M_1, M_2, Y^n) \right) \\
&\geq \sum_{i=1}^n \frac{1}{2} \log \frac{|K_{\mathbf{X}}| N}{|\hat{K}_i| (\mu^T \tilde{K}_i \mu + N)} \\
&\geq \frac{n}{2} \log \frac{|K_{\mathbf{X}}| N}{|\hat{K}| (\mu^T \tilde{K} \mu + N)} \\
&\geq \frac{n}{2} \log \frac{|K_{\mathbf{X}}| N}{|\hat{K}| (\mu^T K(\theta) \mu + N)},
\end{aligned}
\tag{12.7}
$$

where $\hat{K}_i = K_{\mathbf{X}(i)|M_1, M_2, Y^n}$, $\hat{K} = (1/n) \sum_{i=1} \hat{K}_i$, and $K(\theta)$ is defined in (12.5).

We now take $\mu = (1/\sqrt{D_1}, 1/\sqrt{D_2})$ and $N = (1 - \rho^2)/(\rho \sqrt{D_1 D_2})$. Then $\mu^T K(\theta) \mu = 2(1 + \theta)$ and it can be readily shown that $X_{1i} \to Y_i \to X_{2i}$ form a Markov chain. Furthermore, we can upper bound $|\hat{K}|$ as follows.

Lemma 12.6. $|\hat{K}| \leq D_1 D_2 N^2 (1 + \theta)^2 / (2(1 + \theta) + N)^2.$

This can be shown by noting that $\hat{K}$ is diagonal, which follows since $X_{1i} \to Y_i \to X_{2i}$ form a Markov chain, and by establishing the matrix inequality $\hat{K} \preceq (\tilde{K}^{-1} + (1/N) \mu \mu^T)^{-1}$ using results from estimation theory. The details are given in Appendix 12D.

Using the above lemma, the bound in (12.7), and the definitions of μ and N establishes the second lower bound on the sum-rate

$$
R_1 + R_2 \geq R''(\theta) = \mathsf{R}\left(\frac{|K_{\mathbf{X}}| (2(1 + \theta) + N)}{N D_1 D_2 (1 + \theta)^2} \right) = \mathsf{R}\left(\frac{2\rho \sqrt{D_1 D_2}(1 + \theta) + (1 - \rho^2)}{D_1 D_2 (1 + \theta)^2} \right).
\tag{12.8}
$$

Finally, combining the cooperative and μ-sum lower bounds in (12.6) and (12.8), we have

$$
R_1 + R_2 \geq \min_\theta \max\{R'(\theta), R''(\theta)\}.
$$

These two bounds are plotted in Figure 12.6. It can be easily checked that $R'(\theta) = R''(\theta)$ at a unique point

$$
\theta = \theta^* = \frac{\sqrt{(1 - \rho^2)^2 + 4\rho^2 D_1 D_2} - (1 - \rho^2)}{2\rho \sqrt{D_1 D_2}},
$$

which is exactly what we chose in the proof of achievability; see (12.4). Finally, since $R'(\theta)$

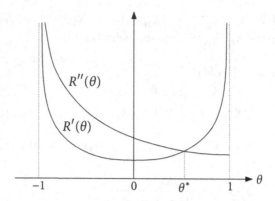

Figure 12.6. Plot of the two bounds on the sum-rate.

is increasing on $[\theta^*, 1)$ and $R''(\theta)$ is decreasing on $(-1, 1)$, we can conclude that

$$\min_{\theta} \max\{R'(\theta), R''(\theta)\} = R'(\theta^*) = R''(\theta^*).$$

Hence

$$R_1 + R_2 \geq R'(\theta^*) = R''(\theta^*) = \mathsf{R}\left(\frac{(1-\rho^2)\phi(D_1, D_2)}{2D_1 D_2}\right).$$

This completes the proof of Theorem 12.3.

Remark 12.2. The choice of the auxiliary random variable Y such that $X_1 \rightarrow Y \rightarrow X_2$ form a Markov chain captures *common information* between X_1 and X_2. A similar auxiliary random variable will be used in the converse proof for quadratic Gaussian multiple descriptions in Section 13.4; see also Section 14.2 for a detailed discussion of common information.

12.4   QUADRATIC GAUSSIAN CEO PROBLEM

The CEO (chief executive or estimation officer) problem is a distributed source coding problem in which the objective is to reconstruct a source from coded noisy observations rather than reconstructing the noisy observations themselves. This problem is motivated by distributed estimation and detection settings, such as a tracker following the location of a moving object over time using noisy measurements from multiple sensors, or a CEO making a decision based on several briefings from subordinates; see Chapter 21 for more discussion on communication for computing. We consider the special case depicted in Figure 12.7. A WGN(P) source X is observed through a Gaussian broadcast channel $Y_j = X + Z_j$, $j = 1, 2$, where $Z_1 \sim \mathrm{N}(0, N_1)$ and $Z_2 \sim \mathrm{N}(0, N_2)$ are noise components, independent of each other and of X. The observation sequences Y_j^n, $j = 1, 2$, are separately encoded with the goal of finding an estimate $\hat{X}^n$ of X^n with mean squared error distortion D.

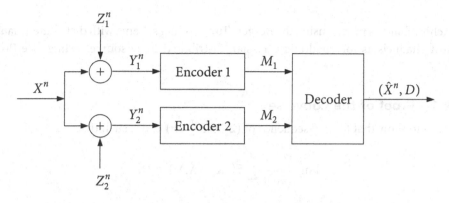

Figure 12.7. Quadratic Gaussian CEO problem.

A $(2^{nR_1}, 2^{nR_2}, n)$ code for the CEO problem consists of

* two encoders, where encoder 1 assigns an index $m_1(y_1^n) \in [1 : 2^{nR_1})$ to each sequence y_1^n and encoder 2 assigns an index $m_2(y_2^n) \in [1 : 2^{nR_2})$ to each sequence y_2^n, and

* a decoder that assigns an estimate $\hat{x}^n$ to each index pair $(m_1, m_2) \in [1 : 2^{nR_1}) \times [1 : 2^{nR_2})$.

A rate–distortion triple (R_1, R_2, D) is said to be *achievable* if there exists a sequence of $(2^{nR_1}, 2^{nR_2}, n)$ codes with

$$\limsup_{n \to \infty} \mathsf{E}\Big(\frac{1}{n}\sum_{i=1}^{n}(X_i - \hat{X}_i)^2\Big) \le D.$$

The rate–distortion region $\mathscr{R}_{\mathrm{CEO}}(D)$ for the quadratic Gaussian CEO problem is the closure of the set of all rate pairs (R_1, R_2) such that (R_1, R_2, D) is achievable.

This problem is closely related to the quadratic Gaussian distributed source coding problem and the rate–distortion function is given by the following.

Theorem 12.4. The rate–distortion region $\mathscr{R}_{\mathrm{CEO}}(D)$ for the quadratic Gaussian CEO problem is the set of rate pairs (R_1, R_2) such that

$$R_1 \ge r_1 + \mathsf{R}\left(\frac{1}{D}\left(\frac{1}{P} + \frac{1 - 2^{-2r_2}}{N_2}\right)^{-1}\right),$$

$$R_2 \ge r_2 + \mathsf{R}\left(\frac{1}{D}\left(\frac{1}{P} + \frac{1 - 2^{-2r_1}}{N_1}\right)^{-1}\right),$$

$$R_1 + R_2 \ge r_1 + r_2 + \mathsf{R}\left(\frac{P}{D}\right)$$

for some $r_1, r_2 \ge 0$ that satisfy the condition

$$D \ge \left(\frac{1}{P} + \frac{1 - 2^{-2r_1}}{N_1} + \frac{1 - 2^{-2r_2}}{N_2}\right)^{-1}.$$

Achievability is proved using the Berger–Tung coding scheme with distributed Gaussian test channels as for quadratic Gaussian distributed lossy source coding; see Problem 12.6.

12.4.1   Proof of the Converse

We need to show that for any sequence of $(2^{nR_1}, 2^{nR_2}, n)$ codes such that

$$\limsup_{n \to \infty} \frac{1}{n} \sum_{i=1}^{n} \mathsf{E}((X_i - \hat{X}_i)^2) \le D,$$

the rate pair $(R_1, R_2) \in \mathscr{R}_{\mathrm{CEO}}(D)$. Let $\mathcal{J}$ be a subset of $\{1, 2\}$ and $\mathcal{J}^c$ be its complement. Consider

$$
\begin{aligned}
\sum_{j \in \mathcal{J}} nR_j &\ge H(M(\mathcal{J})) \\
&\ge H(M(\mathcal{J})|M(\mathcal{J}^c)) \\
&= I(Y^n(\mathcal{J}); M(\mathcal{J})|M(\mathcal{J}^c)) \\
&= I(Y^n(\mathcal{J}), X^n; M(\mathcal{J})|M(\mathcal{J}^c)) \\
&= I(X^n; M(\mathcal{J})|M(\mathcal{J}^c)) + I(Y^n(\mathcal{J}); M(\mathcal{J})|X^n) \\
&= I(X^n; M(\mathcal{J}), M(\mathcal{J}^c)) - I(X^n; M(\mathcal{J}^c)) + \sum_{j \in \mathcal{J}} I(Y_j^n; M_j|X^n) \\
&\ge [I(X^n; \hat{X}^n) - I(X^n; M(\mathcal{J}^c))]^+ + \sum_{j \in \mathcal{J}} I(Y_j^n; M_j|X^n) \\
&\ge \left[\frac{n}{2} \log\left(\frac{P}{D}\right) - I(X^n; M(\mathcal{J}^c)) \right]^+ + \sum_{j \in \mathcal{J}} nr_j,
\end{aligned}
\tag{12.9}
$$

where we define

$$r_j = \frac{1}{n} I(Y_j^n; M_j|X^n), \quad j = 1, 2.$$

In order to upper bound $I(X^n; M(\mathcal{J}^c)) = h(X^n) - h(X^n|M(\mathcal{J}^c)) = (n/2)\log(2\pi e P) - h(X^n|M(\mathcal{J}^c))$, or equivalently, to lower bound $h(X^n|M(\mathcal{J}^c))$, let $\tilde{X}_i$ be the MMSE estimate of X_i given $Y_i(\mathcal{J}^c)$. Then, by the orthogonality principle in Appendix B,

$$\tilde{X}_i = \sum_{j \in \mathcal{J}^c} \frac{\tilde{N}}{N_j} Y_{ji} = \sum_{j \in \mathcal{J}^c} \frac{\tilde{N}}{N_j}(X_i + Z_{ji}),\tag{12.10}$$

$$X_i = \tilde{X}_i + \tilde{Z}_i,\tag{12.11}$$

where $\{\tilde{Z}_i\}$ is a WGN process with average power $\tilde{N} = 1/(1/P + \sum_{j \in \mathcal{J}^c} 1/N_j)$, independent of $\{Y_i(\mathcal{J}^c)\}$. By the conditional EPI and (12.11), we have

$$2^{2h(X^n|M(\mathcal{J}^c))/n} \ge 2^{2h(\tilde{X}^n|M(\mathcal{J}^c))/n} + 2^{2h(\tilde{Z}^n|M(\mathcal{J}^c))/n}$$

$$= 2^{2[h(\tilde{X}^n|X^n, M(\mathcal{J}^c)) + I(\tilde{X}^n; X^n|M(\mathcal{J}^c))]/n} + 2\pi e \tilde{N}.$$

We now lower bound each term in the exponent. First, by (12.10) and the definition of r_j,

$$2^{2h(\tilde{X}^n|X^n,M(\mathcal{J}^c))/n} \geq \sum_{j\in\mathcal{J}^c} \frac{\tilde{N}^2}{N_j^2} 2^{2h(Y_j^n|X^n,M(\mathcal{J}^c))/n}$$

$$= \sum_{j\in\mathcal{J}^c} \frac{\tilde{N}^2}{N_j^2} 2^{2h(Y_j^n|X^n,M_j)/n}$$

$$= \sum_{j\in\mathcal{J}^c} \frac{\tilde{N}^2}{N_j^2} 2^{2(h(Y_j^n|X^n)-I(Y_j^n;M_j|X^n))/n}$$

$$= \sum_{j\in\mathcal{J}^c} \frac{\tilde{N}^2}{N_j}(2\pi e)2^{-2r_j}.$$

For the second term, since $X^n = \tilde{X}^n + \tilde{Z}^n$ and $\tilde{Z}^n$ is conditionally independent of $Y^n(\mathcal{J}^c)$ and thus of $M(\mathcal{J}^c)$,

$$2^{2I(X^n;\tilde{X}^n|M(\mathcal{J}^c))/n} = 2^{2[h(X^n|M(\mathcal{J}^c))-h(X^n|\tilde{X}^n,M(\mathcal{J}^c))]/n}$$

$$= \frac{1}{2\pi e\tilde{N}} 2^{2h(X^n|M(\mathcal{J}^c))/n}.$$

Therefore

$$2^{2h(X^n|M(\mathcal{J}^c))/n} \geq \left(\sum_{j\in\mathcal{J}^c} \frac{\tilde{N}^2}{N_j}(2\pi e)2^{-2r_j}\right)\left(\frac{1}{2\pi e\tilde{N}} 2^{2h(X^n|M(\mathcal{J}^c))/n}\right) + 2\pi e\tilde{N}$$

$$= \sum_{j\in\mathcal{J}^c} \frac{\tilde{N}}{N_j} 2^{-2r_j} 2^{2h(X^n|M(\mathcal{J}^c))/n} + 2\pi e\tilde{N},$$

or equivalently,

$$2^{-2h(X^n|M(\mathcal{J}^c))/n} \leq \frac{1 - \sum_{j\in\mathcal{J}^c} \frac{\tilde{N}}{N_j}2^{-2r_j}}{2\pi e\tilde{N}}$$

and

$$2^{2I(X^n;M(\mathcal{J}^c))/n} \leq \frac{P}{\tilde{N}} - \sum_{j\in\mathcal{J}^c} \frac{P}{N_j} 2^{-2r_j}$$

$$= 1 + \sum_{j\in\mathcal{J}^c} \frac{P}{N_j}(1 - 2^{-2r_j}).$$

Thus, continuing the lower bound in (12.9), we have

$$\sum_{j\in\mathcal{J}} R_j \geq \left[\frac{1}{2}\log\left(\frac{P}{D}\right) - \frac{1}{2}\log\left(1 + \sum_{j\in\mathcal{J}^c} \frac{P}{N_j}(1 - 2^{-2r_j})\right)\right]^+ + \sum_{j\in\mathcal{J}} r_j$$

$$= \left[\frac{1}{2}\log\left(\frac{1}{D}\right) - \frac{1}{2}\log\left(\frac{1}{P} + \sum_{j\in\mathcal{J}^c} \frac{1}{N_j}(1 - 2^{-2r_j})\right)\right]^+ + \sum_{j\in\mathcal{J}} r_j.$$

Substituting $\mathcal{J} = \{1, 2\}, \{1\}, \{2\}$, and $\emptyset$ establishes the four inequalities in Theorem 12.4.

12.5* SUBOPTIMALITY OF BERGER–TUNG CODING

In previous sections we showed that the Berger–Tung coding scheme is optimal in several special cases. The scheme, however, does not perform well when the sources have a *common part*; see Section 14.1 for a formal definition. In this case, the encoders can send the description of the common part cooperatively to reduce the rates.

Proposition 12.1. Let (X_1, X_2) be a 2-DMS such that X_2 is a function of X_1, and $d_2 \equiv 0$, that is, the decoder is interested in reconstructing only X_1. Then the rate–distortion region $\mathscr{R}(D_1)$ for distributed lossy source coding is the set of rate pairs (R_1, R_2) such that

$$R_1 \geq I(X_1; \hat{X}_1 | V),$$
$$R_2 \geq I(X_2; V)$$

for some conditional pmf $p(v|x_2)p(\hat{x}_1|x_1, v)$ with $|\mathcal{V}| \leq |\mathcal{X}_2| + 2$ that satisfy the constraint $E(d_1(X_1, \hat{X}_1)) \leq D_1$.

The converse follows by the Berger–Tung outer bound. To prove achievability, encoder 2 describes X_2 with V. Now that encoder 1 has access to X_2 and thus to V, it can describe X_1 with $\hat{X}_1$ conditioned on V. It can be shown that this scheme achieves the rate–distortion region in the proposition.

To illustrate this result, consider the following.

Example 12.1. Let $(\tilde{X}, \tilde{Y})$ be a DSBS(p), $p \in (0, 1/2)$, $X_1 = (\tilde{X}, \tilde{Y})$, $X_2 = \tilde{Y}$, $d_1(x_1, \hat{x}_1) = d(\tilde{x}, \hat{x}_1)$ be a Hamming distortion measure, and $d_2(x_2, \hat{x}_2) \equiv 0$. We consider the minimum achievable sum-rate for distortion D_1. By taking $p(v|\tilde{y})$ to be a BSC(α), the region in Proposition 12.1 can be further simplified to the set of rate pairs (R_1, R_2) such that

$$R_1 > H(\alpha * p) - H(D_1),$$
$$R_2 > 1 - H(\alpha) \tag{12.12}$$

for some $\alpha \in [0, 1/2]$ that satisfies the constraint $H(\alpha * p) - H(D_1) \geq 0$.

It can be shown that the Berger–Tung inner bound is contained in the above region. In fact, noting that $R_{\text{SI-ED}}(D_1) < R_{\text{SI-D}}(D_1)$ for a DSBS($\alpha * p$) and using Mrs. Gerber's lemma, it can be shown that a boundary point of the region in (12.12) lies strictly outside the Berger–Tung inner bound. Thus, the Berger–Tung coding scheme is suboptimal in general.

Remark 12.3 (Berger–Tung coding with common part). Using the common part of a general 2-DMS, the cooperative coding scheme in Proposition 12.1 and the Berger–Tung coding scheme can be combined to yield a tighter inner bound on the rate–distortion region for distributed lossy source coding.

SUMMARY

- Lossy distributed source coding setup

- Rate–distortion region

- Berger–Tung inner bound:

  - Distributed compress–bin

  - Includes Wyner–Ziv and Slepian–Wolf as special cases

  - Not tight in general

- Markov lemma

- Mutual packing lemma

- Quadratic Gaussian distributed source coding:

  - Berger–Tung coding is optimal

  - Identification of a common-information random variable in the proof of the μ-sum lower bound

  - Use of the EPI in the proof of the converse

  - Use of MMSE estimation results in the proof of the converse

- Berger–Tung coding is optimal for the quadratic Gaussian CEO problem

BIBLIOGRAPHIC NOTES

Theorems 12.1 and 12.2 are due to Berger (1978) and Tung (1978). Tung (1978) established the Markov lemma and also showed that the Berger–Tung outer bound is not tight in general. A strictly improved outer bound was recently established by Wagner and Anantharam (2008). Berger and Yeung (1989) established the rate–distortion region in (12.1).

The quadratic Gaussian distributed source coding problem was studied by Oohama (1997), who showed that $\mathscr{R}(D_1, D_2) \subseteq \mathscr{R}_1(D_1) \cap \mathscr{R}_2(D_2)$. Theorem 12.3 was established by Wagner, Tavildar, and Viswanath (2008). Our proof of the converse combines key ideas from the proofs in Wagner, Tavildar, and Viswanath (2008) and Wang, Chen, and Wu (2010). The CEO problem was first introduced by Berger, Zhang, and Viswanathan (1996). The quadratic Gaussian case was studied by Viswanathan and Berger (1997) and Oohama (1998). Theorem 12.4 is due to Oohama (2005) and Prabhakaran, Tse, and Ramchandran (2004). The rate–distortion region for quadratic Gaussian distributed source coding with more than two sources is known for sources satisfying a certain tree-structured Markov condition (Tavildar, Viswanath, and Wagner 2006). In addition, the minimum sum-rate for quadratic Gaussian distributed lossy source coding with more than two sources is known in several special cases (Yang, Zhang, and Xiong 2010). The rate–distortion region for the Gaussian CEO problem can be extended to more than two encoders (Oohama 2005, Prabhakaran, Tse, and Ramchandran 2004).

Proposition 12.1 was established by Kaspi and Berger (1982). Example 12.1 is due to Wagner, Kelly, and Altuğ (2011), who also showed that the Berger–Tung coding scheme is suboptimal even for sources without a common part.

PROBLEMS

12.1. Prove the mutual packing lemma.

12.2. Show that the Berger–Tung inner bound can be alternatively expressed as the set of rate pairs (R_1, R_2) such that

$$R_1 > I(X_1, X_2; U_1|U_2, Q),$$
$$R_2 > I(X_1, X_2; U_2|U_1, Q),$$
$$R_1 + R_2 > I(X_1, X_2; U_1, U_2|Q)$$

for some conditional pmf $p(q)p(u_1|x_1, q)p(u_2|x_2, q)$ and functions $\hat{x}_1(u_1, u_2, q)$, $\hat{x}_2(u_1, u_2, q)$ that satisfy the constraints $D_j \geq E(d_j(X_j, \hat{X}_j))$, $j = 1, 2$.

12.3. Prove the Berger–Tung outer bound.

12.4. *Lossy source coding with coded side information only at the decoder.* Let (X, Y) be a 2-DMS and $d(x, \hat{x})$ be a distortion measure. Consider the lossy source coding setup depicted in Figure 12.8, where coded side information of Y is available at the decoder.

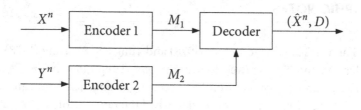

Figure 12.8. Lossy source coding with coded side information.

(a) Let $\mathscr{R}(D)$ be the set of rate pairs (R_1, R_2) such that

$$R_1 > I(X; U|V, Q),$$
$$R_2 > I(Y; V|Q)$$

for some conditional pmf $p(q)p(u|x, q)p(v|y, q)$ and function $\hat{x}(u, v, q)$ such that $E(d(X, \hat{X})) \leq D$. Show that any rate pair (R_1, R_2) is achievable.

(b) Consider the Berger–Tung inner bound with $(X_1, X_2) = (X, Y)$, $d_1 = d$, and $d_2 \equiv 0$. Let $\mathscr{R}'(D)$ denote the resulting region. Show that $\mathscr{R}(D) = \mathscr{R}'(D)$.

12.5. *Distributed lossy source coding with one distortion measure.* Consider the distributed lossy source coding setup where X_2 is to be recovered losslessly and X_1 is to be reconstructed with distortion D_1. Find the rate–distortion region.
Remark: This result is due to Berger and Yeung (1989).

12.6. *Achievability for the quadratic Gaussian CEO problem.* Show that the Berger–Tung inner bound with the single distortion constraint $E((X - \hat{X})^2) \le D$ reduces to the rate–distortion region $\mathcal{R}(D)$ in Theorem 12.4.

(a) Let $\mathcal{R}'(D)$ be the set of rate pairs (R_1, R_2) that satisfy the first three inequalities in Theorem 12.4 and the last one with equality. Show that $\mathcal{R}'(D) = \mathcal{R}(D)$. (Hint: Consider three cases $1/P + (1 - 2^{-2r_1})/N_1 > 1/D, 1/P + (1 - 2^{-2r_2})/N_2 > 1/D$, and otherwise.)

(b) Let $U_j = Y_j + V_j$, where $V_j \sim N(0, \tilde{N}_j)$ for $j = 1, 2$, and $\hat{X} = E(X|U_1, U_2)$. Show that
$$E[(X - \hat{X})^2] = (1/P + (1 - 2^{-2r_1})/N_1 + (1 - 2^{-2r_2})/N_2)^{-1},$$
where $r_j = (1/2)\log(1 + N_j/\tilde{N}_j)$, $j = 1, 2$.

(c) Complete the proof by evaluating the mutual information terms in the Berger–Tung inner bound.

APPENDIX 12A   PROOF OF THE MARKOV LEMMA

By the union of events bound,
$$P\{(x^n, y^n, Z^n) \notin \mathcal{T}_\epsilon^{(n)}\} \le \sum_{x,y,z} P\{|\pi(x, y, z|x^n, y^n, Z^n) - p(x, y, z)| \ge \epsilon p(x, y, z)\}.$$

Hence it suffices to show that
$$P(\mathcal{E}(x, y, z)) = P\{|\pi(x, y, z|x^n, y^n, Z^n) - p(x, y, z)| \ge \epsilon p(x, y, z)\}$$

tends to zero as $n \to \infty$ for each $(x, y, z) \in \mathcal{X} \times \mathcal{Y} \times \mathcal{Z}$.
    Given (x, y, z), define the sets
$$\mathcal{A}_1 = \{z^n : (y^n, z^n) \in \mathcal{T}_{\epsilon'}^{(n)}(Y, Z)\},$$
$$\mathcal{A}_2 = \{z^n : |\pi(x, y, z|x^n, y^n, z^n) - p(x, y, z)| \ge \epsilon p(x, y, z)\}.$$

Then $P(\mathcal{E}(x, y, z)) = P\{Z^n \in \mathcal{A}_2\} \le P\{Z^n \in \mathcal{A}_1^c\} + P\{Z^n \in \mathcal{A}_1 \cap \mathcal{A}_2\}$.
    We bound each term. By the first condition on $p(z^n|y^n)$, the first term tends to zero as $n \to \infty$. For the second term, let
$$\mathcal{B}_1(n_{yz}) = \mathcal{A}_1 \cap \{z^n : \pi(y, z|y^n, z^n) = n_{yz}/n\} \quad \text{for } n_{yz} \in [0:n].$$

Then, for n sufficiently large,

$$P\{Z^n \in \mathcal{A}_1 \cap \mathcal{A}_2\} \leq \sum_{n_{yz}} |\mathcal{B}_1(n_{yz}) \cap \mathcal{A}_2| \, 2^{-n(H(Z|Y)-\delta(\epsilon'))},$$

where the summation is over all n_{yz} such that $|n_{yz} - np(y,z)| \leq \epsilon' np(y,z)$ and the inequality follows by the second condition on $p(z^n|y^n)$.

Let $n_{xy} = n\pi(x,y|x^n, y^n)$ and $n_y = n\pi(y|y^n)$. Let K be a hypergeometric random variable that represents the number of red balls in a sequence of n_{yz} draws without replacement from a bag of n_{xy} red balls and $n_y - n_{xy}$ blue balls. Then

$$P\{K = k\} = \frac{\binom{n_{xy}}{k}\binom{n_y - n_{xy}}{n_{yz} - k}}{\binom{n_y}{n_{yz}}} = \frac{|\mathcal{B}_1(n_{yz}) \cap \mathcal{B}_2(k)|}{|\mathcal{B}_1(n_{yz})|},$$

where

$$\mathcal{B}_2(k) = \{z^n : \pi(x,y,z|x^n, y^n, z^n) = k/n\}$$

for $k \in [0 : \min\{n_{xy}, n_{yz}\}]$. Thus

$$|\mathcal{B}_1(n_{yz}) \cap \mathcal{A}_2| = |\mathcal{B}_1(n_{yz})| \sum_{k:|k-np(x,y,z)|\geq\epsilon np(x,y,z)} \frac{|\mathcal{B}_1(n_{yz}) \cap \mathcal{B}_2(k)|}{|\mathcal{B}_1(n_{yz})|}$$

$$\leq |\mathcal{A}_1| \sum_{k:|k-np(x,y,z)|\geq\epsilon np(x,y,z)} P\{K = k\}$$

$$\leq 2^{n(H(Z|Y)+\delta(\epsilon'))} P\{|K - np(x,y,z)| \geq \epsilon np(x,y,z)\}.$$

Note that from the conditions on n_{xy}, n_{yz}, and n_y, $E(K) = n_{xy}n_{yz}/n_y$ satisfies the bounds $n(1 - \delta(\epsilon'))p(x,y,z) \leq E(K) \leq n(1 + \delta(\epsilon'))p(x,y,z)$. Now let $K' \sim \mathrm{Binom}(n_{xy}, n_{yz}/n_y)$ be a binomial random variable with the same mean $E(K') = E(K) = n_{xy}n_{yz}/n_y$. Then it can be shown (Uhlmann 1966, Orlitsky and El Gamal 1990) that $n(1 - \epsilon)p(x,y,z) \leq E(K) \leq n(1 + \epsilon)p(x,y,z)$ (which holds if ϵ' is sufficiently small) implies that

$$P\{|K - np(x,y,z)| \geq \epsilon np(x,y,z)\} \leq P\{|K' - np(x,y,z)| \geq \epsilon np(x,y,z)\}$$

$$\leq 2e^{-np(x,y,z)(\epsilon-\delta(\epsilon'))^2/(3(1+\delta(\epsilon')))},$$

where the last inequality follows by the Chernoff bound in Appendix B. Combining the above inequalities, we have for n sufficiently large

$$P\{Z^n \in \mathcal{A}_1 \cap \mathcal{A}_2\} \leq \sum_{n_{yz}} 2^{2n\delta(\epsilon')} 2e^{-np(x,y,z)(\epsilon-\delta(\epsilon'))^2/(3(1+\delta(\epsilon')))}$$

$$\leq (n+1)2^{2n\delta(\epsilon')} 2e^{-np(x,y,z)(\epsilon-\delta(\epsilon'))^2/(3(1+\delta(\epsilon')))},$$

which tends to zero as $n \to \infty$ if ϵ' is sufficiently small.

APPENDIX 12B   PROOF OF LEMMA 12.3

For every $u_2^n \in \mathcal{T}_{\epsilon'}^{(n)}(U_2|x_2^n)$,

$$
\begin{aligned}
\mathsf{P}\{U_2^n(L_2) = u_2^n \mid X_2^n = x_2^n\} &= \mathsf{P}\{U_2^n(L_2) = u_2^n, U_2^n(L_2) \in \mathcal{T}_{\epsilon'}^{(n)}(U_2|x_2^n) \mid X_2^n = x_2^n\} \\
&= \mathsf{P}\{U_2^n(L_2) \in \mathcal{T}_{\epsilon'}^{(n)}(U_2|x_2^n) \mid X_2^n = x_2^n\} \\
&\quad \cdot \mathsf{P}\{U_2^n(L_2) = u_2^n \mid X_2^n = x_2^n, U_2^n(L_2) \in \mathcal{T}_{\epsilon'}^{(n)}(U_2|x_2^n)\} \\
&\le \mathsf{P}\{U_2^n(L_2) = u_2^n \mid X_2^n = x_2^n, U_2^n(L_2) \in \mathcal{T}_{\epsilon'}^{(n)}(U_2|x_2^n)\} \\
&= \sum_{l_2} \mathsf{P}\{U_2^n(L_2) = u_2^n, L_2 = l_2 \mid X_2^n = x_2^n, U_2^n(L_2) \in \mathcal{T}_{\epsilon'}^{(n)}(U_2|x_2^n)\} \\
&= \sum_{l_2} \mathsf{P}\{L_2 = l_2 \mid X_2^n = x_2^n, U_2^n(L_2) \in \mathcal{T}_{\epsilon'}^{(n)}(U_2|x_2^n)\} \\
&\quad \cdot \mathsf{P}\{U_2^n(l_2) = u_2^n \mid X_2^n = x_2^n, U_2^n(l_2) \in \mathcal{T}_{\epsilon'}^{(n)}(U_2|x_2^n), L_2 = l_2\} \\
&\overset{(a)}{=} \sum_{l_2} \mathsf{P}\{L_2 = l_2 \mid X_2^n = x_2^n, U_2^n(L_2) \in \mathcal{T}_{\epsilon'}^{(n)}(U_2|x_2^n)\} \\
&\quad \cdot \mathsf{P}\{U_2^n(l_2) = u_2^n \mid U_2^n(l_2) \in \mathcal{T}_{\epsilon'}^{(n)}(U_2|x_2^n)\} \\
&\overset{(b)}{\le} \sum_{l_2} \mathsf{P}\{L_2 = l_2 \mid X_2^n = x_2^n, U_2^n(L_2) \in \mathcal{T}_{\epsilon'}^{(n)}(U_2|x_2^n)\} \\
&\quad \cdot 2^{-n(H(U_2|X_2)-\delta(\epsilon'))} \\
&= 2^{-n(H(U_2|X_2)-\delta(\epsilon'))},
\end{aligned}
$$

where (a) follows since $U_2^n(l_2)$ is independent of X_2^n and $U_2^n(l_2')$ for $l_2' \ne l_2$ and is conditionally independent of L_2 given X_2^n and the indicator variables of the events $\{U_2^n(l_2) \in \mathcal{T}_{\epsilon'}^{(n)}(U_2|x_2^n)\}$, $l_2 \in [1:2^{n\tilde{R}_2}]$, which implies that the event $\{U_2^n(l_2) = u_2^n\}$ is conditionally independent of $\{X_2^n = x_2^n, L_2 = l_2\}$ given $\{U_2^n(l_2) \in \mathcal{T}_{\epsilon'}^{(n)}(U_2|x_2^n)\}$. Step (b) follows from the properties of typical sequences. Similarly, for every $u_2^n \in \mathcal{T}_{\epsilon'}^{(n)}(U_2|x_2^n)$ and n sufficiently large,

$$
\mathsf{P}\{U_2^n(L_2) = u_2^n \mid X_2^n = x_2^n\} \ge (1 - \epsilon')2^{-n(H(U_2|X_2)+\delta(\epsilon'))}.
$$

This completes the proof of Lemma 12.3.

APPENDIX 12C   PROOF OF LEMMA 12.4

Since $D_2 \ge 1 - \rho^2 + \rho^2 D_1$,

$$
\mathcal{R}_{12}(D_1, D_2) \supseteq \mathcal{R}_{12}(D_1, 1 - \rho^2 + \rho^2 D_1) = \{(R_1, R_2): R_1 + R_2 \ge R(1/D_1)\}.
$$

If $(R_1, R_2) \in \mathcal{R}_1(D_1)$, then

$$
R_1 + R_2 \ge g(R_2, D_1) + R_2 \ge g(0, D_1) = R(1/D_1),
$$

since $g(R_2, D_1) + R_2$ is an increasing function of R_2. Thus, $\mathcal{R}_1(D_1) \subseteq \mathcal{R}_{12}(D_1, D_2)$.

Also note that the rate regions $\mathcal{R}_1(D_1)$ and $\mathcal{R}_2(D_2)$ can be expressed as

$$\mathcal{R}_1(D_1) = \left\{ (R_1, R_2) \colon R_1 \geq \mathsf{R}\left(\frac{1 - \rho^2 + \rho^2 2^{-2R_2}}{D_1} \right) \right\},$$

$$\mathcal{R}_2(D_2) = \left\{ (R_1, R_2) \colon R_1 \geq \mathsf{R}\left(\frac{\rho^2}{D_2 2^{2R_2} - 1 + \rho^2} \right) \right\}.$$

But $D_2 \geq 1 - \rho^2 + \rho^2 D_1$ implies that

$$\frac{\rho^2}{D_2 2^{2R_2} - 1 + \rho^2} \leq \frac{1 - \rho^2 + \rho^2 2^{-2R_2}}{D_1}.$$

Thus, $\mathcal{R}_1(D_1) \subseteq \mathcal{R}_2(D_2)$.

APPENDIX 12D PROOF OF LEMMA 12.6

The proof has three steps. First, we show that $\hat{K} \geq 0$ is diagonal. Second, we prove the matrix inequality $\hat{K} \leq (\tilde{K}^{-1} + (1/N)\boldsymbol{\mu}\boldsymbol{\mu}^T)^{-1}$. Since $\tilde{K} \leq K(\theta)$, it can be further shown by the matrix inversion lemma in Appendix B that

$$\hat{K} \leq (K^{-1}(\theta) + (1/N)\boldsymbol{\mu}\boldsymbol{\mu}^T)^{-1} = \begin{bmatrix} (1 - \alpha)D_1 & (\theta - \alpha)\sqrt{D_1 D_2} \\ (\theta - \alpha)\sqrt{D_1 D_2} & (1 - \alpha)D_2 \end{bmatrix},$$

where $\alpha = (1 + \theta)^2/(2(1 + \theta) + N)$. Finally, combining the above matrix inequality with the fact that $\hat{K}$ is diagonal, we show that

$$|\hat{K}| \leq D_1 D_2 (1 + \theta - 2\alpha)^2 = \frac{D_1 D_2 N^2 (1 + \theta)^2}{(2(1 + \theta) + N)^2}.$$

Step 1. Since $M_1 \to X_1^n \to Y^n \to X_2^n \to M_2$ form a Markov chain,

$$\mathsf{E}\left[(X_{1i} - \mathsf{E}(X_{1i}|Y^n, M_1, M_2))(X_{2i'} - \mathsf{E}(X_{2i'}|Y^n, M_1, M_2)) \right]$$
$$= \mathsf{E}\left[(X_{1i} - \mathsf{E}(X_{1i}|Y^n, X_{2i'}, M_1, M_2))(X_{2i'} - \mathsf{E}(X_{2i'}|Y^n, M_1, M_2)) \right] = 0$$

for all $i, i' \in [1 : n]$. Thus

$$\hat{K} = \frac{1}{n} \sum_{i=1}^n K_{X(i)|Y^n, M_1, M_2} = \mathrm{diag}(\beta_1, \beta_2)$$

for some $\beta_1, \beta_2 > 0$.

Step 2. Let $\tilde{Y}(i) = (Y_i, \hat{X}(i))$, where $\hat{X}(i) = \mathsf{E}(X(i)|M_1, M_2)$ for $i \in [1 : n]$, and

$$A = \left(\frac{1}{n} \sum_{i=1}^n K_{X(i), \tilde{Y}(i)} \right) \left(\frac{1}{n} \sum_{i=1}^n K_{\tilde{Y}(i)} \right)^{-1}.$$

Then

$$\frac{1}{n}\sum_{i=1}^{n}K_{\mathbf{X}(i)|\mathbf{Y}^n,M_1,M_2} \stackrel{(a)}{\leq} \frac{1}{n}\sum_{i=1}^{n}K_{\mathbf{X}(i)-A\hat{\mathbf{Y}}(i)}$$

$$\stackrel{(b)}{=} \left(\frac{1}{n}\sum_{i=1}^{n}K_{\mathbf{X}(i)}\right) - \left(\frac{1}{n}\sum_{i=1}^{n}K_{\mathbf{X}(i)\hat{\mathbf{Y}}(i)}\right)\left(\frac{1}{n}\sum_{i=1}^{n}K_{\hat{\mathbf{Y}}(i)}\right)^{-1}\left(\frac{1}{n}\sum_{i=1}^{n}K_{\hat{\mathbf{Y}}(i)\mathbf{X}(i)}\right)$$

$$\stackrel{(c)}{=} K_{\mathbf{X}} - \begin{bmatrix}K_{\mathbf{X}}\mu & K_{\mathbf{X}} - \tilde{K}\end{bmatrix}\begin{bmatrix}\mu^T K_{\mathbf{X}}\mu + N & \mu^T(K_{\mathbf{X}} - \tilde{K}) \\ (K_{\mathbf{X}} - \tilde{K})\mu & K_{\mathbf{X}} - \tilde{K}\end{bmatrix}^{-1}\begin{bmatrix}\mu^T K_{\mathbf{X}} \\ K_{\mathbf{X}} - \tilde{K}\end{bmatrix}$$

$$\stackrel{(d)}{=} (\tilde{K}^{-1} + (1/N)\mu\mu^T)^{-1}$$

$$\stackrel{(e)}{\leq} (K^{-1}(\theta) + (1/N)\mu\mu^T)^{-1},$$

where (a) follows by the optimality of the MMSE estimate $E(\mathbf{X}(i)|\mathbf{Y}^n, M_1, M_2)$ (compared to the estimate $A\hat{\mathbf{Y}}(i)$), (b) follows by the definition of the matrix A, (c) follows since $(1/n)\sum_{i=1}^{n}K_{\hat{\mathbf{X}}(i)} = K_{\mathbf{X}} - \tilde{K}$, (d) follows by the matrix inversion lemma in Appendix B, and (e) follows since $\tilde{K} \preceq K(\theta)$. Substituting for μ and N, we have shown

$$\hat{K} = \text{diag}(\beta_1, \beta_2) \preceq \begin{bmatrix}(1 - \alpha)D_1 & (\theta - \alpha)\sqrt{D_1 D_2} \\ (\theta - \alpha)\sqrt{D_1 D_2} & (1 - \alpha)D_2\end{bmatrix},$$

where $\alpha = (1 + \theta)^2/(2(1 + \theta) + N)$.

Step 3. We first note by simple algebra that if $b_1, b_2 \geq 0$ and

$$\begin{bmatrix}b_1 & 0 \\ 0 & b_2\end{bmatrix} \preceq \begin{bmatrix}a & c \\ c & a\end{bmatrix},$$

then $b_1 b_2 \leq (a - c)^2$. Now let $\Lambda = \text{diag}(1/\sqrt{D_1}, -1/\sqrt{D_2})$. Then

$$\Lambda \, \text{diag}(\beta_1, \beta_2)\Lambda \preceq \Lambda \begin{bmatrix}(1 - \alpha)D_1 & (\theta - \alpha)\sqrt{D_1 D_2} \\ (\theta - \alpha)\sqrt{D_1 D_2} & (1 - \alpha)D_2\end{bmatrix}\Lambda = \begin{bmatrix}1 - \alpha & \alpha - \theta \\ \alpha - \theta & 1 - \alpha\end{bmatrix}.$$

Therefore

$$\frac{\beta_1 \beta_2}{D_1 D_2} \leq ((1 - \alpha) - (\alpha - \theta))^2,$$

or equivalently,

$$\beta_1 \beta_2 \leq D_1 D_2 (1 + \theta - 2\alpha)^2.$$

Plugging in α and simplifying, we obtain the desired inequality.

CHAPTER 13

Multiple Description Coding

We consider the problem of generating two descriptions of a source such that each description by itself can be used to reconstruct the source with some desired distortion and the two descriptions together can be used to reconstruct the source with a lower distortion. This problem is motivated by the need to efficiently communicate multimedia content over networks such as the Internet. Consider the following two scenarios:

- Path diversity: Suppose we wish to send a movie to a viewer over a network that suffers from data loss and delays. We can send multiple copies of the same description of the movie to the viewer via different paths in the network. Such replication, however, is inefficient and the viewer does not benefit from receiving more than one copy of the description. Multiple description coding provides a more efficient means to achieve such "path diversity." We generate multiple descriptions of the movie, so that if the viewer receives only one of them, the movie can be reconstructed with some acceptable quality, and if the viewer receives two of them, the movie can be reconstructed with a higher quality and so on.

- Successive refinement: Suppose we wish to send a movie with different levels of quality to different viewers. We can send a separate description of the movie to each viewer. These descriptions, however, are likely to have significant overlaps. Successive refinement, which is a special case of multiple description coding, provides a more efficient way to distribute the movie. The idea is to send the lowest quality description and successive refinements of it (instead of additional full descriptions). Each viewer then uses the lowest quality description and some of the successive refinements to reconstruct the movie at her desired level of quality.

The optimal scheme for generating multiple descriptions is not known in general. We present the El Gamal–Cover coding scheme for generating two descriptions that are individually good but still carry additional information about the source when combined together. The proof of achievability uses the multivariate covering lemma in Section 8.4. We show that this scheme is optimal for the quadratic Gaussian case. The key to the converse is the identification of a common-information random variable. We then present an improvement on the El Gamal–Cover scheme by Zhang and Berger that involves sending an additional common description. Finally, we briefly discuss extensions of these results to more than two descriptions. We will continue the discussion of multiple description coding in Chapter 20.

13.1   MULTIPLE DESCRIPTION CODING FOR A DMS

Consider the multiple description coding setup for a DMS X and three distortion measures $d_j(x, \hat{x}_j)$, $j = 0, 1, 2$, depicted in Figure 13.1. Each encoder generates a description of X so that decoder 1 that receives only description M_1 can reconstruct X with distortion D_1, decoder 2 that receives only description M_2 can reconstruct X with distortion D_2, and decoder 0 that receives both descriptions can reconstruct X with distortion D_0. We wish to find the optimal tradeoff between the description rate pair (R_1, R_2) and the distortion triple (D_0, D_1, D_2).

A $(2^{nR_1}, 2^{nR_2}, n)$ *multiple description code* consists of

- two encoders, where encoder 1 assigns an index $m_1(x^n) \in [1 : 2^{nR_1})$ and encoder 2 assigns an index $m_2(x^n) \in [1 : 2^{nR_2})$ to each sequence $x^n \in \mathcal{X}^n$, and

- three decoders, where decoder 1 assigns an estimate $\hat{x}_1^n$ to each index m_1, decoder 2 assigns an estimate $\hat{x}_2^n$ to each index m_2, and decoder 0 assigns an estimate $\hat{x}_0^n$ to each index pair (m_1, m_2).

A rate–distortion quintuple $(R_1, R_2, D_0, D_1, D_2)$ is said to be achievable (and a rate pair (R_1, R_2) is said to be achievable for distortion triple (D_0, D_1, D_2)) if there exists a sequence of $(2^{nR_1}, 2^{nR_2}, n)$ codes with

$$\limsup_{n \to \infty} \mathsf{E}(d_j(X^n, \hat{X}_j^n)) \le D_j, \quad j = 0, 1, 2.$$

The *rate–distortion region* $\mathcal{R}(D_0, D_1, D_2)$ for multiple description coding is the closure of the set of rate pairs (R_1, R_2) such that $(R_1, R_2, D_0, D_1, D_2)$ is achievable.

The rate–distortion region for multiple description coding is not known in general. The difficulty is that two good individual descriptions must be close to the source and so must be highly dependent. Thus the second description contributes little extra information beyond the first one. At the same time, to obtain a better reconstruction by combining two descriptions, they must be far apart and so must be highly independent. Two independent descriptions, however, cannot be individually good in general.

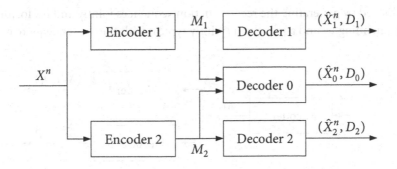

Figure 13.1. Multiple description coding setup.

13.2   SIMPLE SPECIAL CASES

First consider the following special cases of the general multiple description problem.

No combined reconstruction. Suppose $D_0 = \infty$, that is, decoder 0 does not exist as depicted in Figure 13.2. Then the rate–distortion region for distortion pair (D_1, D_2) is the set of rate pairs (R_1, R_2) such that

$$
\begin{aligned}
R_1 &\geq I(X; \hat{X}_1), \\
R_2 &\geq I(X; \hat{X}_2),
\end{aligned}
\tag{13.1}
$$

for some conditional pmf $p(\hat{x}_1|x)p(\hat{x}_2|x)$ that satisfies the constraints $\mathsf{E}(d_j(X, \hat{X}_j)) \leq D_j$, $j = 1, 2$. This rate–distortion region is achieved by generating two reconstruction codebooks independently and performing joint typicality encoding separately to generate each description.

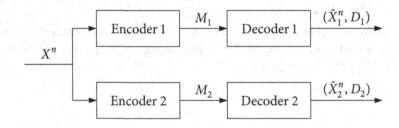

Figure 13.2. Multiple description coding with no combined reconstruction.

Single description with two reconstructions. Suppose that $R_2 = 0$, that is, one description is used to generate two reconstructions $\hat{X}_1^n$ and $\hat{X}_0^n$ as depicted in Figure 13.3. Then the rate–distortion function is

$$
R(D_0, D_1) = \min_{p(\hat{x}_0, \hat{x}_1|x):\, \mathsf{E}(d_j(X, \hat{X}_j)) \leq D_j,\, j=0,1} I(X; \hat{X}_0, \hat{X}_1),
\tag{13.2}
$$

and is achieved by generating the reconstruction sequences jointly and performing joint typicality encoding to find a pair that is jointly typical with the source sequence.

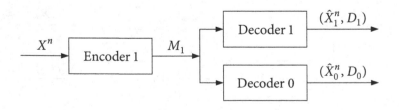

Figure 13.3. Single description with two reconstructions.

Combined description only. Now suppose $D_1 = D_2 = \infty$, that is, decoders 1 and 2 do not exist, as depicted in Figure 13.4. Then the rate–distortion region for distortion D_0 is the set of rate pairs (R_1, R_2) such that

$$R_1 + R_2 \geq I(X; \hat{X}_0) \tag{13.3}$$

for some conditional pmf $p(\hat{x}_0|x)$ that satisfies the constraint $E(d_0(X, \hat{X}_0)) \leq D_0$. This rate–distortion region is achieved by rate splitting. We generate a single description with rate $R_1 + R_2$ as in point-to-point lossy source coding and divide it into two independent indices with rates R_1 and R_2, respectively.

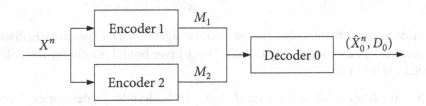

Figure 13.4. Multiple descriptions with combined description only.

The optimality of the above rate regions follows by noting that each region coincides with the following simple outer bound.

Outer bound. Fix a distortion triple (D_0, D_1, D_2). Following similar steps to the converse proof of the lossy source coding theorem, we can readily establish an outer bound on the rate–distortion region that consists of all rate pairs (R_1, R_2) such that

$$\begin{aligned} R_1 &\geq I(X; \hat{X}_1), \\ R_2 &\geq I(X; \hat{X}_2), \\ R_1 + R_2 &\geq I(X; \hat{X}_0, \hat{X}_1, \hat{X}_2) \end{aligned} \tag{13.4}$$

for some conditional pmf $p(\hat{x}_0, \hat{x}_1, \hat{x}_2|x)$ that satisfies the constraints $E(d_j(X, \hat{X}_j)) \leq D_j$, $j = 0, 1, 2$. This outer bound, however, is not tight in general.

13.3   EL GAMAL–COVER INNER BOUND

As we have seen, generating the reconstructions independently and performing joint typicality encoding separately is optimal when decoder 0 does not exist. When it does, however, this scheme is inefficient because three separate descriptions are sent. At the other extreme, generating the reconstructions jointly is optimal when either decoder 1 or 2 does not exist, but is inefficient when both exist because the same description is sent twice. The following inner bound is achieved by generating two descriptions that are individually good, but different enough to provide additional information to decoder 0.

Theorem 13.1 (El Gamal–Cover Inner Bound). Let X be a DMS and $d_j(x, \hat{x}_j)$, $j = 0, 1, 2$, be three distortion measures. A rate pair (R_1, R_2) is achievable with distortion triple (D_0, D_1, D_2) for multiple description coding if

$$R_1 > I(X; \hat{X}_1 | Q),$$
$$R_2 > I(X; \hat{X}_2 | Q),$$
$$R_1 + R_2 > I(X; \hat{X}_0, \hat{X}_1, \hat{X}_2 | Q) + I(\hat{X}_1; \hat{X}_2 | Q)$$

for some conditional pmf $p(q)p(\hat{x}_0, \hat{x}_1, \hat{x}_2 | x, q)$ with $|Q| \le 6$ such that $E(d_j(X, \hat{X}_j)) \le D_j$, $j = 0, 1, 2$.

It is easy to verify that the above inner bound is tight for all the special cases discussed in Section 13.2. In addition, the El Gamal–Cover inner bound is tight for the following nontrivial special cases:

- Successive refinement: In this case $d_2 \equiv 0$, that is, decoder 2 does not exist and the reconstruction at decoder 0 is a refinement of the reconstruction at decoder 1. We discuss this case in detail in Section 13.5.

- Semideterministic distortion measure: Consider the case where $d_1(x, \hat{x}_1) = 0$ if $\hat{x}_1 = g(x)$ and $d_1(x, \hat{x}_1) = 1$, otherwise, and $D_1 = 0$. In other words, $\hat{X}_1$ recovers some function $g(X)$ losslessly.

- Quadratic Gaussian: Let $X \sim N(0, 1)$ and each d_j, $j = 0, 1, 2$, be the squared error (quadratic) distortion measure.

In addition to these special cases, the El Gamal–Cover inner bound is tight when there is *no excess rate*, that is, when the rate pair (R_1, R_2) satisfies the condition $R_1 + R_2 = R(D_0)$, where $R(D_0)$ is the rate–distortion function for X with the distortion measure d_0 evaluated at D_0. It can be shown that $\mathcal{R}(D_1, D_2) \cap \{(R_1, R_2): R_1 + R_2 = R(D_0)\}$ is contained in the closure of the El Gamal–Cover inner bound for each distortion triple (D_0, D_1, D_2). The no-excess rate case is illustrated in the following.

Example 13.1 (Multiple descriptions of a Bernoulli source). Consider the multiple description coding problem for a Bern($1/2$) source and Hamming distortion measures d_0, d_1, and d_2. Suppose that there is no excess rate with $R_1 = R_2 = 1/2$ and $D_0 = 0$. Symmetry suggests that $D_1 = D_2 = D$. Define

$$D_{\min} = \inf\{D: (R_1, R_2) = (1/2, 1/2) \text{ is achievable for distortion triple } (0, D, D)\}.$$

Note that $D_0 = 0$ requires the two descriptions, each at rate $1/2$, to be independent. The lossy source coding theorem in this case gives $R = 1/2 \ge 1 - H(D)$ is necessary, i.e., $D_{\min} \ge 0.11$.

First consider the following simple scheme. We split the source sequence x^n into two equal-length subsequences (e.g., the odd and the even entries). Encoder 1 sends the index of the first subsequence and encoder 2 sends the index of the second. This scheme achieves individual distortions $D_1 = D_2 = D = 1/4$; hence $D_{\min} \le D = 1/4$. Can we do better?

Consider the El Gamal–Cover inner bound with $X = X_1 \cdot X_2$, where $\hat{X}_1$ and $\hat{X}_2$ are independent Bern($1/\sqrt{2}$) random variables, and $\hat{X}_0 = X$. Thus, $X \sim$ Bern($1/2$) as it should be. Substituting in the inner bound rate constraints, we obtain

$$R_1 \ge I(X; \hat{X}_1) = H(X) - H(X|\hat{X}_1) = 1 - \frac{1}{\sqrt{2}} H\left(\frac{1}{\sqrt{2}}\right) \approx 0.383,$$

$$R_2 \ge I(X; \hat{X}_2) \approx 0.383,$$

$$R_1 + R_2 \ge I(X; \hat{X}_0, \hat{X}_1, \hat{X}_2) + I(\hat{X}_1; \hat{X}_2)$$
$$= H(X) - H(X|\hat{X}_0, \hat{X}_1, \hat{X}_2) + I(\hat{X}_1; \hat{X}_2) = 1 - 0 + 0 = 1.$$

The expected distortions are

$$E(d(X, \hat{X}_1)) = P\{\hat{X}_1 \ne X\}$$

$$= P\{\hat{X}_1 = 0, X = 1\} + P\{\hat{X}_1 = 1, X = 0\} = \frac{\sqrt{2} - 1}{2},$$

$$E(d(X, \hat{X}_2)) = \frac{\sqrt{2} - 1}{2},$$

$$E(d(X, \hat{X}_0)) = 0.$$

Thus, the rate pair $(R_1, R_2) = (1/2, 1/2)$ is achievable with distortion triple $(D_1, D_2, D_0) = ((\sqrt{2} - 1)/2, (\sqrt{2} - 1)/2, 0)$. It turns out that $D_{\min} = (\sqrt{2} - 1)/2 \approx 0.207$ is indeed optimal.

The El Gamal–Cover inner bound is not tight in general when there is excess rate even for the Bern($1/2$) source and Hamming distortion measures as discussed in Section 13.6.

13.3.1   Proof of the El Gamal–Cover Inner Bound

The idea of the proof is to generate two descriptions from which we can obtain arbitrarily correlated reconstructions that satisfy the distortion constraints individually and jointly. The proof uses the multivariate covering lemma in Section 8.6. We prove achievability for $|\mathcal{Q}| = 1$; the rest of the proof follows by time sharing.

Rate splitting. Divide index M_j, $j = 1, 2$, into two independent indices M_{0j} at rate R_{0j} and M_{jj} at rate R_{jj}. Thus, $R_j = R_{0j} + R_{jj}$ for $j = 1, 2$. Define $R_0 = R_{01} + R_{02}$.

Codebook generation. Fix a conditional pmf $p(\hat{x}_0, \hat{x}_1, \hat{x}_2 | x)$ such that $E(d_j(X, \hat{X}_j)) \le D_j/(1 + \epsilon)$, $j = 0, 1, 2$. For $j = 1, 2$, randomly and independently generate $2^{nR_{jj}}$ sequences $\hat{x}_j^n(m_{jj})$, $m_{jj} \in [1 : 2^{nR_{jj}}]$, each according to $\prod_{i=1}^{n} p_{\hat{X}_j}(\hat{x}_{ji})$. For each $(m_{11}, m_{22}) \in [1 : 2^{nR_{11}}] \times [1 : 2^{nR_{22}}]$, randomly and conditionally independently generate 2^{nR_0} sequences $\hat{x}_0^n(m_{11}, m_{22}, m_0)$, $m_0 \in [1 : 2^{nR_0}]$, each according to $\prod_{i=1}^{n} p_{\hat{X}_0|\hat{X}_1, \hat{X}_2}(\hat{x}_{0i}|\hat{x}_{1i}(m_{11}), \hat{x}_{2i}(m_{22}))$.

Encoding. Upon observing x^n, the encoder finds an index triple (m_{11}, m_{22}, m_0) such that

$(x^n, \hat{x}_1^n(m_{11}), \hat{x}_2^n(m_{22}), \hat{x}_0^n(m_{11}, m_{22}, m_0)) \in T_\epsilon^{(n)}$. If no such triple exists, the encoder sets $(m_{11}, m_{22}, m_0) = (1, 1, 1)$. It then represents $m_0 \in [1 : 2^{nR_0}]$ by $(m_{01}, m_{02}) \in [1 : 2^{nR_{01}}] \times [1 : 2^{nR_{02}}]$ and sends (m_{11}, m_{01}) to decoder 1, (m_{22}, m_{02}) to decoder 2, and (m_{11}, m_{22}, m_0) to decoder 0.

Decoding. Given the index pair (m_{11}, m_{01}), decoder 1 declares $\hat{x}_1^n(m_{11})$ as its reconstruction of x^n. Similarly, given (m_{22}, m_{02}), decoder 2 declares $\hat{x}_2^n(m_{22})$ as its reconstruction of x^n, and given (m_{11}, m_{22}, m_0), decoder 0 declares $\hat{x}_0^n(m_{11}, m_{22}, m_0)$ as its reconstruction of x^n.

Analysis of expected distortion. Let (M_{11}, M_{22}, M_0) denote the index triple for the reconstruction codeword triple $(\hat{X}_1^n, \hat{X}_2^n, \hat{X}_0^n)$. Define the "error" event

$$\mathcal{E} = \{(X^n, \hat{X}_1^n(M_{11}), \hat{X}_2^n(M_{22}), \hat{X}_0^n(M_{11}, M_{22}, M_0)) \notin T_\epsilon^{(n)}\}$$

and consider the following events:

$$\mathcal{E}_1 = \{(X^n, \hat{X}_1^n(m_{11}), \hat{X}_2^n(m_{22})) \notin T_\epsilon^{(n)} \text{ for all } (m_{11}, m_{22}) \in [1 : 2^{nR_{11}}] \times [1 : 2^{nR_{22}}]\},$$

$$\mathcal{E}_2 = \{(X^n, \hat{X}_1^n(M_{11}), \hat{X}_2^n(M_{22}), \hat{X}_0^n(M_{11}, M_{22}, m_0)) \notin T_\epsilon^{(n)} \text{ for all } m_0 \in [1 : 2^{nR_0}]\}.$$

Then by the union of events bound,

$$P(\mathcal{E}) \leq P(\mathcal{E}_1) + P(\mathcal{E}_1^c \cap \mathcal{E}_2).$$

By the multivariate covering lemma for the 3-DMS $(X, \hat{X}_1, \hat{X}_2)$ with $r_3 = 0$, $P(\mathcal{E}_1)$ tends to zero as $n \to \infty$ if

$$R_{11} > I(X; \hat{X}_1) + \delta(\epsilon),$$
$$R_{22} > I(X; \hat{X}_2) + \delta(\epsilon), \tag{13.5}$$
$$R_{11} + R_{22} > I(X; \hat{X}_1, \hat{X}_2) + I(\hat{X}_1; \hat{X}_2) + \delta(\epsilon).$$

By the covering lemma, $P(\mathcal{E}_1^c \cap \mathcal{E}_2)$ tends to zero as $n \to \infty$ if $R_0 > I(X; \hat{X}_0 | \hat{X}_1, \hat{X}_2) + \delta(\epsilon)$. Eliminating R_{11}, R_{22}, and R_0 by the Fourier–Motzkin procedure in Appendix D, it can be shown that $P(\mathcal{E})$ tends to zero as $n \to \infty$ if

$$R_1 > I(X; \hat{X}_1) + \delta(\epsilon),$$
$$R_2 > I(X; \hat{X}_2) + \delta(\epsilon), \tag{13.6}$$
$$R_1 + R_2 > I(X; \hat{X}_0, \hat{X}_1, \hat{X}_2) + I(\hat{X}_1; \hat{X}_2) + 2\delta(\epsilon).$$

Now by the law of total expectation and the typical average lemma, the asymptotic distortions averaged over random codebooks are bounded as

$$\limsup_{n \to \infty} E(d_j(X^n, \hat{X}_j^n)) \leq D_j, \quad j = 0, 1, 2,$$

if the inequalities on (R_1, R_2) in (13.6) are satisfied. Using time sharing establishes the achievability of every rate pair (R_1, R_2) that satisfies the inequalities in the theorem for some conditional pmf $p(q)p(\hat{x}_0, \hat{x}_1, \hat{x}_2 | x, q)$ such that $E(d_j(X, \hat{X}_j)) < D_j, j = 0, 1, 2$. Finally, using the continuity of mutual information completes the proof with nonstrict distortion inequalities.

13.4 QUADRATIC GAUSSIAN MULTIPLE DESCRIPTION CODING

Consider the multiple description coding problem for a WGN(P) source X and squared error distortion measures d_0, d_1, and d_2. Without loss of generality, we assume that $0 < D_0 \leq D_1, D_2 \leq P$. The El Gamal–Cover inner bound is tight in this case.

Theorem 13.2. The rate–distortion region $\mathcal{R}(D_0, D_1, D_2)$ for multiple description coding of a WGN(P) source X and squared error distortion measures is the set of rate pairs (R_1, R_2) such that

$$R_1 \geq \mathsf{R}\left(\frac{P}{D_1}\right),$$

$$R_2 \geq \mathsf{R}\left(\frac{P}{D_2}\right),$$

$$R_1 + R_2 \geq \mathsf{R}\left(\frac{P}{D_0}\right) + \Delta(P, D_0, D_1, D_2),$$

where $\Delta = \Delta(P, D_0, D_1, D_2)$ is

$$\Delta = \begin{cases} 0 & \text{if } D_0 \leq D_1 + D_2 - P, \\ \mathsf{R}((PD_0)/(D_1 D_2)) & \text{if } D_0 \geq 1/(1/D_1 + 1/D_2 - 1/P), \\ \mathsf{R}\left(\dfrac{(P-D_0)^2}{(P-D_0)^2 - \left(\sqrt{(P-D_1)(P-D_2)} - \sqrt{(D_1-D_0)(D_2-D_0)}\right)^2}\right) & \text{otherwise.} \end{cases}$$

13.4.1 Proof of Achievability

We set $(X, \hat{X}_0, \hat{X}_1, \hat{X}_2)$ to be jointly Gaussian with

$$\hat{X}_j = \left(1 - \frac{D_j}{P}\right)(X + Z_j), \quad j = 0, 1, 2,$$

where (Z_0, Z_1, Z_2) is a zero-mean Gaussian random vector independent of X with covariance matrix

$$K = \begin{bmatrix} N_0 & N_0 & N_0 \\ N_0 & N_1 & N_0 + \rho\sqrt{(N_1 - N_0)(N_2 - N_0)} \\ N_0 & N_0 + \rho\sqrt{(N_1 - N_0)(N_2 - N_0)} & N_2 \end{bmatrix},$$

and

$$N_j = \frac{PD_j}{P - D_j}, \quad j = 0, 1, 2.$$

Note that $X \to \hat{X}_0 \to (\hat{X}_1, \hat{X}_2)$ form a Markov chain for all $\rho \in [-1, 1]$.

We divide the rest of the proof into two parts.

High distortion: $D_1 + D_2 \geq P + D_0$. Note that by relaxing the simple outer bound in Section 13.2, any achievable rate pair (R_1, R_2) must satisfy the inequalities

$$R_1 \geq \mathsf{R}\left(\frac{P}{D_1}\right),$$

$$R_2 \geq \mathsf{R}\left(\frac{P}{D_2}\right),$$

$$R_1 + R_2 \geq \mathsf{R}\left(\frac{P}{D_0}\right).$$

Surprisingly these rates are achievable under the high distortion condition $D_1 + D_2 \geq P + D_0$. Under this condition and the standing assumption $0 < D_0 \leq D_1, D_2 \leq P$, it can be easily verified that $(N_1 - N_0)(N_2 - N_0) \geq (P + N_0)^2$. Thus, there exists $\rho \in [-1, 1]$ such that $N_0 + \rho\sqrt{(N_1 - N_0)(N_2 - N_0)} = -P$. This shows that $\hat{X}_1$ and $\hat{X}_2$ can be made independent of each other, while achieving $\mathsf{E}(d(X, \hat{X}_j)) = D_j$, $j = 0, 1, 2$, which proves the achievability of the simple outer bound.

Low distortion: $D_1 + D_2 < P + D_0$. Under this low distortion condition, the consequent dependence of the descriptions causes an increase in the total description rate $R_1 + R_2$ beyond $\mathsf{R}(P/D_0)$. Consider the above choice of $(\hat{X}_0, \hat{X}_1, \hat{X}_2)$ along with $\rho = -1$. Then it can be shown by a little algebra that

$$I(\hat{X}_1; \hat{X}_2) = \Delta(P, D_0, D_1, D_2),$$

which proves that the rate region in Theorem 13.2 is achievable.

To complete the proof of achievability, we use the discretization procedure described in Section 3.8.

13.4.2   Proof of the Converse

We only need to consider the low distortion case, where $D_0 + P > D_1 + D_2$. Again the inequalities $R_j \geq \mathsf{R}(P/D_j)$, $j = 1, 2$, follow immediately by the lossy source coding theorem. To bound the sum-rate, let $Y_i = X_i + Z_i$, where $\{Z_i\}$ is a WGN(N) process independent of $\{X_i\}$. Then

$$
\begin{aligned}
n(R_1 + R_2) &\geq H(M_1, M_2) + I(M_1; M_2) \\
&= I(X^n; M_1, M_2) + I(Y^n, M_1; M_2) - I(Y^n; M_2 | M_1) \\
&\geq I(X^n; M_1, M_2) + I(Y^n; M_2) - I(Y^n; M_2 | M_1) \\
&= I(X^n; M_1, M_2) + I(Y^n; M_2) + I(Y^n; M_1) - I(Y^n; M_1, M_2) \\
&= \big(I(X^n; M_1, M_2) - I(Y^n; M_1, M_2)\big) + I(Y^n; M_1) + I(Y^n; M_2). \qquad (13.7)
\end{aligned}
$$

We first lower bound the second and third terms. Consider

$$I(Y^n; M_1) \geq \sum_{i=1}^{n} I(Y_i; \hat{X}_{1i})$$

$$\geq \sum_{i=1}^{n}\big(h(Y_i) - h(Y_i - \hat{X}_{1i})\big)$$

$$\geq \sum_{i=1}^{n} \frac{1}{2} \log \left(\frac{P+N}{\mathsf{E}((Y_i - \hat{X}_{1i})^2)} \right)$$

$$\overset{(a)}{\geq} \frac{n}{2} \log \left(\frac{P+N}{D_1 + N} \right),$$

where (a) follows by Jensen's inequality and the fact that

$$\frac{1}{n} \sum_{i=1}^{n} \mathsf{E}((Y_i - \hat{X}_{1i})^2) \leq \frac{1}{n} \sum_{i=1}^{n} \mathsf{E}((X_i - \hat{X}_{1i})^2) + N \leq D_1 + N.$$

Similarly

$$I(Y^n; M_2) \geq \frac{n}{2} \log \left(\frac{P+N}{D_2 + N} \right).$$

Next, we lower bound the first term in (13.7). By the conditional EPI,

$$h(Y^n | M_1, M_2) \geq \frac{n}{2} \log \left(2^{2h(X^n | M_1, M_2)/n} + 2^{2h(Z^n | M_1, M_2)/n} \right)$$

$$= \frac{n}{2} \log \left(2^{2h(X^n | M_1, M_2)/n} + 2\pi eN \right).$$

Since $h(X^n | M_1, M_2) \leq h(X^n | \hat{X}_0^n) \leq (n/2) \log(2\pi e D_0)$, we have

$$h(Y^n | M_1, M_2) - h(X^n | M_1, M_2) \geq \frac{n}{2} \log \left(1 + \frac{2\pi eN}{2^{2h(X^n | M_1, M_2)/n}} \right)$$

$$\geq \frac{n}{2} \log \left(1 + \frac{N}{D_0} \right).$$

Hence

$$I(X^n; M_1, M_2) - I(Y^n; M_1, M_2) \geq \frac{n}{2} \log \left(\frac{P(D_0 + N)}{(P+N)D_0} \right).$$

Combining these inequalities and continuing with the lower bound on the sum-rate, we have

$$R_1 + R_2 \geq \frac{1}{2} \log \left(\frac{P}{D_0} \right) + \frac{1}{2} \log \left(\frac{(P+N)(D_0 + N)}{(D_1 + N)(D_2 + N)} \right).$$

Finally we maximize this sum-rate bound over $N \geq 0$ by taking

$$N = \left[\frac{D_1 D_2 - D_0 P + \sqrt{(D_1 - D_0)(D_2 - D_0)(P - D_1)(P - D_2)}}{P + D_0 - D_1 - D_2} \right]^+, \tag{13.8}$$

which yields the desired inequality

$$R_1 + R_2 \geq R \left(\frac{P}{D_0} \right) + \Delta(P, D_0, D_1, D_2).$$

This completes the proof of Theorem 13.2.

Remark 13.1. It can be shown that if $(R_1, R_2) \in \mathcal{R}(D_0, D_1, D_2)$ and $D_0 \leq D_1 + D_2 - P$, then $(R_1, R_2) \in \mathcal{R}(D_0, D_1^*, D_2^*)$ for some $D_1^* \leq D_1$ and $D_2^* \leq D_2$ such that $D_0 = D_1^* + D_2^* - P$. Also, if $(R_1, R_2) \in \mathcal{R}(D_0, D_1, D_2)$ and $D_0 \geq 1/(1/D_1 + 1/D_2 - 1/P)$, then $(R_1, R_2) \in \mathcal{R}(D_0^*, D_1, D_2)$, where $D_0^* = 1/(1/D_1 + 1/D_2 - 1/P) \leq D_0$. Note that these two cases (high distortion and low distortion with $D_0 \geq 1/(1/D_1 + 1/D_2 - 1/P))$ correspond to $N = \infty$ and $N = 0$ in (13.8), respectively.

Remark 13.2. It can be shown that the optimal Y satisfies the Markov chain relationship $\hat{X}_1 \to Y \to \hat{X}_2$, where $\hat{X}_1$ and $\hat{X}_2$ are the reconstructions specified in the proof of achievability. In addition, the optimal Y minimizes $I(\hat{X}_1, \hat{X}_2; Y)$ over all choices of the form $Y = X + Z$ such that $\hat{X}_1 \to Y \to \hat{X}_2$, where Z is independent of $(X, \hat{X}_1, \hat{X}_2)$. The minimized mutual information can be viewed as "common information" between the optimal reconstructions for the given distortion triple; see Section 14.2 for a detailed discussion of common information.

13.5 SUCCESSIVE REFINEMENT

Consider the special case of multiple description coding depicted in Figure 13.5, where decoder 2 does not exist, or equivalently, $d_2(x, x_2) \equiv 0$. Since there is no standalone decoder for the second description M_2, there is no longer a tension between the two descriptions. As such, the second description can be viewed as a refinement of the first description that helps decoder 0 achieve a lower distortion. However, a tradeoff still exists between the two descriptions because if the first description is optimal for decoder 1, that is, if it achieves the rate–distortion function for d_1, the first and second descriptions combined may not be optimal for decoder 0.

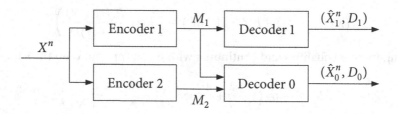

Figure 13.5. Successive refinement.

The El Gamal–Cover inner bound is tight for this case.

Theorem 13.3. The successive refinement rate–distortion region $\mathcal{R}(D_0, D_1)$ for a DMS X and distortion measures d_0 and d_1 is the set of rate pairs (R_1, R_2) such that

$$R_1 \geq I(X; \hat{X}_1),$$
$$R_1 + R_2 \geq I(X; \hat{X}_0, \hat{X}_1)$$

for some conditional pmf $p(\hat{x}_0, \hat{x}_1 | x)$ that satisfies the constraints $\mathrm{E}(d_j(X, \hat{X}_j)) \leq D_j$, $j = 0, 1$.

The proof of achievability follows immediately from the proof of Theorem 13.1 by setting $Q = \hat{X}_2 = \emptyset$. The proof of the converse is also straightforward.

13.5.1 Successively Refinable Sources

Consider the successive refinement of a source under a common distortion measure $d_1 = d_0 = d$. If a rate pair (R_1, R_2) is achievable with distortion pair (D_0, D_1) for $D_0 \le D_1$, then we must have

$$R_1 \ge R(D_1),$$
$$R_1 + R_2 \ge R(D_0),$$

where $R(D) = \min_{p(\hat{x}|x):E(d(X,\hat{X}))\le D} I(X; \hat{X})$ is the rate–distortion function for a single description.

In some cases, $(R_1, R_2) = (R(D_1), R(D_0) - R(D_1))$ is actually achievable for all $D_0 \le D_1$ and there is no loss of optimality in describing the source successively by a coarse description and a refinement of it. Such a source is referred to as *successively refinable*.

Example 13.2 (Bernoulli source with Hamming distortion). A Bern(p) source X is successively refinable under Hamming distortion measures d_0 and d_1. This is shown by considering a cascade of backward binary symmetric test channels

$$X = \hat{X}_0 \oplus Z_0 = (\hat{X}_1 \oplus Z_1) \oplus Z_0,$$

where $Z_0 \sim$ Bern(D_0) and $Z_1 \sim$ Bern(D') such that $D' * D_0 = D_1$. Note that $\hat{X}_1 \to \hat{X}_0 \to X$ form a Markov chain.

Example 13.3 (Gaussian source with squared error distortion). Consider a WGN(P) source with squared error distortion measures. We can achieve $R_1 = R(P/D_1)$ and $R_1 + R_2 = R(P/D_0)$ (or equivalently, $R_2 = R(D_1/D_0)$) simultaneously by taking a cascade of backward Gaussian test channels

$$X = \hat{X}_0 + Z_0 = (\hat{X}_1 + Z_1) + Z_0,$$

where $\hat{X}_1 \sim$ N$(0, P - D_1)$, $Z_1 \sim$ N$(0, D_1 - D_0)$, and $Z_0 \sim$ N$(0, D_0)$ are mutually independent and $\hat{X}_0 = \hat{X}_1 + Z_1$. Again $\hat{X}_1 \to \hat{X}_0 \to X$ form a Markov chain.

Successive refinability of the Gaussian source can be shown more directly by the following heuristic argument. By the quadratic Gaussian source coding theorem in Section 3.6, using rate R_1, the source sequence X^n can be reconstructed using $\hat{X}_1^n$ with error $Y^n = X^n - \hat{X}_1^n$ and distortion $D_1 = (1/n) \sum_{i=1}^{n} E(Y_i^2) \le P2^{-2R_1}$. Then, using rate R_2, the error sequence Y^n can be reconstructed using $\hat{X}_0^n$ with distortion $D_0 \le D_1 2^{-2R_2} \le P2^{-2(R_1+R_2)}$. Subsequently, the error of the error, the error of the error of the error, and so on can be successively described, providing further refinement of the source. Each stage represents a quadratic Gaussian source coding problem. Thus, successive refinement for a WGN source is, in a sense, dual to successive cancellation for the Gaussian MAC and BC.

In general, we can establish the following result on successive refinability.

Proposition 13.1. A DMS X is successively refinable under a common distortion measure $d = d_0 = d_1$ iff for every $D_0 \leq D_1$ there exists a conditional pmf $p(\hat{x}_0, \hat{x}_1|x) = p(\hat{x}_0|x)p(\hat{x}_1|\hat{x}_0)$ such that $p(\hat{x}_0|x)$ and $p(\hat{x}_1|x)$ attain the rate–distortion functions $R(D_0)$ and $R(D_1)$, respectively; in other words, $X \to \hat{X}_0 \to \hat{X}_1$ form a Markov chain, and

$$E(d(X, \hat{X}_0)) \leq D_0,$$
$$E(d(X, \hat{X}_1)) \leq D_1,$$
$$I(X; \hat{X}_0) = R(D_0),$$
$$I(X; \hat{X}_1) = R(D_1).$$

13.6   ZHANG–BERGER INNER BOUND

The Zhang–Berger inner bound extends the El Gamal–Cover inner bound by adding a *common* description.

Theorem 13.4 (Zhang–Berger Inner Bound). Let X be a DMS and $d_j(x, \hat{x}_j)$, $j = 0, 1, 2$, be three distortion measures. A rate pair (R_1, R_2) is achievable for distortion triple (D_0, D_1, D_2) if

$$R_1 > I(X; \hat{X}_1, U),$$
$$R_2 > I(X; \hat{X}_2, U),$$
$$R_1 + R_2 > I(X; \hat{X}_0, \hat{X}_1, \hat{X}_2|U) + 2I(U; X) + I(\hat{X}_1; \hat{X}_2|U)$$

for some conditional pmf $p(u, \hat{x}_0, \hat{x}_1, \hat{x}_2|x)$ with $|\mathcal{U}| \leq |\mathcal{X}| + 5$ such that $E(d_j(X, \hat{X}_j)) \leq D_j$, $j = 0, 1, 2$.

Note that the Zhang–Berger inner bound is convex. It is also easy to see that it contains the El Gamal–Cover inner bound (by taking $U = \emptyset$). This containment can be strict. Consider a Bern($1/2$) source and Hamming distortion measures. If $(D_0, D_1, D_2) = (0, 0.1, 0.1)$, the El Gamal–Cover coding scheme cannot achieve (R_1, R_2) such that $R_1 + R_2 \leq 1.2564$. By comparison, the Zhang–Berger inner bound contains a rate pair (R_1, R_2) with $R_1 + R_2 = 1.2057$. It is not known, however, if this inner bound is tight in general.

Proof of Theorem 13.4 (outline). We first generate a common description of the source and then refine it using the El Gamal–Cover coding scheme to obtain the two descriptions. The common description is sent by both encoders and each encoder sends a different refinement in addition. In the following, we describe the coding scheme in more detail.

Divide each index M_j, $j = 1, 2$, into a common index M_0 at rate R_0 and independent private indices M_{0j} at rate R_{0j} and M_{jj} at rate R_{jj}. Thus $R_j = R_0 + R_{jj} + R_{0j}$, $j = 1, 2$. Fix

$p(u, \hat{x}_1, \hat{x}_2, \hat{x}_0 | x)$ such that $\mathsf{E}(d(X, \hat{X}_j)) \le D_j/(1 + \epsilon)$, $j = 0, 1, 2$. Randomly and independently generate 2^{nR_0} sequences $u^n(m_0)$, $m_0 \in [1 : 2^{nR_0}]$, each according to $\prod_{i=1}^{n} p_U(u_i)$. For each $m_0 \in [1 : 2^{nR_0}]$, randomly and conditional independently generate $2^{nR_{11}}$ sequences $\hat{x}_1^n(m_0, m_{11})$, $m_{11} \in [1 : 2^{nR_{11}}]$, each according to $\prod_{i=1}^{n} p_{\hat{X}_1 | U}(\hat{x}_{1i} | u_i(m_0))$, and $2^{nR_{22}}$ sequences $\hat{x}_2^n(m_0, m_{22})$, $m_{22} \in [1 : 2^{nR_{22}}]$, each according to $\prod_{i=1}^{n} p_{\hat{X}_2 | U}(\hat{x}_{2i} | u_i(m_0))$. For each $(m_0, m_{11}, m_{22}) \in [1 : 2^{nR_0}] \times [1 : 2^{nR_{11}}] \times [1 : 2^{nR_{22}}]$, randomly and conditionally independently generate $2^{n(R_{01} + R_{02})}$ sequences $\hat{x}_0^n(m_0, m_{11}, m_{22}, m_{01}, m_{02})$, $(m_{01}, m_{02}) \in [1 : 2^{nR_{01}}] \times [1 : 2^{nR_{02}}]$, each according to $\prod_{i=1}^{n} p_{\hat{X}_0 | U, \hat{X}_1, \hat{X}_2}(\hat{x}_{0i} | u_i(m_0), \hat{x}_{1i}(m_0, m_{11}), \hat{x}_{2i}(m_0, m_{22}))$.

Upon observing x^n, the encoders find an index tuple $(m_0, m_{11}, m_{22}, m_{01}, m_{02})$ such that $(x^n, u^n(m_0), \hat{x}_1^n(m_0, m_{11}), \hat{x}_2^n(m_0, m_{22}), \hat{x}_0^n(m_0, m_{11}, m_{22}, m_{01}, m_{02})) \in \mathcal{T}_\epsilon^{(n)}$. Encoder 1 sends (m_0, m_{11}, m_{01}) and encoder 2 sends (m_0, m_{22}, m_{02}). By the covering lemma and the multivariate covering lemma, it can be shown that a rate tuple $(R_0, R_{11}, R_{22}, R_{01}, R_{02})$ is achievable if

$$R_0 > I(X; U) + \delta(\epsilon),$$
$$R_{11} > I(X; \hat{X}_1 | U) + \delta(\epsilon),$$
$$R_{22} > I(X; \hat{X}_2 | U) + \delta(\epsilon),$$
$$R_{11} + R_{22} > I(X; \hat{X}_1, \hat{X}_2 | U) + I(\hat{X}_1; \hat{X}_2 | U) + \delta(\epsilon),$$
$$R_{01} + R_{02} > I(X; \hat{X}_0 | U, \hat{X}_1, \hat{X}_2) + \delta(\epsilon)$$

for some conditional pmf $p(u, \hat{x}_0, \hat{x}_1, \hat{x}_2 | x)$ such that $\mathsf{E}(d_j(X, \hat{X}_j)) < D_j$, $j = 0, 1, 2$. Finally, using Fourier–Motzkin elimination and the continuity of mutual information, we arrive at the inequalities in the theorem.

Remark 13.3 (An equivalent characterization). We show that the Zhang–Berger inner bound is equivalent to the set of rate pairs (R_1, R_2) such that

$$R_1 > I(X; U_0, U_1),$$
$$R_2 > I(X; U_0, U_2), \tag{13.9}$$
$$R_1 + R_2 > I(X; U_1, U_2 | U_0) + 2I(U_0; X) + I(U_1; U_2 | U_0)$$

for some conditional pmf $p(u_0, u_1, u_2 | x)$ and functions $\hat{x}_0(u_0, u_1, u_2)$, $\hat{x}_1(u_0, u_1)$, and $\hat{x}_2(u_0, u_2)$ that satisfy the constraints

$$\mathsf{E}[d_0(X, \hat{x}_0(U_0, U_1, U_2))] \le D_0,$$
$$\mathsf{E}[d_1(X, \hat{x}_1(U_0, U_1))] \le D_1,$$
$$\mathsf{E}[d_2(X, \hat{x}_2(U_0, U_2))] \le D_2.$$

Clearly region (13.9) is contained in the Zhang–Berger inner bound in Theorem 13.4. We now show that region (13.9) contains the Zhang–Berger inner bound. Consider the following corner point of the Zhang–Berger inner bound:

$$R_1 = I(X; \hat{X}_1, U),$$
$$R_2 = I(X; \hat{X}_0, \hat{X}_2 | \hat{X}_1, U) + I(X; U) + I(\hat{X}_1; \hat{X}_2 | U)$$

for some conditional pmf $p(u, \hat{x}_0, \hat{x}_1, \hat{x}_2 | x)$. By the functional representation lemma in Appendix B, there exists a random variable W independent of $(U, \hat{X}_1, \hat{X}_2)$ such that $\hat{X}_0$ is a function of $(U, \hat{X}_1, \hat{X}_2, W)$ and $X \to (U, \hat{X}_0, \hat{X}_1, \hat{X}_2) \to W$. Let $U_0 = U$, $U_1 = \hat{X}_1$, and $U_2 = (W, \hat{X}_2)$. Then

$$R_1 = I(X; \hat{X}_1, U) = I(X; U_0, U_1),$$
$$R_2 = I(X; \hat{X}_0, \hat{X}_2 | \hat{X}_1, U) + I(X; U) + I(\hat{X}_1; \hat{X}_2 | U)$$
$$= I(X; \hat{X}_2, W | \hat{X}_1, U_0) + I(X; U_0) + I(\hat{X}_1; \hat{X}_2, W | U_0)$$
$$= I(X; U_2 | U_0, U_1) + I(X; U_0) + I(U_1; U_2 | U_0)$$

and $\hat{X}_1$, $\hat{X}_2$, and $\hat{X}_0$ are functions of U_1, U_2, and (U_0, U_1, U_2), respectively. Hence, (R_1, R_2) is also a corner point of the region in (13.9). The rest of the proof follows by time sharing.

SUMMARY

- Multiple description coding setup

- El Gamal–Cover inner bound:

  - Generation of two descriptions that are individually good and jointly better

  - Use of the multivariate covering lemma

- Quadratic Gaussian multiple description coding:

  - El Gamal–Cover inner bound is tight

  - Individual rate–distortion functions can be achieved at high distortion

  - Identification of a common-information random variable in the proof of the converse

- Successive refinement

- Zhang–Berger inner bound adds a common description to El Gamal–Cover coding

- **Open problems:**

  13.1. Is the Zhang–Berger inner bound tight?

  13.2. What is the multiple description rate–distortion region for a Bern(1/2) source and Hamming distortion measures?

BIBLIOGRAPHIC NOTES

The multiple description coding problem was formulated by A. Gersho and H. .S. Witsenhausen, and initially studied by Witsenhausen (1980), Wolf, Wyner, and Ziv (1980), and Witsenhausen and Wyner (1981). El Gamal and Cover (1982) established Theorem 13.1

and evaluated the inner bound for the quadratic Gaussian case in Theorem 13.2. Ozarow (1980) proved the converse of Theorem 13.2. Chen, Tian, Berger, and Hemami (2006) proposed a successive quantization scheme for the El Gamal–Cover inner bound. Berger and Zhang (1983) considered the Bernoulli source with Hamming distortion measures in Example 13.1 and showed that the El Gamal–Cover inner bound is tight when there is no excess rate. Ahlswede (1985) established the optimality for the general no excess rate case. The rate–distortion region for the semideterministic case is due to Fu and Yeung (2002).

Theorem 13.3 is due to Equitz and Cover (1991) and Rimoldi (1994). Proposition 13.1 and Examples 13.2 and 13.3 also appeared in Equitz and Cover (1991). The inner bound in Theorem 13.4 is due to Venkataramani, Kramer, and Goyal (2003). The equivalence to the original characterization in (13.9) by Zhang and Berger (1987) is due to Wang, Chen, Zhao, Cuff, and Permuter (2011). Venkataramani, Kramer, and Goyal (2003) extended the El Gamal–Cover and Zhang–Berger coding schemes to k descriptions and 2^{k-1} decoders. This extension is optimal in several special cases, including k-level successive refinement (Equitz and Cover 1991) and quadratic Gaussian multiple descriptions with individual decoders, each of which receives its own description, and a central decoder that receives all descriptions (Chen 2009). It is not known, however, if these extensions are optimal in general (Puri, Pradhan, and Ramchandran 2005).

PROBLEMS

13.1. Establish the rate–distortion region for no combined reconstruction in (13.1).

13.2. Establish the outer bound on the rate–distortion region in (13.4).

13.3. Consider the multiple description coding setup, where $d_1(x, \hat{x}_1) = 0$ if $\hat{x}_1 = g(x)$ and $d_1(x, \hat{x}_1) = 1$, otherwise, and $D_1 = 0$. Show that the El Gamal–Cover inner bound is tight for this case.

13.4. Complete the details of the achievability proof of Theorem 13.2.

13.5. *Gaussian auxiliary random variable.* Consider a jointly Gaussian triple $(X, \hat{X}_1, \hat{X}_2)$ that attains the rate–distortion region for quadratic Gaussian multiple description in Theorem 13.2. Let $Y = X + Z$, where $Z \sim N(0, N)$ is independent of $(X, \hat{X}_1, \hat{X}_2)$ and N is given by (13.8).

(a) Show that $\hat{X}_1 \rightarrow Y \rightarrow \hat{X}_2$ form a Markov chain.

(b) Let $\tilde{Y} = X + \tilde{Z}$ such that $\tilde{Z}$ is independent of $(X, \hat{X}_1, \hat{X}_2)$ and $E(\tilde{Z}^2) = N$. Show that $I(\hat{X}_1, \hat{X}_2; Y) \leq I(\hat{X}_1, \hat{X}_2; \tilde{Y})$.

13.6. Prove the converse for Theorem 13.3.

13.7. Prove the sufficient and necessary condition for successive refinability in Proposition 13.1.

CHAPTER 14

Joint Source–Channel Coding

In Chapters 4 through 9, we studied reliable communication of independent messages over noisy single-hop networks (channel coding), and in Chapters 10 through 13, we studied the dual setting of reliable communication of uncompressed sources over noiseless single-hop networks (source coding). These settings are special cases of the more general information flow problem of reliable communication of uncompressed sources over noisy single-hop networks. As we have seen in Section 3.9, separate source and channel coding is asymptotically sufficient for communicating a DMS over a DMC. Does such separation hold in general for communicating a k-DMS over a DM single-hop network?

In this chapter, we show that such separation does not hold in general. Thus in some multiuser settings it is advantageous to perform joint source–channel coding. We demonstrate this breakdown in separation through examples of lossless communication of a 2-DMS over a DM-MAC and over a DM-BC.

For the DM-MAC case, we show that joint source–channel coding can help communication by utilizing the correlation between the sources to induce statistical cooperation between the transmitters. We present a joint source–channel coding scheme that outperforms separate source and channel coding. We then show that this scheme can be improved when the sources have a common part, that is, a source that both senders can agree on with probability one.

For the DM-BC case, we show that joint source–channel coding can help communication by utilizing the statistical compatibility between the sources and the channel. We first consider a separate source and channel coding scheme based on the Gray–Wyner source coding system and Marton's channel coding scheme. The optimal rate–region for the Gray–Wyner system naturally leads to several definitions of common information between correlated sources. We then describe a joint source–channel coding scheme that outperforms the separate Gray–Wyner and Marton coding scheme.

Finally, we present a general single-hop network that includes as special cases many of the multiuser source and channel settings we discussed in previous chapters. We describe a hybrid source–channel coding scheme for this network.

14.1 LOSSLESS COMMUNICATION OF A 2-DMS OVER A DM-MAC

Consider the multiple access communication system depicted in Figure 14.1, where a 2-DMS (U_1, U_2) is to be communicated losslessly over a 2-sender DM-MAC $p(y|x_1, x_2)$.

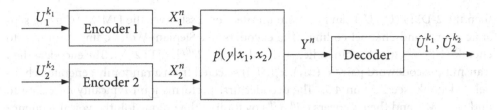

Figure 14.1. Communication of a 2-DMS over a 2-sender DM-MAC.

A $(|\mathcal{U}_1|^{k_1}, |\mathcal{U}_2|^{k_2}, n)$ joint source–channel code of rate pair $(r_1, r_2) = (k_1/n, k_2/n)$ for this setup consists of

- two encoders, where encoder $j = 1, 2$ assigns a sequence $x_j^n(u_j^{k_j}) \in \mathcal{X}_j^n$ to each sequence $u_j^{k_j} \in \mathcal{U}_j^{k_j}$, and

- a decoder that assigns an estimate $(\hat{u}_1^{k_1}, \hat{u}_2^{k_2}) \in \mathcal{U}_1^{k_1} \times \mathcal{U}_2^{k_2}$ to each sequence $y^n \in \mathcal{Y}^n$.

The probability of error is defined as $P_e^{(n)} = P\{(\hat{U}_1^{k_1}, \hat{U}_2^{k_2}) \neq (U_1^{k_1}, U_2^{k_2})\}$. We say that the sources are communicated losslessly over the DM-MAC if there exists a sequence of $(|\mathcal{U}_1|^{k_1}, |\mathcal{U}_2|^{k_2}, n)$ codes such that $\lim_{n \to \infty} P_e^{(n)} = 0$. The problem is to find the necessary and sufficient condition for lossless communication. For simplicity, we assume henceforth the rates $r_1 = r_2 = 1$ symbol/transmission.

First consider the following sufficient condition for separate source and channel coding. We know that the capacity region $\mathscr{C}$ of the DM-MAC is the set of rate pairs (R_1, R_2) such that

$$R_1 \leq I(X_1; Y | X_2, Q),$$
$$R_2 \leq I(X_2; Y | X_1, Q),$$
$$R_1 + R_2 \leq I(X_1, X_2; Y | Q)$$

for some pmf $p(q)p(x_1|q)p(x_2|q)$. We also know from the Slepian–Wolf theorem that the optimal rate region $\mathscr{R}^*$ for distributed lossless source coding is the set of rate pairs (R_1, R_2) such that

$$R_1 \geq H(U_1 | U_2),$$
$$R_2 \geq H(U_2 | U_1),$$
$$R_1 + R_2 \geq H(U_1, U_2).$$

Hence, if the intersection of the interiors of $\mathscr{C}$ and $\mathscr{R}^*$ is not empty, that is, there exists a pmf $p(q)p(x_1|q)p(x_2|q)$ such that

$$H(U_1 | U_2) < I(X_1; Y | X_2, Q),$$
$$H(U_2 | U_1) < I(X_2; Y | X_1, Q), \tag{14.1}$$
$$H(U_1, U_2) < I(X_1, X_2; Y | Q),$$

then the 2-DMS (U_1, U_2) can be communicated losslessly over the DM-MAC using separate source and channel coding. The encoders use Slepian–Wolf coding (binning) to
encode (U_1^n, U_2^n) into the bin indices $(M_1, M_2) \in [1 : 2^{nR_1}] \times [1 : 2^{nR_2}]$. The encoders then
transmit the codeword pair $(x_1^n(M_1), x_2^n(M_2))$ selected from a randomly generated channel codebook; see Section 4.5. The decoder first performs joint typicality decoding to
find (M_1, M_2) and then recovers (U_1^n, U_2^n) by finding the unique jointly typical sequence
pair in the product bin with index pair (M_1, M_2). Since the rate pair (R_1, R_2) satisfies
the conditions for both lossless source coding and reliable channel coding, the end-to-
end probability of error tends to zero as $n \to \infty$. Note that although the joint pmf on
(M_1, M_2) is not necessarily uniform, the message pair can still be reliably transmitted to
the receiver if $(R_1, R_2) \in \mathscr{C}$ (see Problem 4.13).

Consider the following examples for which this sufficient condition for separate source
and channel coding is also necessary.

Example 14.1 (MAC with orthogonal components). Let (U_1, U_2) be an arbitrary 2-
DMS and $p(y|x_1, x_2) = p(y_1|x_1)p(y_2|x_2)$ be a DM-MAC with output $Y = (Y_1, Y_2)$ that
consists of two separate DMCs, $p(y_1|x_1)$ with capacity C_1 and $p(y_2|x_2)$ with capacity C_2.
The sources can be communicated losslessly over this MAC if
$$H(U_1|U_2) < C_1,$$
$$H(U_2|U_1) < C_2,$$
$$H(U_1, U_2) < C_1 + C_2.$$
Conversely, if one of the following inequalities is satisfied:
$$H(U_1|U_2) > C_1,$$
$$H(U_2|U_1) > C_2,$$
$$H(U_1, U_2) > C_1 + C_2,$$
then the sources cannot be communicated losslessly over the channel. Thus source–
channel separation holds for this case.

Example 14.2 (Independent sources). Let U_1 and U_2 be independent sources with en-
tropies $H(U_1)$ and $H(U_2)$, respectively, and $p(y|x_1, x_2)$ be an arbitrary DM-MAC. Source
channel separation again holds in this case. That is, the sources can be communicated
losslessly over the DM-MAC by separate source and channel coding if
$$H(U_1) < I(X_1; Y|X_2, Q),$$
$$H(U_2) < I(X_2; Y|X_1, Q),$$
$$H(U_1) + H(U_2) < I(X_1, X_2; Y|Q),$$
for some pmf $p(q)p(x_1|q)p(x_2|q)$, and the converse holds in general.

Does source–channel separation hold in general for lossless communication of an ar-
bitrary 2-DMS (U_1, U_2) over an arbitrary DM-MAC $p(y|x_1, x_2)$? In other words, is it al-
ways the case that if the intersection of $\mathscr{R}^*$ and $\mathscr{C}$ is empty for a 2-DMS and a DM-MAC,

then the 2-DMS cannot be communicated losslessly over the DM-MAC? To answer this question, consider the following.

Example 14.3. Let (U_1, U_2) be a 2-DMS with $\mathcal{U}_1 = \mathcal{U}_2 = \{0, 1\}$, $p_{U_1,U_2}(0, 0) = p_{U_1,U_2}(0, 1) = p_{U_1,U_2}(1, 1) = 1/3$, and $p_{U_1,U_2}(1, 0) = 0$. Let $p(y|x_1, x_2)$ be a binary erasure MAC with $\mathcal{X}_1 = \mathcal{X}_2 = \{0, 1\}$, $\mathcal{Y} = \{0, 1, 2\}$, and $Y = X_1 + X_2$ (see Example 4.2). The optimal rate region $\mathscr{R}^*$ for this 2-DMS and the capacity region of the binary erasure MAC are sketched in Figure 14.2. Note that the intersection of these two regions is empty since $H(U_1, U_2) = \log 3 = 1.585$ and $\max_{p(x_1)p(x_2)} I(X_1, X_2; Y) = 1.5$. Hence, $H(U_1, U_2) > \max_{p(x_1)p(x_2)} I(X_1, X_2; Y)$ and (U_1, U_2) cannot be communicated losslessly over the erasure DM-MAC using separate source and channel coding.

Now consider an uncoded transmission scheme in which the encoders transmit $X_{1i} = U_{1i}$ and $X_{2i} = U_{2i}$ in time $i \in [1 : n]$. It is easy to see that this scheme achieves *error-free* communication! Thus using separate source and channel coding for sending a 2-DMS over a DM-MAC is *not* optimal in general.

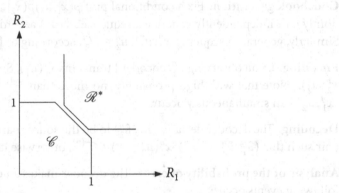

Figure 14.2. Separate source and channel coding fails since $\mathscr{R}^* \cap \mathscr{C} = \emptyset$.

A general necessary and sufficient condition for lossless communication of a 2-DMS over a DM-MAC is not known. In the following we present joint source–channel coding schemes that include as special cases the aforementioned separate source and channel coding scheme and the uncoded transmission scheme in Example 14.3.

14.1.1   A Joint Source–Channel Coding Scheme

We establish the following sufficient condition for lossless communication of a 2-DMS over a DM-MAC.

Theorem 14.1. A 2-DMS (U_1, U_2) can be communicated losslessly over a DM-MAC $p(y|x_1, x_2)$ at rates $r_1 = r_2 = 1$ if

$$H(U_1|U_2) < I(X_1; Y|U_2, X_2, Q),$$
$$H(U_2|U_1) < I(X_2; Y|U_1, X_1, Q),$$

$$H(U_1, U_2) < I(X_1, X_2; Y|Q)$$

for some conditional pmf $p(q, x_1, x_2 | u_1, u_2) = p(q)p(x_1 | u_1, q)p(x_2 | u_2, q)$ with $|\mathcal{Q}| \leq 3$.

This theorem recovers the following as special cases:

- Separate source and channel coding: We set $p(x_1 | u_1, q)p(x_2 | u_2, q) = p(x_1 | q)p(x_2 | q)$, that is, (X_1, X_2, Q) is independent of (U_1, U_2). Then, the set of inequalities in the theorem simplifies to (14.1).

- Example 14.3: Set $Q = \emptyset$, $X_1 = U_1$, and $X_2 = U_2$.

14.1.2 Proof of Theorem 14.1

We establish achievability for $|\mathcal{Q}| = 1$; the rest of the proof follows by time sharing.

Codebook generation. Fix a conditional pmf $p(x_1 | u_1)p(x_2 | u_2)$. For each $u_1^n \in \mathcal{U}_1^n$, randomly and independently generate a sequence $x_1^n(u_1^n)$ according to $\prod_{i=1}^n p_{X_1 | U_1}(x_{1i} | u_{1i})$. Similarly, generate a sequence $x_2^n(u_2^n)$, $u_2^n \in \mathcal{U}_2^n$, according to $\prod_{i=1}^n p_{X_2 | U_2}(x_{2i} | u_{2i})$.

Encoding. Upon observing u_1^n, encoder 1 transmits $x_1^n(u_1^n)$. Similarly encoder 2 transmits $x_2^n(u_2^n)$. Note that with high probability, no more than $2^{n(H(U_1, U_2) + \delta(\epsilon))}$ codeword pairs (x_1^n, x_2^n) can simultaneously occur.

Decoding. The decoder declares $(\hat{u}_1^n, \hat{u}_2^n)$ to be the source pair estimate if it is the unique pair such that $(\hat{u}_1^n, \hat{u}_2^n, x_1^n(\hat{u}_1^n), x_2^n(\hat{u}_2^n), y^n) \in \mathcal{T}_\epsilon^{(n)}$; otherwise it declares an error.

Analysis of the probability of error. The decoder makes an error iff one or more of the following events occur:

$$\mathcal{E}_1 = \{(U_1^n, U_2^n, X_1^n(U_1^n), X_2^n(U_2^n), Y^n) \notin \mathcal{T}_\epsilon^{(n)}\},$$
$$\mathcal{E}_2 = \{(\tilde{u}_1^n, U_2^n, X_1^n(\tilde{u}_1^n), X_2^n(U_2^n), Y^n) \in \mathcal{T}_\epsilon^{(n)} \text{ for some } \tilde{u}_1^n \neq U_1^n\},$$
$$\mathcal{E}_3 = \{(U_1^n, \tilde{u}_2^n, X_1^n(U_1^n), X_2^n(\tilde{u}_2^n), Y^n) \in \mathcal{T}_\epsilon^{(n)} \text{ for some } \tilde{u}_2^n \neq U_2^n\},$$
$$\mathcal{E}_4 = \{(\tilde{u}_1^n, \tilde{u}_2^n, X_1^n(\tilde{u}_1^n), X_2^n(\tilde{u}_2^n), Y^n) \in \mathcal{T}_\epsilon^{(n)} \text{ for some } \tilde{u}_1^n \neq U_1^n, \tilde{u}_2^n \neq U_2^n\}.$$

Thus, the average probability of error is upper bounded as

$$\mathsf{P}(\mathcal{E}) \leq \mathsf{P}(\mathcal{E}_1) + \mathsf{P}(\mathcal{E}_2) + \mathsf{P}(\mathcal{E}_3) + \mathsf{P}(\mathcal{E}_4).$$

By the LLN, $\mathsf{P}(\mathcal{E}_1)$ tends to zero as $n \to \infty$. Next consider the second term. By the union of events bound,

$$\mathsf{P}(\mathcal{E}_2) = \sum_{u_1^n} p(u_1^n) \, \mathsf{P}\{(\tilde{u}_1^n, U_2^n, X_1^n(\tilde{u}_1^n), X_2^n(U_2^n), Y^n) \in \mathcal{T}_\epsilon^{(n)} \text{ for some } \tilde{u}_1^n \neq u_1^n \,|\, U_1^n = u_1^n\}$$

$$\leq \sum_{u_1^n} p(u_1^n) \sum_{\tilde{u}_1^n \neq u_1^n} \mathsf{P}\{(\tilde{u}_1^n, U_2^n, X_1^n(\tilde{u}_1^n), X_2^n(U_2^n), Y^n) \in \mathcal{T}_\epsilon^{(n)} \,|\, U_1^n = u_1^n\}.$$

Now conditioned on $\{U_1^n = u_1^n\}$, $(U_2^n, X_1^n(\tilde{u}_1^n), X_2^n(U_2^n), Y^n) \sim p(u_2^n, x_2^n, y^n | u_1^n)p(x_1^n | \tilde{u}_1^n) = \prod_{i=1}^{n} p_{U_2,X_2,Y|U_1}(u_{2i}, x_{2i}, y_i | u_{1i})p_{X_1|U_1}(x_{1i} | \tilde{u}_{1i})$ for all $\tilde{u}_1^n \neq u_1^n$. Thus

$$P(\mathcal{E}_2) \leq \sum_{u_1^n} p(u_1^n) \sum_{\substack{\tilde{u}_1^n \neq u_1^n \\ (\tilde{u}_1^n, u_2^n, x_1^n, x_2^n, y^n) \in T_\epsilon^{(n)}}} p(u_2^n, x_2^n, y^n | u_1^n)p(x_1^n | \tilde{u}_1^n)$$

$$= \sum_{(\tilde{u}_1^n, u_2^n, x_1^n, x_2^n, y^n) \in T_\epsilon^{(n)}} \sum_{u_1^n \neq \tilde{u}_1^n} p(u_1^n, u_2^n, x_2^n, y^n)p(x_1^n | \tilde{u}_1^n)$$

$$\leq \sum_{(\tilde{u}_1^n, u_2^n, x_1^n, x_2^n, y^n) \in T_\epsilon^{(n)}} \sum_{u_1^n} p(u_1^n, u_2^n, x_2^n, y^n)p(x_1^n | \tilde{u}_1^n)$$

$$= \sum_{(\tilde{u}_1^n, u_2^n, x_1^n, x_2^n, y^n) \in T_\epsilon^{(n)}} p(u_2^n, x_2^n, y^n)p(x_1^n | \tilde{u}_1^n)$$

$$= \sum_{(u_2^n, x_2^n, y^n) \in T_\epsilon^{(n)}} p(u_2^n, x_2^n, y^n) \sum_{(\tilde{u}_1^n, x_1^n) \in T_\epsilon^{(n)}(U_1, X_1 | u_2^n, x_2^n, y^n)} p(x_1^n | \tilde{u}_1^n)$$

$$\leq \sum_{(\tilde{u}_1^n, x_1^n) \in T_\epsilon^{(n)}(U_1, X_1 | u_2^n, x_2^n, y^n)} p(x_1^n | \tilde{u}_1^n)$$

$$\leq 2^{n(H(U_1, X_1 | U_2, X_2, Y) - H(X_1 | U_1) + 2\delta(\epsilon))}.$$

Collecting the entropy terms, we have

$$H(U_1, X_1 | U_2, X_2, Y) - H(X_1 | U_1)$$
$$= H(U_1, X_1 | U_2, X_2, Y) - H(U_1, X_1 | U_2, X_2) - H(X_1 | U_1) + H(U_1, X_1 | U_2, X_2)$$
$$\overset{(a)}{=} -I(U_1, X_1; Y | U_2, X_2) + H(U_1 | U_2)$$
$$\overset{(b)}{=} -I(X_1; Y | U_2, X_2) + H(U_1 | U_2),$$

where (a) follows since $X_1 \rightarrow U_1 \rightarrow U_2 \rightarrow X_2$ form a Markov chain and (b) follows since $(U_1, U_2) \rightarrow (X_1, X_2) \rightarrow Y$ form a Markov chain. Thus $P(\mathcal{E}_2)$ tends to zero as $n \rightarrow \infty$ if $H(U_1 | U_2) < I(X_1; Y | U_2, X_2) - 2\delta(\epsilon)$. Similarly, $P(\mathcal{E}_3)$ and $P(\mathcal{E}_4)$ tend to zero as $n \rightarrow \infty$ if $H(U_2 | U_1) < I(X_2; Y | U_1, X_1) - 2\delta(\epsilon)$ and $H(U_1, U_2) < I(X_1, X_2; Y) - 3\delta(\epsilon)$. This completes the proof of Theorem 14.1.

Suboptimality of the coding scheme. The coding scheme used in the above proof is not optimal in general. Suppose $U_1 = U_2 = U$. Then Theorem 14.1 reduces to the sufficient condition

$$H(U) < \max_{p(q)p(x_1|q,u)p(x_2|q,u)} I(X_1, X_2; Y | Q)$$

$$= \max_{p(x_1|u)p(x_2|u)} I(X_1, X_2; Y). \tag{14.2}$$

However, since both senders observe the same source, they can first encode the source losslessly at rate $H(U)$ and then transmit the source description using cooperative channel coding; see Problem 4.9. Thus, the source can be communicated losslessly if

$$H(U) < \max_{p(x_1, x_2)} I(X_1, X_2; Y),$$

which is a less stringent condition than that in (14.2). Hence, when U_1 and U_2 have a *common part*, we can improve upon the joint source–channel coding scheme for Theorem 14.1. In the following subsection, we formally define the common part between two correlated sources. Subsequently, we present separate and joint source–channel coding schemes that incorporate this common part.

14.1.3 Common Part of a 2-DMS

Let (U_1, U_2) be a pair of random variables. Arrange $p(u_1, u_2)$ in a block diagonal form with the maximum possible number k of nonzero blocks, as shown in Figure 14.3. The *common part* between U_1 and U_2 is the random variable U_0 that takes the value u_0 if (U_1, U_2) is in block $u_0 \in [1 : k]$. Note that U_0 can be determined by U_1 or U_2 alone.

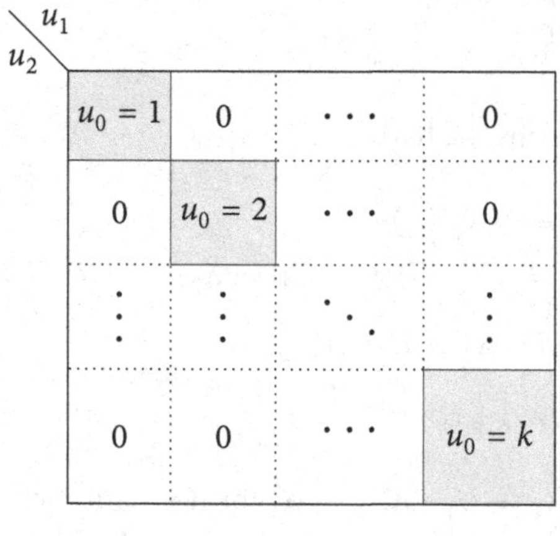

Figure 14.3. Block diagonal arrangement of the joint pmf $p(u_1, u_2)$.

Formally, let $g_1 : \mathcal{U}_1 \to [1 : k]$ and $g_2 : \mathcal{U}_2 \to [1 : k]$ be two functions with the largest integer k such that $P\{g_1(U_1) = u_0\} > 0, P\{g_2(U_2) = u_0\} > 0$ for $u_0 \in [1 : k]$ and $P\{g_1(U_1) = g_2(U_2)\} = 1$. The common part between U_1 and U_2 is defined as $U_0 = g_1(U_1) = g_2(U_2)$, which is unique up to relabeling of the symbols.

To better understand this definition, consider the following.

Example 14.4. Let (U_1, U_2) be a pair of random variables with the joint pmf in Table 14.1. Here $k = 2$ and the common part U_0 has the pmf $p_{U_0}(1) = 0.7$ and $p_{U_0}(2) = 0.3$.

Now let (U_1, U_2) be a 2-DMS. What is the common part between the sequences U_1^n and U_2^n? It turns out that this common part is always U_0^n (up to relabeling). Thus we say that U_0 is the common part of the 2-DMS (U_1, U_2).

	$u_0 = 1$		$u_0 = 2$	
u_2 \ u_1	1	2	3	4
$u_0 = 1$ 1	0.1	0.2	0	0
$u_0 = 1$ 2	0.1	0.1	0	0
3	0.1	0.1	0	0
$u_0 = 2$ 4	0	0	0.2	0.1

Table 14.1. Joint pmf for Example 14.4.

14.1.4   Three-Index Separate Source and Channel Coding Scheme

Taking the common part into consideration, we can generalize the 2-index separate source and channel coding scheme discussed earlier in this section into a 3-index scheme. Source coding is performed by encoding U_1^n into an index pair (M_0, M_1) and U_2^n into an index pair (M_0, M_2) such that (U_1^n, U_2^n) can be losslessly recovered from the index triple (M_0, M_1, M_2) as depicted in Figure 14.4.

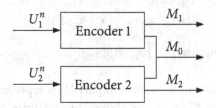

Figure 14.4. Source encoding setup for the 3-index separate source and channel coding scheme. The 2-DMS can be losslessly recovered from (M_0, M_1, M_2).

Since M_0 must be a function only of U_0^n, it can be easily shown that the optimal rate region $\mathscr{R}^*$ is the set of rate triples (R_0, R_1, R_2) such that

$$
\begin{aligned}
R_1 &\geq H(U_1 | U_2), \\
R_2 &\geq H(U_2 | U_1), \\
R_1 + R_2 &\geq H(U_1, U_2 | U_0), \\
R_0 + R_1 + R_2 &\geq H(U_1, U_2).
\end{aligned}
\tag{14.3}
$$

At the same time, the capacity region $\mathscr{C}$ for a DM-MAC $p(y|x_1, x_2)$ with a common message (see Problem 5.19) is the set of rate triples (R_0, R_1, R_2) such that

$$
\begin{aligned}
R_1 &\leq I(X_1; Y | X_2, W), \\
R_2 &\leq I(X_2; Y | X_1, W),
\end{aligned}
$$

$$R_1 + R_2 \leq I(X_1, X_2; Y | W),$$
$$R_0 + R_1 + R_2 \leq I(X_1, X_2; Y)$$

for some pmf $p(w)p(x_1|w)p(x_2|w)$, where $|\mathcal{W}| \leq \min\{|\mathcal{X}_1| \cdot |\mathcal{X}_2| + 2, |\mathcal{Y}| + 3\}$. Hence, if the intersection of the interiors of $\mathcal{R}^*$ and $\mathcal{C}$ is not empty, separate source and channel coding using three indices can be used to communicate the 2-DMS losslessly over the DM-MAC. Note that this coding scheme is not optimal in general as already shown in Example 14.3.

14.1.5   A Joint Source–Channel Coding Scheme with Common Part

By generalizing the coding schemes in Sections 14.1.1 and 14.1.4, we obtain the following sufficient condition for lossless communication of a 2-DMS over a DM-MAC.

Theorem 14.2. A 2-DMS (U_1, U_2) with common part U_0 can be communicated losslessly over a DM-MAC $p(y|x_1, x_2)$ if

$$H(U_1 | U_2) < I(X_1; Y | X_2, U_2, W),$$
$$H(U_2 | U_1) < I(X_2; Y | X_1, U_1, W),$$
$$H(U_1, U_2 | U_0) < I(X_1, X_2; Y | U_0, W),$$
$$H(U_1, U_2) < I(X_1, X_2; Y)$$

for some conditional pmf $p(w)p(x_1|u_1, w)p(x_2|u_2, w)$.

In this sufficient condition, the common part U_0 is represented by the independent auxiliary random variable W, which is chosen to maximize cooperation between the senders.

Remark 14.1. Although the auxiliary random variable W represents the common part U_0, there is no benefit in making it statistically correlated with U_0. This is a consequence of Shannon's source–channel separation theorem in Section 3.9.

Remark 14.2. The above sufficient condition does not change by introducing a time-sharing random variable Q.

Proof of Theorem 14.2 (outline). For each u_0^n, randomly and independently generate $w^n(u_0^n)$ according to $\prod_{i=1}^n p_W(w_i)$. For each (u_0^n, u_1^n), randomly and independently generate $x_1^n(u_0^n, u_1^n)$ according to $\prod_{i=1}^n p_{X_1|U_1, W}(x_{1i}|u_{1i}, w_i(u_0^n))$. Similarly, for (u_0^n, u_2^n), randomly and independently generate $x_2^n(u_0^n, u_2^n)$. The decoder declares $(\hat{u}_0^n, \hat{u}_1^n, \hat{u}_2^n)$ to be the estimate of the sources if it is the unique triple such that $(\hat{u}_0^n, \hat{u}_1^n, \hat{u}_2^n, w^n(\hat{u}_0^n), x_1^n(\hat{u}_0^n, \hat{u}_1^n),$ $x_2^n(\hat{u}_0^n, \hat{u}_2^n), y^n) \in \mathcal{T}_\epsilon^{(n)}$ (this automatically implies that $\hat{u}_0^n$ is the common part of $\hat{u}_1^n$ and $\hat{u}_2^n$). Following the steps in the proof of the previous coding scheme, it can be shown that the above inequalities are sufficient for the probability of error to tend to zero as $n \to \infty$.

Remark 14.3. The above coding scheme is not optimal in general either.

14.2   LOSSLESS COMMUNICATION OF A 2-DMS OVER A DM-BC

Now consider the broadcast communication system depicted in Figure 14.5, where a 2-DMS (U_1, U_2) is to be communicated losslessly over a 2-receiver DM-BC $p(y_1, y_2|x)$. The definitions of a code, probability of error, and lossless communication for this setup are along the same lines as those for the MAC case. As before, assume rates $r_1 = r_2 = 1$ symbol/transmission.

Since the private-message capacity region of the DM-BC is not known in general (see Chapter 8), the necessary and sufficient condition for lossless communication of a 2-DMS over a DM-BC is not known even when the sources are independent. We will show nevertheless that separation does not hold in general for sending a 2-DMS over a DM-BC.

Consider the following separate source and channel coding scheme for this setup. The encoder first assigns an index triple $(M_0, M_1, M_2) \in [1 : 2^{nR_0}] \times [1 : 2^{nR_1}] \times [1 : 2^{nR_2}]$ to the source sequence pair (U_1^n, U_2^n) such that U_1^n can be recovered losslessly from the pair of indices (M_0, M_1) and U_2^n can be recovered losslessly from the pair of indices (M_0, M_2). The encoder then transmits a codeword $x^n(M_0, M_1, M_2)$ from a channel codebook. Decoder 1 first decodes for (M_0, M_1) and then recovers U_1^n. Similarly decoder 2 first decodes for (M_0, M_2) and then recovers U_2^n. The source coding part of this scheme is discussed in the following subsection.

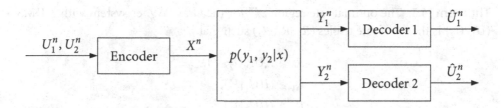

Figure 14.5. Communication of a 2-DMS over a 2-receiver DM-BC.

14.2.1   Gray–Wyner System

The Gray–Wyner system depicted in Figure 14.6 is a distributed lossless source coding setup in which a 2-DMS (U_1, U_2) is described by three encoders so that decoder 1, who receives the descriptions M_0 and M_1, can losslessly recover U_1^n and decoder 2, who receives the descriptions M_0 and M_2, can losslessly recover U_2^n. We wish to find the optimal rate region for this distributed lossless source coding setup.

A $(2^{nR_0}, 2^{nR_1}, 2^{nR_2}, n)$ code for the Gray–Wyner system consists of

- three encoders, where encoder $j = 0, 1, 2$ assigns the index $m_j(u_1^n, u_2^n) \in [1 : 2^{nR_j})$ to each sequence pair $(u_1^n, u_2^n) \in \mathcal{U}_1^n \times \mathcal{U}_2^n$, and

- two decoders, where decoder 1 assigns an estimate $\hat{u}_1^n(m_0, m_1)$ to each index pair

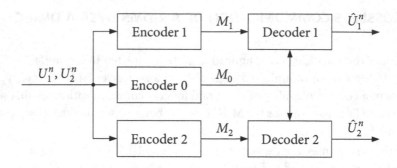

Figure 14.6. Gray–Wyner system.

$(m_0, m_1) \in [1 : 2^{nR_0}) \times [1 : 2^{nR_1})$ and decoder 2 assigns an estimate $\hat{u}_2^n(m_0, m_2)$ to each index pair $(m_0, m_2) \in [1 : 2^{nR_0}) \times [1 : 2^{nR_2})$.

The probability of error is defined as $P_e^{(n)} = \mathsf{P}\{(\hat{U}_1^n, \hat{U}_2^n) \neq (U_1^n, U_2^n)\}$. A rate triple (R_0, R_1, R_2) is said to be achievable if there exists a sequence of $(2^{nR_0}, 2^{nR_1}, 2^{nR_2}, n)$ codes such that $\lim_{n \to \infty} P_e^{(n)} = 0$. The optimal rate region $\mathscr{R}^*$ for the Gray–Wyner system is the closure of the set of achievable rate triples.

The optimal rate region for the Gray–Wyner system is given in the following.

Theorem 14.3. The optimal rate region $\mathscr{R}^*$ for the Gray–Wyner system with 2-DMS (U_1, U_2) is the set of rate triples (R_0, R_1, R_2) such that

$$R_0 \geq I(U_1, U_2; V),$$
$$R_1 \geq H(U_1 | V),$$
$$R_2 \geq H(U_2 | V)$$

for some conditional pmf $p(v|u_1, u_2)$ with $|\mathcal{V}| \leq |\mathcal{U}_1| \cdot |\mathcal{U}_2| + 2$.

The optimal rate region has the following extreme points:

- $R_0 = 0$: By taking $V = \emptyset$, the region reduces to $R_1 \geq H(U_1)$ and $R_2 \geq H(U_2)$.

- $R_1 = 0$: By taking $V = U_1$, the region reduces to $R_0 \geq H(U_1)$ and $R_2 \geq H(U_2 | U_1)$.

- $R_2 = 0$: By taking $V = U_2$, the region reduces to $R_0 \geq H(U_2)$ and $R_1 \geq H(U_1 | U_2)$.

- $(R_1, R_2) = (0, 0)$: By taking $V = (U_1, U_2)$, the region reduces to $R_0 \geq H(U_1, U_2)$.

Proof of Theorem 14.3. To prove achievability, we use joint typicality encoding to find a $v^n(m_0)$, $m_0 \in [1 : 2^{nR_0}]$, jointly typical with (u_1^n, u_2^n). The index m_0 is sent to both decoders. Given $v^n(m_0)$, we assign indices $m_1 \in [1 : 2^{nR_1}]$ and $m_2 \in [1 : 2^{nR_2}]$ to the sequences in $\mathcal{T}_\epsilon^{(n)}(U_1 | v^n(m_0))$ and $\mathcal{T}_\epsilon^{(n)}(U_2 | v^n(m_0))$, respectively, and send them to decoders 1 and 2, respectively. For the proof of the converse, we use standard arguments with the auxiliary random variable identification $V_i = (M_0, U_1^{i-1}, U_2^{i-1})$. The cardinality bound on $\mathcal{V}$ can be proved using the convex cover method in Appendix C.

14.2.2   Common Information

A rate triple (R_0, R_1, R_2) in the optimal rate region for the Gray–Wyner system must satisfy the inequalities

$$R_0 + R_1 \geq H(U_1),$$
$$R_0 + R_2 \geq H(U_2),$$
$$R_0 + R_1 + R_2 \geq H(U_1, U_2).$$

As seen from the extreme points above, the first two inequalities can be simultaneously tight while the third can be individually tight. Interestingly, the values of the common rate R_0 corresponding to these two cases lead to several notions of *common information*.

- **Gács–Körner–Witsenhausen common information.** When $R_0 + R_1 = H(U_1)$ and $R_0 + R_2 = H(U_2)$, the maximum common rate R_0 is the entropy $H(U_0)$ of the common part between U_1 and U_2 (as defined in Section 14.1.3), denoted by $K(U_1; U_2)$.

- **Mutual information.** The minimum sum-rate is $H(U_1, U_2)$. With no common rate, i.e., $R_0 = 0$, the minimum sum-rate jumps to $H(U_1) + H(U_2)$. The difference between these two sum-rates is the mutual information $I(U_1; U_2)$ and represents the value of having a common link.

- **Wyner's common information.** When $R_0 + R_1 + R_2 = H(U_1, U_2)$, the minimum common rate R_0 is

$$J(U_1; U_2) = \min I(U_1, U_2; V), \tag{14.4}$$

where the minimum is over all conditional pmfs $p(v|u_1, u_2)$ with $|\mathcal{V}| \leq |\mathcal{U}_1| \cdot |\mathcal{U}_2|$ such that $I(U_1; U_2|V) = 0$, i.e., $U_1 \to V \to U_2$. Recall that this Markov structure appeared in the converse proofs for the quadratic Gaussian distributed source coding and multiple description coding problems in Sections 12.3 and 13.4, respectively.

The above three quantities represent common information between the random variables U_1 and U_2 in different contexts. The Gács–Körner–Witsenhausen common information $K(X; Y)$ captures the amount of common randomness that can be extracted by knowing U_1 and U_2 separately. In comparison, Wyner's common information captures the amount of common randomness that is needed to generate U_1 and U_2 separately. Mutual information, as we have seen in the Slepian–Wolf theorem, captures the amount of information about U_1 provided by observing U_2 and vice versa.

In general, it can be easily shown that

$$0 \leq K(U_1; U_2) \leq I(U_1; U_2) \leq J(U_1; U_2) \leq H(U_1, U_2), \tag{14.5}$$

and these inequalities can be strict. Furthermore, $K(U_1; U_2) = I(U_1; U_2) = J(U_1; U_2)$ iff $U_1 = (V, V_1)$ and $U_2 = (V, V_2)$ for some pmf $p(v_1)p(v|v_1)p(v_2|v)$.

Example 14.5. Let (U_1, U_2) be a DSBS(p), $p \in [0, 1/2]$. Then it can be easily shown that $J(U_1; U_2) = 1 + H(p) - 2H(\alpha)$, where $\alpha \star \alpha = p$. The minimum in (14.4) is attained by setting $V \sim \text{Bern}(1/2)$, $V_1 \sim \text{Bern}(\alpha)$, and $V_2 \sim \text{Bern}(\alpha)$ to be mutually independent and $U_j = V \oplus V_j$, $j = 1, 2$.

Example 14.6. Let (U_1, U_2) be binary with $p(0, 0) = p(0, 1) = p(1, 1) = 1/3$. Then it can be shown that $J(U_1; U_2) = 2/3$, which is attained by setting $V \sim \text{Bern}(1/2)$, and $U_1 = 0$, $U_2 \sim \text{Bern}(2/3)$ if $V = 0$, and $U_1 \sim \text{Bern}(1/3)$, $U_2 = 1$ if $V = 1$.

14.2.3 A Separate Source–Channel Coding Scheme

We return to the discussion on sending a 2-DMS over a DM-BC using separate source and channel coding. Recall that Marton's inner bound in Section 8.4 is the best-known inner bound on the capacity region of the DM-BC. Denote this inner bound as $\mathscr{R} \subseteq \mathscr{C}$. Then, a 2-DMS can be communicated losslessly over a DM-BC if the intersection of the interiors of Marton's inner bound $\mathscr{R}$ and the optimal rate region $\mathscr{R}^*$ for the Gray–Wyner system is not empty, that is, if

$$I(U_1, U_2; V) + H(U_1|V) < I(W_0, W_1; Y_1),$$
$$I(U_1, U_2; V) + H(U_2|V) < I(W_0, W_2; Y_2),$$
$$I(U_1, U_2; V) + H(U_1|V) + H(U_2|V) < I(W_0, W_1; Y_1) + I(W_2; Y_2|W_0) - I(W_1; W_2|W_0),$$
$$I(U_1, U_2; V) + H(U_1|V) + H(U_2|V) < I(W_1; Y_1|W_0) + I(W_0, W_2; Y_2) - I(W_1; W_2|W_0),$$
$$2I(U_1, U_2; V) + H(U_1|V) + H(U_2|V) < I(W_0, W_1; Y_1) + I(W_0, W_2; Y_2) - I(W_1; W_2|W_0)$$

$$(14.6)$$

for some pmfs $p(v|u_1, u_2)$ and $p(w_0, w_1, w_2)$, and function $x(w_0, w_1, x_2)$. This separate source–channel coding scheme is optimal for some classes of sources and channels.

- More capable BC: Suppose that Y_1 is more capable than Y_2, i.e., $I(X; Y_1) \geq I(X; Y_2)$ for all $p(x)$. Then the 2-DMS (U_1, U_2) can be communicated losslessly if

$$H(U_1, U_2) < I(X; Y_1),$$
$$H(U_1, U_2) < I(X; Y_1|W) + I(W; Y_2),$$
$$H(U_2) < I(W; Y_2)$$

  for some pmf $p(w, x)$.

- Nested sources: Suppose that $U_1 = (V_1, V_2)$ and $U_2 = V_2$ for some $(V_1, V_2) \sim p(v_1, v_2)$. Then the 2-DMS (U_1, U_2) can be communicated losslessly if

$$H(V_1, V_2) = H(U_1) < I(X; Y_1),$$
$$H(V_1, V_2) = H(U_1) < I(X; Y_1|W) + I(W; Y_2),$$
$$H(V_2) = H(U_2) < I(W; Y_2)$$

  for some pmf $p(w, x)$.

In both cases, achievability follows by representing (U_1^n, U_2^n) by a message pair (M_1, M_2) at rates $R_2 = H(U_2)$ and $R_1 = H(U_1|U_2)$, respectively, and using superposition coding. The converse proofs are essentially the same as the converse proofs for the more capable BC and degraded message sets BC, respectively.

Source–channel separation is not optimal in general, however, as demonstrated in the following.

Example 14.7. Consider the 2-DMS (U_1, U_2) with $\mathcal{U}_1 = \mathcal{U}_2 = \{0, 1\}$ and $p_{U_1, U_2}(0, 0) = p_{U_1, U_2}(0, 1) = p_{U_1, U_2}(1, 1) = 1/3$ and the Blackwell channel in Example 8.2 defined by $\mathcal{X} = \{0, 1, 2\}$, $\mathcal{Y}_1 = \mathcal{Y}_2 = \{0, 1\}$, and $p_{Y_1, Y_2|X}(0, 0|0) = p_{Y_1, Y_2|X}(0, 1|1) = p_{Y_1, Y_2|X}(1, 1|2) = 1$. The capacity region of this channel is contained in the set of rate triples (R_0, R_1, R_2) such that

$$R_0 + R_1 \le 1,$$
$$R_0 + R_2 \le 1,$$
$$R_0 + R_1 + R_2 \le \log 3.$$

However, as we found in Example 14.6, the sources require $R_0 \ge J(U_1; U_2) = 2/3$ when $R_0 + R_1 + R_2 = \log 3$, or equivalently, $2R_0 + R_1 + R_2 \ge \log 3 + 2/3 = 2.252$, which implies that $R_0 + R_1 \ge 1.126$ or $R_0 + R_2 \ge 1.126$.

Hence, the intersection of the optimal rate region $\mathcal{R}^*$ for the Gray–Wyner system and the capacity region $\mathcal{C}$ is empty and this 2-DMS cannot be communicated losslessly over the Blackwell channel using separate source and channel coding.

By contrast, setting $X = U_1 + U_2$ achieves error-free transmission since Y_1 and Y_2 uniquely determine U_1 and U_2, respectively. Thus joint source–channel coding can strictly outperform separate source and channel coding for sending a 2-DMS over a DM-BC.

14.2.4   A Joint Source–Channel Coding Scheme

We describe a general joint source–channel coding scheme that improves upon separate Gray–Wyner source coding and Marton's channel coding.

Theorem 14.4. A 2-DMS (U_1, U_2) can be communicated losslessly over a DM-BC $p(y_1, y_2|x)$ if

$$H(U_1|U_2) < I(U_1, W_0, W_1; Y_1) - I(U_1, W_0, W_1; U_2),$$
$$H(U_2|U_1) < I(U_2, W_0, W_2; Y_2) - I(U_2, W_0, W_2; U_1),$$
$$H(U_1, U_2) < I(U_1, W_0, W_1; Y_1) + I(U_2, W_2; Y_2|W_0) - I(U_1, W_1; U_2, W_2|W_0),$$
$$H(U_1, U_2) < I(U_1, W_1; Y_1|W_0) + I(U_2, W_0, W_2; Y_2) - I(U_1, W_1; U_2, W_2|W_0),$$
$$H(U_1, U_2) < I(U_1, W_0, W_1; Y_1) + I(U_2, W_0, W_2; Y_2) - I(U_1, W_1; U_2, W_2|W_0)$$
$$\qquad - I(U_1, U_2; W_0)$$

for some conditional pmf $p(w_0, w_1, w_2|u_1, u_2)$ and function $x(u_1, u_2, w_0, w_1, w_2)$.

This theorem recovers the following as special cases:

- Separate source and channel coding: We set $p(w_0, w_1, w_2 | u_0, u_1) = p(w_0, w_1, w_2)$ and $x(u_1, u_2, w_0, w_1, w_2) = x(w_0, w_1, w_2)$, i.e., (W_0, W_1, W_2, X) is independent of (U_1, U_2). Then, the set of inequalities in the theorem simplifies to (14.6).

- Example 14.7: Set $W_0 = \emptyset$, $W_1 = U_1$, $W_2 = U_2$, $X = U_1 + U_2$.

Remark 14.4. The sufficient condition in Theorem 14.4 does not improve by time sharing.

14.2.5   Proof of Theorem 14.4

Codebook generation.   Fix a conditional pmf $p(w_0, w_1, w_2 | u_1, u_2)$ and function $x(u_1, u_2, w_0, w_1, w_2)$. Randomly and independently generate 2^{nR_0} sequences $w_0^n(m_0)$, $m_0 \in [1 : 2^{nR_0}]$, each according to $\prod_{i=1}^{n} p_{W_0}(w_{0i})$. For each $u_1^n \in \mathcal{U}_1^n$ and $m_0 \in [1 : 2^{nR_0}]$, randomly and independently generate 2^{nR_1} sequences $w_1^n(u_1^n, m_0, m_1)$, $m_1 \in [1 : 2^{nR_1}]$, each according to $\prod_{i=1}^{n} p_{W_1 | U_1, W_0}(w_{1i} | u_{1i}, w_{0i}(m_0))$. Similarly, for each $u_2^n \in \mathcal{U}_2^n$ and $m_0 \in [1 : 2^{nR_0}]$, randomly and independently generate 2^{nR_2} sequences $w_2^n(u_2^n, m_0, m_2)$, $m_2 \in [1 : 2^{nR_2}]$, each according to $\prod_{i=1}^{n} p_{W_2 | U_2, W_0}(w_{2i} | u_{2i}, w_{0i}(m_0))$.

Encoding.   For each sequence pair (u_1^n, u_2^n), choose a triple $(m_0, m_1, m_2) \in [1 : 2^{nR_0}] \times [1 : 2^{nR_1}] \times [1 : 2^{nR_2}]$ such that $(u_1^n, u_2^n, w_0^n(m_0), w_1^n(u_1^n, m_0, m_1), w_2^n(u_2^n, m_0, m_2)) \in \mathcal{T}_{\epsilon'}^{(n)}$. If there is no such triple, choose $(m_0, m_1, m_2) = (1, 1, 1)$. Then the encoder transmits $x_i = x(u_{1i}, u_{2i}, w_{0i}(m_0), w_{1i}(u_1^n, m_0, m_1), w_{2i}(u_2^n, m_0, m_2))$ for $i \in [1 : n]$.

Decoding.   Let $\epsilon > \epsilon'$. Decoder 1 declares $\hat{u}_1^n$ to be the estimate of u_1^n if it is the unique sequence such that $(\hat{u}_1^n, w_0^n(m_0), w_1^n(\hat{u}_1^n, m_0, m_1), y_1^n) \in \mathcal{T}_\epsilon^{(n)}$ for some $(m_0, m_1) \in [1 : 2^{nR_0}] \times [1 : 2^{nR_1}]$. Similarly, decoder 2 declares $\hat{u}_2^n$ to be the estimate of u_2^n if it is the unique sequence such that $(\hat{u}_2^n, w_0^n(m_0), w_2^n(\hat{u}_2^n, m_0, m_2), y_2^n) \in \mathcal{T}_\epsilon^{(n)}$ for some $(m_0, m_2) \in [1 : 2^{nR_0}] \times [1 : 2^{nR_2}]$.

Analysis of the probability of error.   Assume (M_0, M_1, M_2) is selected at the encoder. Then decoder 1 makes an error only if one or more of the following events occur:

$$\mathcal{E}_0 = \{(U_1^n, U_2^n, W_0^n(m_0), W_1^n(U_1^n, m_0, m_1), W_2^n(U_2^n, m_0, m_2)) \notin \mathcal{T}_{\epsilon'}^{(n)}$$
$$\text{for all } m_0, m_1, m_2\},$$

$$\mathcal{E}_{11} = \{(U_1^n, W_0^n(M_0), W_1^n(U_1^n, M_0, M_1), Y_1^n) \notin \mathcal{T}_\epsilon^{(n)}\},$$

$$\mathcal{E}_{12} = \{(\tilde{u}_1^n, W_0^n(M_0), W_1^n(\tilde{u}_1^n, M_0, m_1), Y_1^n) \in \mathcal{T}_\epsilon^{(n)} \text{ for some } \tilde{u}_1^n \neq U_1^n, m_1\},$$

$$\mathcal{E}_{13} = \{(\tilde{u}_1^n, W_0^n(m_0), W_1^n(\tilde{u}_1^n, m_0, m_1), Y_1^n) \in \mathcal{T}_\epsilon^{(n)} \text{ for some } \tilde{u}_1^n \neq U_1^n, m_0 \neq M_0, m_1\}.$$

Thus the probability of error $\mathsf{P}(\mathcal{E}_1)$ for decoder 1 is upper bounded as

$$\mathsf{P}(\mathcal{E}_1) \le \mathsf{P}(\mathcal{E}_0) + \mathsf{P}(\mathcal{E}_0^c \cap \mathcal{E}_{11}) + \mathsf{P}(\mathcal{E}_{12}) + \mathsf{P}(\mathcal{E}_{13}).$$

The first term tends to zero by the following variant of the multivariate covering lemma in Section 8.6.

Lemma 14.1. The probability $P(\mathcal{E}_0)$ tends to zero as $n \to \infty$ if

$$R_0 > I(U_1, U_2; W_0) + \delta(\epsilon'),$$

$$R_0 + R_1 > I(U_1, U_2; W_0) + I(U_2; W_1 | U_1, W_0) + \delta(\epsilon'),$$

$$R_0 + R_2 > I(U_1, U_2; W_0) + I(U_1; W_2 | U_2, W_0) + \delta(\epsilon'),$$

$$R_0 + R_1 + R_2 > I(U_1, U_2; W_0) + I(U_2; W_1 | U_1, W_0) + I(U_1, W_1; W_2 | U_2, W_0) + \delta(\epsilon').$$

The proof of this lemma is given in Appendix 14A.

By the conditional typicality lemma, $P(\mathcal{E}_0^c \cap \mathcal{E}_{11})$ tends to zero as $n \to \infty$. Following steps similar to the DM-MAC joint source–channel coding, it can be shown that $P(\mathcal{E}_{12})$ tends to zero as $n \to \infty$ if $H(U_1) + R_1 < I(U_1, W_1; Y_1 | W_0) + I(U_1; W_0) - \delta(\epsilon)$, and $P(\mathcal{E}_{13})$ tends to zero as $n \to \infty$ if $H(U_1) + R_0 + R_1 < I(U_1, W_0, W_1; Y_1) + I(U_1; W_0) - \delta(\epsilon)$.

Similarly, the probability of error for decoder 2 tends to zero as $n \to \infty$ if $H(U_2) + R_2 < I(U_2, W_2; Y_2 | W_0) + I(U_2; W_0) - \delta(\epsilon)$ and $H(U_2) + R_0 + R_2 < I(U_2, W_0, W_2; Y_2) + I(U_2; W_0) - \delta(\epsilon)$. The rest of the proof follows by combining the above inequalities and eliminating (R_0, R_1, R_2) by the Fourier–Motzkin procedure in Appendix D.

14.3 A GENERAL SINGLE-HOP NETWORK

We end our discussion of single-hop networks with a general network model that includes many of the setups we studied in previous chapters. Consider the 2-sender 2-receiver communication system with general source transmission demand depicted in Figure 14.7. Let (U_1, U_2) be a 2-DMS with common part U_0, $p(y_1, y_2 | x_1, x_2)$ be a DM single-hop network, and $d_{11}(u_1, \hat{u}_{11})$, $d_{12}(u_1, \hat{u}_{12})$, $d_{21}(u_2, \hat{u}_{21})$, $d_{22}(u_2, \hat{u}_{22})$ be four distortion measures. For simplicity, assume transmission rates $r_1 = r_2 = 1$ symbol/transmission. Sender 1 observes the source sequence U_1^n and sender 2 observes the source sequence U_2^n. Receiver 1 wishes to reconstruct (U_1^n, U_2^n) with distortions (D_{11}, D_{21}) and receiver 2 wishes to reconstruct (U_1^n, U_2^n) with distortions (D_{12}, D_{22}). We wish to determine the necessary and sufficient condition for sending the sources within prescribed distortions.

This general network includes the following special cases we discussed earlier.

- Lossless communication of a 2-DMS over a DM-MAC: Assume that $Y_2 = \emptyset$, d_{11} and d_{21} are Hamming distortion measures, and $D_{11} = D_{21} = 0$. As we have seen, this setup

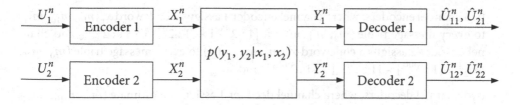

Figure 14.7. A general single-hop communication network.

in turn includes as special cases communication of independent and common messages over a DM-MAC in Problem 5.19 and distributed lossless source coding in Chapter 10.

- Lossy communication of a 2-DMS over a DM-MAC: Assume $Y_2 = \emptyset$ and relabel d_{11} as d_1 and d_{21} as d_2. This setup includes distributed lossy source coding discussed in Chapter 12 as a special case.

- Lossless communication of a 2-DMS over a DM-BC: Assume that $X_2 = U_2 = \emptyset$, $U_1 = (V_1, V_2)$, d_{11} and d_{21} are Hamming distortion measures on V_1 and V_2, respectively, and $D_{11} = D_{21} = 0$. As we have seen, this setup includes sending private and common messages over a DM-BC in Chapter 8 and the Gray–Wyner system in Section 14.2.1 as special cases.

- Lossy communication of a 2-DMS over a DM-BC: Assume that $X_2 = U_2 = \emptyset$, and relabel d_{11} as d_1 and d_{21} as d_2. This setup includes several special cases of the multiple-description coding problem in Chapter 13 such as successive refinement.

- Interference channel: Assume that U_1 and U_2 are independent, d_{11} and d_{22} are Hamming distortion measures, and $D_{11} = D_{22} = 0$. This yields the DM-IC in Chapter 6.

14.3.1   Separate Source and Channel Coding Scheme

We define separate source and channel coding for this general single-hop network as follows. A $(2^{nR_0}, 2^{nR_{10}}, 2^{nR_{11}}, 2^{nR_{20}}, 2^{nR_{22}}, n)$ source code consists of

- two source encoders, where source encoder 1 assigns an index triple $(m_0, m_{10}, m_{11}) \in [1 : 2^{nR_0}] \times [1 : 2^{nR_{10}}] \times [1 : 2^{nR_{11}}]$ to every u_1^n and source encoder 2 assigns an index triple $(m_0, m_{20}, m_{22}) \in [1 : 2^{nR_0}] \times [1 : 2^{nR_{20}}] \times [1 : 2^{nR_{22}}]$ to every u_2^n (here m_0 is a common index that is a function only of the common part u_0^n), and

- two source decoders, where source decoder 1 assigns an estimate $(\hat{u}_{11}^n, \hat{u}_{21}^n)$ to every index quadruple $(m_0, m_{10}, m_{11}, m_{20})$ and source decoder 2 assigns an estimate $(\hat{u}_{12}^n, \hat{u}_{22}^n)$ to every index quadruple $(m_0, m_{10}, m_{20}, m_{22})$.

Achievability and the rate–distortion region $\mathcal{R}(D_{11}, D_{12}, D_{21}, D_{22})$ are defined as for other lossy source coding problems. A $(2^{nR_0}, 2^{nR_{10}}, 2^{nR_{11}}, 2^{nR_{20}}, 2^{nR_{22}}, n)$ channel code consists of

- five message sets $[1 : 2^{nR_0}]$, $[1 : 2^{nR_{10}}]$, $[1 : 2^{nR_{11}}]$, $[1 : 2^{nR_{20}}]$, and $[1 : 2^{nR_{22}}]$,

- two channel encoders, where channel encoder 1 assigns a codeword $x_1^n(m_0, m_{10}, m_{11})$ to every message triple $(m_0, m_{10}, m_{11}) \in [1 : 2^{nR_0}] \times [1 : 2^{nR_{10}}] \times [1 : 2^{nR_{11}}]$ and channel encoder 2 assigns a codeword $x_2^n(m_0, m_{20}, m_{22})$ to every message triple $(m_0, m_{20}, m_{22}) \in [1 : 2^{nR_0}] \times [1 : 2^{nR_{20}}] \times [1 : 2^{nR_{22}}]$, and

- two channel decoders, where channel decoder 1 assigns an estimate $(\hat{m}_{01}, \hat{m}_{101}, \hat{m}_{11}, \hat{m}_{201})$ to every received sequence y_1^n and channel decoder 2 assigns an estimate $(\hat{m}_{02}, \hat{m}_{102}, \hat{m}_{202}, \hat{m}_{22})$ to every received sequence y_2^n.

The average probability of error, achievability, and the capacity region $\mathscr{C}$ are defined as for other channel coding settings.

The sources can be communicated over the channel with distortion quadruple $(D_{11}, D_{12}, D_{21}, D_{22})$ using separate source and channel coding if the intersection of the interiors of $\mathscr{R}(D_{11}, D_{12}, D_{21}, D_{22})$ and $\mathscr{C}$ is nonempty. As we have already seen, source–channel separation does not hold in general, that is, there are cases where this intersection is empty, yet the sources can be still communicated over the channel as specified.

14.3.2*   A Hybrid Source–Channel Coding Scheme

In separate source and channel coding, channel codewords are conditionally independent of the source sequences given the descriptions (indices). Hence, the correlation between the sources is not utilized in channel coding. The hybrid source–channel coding scheme we discuss here captures this correlation in channel coding, while utilizing known lossy source coding and channel coding schemes. Each sender first performs source encoding on its source sequences. It then maps the resulting codewords and the source sequence symbol-by-symbol into a channel input sequence and transmits it. Each receiver performs channel decoding for the codewords generated through source encoding and then maps the codeword estimates and the received sequence symbol-by-symbol into reconstructions of the desired source sequences.

For simplicity of presentation, we describe this scheme only for the special case of lossy communication of a 2-DMS over a DM-MAC.

Proposition 14.1. Let (U_1, U_2) be a 2-DMS and $d_1(u_1, \hat{u}_1)$, $d_2(u_2, \hat{u}_2)$ be two distortion measures. The 2-DMS (U_1, U_2) can be communicated over a DM-MAC $p(y|x_1, x_2)$ with distortion pair (D_1, D_2) if

$$I(U_1; V_1|Q) < I(V_1; Y, V_2|Q),$$
$$I(U_2; V_2|Q) < I(V_2; Y, V_1|Q),$$
$$I(U_1; V_1|Q) + I(U_2; V_2|Q) < I(V_1, V_2; Y|Q) + I(V_1; V_2|Q)$$

for some conditional pmf $p(q, v_1, v_2|u_1, u_2) = p(q)p(v_1|u_1, q)p(v_2|u_2, q)$ and functions $x_1(u_1, v_1, q), x_2(u_2, v_2, q), \hat{u}_1(v_1, v_2, y, q)$, and $\hat{u}_2(v_1, v_2, y, q)$ such that $\mathsf{E}(d_j(U_j, \hat{U}_j)) \leq D_j, j = 1, 2$.

Proof outline. The coding scheme used to prove this proposition is depicted in Figure 14.8. For simplicity, let $Q = \emptyset$. Fix a conditional pmf $p(v_1|u_1)p(v_2|u_2)$ and functions $x_1(u_1, v_1), x_2(u_2, v_2), \hat{u}_1(v_1, v_2, y)$, and $\hat{u}_2(v_1, v_2, y)$. For $j = 1, 2$, randomly and independently generate 2^{nR_j} sequences $v_j^n(m_j), m_j \in [1 : 2^{nR_j}]$, each according to $\prod_{i=1}^{n} p_{V_j}(v_{ji})$. Given u_j^n, encoder $j = 1, 2$ finds an index $m_j \in [1 : 2^{nR_j}]$ such that $(u_j^n, v_j^n(m_j)) \in \mathcal{T}_{\epsilon'}^{(n)}$. By the covering lemma, the probability of error for this joint typicality encoding step tends

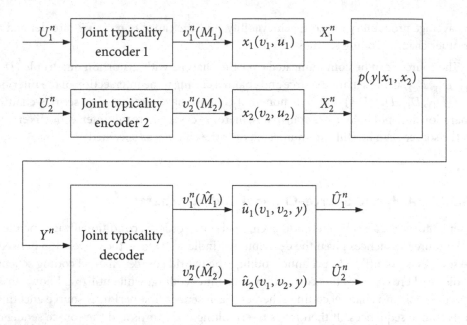

Figure 14.8. Hybrid source-channel coding for communicating a 2-DMS over a DM-MAC.

to zero as $n \to \infty$ if

$$R_1 > I(U_1; V_1) + \delta(\epsilon'),$$
$$R_2 > I(U_2; V_2) + \delta(\epsilon').$$

Encoder $j = 1, 2$ then transmits $x_{ji} = x_j(u_{ji}, v_{ji}(m_j))$ for $i \in [1 : n]$. Upon receiving y^n, the decoder finds the unique index pair $(\hat{m}_1, \hat{m}_2)$ such that $(v_1^n(\hat{m}_1), v_2^n(\hat{m}_2), y^n) \in T_\epsilon^{(n)}$. It can be shown that the probability of error for this joint typicality decoding step tends to zero as $n \to \infty$ if

$$R_1 < I(V_1; Y, V_2) - \delta(\epsilon),$$
$$R_2 < I(V_2; Y, V_1) - \delta(\epsilon),$$
$$R_1 + R_2 < I(V_1, V_2; Y) + I(V_1; V_2) - \delta(\epsilon).$$

The decoder then sets the reconstruction sequences as $\hat{u}_{ji} = \hat{u}_j(v_{1i}(\hat{m}_1), v_{2i}(\hat{m}_2), y_i)$, $i \in [1 : n]$, for $j = 1, 2$. Eliminating R_1 and R_2 and following similar arguments to the achievability proof for distributed lossy source coding completes the proof.

Remark 14.5. By setting $V_j = (U_j, X_j)$ and $\hat{U}_j = U_j$, $j = 1, 2$, Proposition 14.1 reduces to Theorem 14.1.

Remark 14.6. Due to the dependence between the codebook $\{U_j^n(m_j): m_j \in [1 : 2^{nR_j}]\}$ and the index M_j, $j = 1, 2$, the analysis of the probability error for joint typicality decoding requires nontrivial extensions of the packing lemma and the proof of achievability for the DM-MAC.

Remark 14.7. This hybrid coding scheme can be readily extended to the case of sources with a common part. It can be extended also to the general single-hop network depicted in Figure 14.7 by utilizing the source coding and channel coding schemes discussed in previous chapters.

SUMMARY

- Source–channel separation does not hold in general for communicating correlated sources over multiuser channels

- Joint source–channel coding schemes that utilize the correlation between the sources for cooperative transmission

- Common part of a 2-DMS

- Gray–Wyner system

- Notions of common information:

    - Gács–Körner–Witsenhausen common information $K(X; Y)$

    - Wyner's common information $J(X; Y)$

    - Mutual information $I(X; Y)$

    - $K(X; Y) \leq I(X; Y) \leq J(X; Y)$

- Joint Gray–Wyner–Marton coding for lossless communication of a 2-DMS over a DM-BC

- Hybrid source–channel coding scheme for a general single-hop network

BIBLIOGRAPHIC NOTES

The joint source–channel coding schemes for sending a 2-DMS over a DM-MAC in Theorems 14.1 and 14.2 are due to Cover, El Gamal, and Salehi (1980), who also showed via Example 14.3 that source–channel separation does not always hold. The definition of a common part of a 2-DMS and its characterization are due to Gács and Körner (1973) and Witsenhausen (1975). Dueck (1981a) showed via an example that the coding scheme used in the proof of Theorem 14.2, which utilizes the common part, is still suboptimal.

Theorem 14.3 is due to Gray and Wyner (1974), who also established the rate–distortion region for the lossy case. The definitions of common information and their properties can be found in Wyner (1975a). Examples 14.5 and 14.6 are due to Wyner (1975a) and Witsenhausen (1976a). Theorem 14.4 was established by Han and Costa (1987); see also Kramer and Nair (2009). The proof in Section 14.2.5 is due to Minero and Kim (2009). The hybrid source–channel coding scheme in Section 14.3.2 was proposed by Lim, Minero, and Kim (2010), who also established Proposition 14.1.

PROBLEMS

14.1. Establish the necessarily condition for lossless communication of an arbitrary 2-DMS (U_1, U_2) over a DM-MAC with orthogonal components $p(y_1|x_1)p(y_2|x_2)$ in Example 14.1.

14.2. Consider the 3-index lossless source coding setup in Section 14.1.4. Show that the optimal rate region is given by (14.3).

14.3. Provide the details of the proof of Theorem 14.2.

14.4. Consider the sufficient condition for lossless communication of a 2-DMS (U_1, U_2) over a DM-MAC $p(y|x_1, x_2)$ in Theorem 14.2. Show that the condition does not change by considering conditional pmfs $p(w|u_0)p(x_1|u_1, w)p(x_2|u_2, w)$. Hence, joint source–channel coding of the common part U_0^n via the codeword W^n does not help.

14.5. Provide the details of the proof of Theorem 14.3.

14.6. Show that every rate triple (R_0, R_1, R_2) in the optimal rate region $\mathcal{R}^*$ of the Gray–Wyner system must satisfy the inequalities

$$R_0 + R_1 \geq H(U_1),$$
$$R_0 + R_2 \geq H(U_2),$$
$$R_0 + R_1 + R_2 \geq H(U_1, U_2).$$

14.7. *Separate source and channel coding over a DM-BC.* Consider the sufficient condition for lossless communication of a 2-DMS over a DM-BC via separate source and channel coding in (14.6).

(a) Show that, when specialized to a noiseless BC, the condition simplifies to the set of rate triples (R_0, R_1, R_2) such that

$$R_0 + R_1 \geq I(U_1, U_2; V) + H(U_1|V),$$
$$R_0 + R_2 \geq I(U_1, U_2; V) + H(U_2|V),$$
$$R_0 + R_1 + R_2 \geq I(U_1, U_2; V) + H(U_1|V) + H(U_2|V)$$

for some conditional pmf $p(v|u_1, u_2)$.

(b) Show that the above region is equivalent to the optimal rate region for the Gray–Wyner system in Theorem 14.3.

14.8. *Common information.* Consider the optimal rate region $\mathcal{R}^*$ of the Gray–Wyner system in Theorem 14.3.

(a) Complete the derivations of Gács–Körner–Witsenhausen common information and Wyner's common information as extreme points of $\mathcal{R}^*$.

(b) Show that the three measures of common information satisfy the inequalities

$$0 \leq K(U_1; U_2) \leq I(U_1; U_2) \leq J(U_1; U_2) \leq H(U_1, U_2).$$

(c) Show that $K(U_1; U_2) = I(U_1; U_2) = J(U_1; U_2)$ iff $U_1 = (V, V_1)$ and $U_2 = (V, V_2)$ for some $(V, V_1, V_2) \sim p(v)p(v_1|v)p(v_2|v)$.

14.9. *Lossy Gray–Wyner system.* Consider the Gray–Wyner system in Section 14.2.1 for a 2-DMS (U_1, U_2) and two distortion measures d_1 and d_2. The sources are to be reconstructed with prescribed distortion pair (D_1, D_2). Show that the rate–distortion region $\mathscr{R}(D_1, D_2)$ is the set of rate pairs (R_1, R_2) such that

$$R_0 \geq I(U_1, U_2; V),$$
$$R_1 \geq I(U_1; \hat{U}_1|V),$$
$$R_2 \geq I(U_2; \hat{U}_2|V)$$

for some conditional pmf $p(v|u_1, u_2)p(\hat{u}_1|u_1, v)p(\hat{u}_2|u_2, v)$ that satisfy the constraints $\mathsf{E}(d_j(U_j, \hat{U}_j)) \leq D_j$, $j = 1, 2$.

14.10. *Nested sources over a DM-MAC.* Let (U_1, U_2) be a 2-DMS with common part $U_0 = U_2$. We wish to send this 2-DMS over a DM-MAC $p(y|x_1, x_2)$ at rates $r_1 = r_2 = r$ symbol/transmission. Show that source–channel separation holds for this setting. Remark: This problem was studied by De Bruyn, Prelov, and van der Meulen (1987).

14.11. *Nested sources over a DM-BC.* Consider the nested 2-DMS (U_1, U_2) in the previous problem. We wish to communicated this 2-DMS over a DM-BC $p(y_1, y_2|x)$ at rates $r_1 = r_2 = r$. Show that source–channel separation holds again for this setting.

14.12. *Lossy communication of a Gaussian source over a Gaussian BC.* Consider a Gaussian broadcast channel $Y_1 = X + Z_1$ and $Y_2 = X + Z_2$, where $Z_1 \sim N(0, N_1)$ and $Z_2 \sim N(0, N_2)$ are noise components with $N_2 > N_1$. Assume average power constraint P on X. We wish to communicate a WGN(P) source U with mean squared error distortions D_1 to Y_1 and D_2 to Y_2 at rate $r = 1$ symbol/transmission.

(a) Find the minimum achievable individual distortions D_1 and D_2 in terms of P, N_1, and N_2.

(b) Suppose we use separate source and channel coding by first using successive refinement coding for the quadratic Gaussian source in Example 13.3 and then using optimal Gaussian BC codes for independent messages. Characterize the set of achievable distortion pairs (D_1, D_2) using this scheme.

(c) Now suppose we send the source with no coding, i.e., set $X_i = U_i$ for $i \in [1:n]$, and use the linear MMSE estimate $\hat{U}_{1i}$ at Y_1 and $\hat{U}_{2i}$ at Y_2. Characterize the set of achievable distortion pairs (D_1, D_2) using this scheme.

(d) Does source–channel separation hold for communicating a Gaussian source over a Gaussian BC with squared error distortion measure?

APPENDIX 14A   PROOF OF LEMMA 14.1

The proof follows similar steps to the mutual covering lemma in Section 8.3. For each $(u_1^n, u_2^n) \in \mathcal{T}_{\epsilon'}^{(n)}(U_1, U_2)$, define

$$\mathcal{A}(u_1^n, u_2^n) = \{(m_0, m_1, m_2) \in [1 : 2^{nR_0}] \times [1 : 2^{nR_1}] \times [1 : 2^{nR_2}] :$$
$$(u_1^n, u_2^n, W_0^n(m_0), W_1^n(u_1^n, m_0, m_1), W_2^n(u_2^n, m_0, m_2)) \in \mathcal{T}_{\epsilon'}^{(n)}\}.$$

Then

$$P(\mathcal{E}_0) \le P\{(U_1^n, U_2^n) \notin \mathcal{T}_{\epsilon'}^{(n)}\} + \sum_{(u_1^n, u_2^n) \in \mathcal{T}_{\epsilon'}^{(n)}} p(u_1^n, u_2^n) P\{|\mathcal{A}(u_1^n, u_2^n)| = 0\}.$$

By the LLN, the first term tends to zero as $n \to \infty$. To bound the second term, recall from the proof of the mutual covering lemma that

$$P\{|\mathcal{A}(u_1^n, u_2^n)| = 0\} \le \frac{\mathrm{Var}(|\mathcal{A}(u_1^n, u_2^n)|)}{(E(|\mathcal{A}(u_1^n, u_2^n)|))^2}.$$

Now, define the indicator function

$$E(m_0, m_1, m_2) = \begin{cases} 1 & \text{if } (u_1^n, u_2^n, W_0^n(m_0), W_1^n(u_1^n, m_0, m_1), W_2^n(u_2^n, m_0, m_2)) \in \mathcal{T}_{\epsilon'}^{(n)}, \\ 0 & \text{otherwise} \end{cases}$$

for each (m_0, m_1, m_2). We can then write

$$|\mathcal{A}(u_1^n, u_2^n)| = \sum_{m_0, m_1, m_2} E(m_0, m_1, m_2),$$

Let

$$p_1 = E[E(1, 1, 1)]$$
$$= P\{(u_1^n, u_2^n, W_0^n(m_0), W_1^n(u_1^n, m_0, m_1), W_2^n(u_2^n, m_0, m_2)) \in \mathcal{T}_{\epsilon'}^{(n)}\},$$
$$p_2 = E[E(1, 1, 1)E(1, 2, 1)],$$
$$p_3 = E[E(1, 1, 1)E(1, 1, 2)],$$
$$p_4 = E[E(1, 1, 1)E(1, 2, 2)],$$
$$p_5 = E[E(1, 1, 1)E(2, 1, 1)] = E[E(1, 1, 1)E(2, 1, 2)]$$
$$= E[E(1, 1, 1)E(2, 2, 1)] = E[E(1, 1, 1)E(2, 2, 2)] = p_1^2.$$

Then

$$E(|\mathcal{A}(u_1^n, u_2^n)|) = \sum_{m_0, m_1, m_2} E[E(m_0, m_1, m_2)] = 2^{n(R_0 + R_1 + R_2)} p_1$$

and

$$E(|\mathcal{A}(u_1^n, u_2^n)|^2) = \sum_{m_0, m_1, m_2} E[E(m_0, m_1, m_2)]$$

$$+ \sum_{m_0, m_1, m_2} \sum_{m_1' \neq m_1} E[E(m_0, m_1, m_2)E(m_0, m_1', m_2)]$$

$$+ \sum_{m_0, m_1, m_2} \sum_{m_2' \neq m_2} E[E(m_0, m_1, m_2)E(m_0, m_1, m_2')]$$

$$+ \sum_{m_0, m_1, m_2} \sum_{m_1' \neq m_1, m_2' \neq m_2} E[E(m_0, m_1, m_2)E(m_0, m_1', m_2')]$$

$$+ \sum_{m_0, m_1, m_2} \sum_{m_0' \neq m_0, m_1', m_2'} E[E(m_0, m_1, m_2)E(m_0', m_1', m_2')]$$

$$\leq 2^{n(R_0 + R_1 + R_2)} p_1 + 2^{n(R_0 + 2R_1 + R_2)} p_2 + 2^{n(R_0 + R_1 + 2R_2)} p_3$$

$$+ 2^{n(R_0 + 2R_1 + 2R_2)} p_4 + 2^{2n(R_0 + R_1 + R_2)} p_5.$$

Hence

$$\text{Var}(|\mathcal{A}(u_1^n, u_2^n)|) \leq 2^{n(R_0 + R_1 + R_2)} p_1 + 2^{n(R_0 + 2R_1 + R_2)} p_2 + 2^{n(R_0 + R_1 + 2R_2)} p_3 + 2^{n(R_0 + 2R_1 + 2R_2)} p_4.$$

Now by the joint typicality lemma, we have

$$p_1 \geq 2^{-n(I(U_1, U_2; W_0) + I(U_2; W_1 | U_1, W_0) + I(U_1, W_1; W_2 | U_2, W_0) + \delta(\epsilon'))},$$

$$p_2 \leq 2^{-n(I(U_1, U_2; W_0) + 2I(U_2, W_2; W_1 | U_1, W_0) + I(U_1; W_2 | U_2, W_0) - \delta(\epsilon'))},$$

$$p_3 \leq 2^{-n(I(U_1, U_2; W_0) + I(U_2; W_1 | U_1, W_0) + 2I(U_1, W_1, U_1; W_2 | U_2, W_0) - \delta(\epsilon'))},$$

$$p_4 \leq 2^{-n(I(U_1, U_2; W_0) + 2I(U_2; W_1 | U_1, W_0) + 2I(U_1, W_1; W_2 | U_2, W_0) - \delta(\epsilon'))}.$$

Hence

$$\frac{\text{Var}(|\mathcal{A}(u_1^n, u_2^n)|)}{(E(|\mathcal{A}(u_1^n, u_2^n)|))^2} \leq 2^{-n(R_0 + R_1 + R_2 - I(U_1, U_2; W_0) - I(U_2; W_1 | U_1, W_0) - I(U_1, W_1; W_2 | U_2, W_0) - \delta(\epsilon'))}$$

$$+ 2^{-n(R_0 + R_2 - I(U_1, U_2; W_0) - I(U_1; W_2 | U_2, W_0) - 3\delta(\epsilon'))}$$

$$+ 2^{-n(R_0 + R_1 - I(U_1, U_2; W_0) - I(U_2; W_1 | U_1, W_0) - 3\delta(\epsilon'))}$$

$$+ 2^{-n(R_0 - I(U_1, U_2; W_0) - 3\delta(\epsilon'))}.$$

Therefore, $P\{|\mathcal{A}(u_1^n, u_2^n)| = 0\}$ tends to zero as $n \to \infty$ if

$$R_0 > I(U_1, U_2; W_0) + 3\delta(\epsilon'),$$

$$R_0 + R_1 > I(U_1, U_2; W_0) + I(U_2; W_1 | U_1, W_0) + 3\delta(\epsilon'),$$

$$R_0 + R_2 > I(U_1, U_2; W_0) + I(U_1; W_2 | U_2, W_0) + 3\delta(\epsilon'),$$

$$R_0 + R_1 + R_2 > I(U_1, U_2; W_0) + I(U_2; W_1 | U_1, W_0) + I(U_1, W_1; W_2 | U_2, W_0) + \delta(\epsilon').$$

This completes the proof of Lemma 14.1.

PART III

MULTIHOP NETWORKS

CHAPTER 15

Graphical Networks

So far we have studied single-hop networks in which each node is either a sender or a receiver. In this chapter, we begin the discussion of multihop networks, where some nodes can act as both senders and receivers and hence communication can be performed over multiple rounds. We consider the limits on communication of independent messages over networks modeled by a weighted directed acyclic graph. This network model represents, for example, a wired network or a wireless mesh network operated in time or frequency division, where the nodes may be servers, handsets, sensors, base stations, or routers. The edges in the graph represent point-to-point communication links that use channel coding to achieve close to error-free communication at rates below their respective capacities. We assume that each node wishes to communicate a message to other nodes over this graphical network. The nodes can also act as relays to help other nodes communicate their messages. What is the capacity region of this network?

Although communication over such a graphical network is not hampered by noise or interference, the conditions on optimal information flow are not known in general. The difficulty arises in determining the optimal relaying strategies when several messages are to be sent to different destination nodes.

We first consider the graphical multicast network, where a source node wishes to communicate a message to a set of destination nodes. We establish the cutset upper bound on the capacity and show that it is achievable error-free via routing when there is only one destination, leading to the celebrated max-flow min-cut theorem. When there are multiple destinations, routing alone cannot achieve the capacity, however. We show that the cutset bound is still achievable, but using more sophisticated coding at the relays. The proof of this result involves linear network coding in which the relays perform simple linear operations over a finite field.

We then consider graphical networks with multiple independent messages. We show that the cutset bound is tight when the messages are to be sent to the same set of destination nodes (multimessage multicast), and is achieved again error-free using linear network coding. When each message is to be sent to a different set of destination nodes, however, neither the cutset bound nor linear network coding is optimal in general.

The aforementioned capacity results can be extended to networks with broadcasting and cycles. In Chapter 18, we present a coding scheme that extends network coding to general networks. This noisy network coding scheme yields an alternative achievability proof of the network coding theorem that applies to graphical networks with and without cycles.

15.1 GRAPHICAL MULTICAST NETWORK

Consider a *multicast network* modeled by a weighted directed acyclic graph $\mathcal{G} = (\mathcal{N}, \mathcal{E}, \mathcal{C})$ as depicted in Figure 15.1. Here $\mathcal{N} = [1:N]$ is the set of nodes, $\mathcal{E} \subset \mathcal{N} \times \mathcal{N}$ is the set of edges, and $\mathcal{C} = \{C_{jk}: (j,k) \in \mathcal{E}\}$ is the set of edge weights. Each node represents a sender–receiver pair and each edge (j,k) represents a noiseless communication link with capacity C_{jk}. Source node 1 wishes to communicate a message $M \in [1:2^{nR}]$ to a set of destination nodes $\mathcal{D} \in \mathcal{N}$. Each node $k \in [2:N]$ can also act as a relay to help the source node communicate its message to the destination nodes. Note that in addition to being noiseless, this network model does not allow for broadcasting or interference. However, we do not assume any constraints on the functions performed by the nodes; hence general relaying operations are allowed.

A $(2^{nR}, n)$ code for a graphical multicast network $\mathcal{G} = (\mathcal{N}, \mathcal{E}, \mathcal{C})$ consists of

- a message set $[1:2^{nR}]$,

- a source encoder that assigns an index $m_{1j}(m) \in [1:2^{nC_{1j}}]$ to each message $m \in [1:2^{nR}]$ for each edge $(1,j) \in \mathcal{E}$,

- a set of relay encoders, where encoder $k \in [2:N-1]$ assigns an index $m_{kl} \in [1:2^{nC_{kl}}]$ to each received index tuple $(m_{jk}: (j,k) \in \mathcal{E})$ for each $(k,l) \in \mathcal{E}$, and

- a set of decoders, where decoder $k \in \mathcal{D}$ assigns an estimate $\hat{m}_k \in [1:2^{nR}]$ or an error message e to every received index tuple $(m_{jk}: (j,k) \in \mathcal{E})$.

These coding operations are illustrated in Figure 15.2.

We assume that the message M is uniformly distributed over $[1:2^{nR}]$. The average

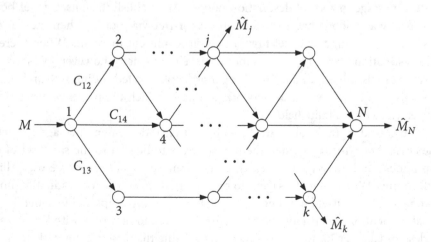

Figure 15.1. Graphical multicast network.

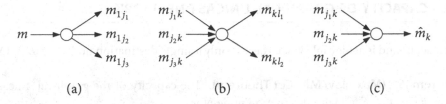

Figure 15.2. Coding operations at the nodes: (a) source encoder, (b) relay encoder, (c) decoder.

probability of error is defined as $P_e^{(n)} = P\{\hat{M}_j \neq M$ for some $j \in \mathcal{D}\}$. A rate R is said to be achievable if there exists a sequence of $(2^{nR}, n)$ codes such that $\lim_{n\to\infty} P_e^{(n)} = 0$. The capacity (maximum flow) of the graphical multicast network is the supremum of the set of achievable rates.

First consider the following upper bound on the capacity.

Cutset bound. For destination node $j \in \mathcal{D}$, define a *cut* $(\mathcal{S}, \mathcal{S}^c)$ as a partition of the set of nodes $\mathcal{N}$ such that source node $1 \in \mathcal{S}$ and destination node $j \in \mathcal{S}^c$. The capacity of the cut is defined as

$$C(\mathcal{S}) = \sum_{\substack{(k,l)\in\mathcal{E} \\ k\in\mathcal{S},\, l\in\mathcal{S}^c}} C_{kl},$$

that is, the sum of the capacities of the edges from the subset of nodes $\mathcal{S}$ to its complement $\mathcal{S}^c$. Intuitively, the capacity of this network cannot exceed the smallest cut capacity $C(\mathcal{S})$ of every cut $(\mathcal{S}, \mathcal{S}^c)$ and every destination node $j \in \mathcal{D}$. We formalize this statement in the following.

Theorem 15.1 (Cutset Bound for the Graphical Multicast Network). The capacity of the graphical multicast network $\mathcal{G} = (\mathcal{N}, \mathcal{E}, C)$ with destination set $\mathcal{D}$ is upper bounded as

$$C \leq \min_{j\in\mathcal{D}} \ \min_{\substack{\mathcal{S}\subset\mathcal{N} \\ 1\in\mathcal{S},\, j\in\mathcal{S}^c}} C(\mathcal{S}).$$

Proof. To establish this cutset upper bound, consider a cut $(\mathcal{S}, \mathcal{S}^c)$ such that $1 \in \mathcal{S}$ and $j \in \mathcal{S}^c$ for some $j \in \mathcal{D}$. Then for every $(2^{nR}, n)$ code, $\hat{M}_j$ is a function of $M(\mathcal{N}, \mathcal{S}^c) = (M_{kl}: (k,l) \in \mathcal{E}, l \in \mathcal{S}^c)$, which in turn is a function of $M(\mathcal{S}, \mathcal{S}^c) = (M_{kl}: (k,l) \in \mathcal{E}, k \in \mathcal{S}, l \in \mathcal{S}^c)$. Thus by Fano's inequality,

$$nR \leq I(M; \hat{M}_j) + n\epsilon_n$$
$$\leq H(\hat{M}_j) + n\epsilon_n$$
$$\leq H(M(\mathcal{S}, \mathcal{S}^c)) + n\epsilon_n$$
$$\leq nC(\mathcal{S}) + n\epsilon_n.$$

Repeating this argument over all cuts and all destination nodes completes the proof of the cutset bound.

15.2   CAPACITY OF GRAPHICAL UNICAST NETWORK

The cutset bound is achievable when there is only a single destination node, say, $\mathcal{D} = \{N\}$.

Theorem 15.2 (Max-Flow Min-Cut Theorem). The capacity of the graphical unicast network $\mathcal{G} = (\mathcal{N}, \mathcal{E}, \mathcal{C})$ with destination node N is

$$C = \min_{\substack{\mathcal{S} \subset \mathcal{N} \\ 1 \in \mathcal{S}, N \in \mathcal{S}^c}} C(\mathcal{S}).$$

Note that it suffices to take the minimum over *connected* cuts $(\mathcal{S}, \mathcal{S}^c)$, that is, if $k \in \mathcal{S}$ and $k \neq 1$, then $(j, k) \in \mathcal{E}$ for some $j \in \mathcal{S}$, and if $j \in \mathcal{S}^c$ and $j \neq N$, then $(j, k) \in \mathcal{E}$ for some $k \in \mathcal{S}^c$. This theorem is illustrated in the following.

Example 15.1. Consider the graphical unicast network depicted in Figure 15.3. The capacity of this network is $C = 3$, with the minimum cut $\mathcal{S} = \{1, 2, 3, 5\}$, and is achieved by routing 1 bit along the path $1 \rightarrow 2 \rightarrow 4 \rightarrow 6$ and 2 bits along the path $1 \rightarrow 3 \rightarrow 5 \rightarrow 6$.

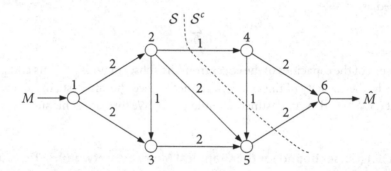

Figure 15.3. Graphical unicast network for Example 15.1.

To prove the max-flow min-cut theorem, we only need to establish achievability, since the converse follows by the cutset bound.

Proof of achievability. Assume without loss of generality that every node lies on at least one path from node 1 to node N. Suppose we allocate rate $r_{jk} \leq C_{jk}$ to each edge $(j, k) \in \mathcal{E}$ such that

$$\sum_{j:(j,k)\in\mathcal{E}} r_{jk} = \sum_{l:(k,l)\in\mathcal{E}} r_{kl}, \quad k \neq 1, N,$$

$$\sum_{k:(1,k)\in\mathcal{E}} r_{1k} = \sum_{j:(j,N)\in\mathcal{E}} r_{jN} = R,$$

that is, the total incoming information rate at each node is equal to the total outgoing information rate from it. Then it is straightforward to check that the rate R is achievable by splitting the message into multiple messages and routing them according to the rate allocation r_{jk} as in commodity flow. Hence, the optimal rate allocation can be found by solving the optimization problem

$$\text{maximize } R$$
$$\text{subject to } 0 \le r_{jk} \le C_{jk},$$
$$\sum_j r_{jk} = \sum_l r_{kl}, \quad k \ne 1, N,$$
$$\sum_k r_{1k} = R,$$
$$\sum_j r_{jN} = R.$$

(15.1)

This is a linear program (LP) and the optimal solution can be found by solving its dual problem, which is another LP (see Appendix E)

$$\text{minimize } \sum_{j,k} \lambda_{jk} C_{jk}$$
$$\text{subject to } \lambda_{jk} \ge 0,$$
$$v_j - v_k = \lambda_{jk},$$
$$v_1 - v_N = 1$$

(15.2)

with variables $\lambda_{jk}, (j, k) \in \mathcal{E}$ (weights for the link capacities) and $v_j, j \in \mathcal{N}$ (differences of the weights). Since the minimum depends on $v_j, j \in [1 : N]$, only through their differences, the dual LP is equivalent to

$$\text{minimize } \sum_{j,k} \lambda_{jk} C_{jk}$$
$$\text{subject to } \lambda_{jk} \ge 0,$$
$$v_j - v_k = \lambda_{jk},$$
$$v_1 = 1,$$
$$v_N = 0.$$

(15.3)

Now it can be shown (see Problem 15.4) that the set of feasible solutions to (15.3) is a polytope with extreme points of the form

$$v_j = \begin{cases} 1 & \text{if } j \in \mathcal{S}, \\ 0 & \text{otherwise,} \end{cases}$$
$$\lambda_{jk} = \begin{cases} 1 & \text{if } j \in \mathcal{S}, k \in \mathcal{S}^c, (j, k) \in \mathcal{E}, \\ 0 & \text{otherwise} \end{cases}$$

(15.4)

for some $\mathcal{S}$ such that $1 \in \mathcal{S}, N \in \mathcal{S}^c$, and $\mathcal{S}$ and $\mathcal{S}^c$ are each connected. Furthermore, it can be readily checked that the value of (15.3) at each of these extreme solutions corresponds to $C(S)$. Since the minimum of a linear (or convex in general) function is attained by an extreme point of the feasible set, the minimum of (15.3) and, in turn, of (15.2) is equal to $\min C(S)$, where the minimum is over all (connected) cuts. Finally, since both the primal and dual optimization problems satisfy *Slater's condition* (see Appendix E), the dual optimum is equal to the primal optimum, i.e., $\max R = \min C(S)$. This completes the proof of the max-flow min-cut theorem.

Remark 15.1. The capacity of a unicast graphical network is achieved error-free using routing. Hence, information in such a network can be treated as water flowing in pipes or a commodity transported over a network of roads.

Remark 15.2. The max-flow min-cut theorem continues to hold for networks with cycles and delays (see Chapter 18), and capacity is achieved error-free using routing. For networks with cycles, there is always an optimal routing that does not involve any cycle.

15.3 CAPACITY OF GRAPHICAL MULTICAST NETWORK

The cutset bound turns out to be achievable also when the network has more than one destination. Unlike the unicast case, however, it is *not* always achievable using only routing as demonstrated in the following.

Example 15.2 (Butterfly network). Consider the graphical multicast network depicted in Figure 15.4 with $C_{jk} = 1$ for all (j, k) and $\mathcal{D} = \{6, 7\}$. It is easy to see that the cutset upper bound for this network is $C \leq 2$. To send the message M via routing, we split it into two independent messages M_1 at rate R_1 and M_2 at rate R_2 with $R = R_1 + R_2$. If each relay node k simply forwards its incoming messages (i.e., $\sum_j r_{jk} = \sum_l r_{kl}$ for each relay node k), then it can be easily seen that $R_1 + R_2$ cannot exceed 1 bit per transmission with link $(4, 5)$ being the main bottleneck. Even if we allow the relay nodes to forward multiple copies of its incoming messages, we must still have $R_1 + R_2 \leq 1$.

Surprisingly, if we allow simple encoding operations at the relay nodes (network cod-

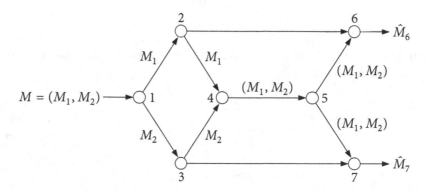

Figure 15.4. Optimal routing for the butterfly network.

ing), we can achieve the cutset bound. Let $R_1 = R_2 = 1$. As illustrated in Figure 15.5, relay nodes 2, 3, and 5 forward multiple copies of their incoming messages, and relay node 4 sends the modulo-2 sum of its incoming messages. Then both destination nodes 6 and 7 can recover (M_1, M_2) error-free, achieving the cutset upper bound. This simple example shows that treating information as a physical commodity is not optimal in general.

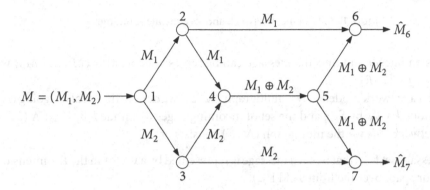

Figure 15.5. Network coding for the butterfly network.

The above *network coding* idea can be generalized to achieve the cutset bound for arbitrary graphical multicast networks.

Theorem 15.3 (Network Coding Theorem). The capacity of the graphical multicast network $\mathcal{G} = (\mathcal{N}, \mathcal{E}, \mathcal{C})$ with destination set $\mathcal{D}$ is

$$C = \min_{j \in \mathcal{D}} \quad \min_{\substack{\mathcal{S} \subset \mathcal{N} \\ 1 \in \mathcal{S}, j \in \mathcal{S}^c}} C(\mathcal{S}).$$

As in the butterfly network example, capacity can be achieved error-free using simple linear operations at the relay nodes.

Remark 15.3. The network coding theorem can be extended to networks with broadcasting (that is, networks modeled by a *hypergraph*), cycles, and delays. The proof of the extension to networks with cycles is sketched in the Bibliographic Notes. In Section 18.3, we show that network coding and these extensions are special cases of the noisy network coding scheme.

15.3.1   Linear Network Coding

For simplicity, we first consider a multicast network with integer link capacities, represented by a directed acyclic *multigraph* $\mathcal{G} = (\mathcal{N}, \mathcal{E})$ with links of the same 1-bit capacity as depicted in Figure 15.6. Hence, each link of the multigraph $\mathcal{G}$ can carry n bits of information (a symbol from the finite field $\mathbb{F}_{2^n}$) per n-transmission block. Further, we assume

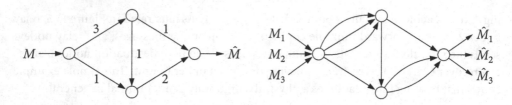

Figure 15.6. Graphical network and its equivalent multigraph.

that R is an integer and thus the message can be represented as $M = (M_1, \ldots, M_R)$ with $M_j \in \mathbb{F}_{2^n}$, $j \in [1 : R]$.

Given a network modeled by a multigraph $(\mathcal{N}, \mathcal{E})$, we denote the set of outgoing edges from a node $k \in \mathcal{N}$ by $\mathcal{E}_{k\rightarrow}$ and the set of incoming edges to a node k by $\mathcal{E}_{\rightarrow k}$. A $(2^{nR}, n)$ *linear network code* for the multigraph $(\mathcal{N}, \mathcal{E})$ consists of

- a message set $\mathbb{F}_{2^n}^{R}$ (that is, each message is represented by a vector in the R-dimensional vector space over the finite field $\mathbb{F}_{2^n}$),

- a linear source encoder that assigns an index tuple $m(\mathcal{E}_{1\rightarrow}) = \{m_e \in \mathbb{F}_{2^n} : e \in \mathcal{E}_{1\rightarrow}\}$ to each $(m_1, \ldots, m_R) \in \mathbb{F}_{2^n}^{R}$ via a linear transformation (with coefficients in $\mathbb{F}_{2^n}$),

- a set of linear relay encoders, where encoder $k \in [2 : N - 1]$ assigns an index tuple $m(\mathcal{E}_{k\rightarrow})$ to each $m(\mathcal{E}_{\rightarrow k})$ via a linear transformation, and

- a set of linear decoders, where decoder $j \in \mathcal{D}$ assigns $\hat{m}_j^R$ to each $m(\mathcal{E}_{\rightarrow j})$ via a linear transformation.

These linear network coding operations are illustrated in Figure 15.7.

Note that for each destination node $j \in \mathcal{D}$, a linear network code induces a linear transformation

$$\hat{m}_j^R = A_j(\alpha)\, m^R$$

for some $A_j(\alpha) \in \mathbb{F}_{2^n}^{R \times R}$, where α is a vector of linear encoding/decoding coefficients with elements taking values in $\mathbb{F}_{2^n}$. A rate R is said to be achievable error-free if there exist an integer n and a vector α such that $A_j(\alpha) = I_R$ for every $j \in \mathcal{D}$. Note that any invertible $A_j(\alpha)$ suffices since the decoder can multiply $\hat{m}_j^R$ by $A_j^{-1}(\alpha)$ to recover m. The resulting decoder is still linear (with a different α).

Example 15.3. Consider the 4-node network depicted in Figure 15.8. A linear network code with $R = 2$ induces the linear transformation

$$\begin{bmatrix} \hat{m}_1 \\ \hat{m}_2 \end{bmatrix} = \begin{bmatrix} \alpha_9 & \alpha_{10} \\ \alpha_{11} & \alpha_{12} \end{bmatrix} \begin{bmatrix} \alpha_6 & 0 \\ \alpha_5\alpha_8 & \alpha_7 \end{bmatrix} \begin{bmatrix} \alpha_1 & \alpha_2 \\ \alpha_3 & \alpha_4 \end{bmatrix} \begin{bmatrix} m_1 \\ m_2 \end{bmatrix} = A(\alpha) \begin{bmatrix} m_1 \\ m_2 \end{bmatrix}.$$

If $A(\alpha)$ is invertible, then the rate $R = 2$ can be achieved error-free by substituting the matrix

$$\begin{bmatrix} \alpha_9 & \alpha_{10} \\ \alpha_{11} & \alpha_{12} \end{bmatrix}$$

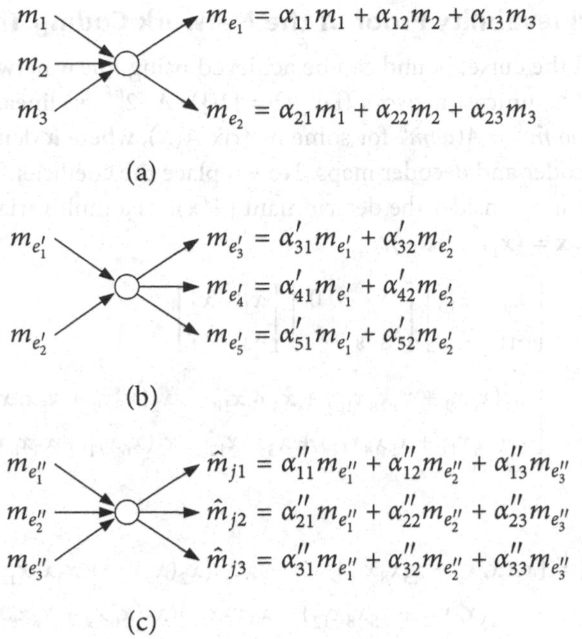

m_1
m_2
m_3
$m_{e_1} = \alpha_{11}m_1 + \alpha_{12}m_2 + \alpha_{13}m_3$
$m_{e_2} = \alpha_{21}m_1 + \alpha_{22}m_2 + \alpha_{23}m_3$
(a)
$m_{e'_1}$
$m_{e'_2}$
$m_{e'_3} = \alpha'_{31}m_{e'_1} + \alpha'_{32}m_{e'_2}$
$m_{e'_4} = \alpha'_{41}m_{e'_1} + \alpha'_{42}m_{e'_2}$
$m_{e'_5} = \alpha'_{51}m_{e'_1} + \alpha'_{52}m_{e'_2}$
(b)
$m_{e''_1}$
$m_{e''_2}$
$m_{e''_3}$
$\hat{m}_{j1} = \alpha''_{11}m_{e''_1} + \alpha''_{12}m_{e''_2} + \alpha''_{13}m_{e''_3}$
$\hat{m}_{j2} = \alpha''_{21}m_{e''_1} + \alpha''_{22}m_{e''_2} + \alpha''_{23}m_{e''_3}$
$\hat{m}_{j3} = \alpha''_{31}m_{e''_1} + \alpha''_{32}m_{e''_2} + \alpha''_{33}m_{e''_3}$
(c)

Figure 15.7. Linear network code: (a) source encoder, (b) relay encoder, (c) decoder.

m_1
m_2
1
2
3
4
$\alpha_1 m_1 + \alpha_2 m_2$
$\alpha_3 m_1 + \alpha_4 m_2$
$\alpha_5 m_{12}$
$\alpha_6 m_{12}$
$\alpha_7 m_{13} + \alpha_8 m_{23}$
$\hat{m}_1 = \alpha_9 m_{24} + \alpha_{10} m_{34}$
$\hat{m}_2 = \alpha_{11} m_{24} + \alpha_{12} m_{34}$

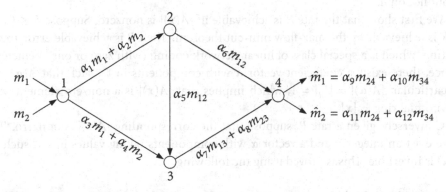

Figure 15.8. Linear network code for Example 15.3.

at the decoder with

$$A^{-1}(\alpha)\begin{bmatrix} \alpha_9 & \alpha_{10} \\ \alpha_{11} & \alpha_{12} \end{bmatrix}.$$

Remark 15.4. A general network with noninteger link capacities can be approximated by a multigraph with links of the same n/k bit capacities (each link carries n information bits per k transmissions). A $(2^{nR}, k)$ linear code with rate nR/k is defined as before with the R-dimensional vector space over $\mathbb{F}_{2^n}$.

15.3.2 Achievability Proof of the Network Coding Theorem

We show that the cutset bound can be achieved using linear network coding. First consider a graphical unicast network (i.e., $\mathcal{D} = \{N\}$). A $(2^{nR}, n)$ linear code induces a linear transformation $\hat{m}^R = A(\boldsymbol{\alpha})m^R$ for some matrix $A(\boldsymbol{\alpha})$, where $\boldsymbol{\alpha}$ denotes the coefficients in the linear encoder and decoder maps. Now replace the coefficients $\boldsymbol{\alpha}$ with an indeterminate vector $\mathbf{x}$ and consider the determinant $|A(\mathbf{x})|$ as a multivariate polynomial in $\mathbf{x}$. In Example 15.3, $\mathbf{x} = (x_1, \ldots, x_{12})$,

$$A(\mathbf{x}) = \begin{bmatrix} x_9 & x_{10} \\ x_{11} & x_{12} \end{bmatrix} \begin{bmatrix} x_6 & 0 \\ x_5 x_8 & x_7 \end{bmatrix} \begin{bmatrix} x_1 & x_2 \\ x_3 & x_4 \end{bmatrix}$$

$$= \begin{bmatrix} x_1(x_6 x_9 + x_5 x_8 x_{10}) + x_3 x_7 x_{10} & x_2(x_6 x_9 + x_5 x_8 x_{10}) + x_4 x_7 x_{10} \\ x_1(x_6 x_{11} + x_5 x_8 x_{12}) + x_3 x_7 x_{12} & x_2(x_6 x_{11} + x_5 x_8 x_{12}) + x_4 x_7 x_{12} \end{bmatrix},$$

and

$$|A(\mathbf{x})| = (x_1(x_6 x_9 + x_5 x_8 x_{10}) + x_3 x_7 x_{10})(x_2(x_6 x_{11} + x_5 x_8 x_{12}) + x_4 x_7 x_{12})$$
$$- (x_1(x_6 x_{11} + x_5 x_8 x_{12}) + x_3 x_7 x_{12})(x_2(x_6 x_9 + x_5 x_8 x_{10}) + x_4 x_7 x_{10}).$$

In general, $|A(\mathbf{x})|$ is a polynomial in $\mathbf{x}$ with binary coefficients, that is, $|A(\mathbf{x})| \in \mathbb{F}_2[\mathbf{x}]$, the polynomial ring over $\mathbb{F}_2$. This polynomial depends on the network topology and the rate R, but not on n.

We first show that the rate R is achievable iff $|A(\mathbf{x})|$ is nonzero. Suppose $R \le C$, that is, R is achievable by the max-flow min-cut theorem. Then R is achievable error-free via routing, which is a special class of linear network coding (with zero or one coefficients). Hence, there exists a coefficient vector $\boldsymbol{\alpha}$ with components in $\mathbb{F}_2$ such that $A(\boldsymbol{\alpha}) = I_R$. In particular, $|A(\boldsymbol{\alpha})| = |I_R| = 1$, which implies that $|A(\mathbf{x})|$ is a nonzero element of the polynomial ring $\mathbb{F}_2[\mathbf{x}]$.

Conversely, given a rate R, suppose that the corresponding $|A(\mathbf{x})|$ is nonzero. Then there exist an integer n and a vector $\boldsymbol{\alpha}$ with components taking values in $\mathbb{F}_{2^n}$ such that $A(\boldsymbol{\alpha})$ is invertible. This is proved using the following.

Lemma 15.1. If $P(\mathbf{x})$ is a nonzero polynomial over $\mathbb{F}_2$, then there exist an integer n and a vector $\boldsymbol{\alpha}$ with components taking values in $\mathbb{F}_{2^n}$ such that $P(\boldsymbol{\alpha}) \ne 0$.

For example, $x^2 + x = 0$ for all $x \in \mathbb{F}_2$, but $x^2 + x \ne 0$ for some $x \in \mathbb{F}_4$. The proof of this lemma is given in Appendix 15A.

Now by Lemma 15.1, $|A(\boldsymbol{\alpha})|$ is nonzero, that is, $A(\boldsymbol{\alpha})$ is invertible and hence R is achievable.

Next consider the multicast network with destination set $\mathcal{D}$. If $R \le C$, then by the above argument, $|A_j(\mathbf{x})|$ is a nonzero polynomial over $\mathbb{F}_2$ for every $j \in \mathcal{D}$. Since the polynomial ring $\mathbb{F}_2[\mathbf{x}]$ is an integral domain, that is, the product of two nonzero elements is always nonzero (Lidl and Niederreiter 1997, Theorem 1.51), the product $\prod_{j \in \mathcal{D}} |A_j(\mathbf{x})|$ is also nonzero. As before, this implies that there exist n and $\boldsymbol{\alpha}$ with elements in $\mathbb{F}_{2^n}$ such that $|A_j(\alpha)|$ is nonzero for all $j \in \mathcal{D}$, that is, $A_j^{-1}(\alpha)$ is invertible for every $j \in \mathcal{D}$. This completes the proof of achievability.

Remark 15.5. It can be shown that block length $n \le \lceil \log(|\mathcal{D}|R + 1) \rceil$ suffices in the above proof.

Remark 15.6. The achievability proof of the network coding theorem via linear network coding readily extends to broadcasting. It can be also extended to networks with cycles by using convolutional codes.

15.4 GRAPHICAL MULTIMESSAGE NETWORK

We now consider the more general problem of communicating multiple independent messages over a network. As before, we model the network by a directed acyclic graph $\mathcal{G} = (\mathcal{N}, \mathcal{E}, \mathcal{C})$ as depicted in Figure 15.9. Assume that the nodes are ordered so that there is no path from node k to node j if $j < k$. Each node $j \in [1 : N - 1]$ wishes to send a message M_j to a set $\mathcal{D}_j \subseteq [j + 1 : N]$ of destination nodes. This setting includes the case where only a subset of nodes are sending messages by taking $M_j = \emptyset$ ($R_j = 0$) for each nonsource node j.

A $(2^{nR_1}, \dots, 2^{nR_{N-1}}, n)$ code for the graphical multimessage network $\mathcal{G} = (\mathcal{N}, \mathcal{E}, \mathcal{C})$ consists of

- message sets $[1 : 2^{nR_1}], \dots, [1 : 2^{nR_{N-1}}]$,

- a set of encoders, where encoder $k \in [1 : N - 1]$ assigns an index $m_{kl} \in [1 : 2^{nC_{kl}}]$ to each received index tuple $(m_{jk} : (j, k) \in \mathcal{E})$ and its own message m_k for each $(k, l) \in \mathcal{E}$, and

- a set of decoders, where decoder $l \in [2 : N]$ assigns an estimate $\hat{m}_{jl}$ or an error message e to each received index tuple $(m_{kl} : (k, l) \in \mathcal{E})$ for j such that $l \in \mathcal{D}_j$.

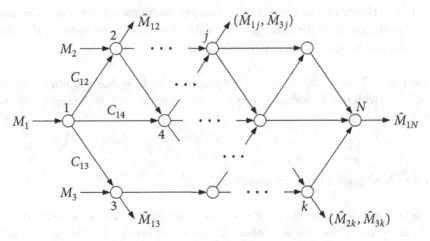

Figure 15.9. Graphical multimessage network.

Assume that the message tuple $(M_1, \ldots, M_{N-1})$ is uniformly distributed over $[1 : 2^{nR_1}] \times \cdots \times [1 : 2^{nR_{N-1}}]$. The average probability of error is defined as $P_e^{(n)} = \mathrm{P}\{\hat{M}_{jk} \neq M_j$ for some $j \in [1 : N-1], k \in \mathcal{D}_j\}$. A rate tuple $(R_1, \ldots, R_{N-1})$ is said to be achievable if there exists a sequence of $(2^{nR_1}, \ldots, 2^{nR_{N-1}}, n)$ codes such that $\lim_{n \to \infty} P_e^{(n)} = 0$. The capacity region of the graphical network is the closure of the set of achievable rate tuples.

The capacity region for the multimessage setting is not known in general. By extending the proof of the cutset bound for the multicast case, we can obtain the following outer bound on the capacity region.

Theorem 15.4 (Cutset Bound for the Graphical Multimessage Network). If the rate tuple $(R_1, \ldots, R_{N-1})$ is achievable for the graphical multimessage network $\mathcal{G} = (\mathcal{N}, \mathcal{E}, \mathcal{C})$ with destination sets $(\mathcal{D}_1, \ldots, \mathcal{D}_{N-1})$, then it must satisfy the inequality

$$\sum_{j \in \mathcal{S} : \mathcal{D}_j \cap \mathcal{S}^c \neq \emptyset} R_j \leq C(\mathcal{S})$$

for all $\mathcal{S} \subset \mathcal{N}$ such that $\mathcal{D}_j \cap \mathcal{S}^c \neq \emptyset$ for some $j \in [1 : N-1]$.

Remark 15.7. As we show in Chapter 18, this cutset outer bound continues to hold even when the network has cycles and allows for broadcasting and interaction between the nodes.

In the following, we consider two special classes of multimessage networks.

15.4.1   Graphical Multimessage Multicast Network

When $R_2 = \cdots = R_{N-1} = 0$, the network reduces to a multicast network and the cutset bound is tight. More generally, let $[1 : k]$, for some $k \leq N-1$, be the set of source nodes and assume that the sets of destination nodes are the same for every source, i.e., $\mathcal{D}_j = \mathcal{D}$ for $j \in [1 : k]$. Hence in this general multimessage multicast setting, every destination node in $\mathcal{D}$ is to recover all the messages. The cutset bound is again tight for this class of networks and is achieved via linear network coding.

Theorem 15.5. The capacity region of the graphical multimessage multicast network $\mathcal{G} = (\mathcal{N}, \mathcal{E}, \mathcal{C})$ with source nodes $[1 : k]$ and destination nodes $\mathcal{D}$ is the set of rate tuples $(R_1, \ldots, R_k)$ such that

$$\sum_{j \in \mathcal{S}} R_j \leq C(\mathcal{S})$$

for all $\mathcal{S} \subset \mathcal{N}$ with $[1 : k] \cap \mathcal{S} \neq \emptyset$ and $\mathcal{D} \cap \mathcal{S}^c \neq \emptyset$.

It can be easily checked that for $k = 1$, this theorem reduces to the network coding theorem. The proof of the converse follows by the cutset bound. For the proof of achievability, we use linear network coding.

Proof of achievability. Since the messages are to be sent to the same set of destination nodes, we can treat them as a single large message rather than distinct nonexchangeable quantities. This key observation makes it straightforward to reduce the multimessage multicast problem to a single-message multicast one as follows.

Consider the augmented network G' depicted in Figure 15.10, where an auxiliary node 0 is connected to every source node $j \in [1 : k]$ by an edge $(0, j)$ of capacity R_j. Suppose that the auxiliary node 0 wishes to communicate a message $M_0 \in [1 : 2^{nR_0}]$ to $\mathcal{D}$. Then, by the achievability proof of the network coding theorem for the multicast network, linear network coding achieves the cutset bound for the augmented network G' and hence its capacity is

$$C_0 = \min C(\mathcal{S}'),$$

where the minimum is over all $\mathcal{S}'$ such that $0 \in \mathcal{S}'$ and $\mathcal{D} \cap \mathcal{S}'^c \neq \emptyset$. Now if the rate tuple $(R_1, \ldots, R_k)$ satisfies the rate constraints in Theorem 15.5, then it can be easily shown that

$$C_0 = \sum_{j=1}^{k} R_j.$$

Hence, there exist an integer n and a coefficient vector α such that for every $j \in \mathcal{D}$, the linear transformation $A'_j(\alpha)$ induced by the corresponding linear network code is invertible. But from the structure of the augmented graph G', the linear transformation can be factored as $A'_j(\alpha) = A_j(\alpha)B_j(\alpha)$, where $B_j(\alpha)$ is a square matrix that encodes $m_0 \in \mathbb{F}_{2^n}^{C_0}$ into $(m_1, \ldots, m_k) \in \mathbb{F}_{2^n}^{C_0}$. Since $A'_j(\alpha)$ is invertible, both $A_j(\alpha)$ and $B_j(\alpha)$ must be invertible as well. Therefore, each destination node $j \in \mathcal{D}$ can recover $(m_1, \ldots, m_k)$ as well as m_0, which establishes the achievability of the rate tuple $(R_1, \ldots, R_k)$.

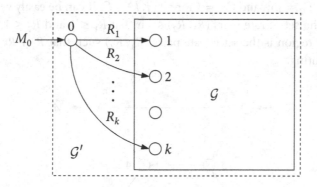

Figure 15.10. Augmented single-message multicast network.

15.4.2   Graphical Multiple-Unicast Network

Consider a multiple-unicast network, where each source j has a single destination, i.e., $|\mathcal{D}_j| = 0$ or 1 for all $j \in [1 : N - 1]$. If the operations at the nodes are restricted to routing, then the problem reduces to the well-studied *multicommodity flow*. The necessary and sufficient conditions for optimal multicommodity flow can be found by linear programming as for the max-flow min-cut theorem. This provides an inner bound on the capacity region of the multiple-unicast network, which is not optimal in general. More generally, each node can perform linear network coding.

Example 15.4. Consider the 2-unicast butterfly network depicted in Figure 15.11, where $R_3 = \cdots = R_6 = 0$, $\mathcal{D}_1 = \{6\}$, $\mathcal{D}_2 = \{5\}$, and $C_{jk} = 1$ for all $(j, k) \in \mathcal{E}$. By the cutset bound, we must have $R_1 \le 1$ and $R_2 \le 1$ (by setting $\mathcal{S} = \{1, 3, 4, 5\}$ and $\mathcal{S} = \{2, 3, 4, 6\}$, respectively), which is achievable via linear network coding. In comparison, routing can achieve at most $R_1 + R_2 \le 1$, because of the bottleneck edge $(3, 4)$.

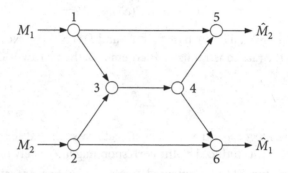

Figure 15.11. Two-unicast butterfly network.

The cutset bound is not tight in general, however, as demonstrated in the following.

Example 15.5. Consider the 2-unicast network depicted in Figure 15.12, where $R_3 = \cdots = R_6 = 0$, $\mathcal{D}_1 = \{6\}$, $\mathcal{D}_2 = \{5\}$, and $C_{jk} = 1$ for all $(j, k) \in \mathcal{E}$. It can be easily verified that the cutset bound is the set of rate pairs (R_1, R_2) such that $R_1 \le 1$ and $R_2 \le 1$. By comparison, the capacity region is the set of rate pairs (R_1, R_2) such that $R_1 + R_2 \le 1$, which is achievable via routing.

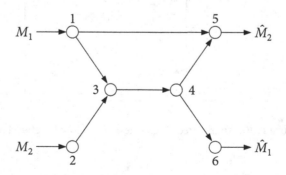

Figure 15.12. Two-unicast network for which the cutset bound is not tight.

The capacity region of the multiple-unicast network is not known in general.

SUMMARY

- Cutset bounds on the capacity of graphical networks

- Max-flow min-cut theorem for graphical unicast networks

- Routing alone does not achieve the capacity of general graphical networks

- Network coding theorem for graphical multicast networks

- Linear network coding achieves the capacity of graphical multimessage multicast networks (error-free and with finite block length)

- **Open problems:**

  15.1. What is the capacity region of the general graphical 2-unicast network?

  15.2. Does linear network coding achieve the capacity region of graphical multiple-unicast networks?

  15.3. Is the average probability of error capacity region of the general graphical multimessage network equal to its maximal probability of error capacity region?

BIBLIOGRAPHIC NOTES

The max-flow min-cut theorem was established by Ford and Fulkerson (1956); see also Elias, Feinstein, and Shannon (1956). The capacity and optimal routing can be found by the Ford–Fulkerson algorithm, which is constructive and more efficient than standard linear programming algorithms such as the simplex and interior point methods.

The butterfly network in Example 15.2 and the network coding theorem are due to Ahlswede, Cai, Li, and Yeung (2000). The proof of the network coding theorem in their paper uses random coding. The source and relay encoding and the decoding mappings are randomly and independently generated, each according to a uniform pmf. The key step in the proof is to show that if the rate R is less than the cutset bound, then the end-to-end mapping is one-to-one with high probability. This proof was extended to cyclic networks by constructing a time-expanded acyclic network as illustrated in Figure 15.13. The nodes in the original network are replicated b times and auxiliary source and destination nodes are added as shown in the figure. An edge is drawn between two nodes in consecutive levels of the time-expanded network if the nodes are connected by an edge in the original network. The auxiliary source node is connected to each copy of the original source node and each copy of a destination node is connected to the corresponding auxiliary destination node. Note that this time-expanded network is always acyclic. Hence, we can use the random coding scheme for the acyclic network, which implies that the same message is in effect sent over b transmission blocks using independent mappings. The key step is to show that for sufficiently large b, the cutset bound for the time-expanded network is roughly b times the capacity of the original network.

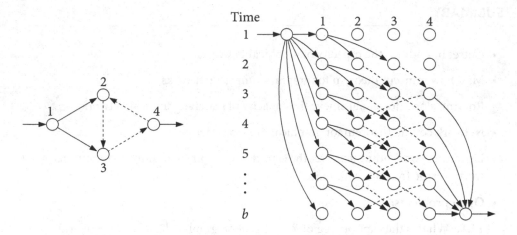

Figure 15.13. A cyclic network and its time-expanded acyclic network. The cycle $2 \to 3 \to 4 \to 2$ in the original network is unfolded to the paths $2(t) \to 3(t+1) \to 4(t+2) \to 2(t+3)$ in the time-expanded network.

The network coding theorem was subsequently proved using linear coding by Li, Yeung, and Cai (2003) and Koetter and Médard (2003). The proof of achievability in Section 15.3.2 follows the latter, who also extended the result to cyclic networks using convolutional codes. Jaggi, Sanders, Chou, Effros, Egner, Jain, and Tolhuizen (2005) developed a polynomial-time algorithm for finding a vector α that makes $A_j(\alpha)$ invertible for each $j \in \mathcal{D}$. In comparison, finding the optimal routing is the same as packing Steiner trees (Chou, Wu, and Jain 2003), which is an NP-complete problem. For sufficiently large n, a *randomly* generated linear network code achieves the capacity with high probability. This random linear network coding can be used as both a construction tool and a method of attaining robustness to link failures and network topology changes (Ho, Médard, Koetter, Karger, Effros, Shi, and Leong 2006, Chou, Wu, and Jain 2003).

Multicommodity flow was first studied by Hu (1963), who established a version of max-flow min-cut theorem for undirected networks with two commodities. Extensions to (unicast) networks with more than two commodities can be found, for example, in Schrijver (2003). The *routing* capacity region of a general multimessage network was established by Cannons, Dougherty, Freiling, and Zeger (2006). The capacity of the multimessage multicast network in Theorem 15.5 is due to Ahlswede, Cai, Li, and Yeung (2000). Dougherty, Freiling, and Zeger (2005) showed via an ingenious counterexample that unlike the multicast case, linear network coding fails to achieve the capacity region of a general graphical multimessage network error-free. This counterexample hinges on a deep connection between linear network coding and matroid theory; see Dougherty, Freiling, and Zeger (2011) and the references therein. Network coding has attracted much attention from researchers in coding theory, wireless communication, and networking, in addition to information theorists. Comprehensive treatments of network coding and its extensions and applications can be found in Yeung, Li, Cai, and Zhang (2005a,b), Fragouli and Soljanin (2007a,b), Yeung (2008), and Ho and Lun (2008).

PROBLEMS

15.1. Show that it suffices to take the minimum over connected cuts in the max-flow min-cut theorem.

15.2. Show that it suffices to take $n \leq \lceil \log(|\mathcal{D}|R + 1) \rceil$ in the achievability proof of the network coding theorem in Section 15.3.2.

15.3. Prove the cutset bound for the general graphical multimessage network in Theorem 15.4.

15.4. *Duality for the max-flow min-cut theorem.* Consider the optimization problem in (15.1).

(a) Verify that (15.1) and (15.2) are dual to each other by rewriting them in the standard forms of the LP and the dual LP in Appendix E.

(b) Show that the set of feasible solutions to (15.3) is characterized by $(v_1, \ldots, v_N)$ such that $v_j \in [0, 1]$, $v_1 = 1$, $v_N = 0$, and $v_j \geq v_k$ if j precedes k. Verify that the solution in (15.4) is feasible.

(c) Using part (b) and the fact that every node lies on some path from node 1 to node N, show that every feasible solution to (15.3) is a convex combination of extreme points of the form (15.4).

15.5. *Hypergraphical network.* Consider a wireless network modeled by a weighted directed acyclic hypergraph $\mathcal{H} = (\mathcal{N}, \mathcal{E}, \mathcal{C})$, where $\mathcal{E}$ now consists of a set of hyperedges $(j, \mathcal{N}_j)$ with capacity $C_{j\mathcal{N}_j}$. Each hyperedge models a noiseless broadcast channel from node j to a set of receiver nodes $\mathcal{N}_j$. Suppose that source node 1 wishes to communicate a message $M \in [1 : 2^{nR}]$ to a set of destination nodes $\mathcal{D}$.

(a) Generalize the cutset bound for the graphical multicast network to establish an upper bound on the capacity of the hypergraphical multicast network.

(b) Show that the cutset bound in part (a) is achievable via linear network coding.

15.6. *Multimessage network.* Consider the network depicted in Figure 15.14, where $R_3 = \cdots = R_6 = 0$, $\mathcal{D}_1 = \{4, 5\}$, $\mathcal{D}_2 = \{6\}$, and $C_{jk} = 1$ for all $(j, k) \in \mathcal{E}$.

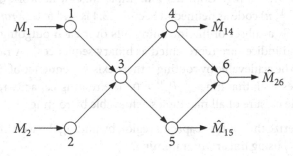

Figure 15.14. Multimessage network for which the cutset bound is not tight.

(a) Show that the cutset bound is the set of rate pairs (R_1, R_2) such that $R_1, R_2 \leq 1$.

(b) Show that the capacity region is the set of rate pairs (R_1, R_2) such that $2R_1 + R_2 \leq 2$ and $R_2 \leq 1$. (Hint: For the proof of the converse, use Fano's inequality and the data processing inequality to show that $nR_1 \leq I(\hat{M}_{14}; \hat{M}_{15}) - n\epsilon_n \leq I(M_{34}; M_{35}) - n\epsilon_n$, and $n(R_1 + R_2) \leq H(M_{34}, M_{35}) - n\epsilon_n$.)

Remark: A similar example appeared in Yeung (2008, Section 21.1) and the hint is due to G. Kramer.

15.7.   *Triangular cyclic network.* Consider the 3-node graphical multiple-unicast network in Figure 15.15, where $C_{12} = C_{23} = C_{31} = 1$. Node $j = 1, 2, 3$ wishes to communicate a message $M_j \in [1 : 2^{nR_j}]$ to its predecessor node.

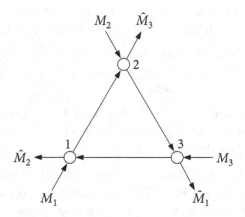

Figure 15.15. Triangular cyclic network.

(a) Find the cutset bound.

(b) Show that the capacity region is the set of rate triples (R_1, R_2, R_3) such that $R_j + R_k \leq 1$ for $j, k = 1, 2, 3$ with $j \neq k$.

Remark: This problem was studied by Kramer and Savari (2006), who developed the edge-cut outer bound.

15.8.   *Multiple-unicast routing.* Consider a multiple-unicast network $\mathcal{G} = (\mathcal{N}, \mathcal{E}, \mathcal{C})$. A $(2^{nR_1}, \ldots, 2^{nR_k}, n)$ code as defined in Section 15.4 is said to be a *routing* code if each encoder sends a subset of the incoming bits over each outgoing edge (when the messages and indices are represented as binary sequences). A rate tuple $(R_1, \ldots, R_k)$ is said to be achievable by routing if there exists a sequence of $(2^{nR_1}, \ldots, 2^{nR_k}, n)$ routing codes such that $\lim_{n \to \infty} P_e^{(n)} = 0$. The routing capacity region of the network $\mathcal{G}$ is the closure of all rate tuples achievable by routing.

(a) Characterize the routing capacity region by finding the weighted sum-capacity $\sum_{j=1}^{k} \lambda_j R_j$ using linear programming.

(b) Show that the routing capacity region is achieved via *forwarding*, that is, no duplication of the same information is needed. Thus, for routing over unicast networks, information from different sources can be treated as physical commodities.

(c) Show that the routing capacity region for average probability of error is the same as that for maximal probability of error.

APPENDIX 15A   PROOF OF LEMMA 15.1

Let $P(x_1, \ldots, x_k)$ be a nonzero polynomial in $\mathbb{F}_2[x_1, \ldots, x_k]$. We show that if n is sufficiently large, then there exist $\alpha_1, \ldots, \alpha_k \in \mathbb{F}_{2^n}$ such that $P(\alpha_1, \ldots, \alpha_k) \neq 0$. First, suppose $k = 1$ and recall the fact that the number of roots of a (single-variable) polynomial $P(x_1) \in \mathbb{F}[x_1]$ cannot exceed its degree for any field $\mathbb{F}$ (Lidl and Niederreiter 1997, Theorem 1.66). Hence by treating $P(x_1)$ as a polynomial over $\mathbb{F}_{2^n}$, there exists an element $\alpha_1 \in \mathbb{F}_{2^n}$ with $P(\alpha_1) \neq 0$, if 2^n is strictly larger than the degree of the polynomial.

We proceed by induction on the number of variables k. Express the polynomial as

$$P(x_1, \ldots, x_k) = \sum_{j=0}^{d} P_j(x_2, \ldots, x_k) x_1^j.$$

Since $P(x_1, \ldots, x_k) \neq 0$, $P_j(x_2, \ldots, x_k) \neq 0$ for some j. Then by the induction hypothesis, if n is sufficiently large, there exist $\alpha_2, \ldots, \alpha_k \in \mathbb{F}_{2^n}$ such that $P_j(\alpha_2, \ldots, \alpha_k) \neq 0$ for some j. But this implies that $P(x_1, \alpha_2, \ldots, \alpha_k) \in \mathbb{F}_{2^n}$ is nonzero. Hence, using the aforementioned fact on the number of roots of single-variable polynomials, we conclude that $P(\alpha_1, \alpha_2, \ldots, \alpha_k)$ is nonzero for some $\alpha_1 \in \mathbb{F}_{2^n}$ if n is sufficiently large.

CHAPTER 16

Relay Channels

In this chapter, we begin our discussion of communication over general multihop networks. We study the 3-node relay channel, which is a model for point-to-point communication with the help of a relay, such as communication between two base stations through both a terrestrial link and a satellite, or between two nodes in a mesh network with an intermediate node acting as a relay.

The capacity of the relay channel is not known in general. We establish a cutset upper bound on the capacity and discuss several coding schemes that are optimal in some special cases. We first discuss the following two extreme schemes.

- Direct transmission: In this simple scheme, the relay is not actively used in the communication.

- Decode–forward: In this multihop scheme, the relay plays a central role in the communication. It decodes for the message and coherently cooperates with the sender to communicate it to the receiver. This scheme involves the new techniques of block Markov coding, backward decoding, and the use of binning in channel coding.

We observe that direct transmission can outperform decode–forward when the channel from the sender to the relay is weaker than that to the receiver. This motivates the development of the following two schemes.

- Partial decode–forward: Here the relay recovers only part of the message and the rest of the message is recovered only by the receiver. We show that this scheme is optimal for a class of semideterministic relay channels and for relay channels with orthogonal sender components.

- Compress–forward: In this scheme, the relay does not attempt to recover the message. Instead, it uses Wyner–Ziv coding with the receiver's sequence acting as side information, and forwards the bin index. The receiver then decodes for the bin index, finds the corresponding reconstruction of the relay received sequence, and uses it together with its own sequence to recover the message. Compress–forward is shown to be optimal for a class of deterministic relay channels and for a modulo-2 sum relay channel example whose capacity turns out to be strictly lower than the cutset bound.

Motivated by wireless networks, we study the following three Gaussian relay channel models.

- Full-duplex Gaussian RC: The capacity for this model is not known for any set of nonzero channel parameter values. We evaluate and compare the cutset upper bound and the decode–forward and compress–forward lower bounds. We show that the partial decode–forward lower bound reduces to the largest of the rates for direct transmission and decode–forward.

- Half-duplex Gaussian RC with sender frequency division: In contrast to the full-duplex RC, we show that partial decode–forward is optimal for this model.

- Half-duplex Gaussian RC with receiver frequency division: We show that the cutset bound coincides with the decode–forward bound for a range of channel parameter values. We then present the amplify–forward coding scheme in which the relay sends a scaled version of its previously received signal. We generalize amplify–forward to linear relaying functions that are weighted sums of past received signals and establish a single-letter characterization of the capacity with linear relaying.

In the last section of this chapter, we study the effect of relay lookahead on capacity. In the relay channel setup, we assume that the relaying functions depend only on past received relay symbols; hence they are strictly causal. Here we allow the relaying functions to depend with some lookahead on the relay received sequence. We study two extreme lookahead models—the noncausal relay channel and the causal relay channel. We present upper and lower bounds on the capacity for these two models that are tight in some cases. In particular, we show that the cutset bound for the strictly causal relay channel does not hold for the causal relay channel. We further show that simple instantaneous relaying can be optimal and achieves higher rates than the cutset bound for the strictly causal relay channel. We then extend these results to the (full-duplex) Gaussian case. For the noncausal Gaussian RC, we show that capacity is achieved via noncausal decode–forward if the channel from the sender to the relay is sufficiently strong, while for the causal Gaussian RC, we show that capacity is achieved via instantaneous amplify–forward if the channel from the sender to the relay is sufficiently weaker than the other two channels. These results are in sharp contrast to the strictly causal case for which capacity is not known for any nonzero channel parameter values.

16.1   DISCRETE MEMORYLESS RELAY CHANNEL

Consider the 3-node point-to-point communication system with a relay depicted in Figure 16.1. The sender (node 1) wishes to communicate a message M to the receiver (node 3) with the help of the relay (node 2). We first consider the *discrete memoryless relay channel* (DM-RC) model $(\mathcal{X}_1 \times \mathcal{X}_2, p(y_2, y_3 | x_1, x_2), \mathcal{Y}_2 \times \mathcal{Y}_3)$ that consists of four finite sets $\mathcal{X}_1$, $\mathcal{X}_2, \mathcal{Y}_2, \mathcal{Y}_3$, and a collection of conditional pmfs $p(y_2, y_3 | x_1, x_2)$ on $\mathcal{Y}_2 \times \mathcal{Y}_3$.

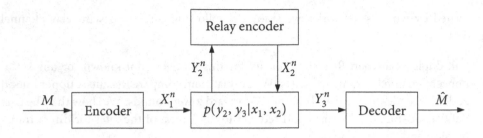

Figure 16.1. Point-to-point communication system with a relay.

A $(2^{nR}, n)$ code for the DM-RC consists of

- a message set $[1 : 2^{nR}]$,

- an encoder that assigns a codeword $x_1^n(m)$ to each message $m \in [1 : 2^{nR}]$,

- a relay encoder that assigns a symbol $x_{2i}(y_2^{i-1})$ to each past received sequence $y_2^{i-1} \in \mathcal{Y}_2^{i-1}$ for each time $i \in [1 : n]$, and

- a decoder that assigns an estimate $\hat{m}$ or an error message e to each received sequence $y_3^n \in \mathcal{Y}_3^n$.

The channel is memoryless in the sense that the current received symbols (Y_{2i}, Y_{3i}) and the message and past symbols $(m, X_1^{i-1}, X_2^{i-1}, Y_2^{i-1}, Y_3^{i-1})$ are conditionally independent given the current transmitted symbols (X_{1i}, X_{2i}).

We assume that the message M is uniformly distributed over the message set. The average probability of error is defined as $P_e^{(n)} = P\{\hat{M} \neq M\}$. A rate R is said to be *achievable* for the DM-RC if there exists a sequence of $(2^{nR}, n)$ codes such that $\lim_{n\to\infty} P_e^{(n)} = 0$. The *capacity* C of the DM-RC is the supremum of all achievable rates.

The capacity of the DM-RC is not known in general. We discuss upper and lower bounds on the capacity that are tight for some special classes of relay channels.

16.2 CUTSET UPPER BOUND ON THE CAPACITY

The following upper bound is motivated by the cutset bounds for graphical networks we discussed in Chapter 15.

Theorem 16.1 (Cutset Bound for the DM-RC). The capacity of the DM-RC is upper bounded as

$$C \leq \max_{p(x_1, x_2)} \min\{I(X_1, X_2; Y_3), I(X_1; Y_2, Y_3 | X_2)\}.$$

The terms under the minimum in the bound can be interpreted as *cooperative* multiple access and broadcast bounds as illustrated in Figure 16.2—the sender cannot transmit information at a higher rate than if both senders or both receivers fully cooperate.

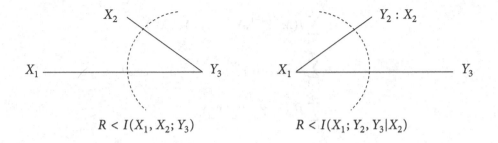

$$R < I(X_1, X_2; Y_3)$$

$$R < I(X_1; Y_2, Y_3 | X_2)$$

Cooperative multiple access bound   Cooperative broadcast bound

Figure 16.2. Min-cut interpretation of the cutset bound.

The cutset bound is tight for many classes of DM-RCs with known capacity. However, it is not tight in general as shown via an example in Section 16.7.3.

Proof of Theorem 16.1. By Fano's inequality,

$$nR = H(M) = I(M; Y_3^n) + H(M|Y_3^n) \le I(M; Y_3^n) + n\epsilon_n,$$

where ϵ_n tends to zero as $n \to \infty$. We now show that

$$I(M; Y_3^n) \le \min\left\{ \sum_{i=1}^{n} I(X_{1i}, X_{2i}; Y_{3i}), \sum_{i=1}^{n} I(X_{1i}; Y_{2i}, Y_{3i} | X_{2i}) \right\}.$$

To establish the first inequality, consider

$$I(M; Y_3^n) = \sum_{i=1}^{n} I(M; Y_{3i} | Y_3^{i-1})$$

$$\le \sum_{i=1}^{n} I(M, Y_3^{i-1}; Y_{3i})$$

$$\le \sum_{i=1}^{n} I(X_{1i}, X_{2i}, M, Y_3^{i-1}; Y_{3i})$$

$$= \sum_{i=1}^{n} I(X_{1i}, X_{2i}; Y_{3i}).$$

To establish the second inequality, consider

$$I(M; Y_3^n) \le I(M; Y_2^n, Y_3^n)$$

$$= \sum_{i=1}^{n} I(M; Y_{2i}, Y_{3i} | Y_2^{i-1}, Y_3^{i-1})$$

$$\overset{(a)}{=} \sum_{i=1}^{n} I(M; Y_{2i}, Y_{3i} | Y_2^{i-1}, Y_3^{i-1}, X_{2i})$$

$$\leq \sum_{i=1}^{n} I(M, Y_2^{i-1}, Y_3^{i-1}; Y_{2i}, Y_{3i} | X_{2i})$$

$$= \sum_{i=1}^{n} I(X_{1i}, M, Y_2^{i-1}, Y_3^{i-1}; Y_{2i}, Y_{3i} | X_{2i})$$

$$= \sum_{i=1}^{n} I(X_{1i}; Y_{2i}, Y_{3i} | X_{2i}),$$

where (a) follows since X_{2i} is a function of Y_2^{i-1}. Finally, let $Q \sim \text{Unif}[1:n]$ be independent of $(X_1^n, X_2^n, Y_2^n, Y_3^n)$ and set $X_1 = X_{1Q}, X_2 = X_{2Q}, Y_2 = Y_{2Q}, Y_3 = Y_{3Q}$. Since $Q \to (X_1, X_2) \to (Y_2, Y_3)$, we have

$$\sum_{i=1}^{n} I(X_{1i}, X_{2i}; Y_{3i}) = nI(X_1, X_2; Y_3 | Q) \leq nI(X_1, X_2; Y_3),$$

$$\sum_{i=1}^{n} I(X_{1i}; Y_{2i}, Y_{3i} | X_{2i}) = nI(X_1; Y_2, Y_3 | X_2, Q) \leq nI(X_1; Y_2, Y_3 | X_2).$$

Thus

$$R \leq \min\{I(X_1, X_2; Y_3), I(X_1; Y_2, Y_3 | X_2)\} + \epsilon_n.$$

Taking $n \to \infty$ completes the proof of the cutset bound.

16.3   DIRECT-TRANSMISSION LOWER BOUND

One simple coding scheme for the relay channel is to fix the relay transmission at the most favorable symbol to the channel from the sender to the receiver and to communicate the message directly using optimal point-to-point channel coding. The capacity of the relay channel is thus lower bounded by the capacity of the resulting DMC as

$$C \geq \max_{p(x_1), x_2} I(X_1; Y_3 | X_2 = x_2). \tag{16.1}$$

This bound is tight for the class of *reversely degraded* DM-RCs in which

$$p(y_2, y_3 | x_1, x_2) = p(y_3 | x_1, x_2)p(y_2 | y_3, x_2),$$

that is, $X_1 \to Y_3 \to Y_2$ form a Markov chain conditioned on X_2. The proof of the converse follows by the cutset bound and noting that $I(X_1; Y_2, Y_3 | X_2) = I(X_1; Y_3 | X_2)$.

When the relay channel is not reversely degraded, however, we can achieve better rates by actively using the relay.

16.4   DECODE–FORWARD LOWER BOUND

At the other extreme of direct transmission, the decode–forward coding scheme relies heavily on the relay to help communicate the message to the receiver. We develop this scheme in three steps. In the first two steps, we use a multihop relaying scheme in which the receiver treats the transmission from the sender as noise. The decode–forward scheme improves upon this multihop scheme by having the receiver decode also for the information sent directly by the sender.

16.4.1   Multihop Lower Bound

In the multihop relaying scheme, the relay recovers the message received from the sender in each block and retransmits it in the following block. This gives the lower bound on the capacity of the DM-RC

$$C \geq \max_{p(x_1)p(x_2)} \min\{I(X_2; Y_3), I(X_1; Y_2|X_2)\}. \tag{16.2}$$

It is not difficult to show that this lower bound is tight when the DM-RC consists of a cascade of two DMCs, i.e., $p(y_2, y_3|x_1, x_2) = p(y_2|x_1)p(y_3|x_2)$. In this case the capacity expression simplifies to

$$
\begin{aligned}
C &= \max_{p(x_1)p(x_2)} \min\{I(X_2; Y_3), I(X_1; Y_2|X_2)\} \\
&= \max_{p(x_1)p(x_2)} \min\{I(X_2; Y_3), I(X_1; Y_2)\} \\
&= \min\Big\{\max_{p(x_2)} I(X_2; Y_3), \max_{p(x_1)} I(X_1; Y_2)\Big\}.
\end{aligned}
$$

Achievability of the multihop lower bound uses b transmission blocks, each consisting of n transmissions, as illustrated in Figure 16.3. A sequence of $(b - 1)$ messages M_j, $j \in [1 : b - 1]$, each selected independently and uniformly over $[1 : 2^{nR}]$, is sent over these b blocks, We assume $m_b = 1$ by convention. Note that the average rate over the b blocks is $R(b - 1)/b$, which can be made as close to R as desired.

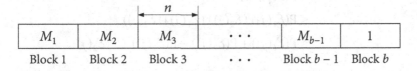

M_1	M_2	M_3	$\cdots$	M_{b-1}	1
Block 1	Block 2	Block 3	$\cdots$	Block $b - 1$	Block b

Figure 16.3. Multiple transmission blocks used in the multihop scheme.

Codebook generation. Fix the product pmf $p(x_1)p(x_2)$ that attains the multihop lower bound in (16.2). Randomly and independently generate a codebook for each block. For each $j \in [1 : b]$, randomly and independently generate 2^{nR} sequences $x_1^n(m_j)$, $m_j \in [1 : 2^{nR}]$, each according to $\prod_{i=1}^n p_{X_1}(x_{1i})$. Similarly, generate 2^{nR} sequences $x_2^n(m_{j-1})$, $m_{j-1} \in [1 : 2^{nR}]$, each according to $\prod_{i=1}^n p_{X_2}(x_{2i})$. This defines the codebook

$$\mathcal{C}_j = \{(x_1^n(m_j), x_2^n(m_{j-1})): m_{j-1}, m_j \in [1 : 2^{nR}]\}, \quad j \in [1 : b].$$

The codebooks are revealed to all parties.

Encoding. Let $m_j \in [1 : 2^{nR}]$ be the new message to be sent in block j. The encoder transmits $x_1^n(m_j)$ from codebook C_j.

Relay encoding. By convention, let $\tilde{m}_0 = 1$. At the end of block j, the relay finds the unique message $\tilde{m}_j$ such that $(x_1^n(\tilde{m}_j), x_2^n(\tilde{m}_{j-1}), y_2^n(j)) \in T_\epsilon^{(n)}$. In block $j + 1$, it transmits $x_2^n(\tilde{m}_j)$ from codebook C_{j+1}.

Since the relay codeword transmitted in a block depends statistically on the message transmitted in the previous block, we refer to this scheme as *block Markov coding*.

Decoding. At the end of block $j + 1$, the receiver finds the unique message $\hat{m}_j$ such that $(x_2^n(\hat{m}_j), y_3^n(j + 1)) \in T_\epsilon^{(n)}$.

Analysis of the probability of error. We analyze the probability of decoding error for the message M_j averaged over codebooks. Assume without loss of generality that $M_j = 1$. Let $\tilde{M}_j$ be the relay message estimate at the end of block j. Since

$$\{\hat{M}_j \neq 1\} \subseteq \{\tilde{M}_j \neq 1\} \cup \{\hat{M}_j \neq \tilde{M}_j\},$$

the decoder makes an error only if one or more of the following events occur:

$$\tilde{\mathcal{E}}_1(j) = \{(X_1^n(1), X_2^n(\tilde{M}_{j-1}), Y_2^n(j)) \notin T_\epsilon^{(n)}\},$$
$$\tilde{\mathcal{E}}_2(j) = \{(X_1^n(m_j), X_2^n(\tilde{M}_{j-1}), Y_2^n(j)) \in T_\epsilon^{(n)} \text{ for some } m_j \neq 1\},$$
$$\mathcal{E}_1(j) = \{(X_2^n(\tilde{M}_j), Y_3^n(j + 1)) \notin T_\epsilon^{(n)}\},$$
$$\mathcal{E}_2(j) = \{(X_2^n(m_j), Y_3^n(j + 1)) \in T_\epsilon^{(n)} \text{ for some } m_j \neq \tilde{M}_j\}.$$

Thus, the probability of error is upper bounded as

$$\begin{aligned}
P(\mathcal{E}(j)) &= P\{\hat{M}_j \neq 1\} \\
&\leq P(\tilde{\mathcal{E}}_1(j) \cup \tilde{\mathcal{E}}_2(j) \cup \mathcal{E}_1(j) \cup \mathcal{E}_2(j)) \\
&\leq P(\tilde{\mathcal{E}}_1(j)) + P(\tilde{\mathcal{E}}_2(j)) + P(\mathcal{E}_1(j)) + P(\mathcal{E}_2(j)),
\end{aligned}$$

where the first two terms upper bound $P\{\tilde{M}_j \neq 1\}$ and the last two terms upper bound $P\{\hat{M}_j \neq \tilde{M}_j\}$.

Now, by the independence of the codebooks, the relay message estimate $\tilde{M}_{j-1}$, which is a function of $Y_2^n(j - 1)$ and codebook C_{j-1}, is independent of the codewords $X_1^n(m_j)$, $X_2^n(m_{j-1})$, $m_j, m_{j-1} \in [1 : 2^{nR}]$, from codebook C_j. Hence, by the LLN, $P(\tilde{\mathcal{E}}_1(j))$ tends to zero as $n \to \infty$, and by the packing lemma, $P(\tilde{\mathcal{E}}_2(j))$ tends to zero as $n \to \infty$ if $R < I(X_1; Y_2|X_2) - \delta(\epsilon)$. Similarly, by the independence of the codebooks and the LLN, $P(\mathcal{E}_1(j))$ tends to zero as $n \to \infty$, and by the same independence and the packing lemma, $P(\mathcal{E}_2(j))$ tends to zero as $n \to \infty$ if $R < I(X_2; Y_3) - \delta(\epsilon)$. Thus we have shown that under the given constraints on the rate, $P\{\hat{M}_j \neq M_j\}$ tends to zero as $n \to \infty$ for each $j \in [1 : b - 1]$. This completes the proof of the multihop lower bound.

16.4.2 Coherent Multihop Lower Bound

The multihop scheme discussed in the previous subsection can be improved by having the sender and the relay *coherently* cooperate in transmitting their codewords. With this improvement, we obtain the lower bound on the capacity of the DM-RC

$$C \geq \max_{p(x_1, x_2)} \min\{I(X_2; Y_3), I(X_1; Y_2 | X_2)\}. \tag{16.3}$$

Again we use a block Markov coding scheme in which a sequence of $(b-1)$ i.i.d. messages M_j, $j \in [1 : b-1]$, is sent over b blocks each consisting of n transmissions.

Codebook generation. Fix the pmf $p(x_1, x_2)$ that attains the lower bound in (16.3). For $j \in [1 : b]$, randomly and independently generate 2^{nR} sequences $x_2^n(m_{j-1})$, $m_{j-1} \in [1 : 2^{nR}]$, each according to $\prod_{i=1}^n p_{X_2}(x_{2i})$. For each $m_{j-1} \in [1 : 2^{nR}]$, randomly and conditionally independently generate 2^{nR} sequences $x_1^n(m_j | m_{j-1})$, $m_j \in [1 : 2^{nR}]$, each according to $\prod_{i=1}^n p_{X_1 | X_2}(x_{1i} | x_{2i}(m_{j-1}))$. This defines the codebook

$$\mathcal{C}_j = \{(x_1^n(m_j | m_{j-1}), x_2^n(m_{j-1})): m_{j-1}, m_j \in [1 : 2^{nR}]\}, \quad j \in [1 : b].$$

The codebooks are revealed to all parties.

Encoding and decoding are explained with the help of Table 16.1.

Block	1	2	3	$\cdots$	$b-1$	b
X_1	$x_1^n(m_1\|1)$	$x_1^n(m_2\|m_1)$	$x_1^n(m_3\|m_2)$	$\cdots$	$x_1^n(m_{b-1}\|m_{b-2})$	$x_1^n(1\|m_{b-1})$
Y_2	$\tilde{m}_1$	$\tilde{m}_2$	$\tilde{m}_3$	$\cdots$	$\tilde{m}_{b-1}$	$\emptyset$
X_2	$x_2^n(1)$	$x_2^n(\tilde{m}_1)$	$x_2^n(\tilde{m}_2)$	$\cdots$	$x_2^n(\tilde{m}_{b-2})$	$x_2^n(\tilde{m}_{b-1})$
Y_3	$\emptyset$	$\hat{m}_1$	$\hat{m}_2$	$\cdots$	$\hat{m}_{b-2}$	$\hat{m}_{b-1}$

Table 16.1. Encoding and decoding for the coherent multihop lower bound.

Encoding. Let $m_j \in [1 : 2^{nR}]$ be the message to be sent in block j. The encoder transmits $x_1^n(m_j | m_{j-1})$ from codebook $\mathcal{C}_j$, where $m_0 = m_b = 1$ by convention.

Relay encoding. By convention, let $\tilde{m}_0 = 1$. At the end of block j, the relay finds the unique message $\tilde{m}_j$ such that $(x_1^n(\tilde{m}_j | \tilde{m}_{j-1}), x_2^n(\tilde{m}_{j-1}), y_2^n(j)) \in \mathcal{T}_\epsilon^{(n)}$. In block $j + 1$, it transmits $x_2^n(\tilde{m}_j)$ from codebook $\mathcal{C}_{j+1}$.

Decoding. At the end of block $j + 1$, the receiver finds the unique message $\hat{m}_j$ such that $(x_2^n(\hat{m}_j), y_3^n(j + 1)) \in \mathcal{T}_\epsilon^{(n)}$.

Analysis of the probability of error. We analyze the probability of decoding error for M_j averaged over codebooks. Assume without loss of generality that $M_{j-1} = M_j = 1$. Let $\tilde{M}_j$

be the relay message estimate at the end of block j. As before, the decoder makes an error only if one or more of the following events occur:

$$\tilde{\mathcal{E}}(j) = \{\tilde{M}_j \neq 1\},$$
$$\mathcal{E}_1(j) = \{(X_2^n(\tilde{M}_j), Y_3^n(j+1)) \notin \mathcal{T}_{\epsilon}^{(n)}\},$$
$$\mathcal{E}_2(j) = \{(X_2^n(m_j), Y_3^n(j+1)) \in \mathcal{T}_{\epsilon}^{(n)} \text{ for some } m_j \neq \tilde{M}_j\}.$$

Thus, the probability of error is upper bounded as

$$P(\mathcal{E}(j)) = P\{\hat{M}_j \neq 1\} \leq P(\tilde{\mathcal{E}}(j) \cup \mathcal{E}_1(j) \cup \mathcal{E}_2(j)) \leq P(\tilde{\mathcal{E}}(j)) + P(\mathcal{E}_1(j)) + P(\mathcal{E}_2(j)).$$

Following the same steps to the analysis of the probability of error for the (noncoherent) multihop scheme, the last two terms, $P(\mathcal{E}_1(j))$ and $P(\mathcal{E}_2(j))$, tend to zero as $n \to \infty$ if $R < I(X_2; Y_3) - \delta(\epsilon)$. To upper bound the first term $P(\tilde{\mathcal{E}}(j))$, define

$$\tilde{\mathcal{E}}_1(j) = \{(X_1^n(1|\tilde{M}_{j-1}), X_2^n(\tilde{M}_{j-1}), Y_2^n(j)) \notin \mathcal{T}_{\epsilon}^{(n)}\},$$
$$\tilde{\mathcal{E}}_2(j) = \{(X_1^n(m_j|\tilde{M}_{j-1}), X_2^n(\tilde{M}_{j-1}), Y_2^n(j)) \in \mathcal{T}_{\epsilon}^{(n)} \text{ for some } m_j \neq 1\}.$$

Then

$$P(\tilde{\mathcal{E}}(j)) \leq P(\tilde{\mathcal{E}}(j-1) \cup \tilde{\mathcal{E}}_1(j) \cup \tilde{\mathcal{E}}_2(j))$$
$$\leq P(\tilde{\mathcal{E}}(j-1)) + P(\tilde{\mathcal{E}}_1(j) \cap \tilde{\mathcal{E}}^c(j-1)) + P(\tilde{\mathcal{E}}_2(j)).$$

Consider the second term

$$P(\tilde{\mathcal{E}}_1(j) \cap \tilde{\mathcal{E}}^c(j-1)) = P\{(X_1^n(1|\tilde{M}_{j-1}), X_2^n(\tilde{M}_{j-1}), Y_2^n(j)) \notin \mathcal{T}_{\epsilon}^{(n)}, \tilde{M}_{j-1} = 1\}$$
$$\leq P\{(X_1^n(1|1), X_2^n(1), Y_2^n(j)) \notin \mathcal{T}_{\epsilon}^{(n)} \mid \tilde{M}_{j-1} = 1\},$$

which, by the independence of the codebooks and the LLN, tends to zero as $n \to \infty$. By the packing lemma, $P(\tilde{\mathcal{E}}_2(j))$ tends to zero as $n \to \infty$ if $R < I(X_1; Y_2|X_2) - \delta(\epsilon)$. Note that $\tilde{M}_0 = 1$ by definition. Hence, by induction, $P(\tilde{\mathcal{E}}(j))$ tends to zero as $n \to \infty$ for every $j \in [1:b-1]$. Thus we have shown that under the given constraints on the rate, $P\{\hat{M}_j \neq M_j\}$ tends to zero as $n \to \infty$ for every $j \in [1:b-1]$. This completes the proof of achievability of the coherent multihop lower bound in (16.3).

16.4.3 Decode–Forward Lower Bound

The coherent multihop scheme can be further improved by having the receiver decode simultaneously for the messages sent by the sender and the relay. This leads to the following.

Theorem 16.2 (Decode–Forward Lower Bound). The capacity of the DM-RC is lower bounded as

$$C \geq \max_{p(x_1, x_2)} \min\{I(X_1, X_2; Y_3), I(X_1; Y_2|X_2)\}.$$

Note that the main difference between this bound and the cutset bound is that the latter includes Y_3 in the second mutual information term.

This lower bound is tight when the DM-RC is *degraded*, i.e.,

$$p(y_2, y_3 | x_1, x_2) = p(y_2 | x_1, x_2) p(y_3 | y_2, x_2).$$

The proof of the converse for this case follows by the cutset upper bound in Theorem 16.1 since the degradedness of the channel implies that $I(X_1; Y_2, Y_3 | X_2) = I(X_1; Y_2 | X_2)$. We illustrate this capacity result in the following.

Example 16.1 (Sato relay channel).   Consider the degraded DM-RC with $\mathcal{X}_1 = \mathcal{Y}_2 = \mathcal{Y}_3 = \{0, 1, 2\}$, $\mathcal{X}_2 = \{0, 1\}$, and $Y_2 = X_1$ as depicted in Figure 16.4. With direct transmission, $R_0 = 1$ bits/transmission can be achieved by setting $X_2 = 0$ (or 1). By comparison, using the optimal first-order Markov relay function $x_{2i}(y_{2,i-1})$ yields $R_1 = 1.0437$, and using the optimal second-order Markov relay function $x_{2i}(y_{2,i-1}, y_{2,i-2})$ yields $R_2 = 1.0549$. Since the channel is degraded, the capacity coincides with the decode–forward lower bound in Theorem 16.2. Evaluating this bound yields $C = 1.1619$.

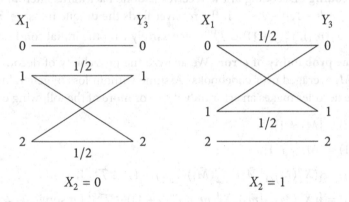

Figure 16.4. Sato relay channel.

16.4.4   Proof via Backward Decoding

Again we consider b transmission blocks, each consisting of n transmissions, and use a block Markov coding scheme. A sequence of $(b - 1)$ i.i.d. messages $M_j \in [1 : 2^{nR}]$, $j \in [1 : b - 1]$, is to be sent over the channel in nb transmissions.

Codebook generation.   Fix the pmf $p(x_1, x_2)$ that attains the lower bound. As in the coherent multihop scheme, we randomly and independently generate codebooks $\mathcal{C}_j = \{(x_1^n(m_j | m_{j-1}), x_2^n(m_{j-1})): m_{j-1}, m_j \in [1 : 2^{nR}]\}$, $j \in [1 : b]$.

Encoding and backward decoding are explained with the help of Table 16.2.

Encoding.   Encoding is again the same as in the coherent multihop scheme. To send m_j in block j, the encoder transmits $x_1^n(m_j | m_{j-1})$ from codebook $\mathcal{C}_j$, where $m_0 = m_b = 1$ by convention.

Block	1	2	3	$\cdots$	$b-1$	b
X_1	$x_1^n(m_1\|1)$	$x_1^n(m_2\|m_1)$	$x_1^n(m_3\|m_2)$	$\cdots$	$x_1^n(m_{b-1}\|m_{b-2})$	$x_1^n(1\|m_{b-1})$
Y_2	$\tilde{m}_1 \rightarrow$	$\tilde{m}_2 \rightarrow$	$\tilde{m}_3 \rightarrow$	$\cdots$	$\tilde{m}_{b-1}$	$\emptyset$
X_2	$x_2^n(1)$	$x_2^n(\tilde{m}_1)$	$x_2^n(\tilde{m}_2)$	$\cdots$	$x_2^n(\tilde{m}_{b-2})$	$x_2^n(\tilde{m}_{b-1})$
Y_3	$\emptyset$	$\tilde{m}_1$	$\leftarrow \hat{m}_2$	$\cdots$	$\leftarrow \hat{m}_{b-2}$	$\leftarrow \hat{m}_{b-1}$

Table 16.2. Encoding and backward decoding for the decode–forward lower bound.

Relay encoding. Relay encoding is also the same as in the coherent multihop scheme. By convention, let $\tilde{m}_0 = 1$. At the end of block j, the relay finds the unique message $\tilde{m}_j$ such that $(x_1^n(\tilde{m}_j|\tilde{m}_{j-1}), x_2^n(\tilde{m}_{j-1}), y_2^n(j)) \in T_\epsilon^{(n)}$. In block $j + 1$, it transmits $x_2^n(\tilde{m}_j)$ from codebook C_{j+1}.

Backward decoding. Decoding at the receiver is done backwards after all b blocks are received. For $j = b - 1, b - 2, \ldots, 1$, the receiver finds the unique message $\hat{m}_j$ such that $(x_1^n(\hat{m}_{j+1}|\hat{m}_j), x_2^n(\hat{m}_j), y_3^n(j + 1)) \in T_\epsilon^{(n)}$, successively with the initial condition $\hat{m}_b = 1$.

Analysis of the probability of error. We analyze the probability of decoding error for the message M_j averaged over codebooks. Assume without loss of generality that $M_j = M_{j+1} = 1$. The decoder makes an error only if one or more of the following events occur:

$$\tilde{\mathcal{E}}(j) = \{\tilde{M}_j \neq 1\},$$
$$\mathcal{E}(j + 1) = \{\hat{M}_{j+1} \neq 1\},$$
$$\mathcal{E}_1(j) = \{(X_1^n(\hat{M}_{j+1}|\tilde{M}_j), X_2^n(\tilde{M}_j), Y_3^n(j + 1)) \notin T_\epsilon^{(n)}\},$$
$$\mathcal{E}_2(j) = \{(X_1^n(\hat{M}_{j+1}|m_j), X_2^n(m_j), Y_3^n(j + 1)) \in T_\epsilon^{(n)} \text{ for some } m_j \neq \tilde{M}_j\}.$$

Thus the probability of error is upper bounded as

$$P(\mathcal{E}(j)) = P\{\hat{M}_j \neq 1\}$$
$$\leq P(\tilde{\mathcal{E}}(j) \cup \mathcal{E}(j + 1) \cup \mathcal{E}_1(j) \cup \mathcal{E}_2(j))$$
$$\leq P(\tilde{\mathcal{E}}(j)) + P(\mathcal{E}(j + 1)) + P(\mathcal{E}_1(j) \cap \tilde{\mathcal{E}}^c(j) \cap \mathcal{E}^c(j + 1)) + P(\mathcal{E}_2(j)).$$

Following the same steps as in the analysis of the probability of error for the coherent multihop scheme, the first term $P(\tilde{\mathcal{E}}(j))$ tends to zero as $n \rightarrow \infty$ if $R < I(X_1; Y_2|X_2) - \delta(\epsilon)$. The third term is upper bounded as

$$P(\mathcal{E}_1(j) \cap \tilde{\mathcal{E}}^c(j) \cap \mathcal{E}^c(j + 1))$$
$$= P(\mathcal{E}_1(j) \cap \{\hat{M}_{j+1} = 1\} \cap \{\tilde{M}_j = 1\})$$
$$= P\{(X_1^n(1|1), X_2^n(1), Y_3^n(j + 1)) \notin T_\epsilon^{(n)}, \hat{M}_{j+1} = 1, \tilde{M}_j = 1\}$$
$$\leq P\{(X_1^n(1|1), X_2^n(1), Y_3^n(j + 1)) \notin T_\epsilon^{(n)} | \tilde{M}_j = 1\},$$

which, by the independence of the codebooks and the LLN, tends to zero as $n \to \infty$. By the same independence and the packing lemma, the fourth term $P(\mathcal{E}_2(j))$ tends to zero as $n \to \infty$ if $R < I(X_1, X_2; Y_3) - \delta(\epsilon)$. Finally for the second term $P(\mathcal{E}(j+1))$, note that $\hat{M}_b = M_b = 1$. Hence, by induction, $P\{\hat{M}_j \neq M_j\}$ tends to zero as $n \to \infty$ for every $j \in [1:b-1]$ if the given constraints on the rate are satisfied. This completes the proof of the decode–forward lower bound.

16.4.5   Proof via Binning

The excessive delay of backward decoding can be alleviated by using *binning* or *sliding window decoding*. We describe the binning scheme here. The proof using sliding window decoding will be given in Chapter 18.

In the binning scheme, the sender and the relay cooperatively send the bin index L_j of the message M_j (instead of the message itself) in block $j + 1$ to help the receiver recover M_j.

Codebook generation. Fix the pmf $p(x_1, x_2)$ that attains the lower bound. Let $0 \le R_2 \le R$. For each $j \in [1:b]$, randomly and independently generate 2^{nR_2} sequences $x_2^n(l_{j-1})$, $l_{j-1} \in [1:2^{nR_2}]$, each according to $\prod_{i=1}^{n} p_{X_2}(x_{2i})$. For each $l_{j-1} \in [1:2^{nR_2}]$, randomly and conditionally independently generate 2^{nR} sequences $x_1^n(m_j|l_{j-1})$, $m_j \in [1:2^{nR}]$, each according to $\prod_{i=1}^{n} p_{X_1|X_2}(x_{1i}|x_{2i}(l_{j-1}))$. This defines the codebook

$$C_j = \{(x_1^n(m_j|l_{j-1}), x_2^n(l_{j-1})): m_j \in [1:2^{nR}], l_{j-1} \in [1:2^{nR_2}]\}, \quad j \in [1:b].$$

Partition the set of messages into 2^{nR_2} equal size bins $\mathcal{B}(l) = [(l-1)2^{n(R-R_2)} + 1 : l2^{n(R-R_2)}]$, $l \in [1:2^{nR_2}]$. The codebooks and bin assignments are revealed to all parties. Encoding and decoding are explained with the help of Table 16.3.

Block	1	2	3	$\cdots$	$b-1$	b
X_1	$x_1^n(m_1|1)$	$x_1^n(m_2|l_1)$	$x_1^n(m_3|l_2)$	$\cdots$	$x_1^n(m_{b-1}|l_{b-2})$	$x_1^n(1|l_{b-1})$
Y_2	$\tilde{m}_1, \tilde{l}_1$	$\tilde{m}_2, \tilde{l}_2$	$\tilde{m}_3, \tilde{l}_3$	$\cdots$	$\tilde{m}_{b-1}, \tilde{l}_{b-1}$	$\emptyset$
X_2	$x_2^n(1)$	$x_2^n(\tilde{l}_1)$	$x_2^n(\tilde{l}_2)$	$\cdots$	$x_2^n(\tilde{l}_{b-2})$	$x_2^n(\tilde{l}_{b-1})$
Y_3	$\emptyset$	$\hat{l}_1, \hat{m}_1$	$\hat{l}_2, \hat{m}_2$	$\cdots$	$\hat{l}_{b-2}, \hat{m}_{b-2}$	$\hat{l}_{b-1}, \hat{m}_{b-1}$

Table 16.3. Encoding and decoding of the binning scheme for decode–forward.

Encoding. Let $m_j \in [1:2^{nR}]$ be the new message to be sent in block j and assume that $m_{j-1} \in \mathcal{B}(l_{j-1})$. The encoder transmits $x_1^n(m_j|l_{j-1})$ from codebook C_j, where $l_0 = m_b = 1$ by convention.

Relay encoding. At the end of block j, the relay finds the unique message $\tilde{m}_j$ such that

$(x_1^n(\tilde{m}_j|\tilde{l}_{j-1}), x_2^n(\tilde{l}_{j-1}), y_2^n(j)) \in \mathcal{T}_\epsilon^{(n)}$. If $\tilde{m}_j \in \mathcal{B}(\tilde{l}_j)$, the relay transmits $x_2^n(\tilde{l}_j)$ from code-book $\mathcal{C}_{j+1}$ in block $j + 1$, where by convention $\tilde{l}_0 = 1$.

Decoding. At the end of block $j + 1$, the receiver finds the unique index $\hat{l}_j$ such that $(x_2^n(\hat{l}_j), y_3^n(j+1)) \in \mathcal{T}_\epsilon^{(n)}$. It then finds the unique message $\hat{m}_j$ such that $(x_1^n(\hat{m}_j|\hat{l}_{j-1}), x_2^n(\hat{l}_{j-1}), y_3^n(j)) \in \mathcal{T}_\epsilon^{(n)}$ and $\hat{m}_j \in \mathcal{B}(\hat{l}_j)$.

Analysis of the probability of error. We analyze the probability of decoding error for the message M_j averaged over codebooks. Assume without loss of generality that $M_j = L_{j-1} = L_j = 1$ and let $\tilde{L}_j$ be the relay estimate of L_j. Then the decoder makes an error only if one or more of the following events occur:

$$\tilde{\mathcal{E}}(j - 1) = \{\tilde{L}_{j-1} \neq 1\},$$
$$\mathcal{E}_1(j - 1) = \{\hat{L}_{j-1} \neq 1\},$$
$$\mathcal{E}_1(j) = \{\hat{L}_j \neq 1\},$$
$$\mathcal{E}_2(j) = \{(X_1^n(1|\hat{L}_{j-1}), X_2^n(\hat{L}_{j-1}), Y_3^n(j)) \notin \mathcal{T}_\epsilon^{(n)}\},$$
$$\mathcal{E}_3(j) = \{(X_1^n(m_j|\hat{L}_{j-1}), X_2^n(\hat{L}_{j-1}), Y_3^n(j)) \in \mathcal{T}_\epsilon^{(n)} \text{ for some } m_j \neq 1, m_j \in \mathcal{B}(\hat{L}_j)\}.$$

Thus the probability of error is upper bounded as

$$\begin{aligned}
P(\mathcal{E}(j)) &= P\{\hat{M}_j \neq 1\} \\
&\leq P(\tilde{\mathcal{E}}(j - 1) \cup \mathcal{E}_1(j - 1) \cup \mathcal{E}_1(j) \cup \mathcal{E}_2(j) \cup \mathcal{E}_3(j)) \\
&\leq P(\tilde{\mathcal{E}}(j - 1)) + P(\mathcal{E}_1(j)) + P(\mathcal{E}_1(j - 1)) \\
&\quad + P(\mathcal{E}_2(j) \cap \tilde{\mathcal{E}}^c(j - 1) \cap \mathcal{E}_1^c(j - 1) \cap \mathcal{E}_1^c(j)) \\
&\quad + P(\mathcal{E}_3(j) \cap \tilde{\mathcal{E}}^c(j - 1) \cap \mathcal{E}_1^c(j - 1) \cap \mathcal{E}_1^c(j)).
\end{aligned}$$

Following similar steps to the analysis of the error probability for the coherent multihop scheme with $\tilde{L}_{j-1}$ replacing $\tilde{M}_{j-1}$, the first term $P(\tilde{\mathcal{E}}(j - 1))$ tends to zero as $n \to \infty$ if $R_2 < I(X_1; Y_2|X_2) - \delta(\epsilon)$. Again following similar steps to the analysis of the error probability for the coherent multihop scheme, the second and third terms, $P(\mathcal{E}_1(j))$ and $P(\mathcal{E}_1(j - 1))$, tend to zero as $n \to \infty$ if $R_2 < I(X_2; Y_3) - \delta(\epsilon)$. The fourth term is upper bounded as

$$\begin{aligned}
&P(\mathcal{E}_2(j) \cap \tilde{\mathcal{E}}^c(j - 1) \cap \mathcal{E}_1^c(j - 1) \cap \mathcal{E}_1^c(j)) \\
&= P(\mathcal{E}_2(j) \cap \{\tilde{L}_{j-1} = 1\} \cap \{\hat{L}_{j-1} = 1\} \cap \{\hat{L}_j = 1\}) \\
&\leq P\{(X_1^n(1|1), X_2^n(1), Y_3^n(j)) \notin \mathcal{T}_\epsilon^{(n)} \mid \tilde{L}_{j-1} = 1\},
\end{aligned}$$

which, by the independence of the codebooks and the LLN, tends to zero as $n \to \infty$. The last term is upper bounded as

$$\begin{aligned}
&P(\mathcal{E}_3(j) \cap \tilde{\mathcal{E}}^c(j - 1) \cap \mathcal{E}_1^c(j - 1) \cap \mathcal{E}_1^c(j)) \\
&= P(\mathcal{E}_3(j) \cap \{\tilde{L}_{j-1} = 1\} \cap \{\hat{L}_{j-1} = 1\} \cap \{\hat{L}_j = 1\}) \\
&\leq P\{(X_1^n(m_j|1), X_2^n(1), Y_3^n(j)) \in \mathcal{T}_\epsilon^{(n)} \text{ for some } m_j \neq 1, m_j \in \mathcal{B}(1) \mid \tilde{L}_{j-1} = 1\},
\end{aligned}$$

which, by the same independence and the packing lemma, tends to zero as $n \to \infty$ if $R - R_2 < I(X_1; Y_3|X_2) - \delta(\epsilon)$. Combining the bounds and eliminating R_2, we have shown that $P\{\hat{M}_j \neq M_j\}$ tends to zero as $n \to \infty$ for each $j \in [1 : b - 1]$ if $R < I(X_1; Y_2|X_2) - \delta(\epsilon)$ and $R < I(X_1; Y_3|X_2) + I(X_2; Y_3) - 2\delta(\epsilon) = I(X_1, X_2; Y_3) - 2\delta(\epsilon)$. This completes the proof of the decode–forward lower bound using binning.

16.5 GAUSSIAN RELAY CHANNEL

Consider the Gaussian relay channel depicted in Figure 16.5, which is a simple model for wireless point-to-point communication with a relay. The channel outputs corresponding to the inputs X_1 and X_2 are

$$Y_2 = g_{21}X_1 + Z_2,$$
$$Y_3 = g_{31}X_1 + g_{32}X_2 + Z_3,$$

where g_{21}, g_{31}, and g_{32} are channel gains, and $Z_2 \sim N(0, 1)$ and $Z_3 \sim N(0, 1)$ are independent noise components. Assume average power constraint P on each of X_1 and X_2. Since the relay can both send X_2 and receive Y_2 at the same time, this model is sometimes referred to as the *full-duplex* Gaussian RC, compared to the half-duplex models we discuss in Sections 16.6.3 and 16.8.

We denote the SNR of the direct channel by $S_{31} = g_{31}^2 P$, the SNR of the channel from the sender to the relay receiver by $S_{21} = g_{21}^2 P$, and the SNR of the channel from the relay to the receiver by $S_{32} = g_{32}^2 P$. Note that under this model, the RC cannot be degraded or reversely degraded. In fact, the capacity is not known for any $S_{21}, S_{31}, S_{32} > 0$.

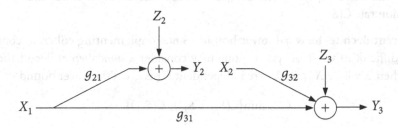

Figure 16.5. Gaussian relay channel.

16.5.1 Upper and Lower Bounds on the Capacity of the Gaussian RC

We evaluate the upper and lower bounds we discussed in the previous sections.

Cutset upper bound. The proof of the cutset bound in Theorem 16.1 applies to arbitrary alphabets. By optimizing the bound subject to the power constraints, we can show that it is attained by jointly Gaussian (X_1, X_2) (see Appendix 16A) and simplifies to

$$C \leq \max_{0 \leq \rho \leq 1} \min\{C(S_{31} + S_{32} + 2\rho\sqrt{S_{31}S_{32}}), C((1 - \rho^2)(S_{31} + S_{21}))\}$$

$$= \begin{cases} C\left(\left(\sqrt{S_{21}S_{32}} + \sqrt{S_{31}(S_{31} + S_{21} - S_{32})}\right)^2 / (S_{31} + S_{21})\right) & \text{if } S_{21} \geq S_{32}, \\ C(S_{31} + S_{21}) & \text{otherwise.} \end{cases} \quad (16.4)$$

Direct-transmission lower bound. It is straightforward to see that the lower bound in (16.1) yields the lower bound

$$C \geq C(S_{31}).$$

Multihop lower bound. Consider the multihop lower bound in (16.2) subject to the power constraints. The distributions on the inputs X_1 and X_2 that optimize the bound are not known in general. Assuming X_1 and X_2 to be Gaussian, we obtain the lower bound

$$C \geq \min\{C(S_{21}), \, C(S_{32}/(S_{31} + 1))\}. \quad (16.5)$$

To prove achievability of this bound, we extend the multihop achievability to the case with input costs and use the discretization procedure in Section 3.4.

Decode–forward lower bound. Maximizing the decode–forward lower bound in Theorem 16.2 subject to the power constraints yields

$$C \geq \max_{0 \leq \rho \leq 1} \min\{C(S_{31} + S_{32} + 2\rho\sqrt{S_{31}S_{32}}), \, C(S_{21}(1 - \rho^2))\}$$

$$= \begin{cases} C\left(\left(\sqrt{S_{31}(S_{21} - S_{32})} + \sqrt{S_{32}(S_{21} - S_{31})}\right)^2 / S_{21}\right) & \text{if } S_{21} \geq S_{31} + S_{32}, \\ C(S_{21}) & \text{otherwise.} \end{cases} \quad (16.6)$$

Achievability follows by setting $X_2 \sim N(0, P)$ and $X_1 = \rho X_2 + X_1'$, where $X_1' \sim N(0, (1 - \rho^2)P)$ is independent of X_2 and carries the new message to be recovered first by the relay. Note that when $S_{21} < S_{31}$, the decode–forward rate becomes lower than the direct transmission rate $C(S_{31})$.

Noncoherent decode–forward lower bound. Since implementing coherent communication is difficult in wireless systems, one may consider a *noncoherent* decode–forward coding scheme, where X_1 and X_2 are independent. This gives the lower bound

$$C \geq \min\{C(S_{31} + S_{32}), \, C(S_{21})\}. \quad (16.7)$$

This scheme uses the same codebook generation and encoding steps as the (noncoherent) multihop scheme, but achieves a higher rate by performing simultaneous decoding.

16.6   PARTIAL DECODE–FORWARD LOWER BOUND

In decode–forward, the relay fully recovers the message, which is optimal for the degraded RC because the relay receives a strictly better version of X_1 than the receiver. But in some cases (e.g., the Gaussian RC with $S_{21} < S_{31}$), the channel to the relay can be a bottleneck and decode–forward can be strictly worse than direct transmission. In *partial decode–forward*, the relay recovers only part of the message. This yields a tighter lower bound on the capacity than both decode–forward and direct transmission.

Theorem 16.3 (Partial Decode–Forward Lower Bound). The capacity of the DM-RC is lower bounded as

$$C \geq \max_{p(u,x_1,x_2)} \min\{I(X_1, X_2; Y_3),\ I(U; Y_2|X_2) + I(X_1; Y_3|X_2, U)\},$$

where $|\mathcal{U}| \leq |\mathcal{X}_1| \cdot |\mathcal{X}_2|$.

Note that if we set $U = X_1$, this lower bound reduces to the decode–forward lower bound, and if we set $U = \emptyset$, it reduces to the direct-transmission lower bound.

Proof outline. We use block Markov coding and backward decoding. Divide the message M_j, $j \in [1 : b - 1]$, into two independent messages M_j' at rate R' and M_j'' at rate R''. Hence $R = R' + R''$. Fix the pmf $p(u, x_1, x_2)$ that attains the lower bound and randomly generate an independent codebook

$$\mathcal{C}_j = \{(u^n(m_j'|m_{j-1}'), x_1^n(m_j', m_j''|m_{j-1}'), x_2^n(m_{j-1}')):\ m_{j-1}', m_j' \in [1 : 2^{nR'}],\ m_j'' \in [1 : 2^{nR''}]\}$$

for each block $j \in [1 : b]$. The sender and the relay cooperate to communicate m_j' to the receiver. The relay recovers m_j' at the end of block j using joint typicality decoding (with $u^n(m_j'|m_{j-1}')$ replacing $x_1^n(m_j|m_{j-1})$ in decode–forward). The probability of error for this step tends to zero as $n \to \infty$ if $R' < I(U; Y_2|X_2) - \delta(\epsilon)$. After receiving all the blocks, the messages m_j', $j \in [1 : b - 1]$, are first recovered at the receiver using backward decoding (with $(u^n(m_{j+1}'|m_j'), x_2^n(m_j'))$ replacing $(x_1^n(m_{j+1}|m_j), x_2^n(m_j))$ in decode–forward). The probability of error for this step tends to zero as $n \to \infty$ if $R' < I(U, X_2; Y_3) - \delta(\epsilon)$. The receiver then finds the unique message m_j'', $j \in [1 : b - 1]$, such that $(u^n(m_j'|m_{j-1}'), x_1^n(m_j', m_j''|m_{j-1}'), x_2^n(m_{j-1}'), y_3^n(j)) \in \mathcal{T}_\epsilon^{(n)}$. The probability of error for this step tends to zero as $n \to \infty$ if $R'' < I(X_1; Y_3|U, X_2) - \delta(\epsilon)$. Eliminating R' and R'' from the rate constraints establishes the partial decode–forward lower bound in Theorem 16.3.

The partial decode–forward scheme is optimal in some special cases.

16.6.1   Semideterministic DM-RC

Suppose that Y_2 is a function of (X_1, X_2), i.e., $Y_2 = y_2(X_1, X_2)$. Then, the capacity of this semideterministic DM-RC is

$$C = \max_{p(x_1,x_2)} \min\{I(X_1, X_2; Y_3),\ H(Y_2|X_2) + I(X_1; Y_3|X_2, Y_2)\}. \tag{16.8}$$

Achievability follows by setting $U = Y_2$ in the partial decode–forward lower bound in Theorem 16.3, which is feasible since Y_2 is a function of (X_1, X_2). The converse follows by the cutset bound in Theorem 16.1.

16.6.2   Relay Channel with Orthogonal Sender Components

The relay channel with orthogonal components is motivated by the fact that in many wireless communication systems the relay cannot send and receive in the same time slot or in the same frequency band. The relay channel model can be specialized to accommodate this constraint by assuming orthogonal sender or receiver components. Here we consider the DM-RC with *orthogonal sender components* depicted in Figure 16.6, where $X_1 = (X_1', X_1'')$ and $p(y_2, y_3|x_1, x_2) = p(y_3|x_1', x_2)p(y_2|x_1'', x_2)$. The relay channel with orthogonal receiver components is discussed in Section 16.7.3. It turns out that the capacity is known for this case.

Proposition 16.1. The capacity of the DM-RC with orthogonal sender components is

$$C = \max_{p(x_2)p(x_1'|x_2)p(x_1''|x_2)} \min\{I(X_1', X_2; Y_3), I(X_1''; Y_2|X_2) + I(X_1'; Y_3|X_2)\}.$$

The proof of achievability uses partial decode–forward with $U = X_1''$. The proof of the converse follows by the cutset bound.

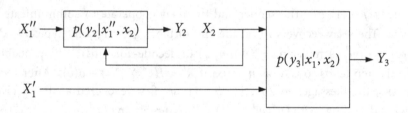

Figure 16.6. Relay channel with orthogonal sender components.

16.6.3   SFD Gaussian Relay Channel

We consider the Gaussian counterpart of the relay channel with orthogonal sender components depicted in Figure 16.7, which we refer to as the *sender frequency-division* (SFD) Gaussian RC. In this half-duplex model, the channel from the sender to the relay uses a separate frequency band. More specifically, in this model $X_1 = (X_1', X_1'')$ and

$$Y_2 = g_{21}X_1'' + Z_2,$$
$$Y_3 = g_{31}X_1' + g_{32}X_2 + Z_3,$$

where g_{21}, g_{31}, and g_{32} are channel gains, and $Z_2 \sim N(0, 1)$ and $Z_3 \sim N(0, 1)$ are independent noise components. Assume average power constraint P on each of $X_1 = (X_1', X_1'')$ and X_2. The capacity of the SFD Gaussian RC is

$$C = \max_{0 \leq \alpha, \rho \leq 1} \min\{C(\alpha S_{31} + S_{32} + 2\rho\sqrt{\alpha S_{31}S_{32}}), C(\tilde{\alpha}S_{21}) + C(\alpha(1 - \rho^2)S_{31})\}. \tag{16.9}$$

Achievability is proved by extending the partial decode–forward lower bound in Theorem 16.3 to the case with input cost constraints and using the discretization procedure in

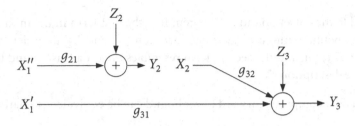

Figure 16.7. Sender frequency-division Gaussian relay channel.

Section 3.4 with $U = X_1'' \sim \mathrm{N}(0, \bar{\alpha}P)$, $X_1' \sim \mathrm{N}(0, \alpha P)$, and $X_2 \sim \mathrm{N}(0, P)$, where (X_1', X_2) is jointly Gaussian with correlation coefficient ρ and is independent of X_1''. The converse follows by showing that the capacity in Proposition 16.1 is attained by the same choice of (X_1', X_1'', X_2).

Remark 16.1. It can be readily verified that direct transmission achieves $C(S_{31})$ (corresponding to $\alpha = 1$ in (16.9)), while decode–forward achieves $\min\{C(S_{21}), C(S_{32})\}$ (corresponding to $\alpha = 0$ in (16.9)). Both of these coding schemes are strictly suboptimal in general.

Remark 16.2. By contrast, it can be shown that the partial decode–forward lower bound for the (full-duplex) Gaussian RC in Section 16.5 is equal to the maximum of the direct-transmission and decode–forward lower bounds; see Appendix 16B. As such, partial decode–forward does not offer any rate improvement over these simpler schemes for the full-duplex case.

16.7 COMPRESS–FORWARD LOWER BOUND

In the (partial) decode–forward coding scheme, the relay recovers the entire message (or part of it). If the channel from the sender to the relay is weaker than the direct channel to the receiver, this requirement can reduce the rate below that for direct transmission in which the relay is not used at all. In the *compress–forward* coding scheme, the relay helps communication by sending a description of its received sequence to the receiver. Because this description is correlated with the received sequence, Wyner–Ziv coding is used to reduce the rate needed to communicate it to the receiver. This scheme achieves the following lower bound.

Theorem 16.4 (Compress–Forward Lower Bound). The capacity of the DM-RC is lower bounded as

$$C \geq \max \min\{I(X_1, X_2; Y_3) - I(Y_2; \hat{Y}_2 | X_1, X_2, Y_3), I(X_1; \hat{Y}_2, Y_3 | X_2)\},$$

where the maximum is over all conditional pmfs $p(x_1)p(x_2)p(\hat{y}_2 | x_2, y_2)$ with $|\hat{\mathcal{Y}}_2| \leq |\mathcal{X}_2| \cdot |\mathcal{Y}_2| + 1$.

Compared to the cutset bound in Theorem 16.1, the first term in the minimum is the multiple access bound without coherent cooperation (X_1 and X_2 are independent) and with a subtracted term, and the second term resembles the broadcast bound but with Y_2 replaced by the description $\hat{Y}_2$.

Remark 16.3. The compress–forward lower bound can be equivalently characterized as

$$C \geq \max I(X_1; \hat{Y}_2, Y_3 | X_2), \qquad (16.10)$$

where the maximum is over all conditional pmfs $p(x_1)p(x_2)p(\hat{y}_2|x_2, y_2)$ such that

$$I(X_2; Y_3) \geq I(Y_2; \hat{Y}_2 | X_2, Y_3).$$

We establish this equivalence in Appendix 16C.

Remark 16.4. The bound (before maximization) is not in general convex in $p(x_1)p(x_2)$ $p(\hat{y}_2|x_2, y_2)$ and hence the compress–forward scheme can be improved using coded time sharing to yield the lower bound

$$C \geq \max \min\{I(X_1, X_2; Y_3|Q) - I(Y_2; \hat{Y}_2|X_1, X_2, Y_3, Q), I(X_1; \hat{Y}_2, Y_3|X_2, Q)\},$$

where the maximum is over all conditional pmfs $p(q)p(x_1|q)p(x_2|q)p(\hat{y}_2|x_2, y_2, q)$ with $|\hat{\mathcal{Y}}_2| \leq |\mathcal{X}_2| \cdot |\mathcal{Y}_2| + 1$ and $|\mathcal{Q}| \leq 2$.

16.7.1   Proof of the Compress–Forward Lower Bound

Again a block Markov coding scheme is used to communicate $(b - 1)$ i.i.d. messages in b blocks. At the end of block j, a reconstruction sequence $\hat{y}_2^n(j)$ of $y_2^n(j)$ conditioned on $x_2^n(j)$ (which is known to both the relay and the receiver) is chosen by the relay. Since the receiver has side information $y_3^n(j)$ about $\hat{y}_2^n(j)$, we use binning as in Wyner–Ziv coding to reduce the rate necessary to send $\hat{y}_2^n(j)$. The bin index is sent to the receiver in block $j + 1$ via $x_2^n(j + 1)$. At the end of block $j + 1$, the receiver decodes for $x_2^n(j + 1)$. It then uses $y_3^n(j)$ and $x_2^n(j)$ to decode for $\hat{y}_2^n(j)$ and $x_1^n(j)$ simultaneously. We now give the details of the proof.

Codebook generation. Fix the conditional pmf $p(x_1)p(x_2)p(\hat{y}_2|y_2, x_2)$ that attains the lower bound. Again, we randomly generate an independent codebook for each block. For $j \in [1 : b]$, randomly and independently generate 2^{nR} sequences $x_1^n(m_j)$, $m_j \in [1 : 2^{nR}]$, each according to $\prod_{i=1}^{n} p_{X_1}(x_{1i})$. Randomly and independently generate $2^{n\tilde{R}_2}$ sequences $x_2^n(l_{j-1})$, $l_{j-1} \in [1 : 2^{nR_2}]$, each according to $\prod_{i=1}^{n} p_{X_2}(x_{2i})$. For each $l_{j-1} \in [1 : 2^{nR_2}]$, randomly and conditionally independently generate $2^{n\tilde{R}_2}$ sequences $\hat{y}_2^n(k_j|l_{j-1})$, $k_j \in [1 : 2^{n\tilde{R}_2}]$, each according to $\prod_{i=1}^{n} p_{\hat{Y}_2|X_2}(\hat{y}_{2i}|x_{2i}(l_{j-1}))$. This defines the codebook

$$\mathcal{C}_j = \{(x_1^n(m_j), x_2^n(l_{j-1}), \hat{y}_2^n(k_j|l_{j-1})): m_j \in [1 : 2^{nR}], l_{j-1} \in [1 : 2^{nR_2}], k_j \in [1 : 2^{n\tilde{R}_2}]\}.$$

Partition the set $[1 : 2^{n\tilde{R}_2}]$ into 2^{nR_2} equal size bins $\mathcal{B}(l_j)$, $l_j \in [1 : 2^{nR_2}]$. The codebook and bin assignments are revealed to all parties.

Encoding and decoding are explained with the help of Table 16.4.

Block	1	2	3	$\cdots$	$b-1$	b
X_1	$x_1^n(m_1)$	$x_1^n(m_2)$	$x_1^n(m_3)$	$\cdots$	$x_1^n(m_{b-1})$	$x_1^n(1)$
Y_2	$\hat{y}_2^n(k_1\|1), l_1$	$\hat{y}_2^n(k_2\|l_1), l_2$	$\hat{y}_2^n(k_3\|l_2), l_3$	$\cdots$	$\hat{y}_2^n(k_{b-1}\|l_{b-2}), l_{b-1}$	$\emptyset$
X_2	$x_2^n(1)$	$x_2^n(l_1)$	$x_2^n(l_2)$	$\cdots$	$x_2^n(l_{b-2})$	$x_2^n(l_{b-1})$
Y_3	$\emptyset$	$\hat{l}_1, \hat{k}_1,$ $\hat{m}_1$	$\hat{l}_2, \hat{k}_2,$ $\hat{m}_2$	$\cdots$	$\hat{l}_{b-2}, \hat{k}_{b-2},$ $\hat{m}_{b-2}$	$\hat{l}_{b-1}, \hat{k}_{b-1},$ $\hat{m}_{b-1}$

Table 16.4. Encoding and decoding for the compress–forward lower bound.

Encoding. Let $m_j \in [1:2^{nR}]$ be the message to be sent in block j. The encoder transmits $x_1^n(m_j)$ from codebook $\mathcal{C}_j$, where $m_b = 1$ by convention.

Relay encoding. By convention, let $l_0 = 1$. At the end of block j, the relay finds an index k_j such that $(y_2^n(j), \hat{y}_2^n(k_j|l_{j-1}), x_2^n(l_{j-1})) \in \mathcal{T}_{\epsilon'}^{(n)}$. If there is more than one such index, it selects one of them uniformly at random. If there is no such index, it selects an index from $[1:2^{n\tilde{R}_2}]$ uniformly at random. In block $j+1$ the relay transmits $x_2^n(l_j)$, where l_j is the bin index of k_j.

Decoding. Let $\epsilon > \epsilon'$. At the end of block $j+1$, the receiver finds the unique index $\hat{l}_j$ such that $(x_2^n(\hat{l}_j), y_3^n(j+1)) \in \mathcal{T}_{\epsilon}^{(n)}$. It then finds the unique message $\hat{m}_j$ such that $(x_1^n(\hat{m}_j), x_2^n(\hat{l}_{j-1}), \hat{y}_2^n(\hat{k}_j|\hat{l}_{j-1}), y_3^n(j)) \in \mathcal{T}_{\epsilon}^{(n)}$ for some $\hat{k}_j \in \mathcal{B}(\hat{l}_j)$.

Analysis of the probability of error. We analyze the probability of decoding error for the message M_j averaged over codebooks. Assume without loss of generality that $M_j = 1$ and let L_{j-1}, L_j, K_j denote the indices chosen by the relay in block j. Then the decoder makes an error only if one or more of the following events occur:

$$\tilde{\mathcal{E}}(j) = \{(X_2^n(L_{j-1}), \hat{Y}_2^n(k_j|L_{j-1}), Y_2^n(j)) \notin \mathcal{T}_{\epsilon'}^{(n)} \text{ for all } k_j \in [1:2^{n\tilde{R}_2}]\},$$

$$\mathcal{E}_1(j-1) = \{\hat{L}_{j-1} \neq L_{j-1}\},$$

$$\mathcal{E}_1(j) = \{\hat{L}_j \neq L_j\},$$

$$\mathcal{E}_2(j) = \{(X_1^n(1), X_2^n(\hat{L}_{j-1}), \hat{Y}_2^n(K_j|\hat{L}_{j-1}), Y_3^n(j)) \notin \mathcal{T}_{\epsilon}^{(n)}\},$$

$$\mathcal{E}_3(j) = \{(X_1^n(m_j), X_2^n(\hat{L}_{j-1}), \hat{Y}_2^n(K_j|\hat{L}_{j-1}), Y_3^n(j)) \in \mathcal{T}_{\epsilon}^{(n)} \text{ for some } m_j \neq 1\},$$

$$\mathcal{E}_4(j) = \{(X_1^n(m_j), X_2^n(\hat{L}_{j-1}), \hat{Y}_2^n(\hat{k}_j|\hat{L}_{j-1}), Y_3^n(j)) \in \mathcal{T}_{\epsilon}^{(n)}$$
$$\text{for some } \hat{k}_j \in \mathcal{B}(\hat{L}_j), \hat{k}_j \neq K_j, m_j \neq 1\}.$$

Thus the probability of error is upper bounded as

$$P(\mathcal{E}(j)) = P\{\hat{M}_j \neq 1\}$$
$$\leq P(\tilde{\mathcal{E}}(j)) + P(\mathcal{E}_1(j-1)) + P(\mathcal{E}_1(j)) + P(\mathcal{E}_2(j) \cap \tilde{\mathcal{E}}^c(j) \cap \mathcal{E}_1^c(j-1))$$
$$+ P(\mathcal{E}_3(j)) + P(\mathcal{E}_4(j) \cap \mathcal{E}_1^c(j-1) \cap \mathcal{E}_1^c(j)).$$

By independence of the codebooks and the covering lemma, the first term $P(\tilde{\mathcal{E}}(j))$ tends to zero as $n \to \infty$ if $\tilde{R}_2 > I(\hat{Y}_2; Y_2|X_2) + \delta(\epsilon')$. Following the analysis of the error probability in the multihop coding scheme, the next two terms $P(\mathcal{E}_1(j-1)) = P\{\hat{L}_{j-1} \neq L_{j-1}\}$ and $P(\mathcal{E}_1(j)) = P\{\hat{L}_j \neq L_j\}$ tend to zero as $n \to \infty$ if $R_2 < I(X_2; Y_3) - \delta(\epsilon)$. The fourth term is upper bounded as

$$P(\mathcal{E}_2(j) \cap \tilde{\mathcal{E}}^c(j) \cap \mathcal{E}_1^c(j-1)) \leq P\{(X_1^n(1), X_2^n(L_{j-1}), \hat{Y}_2^n(K_j|L_{j-1}), Y_3^n(j)) \notin \mathcal{T}_\epsilon^{(n)} \mid \tilde{\mathcal{E}}^c(j)\},$$

which, by the independence of the codebooks and the conditional typicality lemma, tends to zero as $n \to \infty$. By the same independence and the packing lemma, $P(\mathcal{E}_3(j))$ tends to zero as $n \to \infty$ if $R < I(X_1; X_2, \hat{Y}_2, Y_3) + \delta(\epsilon) = I(X_1; \hat{Y}_2, Y_3|X_2) + \delta(\epsilon)$. Finally, following similar steps as in Lemma 11.1, the last term is upper bounded as

$$P(\mathcal{E}_4(j) \cap \mathcal{E}_1^c(j-1) \cap \mathcal{E}_1^c(j)) \leq P\{(X_1^n(m_j), X_2^n(L_{j-1}), \hat{Y}_2^n(\hat{k}_j|L_{j-1}), Y_3^n(j)) \in \mathcal{T}_\epsilon^{(n)}$$
$$\text{for some } \hat{k}_j \in \mathcal{B}(L_j), \hat{k}_j \neq K_j, m_j \neq 1\}$$
$$\leq P\{(X_1^n(m_j), X_2^n(L_{j-1}), \hat{Y}_2^n(\hat{k}_j|L_{j-1}), Y_3^n(j)) \in \mathcal{T}_\epsilon^{(n)}$$
$$\text{for some } \hat{k}_j \in \mathcal{B}(1), m_j \neq 1\}, \tag{16.11}$$

which, by the independence of the codebooks, the joint typicality lemma (twice), and the union of events bound, tends to zero as $n \to \infty$ if $R + \tilde{R}_2 - R_2 < I(X_1; Y_3|X_2) + I(\hat{Y}_2; X_1, Y_3|X_2) - \delta(\epsilon)$. Combining the bounds and eliminating R_2 and $\tilde{R}_2$, we have shown that $P\{\hat{M}_j \neq M_j\}$ tends to zero as $n \to \infty$ for every $j \in [1 : b-1]$ if

$$R < I(X_1, X_2; Y_3) + I(\hat{Y}_2; X_1, Y_3|X_2) - I(\hat{Y}_2; Y_2|X_2) - 2\delta(\epsilon) - \delta(\epsilon')$$
$$\overset{(a)}{=} I(X_1, X_2; Y_3) + I(\hat{Y}_2; X_1, Y_3|X_2) - I(\hat{Y}_2; X_1, Y_2, Y_3|X_2) - \delta'(\epsilon)$$
$$= I(X_1, X_2; Y_3) - I(\hat{Y}_2; Y_2|X_1, X_2, Y_3) - \delta'(\epsilon),$$

where (a) follows since $\hat{Y}_2 \to (X_2, Y_2) \to (X_1, Y_3)$ form a Markov chain. This completes the proof of the compress–forward lower bound.

Remark 16.5. There are several other coding schemes that achieve the compress–forward lower bound, most notably, the noisy network coding scheme described in Section 18.3.

16.7.2 Compress–Forward for the Gaussian RC

The conditional distribution $F(x_1)F(x_2)F(\hat{y}_2|y_2, x_2)$ that attains the compress–forward lower bound in Theorem 16.4 is not known for the Gaussian RC in general. Let $X_1 \sim N(0, P)$, $X_2 \sim N(0, P)$, and $Z \sim N(0, N)$ be mutually independent and $\hat{Y}_2 = Y_2 + Z$ (see

Figure 16.8). Substituting in the compress–forward lower bound in Theorem 16.4 and optimizing over N, we obtain the lower bound

$$C \geq C\left(S_{31} + \frac{S_{21}S_{32}}{S_{31} + S_{21} + S_{32} + 1}\right). \tag{16.12}$$

This bound becomes tight as S_{32} tends to infinity. When S_{21} is small, the bound can be improved via time sharing on the sender side.

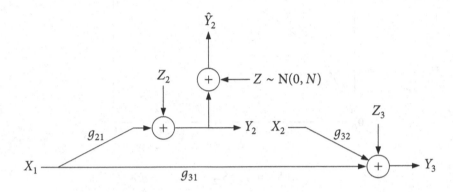

Figure 16.8. Compress–forward for the Gaussian RC.

Figure 16.9 compares the cutset bound, the decode–forward lower bound, and the compress–forward lower bound as a function of S_{31} for different values of S_{21} and S_{32}. Note that in general compress–forward outperforms decode–forward when the channel from the sender to the relay is weaker than that to the receiver, i.e., $S_{21} < S_{31}$, or when the channel from the relay to the receiver is sufficiently strong, specifically if $S_{32} \geq S_{21}(1 + S_{21})/S_{31} - (1 + S_{31})$. Decode–forward outperforms compress–forward (when the latter is evaluated using Gaussian distributions) in other regimes. In general, it can be shown that both decode–forward and compress–forward achieve rates within half a bit of the cutset bound.

16.7.3 Relay Channel with Orthogonal Receiver Components

As a dual model to the DM-RC with orthogonal sender components (see Section 16.6.2), consider the DM-RC with orthogonal receiver components depicted in Figure 16.10. Here $Y_3 = (Y_3', Y_3'')$ and $p(y_2, y_3|x_1, x_2) = p(y_3', y_2|x_1)p(y_3''|x_2)$, decoupling the broadcast channel from the sender to the relay and the receiver from the direct channel from the relay to the receiver.

The capacity of the DM-RC with orthogonal receiver components is not known in general. The cutset bound in Theorem 16.1 simplifies to

$$C \leq \max_{p(x_1)p(x_2)} \min\{I(X_1; Y_3') + I(X_2; Y_3''), I(X_1; Y_2, Y_3')\}.$$

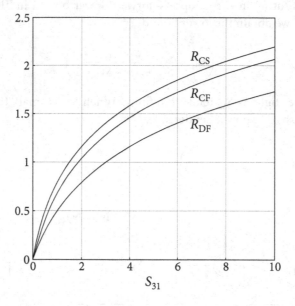

(a) $S_{21} = S_{31}$, $S_{32} = 4S_{31}$.

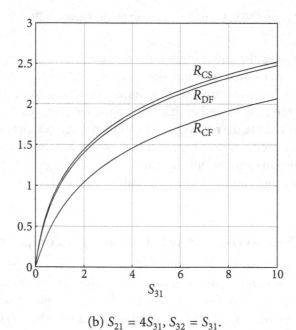

(b) $S_{21} = 4S_{31}$, $S_{32} = S_{31}$.

Figure 16.9. Comparison of the cutset bound R_{CS}, the decode–forward lower bound R_{DF}, and the compress–forward lower bound R_{CF} on the capacity of the Gaussian relay channel.

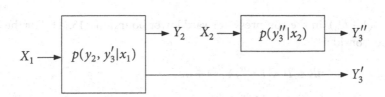

Figure 16.10. Relay channel with orthogonal receiver components.

Let $C_0 = \max_{p(x_2)} I(X_2; Y_3'')$ denote the capacity of the channel from the relay to the receiver. Then the cutset bound can be expressed as

$$C \le \max_{p(x_1)} \min\{I(X_1; Y_3') + C_0, \ I(X_1; Y_2, Y_3')\}. \tag{16.13}$$

By comparison, the compress–forward lower bound in Theorem 16.4 simplifies to

$$C \ge \max_{p(x_1)p(\hat{y}_2|y_2)} \min\{I(X_1; Y_3') - I(Y_2; \hat{Y}_2|X_1, Y_3') + C_0, \ I(X_1; \hat{Y}_2, Y_3')\}. \tag{16.14}$$

These two bounds coincide for the deterministic relay channel with orthogonal receiver components where Y_2 is a function of (X_1, Y_3'). The proof follows by setting $\hat{Y}_2 = Y_2$ in the compress–forward lower bound (16.14) and using the fact that $H(Y_2|X_1, Y_3') = 0$. Note that in general, the capacity itself depends on $p(y_3''|x_2)$ only through C_0.

The following example shows that the cutset bound is not tight in general.

Example 16.2 (Modulo-2 sum relay channel). Consider the DM-RC with orthogonal receiver components depicted in Figure 16.11, where

$$Y_3' = X_1 \oplus Z_3,$$
$$Y_2 = Z_2 \oplus Z_3,$$

and $Z_2 \sim \text{Bern}(p)$ and $Z_3 \sim \text{Bern}(1/2)$ are independent of each other and of X_1.

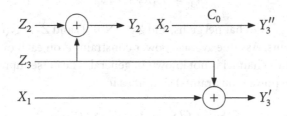

Figure 16.11. Modulo-2 sum relay channel.

For $C_0 \in [0, 1]$, the capacity of this relay channel is

$$C = 1 - H(p * H^{-1}(1 - C_0)),$$

where $H^{-1}(v) \in [0, 1/2]$ is the inverse of the binary entropy function. The proof of achievability follows by setting $\hat{Y}_2 = Y_2 \oplus V$, where $V \sim \text{Bern}(\alpha)$ is independent of (X_1, Z_2, Z_3)

and $\alpha = H^{-1}(1 - C_0)$, in the compress–forward lower bound in (16.14). For the proof of the converse, consider

$$nR \leq I(X_1^n; Y_3'^n, Y_3''^n) + n\epsilon_n$$

$$\overset{(a)}{=} I(X_1^n; Y_3'^n | Y_3''^n) + n\epsilon_n$$

$$\leq n - H(Y_3'^n | X_1^n, Y_3''^n) + n\epsilon_n$$

$$= n - H(Z_3^n | Y_3''^n) + n\epsilon_n$$

$$\overset{(b)}{\leq} n - nH(p * H^{-1}(H(Y_2^n | Y_3''^n)/n)) + n\epsilon_n$$

$$\overset{(c)}{\leq} n - nH(p * H^{-1}(1 - C_0)) + n\epsilon_n,$$

where (a) follows by the independence of X_1^n and $(Z_2^n, Z_3^n, X_2^n, Y_3''^n)$, (b) follows by the vector MGL with $Z_3^n = Y_2^n \oplus Z_2^n$, which yields $H(Z_3^n | Y_3''^n) \geq nH(p * H^{-1}(H(Y_2^n | Y_3''^n)/n))$, and (c) follows since $nC_0 \geq I(X_2^n; Y_3''^n) \geq I(Y_2^n; Y_3''^n) = n - H(Y_2^n | Y_3''^n)$. Note that the cutset bound in (16.13) simplifies to $\min\{1 - H(p), C_0\}$, which is strictly larger than the capacity if $p \neq 1/2$ and $1 - H(p) \leq C_0$. Hence, the cutset bound is not tight in general.

16.8   RFD GAUSSIAN RELAY CHANNEL

The receiver frequency-division (RFD) Gaussian RC depicted in Figure 16.12 has orthogonal receiver components. In this half-duplex model, the channel from the relay to the receiver uses a different frequency band from the broadcast channel from the sender to the relay and the receiver. More specifically, in this model $Y_3 = (Y_3', Y_3'')$ and

$$Y_2 = g_{21}X_1 + Z_2,$$

$$Y_3' = g_{31}X_1 + Z_3',$$

$$Y_3'' = g_{32}X_2 + Z_3'',$$

where g_{21}, g_{31}, and g_{32} are channel gains, and $Z_2 \sim N(0, 1)$ and $Z_3 \sim N(0, 1)$ are independent noise components. Assume average power constraint P on each of X_1 and X_2.

The capacity of this channel is not known in general. The cutset upper bound in Theorem 16.1 (under the power constraints) simplifies to

$$C \leq \begin{cases} C(S_{31}) + C(S_{32}) & \text{if } S_{21} \geq S_{32}(S_{31} + 1), \\ C(S_{31} + S_{21}) & \text{otherwise.} \end{cases} \tag{16.15}$$

The decode–forward lower bound in Theorem 16.2 simplifies to

$$C \geq \begin{cases} C(S_{31}) + C(S_{32}) & \text{if } S_{21} \geq S_{31} + S_{32}(S_{31} + 1), \\ C(S_{21}) & \text{otherwise.} \end{cases} \tag{16.16}$$

When $S_{21} \geq S_{31} + S_{32}(S_{31} + 1)$, the bounds in (16.15) and (16.16) coincide and the capacity

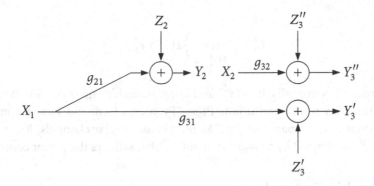

Figure 16.12. Receiver frequency-division Gaussian relay channel.

$C = C(S_{31}) + C(S_{32})$ is achieved by decode–forward. If $S_{21} \leq S_{31}$, the decode–forward lower bound is worse than the direct-transmission lower bound. As in the full-duplex case, the partial decode–forward lower bound reduces to the maximum of the direct-transmission and decode–forward lower bounds, which is in sharp contrast to the sender frequency-division Gaussian RC; see Remarks 16.1 and 16.2. The compress–forward lower bound in Theorem 16.4 with $X_1 \sim N(0, P)$, $X_2 \sim N(0, P)$, and $Z \sim N(0, N)$, independent of each other, and $\hat{Y}_2 = Y_2 + Z$, simplifies (after optimizing over N) to

$$C \geq C\left(S_{31} + \frac{S_{21}S_{32}(S_{31} + 1)}{S_{21} + (S_{31} + 1)(S_{32} + 1)}\right). \tag{16.17}$$

This bound becomes asymptotically tight as either S_{31} or S_{32} approaches infinity. At low S_{21}, that is, low SNR for the channel from the sender to the relay, compress–forward outperforms both direct transmission and decode–forward. Furthermore, the compress–forward rate can be improved via time sharing at the sender, that is, by having the sender transmit at power P/α for a fraction $\alpha \in [0, 1]$ of the time and at zero power for the rest of the time.

16.8.1   Linear Relaying for RFD Gaussian RC

Consider the RFD Gaussian RC with the relaying functions restricted to being linear combinations of past received symbols. Note that under the orthogonal receiver components assumption, we can eliminate the delay in relay encoding simply by relabeling the transmission time for the channel from X_2 to Y_3''. Hence, we equivalently consider relaying functions of the form $x_{2i} = \sum_{j=1}^{i} a_{ij} y_{2j}$, $i \in [1:n]$, or in vector notation of the form $X_2^n = AY_2^n$, where the A is an $n \times n$ lower triangular matrix. This scheme reduces the relay channel to a point-to-point Gaussian channel with input X_1^n and output $Y_3^n = (Y_3'^n, Y_3''^n)$. Note that linear relaying is considerably simpler to implement in practice than decode–forward and compress–forward. It turns out that its performance also compares well with these more complex schemes under certain high SNR conditions.

The capacity with linear relaying, C_L, is characterized by the multiletter expression

$$C_L = \lim_{k \to \infty} C_L^{(k)},$$

where

$$C_L^{(k)} = \sup_{F(x_1^k),\,A} \frac{1}{k} I(X_1^k; Y_3^k)$$

and the supremum is over all cdfs $F(x_1^k)$ and lower triangular matrices A that satisfy the sender and the relay power constraints P; see Problems 16.10 and 16.12 for multiletter characterizations of the capacity of the DM and Gaussian relay channels. It can be easily shown that $C_L^{(k)}$ is attained by a Gaussian input X_1^k that satisfies the power constraint.

16.8.2   Amplify–Forward

Consider $C_L^{(1)}$, which is the maximum rate achieved via a simple *amplify–forward* relaying scheme. It can be shown that $C_L^{(1)}$ is attained by $X_1 \sim N(0, P)$ and $X_2 = \sqrt{P/(S_{21} + 1)}\, Y_2$. Therefore

$$C_L^{(1)} = C\left(S_{31} + \frac{S_{21} S_{32}}{S_{21} + S_{32} + 1}\right). \tag{16.18}$$

This rate approaches the capacity as $S_{32} \to \infty$.

The amplify–forward rate $C_L^{(1)}$ is not convex in P. In fact, it is concave for small P and convex for large P. Hence the rate can be improved by time sharing between direct transmission and amplify–forward. Assuming amplify–forward is performed α of the time and the relay transmits at power P/α during this time, we can achieve the improved rate

$$\max_{0<\alpha,\beta\leq 1} \left(\bar{\alpha}\, C\left(\frac{\bar{\beta} S}{\bar{\alpha}}\right) + \alpha\, C\left(\frac{(\beta/\alpha)(S_{31} + S_{21} S_{32})}{\beta S_{21} + S_{32} + \alpha}\right)\right). \tag{16.19}$$

Figure 16.13 compares the cutset bound to the decode–forward, compress–forward, and amplify–forward lower bounds for different SNRs. Note that compress–forward outperforms amplify–forward in general, but is significantly more complex to implement.

16.8.3*   Linear Relaying Capacity of the RFD Gaussian RC

We can establish the following single-letter characterization of the capacity with linear relaying.

Theorem 16.5. The linear relaying capacity of the RFD Gaussian RC is

$$C_L = \max\left[\alpha_0\, C\left(\frac{\beta_0 P}{\alpha_0}\right) + \sum_{j=1}^{4} \alpha_j\, C\left(\frac{\beta_j P}{\alpha_j}\left(1 + \frac{g_{21}^2 g_{32}^2 \eta_j}{1 + g_{32}^2 \eta_j}\right)\right)\right],$$

where the maximum is over $\alpha_j, \beta_j \geq 0$ and $\eta_j > 0$ such that $\sum_{j=0}^{4} \alpha_j = \sum_{j=0}^{4} \beta_j = 1$ and $\sum_{j=1}^{4} \eta_j \left(g_{21}^2 \beta_j + \alpha_j\right) = P$.

Proof outline. Assume without loss of generality that $g_{31} = 1$. Then $C_L^{(k)}$ is the solution

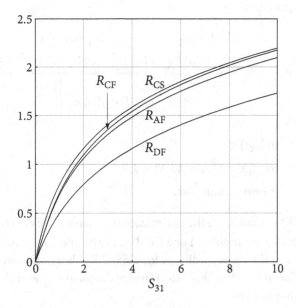

(a) $S_{21} = S_{31}$, $S_{32} = 3S_{31}$.

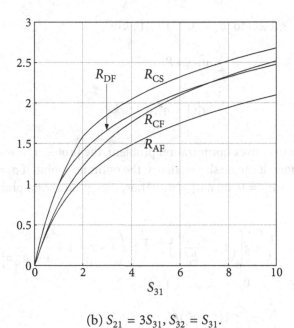

(b) $S_{21} = 3S_{31}$, $S_{32} = S_{31}$.

Figure 16.13. Comparison of the cutset bound R_{CS}, the decode–forward lower bound R_{DF}, the compress–forward lower bound R_{CF}, and the amplify–forward lower bound R_{AF} on the capacity of the RFD Gaussian RC.

to the optimization problem

$$
\text{maximize} \quad \frac{1}{2k} \log \frac{\left\| \begin{bmatrix} I + K_{\mathbf{X}_1} & g_{21}g_{32}K_{\mathbf{X}_1}A^T \\ g_{21}g_{32}AK_{\mathbf{X}_1} & I + g_{21}^2 g_{32}^2 AK_{\mathbf{X}_1}A^T + g_{32}^2 AA^T \end{bmatrix} \right\|}{\left\| \begin{bmatrix} I & 0 \\ 0 & I + g_{32}^2 AA^T \end{bmatrix} \right\|}
$$

subject to $K_{\mathbf{X}_1} \geq 0,$

$\qquad \text{tr}(K_{\mathbf{X}_1}) \leq kP,$

$\qquad \text{tr}(g_{21}^2 K_{\mathbf{X}_1} A^T A + A^T A) \leq kP,$

$\qquad A \text{ lower triangular,}$

where $K_{\mathbf{X}_1} = \mathsf{E}(X_1^k (X_1^k)^T)$ and A are the optimization variables. This is a nonconvex problem in $(K_{\mathbf{X}_1}, A)$ with $k^2 + k$ variables. For a fixed A, the problem is convex in $K_{\mathbf{X}_1}$ and has a water-filling solution. However, finding A for a fixed $K_{\mathbf{X}_1}$ is a nonconvex problem.

Now it can be shown that it suffices to consider diagonal $K_{\mathbf{X}_1}$ and A. Thus, the optimization problem simplifies to

$$
\text{maximize} \quad \frac{1}{2k} \log \prod_{j=1}^{k} \left(1 + \sigma_j \left(1 + \frac{g_{21}^2 g_{32}^2 a_j^2}{1 + g_{32}^2 a_j^2} \right) \right)
$$

subject to $\sigma_j \geq 0, \quad j \in [1:k],$

$$
\sum_{j=1}^{k} \sigma_j \leq kP,
$$

$$
\sum_{j=1}^{k} a_j^2 (1 + g_{21}^2 \sigma_j) \leq kP.
$$

While this is still a nonconvex optimization problem, the problem now involves only $2k$ variables. Furthermore, it can be shown that at the optimum point, if $\sigma_j = 0$, then $a_j = 0$, and conversely if $a_j = a_{j'} = 0$, then $\sigma_j = \sigma_{j'}$. Thus, the optimization problem can be further simplified to

$$
\text{maximize} \quad \frac{1}{2k} \log \left[\left(1 + \frac{k\beta_0 P}{k_0} \right)^{k_0} \prod_{j=k_0+1}^{k} \left(1 + \sigma_j \left(1 + \frac{g_{21}^2 g_{32}^2 a_j^2}{1 + g_{32}^2 a_j^2} \right) \right) \right]
$$

subject to $a_j > 0, \quad j \in [k_0 + 1:k],$

$$
\sum_{j=k_0}^{k} \sigma_j \leq k(1 - \beta_0)P,
$$

$$
\sum_{j=k_0+1}^{k} a_j^2 (1 + g_{21}^2 \sigma_j) \leq kP.
$$

By the KKT condition, it can be shown that at the optimum, there are no more than four distinct nonzero (σ_j, a_j) pairs. Hence, $C_L^{(k)}$ is the solution to the optimization problem

$$\text{maximize} \quad \frac{1}{2k} \log\left[\left(1 + \frac{k\beta_0 P}{k_0}\right)^{k_0} \prod_{j=1}^{4}\left(1 + \sigma_j\left(1 + \frac{g_{21}^2 g_{32}^2 a_j^2}{1 + g_{32}^2 a_j^2}\right)\right)^{k_j}\right]$$

subject to $\quad a_j > 0, \quad j \in [1:4],$

$$\sum_{j=1}^{4} k_j \sigma_j \le k(1 - \beta_0)P,$$

$$\sum_{j=1}^{4} k_j a_j^2 (1 + g_{21}^2 \sigma_j) \le kP,$$

$$\sum_{j=0}^{4} k_j = k,$$

where k_j is a new optimization variable that denotes the number of times the pair (σ_j, a_j) is used during transmission. Taking $k \to \infty$ completes the proof.

Remark 16.6. This is a rare example for which there is no known single-letter mutual information characterization of the capacity, yet we are able to reduce the multiletter characterization *directly* to a computable characterization.

16.9 LOOKAHEAD RELAY CHANNELS

The relay channel can be viewed as a point-to-point communication system with side information about the input X_1 available at the receiver through the relay. As we have seen in Chapters 7 and 11, the degree to which side information can help communication depends on its temporal availability (causal versus noncausal). In our discussion of the relay channel so far, we have assumed that the relaying functions depend only on the past received relay sequence, and hence the side information at the receiver is available *strictly causally*. In this section, we study the relay channel with causal or lookahead relaying functions. The lookahead at the relay may be, for example, the result of a difference between the arrival times of the signal from the sender to the receiver and to the relay. If the signal arrives at both the relay and the receiver at the same time, then we obtain the strictly causal relaying functions assumed so far. If the signal arrives at the receiver later than at the relay, then we effectively have lookahead relaying functions.

To study the effect of lookahead on the capacity of the relay channel, consider the lookahead DM-RC $(\mathcal{X}_1 \times \mathcal{X}_2, p(y_2|x_1)p(y_3|x_1, x_2, y_2), \mathcal{Y}_2 \times \mathcal{Y}_3, l)$ depicted in Figure 16.14. The integer parameter l specifies the amount of relaying lookahead. Note that here we define the lookahead relay channel as $p(y_2|x_1)p(y_3|x_1, x_2, y_2)$, since the conditional pmf $p(y_2, y_3|x_1, x_2)$ depends on the code due to the instantaneous or lookahead dependency of X_2 on Y_2. The channel is memoryless in the sense that $p(y_{2i}|x_1^i, y_2^{i-1}, m) = p_{Y_2|X_1}(y_{2i}|x_{1i})$ and $p(y_{3i}|x_1^i, x_2^i, y_2^i, y_3^{i-1}, m) = p_{Y_3|X_1, X_2, Y_2}(y_{3i}|x_{1i}, x_{2i}, y_{2i})$.

A $(2^{nR}, n)$ code for the lookahead relay channel consists of

- a message set $[1 : 2^{nR}]$,

- an encoder that assigns a codeword $x_1^n(m)$ to each message $m \in [1 : 2^{nR}]$,

- a relay encoder that assigns a symbol $x_{2i}(y_2^{i+l})$ to each sequence y_2^{i+l} for $i \in [1 : n]$, where the symbols that do not have positive time indices or time indices greater than n are arbitrary, and

- a decoder that assigns an estimate $\hat{m}(y_3^n)$ or an error message e to each received sequence y_3^n.

We assume that the message M is uniformly distributed over $[1 : 2^{nR}]$. The definitions of probability of error, achievability, and capacity C_l are as before. Clearly, C_l is monotonically nondecreasing in l. The capacity C_l of the lookahead DM-RC is not known in general for any finite or unbounded l.

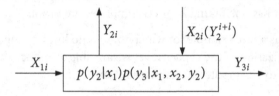

Figure 16.14. Lookahead relay channel.

The DM-RC we studied earlier corresponds to lookahead parameter $l = -1$, or equivalently, a delay of 1. We denote its capacity C_{-1} by C. In the following, we discuss two special cases.

- Noncausal relay channel: For this setup, l is unbounded, that is, the relaying functions can depend on the entire relay received sequence y_2^n. We denote the capacity of the noncausal relay channel by C_∞. The purpose of studying this extreme case is to quantify the limit on the potential gain from allowing lookahead at the relay.

- Causal relay channel: for this setup, the lookahead parameter $l = 0$, that is, the relaying function at time i can depend only on the past and present relay received symbols y_2^i (instead of y_2^{i-1} as in the DM-RC). We investigate the effect of this change on the relaying gain.

16.9.1 Noncausal Relay Channels

The noncausal relay allows for arbitrarily large lookahead relaying functions. Its capacity C_∞ is not known in general. We establish upper and lower bounds on C_∞ that are tight in some cases, and show that C_∞ can be strictly larger than the cutset bound on the capacity C of the (strictly causal) relay channel in Theorem 16.1. Hence C_∞ can be larger than the capacity C itself.

We first consider the following upper bound on C_∞.

> **Theorem 16.6 (Cutset Bound for the Noncausal DM-RC).** The capacity of the noncausal DM-RC $p(y_2|x_1)p(y_3|x_1, x_2, y_2)$ is upper bounded as
>
> $$C_\infty \le \max_{p(x_1)p(u|x_1,y_2),\, x_2(u,y_2)} \min\{I(X_1; Y_2) + I(X_1; Y_3|X_2, Y_2),\, I(U, X_1; Y_3)\}.$$

The proof follows by noting that a $(2^{nR}, n)$ code for the noncausal relay channel induces a joint pmf of the form

$$p(m, x_1^n, x_2^n, y_2^n, y_3^n)$$
$$= 2^{-nR} p(x_1^n|m)\left(\prod_{i=1}^n p_{Y_2|X_1}(y_{2i}|x_{1i})\right) p(x_2^n|y_2^n)\left(\prod_{i=1}^n p_{Y_3|X_1,X_2,Y_2}(y_{3i}|x_{1i}, x_{2i}, y_{2i})\right)$$

and using standard converse proof arguments.

By extending decode–forward to the noncausal case, we can establish the following lower bound.

> **Theorem 16.7 (Noncausal Decode–Forward Lower Bound).** The capacity of the noncausal DM-RC $p(y_2|x_1)p(y_3|x_1, x_2, y_2)$ is lower bounded as
>
> $$C_\infty \ge \max_{p(x_1,x_2)} \min\{I(X_1; Y_2),\, I(X_1, X_2; Y_3)\}.$$

To prove achievability, fix the pmf $p(x_1, x_2)$ that attains the lower bound and randomly and independently generate 2^{nR} sequence pairs $(x_1^n, x_2^n)(m)$, $m \in [1 : 2^{nR}]$, each according to $\prod_{i=1}^n p_{X_1,X_2}(x_{1i}, x_{2i})$. Since the relay knows y_2^n in advance, it decodes for the message m before transmission commences. The probability of error at the relay tends to zero as $n \to \infty$ if $R < I(X_1; Y_2) - \delta(\epsilon)$. The sender and the relay then cooperatively transmit $(x_1^n, x_2^n)(m)$ and the receiver decodes for m. The probability of error at the receiver tends to zero as $n \to \infty$ if $R < I(X_1, X_2; Y_3) - \delta(\epsilon)$. Combining the two conditions completes the proof.

This lower bound is tight in some special cases.

Example 16.3 (Noncausal Sato relay channel). Consider the noncausal version of the Sato relay channel in Example 16.1. We showed that the capacity for the strictly causal case is $C = 1.1619$ and coincides with the cutset bound in Theorem 16.1. Now for the noncausal case, note that by the degradedness of the channel, $I(X_1; Y_3|X_2, Y_2) = 0$, and $p(x_2|x_1, y_2) = p(x_2|x_1)$, since $Y_2 = X_1$. Hence, the noncausal decode–forward lower bound in Theorem 16.7 coincides with the cutset bound in Theorem 16.6 (maximized by setting $U = (X_2, Y_2) = (X_1, X_2)$) and characterizes the capacity C_∞. Optimizing the capacity expression $C_\infty = \max_{p(x_1,x_2)} \min\{I(X_1; Y_2),\, I(X_1, X_2; Y_3)\}$ by setting $p(x_1, x_2)$ as $p_{X_1,X_2}(0, 1) = p_{X_1,X_2}(1, 0) = p_{X_1,X_2}(1, 1) = p_{X_1,X_2}(2, 1) = 1/18$ and $p_{X_1,X_2}(0, 0) = p_{X_1,X_2}(2, 1) = 7/18$, we obtain $C_\infty = \log(9/4) = 1.1699$. Therefore, for this example C_∞ is strictly larger than the cutset bound on the capacity C of the relay channel in Theorem 16.1.

As another example, consider the noncausal version of the Gaussian relay channel in Section 16.5. By evaluating the cutset bound in Theorem 16.6 and the noncausal decode–forward lower bound in Theorem 16.7 with average power constraint P on each of X_1 and X_2, we can show that if $S_{21} \geq S_{31} + S_{32}$, the capacity is

$$C_\infty = C(S_{31} + S_{32} + 2\sqrt{S_{31}S_{32}}). \tag{16.20}$$

As in the Sato relay channel, this capacity is strictly larger than the cutset bound for the Gaussian RC in (16.4).

16.9.2 Causal Relay Channels

We now consider the causal DM-RC, where for each $i \in [1:n]$, the relay encoder assigns a symbol $x_{2i}(y_2^i)$ to every $y_2^i \in \mathcal{Y}_2^i$, that is, the relaying lookahead is $l = 0$. Again the capacity C_0 is not known in general. We provide upper and lower bounds on C_0 and show that rates higher than the cutset bound on the capacity C of the corresponding DM-RC can still be achieved.

Consider the following upper bound on the capacity.

Theorem 16.8 (Cutset Bound for the Causal DM-RC).   The capacity of the causal DM-RC $p(y_2|x_1)p(y_3|x_1, x_2, y_2)$ is upper bounded as

$$C_0 \leq \max_{p(u,x_1), x_2(u,y_2)} \min\{I(X_1; Y_2, Y_3|U), I(U, X_1; Y_3)\},$$

where $|\mathcal{U}| \leq |\mathcal{X}_1| \cdot |\mathcal{X}_2| + 1$.

Note that restricting x_2 to be a function only of u reduces this bound to the cutset bound for the DM-RC in Theorem 16.1. Conversely, this upper bound can be expressed as a cutset bound $C_0 \leq \max_{p(x_1,x_2')} \min\{I(X_1; Y_2, Y_3|X_2'), I(X_1, X_2'; Y_3)\}$ for a DM-RC with relay sender alphabet $\mathcal{X}_2'$ that consists of all mappings $x_2' : \mathcal{Y}_2 \to \mathcal{X}_2$. This is analogous to the capacity expression for the DMC with DM state available causally at the encoder in Section 7.5, which is achieved via the Shannon strategy.

Now we present lower bounds on the capacity of the causal DM-RC. Note that any lower bound on the capacity of the DM-RC, for example, using partial decode–forward or compress–forward, is a lower bound on the capacity of the causal relay channel. We expect, however, that higher rates can be achieved by using the present relay received symbol in addition to its past received symbols.

Instantaneous relaying lower bound. In this simple scheme, the relay transmitted symbol $x_{2i} = x_2(y_{2i})$ for $i \in [1:n]$, that is, x_{2i} is a function only of y_{2i}. This simple scheme yields the lower bound

$$C_0 \geq \max_{p(x_1), x_2(y_2)} I(X_1; Y_3). \tag{16.21}$$

We now show that this simple lower bound can be tight.

Example 16.4 (Causal Sato relay channel). Consider the causal version of the Sato relay channel in Example 16.1. We have shown that $C = 1.1619$ and $C_\infty = \log(9/4) = 1.1699$. Consider the instantaneous relaying lower bound in (16.21) with pmf $(3/9, 2/9, 4/9)$ on X_1 and the function $x_2(y_2) = 0$ if $y_2 = 0$ and $x_2(y_2) = 1$ if $y_2 = 1$ or 2. It can be shown that this choice yields $I(X_1; Y_3) = 1.1699$. Hence the capacity of the causal Sato relay channel is $C_0 = C_\infty = 1.1699$.

This result is not too surprising. Since the channel from the sender to the relay is noiseless, complete cooperation, which requires knowledge of the entire received sequence in advance, can be achieved simply via instantaneous relaying. Since for this example $C_0 > C$ and C coincides with the cutset bound, instantaneous relaying alone can achieve a higher rate than the cutset bound on C!

Causal decode–forward lower bound. The decode–forward lower bound for the DM-RC can be easily extended to incorporate the present received symbol at the relay. This yields the lower bound

$$C_0 \geq \max_{p(u,x_1),\, x_2(u,y_2)} \min\{I(X_1; Y_2|U),\, I(U, X_1; Y_3)\}. \tag{16.22}$$

As for the cutset bound in Theorem 16.8, this bound can be viewed as a decode–forward lower bound for the DM-RC $p(y_2, y_3|x_1, x_2')$, where the relay sender alphabet $\mathcal{X}_2'$ consists of all mappings $x_2': \mathcal{Y}_2 \to \mathcal{X}_2$. Note that the causal decode–forward lower bound coincides with the cutset bound when the relay channel is degraded, i.e., $p(y_3|x_1, x_2, y_2) = p(y_3|x_2, y_2)$.

Now we investigate the causal version of the Gaussian relay channel in the previous subsection and in Section 16.5. Consider the amplify–forward relaying scheme, which is a special case of instantaneous relaying, where the relay in time $i \in [1:n]$ transmits a scaled version of its received signal, i.e., $x_{2i} = ay_{2i}$. To satisfy the relay power constraint, we must have $a^2 \leq P/(g_{21}^2 P + 1)$. The capacity of the resulting equivalent point-to-point Gaussian channel with average received power $(ag_{21}g_{32} + g_{31})^2 P$ and noise power $(a^2 g_{32}^2 + 1)$ yields the lower bound on the capacity of the causal Gaussian relay channel

$$C_0 \geq \mathsf{C}\left(\frac{(ag_{21}g_{32} + g_{31})^2 P}{a^2 g_{32}^2 + 1} \right).$$

Now, it can be shown that if $S_{21}(S_{21} + 1) \leq S_{31}S_{32}$, then this bound is optimized by $a^* = g_{21}/g_{31}g_{32}$ and simplifies to

$$C_0 \geq \mathsf{C}(S_{21} + S_{31}). \tag{16.23}$$

Evaluating the cutset bound on C_∞ in Theorem 16.6 for the Gaussian case with average power constraints yields

$$C_\infty \leq \mathsf{C}(S_{21} + S_{31}), \tag{16.24}$$

if $S_{21} \leq S_{32}$. Thus, if $S_{21} \leq S_{32} \min\{1, S_{31}/(S_{21} + 1)\}$, the bounds in (16.23) and (16.24) coincide and

$$C_0 = C_\infty = C(S_{21} + S_{31}). \tag{16.25}$$

This shows that amplify–forward alone can be *optimal* for both the causal and the non-causal Gaussian relay channels, which is surprising given the extreme simplicity of this relaying scheme. Note that the capacity in the above range of SNR values coincides with the cutset bound on C. This is not the case in general, however, and it can be shown using causal decode–forward that the capacity of the causal Gaussian relay channel can exceed this cutset bound.

Remark 16.7. Combining (16.25) and (16.20), we have shown that the capacity of the non-causal Gaussian RC is known if $S_{21} \geq S_{31} + S_{32}$ or if $S_{21} \leq S_{32} \min\{1, S_{31}/(S_{21} + 1)\}$.

16.9.3 Coherent Cooperation

In studying the relay channel with and without lookahead, we have encountered three different forms of coherent cooperation:

- Decode–forward: Here the relay recovers part or all of the message and the sender and the relay cooperate on communicating the *previous* message. This requires that the relay know only its past received symbols and therefore decode–forward can be implemented for any finite lookahead l.

- Instantaneous relaying: Here the relay sends a function only of its current received symbol. This is possible when the relay has access to the current received symbol, which is the case for every $l \geq 0$.

- Noncausal decode–forward: This scheme is possible only when the relaying functions are noncausal. The relay recovers part or all of the message *before* communication commences and cooperates with the sender to communicate the message to the receiver.

Although instantaneous relaying alone is sometimes optimal (e.g., for the causal Sato relay channel and for a class of causal Gaussian relay channels), utilizing past received symbols at the relay can achieve a higher rate in general.

SUMMARY

- Discrete memoryless relay channel (DM-RC)
- Cutset bound for the relay channel:
  - Cooperative MAC and BC bounds
  - Not tight in general
- Block Markov coding
- Use of multiple independent codebooks

- Decode–forward:

  - Backward decoding

  - Use of binning in channel coding

  - Optimal for the degraded relay channel

- Partial decode–forward:

  - Optimal for the semideterministic relay channel

  - Optimal for the relay channel with orthogonal sender components

- Compress–forward:

  - Optimal for the deterministic relay channel with orthogonal receiver components

  - Optimal for the modulo-2 sum relay channel, which shows that cutset bound is not always tight

- Gaussian relay channel:

  - Capacity is not known for any nonzero channel gains

  - Decode–forward and compress–forward are within 1/2 bit of the cutset bound

  - Partial decode–forward reduces to maximum of decode–forward and direct transmission rates

- SFD Gaussian RC: Capacity is achieved by partial decode–forward

- RFD Gaussian RC:

  - Capacity is known for a range of channel gains and achieved using decode–forward

  - Compress–forward outperforms amplify–forward

  - Linear relaying capacity

- Lookahead relay channels:

  - Cutset bounds for causal and noncausal relay channels

  - Causal and noncausal decode–forward

  - Instantaneous relaying

  - Capacity of noncausal Gaussian RC is known for a large range of channel gains and achieved by noncausal decode–forward/amplify forward

- Coherent cooperation:

  - Cooperation on the previous message (decode–forward)

  - Cooperation on the current symbol (instantaneous relaying)

- Combination of the above (causal decode–forward)

- Cooperation on the current message (noncausal decode–forward)

- **Open problems:**

16.1. Is decode–forward or compress–forward optimal for the Gaussian RC with any nonzero channel gains?

16.2. What joint distribution maximizes the compress–forward lower bound for the Gaussian RC?

16.3. What is the linear relaying capacity of the Gaussian RC?

16.4. Can linear relaying outperform compress–forward?

BIBLIOGRAPHIC NOTES

The relay channel was first introduced by van der Meulen (1971a), who also established the multiletter characterization of the capacity in Problem 16.10. The relay channel was independently motivated by work on packet radio systems at the University of Hawaii in the mid 1970s by Sato (1976) among others. The relay channel in Example 16.1 is due to him. The cutset bound and the decode–forward, partial decode–forward, and compress–forward coding schemes are due to Cover and El Gamal (1979). They also established the capacity of the degraded and reversely degraded relay channels as well as a lower bound that combines compress–forward and partial decode–forward. The proof of decode–forward using binning is due to Cover and El Gamal (1979). Backward decoding is due to Willems and van der Meulen (1985), who used it to establish the capacity region of the DM-MAC with cribbing encoders. The proof of decode–forward using backward decoding is due to Zeng, Kuhlmann, and Buzo (1989).

The capacity of the semideterministic DM-RC is due to El Gamal and Aref (1982). The capacity of the DM and Gaussian relay channels with orthogonal sender components was established by El Gamal and Zahedi (2005). The characterization of the compress–forward lower bound in Theorem 16.4 is due to El Gamal, Mohseni, and Zahedi (2006). They also established the equivalence to the original characterization in (16.10) by Cover and El Gamal (1979). The capacity of the deterministic RC with orthogonal receiver components is due to Kim (2008a), who established it using the hash–forward coding scheme due to Cover and Kim (2007). The modulo-2 sum relay channel example, which shows that the cutset bound can be strictly loose, is due to Aleksic, Razaghi, and Yu (2009). The bounds on the capacity of the receiver frequency-division Gaussian RC in Section 16.8 are due to Høst-Madsen and Zhang (2005), Liang and Veeravalli (2005), and El Gamal, Mohseni, and Zahedi (2006).

The amplify–forward relaying scheme was proposed by Schein and Gallager (2000) for the diamond relay network and subsequently studied for the Gaussian relay channel in Laneman, Tse, and Wornell (2004). The capacity of the RFD Gaussian RC with linear relaying was established by El Gamal, Mohseni, and Zahedi (2006). The results on the lookahead relay channel in Section 16.9 mostly follow El Gamal, Hassanpour, and Mammen (2007). The capacity of the causal relay channel in Example 16.4 is due to Sato (1976). A survey of the literature on the relay channel can found in van der Meulen (2007).

PROBLEMS

16.1. Provide the details of the achievability proof of the partial decode–forward lower bound in Theorem 16.3.

16.2. Provide the details of the converse proof for the capacity of the DM-RC with orthogonal sender components in Proposition 16.1.

16.3. Beginning with the capacity expression for the DM case in Proposition 16.1, establish the capacity of the sender frequency-division Gaussian RC in (16.9).

16.4. Justify the last inequality in (16.11) and show that the upper bound tends to zero as $n \to \infty$ using the joint typicality lemma.

16.5. Show that the capacity of the DM-RC with orthogonal receiver components $p(y_3', y_2|x_1)p(y_3''|x_2)$ depends on $p(y_3''|x_2)$ only through $C_0 = \max_{p(x_2)} I(X_2; Y_3'')$.

16.6. Using the definition of the $(2^{nR}, n)$ code and the memoryless property of the DM lookahead RC $p(y_2|x_1)p(y_3|x_1, x_2, y_2)$ with $l \geq 0$, show that

$$p(y_3^n|x_1^n, x_2^n, y_2^n, m) = \prod_{i=1}^{n} p_{Y_3|X_1,X_2,Y_2}(y_{3i}|x_{1i}, x_{2i}, y_{2i}).$$

16.7. Prove the cutset bound for the noncausal DM-RC in Theorem 16.6.

16.8. Provide the details of the achievability proof for the decode–forward lower bound for the noncausal DM-RC in Theorem 16.7.

16.9. Establish the capacity of the noncausal Gaussian relay channel for $S_{21} \geq S_{31} + S_{32}$ in (16.20).

16.10. *Multiletter characterization of the capacity.* Show that the capacity of the DM-RC $p(y_2, y_3|x_1, x_2)$ is characterized by

$$C = \sup_k C^{(k)} = \lim_{k \to \infty} C^{(k)}, \tag{16.26}$$

where

$$C^{(k)} = \max_{p(x_1^k),\{x_{2i}(y_2^{i-1})\}_{i=1}^k} \frac{1}{k} I(X_1^k; Y_3^k).$$

(Hint: To show that the supremum is equal to the limit, establish the superadditivity of $kC^{(k)}$, i.e., $jC^{(j)} + kC^{(k)} \leq (j+k)C^{(j+k)}$ for all j, k.)

16.11. *Gaussian relay channel.* Consider the Gaussian relay channel with SNRs $S_{21}, S_{31},$ and S_{32}.

(a) Derive the cutset bound in (16.4) starting from the cutset bound in Theorem 16.1 under the average power constraints.

(b) Derive the multihop lower bound in (16.5).

(c) Using jointly Gaussian input distributions, derive an expression for the coherent multihop lower bound for the Gaussian RC.

(d) Derive the decode–forward lower bound in (16.6) starting from the decode–forward lower bound in Theorem 16.2 under average power constraints.

(e) Show that the decode–forward lower bound is within $1/2$ bit of the cutset upper bound.

(f) Derive the noncoherent decode–forward lower bound in (16.7).

(g) Using the Gaussian input distribution on (X_1, X_2) and Gaussian test channel for $\hat{Y}_2$, derive the compress–forward lower bound in (16.12).

(h) Show that the compress–forward lower bound is within $1/2$ bit of the cutset upper bound.

16.12. *A multiletter characterization of the Gaussian relay channel capacity.* Show that the capacity of the Gaussian relay channel with average power constraint P on each of X_1 and X_2 is characterized by

$$C(P) = \lim_{k \to \infty} C^{(k)}(P),$$

where

$$C^{(k)}(P) = \sup_{\substack{F(x_1^k), \{x_{2i}(y_2^{i-1})\}_{i=1}^k \\ \sum_{i=1}^k E(X_{1i}^2) \le kP, \sum_{i=1}^k E(X_{2i}^2) \le kP}} \frac{1}{k} I(X_1^k; Y_3^k).$$

16.13. *Properties of the Gaussian relay channel capacity.* Let $C(P)$ be the capacity of the Gaussian RC with average power constraint P on each of X_1 and X_2.

(a) Show that $C(P) > 0$ if $P > 0$ and $C(P)$ tends to infinity as $P \to \infty$.

(b) Show that $C(P)$ tends to zero as $P \to 0$.

(c) Show that $C(P)$ is concave and strictly increasing in P.

16.14. *Time-division lower bound for the Gaussian RC.* Consider the Gaussian relay channel with SNRs S_{21}, S_{31}, and S_{32}. Suppose that the sender transmits to the receiver for a fraction α_1 of the time, to the relay α_2 of the time, and the relay transmits to the receiver the rest of the time. Find the highest achievable rate using this scheme in terms of the SNRs. Compare this time-division lower bound to the multihop and direct transmission lower bounds.

16.15. *Gaussian relay channel with correlated noise components.* Consider a Gaussian relay channel where $g_{31} = g_{21} = g_{32} = 1$ and the noise components $Z_2 \sim N(0, N_2)$ and $Z_3 \sim N(0, N_3)$ are jointly Gaussian with correlation coefficient ρ. Assume average power constraint P on each of X_1 and X_2. Derive expressions for the cutset bound and the decode–forward lower bound in terms of P, N_1, N_2, and ρ. Under what conditions do these bounds coincide? Interpret the result.

16.16. *RFD-Gaussian relay channel.* Consider the receiver frequency-division Gaussian RC with SNRs S_{21}, S_{31}, and S_{32}.

(a) Derive the cutset bound in (16.15).

(b) Derive the decode–forward lower bound in (16.16). Show that it coincides with the cutset bound when $S_{21} \geq S_{31} + S_{32}(S_{31} + 1)$.

(c) Using Gaussian inputs and test channel, derive the compress–forward lower bound in (16.17).

(d) Consider Gaussian inputs and test channel as in part (c) with time sharing between $X_1 \sim N(0, P/\alpha)$ for a fraction α of the time and $X_1 = 0$ the rest of the time. Derive an expression for the corresponding compress–forward lower bound. Compare the bound to the one without time sharing.

(e) Show that the partial decode–forward lower bound reduces to the maximum of the direct-transmission and decode–forward lower bounds.

(f) Derive the amplify–forward lower bound without time sharing in (16.18) and the one with time sharing in (16.19).

(g) Show that the compress–forward lower bound with or without time sharing is tighter than the amplify–forward lower bound with or without time sharing, respectively.

16.17. *Another modulo-2 sum relay channel.* Consider the DM-RC with orthogonal receiver components $p(y_2, y_3'|x_1)p(y_3''|x_2)$, where

$$Y_3' = Z_2 \oplus Z_3,$$
$$Y_2 = X_1 \oplus Z_2,$$

and $Z_2 \sim \text{Bern}(1/2)$ and $Z_3 \sim \text{Bern}(p)$ are independent of each other and of X_1. Suppose that the capacity of the relay-to-receiver channel $p(y_3''|x_2)$ is $C_0 \in [0, 1]$. Find the capacity of the relay channel.

16.18. *Broadcasting over the relay channel.* Consider the DM-RC $p(y_2, y_3|x_1, x_2)$. Suppose that the message M is to be reliably communicated to both the relay and the receiver. Find the capacity.

16.19. *Partial decode–forward for noncausal relay channels.* Consider the noncausal DM-RC $p(y_2|x_1)p(y_3|x_1, x_2, y_2)$.

(a) By adapting partial decode–forward for the DM-RC to this case, show that the capacity is lower bounded as

$$C_\infty \geq \max_{p(u,x_1,x_2)} \min\{I(X_1, X_2; Y_3), I(U; Y_2) + I(X_1; Y_3|X_2, U)\}.$$

(b) Suppose that $Y_2 = y_2(X_1)$. Using the partial decode–forward lower bound in part (a) and the cutset bound, show that the capacity of this noncausal semideterministic RC is

$$C_\infty = \max_{p(x_1, x_2)} \min\{I(X_1, X_2; Y_3), H(Y_2) + I(X_1; Y_3|X_2, Y_2)\}.$$

(c) Consider the noncausal DM-RC with orthogonal sender components, where $X_1 = (X_1', X_1'')$ and $p(y_2|x_1)p(y_3|x_1, x_2, y_2) = p(y_2|x_1')p(y_3|x_1', x_2)$. Show that the capacity is

$$C_\infty = \max_{p(x_1', x_2)p(x_1'')} \min\{I(X_1', X_2; Y_3), I(X_1''; Y_2) + I(X_1'; Y_3|X_2)\}.$$

16.20. *Partial decode–forward for causal relay channels.* Consider the causal DM-RC $p(y_2|x_1)p(y_3|x_1, x_2, y_2)$.

(a) By adapting partial decode–forward for the DM-RC to this case, show that the capacity is lower bounded as

$$C_0 \geq \max_{p(u, v, x_1), x_2(v, y_2)} \min\{I(V, X_1; Y_3), I(U; Y_2|V) + I(X_1; Y_3|U, V)\},$$

(b) Suppose that $Y_2 = y_2(X_1)$. Using the partial decode–forward lower bound in part (a) and the cutset bound, show that the capacity of this causal semideterministic RC is

$$C_0 = \max_{p(v, x_1), x_2(v, y_2)} \min\{I(V, X_1; Y_3), H(Y_2|V) + I(X_1; Y_3|V, Y_2)\}.$$

16.21. *Instantaneous relaying and compress–forward.* Show how compress–forward can be combined with instantaneous relaying for the causal DM-RC. What is the resulting lower bound on the capacity C_0?

16.22. *Lookahead relay channel with orthogonal receiver components.* Consider the DM-RC with orthogonal receiver components $p(y_3', y_2|x_1)p(y_3''|x_2)$. Show that the capacity with and without lookahead is the same, i.e., $C = C_l = C_\infty$ for all l.

16.23. *MAC with cribbing encoders.* Consider the DM-MAC $p(y|x_1, x_2)$, where sender $j = 1, 2$ wishes to communicate an independent message M_j to the receiver.

(a) Suppose that the codeword from sender 2 is known strictly causally at sender 1, that is, the encoding function at sender 1 is $x_{1i}(m_1, x_2^{i-1})$ at time $i \in [1:n]$. Find the capacity region.

(b) Find the capacity region when the encoding function is $x_{1i}(m_1, x_2^i)$.

(c) Find the capacity region when the encoding function is $x_{1i}(m_1, x_2^n)$.

(d) Find the capacity region when both encoders are "cribbing," that is, the encoding functions are $x_{1i}(m_1, x_2^i)$ and $x_{2i}(m_2, x_1^{i-1})$.

Remark: This problem was studied by Willems and van der Meulen (1985).

APPENDIX 16A   CUTSET BOUND FOR THE GAUSSIAN RC

The cutset bound for the Gaussian RC is given by

$$C \leq \sup_{F(x_1,x_2):E(X_1^2)\leq P,\, E(X_2^2)\leq P} \min\{I(X_1, X_2; Y_3),\ I(X_1; Y_2, Y_3|X_2)\}.$$

We perform the maximization by first establishing an upper bound on the right hand side of this expression and then showing that it is attained by jointly Gaussian (X_1, X_2).

We begin with the first mutual information term. Assume without loss of generality that $E(X_1) = E(X_2) = 0$. Consider

$$\begin{aligned}
I(X_1, X_2; Y_3) &= h(Y_3) - h(Y_3|X_1, X_2) \\
&= h(Y_3) - \frac{1}{2}\log(2\pi e) \\
&\leq \frac{1}{2}\log(E(Y_3^2)) \\
&\leq \frac{1}{2}\log(1 + g_{31}^2\, E(X_1^2) + g_{32}^2\, E(X_2^2) + 2g_{31}g_{32}\, E(X_1 X_2)) \\
&\leq \frac{1}{2}\log(1 + S_{31} + S_{32} + 2\rho\sqrt{S_{31} S_{32}}) \\
&= \mathsf{C}(S_{31} + S_{32} + 2\rho\sqrt{S_{31} S_{32}}),
\end{aligned}$$

where $\rho = E(X_1 X_2)/\sqrt{E(X_1^2)\, E(X_2^2)}$ is the correlation coefficient.

Next we consider the second mutual information term in the cutset bound

$$\begin{aligned}
I(X_1; Y_2, Y_3|X_2) &= h(Y_2, Y_3|X_2) - h(Y_2, Y_3|X_1, X_2) \\
&= h(Y_2, Y_3|X_2) - h(Z_2, Z_3) \\
&= h(Y_3|X_2) + h(Y_2|Y_3, X_2) - \log(2\pi e) \\
&\leq \frac{1}{2}\log(E(\mathrm{Var}(Y_3|X_2))) + \frac{1}{2}\log(E(\mathrm{Var}(Y_2|Y_3, X_2))) \\
&\overset{(a)}{\leq} \frac{1}{2}\log(1 + g_{31}^2(E(X_1^2) - (E(X_1 X_2))^2/E(X_2^2))) \\
&\quad + \frac{1}{2}\log\left(\frac{1 + (g_{21}^2 + g_{31}^2)(E(X_1^2) - (E(X_1 X_2))^2/E(X_2^2))}{1 + g_{31}^2(E(X_1^2) - (E(X_1 X_2))^2/E(X_2^2))}\right) \\
&= \frac{1}{2}\log(1 + (g_{21}^2 + g_{31}^2)(E(X_1^2) - (E(X_1 X_2))^2/E(X_2^2))) \\
&\leq \frac{1}{2}\log(1 + (1 - \rho^2)(S_{21} + S_{31})) \\
&= \mathsf{C}((1 - \rho^2)(S_{21} + S_{31})),
\end{aligned}$$

where (a) follows from the fact that the mean-squared errors of the linear MMSE estimates of Y_3 given X_2 and of Y_2 given (Y_3, X_2) are greater than or equal to the expected conditional variances $E(\mathrm{Var}(Y_3|X_2))$ and $E(\mathrm{Var}(Y_2|Y_3, X_2))$, respectively. Now it is clear that a zero-mean Gaussian (X_1, X_2) with the same power P and correlation coefficient ρ attains the above upper bounds.

APPENDIX 16B   PARTIAL DECODE–FORWARD FOR THE GAUSSIAN RC

It is straightforward to verify that

$$I(X_1, X_2; Y_3) \le C(S_{31} + S_{32} + 2\rho\sqrt{S_{31}S_{32}}),$$

where ρ is the correlation coefficient between X_1 and X_2. Next consider

$$
\begin{aligned}
&I(U; Y_2|X_2) + I(X_1; Y_3|X_2, U) \\
&= h(Y_2|X_2) - h(Y_2|X_2, U) + h(Y_3|X_2, U) - h(Y_3|X_1, X_2, U) \\
&\le \frac{1}{2}\log(2\pi e\, \mathrm{E}[\mathrm{Var}(Y_2|X_2)]) - h(Y_2|X_2, U) + h(Y_3|X_2, U) - \frac{1}{2}\log(2\pi e) \\
&\le \frac{1}{2}\log(1 + g_{21}^2(\mathrm{E}[X_1^2] - (\mathrm{E}[X_1X_2])^2/\mathrm{E}[X_2^2])) - h(Y_2|X_2, U) + h(Y_3|X_2, U) \\
&= C((1 - \rho^2)S_{21}) - h(Y_2|X_2, U) + h(Y_3|X_2, U).
\end{aligned}
$$

We now upper bound $(h(Y_3|X_2, U) - h(Y_2|X_2, U))$. First consider the case $S_{21} > S_{31}$, i.e., $|g_{21}| > |g_{31}|$. In this case

$$
\begin{aligned}
h(Y_2|X_2, U) &= h(g_{21}X_1 + Z_2|X_2, U) \\
&= h(g_{21}X_1 + Z_3|X_2, U) \\
&\ge h(g_{31}X_1 + Z_3|X_2, U) \\
&= h(Y_3|X_2, U).
\end{aligned}
$$

Hence, $I(U; Y_2|X_2) + I(X_1; Y_3|X_2, U) \le C((1 - \rho^2)S_{21})$ and the rate of partial decode–forward reduces to that of decode–forward.

Next consider the case $S_{21} \le S_{31}$. Since

$$\frac{1}{2}\log(2\pi e) \le h(Y_2|X_2, U) \le h(Y_2) = \frac{1}{2}\log(2\pi e(1 + S_{21})),$$

there exists a constant $\beta \in [0, 1]$ such that

$$h(Y_2|X_2, U) = \frac{1}{2}\log(2\pi e(1 + \beta S_{21})).$$

Now consider

$$
\begin{aligned}
h(g_{21}X_1 + Z_2|X_2, U) &= h((g_{21}/g_{31})(g_{31}X_1 + (g_{31}/g_{21})Z_2)\,|\,X_2, U) \\
&= h(g_{31}X_1 + (g_{31}/g_{21})Z_2\,|\,X_2, U) + \log|g_{21}/g_{31}| \\
&\overset{(a)}{=} h(g_{31}X_1 + Z_3' + Z_3''\,|\,X_2, U) + \log|g_{21}/g_{31}| \\
&\overset{(b)}{\ge} \frac{1}{2}\log\left(2^{2h(g_{31}X_1+Z_3'|X_2,U)} + 2^{2h(Z_3''|X_2,U)}\right) + \log|g_{21}/g_{31}| \\
&= \frac{1}{2}\log\left(2^{2h(g_{31}X_1+Z_3'|X_2,U)} + 2\pi e(g_{31}^2/g_{21}^2 - 1)\right) + \log|g_{21}/g_{31}| \\
&= \frac{1}{2}\log\left(2^{2h(Y_3|X_2,U)} + 2\pi e(S_{31}/S_{21} - 1)\right) + \frac{1}{2}\log(S_{21}/S_{31}),
\end{aligned}
$$

where in (a) $Z'_3 \sim N(0, 1)$ and $Z''_3 \sim N(0, g^2_{31}/g^2_{21} - 1)$ are independent, and (b) follows by the entropy power inequality. Since

$$h(g_{21}X_1 + Z_2 \mid X_2, U) = \frac{1}{2}\log(2\pi e(1 + \beta S_{21})),$$

we obtain

$$2\pi e(S_{31}/S_{21} + \beta S_{31}) \geq 2^{2h(Y_3 \mid X_2, U)} + 2\pi e(S_{31}/S_{21} - 1).$$

Thus $h(Y_3 \mid X_2, U) \leq (1/2)\log(2\pi e(1 + \beta S_{31}))$ and

$$h(Y_3 \mid X_2, U) - h(Y_2 \mid X_2, U) \leq \frac{1}{2}\log\left(\frac{1 + \beta S_{31}}{1 + \beta S_{21}}\right)$$

$$\overset{(a)}{\leq} \frac{1}{2}\log\left(\frac{1 + S_{31}}{1 + S_{21}}\right),$$

where (a) follows since if $S_{21} \leq S_{31}$, $(1 + \beta S_{31})/(1 + \beta S_{21})$ is a strictly increasing function of β and attains its maximum when $\beta = 1$. Substituting, we obtain

$$I(U; Y_2 \mid X_2) + I(X_1; Y_3 \mid X_2, U) \leq \frac{1}{2}\log(1 + S_{21}(1 - \rho^2)) + \frac{1}{2}\log\left(\frac{1 + S_{31}}{1 + S_{21}}\right)$$

$$\leq \frac{1}{2}\log(1 + S_{21}) + \frac{1}{2}\log\left(\frac{1 + S_{31}}{1 + S_{21}}\right)$$

$$= C(S_{31}),$$

which is the capacity of the direct channel. Thus the rate of partial decode–forward reduces to that of direct transmission.

APPENDIX 16C   EQUIVALENT COMPRESS–FORWARD LOWER BOUND

Denote the compress–forward lower bound in Theorem 16.4 by

$$R' = \max \min\{I(X_1, X_2; Y_3) - I(Y_2; \hat{Y}_2 \mid X_1, X_2, Y_3), \ I(X_1; \hat{Y}_2, Y_3 \mid X_2)\},$$

where the maximum is over all conditional pmfs $p(x_1)p(x_2)p(\hat{y}_2 \mid x_2, y_2)$, and denote the alternative characterization in (16.10) by

$$R'' = \max I(X_1; \hat{Y}_2, Y_3 \mid X_2),$$

where the maximum is over all conditional pmfs $p(x_1)p(x_2)p(\hat{y}_2 \mid x_2, y_2)$ that satisfy the constraint

$$I(X_2; Y_3) \geq I(Y_2; \hat{Y}_2 \mid X_2, Y_3). \qquad (16.27)$$

We first show that $R'' \leq R'$. For the conditional pmf that attains R'', we have

$$R'' = I(X_1; Y_3, \hat{Y}_2 | X_2)$$
$$= I(X_1; Y_3 | X_2) + I(X_1; \hat{Y}_2 | X_2, Y_3)$$
$$= I(X_1, X_2; Y_3) - I(X_2; Y_3) + I(X_1; \hat{Y}_2 | X_2, Y_3)$$
$$\overset{(a)}{\leq} I(X_1, X_2; Y_3) - I(Y_2; \hat{Y}_2 | X_2, Y_3) + I(X_1; \hat{Y}_2 | X_2, Y_3)$$
$$= I(X_1, X_2; Y_3) - I(X_1, Y_2; \hat{Y}_2 | X_2, Y_3) + I(X_1; \hat{Y}_2 | X_2, Y_3)$$
$$= I(X_1, X_2; Y_3) - I(Y_2; \hat{Y}_2 | X_1, X_2, Y_3),$$

where (a) follows by (16.27).

To show that $R' \leq R''$, note that this is the case if $I(X_1; Y_3, \hat{Y}_2 | X_2) \leq I(X_1, X_2; Y_3) - I(Y_2; \hat{Y}_2 | X_1, X_2, Y_3)$ for the conditional pmf that attains R'. Now assume that at the optimum conditional pmf, $I(X_1; Y_3, \hat{Y}_2 | X_2) > I(X_1, X_2; Y_3) - I(Y_2; \hat{Y}_2 | X_1, X_2, Y_3)$. We show that a higher rate can be achieved. Fix a product pmf $p(x_1)p(x_2)$ and let $\hat{Y}_2' = \hat{Y}_2$ with probability p and $\hat{Y}_2' = \emptyset$ with probability $(1 - p)$. Then $\hat{Y}_2' \to \hat{Y}_2 \to (Y_2, X_2)$ form a Markov chain. Note that the two mutual information terms are continuous in p and that as p increases, the first term decreases and the second increases. Thus there exists p^* such that

$$I(X_1; Y_3, \hat{Y}_2' | X_2) = I(X_1, X_2; Y_3) - I(Y_2; \hat{Y}_2' | X_1, X_2, Y_3)$$

and the rate using $p(\hat{y}_2' | y_2, x_2)$ is larger than that using $p(\hat{y}_2 | y_2, x_2)$. By the above argument, at the optimum conditional pmf,

$$I(X_1, X_2; Y_3) - I(Y_2; \hat{Y}_2 | X_1, X_2, Y_3) = I(X_1; Y_3, \hat{Y}_2 | X_2).$$

Thus

$$I(X_2; Y_3) = I(X_1; Y_3, \hat{Y}_2 | X_2) + I(Y_2; \hat{Y}_2 | X_1, X_2, Y_3) - I(X_1; Y_3 | X_2)$$
$$= I(X_1; \hat{Y}_2 | X_2, Y_3) + I(Y_2; \hat{Y}_2 | X_1, X_2, Y_3)$$
$$= I(X_1, Y_2; \hat{Y}_2 | X_2, Y_3)$$
$$= I(Y_2; \hat{Y}_2 | X_2, Y_3).$$

This completes the proof of the equivalence.

CHAPTER 17

Interactive Channel Coding

The network models we studied so far involve only one-way (feedforward) communication. Many communication systems are inherently interactive, allowing for cooperation through feedback and information exchange over multiway channels. In this chapter, we study the role of feedback in communication and present results on the two-way channel introduced by Shannon as the first multiuser channel. The role of multiway interaction in compression and secure communication will be studied in Chapters 20 and 22, respectively.

As we showed in Section 3.1.1, the capacity of a memoryless point-to-point channel does not increase when noiseless causal feedback is present. Feedback can still benefit point-to-point communication, however, by simplifying coding and improving reliability. The idea is to first send the message uncoded and then to use feedback to iteratively reduce the receiver's error about the message, the error about the error, and so on. We demonstrate this iterative refinement paradigm via the Schalkwijk–Kailath coding scheme for the Gaussian channel and the Horstein and block feedback coding schemes for the binary symmetric channel. We show that the probability of error for the Schalkwijk–Kailath scheme decays double-exponentially in the block length, which is significantly faster than the single-exponential decay of the probability of error without feedback.

We then show that feedback can enlarge the capacity region in multiuser channels. For the multiple access channel, feedback enlarges the capacity region by enabling statistical cooperation between the transmitters. We show that the capacity of the Gaussian MAC with feedback coincides with the outer bound obtained by allowing arbitrary (instead of product) joint input distributions. For the broadcast channel, feedback can enlarge the capacity region by enabling the sender to simultaneously refine both receivers' knowledge about the messages. For the relay channel, we show that the cutset bound is achievable when noiseless causal feedback from the receiver to the relay is allowed. This is in contrast to the case without feedback in which the cutset bound is not achievable in general.

Finally, we discuss the two-way channel, where two nodes wish to exchange their messages interactively over a shared noisy channel. The capacity region of this channel is not known in general. We first establish simple inner and outer bounds on the capacity region. The outer bound is further improved using the new idea of dependence balance. Finally we introduce the notion of directed information and use it to establish a nontrivial multi-letter characterization of the capacity region of the two-way channel.

17.1   POINT-TO-POINT COMMUNICATION WITH FEEDBACK

Consider the point-to-point feedback communication system depicted in Figure 17.1. The sender wishes to communicate a message M to the receiver in the presence of noiseless causal feedback from the receiver, that is, the encoder assigns a symbol $x_i(m, y^{i-1})$ to each message $m \in [1 : 2^{nR}]$ and past received output symbols $y^{i-1} \in \mathcal{Y}^{i-1}$ for $i \in [1 : n]$. Achievability and capacity are defined as for the DMC with no feedback in Section 3.1.

Recall from Section 3.1.1 that feedback does not increase the capacity when the channel is memoryless, e.g., a DMC or a Gaussian channel. Feedback, however, can greatly simplify coding and improve reliability. For example, consider a BEC with erasure probability p. Without feedback, we need to use block error correcting codes to approach the channel capacity $C = (1 - p)$ bits/transmission. With feedback, however, we can achieve capacity by simply retransmitting each bit immediately after it is erased. It can be shown that roughly $n = k/(1 - p)$ transmissions suffice to send k bits of information reliably. Thus, with feedback there is no need for sophisticated error correcting codes.

This simple observation can be extended to other channels with feedback. The basic idea is to first send the message uncoded and then to iteratively refine the receiver's knowledge about it. In the following, we demonstrate this general paradigm of iterative refinement for feedback communication.

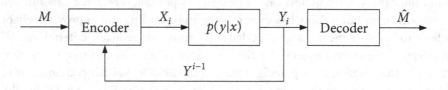

Figure 17.1. Point-to-point feedback communication system.

17.1.1   Schalkwijk–Kailath Coding Scheme for the Gaussian Channel

Consider a Gaussian channel with noiseless causal feedback, where the channel output is $Y = X + Z$ and $Z \sim N(0, 1)$ is the noise. Assume the *expected* average transmission power constraint

$$\sum_{i=1}^{n} \mathsf{E}(x_i^2(m, Y^{i-1})) \le nP, \quad m \in [1 : 2^{nR}]. \tag{17.1}$$

As shown in Section 3.4, the capacity of this channel is $C = C(P)$. We present a simple coding scheme by Schalkwijk and Kailath that achieves any rate $R < C(P)$.

Codebook. Divide the interval $[-\sqrt{P}, \sqrt{P}]$ into 2^{nR} equal-length "message intervals." Represent each message $m \in [1 : 2^{nR}]$ by the midpoint $\theta(m)$ of its interval with distance $\Delta = 2\sqrt{P} \cdot 2^{-nR}$ between neighboring message points.

Encoding. To simplify notation, we assume that transmission begins at time $i = 0$. The encoder first transmits $X_0 = \theta(m)$ at time $i = 0$; hence $Y_0 = \theta(m) + Z_0$. Because of the feedback of Y_0, the encoder can learn the noise $Z_0 = Y_0 - X_0$. The encoder then transmits $X_1 = \gamma_1 Z_0$ at time $i = 1$, where $\gamma_1 = \sqrt{P}$ is chosen so that $\mathsf{E}(X_1^2) = P$. Subsequently, in time $i \in [2:n]$, the encoder forms the MMSE estimate $\mathsf{E}(Z_0|Y^{i-1})$ of Z_0 given Y^{i-1}, and transmits

$$X_i = \gamma_i(Z_0 - \mathsf{E}(Z_0|Y^{i-1})), \tag{17.2}$$

where γ_i is chosen so that $\mathsf{E}(X_i^2) = P$ for $i \in [1:n]$. Hence, $Y_i \sim \mathrm{N}(0, P+1)$ for every $i \in [1:n]$, and the total (expected) power consumption over the $(n+1)$ transmissions is upper bounded as

$$\sum_{i=0}^{n} \mathsf{E}(X_i^2) \le (n+1)P.$$

Decoding. Upon receiving Y^n, the receiver estimates $\theta(m)$ by

$$\hat{\Theta}_n = Y_0 - \mathsf{E}(Z_0|Y^n) = \theta(m) + Z_0 - \mathsf{E}(Z_0|Y^n),$$

and declares that $\hat{m}$ is sent if $\theta(\hat{m})$ is the closest message point to $\hat{\Theta}_n$.

Analysis of the probability of error. Since Z_0 and Z_1 are independent and Gaussian, and $Y_1 = \gamma_1 Z_0 + Z_1$, it follows that $\mathsf{E}(Z_0|Y_1)$ is linear in Y_1. Thus by the orthogonality principle in Appendix B, $X_2 = \gamma_2(Z_0 - \mathsf{E}(Z_0|Y_1))$ is Gaussian and independent of Y_1. Furthermore, since Z_2 is Gaussian and independent of (Y_1, X_2), $Y_2 = X_2 + Z_2$ is also Gaussian and independent of Y_1. In general, for $i \ge 1$, $\mathsf{E}(Z_0|Y^{i-1})$ is linear in Y^{i-1}, and Y_i is Gaussian and independent of Y^{i-1}. Thus the output sequence Y^n is i.i.d. with $Y_i \sim \mathrm{N}(0, P+1)$.

Now we expand $I(Z_0; Y^n)$ in two ways. On the one hand,

$$I(Z_0; Y^n) = \sum_{i=1}^{n} I(Z_0; Y_i|Y^{i-1})$$

$$= \sum_{i=1}^{n} \left(h(Y_i|Y^{i-1}) - h(Y_i|Z_0, Y^{i-1}) \right)$$

$$= \sum_{i=1}^{n} \left(h(Y_i) - h(Z_i|Z_0, Y^{i-1}) \right)$$

$$= \sum_{i=1}^{n} \left(h(Y_i) - h(Z_i) \right)$$

$$= \frac{n}{2} \log(1 + P)$$

$$= n\,C(P).$$

On the other hand,

$$I(Z_0; Y^n) = h(Z_0) - h(Z_0|Y^n) = \frac{1}{2} \log \frac{1}{\mathrm{Var}(Z_0|Y^n)}.$$

Hence, $\mathrm{Var}(Z_0|Y^n) = 2^{-2n\,C(P)}$ and $\hat{\Theta}_n \sim \mathrm{N}(\theta(m), 2^{-2n\,C(P)})$. It is easy to see that the decoder makes an error only if $\hat{\Theta}_n$ is closer to the nearest neighbors of $\theta(m)$ than to $\theta(m)$

itself, that is, if $|\hat{\Theta}_n - \theta(m)| > \Delta/2 = 2^{-nR}\sqrt{P}$ (see Figure 17.2). The probability of error is thus upper bounded as $P_e^{(n)} \le 2Q(2^{n(C(P)-R)}\sqrt{P})$, where

$$Q(x) = \int_x^\infty \frac{1}{\sqrt{2\pi}} e^{-t^2/2}\, dt, \quad x \ge 0.$$

Since $Q(x) \le (1/\sqrt{2\pi})e^{-x^2/2}$ for $x \ge 1$ (Durrett 2010, Theorem 1.2.3), if $R < C(P)$, we have

$$P_e^{(n)} \le \sqrt{\frac{2}{\pi}} \exp\left(-\frac{2^{2n(C(P)-R)}P}{2}\right),$$

that is, the probability of error decays double-exponentially fast in block length n.

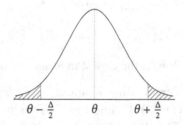

Figure 17.2. Error event for the Schalkwijk–Kailath coding scheme.

Remark 17.1. The Schalkwijk–Kailath encoding rule (17.2) can be interpreted as updating the receiver's knowledge about the initial noise Z_0 (or equivalently the message $\theta(m)$) in each transmission. This encoding rule can be alternatively expressed as

$$
\begin{aligned}
X_i &= \gamma_i(Z_0 - E(Z_0|Y^{i-1})) \\
&= \gamma_i(Z_0 - E(Z_0|Y^{i-2}) + E(Z_0|Y^{i-2}) - E(Z_0|Y^{i-1})) \\
&= \frac{\gamma_i}{\gamma_{i-1}}(X_{i-1} - E(X_{i-1}|Y^{i-1})) \\
&= \frac{\gamma_i}{\gamma_{i-1}}(X_{i-1} - E(X_{i-1}|Y_{i-1})).
\end{aligned}
\tag{17.3}
$$

Thus, the sender iteratively corrects the receiver's error in estimating the previous transmission.

Remark 17.2. Consider the channel with input X_1 and output $\hat{X}_1(Y^n) = E(X_1|Y^n)$. Because the MMSE estimate $\hat{X}_1(Y^n)$ is a linear function of X_1 and Z^n with $I(X_1; \hat{X}_1) = nC(P)$ for Gaussian X_1, the channel from X_1 to $\hat{X}_1$ is equivalent to a Gaussian channel with SNR $2^{2nC(P)} - 1$ (independent of the specific input distribution on X_1). Hence, the Schalkwijk–Kailath scheme transforms n uses of the Gaussian channel with SNR P into a single use of the channel with received SNR $2^{2nC(P)} - 1$. Thus to achieve the capacity, the sender can first send $X_1 = \theta(m)$ and then use the same linear feedback functions as in (17.3) in subsequent transmissions.

Remark 17.3. As another implication of its linearity, the Schalkwijk–Kailath scheme can be used even when the additive noise is not Gaussian. In this case, by the Chebyshev inequality, $P_e^{(n)} \leq P\{|\hat{\Theta}_n - \theta(m)| > 2^{-nR}\sqrt{P}\} \leq 2^{-2n(C(P)-R)}P$, which tends to zero as $n \to \infty$ if $R < C(P)$.

Remark 17.4. The double-exponential decay of the error probability depends crucially on the assumption of an expected power constraint $\sum_{i=1}^{n} E(x_i^2(m, Y^{i-1})) \leq nP$. Under the more stringent *almost-sure* average power constraint

$$P\left\{\sum_{i=1}^{n} x_i^2(m, Y^{i-1}) \leq nP\right\} = 1 \tag{17.4}$$

as assumed in the nonfeedback case, the double-exponential decay is no longer achievable. Nevertheless, Schalkwijk–Kailath coding can still be used with a slight modification to provide a simple constructive coding scheme.

17.1.2*   Horstein's Coding Scheme for the BSC

Consider a BSC(p), where the channel output is $Y = X \oplus Z$ and $Z \sim \text{Bern}(p)$. We present a simple coding scheme by Horstein that achieves any rate $R < 1 - H(p)$. The scheme is illustrated in Figure 17.3.

Codebook. Represent each message $m \in [1 : 2^{nR}]$ by one of 2^{nR} equidistant points $\theta(m) = \alpha + (m-1)2^{-nR} \in [0, 1)$, where the offset $\alpha \in [0, 2^{-nR})$ is to be specified later.

Encoding. Define the encoding map for every $\theta_0 \in [0, 1]$ (not only the message points $\theta(m)$). For each $\theta \in [0, 1)$ and pdf f on $[0, 1)$, define

$$\phi(\theta, f) = \begin{cases} 1 & \text{if } \theta \text{ is greater than the median of } f, \\ 0 & \text{otherwise.} \end{cases}$$

Let $f_0 = f(\theta)$ be the uniform pdf (prior) on $[0, 1)$ (i.e., $\Theta \sim \text{Unif}[0, 1)$). In time $i = 1$, the encoder transmits $x_1 = 1$ if $\theta_0 > 1/2$, and $x_1 = 0$ otherwise. In other words, $x_1 = \phi(\theta_0, f_0)$.

In time $i \in [2 : n]$, upon receiving $Y^{i-1} = y^{i-1}$, the encoder calculates the conditional pdf (posterior) as

$$f_{i-1} = f(\theta|y^{i-1}) = \frac{f(\theta|y^{i-2})p(y_{i-1}|y^{i-2}, \theta)}{p(y_{i-1}|y^{i-2})}$$

$$= f(\theta|y^{i-2}) \cdot \frac{p(y_{i-1}|y^{i-2}, \theta)}{\int p(y_{i-1}|y^{i-2}, \theta')f(\theta'|y^{i-2})d\theta'}$$

$$= \begin{cases} 2\bar{p}f_{i-2} & \text{if } y_{i-1} = \phi(\theta, f_{i-2}), \\ 2p f_{i-2} & \text{otherwise.} \end{cases}$$

The encoder then transmits $x_i = \phi(\theta_0, f_{i-1})$. Note that f_{i-1} is a function of y^{i-1}.

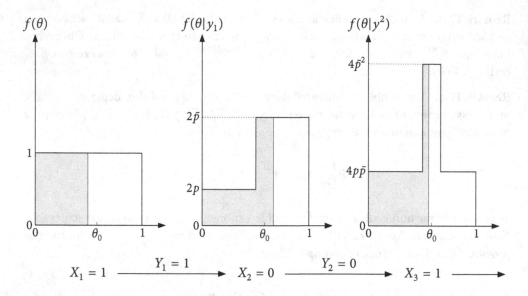

Figure 17.3. Horstein coding scheme. The area of the shaded region under each conditional pdf is equal to 1/2. At time i, $X_i = 0$ is transmitted if θ_0 is in the shaded region and $X_i = 1$ is transmitted, otherwise.

Decoding. Upon receiving $Y^n = y^n$, the decoder uses *maximal posterior interval decoding*. It finds the interval $[\beta, \beta + 2^{-nR})$ of length 2^{-nR} that maximizes the posterior probability

$$\int_{\beta}^{\beta+2^{-nR}} f(\theta|y^n)d\theta.$$

If there is a tie, it chooses the smallest such β. Then it declares that $\hat{m}$ is sent if $\theta(\hat{m}) \in [\beta, \beta + 2^{-nR})$.

Analysis of the probability of error (outline). We first consider the average probability of error for $\Theta \sim \text{Unif}[0, 1)$. Note that $X_1 \sim \text{Bern}(1/2)$ and $Y_1 \sim \text{Bern}(1/2)$. Moreover, for every y^{i-1},

$$P\{X_i = 1 \mid Y^{i-1} = y^{i-1}\} = P\{\phi(\Theta, f_{\Theta|Y^{i-1}}(\theta|y^{i-1})) = 1 \mid Y^{i-1} = y^{i-1}\} = \frac{1}{2}.$$

Thus, $X_i \sim \text{Bern}(1/2)$ is independent of Y^{i-1} and hence $Y_i \sim \text{Bern}(1/2)$ is also independent of Y^{i-1}. Therefore,

$$I(\Theta; Y^n) = H(Y^n) - H(Z^n) = n(1 - H(p)) = nC,$$

which implies that $h(\Theta|Y^n) = -nC$. Moreover, by the LLN and the recursive definitions of the encoding maps and conditional pdfs, it can be shown that $(1/n) \log f(\Theta|Y^n)$ converges to C in probability. These two facts strongly suggest (albeit do not prove) that the probability of error $P\{\Theta \notin [\beta, \beta + 2^{-nR})\}$ tends to zero as $n \to \infty$ if $R < C$. By a more refined analysis of the evolution of $f(\Theta|Y^i)$, $i \in [1:n]$, based on iterated function systems (Diaconis and Freedman 1999, Steinsaltz 1999), it can be shown that the probability of error $E_\Theta(P\{\Theta \notin [\beta, \beta + 2^{-nR})|\Theta\})$ indeed tends to zero as $n \to \infty$ if $R < C$. Therefore, there must exist an $\alpha_n \in [0, 2^{-nR})$ such that

$$P_e^{(n)} = \frac{1}{2^{nR}} \sum_{m=1}^{2^{nR}} P\{\theta \notin [\beta, \beta + 2^{-nR}) \,|\, \Theta = \alpha_n + (m-1)2^{-nR}\}$$

tends to zero as $n \to \infty$ if $R < C$. Thus, with high probability $\theta(M) = \alpha_n + (M-1)2^{-nR}$ is the unique message point within the $[\beta, \beta + 2^{-nR})$ interval.

Posterior matching scheme. The Schalkwijk–Kailath and Horstein coding schemes are special cases of a more general posterior matching coding scheme, which can be applied to any DMC. In this coding scheme, to send a message point $\theta_0 \in [0, 1)$, the encoder transmits

$$X_i = F_X^{-1}(F_{\Theta|Y^{i-1}}(\theta_0|Y^{i-1}))$$

in time $i \in [1:n]$, where $F_X^{-1}(u) = \min\{x: F(x) \geq u\}$ denotes the generalized inverse of the capacity-achieving input cdf $F(x)$ and the iteration begins with the uniform prior on Θ. Note that $X_i \sim F_X(x_i)$ is independent of Y^{i-1}.

Letting $U_i(\theta_0, Y^{i-1}) = F_{\Theta|Y^{i-1}}(\theta_0|Y^{i-1})$, we can express the coding scheme recursively as

$$U_i = F_{U|Y}(U_{i-1}|Y_{i-1}),$$
$$X_i = F_X^{-1}(U_i), \qquad i \in [1:n],$$

where $F_{U|Y}(u|y)$ is the backward channel cdf for $U \sim \text{Unif}[0, 1)$, $X = F_X^{-1}(U)$, and $Y|\{X = x\} \sim p(y|x)$. Thus $\{(U_i, Y_i)\}$ is a Markov process. Using this Markovity and properties of iterated function systems, it can be shown that the maximal posterior interval decoding used in the Horstein scheme achieves the capacity of the DMC (for some relabeling of the input symbols in $\mathcal{X}$).

17.1.3   Block Feedback Coding Scheme for the BSC

In the previous iterative refinement coding schemes, the encoder sends a large amount of uncoded information in the initial transmission and then iteratively refines the receiver's knowledge of this information in subsequent transmissions. Here we present a coding scheme that implements iterative refinement at the block level; the encoder initially transmits an uncoded *block* of information and then refines the receiver's knowledge about it in subsequent blocks.

Again consider a BSC(p), where the channel output is $Y = X \oplus Z$ and $Z \sim \text{Bern}(p)$.

Encoding. To send message $m \in [1:2^n]$, the encoder first transmits the binary representation $X_1 = X^n(m)$ of the message m. Upon receiving the feedback signal Y_1, the encoder compresses the noise sequence $\mathbf{Z}_1 = \mathbf{X}_1 \oplus \mathbf{Y}_1$ losslessly and transmits the binary representation $\mathbf{X}_2(\mathbf{Z}_1)$ of the compression index over the next block of $n(H(p) + \delta(\epsilon))$ transmissions. Upon receiving the resulting feedback signal $\mathbf{Y}_2$, the encoder transmits the corresponding noise index $\mathbf{X}_3(\mathbf{Z}_2)$ over the next $n(H(p) + \delta(\epsilon))^2$ transmissions. Continuing in the same manner, in the j-th block, the encoder transmits the noise index $\mathbf{X}_j(\mathbf{Z}_{j-1})$ over $n(H(p) + \delta(\epsilon))^{j-1}$ transmissions. After k_1 such unequal-length block transmissions, the encoder transmits the compression index of $\mathbf{Z}_{k_1}$ over $nk_2(H(p) + \delta(\epsilon))^{k_1-1}$ transmissions using repetition coding, that is, the encoder transmits the noise index k_2 times without any coding. Note that the rate of this code is $1/(\sum_{j=1}^{k_1}(H(p) + \delta(\epsilon))^{j-1} + k_2(H(p) + \delta(\epsilon))^{k_1})$.

Decoding and analysis of the probability of error. Decoding is performed backwards. The decoder first finds $\mathbf{Z}_{k_1}$ using majority decoding, recovers $\mathbf{Z}_{k_1-1}$ through $\mathbf{X}(\mathbf{Z}_{k_1-1}) = \mathbf{Y}_{k_1} \oplus \mathbf{Z}_{k_1}$, then recovers $\mathbf{Z}_{k_1-2}$, and so on, until $\mathbf{Z}_1$ is recovered. The message is finally recovered through $\mathbf{X}_1(m) = \mathbf{Y}_1 \oplus \mathbf{Z}_1$. By choosing $k_1 = \log(n/\log n)$ and $k_2 = \log^2 n$, it can be shown that the probability of error over all transmission blocks tends to zero while the achieved rate approaches $1/\sum_{j=1}^{\infty}(H(p) + \delta(\epsilon))^{j-1} = 1 - H(p) - \delta(\epsilon)$ as $n \to \infty$.

17.2   MULTIPLE ACCESS CHANNEL WITH FEEDBACK

Consider the multiple access channel with noiseless causal feedback from the receiver to both senders depicted in Figure 17.4. A $(2^{nR_1}, 2^{nR_2}, n)$ code for the DM-MAC with feedback consists of two message sets $[1:2^{nR_j}]$, $j = 1, 2$, two encoders $x_{ji}(m_j, y^{i-1})$, $i \in [1:n]$, $j = 1, 2$, and a decoder $(\hat{m}_1(y^n), \hat{m}_2(y^n))$. The probability of error, achievability, and the capacity region are defined as for the DM-MAC with no feedback in Section 4.1.

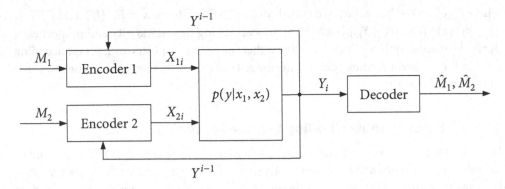

Figure 17.4. Multiple access channel with feedback.

Unlike the point-to-point case, feedback can enlarge the capacity region by inducing statistical cooperation between the two senders. The capacity region with feedback, however, is known only in some special cases, including the Gaussian MAC in Section 17.2.3 and the binary erasure MAC in Example 17.1.

Consider the following outer bound on the capacity region with feedback.

Proposition 17.1. Any achievable rate pair (R_1, R_2) for the DM-MAC $p(y|x_1, x_2)$ with feedback must satisfy the inequalities

$$R_1 \leq I(X_1; Y|X_2),$$
$$R_2 \leq I(X_2; Y|X_1),$$
$$R_1 + R_2 \leq I(X_1, X_2; Y)$$

for some pmf $p(x_1, x_2)$.

Proof. To establish this cooperative outer bound, note that by the memoryless property, $(M_1, M_2, Y^{i-1}) \rightarrow (X_{1i}, X_{2i}) \rightarrow Y_i$ form a Markov chain. Hence, as in the proof of the converse for the DM-MAC in Section 4.5, the following inequalities continue to hold

$$R_1 \leq I(X_{1Q}; Y_Q|X_{2Q}, Q),$$
$$R_2 \leq I(X_{2Q}; Y_Q|X_{1Q}, Q),$$
$$R_1 + R_2 \leq I(X_{1Q}, X_{2Q}; Y_Q|Q).$$

Now, since $Q \rightarrow (X_{1Q}, X_{2Q}) \rightarrow Y_Q$ form a Markov chain, the above inequalities can be further relaxed to

$$R_1 \leq I(X_{1Q}; Y_Q|X_{2Q}),$$
$$R_2 \leq I(X_{2Q}; Y_Q|X_{1Q}),$$
$$R_1 + R_2 \leq I(X_{1Q}, X_{2Q}; Y_Q).$$

Identifying $X_1 = X_{1Q}$, $X_2 = X_{2Q}$, and $Y = Y_Q$ completes the proof of Proposition 17.1.

Remark 17.5. Note that unlike the case with no feedback, X_1 and X_2 are no longer conditionally independent given Q. Hence, the cooperative outer bound is evaluated over all joint pmfs $p(x_1, x_2)$. Later in Section 17.5.2, we find constraints on the set of joint pmfs induced between X_1 and X_2 through feedback, which yield a tighter outer bound in general.

17.2.1   Cover–Leung Inner Bound

We establish the following inner bound on the capacity region of the DM-MAC with feedback.

Theorem 17.1 (Cover–Leung Inner Bound). A rate pair (R_1, R_2) is achievable for the DM-MAC $p(y|x_1, x_2)$ with feedback if

$$R_1 < I(X_1; Y|X_2, U),$$
$$R_2 < I(X_2; Y|X_1, U),$$
$$R_1 + R_2 < I(X_1, X_2; Y)$$

for some pmf $p(u)p(x_1|u)p(x_2|u)$, where $|\mathcal{U}| \leq \min\{|\mathcal{X}_1| \cdot |\mathcal{X}_2| + 1, |\mathcal{Y}| + 2\}$.

Note that this region is convex and has a very similar form to the capacity region with no feedback, which consists of all rate pairs (R_1, R_2) such that

$$R_1 \leq I(X_1; Y | X_2, Q),$$
$$R_2 \leq I(X_2; Y | X_1, Q),$$
$$R_1 + R_2 \leq I(X_1, X_2; Y | Q)$$

for some pmf $p(q)p(x_1|q)p(x_2|q)$, where $|\mathcal{Q}| \leq 2$. The only difference between the two regions is the conditioning in the sum-rate bound.

The above inner bound is tight in some special cases.

Example 17.1 (Binary erasure MAC with feedback). Consider the binary erasure MAC in which the channel input symbols X_1 and X_2 are binary and the channel output symbol $Y = X_1 + X_2$ is ternary. We know that the capacity region without feedback is given by

$$R_1 \leq 1,$$
$$R_2 \leq 1,$$
$$R_1 + R_2 \leq \frac{3}{2}.$$

First we show that this region can be achieved via a simple feedback coding scheme reminiscent of the block feedback coding scheme in Section 17.1.3. We focus on the symmetric rate pair $(R_1, R_2) = (R, R)$.

Suppose that each encoder sends k independent bits uncoded. Then on average $k/2$ bits are erased (that is, $Y = 0 + 1 = 1 + 0 = 1$ is received). Since encoder 1 knows the exact locations of the erasures through feedback, it can retransmit the erased bits over the next $k/2$ transmissions (while encoder 2 sends $X_2 = 0$). The decoder now can recover both messages. Since k bits are sent over $k + k/2$ transmissions, the rate $R = 2/3$ is achievable.

Alternatively, since encoder 2 also knows the $k/2$ erased bits for encoder 1 through feedback, the two encoders can cooperate by each sending half of the $k/2$ erased bits over the following $k/4$ transmissions. These retransmissions result in roughly $k/8$ erased bits, which can be retransmitted again over the following $k/16$ transmissions, and so on. Proceeding recursively, we can achieve the rate $R = k/(k + k/4 + k/16 + \cdots) = 3/4$. This coding scheme can be further improved by inducing further cooperation between the encoders. Again suppose that k independent bits are sent initially. Since both encoders know the $k/2$ erased bits, they can cooperate to send them over the next $k/(2 \log 3)$ transmissions. Hence we can achieve the rate $R = k/(k + k/(2 \log 3)) = 0.7602 > 3/4$.

Now evaluating the Cover–Leung inner bound by taking $X_j = U \oplus V_j$, where $U \sim$ Bern$(1/2)$ and $V_j \sim$ Bern(0.2377), $j = 1, 2$, are mutually independent, we can achieve $R = 0.7911$. This rate can be shown to be optimal.

17.2.2   Proof of the Cover–Leung Inner Bound

The proof of achievability involves superposition coding, block Markov coding, and backward decoding. A sequence of b i.i.d. message pairs $(m_{1j}, m_{2j}) \in [1 : 2^{nR_1}] \times [1 : 2^{nR_2}]$,

$j \in [1:b]$, are sent over b blocks of n transmissions, where by convention, $m_{20} = m_{2b} = 1$. At the end of block $j - 1$, sender 1 recovers $m_{2,j-1}$ and then in block j both senders cooperatively transmit information to the receiver (carried by U) to resolve the remaining uncertainty about $m_{2,j-1}$, superimposed with information about the new message pair (m_{1j}, m_{2j}).

We now provide the details of the proof.

Codebook generation. Fix a pmf $p(u)p(x_1|u)p(x_2|u)$. As in block Markov coding for the relay channel, we randomly and independently generate a codebook for each block. For $j \in [1:b]$, randomly and independently generate 2^{nR_2} sequences $u^n(m_{2,j-1})$, $m_{2,j-1} \in [1:2^{nR_2}]$, each according to $\prod_{i=1}^{n} p_U(u_i)$. For each $m_{2,j-1} \in [1:2^{nR_2}]$, randomly and conditionally independently generate 2^{nR_1} sequences $x_1^n(m_{1j}|m_{2,j-1})$, $m_{1j} \in [1:2^{nR_1}]$, each according to $\prod_{i=1}^{n} p_{X_1|U}(x_{1i}|u_i(m_{2,j-1}))$, and 2^{nR_2} sequences $x_2^n(m_{2j}|m_{2,j-1})$, $m_{2j} \in [1:2^{nR_2}]$, each according to $\prod_{i=1}^{n} p_{X_2|U}(x_{2i}|u_i(m_{2,j-1}))$. This defines the codebook $C_j = \{(u(m_{2,j-1}), x_1^n(m_{1j}|m_{2,j-1}), x_2^n(m_{2j}|m_{2,j-1})) : m_{1j} \in [1:2^{nR_1}], m_{2,j-1}, m_{2j} \in [1:2^{nR_2}]\}$ for $j \in [1:b]$. The codebooks are revealed to all parties.

Encoding and decoding are explained with the help of Table 17.1.

Block	1	2	$\cdots$	$b-1$	b
X_1	$x_1^n(m_{11}\|1)$	$x_1^n(m_{12}\|\tilde{m}_{21})$	$\cdots$	$x_1^n(m_{1,b-1}\|\tilde{m}_{2,b-2})$	$x_1^n(m_{1b}\|\tilde{m}_{2,b-1})$
(X_1, Y)	$\tilde{m}_{21} \rightarrow$	$\tilde{m}_{22} \rightarrow$	$\cdots$	$\tilde{m}_{2,b-1}$	$\emptyset$
X_2	$x_2^n(m_{21}\|1)$	$x_2^n(m_{22}\|m_{21})$	$\cdots$	$x_2^n(m_{2,b-1}\|m_{2,b-2})$	$x_2^n(1\|m_{2,b-1})$
Y	$\hat{m}_{11}$	$\leftarrow \hat{m}_{12}, \hat{m}_{21}$	$\cdots$	$\leftarrow \hat{m}_{1,b-1}, \hat{m}_{2,b-2}$	$\leftarrow \hat{m}_{1b}, \hat{m}_{2,b-1}$

Table 17.1. Encoding and decoding for the Cover–Leung inner bound.

Encoding. Let (m_{1j}, m_{2j}) be the messages to be sent in block j. Encoder 2 transmits $x_2^n(m_{2j}|m_{2,j-1})$ from codebook C_j. At the end of block $j - 1$, encoder 1 finds the unique message $\tilde{m}_{2,j-1}$ such that $(u^n(\tilde{m}_{2,j-2}), x_1^n(m_{1,j-1}|\tilde{m}_{2,j-2}), x_2^n(\tilde{m}_{2,j-1}|\tilde{m}_{2,j-2}), y^n(j - 1)) \in \mathcal{T}_\epsilon^{(n)}$. In block j, encoder 1 transmits $x_1^n(m_{1j}|\tilde{m}_{2,j-1})$ from codebook C_j, where $\tilde{m}_{20} = 1$ by convention.

Decoding. Decoding at the receiver is performed successively backwards after all b blocks are received. For $j = b, \ldots, 1$, the decoder finds the unique message $(\hat{m}_{1j}, \hat{m}_{2,j-1})$ such that $(u^n(\hat{m}_{2,j-1}), x_1^n(\hat{m}_{1j}|\hat{m}_{2,j-1}), x_2^n(\hat{m}_{2j}|\hat{m}_{2,j-1}), y^n(j)) \in \mathcal{T}_\epsilon^{(n)}$, with the initial condition $\hat{m}_{2b} = 1$.

Analysis of the probability of error. We analyze the probability of decoding error for the message pair $(M_{1j}, M_{2,j-1})$ averaged over codebooks. Assume without loss of generality that $M_{1j} = M_{2,j-1} = M_{2j} = 1$. Then the decoder makes an error only if one or more of the following events occur:

$$\tilde{\mathcal{E}}(j) = \{\tilde{M}_{2,j-1} \neq 1\},$$
$$\mathcal{E}(j+1) = \{(\hat{M}_{1,j+1}, \hat{M}_{2j}) \neq (1,1)\},$$
$$\mathcal{E}_1(j) = \{(U^n(1), X_1^n(1|\tilde{M}_{2,j-1}), X_2^n(1|1), Y^n(j)) \notin \mathcal{T}_\epsilon^{(n)}\},$$
$$\mathcal{E}_2(j) = \{(U^n(1), X_1^n(m_{1j}|1), X_2^n(\hat{M}_{2j}|1), Y^n(j)) \in \mathcal{T}_\epsilon^{(n)} \text{ for some } m_{1j} \neq 1\},$$
$$\mathcal{E}_3(j) = \{(U^n(m_{2,j-1}), X_1^n(m_{1j}|m_{2,j-1}), X_2^n(\hat{M}_{2j}|m_{2,j-1}), Y^n(j)) \in \mathcal{T}_\epsilon^{(n)}$$
$$\text{for some } m_{2,j-1} \neq 1, m_{1j}\}.$$

Thus the probability of error is upper bounded as

$$P(\mathcal{E}(j)) = P\{(\hat{M}_{1j}, \hat{M}_{2,j-1}) \neq (1,1)\}$$
$$\leq P(\tilde{\mathcal{E}}(j) \cup \mathcal{E}(j+1) \cup \mathcal{E}_1(j) \cup \mathcal{E}_2(j) \cup \mathcal{E}_3(j))$$
$$\leq P(\tilde{\mathcal{E}}(j)) + P(\mathcal{E}(j+1)) + P(\mathcal{E}_1(j) \cap \tilde{\mathcal{E}}^c(j) \cap \mathcal{E}^c(j+1))$$
$$+ P(\mathcal{E}_2(j) \cap \tilde{\mathcal{E}}^c(j)) + P(\mathcal{E}_3(j) \cap \tilde{\mathcal{E}}^c(j)).$$

Following steps similar to the analysis of the coherent multihop relaying scheme in Section 16.4.2, by independence of the codebooks, the LLN, the packing lemma, and induction, $P(\tilde{\mathcal{E}}(j))$ tends to zero as $n \to \infty$ for all $j \in [1:b]$ if $R_2 < I(X_2; Y, X_1|U) - \delta(\epsilon) = I(X_2; Y|X_1, U) - \delta(\epsilon)$. Also, following steps similar to the analysis of the decode–forward scheme via backward decoding in Section 16.4.4, by the independence of the codebooks, the LLN, the packing lemma, and induction, the last three terms tend to zero as $n \to \infty$ if $R_1 < I(X_1; Y|X_2, U) - \delta(\epsilon)$ and $R_1 + R_2 < I(U, X_1, X_2; Y) - \delta(\epsilon) = I(X_1, X_2; Y) - \delta(\epsilon)$. By induction, the second term $P(\mathcal{E}(j+1))$ tends to zero as $n \to \infty$ and hence $P(\mathcal{E}(j))$ tends to zero as $n \to \infty$ for every $j \in [1:b]$ if the inequalities in the theorem are satisfied. This completes the proof of the Cover–Leung inner bound.

Remark 17.6. The Cover–Leung coding scheme uses only one-sided feedback. As such, it is not surprising that this scheme is suboptimal in general, as we show next.

17.2.3 Gaussian MAC with Feedback

Consider the Gaussian multiple access channel $Y = g_1 X_1 + g_2 X_2 + Z$ with feedback under expected average power constraint P on each of X_1 and X_2 (as defined in (17.1)). It can be shown that the Cover–Leung inner bound simplifies to the inner bound consisting of all (R_1, R_2) such that

$$R_1 < C(\bar{\alpha}_1 S_1),$$
$$R_2 < C(\bar{\alpha}_2 S_2), \tag{17.5}$$
$$R_1 + R_2 < C(S_1 + S_2 + 2\sqrt{\alpha_1 \alpha_2 S_1 S_2})$$

for some $\alpha_1, \alpha_2 \in [0,1]$, where $S_1 = g_1^2 P$ and $S_2 = g_2^2 P$. Achievability follows by setting $X_1 = \sqrt{\alpha_1 P} U + V_1$ and $X_2 = \sqrt{\alpha_2 P} U + V_2$, where $U \sim N(0,1)$, $V_1 \sim N(0, \bar{\alpha}_1 P)$, and $V_2 \sim N(0, \bar{\alpha}_2 P)$ are mutually independent.

It turns out this inner bound is strictly suboptimal.

Theorem 17.2. The capacity region of the Gaussian MAC with feedback is the set of rate pairs (R_1, R_2) such that

$$R_1 \le C((1 - \rho^2)S_1),$$
$$R_2 \le C((1 - \rho^2)S_2),$$
$$R_1 + R_2 \le C(S_1 + S_2 + 2\rho\sqrt{S_1 S_2})$$

for some $\rho \in [0, 1]$.

The converse follows by noting that in order to evaluate the cooperative outer bound in Proposition 17.1, it suffices to consider a zero-mean Gaussian (X_1, X_2) with the same average power P and correlation coefficient ρ.

17.2.4 Achievability Proof of Theorem 17.2

Achievability is proved using an interesting extension of the Schalkwijk–Kailath coding scheme. We first show that the sum-capacity

$$C_{\text{sum}} = \max_{\rho \in [0,1]} \min\{C((1 - \rho^2)S_1) + C((1 - \rho^2)S_2), C(S_1 + S_2 + 2\rho\sqrt{S_1 S_2})\} \quad (17.6)$$

is achievable. Since the first term in the minimum decreases with ρ while the second term increases with ρ, the maximum is attained by a unique $\rho^* \in [0, 1]$ such that

$$C((1 - \rho^{*2})S_1) + C((1 - \rho^{*2})S_2) = C(S_1 + S_2 + 2\rho^*\sqrt{S_1 S_2}).$$

This sum-capacity corresponds to the rate pair

$$R_1 = I(X_1; Y) = I(X_1; Y|X_2) = C((1 - \rho^{*2})S_1),$$
$$R_2 = I(X_2; Y) = I(X_2; Y|X_1) = C((1 - \rho^{*2})S_2) \quad (17.7)$$

attained by a zero-mean Gaussian (X_1, X_2) with the same average power P and correlation coefficient ρ^*. For simplicity of exposition, we only consider the symmetric case $S_1 = S_2 = S = g^2 P$. The general case follows similarly.

Codebook. Divide the interval $[-\sqrt{P(1 - \rho^*)}, \sqrt{P(1 - \rho^*)}]$ into 2^{nR_j}, $j = 1, 2$, message intervals and represent each message $m_j \in [1 : 2^{nR_j}]$ by the midpoint $\theta_j(m_j)$ of an interval with distance $\Delta_j = 2\sqrt{P(1 - \rho^*)} \cdot 2^{-nR_j}$ between neighboring messages.

Encoding. To simplify notation, we assume that transmission commences at time $i = -2$. To send the message pair (m_1, m_2), in the initial 3 transmissions the encoders transmit

$$(X_{1,-2}, X_{2,-2}) = (0, \theta_2(m_2)/\sqrt{1 - \rho^*}),$$
$$(X_{1,-1}, X_{2,-1}) = (\theta_1(m_1)/\sqrt{1 - \rho^*}, 0),$$
$$(X_{1,0}, X_{2,0}) = (0, 0).$$

From the noiseless feedback, encoder 1 knows the initial noise values (Z_{-1}, Z_0) and encoder 2 knows (Z_{-2}, Z_0). Let $U_1 = \sqrt{1 - \rho^*}\, Z_{-1} + \sqrt{\rho^*}\, Z_0$ and $U_2 = \sqrt{1 - \rho^*}\, Z_{-2} + \sqrt{\rho^*}\, Z_0$ be jointly Gaussian with zero means, unit variances, and correlation coefficient ρ^*.

In time $i = 1$, the encoders transmit $X_{11} = \gamma_1 U_1$ and $X_{21} = \gamma_1 U_2$, respectively, where γ_1 is chosen so that $E(X_{11}^2) = E(X_{21}^2) = P$. In time $i \in [2:n]$, the encoders transmit

$$X_{1i} = \gamma_i(X_{1,i-1} - E(X_{1,i-1}|Y_{i-1})),$$
$$X_{2i} = -\gamma_i(X_{2,i-1} - E(X_{2,i-1}|Y_{i-1})),$$

where γ_i, $i \in [2:n]$, is chosen so that $E(X_{1i}^2) = E(X_{2i}^2) = P$ for each i. Such γ_i exists since the "errors" $(X_{1i} - E(X_{1i}|Y_i))$ and $(X_{2i} - E(X_{2i}|Y_i))$ have the same power. Note that the errors are scaled with opposite signs. The total (expected) power consumption for each sender over the $(n + 3)$ transmissions is upper bounded as

$$\sum_{i=-2}^{n} E(X_{ji}^2) \le (n + 1)P \quad \text{for } j = 1, 2.$$

Decoding. The decoder estimates $\theta_1(m_1)$ and $\theta_2(m_2)$ as

$$\hat{\Theta}_{1n} = \sqrt{1 - \rho^*}\, Y_{-1} + \sqrt{\rho^*}\, Y_0 - E(U_1|Y^n) = \theta_1(m_1) + U_1 - E(U_1|Y^n),$$
$$\hat{\Theta}_{2n} = \sqrt{1 - \rho^*}\, Y_{-2} + \sqrt{\rho^*}\, Y_0 - E(U_2|Y^n) = \theta_2(m_2) + U_2 - E(U_2|Y^n).$$

Analysis of the probability of error. Following similar steps as in Remark 17.1, note that we can rewrite X_{1i}, X_{2i} as

$$X_{1i} = \gamma_i'(U_1 - E(U_1|Y^{i-1})),$$
$$X_{2i} = (-1)^{i-1}\gamma_i'(U_2 - E(U_2|Y^{i-1})).$$

Hence X_{1i}, X_{2i}, and Y_i are independent of Y^{i-1}. Now consider

$$I(U_1; Y^n) = \sum_{i=1}^{n} I(U_1; Y_i|Y^{i-1})$$

$$= \sum_{i=1}^{n} h(Y_i) - h(Y_i|U_1, Y^{i-1})$$

$$= \sum_{i=1}^{n} h(Y_i) - h(Y_i|X_{1i}, Y^{i-1})$$

$$\overset{(a)}{=} \sum_{i=1}^{n} h(Y_i) - h(Y_i|X_{1i})$$

$$= \sum_{i=1}^{n} I(X_{1i}; Y_i), \tag{17.8}$$

where (a) follows since X_{1i} and Y_i are independent of Y^{i-1}. Each mutual information term satisfies the following fixed point property.

Lemma 17.1. Let ρ_i be the correlation coefficient between X_{1i} and X_{2i}. Then, $\rho_i = \rho^*$ for all $i \in [1:n]$, where $\rho^* \in [0, 1]$ uniquely satisfies the condition

$$2\,C((1 - \rho^{*2})S) = C(2(1 + \rho^*)S).$$

In other words, the correlation coefficient ρ_i between X_{1i} and X_{2i} stays constant at ρ^* over all $i \in [1:n]$. The proof of this lemma is given in Appendix 17A.

Continuing with the proof of achievability, from (17.7), (17.8), and Lemma 17.1, we have

$$I(U_1; Y^n) = n\,C((1 - \rho^{*2})S),$$

which implies that

$$\hat{\Theta}_{1n} - \theta_1(m_1) = U_1 - \mathrm{E}(U_1|Y^n) \sim \mathrm{N}(0, 2^{-2n\,C((1-\rho^{*2})S)}).$$

Similarly, we have $I(U_2; Y^n) = n\,C((1 - \rho^{*2})S)$ and

$$\hat{\Theta}_{2n} - \theta_2(m_2) = U_2 - \mathrm{E}(U_2|Y^n) \sim \mathrm{N}(0, 2^{-2n\,C((1-\rho^{*2})S)}).$$

Hence, as in the Schalkwijk–Kailath coding scheme, the probability of error $P_e^{(n)}$ tends to zero double-exponentially fast as $n \to \infty$ if $R_1, R_2 < C((1 - \rho^{*2})S)$. This completes the proof of achievability for the sum-capacity in (17.6) when $S_1 = S_2 = S$.

Achieving other points in the capacity region. Other rate pairs in the capacity region can be achieved by combining rate splitting, superposition coding, and successive cancellation decoding with the above feedback coding scheme for the sum-capacity. Encoder 1 divides M_1 into two independent messages M_{10} and M_{11}, and sends M_{10} using a Gaussian random code with power αP_1 and M_{11} using the above feedback coding scheme while treating the codeword for M_{10} as noise. Encoder 2 sends M_2 using the above scheme. The decoder first recovers (M_{11}, M_2) as in the above feedback coding scheme and then recovers M_{10} using successive cancellation decoding.

Now it can be easily shown that the rate triple (R_{10}, R_{11}, R_2) is achievable if

$$R_{10} < C(\alpha S_1),$$

$$R_{11} < C\left(\frac{(1 - (\rho^*)^2)\bar{\alpha}S_1}{1 + \alpha S_1}\right),$$

$$R_2 < C\left(\frac{(1 - (\rho^*)^2)S_2}{1 + \alpha S_1}\right),$$

where $\rho^* \in [0, 1]$ satisfies the condition

$$C\left(\frac{(1 - (\rho^*)^2)\bar{\alpha}S_1}{1 + \alpha S_1}\right) + C\left(\frac{(1 - (\rho^*)^2)S_2}{1 + \alpha S_1}\right) = C\left(\frac{\bar{\alpha}S_1 + S_2 + 2\rho\sqrt{\bar{\alpha}S_1 S_2}}{1 + \alpha S_1}\right).$$

By substituting $R_1 = R_{10} + R_{11}$, taking $\rho = \sqrt{\bar{\alpha}}\rho^*$, and varying $\alpha \in [0, 1]$, we can establish the achievability of any rate pair (R_1, R_2) such that

$$R_1 < C((1 - \rho^2)S_1),$$
$$R_2 < C(S_1 + S_2 + 2\rho\sqrt{S_1 S_2}) - C((1 - \rho^2)S_1)$$

for some $\rho \le \rho^*$. By symmetry, we can show the achievability of any rate pair (R_1, R_2) such that

$$R_1 < C(S_1 + S_2 + 2\rho\sqrt{S_1 S_2}) - C((1 - \rho^2)S_2),$$
$$R_2 < C((1 - \rho^2)S_2)$$

for some $\rho \le \rho^*$. Finally, taking the union over these two regions and noting that the inequalities in the capacity region are inactive for $\rho > \rho^*$ completes the proof of Theorem 17.2.

Remark 17.7. Consider the coding scheme for achieving the sum-capacity in (17.6). Since the covariance matrix $K_{\mathbf{X}_i}$ of (X_{1i}, X_{2i}) is constant over time, the error scaling factor $\gamma_i \equiv \gamma$ is also constant for $i \ge 2$. Let

$$A = \begin{bmatrix} \gamma & 0 \\ 0 & -\gamma \end{bmatrix},$$
$$B = \begin{bmatrix} g & g \end{bmatrix},$$

and consider

$$X_{1i} = \gamma(X_{1,i-1} - E(X_{1,i-1} | Y_{i-1})),$$
$$X_{2i} = -\gamma(X_{2,i-1} - E(X_{2,i-1} | Y_{i-1})).$$

Then $K_{\mathbf{X}_i}$ can be expressed recursively as

$$K_{\mathbf{X}_i} = AK_{\mathbf{X}_{i-1}}A^T - (AK_{\mathbf{X}_{i-1}}B^T)(1 + BK_{\mathbf{X}_{i-1}}B^T)^{-1}(AK_{\mathbf{X}_{i-1}}B^T)^T.$$

Now from the properties of discrete algebraic Riccati equations (Lancaster and Rodman 1995, Chapter 13; Kailath, Sayed, and Hassibi 2000, Section 14.5), it can be shown that

$$\lim_{i \to \infty} K_{\mathbf{X}_i} = K^* = \begin{bmatrix} P & P\rho^* \\ P\rho^* & P \end{bmatrix}$$

for any initial condition $K_{\mathbf{X}_1} > 0$. Equivalently, both $(1/n)I(X_{11}; Y^n)$ and $(1/n)I(X_{21}; Y^n)$ converge to $C((1 - \rho^{*2})S)$ as $n \to \infty$. Thus using the same argument as for the Schalkwijk–Kailath coding scheme (see Remark 17.2), this implies that no initial transmission phase is needed and the same rate pair can be achieved over any additive (non-Gaussian) noise MAC using linear feedback coding.

17.3   BROADCAST CHANNEL WITH FEEDBACK

Consider the broadcast channel with noiseless causal feedback from both receivers depicted in Figure 17.5. The sender wishes to communicate independent messages M_1 and M_2 to receivers 1 and 2, respectively.

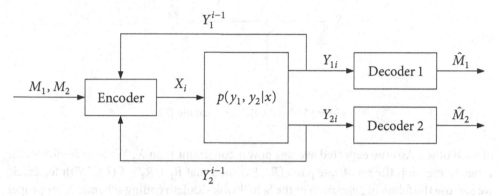

Figure 17.5. Broadcast channel with feedback.

If the broadcast channel is physically degraded, i.e., $p(y_1, y_2 | x) = p(y_1 | x)p(y_2 | y_1)$, then the converse proof in Section 5.4 can be modified to show that feedback does not enlarge the capacity region. Feedback can, however, enlarge the capacity region of broadcast channels in general. We illustrate this fact via the following two examples.

Example 17.2 (Dueck's example). Consider the DM-BC depicted in Figure 17.6 with input $X = (X_0, X_1, X_2)$ and outputs $Y_1 = (X_0, X_1 \oplus Z)$ and $Y_2 = (X_0, X_2 \oplus Z)$, where $\mathcal{X}_0 = \mathcal{X}_1 = \mathcal{X}_2 = \{0, 1\}$ and $Z \sim \text{Bern}(1/2)$. It is easy to show that the capacity region of this channel without feedback is the set of rate pairs (R_1, R_2) such that $R_1 + R_2 \le 1$. By comparison, the capacity region with feedback is the set of rate pairs (R_1, R_2) such that $R_1 \le 1$ and $R_2 \le 1$. The proof of the converse is straightforward. To prove achievability, we show that i.i.d. $\text{Bern}(1/2)$ sequences X_1^n and X_2^n can be sent error-free to receivers 1 and 2, respectively, in $(n + 1)$ transmissions. In the first transmission, we send $(0, X_{11}, X_{21})$ over the channel. Using the feedback link, the sender can determine the noise Z_1. The triple (Z_1, X_{12}, X_{22}) is then sent over the channel in the second transmission, and upon receiving the common information Z_1, the decoders can recover X_{11} and X_{21} perfectly. Thus using the feedback link, the sender and the receivers can recover Z^n and hence each receiver can recover its intended sequence perfectly. Therefore, $(R_1, R_2) = (1, 1)$ is achievable with feedback.

As shown in the above example, feedback can increase the capacity of the DM-BC by letting the encoder broadcast common channel information to all decoders. The following example demonstrates this role of feedback albeit in a less transparent manner.

Example 17.3 (Gaussian BC with feedback). Consider the symmetric Gaussian broadcast channel $Y_j = X + Z_j$, $j = 1, 2$, where $Z_1 \sim \text{N}(0, 1)$ and $Z_2 \sim \text{N}(0, 1)$ are independent

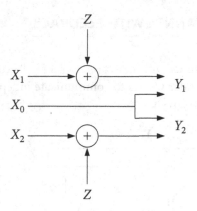

Figure 17.6. DM-BC for Example 17.2.

of each other. Assume expected average power constraint P on X. Without feedback, the capacity region is the set of rate pairs (R_1, R_2) such that $R_1 + R_2 \leq C(P)$. With feedback, we can use the following variation of the Schalkwijk–Kailath coding scheme. After proper initialization, we send $X_i = X_{1i} + X_{2i}$, where

$$X_{1i} = \gamma_i(X_{1,i-1} - E(X_{1,i-1}|Y_{1,i-1})),$$
$$X_{2i} = -\gamma_i(X_{2,i-1} - E(X_{2,i-1}|Y_{2,i-1})),$$

and γ_i is chosen so that $E(X_i^2) \leq P$ for each i. It can be shown that

$$R_1 = R_2 = \frac{1}{2} C\left(\frac{P(1 + \rho^*)/2}{1 + P(1 - \rho^*)/2} \right)$$

is achievable, where ρ^* satisfies the condition

$$\rho^*\left(1 + (P+1)\left(1 + \frac{P(1 - \rho^*)}{2} \right) \right) = \frac{P(P+2)}{2}(1 - \rho^*).$$

For example, when $P = 1$, $(R_1, R_2) = (0.2803, 0.2803)$ is achievable with feedback, which is strictly outside the capacity region without feedback.

Remark 17.8. As shown in Lemma 5.1, the capacity region of the broadcast channel with no feedback depends only on its conditional marginal pmfs. Hence, the capacity region of the Gaussian BC in Example 17.3 is the same as that of its corresponding physically degraded broadcast channel $Y_1 = Y_2 = X + Z$, where $Z \sim N(0, 1)$. The above example shows that this is not the case when feedback is present. Hence, in general the capacity region of a stochastically degraded BC can be strictly larger than that of its physically degraded version.

17.4 RELAY CHANNEL WITH FEEDBACK

Consider the DM-RC $p(y_2, y_3|x_1, x_2)$ with noiseless causal feedback from the receivers to the relay and the sender as depicted in Figure 17.7. Here, for each $i \in [1:n]$, the relay encoder assigns a symbol $x_{2i}(y_2^{i-1}, y_3^{i-1})$ to each pair (y_2^{i-1}, y_3^{i-1}) and the encoder assigns a symbol $x_{1i}(m, y_2^{i-1}, y_3^{i-1})$ to each triple $(m, y_2^{i-1}, y_3^{i-1})$.

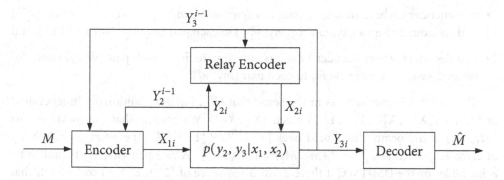

Figure 17.7. Relay channel with feedback.

Unlike the nonfeedback case, we have the following single-letter characterization of the capacity.

Theorem 17.3. The capacity of DM-RC $p(y_2, y_3|x_1, x_2)$ with feedback is

$$C = \max_{p(x_1, x_2)} \min\{I(X_1, X_2; Y_3), I(X_1; Y_2, Y_3|X_2)\}$$

The proof of the converse follows by noting that the cutset bound on the capacity in Theorem 16.1 continues to hold when feedback is present. To prove achievability, note that due to feedback, the relay encoder at time i observes (Y_2^{i-1}, Y_3^{i-1}) instead of only Y_2^{i-1}. Hence, feedback in effect converts the relay channel $p(y_2, y_3|x_1, x_2)$ into a (physically) degraded relay channel $p(y_2', y_3|x_1, x_2)$, where $Y_2' = (Y_2, Y_3)$. Substituting Y_2' in the corresponding decode–forward lower bound

$$C \geq \max_{p(x_1, x_2)} \min\{I(X_1, X_2; Y_3), I(X_1; Y_2'|X_2)\}$$

completes the proof of achievability.

Remark 17.9. To achieve the capacity in Theorem 17.3, we only need feedback from the receiver to the relay. Hence, adding other feedback links does not increase the capacity.

17.5 TWO-WAY CHANNEL

Consider the two-way communication system depicted in Figure 17.8. We assume a *discrete memoryless two-way channel* (DM-TWC) model $(\mathcal{X}_1 \times \mathcal{X}_2, p(y_1, y_2|x_1, x_2), \mathcal{Y}_1 \times \mathcal{Y}_2)$ that consists of four finite sets $\mathcal{X}_1, \mathcal{X}_2, \mathcal{Y}_1, \mathcal{Y}_2$ and a collection of conditional pmfs $p(y_1, y_2|x_1, x_2)$ on $\mathcal{Y}_1 \times \mathcal{Y}_2$. Each node wishes to send a message to the other node.

A $(2^{nR_1}, 2^{nR_2}, n)$ code for the DM-TWC consists of

- two message sets $[1 : 2^{nR_1}]$ and $[1 : 2^{nR_2}]$,

- two encoders, where encoder 1 assigns a symbol $x_{1i}(m_1, y_1^{i-1})$ to each pair (m_1, y_1^{i-1}) and encoder 2 assigns a symbol $x_{2i}(m_2, y_2^{i-1})$ to each pair (m_2, y_2^{i-1}) for $i \in [1 : n]$, and

- two decoders, where decoder 1 assigns an estimate $\hat{m}_2$ to each pair (m_1, y_1^n) and decoder 2 assigns an estimate $\hat{m}_1$ to each pair (m_2, y_2^n).

The channel is memoryless in the sense that (Y_{1i}, Y_{2i}) is conditionally independent of $(M_1, M_2, X_1^{i-1}, X_2^{i-1}, Y_1^{i-1}, Y_2^{i-1})$ given (X_{1i}, X_{2i}). We assume that the message pair (M_1, M_2) is uniformly distributed over $[1 : 2^{nR_1}] \times [1 : 2^{nR_2}]$. The average probability of error is defined as $P_e^{(n)} = P\{(\hat{M}_1, \hat{M}_2) \neq (M_1, M_2)\}$. A rate pair (R_1, R_2) is said to be achievable for the DM-TWC if there exists a sequence of $(2^{nR_1}, 2^{nR_2}, n)$ codes such that $\lim_{n \to \infty} P_e^{(n)} = 0$. The capacity region $\mathscr{C}$ of the DM-TWC is the closure of the set of achievable rate pairs (R_1, R_2).

The capacity region of the DM-TWC is not known in general. The main difficulty is that the two information flows share the same channel, causing interference to each other. In addition, each node has to play the two competing roles of communicating its own message and providing feedback to help the other node.

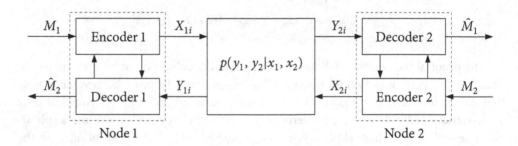

Figure 17.8. Two-way communication system.

17.5.1 Simple Inner and Outer Bounds

By having the encoders completely ignore the received outputs and use randomly and independently generated point-to-point codebooks, we obtain the following inner bound.

Proposition 17.2. A rate pair (R_1, R_2) is achievable for the DM-TWC $p(y_1, y_2 | x_1, x_2)$ if

$$R_1 < I(X_1; Y_2 | X_2, Q),$$
$$R_2 < I(X_2; Y_1 | X_1, Q)$$

for some pmf $p(q)p(x_1|q)p(x_2|q)$.

This inner bound is tight in some special cases, for example, when the channel decomposes into two separate DMCs, i.e., $p(y_1, y_2|x_1, x_2) = p(y_1|x_2)p(y_2|x_1)$. The bound is not tight in general, however, as will be shown in Example 17.5.

Using standard converse techniques, we can establish the following outer bound.

Proposition 17.3. If a rate pair (R_1, R_2) is achievable for the DM-TWC $p(y_1, y_2|x_1, x_2)$, then it must satisfy the inequalities

$$R_1 \le I(X_1; Y_2|X_2),$$
$$R_2 \le I(X_2; Y_1|X_1)$$

for some pmf $p(x_1, x_2)$.

This outer bound is a special case of the cutset bound that will be discussed in Section 18.4 and is not tight in general.

Example 17.4 (Binary multiplier channel (BMC)). Consider the DM-TWC with $\mathcal{X}_1 = \mathcal{X}_2 = \{0, 1\}$ and $Y_1 = Y_2 = Y = X_1 \cdot X_2$ depicted in Figure 17.9.

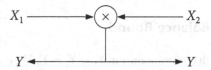

Figure 17.9. Binary multiplier channel.

The inner bound in Proposition 17.2 simplifies to the set of rate pairs (R_1, R_2) such that

$$R_1 < \alpha_2 H(\alpha_1),$$
$$R_2 < \alpha_1 H(\alpha_2),$$

which is attained by $Q = \emptyset$, $X_1 \sim \text{Bern}(\alpha_1)$, and $X_2 \sim \text{Bern}(\alpha_2)$ for some $\alpha_1, \alpha_2 \in [1/2, 1]$. In the other direction, the outer bound in Proposition 17.3 simplifies to the set of rate pairs (R_1, R_2) such that

$$R_1 \le \bar{\alpha}_1 H\left(\frac{\alpha_2}{\bar{\alpha}_1}\right),$$

$$R_2 \le \bar{\alpha}_2 H\left(\frac{\alpha_1}{\bar{\alpha}_2}\right),$$

which is attained by setting $p_{X_1, X_2}(1, 0) = \alpha_1$, $p_{X_1, X_2}(0, 1) = \alpha_2$, and $p_{X_1, X_2}(1, 1) = 1 - \alpha_1 - \alpha_2$ for some $\alpha_1, \alpha_2 \ge 0$ such that $\alpha_1 + \alpha_2 \le 1$. The two bounds are plotted in Figure 17.10.

Note that these bounds lead to the lower and upper bounds $0.6170 \le C_{\text{sym}} \le 0.6942$ on the symmetric capacity $C_{\text{sym}} = \max\{R : (R, R) \in \mathscr{C}\}$.

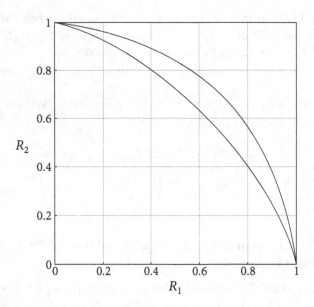

Figure 17.10. Simple inner and outer bounds on the capacity region of the BMC.

17.5.2 Dependence Balance Bound

Consider the DM-TWC with common output $Y_1 = Y_2 = Y$. We can establish the following improved outer bound on the capacity region.

> **Theorem 17.4 (Dependence Balance Bound for DM-TWC with Common Output).**
> If a rate pair (R_1, R_2) is achievable for the DM-TWC with common output $p(y|x_1, x_2)$, then it must satisfy the inequalities
>
> $$R_1 \le I(X_1; Y|X_2, U),$$
> $$R_2 \le I(X_2; Y|X_1, U)$$
>
> for some pmf $p(u, x_1, x_2)$ such that
>
> $$I(X_1; X_2|U) \le I(X_1; X_2|Y, U) \tag{17.9}$$
>
> and $|\mathcal{U}| \le 3$.

   Note that the only difference between this bound and the simple outer bound in Proposition 17.3 is the extra *dependence balance* condition in (17.9), which limits the set of joint pmfs that can be formed through sequential transmissions. This bound is not tight in general, however. For the case of the BMC, the dependence balance condition in (17.9) is inactive (since $U = Y = X_1 \cdot X_2$), which leads to the same upper bound on the symmetric capacity $C_{\text{sym}} \le 0.6942$ obtained by the simple outer bound in Proposition 17.3.

The dependence balance bound, however, can be improved by a genie argument to yield the tighter upper bound $C_{\text{sym}} \leq 0.6463$.

The proof of the dependence balance bound follows by standard converse techniques with the auxiliary random variable identification $U_i = Y^{i-1}$.

Remark 17.10. A similar outer bound can be derived for the DM-MAC with feedback. In this case, we add the inequality

$$R_1 + R_2 \leq I(X_1, X_2; Y|U)$$

to the inequalities in Theorem 17.4. It can be shown that the resulting outer bound is in general tighter than the cooperative outer bound for the DM-MAC with feedback in Proposition 17.1, due to the dependence balance condition in (17.9) on the set of joint pmfs on (X_1, X_2).

17.6   DIRECTED INFORMATION

Just as mutual information between two random sequences X^n and Y^n captures the uncertainty about Y^n reduced by knowing X^n, *directed information* from X^n to Y^n captures the uncertainty about Y^n reduced by *causal* knowledge of X^n. This notion of directed information provides a nontrivial multiletter characterization of the capacity region of the DM-TWC.

More precisely, the directed information from a random sequence X^n to another random sequence Y^n of the same length is defined as

$$I(X^n \to Y^n) = \sum_{i=1}^{n} I(X^i; Y_i|Y^{i-1}) = H(Y^n) - \sum_{i=1}^{n} H(Y_i|Y^{i-1}, X^i).$$

In comparison, the mutual information between X^n and Y^n is

$$I(X^n; Y^n) = \sum_{i=1}^{n} I(X^n; Y_i|Y^{i-1}) = H(Y^n) - \sum_{i=1}^{n} H(Y_i|Y^{i-1}, X^n).$$

The directed information from X^n to Y^n *causally conditioned on* Z^n is similarly defined as

$$I(X^n \to Y^n \| Z^n) = \sum_{i=1}^{n} I(X^i; Y_i|Y^{i-1}, Z^i).$$

As a related notion, the *pmf of X^n causally conditioned on* Y^n is defined as

$$p(x^n \| y^n) = \prod_{i=1}^{n} p(x_i|x^{i-1}, y^i).$$

By convention, the pmf of X^n causally conditioned on $(\emptyset, Y^{n-1})$ is expressed as

$$p(x^n \| y^{n-1}) = \prod_{i=1}^{n} p(x_i|x^{i-1}, y^{i-1}).$$

17.6.1   Multiletter Characterization of the DM-TWC Capacity Region

We use directed information to obtain a multiletter characterization of the capacity region of the DM-TWC with common output.

Theorem 17.5. Let $\mathscr{C}^{(k)}$, $k \geq 1$, be the set of rate pairs (R_1, R_2) such that

$$R_1 \leq \frac{1}{k} I(X_1^k \to Y^k \| X_2^k),$$

$$R_2 \leq \frac{1}{k} I(X_2^k \to Y^k \| X_1^k)$$

for some (causally conditional) pmf $p(x_1^k \| y^{k-1}) p(x_2^k \| y^{k-1})$. Then the capacity region $\mathscr{C}$ of the DM-TWC with common output $p(y|x_1, x_2)$ is the closure of $\bigcup_k \mathscr{C}^{(k)}$.

While this characterization in itself is not computable and thus of little use, each choice of k and $p(x_1^k \| y^{k-1}) p(x_2^k \| y^{k-1})$ leads to an inner bound on the capacity region, as illustrated in the following.

Example 17.5 (BMC revisited). Consider the binary multiplier channel $Y = X_1 \cdot X_2$ in Example 17.4. We evaluate the inequalities in Theorem 17.5 with $U_0 = \emptyset$ and $U_i = u_i(Y^i) \in \{0, 1, 2\}$, $i \in [1:n]$, such that

$$U_i = \begin{cases} 0 & \text{if } (U_{i-1}, Y_i) = (0, 1) \text{ or } (U_{i-1}, Y_i) = (1, 0) \text{ or } U_i = 2, \\ 1 & \text{if } (U_{i-1}, Y_i) = (0, 0), \\ 2 & \text{otherwise.} \end{cases}$$

For $j = 1, 2$, consider the same causally conditional pmf $p(x_j^k \| y^{k-1})$ of the form

$$p(x_{ji} = 1 | u_{i-1} = 0) = \alpha,$$
$$p(x_{ji} = 1 | u_{i-1} = 1, x_{j,i-1} = 1) = \beta,$$
$$p(x_{ji} = 1 | u_{i-1} = 1, x_{j,i-1} = 0) = 1,$$
$$p(x_{ji} = 1 | u_{i-1} = 2, x_{j,i-2} = 1) = 0,$$
$$p(x_{ji} = 1 | u_{i-1} = 2, x_{j,i-2} = 0) = 1$$

for some $\alpha, \beta \in [0, 1]$. Then, by optimizing over α, β, it can be shown that

$$\lim_{k \to \infty} \frac{1}{k} I(X_1^k \to Y^k \| X_2^k) = \lim_{k \to \infty} \frac{1}{k} I(X_2^k \to Y^k \| X_1^k) = 0.6191,$$

and hence $(R_1, R_2) = (0.6191, 0.6191)$ is achievable, which is strictly outside the inner bound in Proposition 17.2.

Remark 17.11. A similar multiletter characterization to Theorem 17.5 can be found for the general DM-TWC $p(y_1, y_2|x_1, x_2)$ and the DM-MAC with feedback.

17.6.2   Proof of the Converse

By Fano's inequality,

$$nR_1 \le I(M_1; Y^n, M_2) + n\epsilon_n$$
$$= I(M_1; Y^n | M_2) + n\epsilon_n$$
$$= \sum_{i=1}^{n} I(M_1; Y_i | M_2, Y^{i-1}) + n\epsilon_n$$
$$= \sum_{i=1}^{n} I(M_1, X_1^i; Y_i | M_2, X_2^i, Y^{i-1}) + n\epsilon_n$$
$$\le \sum_{i=1}^{n} I(X_1^i; Y_i | X_2^i, Y^{i-1}) + n\epsilon_n$$
$$= I(X_1^n \to Y^n \| X_2^n) + n\epsilon_n.$$

Similarly, $nR_2 \le I(X_2^n \to Y^n \| X_1^n) + n\epsilon_n$. Also it can be shown that $I(M_1; M_2 | X_2^i, Y^i) \le I(M_1; M_2 | X_2^{i-1}, Y^{i-i})$ for all $i \in [1:n]$. Hence

$$I(X_1^i; X_{2i} | X_2^{i-1}, Y^{i-1}) \le I(M_1; M_2 | X_2^{i-1}, Y^{i-1}) \le I(M_1; M_2) = 0,$$

or equivalently, $X_{2i} \to (X_2^{i-1}, Y^{i-1}) \to X_1^i$ form a Markov chain. This implies that the joint pmf is of the form $p(x_1^n \| y^{n-1}) p(x_2^n \| y^{n-1})$. Therefore, for any $\epsilon > 0$, $(R_1 - \epsilon, R_2 - \epsilon) \in \mathscr{C}^{(n)}$ for n sufficiently large. This completes the converse proof of Theorem 17.5.

17.6.3   Proof of Achievability

Communication takes place over k *interleaved* blocks, each used for one of k independent message pairs $(M_{1j}, M_{2j}) \in [1:2^{nR_{1j}}] \times [1:2^{nR_{2j}}]$, $j \in [1:k]$. Block j consists of transmission times $j, k + j, 2k + j, \ldots, (n-1)k + j$, as illustrated in Figure 17.11.

For each block, we treat the channel input and output sequences from the previous blocks as causal state information available at each encoder–decoder pair and use the multiplexing technique in Section 7.4.1.

Codebook generation. Fix k and a conditional pmf

$$p(x_1^k \| y^{k-1}) p(x_2^k \| y^{k-1}) = \prod_{j=1}^{k} p(x_{1j} | x_1^{j-1}, y^{j-1}) p(x_{2j} | x_1^{j-1}, y^{j-1}).$$

Let $S_{1j} = (X_1^{j-1}, Y^{j-1})$ and $S_{2j} = (X_2^{j-1}, Y^{j-1})$. Divide the message m_1 into k independent messages $(m_{11}, \ldots, m_{1k})$. Then further divide each message m_{1j} into the messages $(m_{1j}(s_{1j}): s_{1j} \in S_{1j} = \mathcal{X}_1^{j-1} \times \mathcal{Y}^{j-1})$. Thus, $R_1 = \sum_{j=1}^{k} \sum_{s_{1j} \in S_{1j}} R_{1j}(s_{1j})$. For $j \in [1:k]$ and $s_{1j} \in S_{1j}$, randomly and conditionally independently generate $2^{nR_{1j}(s_{1j})}$ sequences $x_{1j}(m_{1j}(s_{1j}), s_{1j})$, $m_{1j}(s_{1j}) \in [1:2^{nR_{1j}(s_{1j})}]$, each according to $\prod_{i=1}^{n} p_{X_{1j}|S_{1j}}(x_{1,(i-1)k+j} | s_{1j})$.

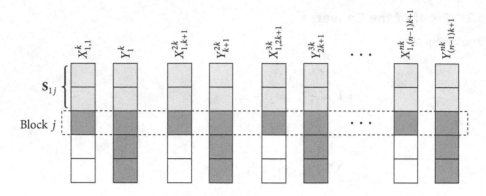

Figure 17.11. Multiletter coding scheme for the DM-TWC.

These sequences form the codebook $C_{1j}(s_{1j})$ for $j \in [1:k]$. Similarly, generate random codebooks $C_{2j}(s_{2j})$, $j \in [1:k]$, $s_{2j} \in S_{2j} = \mathcal{X}_2^{j-1} \times \mathcal{Y}^{j-1}$ for $m_2 = (m_{21}, \dots, m_{2k})$ and $m_{2j} = (m_{2j}(s_{2j})\colon s_{2j} \in S_{2j})$.

Encoding. For each block $j \in [1:k]$ (consisting of transmission times $j, k+j, 2k+j, \dots, (n-1)k+j$), encoder 1 treats the sequences $(\mathbf{x}_1^{j-1}, \mathbf{y}^{j-1}) \in S_{1j}^n$ as causal state information available at both encoder 1 and decoder 2. As in the achievability proof in Section 7.4.1, encoder 1 stores the codewords $\mathbf{x}_{1j}(m_{1j}(s_{1j}), s_{1j})$, $s_{1j} \in \mathcal{X}_1^{j-1} \times \mathcal{Y}^{j-1}$, in FIFO buffers and transmits a symbol from the buffer that corresponds to the state symbol of the given time. Similarly, encoder 2 stores the codewords in FIFO buffers and multiplexes over them according to its state sequence.

Decoding and the analysis of the probability of error. Upon receiving y^{kn}, decoder 2 recovers $m_{11}, \dots, m_{1k}$ successively. For interleaved block j, decoder 2 first forms the state sequence estimate $\hat{s}_{1j} = (\hat{\mathbf{x}}_1^{j-1}, \mathbf{y}^{j-1})$ from the codeword estimates and output sequences from previous blocks, and demultiplexes the output sequence for block j into $|S_{1j}|$ subsequences accordingly. Then it finds the unique index $\hat{m}_{1j}(s_{1j})$ for each subsequence corresponding to the state s_{1j}. Following the same argument as in Section 7.4.1, the probability of error for M_{1j} tends to zero as $n \to \infty$ if previous decoding steps are successful and

$$R_{1j} < I(X_{1j}; X_2^k, Y_j^k | S_{1j}) - \delta(\epsilon)$$

$$\overset{(a)}{=} I(X_{1j}; X_{2,j+1}^k, Y_j^k | X_2^j, X_1^{j-1}, Y^{j-1}) - \delta(\epsilon)$$

$$= \sum_{i=j}^{k} I(X_{1j}; X_{2,i+1}, Y_i | X_2^i, X_1^{j-1}, Y^{i-1}) - \delta(\epsilon)$$

$$\overset{(b)}{=} \sum_{i=j}^{k} I(X_{1j}; Y_i | X_2^i, X_1^{j-1}, Y^{i-1}) - \delta(\epsilon),$$

where (a) and (b) follow since $X_{1j} \to (X_1^{j-1}, Y^{j-1}) \to X_2^j$ and $X_{2,i+1} \to (X_2^i, Y^i) \to X_1^{i+1}$, respectively. Therefore, by induction, the probability of error over all time slots $j \in [1:k]$

tends to zero as $n \to \infty$ if

$$\sum_{j=1}^{k} R_{1j} < \sum_{j=1}^{k} \sum_{i=j}^{k} I(X_{1j}; Y_i | X_2^i, X_1^{j-1}, Y^{i-1}) - k\delta(\epsilon)$$

$$= \sum_{i=1}^{k} \sum_{j=1}^{i} I(X_{1j}; Y_i | X_2^i, X_1^{j-1}, Y^{i-1}) - k\delta(\epsilon)$$

$$= \sum_{i=1}^{k} I(X_1^i; Y_i | X_2^i, Y^{i-1}) - k\delta(\epsilon)$$

$$= I(X_1^k \to Y^k \| X_2^k) - k\delta(\epsilon).$$

Similarly, the probability of error for decoder 1 tends to zero as $n \to \infty$ if

$$\sum_{j=1}^{k} R_{2j} < I(X_2^k \to Y^k \| X_1^k) - k\delta(\epsilon).$$

This completes the achievability proof of Theorem 17.5.

SUMMARY

- Feedback can simplify coding and improve reliability

- Iterative refinement:

  - Schalkwijk–Kailath scheme for the Gaussian channel

  - Horstein's scheme for the BSC

  - Posterior matching scheme for the DMC

  - Block feedback scheme for the DMC

- Feedback can enlarge the capacity region of multiuser channels:

  - Feedback can be used to induce cooperation between the senders

  - Cover–Leung inner bound

  - A combination of the Schalkwijk–Kailath scheme, rate splitting, and superposition coding achieves the capacity region of the 2-sender Gaussian MAC with feedback

  - Feedback can be used also to provide channel information to be broadcast to multiple receivers

  - The cutset bound is achievable for the relay channel with feedback

- Dependence balance bound

- Directed information and the multiletter characterization of the capacity region of the two-way channel

- **Open problems:**

    17.1. What is the sum-capacity of the 3-sender symmetric Gaussian MAC with feed-back?

    17.2. What is the sum-capacity of the 2-receiver symmetric Gaussian BC with feed-back?

    17.3. What is the symmetric capacity of the binary multiplier channel?

BIBLIOGRAPHIC NOTES

The Schalkwijk–Kailath coding scheme appeared in Schalkwijk and Kailath (1966) and Schalkwijk (1966). Pinsker (1968) and Shepp, Wolf, Wyner, and Ziv (1969) showed that under the almost-sure average power constraint in (17.4), the probability of error decays only (single) exponentially fast, even for sending a binary message $M \in \{1, 2\}$. Wyner (1968) provided a simple modification of the Schalkwijk–Kailath scheme that achieves any rate $R < C$ under this more stringent power constraint. Butman (1976) extended the Schalkwijk–Kailath scheme to additive *colored* Gaussian noise channels with feedback, the optimality of which was established by Kim (2010).

The Horstein coding scheme appeared in Horstein (1963). The first rigorous proof that this scheme achieves the capacity of the BSC is due to Shayevitz and Feder (2011), who also developed the posterior matching scheme. The block feedback coding scheme for the BSC in Section 17.1.3 is originally due to Weldon (1963) and has been extended to arbitrary DMCs by Ahlswede (1973) and Ooi and Wornell (1998).

Gaarder and Wolf (1975) first showed that feedback can enlarge the capacity region of DM multiuser channels via the binary erasure MAC ($R = 0.7602$ in Example 17.1). The Cover–Leung inner bound on the capacity region of the DM-MAC with feedback appeared in Cover and Leung (1981). The proof of achievability in Section 17.2.2 follows Zeng, Kuhlmann, and Buzo (1989). The capacity region of the Gaussian MAC with feedback in Theorem 17.2 is due to Ozarow (1984). Carleial (1982) studied the MAC with generalized feedback, where the senders observe noisy versions of the channel output at the receiver; see Kramer (2007) for a discussion of this model.

El Gamal (1978) showed that feedback does not enlarge the capacity region of the physically degraded DM-BC. Example 17.2 is due to Dueck (1980). Ozarow and Leung (1984) extended the Schalkwijk–Kailath coding scheme to the Gaussian BC as described in Example 17.3. Elia (2004) improved the resulting inner bound using control theoretic tools. Cover and El Gamal (1979) established the capacity of the DM-RC with feedback in Theorem 17.3.

Shannon (1961) introduced the DM-TWC and established the inner and outer bounds in Propositions 17.2 and 17.3. The binary multiplier channel (BMC) in Example 17.4 first appeared in the same paper, where it is attributed to D. Blackwell. Dueck (1979) showed via a counterexample that Shannon's inner bound is not tight in general. Hekstra and Willems (1989) established the dependence balance bound and its extensions. They also established the symmetric capacity upper bound $C_{\text{sym}} \leq 0.6463$. The symmetric capacity lower bound in Example 17.5 was originally obtained by Schalkwijk (1982) using a constructive coding scheme in the flavor of the Horstein scheme. This bound was further improved by Schalkwijk (1983) to $C_{\text{sym}} \geq 0.6306$.

The notion of directed information was first introduced by Marko (1973) in a slightly different form. The definition in Section 17.6 is due to Massey (1990), who demonstrated its utility by showing that the capacity of point-to-point channels with memory is upper bounded by the maximum directed information from X^n to Y^n. The causal conditioning notation was developed by Kramer (1999), who also established the multiletter characterization of the capacity region for the DM-TWC in Kramer (2003). Massey's upper bound was shown to be tight for classes of channels with memory by Tatikonda and Mitter (2009), Kim (2008b), and Permuter, Weissman, and Goldsmith (2009). The inner bound on the capacity region of the BMC using directed information in Example 17.5 is due to Ardestanizadeh (2010). Directed information and causally conditional probabilities arise as a canonical answer to many other problems with causality constraints (Permuter, Kim, and Weissman 2011).

PROBLEMS

17.1. Show that the cutset bound on the capacity of the DM-RC in Theorem 16.1 continues to hold when feedback from the receivers to the senders is present.

17.2. Provide the details of the analysis of the probability of error for the Cover–Leung inner bound in Theorem 17.1.

17.3. Show that the Cover–Leung inner bound for the Gaussian MAC simplifies to (17.5).

17.4. Prove achievability of the sum-capacity in (17.6) of the Gaussian MAC with feedback when $S_1 \neq S_2$.

17.5. Provide the details of the achievability proof of every rate pair in the capacity region of the Gaussian MAC.

17.6. Show that feedback does not increase the capacity region of the physically degraded broadcast channel. (Hint: Consider auxiliary random variable identification $U_i = (M_2, Y_1^{i-1}, Y_2^{i-1})$.)

17.7. Show that the simple outer bound on the capacity region of the DM-BC $p(y_1, y_2|x)$ in Figure 5.2 continues to hold with feedback.

17.8. Provide the details of the proof of the dependence balance bound in Theorem 17.4.

17.9. *Semideterministic MAC with feedback.* Show that the Cover–Leung inner bound on the capacity region of the DM-MAC $p(y|x_1, x_2)$ with feedback is tight when X_1 is a function of (X_2, Y).
Remark: This result is due to Willems (1982).

17.10. *Gaussian MAC with noise feedback.* Consider the Gaussian multiple access channel $Y = g_1 X_1 + g_2 X_2 + Z$ in Section 17.2.3. Suppose that there is noiseless causal feedback of the noise Z, instead of Y, to the senders. Thus, encoder $j = 1, 2$ assigns a symbol $x_{ji}(m_j, z^{i-1})$ at time $i \in [1 : N]$. Show that the capacity region of this channel is the set of rate pairs (R_1, R_2) such that

$$R_1 + R_2 \leq C(S_1 + S_2 + 2\sqrt{S_1 S_2}),$$

which is equal to the cooperative capacity region (see Problem 4.9).

17.11. *Directed information.* Prove the following properties of directed information and causally conditional probability distributions:

(a) Chain rule: $p(y^n, x^n) = p(y^n||x^n)p(x^n||y^{n-1})$.

(b) Nonnegativity: $I(X^n \to Y^n) \geq 0$ with equality iff $p(y^n||x^n) = p(y^n)$.

(c) Conservation: $I(X^n; Y^n) = I(X^n \to Y^n) + I(Y_0, Y^{n-1} \to X^n)$, where $Y_0 = \emptyset$.

(d) Comparison to mutual information: $I(X^n \to Y^n) \leq I(X^n; Y^n)$ with equality if there is no feedback, i.e., $p(x_i|x^{i-1}, y^{i-1}) = p(x_i|x^{i-1})$, $i \in [1 : n]$.

17.12. *Compound channel with feedback.* Consider the compound channel $p(y_s|x)$ in Chapter 7 with finite state alphabet S.

(a) Find the capacity of the compound channel with noiseless causal feedback.

(b) Compute the capacity of the binary erasure compound channel with feedback where the erasure probability can take one of the four values $(0, 0.1, 0.2, 0.25)$.

(c) Let C_1 be the capacity of the compound channel without feedback, C_2 be the capacity of the compound channel without feedback but when the sender and code designer know the actual channel (in addition to the decoder), and C_3 be the capacity of the compound channel with feedback. Which of the following statements hold in general?

  (1) $C_1 = C_2$.
  (2) $C_1 = C_3$.
  (3) $C_2 = C_3$.
  (4) $C_1 = C_2 = C_3$.
  (5) $C_1 \leq C_2$.

(d) Consider a compound channel where the capacity of each individual channel $s \in S$ is achieved by the same pmf $p^*(x)$ on $\mathcal{X}$. Answer part (c) for this case.

(e) Consider a compound channel having zero capacity without feedback. Is the capacity with feedback also equal to zero in this case? Prove it or provide a counterexample.

17.13. *Modulo-2 sum two-way channel.* Consider a DM-TWC with $\mathcal{X}_1 = \mathcal{X}_2 = \mathcal{Y}_1 = \mathcal{Y}_2 = \{0, 1\}$ and $Y_1 = Y_2 = X_1 \oplus X_2$. Find its capacity region.

17.14. *Gaussian two-way channel.* Consider the Gaussian two-way channel

$$Y_1 = g_{12}X_2 + Z_1,$$
$$Y_2 = g_{21}X_1 + Z_2,$$

where the noise pair $(Z_1, Z_2) \sim N(0, K)$. Assume average power constraint P on each of X_1 and X_2. Find the capacity region of this channel in terms of the power constraint P, channel gains g_{12} and g_{21}, and the noise covariance matrix K.

17.15. *Common-message feedback capacity of broadcast channels.* Consider the DM-BC $p(y_1, y_2|x)$ with feedback from the receivers. Find the common-message capacity C_F.

17.16. *Broadcast channel with feedback.* Consider the generalization of Dueck's broadcast channel with feedback example as depicted in Figure 17.12, where (Z_1, Z_2) is a DSBS(p). Show that the capacity region with feedback is the set of rate pairs (R_1, R_2) such that

$$R_1 \le 1,$$
$$R_2 \le 1,$$
$$R_1 + R_2 \le 2 - H(p).$$

Remark: This result is due to Shayevitz and Wigger (2010).

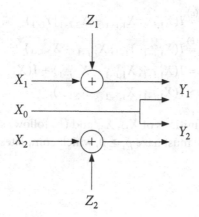

Figure 17.12. Generalized Dueck's example.

17.17. *Channels with state and feedback.* Consider the DMC with DM state $p(y|x, s)p(s)$ in Chapter 7. Suppose that noiseless causal feedback from the receiver is available at the encoder. Show that feedback does not increase the capacity when the state information is available causally or noncausally at the encoder.

17.18. *Delayed feedback.* We considered the capacity region for channels with causal feedback. Now consider the DM-MAC $p(y|x_1, x_2)$ and assume that the feedback has a constant delay d such that encoder $j = 1, 2$ assigns a symbol $x_{ji}(m_j, y^{i-d})$ for time $i \in [1 : N]$. Show that the capacity region is identical to that for the causal case ($d = 1$). (Hint: For example, when $d = 2$, consider 2 parallel DM-MACs corresponding to even and odd transmissions.)

APPENDIX 17A   PROOF OF LEMMA 17.1

We use induction. By the construction of U_1 and U_2, $\rho_1 = \rho^*$. For $i \geq 2$, consider

$$
\begin{aligned}
I(X_{1i}; X_{2i}) &\stackrel{(a)}{=} I(X_{1i}; X_{2i} | Y_{i-1}) \\
&= I(X_{1,i-1}; X_{2,i-1} | Y_{i-1}) \\
&= I(X_{1,i-1}; X_{2,i-1}) + I(X_{1,i-1}; Y_{i-1} | X_{2,i-1}) - I(X_{1,i-1}; Y_{i-1}) \\
&\stackrel{(b)}{=} I(X_{1,i-1}; X_{2,i-1}),
\end{aligned}
$$

where (a) follows since the pair (X_{1i}, X_{2i}) is independent of Y_{i-1}, (b) follows by the induction hypothesis $I(X_{1,i-1}; Y_{i-1} | X_{2,i-1}) = I(X_{1,i-1}; Y_{i-1})$. Hence, we have $\rho_i^2 = \rho_{i-1}^2$.

To show that $\rho_i = \rho_{i-1}$ (same sign), consider

$$
\begin{aligned}
I(X_{1i}; X_{1i} + X_{2i}) &= I(X_{1i}; X_{1i} + X_{2i} | Y_{i-1}) \\
&\stackrel{(a)}{=} I(X_{1,i-1}; X_{1,i-1} - X_{2,i-1} | Y_{i-1}) \\
&\stackrel{(b)}{=} I(X_{1,i-1}, Y_{i-1}; X_{1,i-1} - X_{2,i-1}) \\
&= I(X_{1,i-1}; X_{1,i-1} - X_{2,i-1}) + I(X_{2,i-1}; Y_{i-1} | X_{1,i-1}) \\
&> I(X_{1,i-1}; X_{1,i-1} - X_{2,i-1}),
\end{aligned}
$$

where (a) follows by the definitions of X_{1i}, X_{2i} and (b) follows since $X_{1,i-1} - X_{2,i-1}$ is independent of Y_{i-1}. Hence we must have $\rho_i \geq 0$, which completes the proof of Lemma 17.1.

CHAPTER 18

Discrete Memoryless Networks

We introduce the discrete memoryless network (DMN) as a model for general multihop networks that includes as special cases graphical networks, the DM-RC, the DM-TWC, and the DM single-hop networks with and without feedback. In this chapter, we establish outer and inner bounds on the capacity region of this network.

We first consider the multicast messaging case and establish a cutset upper bound on the capacity. We then extend the decode–forward coding scheme for the relay channel to multicast networks using the new idea of sliding block decoding. The resulting network decode–forward lower bound on the capacity is tight for the physically degraded multicast network. Generalizing compress–forward for the relay channel and network coding for graphical networks, we introduce the noisy network coding scheme. This scheme involves several new ideas. Unlike the compress–forward coding scheme for the relay channel in Section 16.7, where independent messages are sent over multiple blocks, in noisy network coding the same message is sent multiple times using independent codebooks. Furthermore, the relays do not use Wyner–Ziv binning and each decoder performs simultaneous nonunique decoding on the received signals from all the blocks without explicitly decoding for the compression indices. We show that this scheme is optimal for certain classes of deterministic networks and wireless erasure networks. These results also extend the network coding theorem in Section 15.3 to graphical networks with cycles and broadcasting.

We then turn our attention to the general multimessage network. We establish a cutset bound on the multimessage capacity region of the DMN, which generalizes all previous cutset bounds. We also extend the noisy network coding inner bound to multimessage multicast networks and show that it coincides with the cutset bound for the aforementioned classes of deterministic networks and wireless erasure networks. Finally, we show how the noisy network coding scheme can be combined with decoding techniques for the interference channel to establish inner bounds on the capacity region of the general DMN.

18.1 DISCRETE MEMORYLESS MULTICAST NETWORK

Consider the multicast network depicted in Figure 18.1. The network is modeled by an N-node *discrete memoryless network* (DMN) $(\mathcal{X}_1 \times \cdots \times \mathcal{X}_N, p(y_1, \ldots, y_N | x_1, \ldots, x_N), \mathcal{Y}_1 \times \cdots \times \mathcal{Y}_N)$ that consists of N sender–receiver alphabet pairs $(\mathcal{X}_k, \mathcal{Y}_k)$, $k \in [1:N]$,

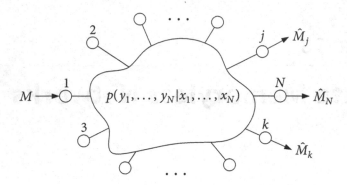

Figure 18.1. Discrete memoryless multicast network.

and a collection of conditional pmfs $p(y_1, \ldots, y_N | x_1, \ldots, x_N)$. Note that the *topology* of this network, that is, which nodes can communicate directly to which other nodes, is defined through the structure of the conditional pmf $p(y^N | x^N)$. In particular, the graphical network discussed in Chapter 15 is a special case of the DMN.

We assume that source node 1 wishes to send a message M to a set of destination nodes $\mathcal{D} \subseteq [2:N]$. We wish to find the capacity of this DMN.

A $(2^{nR}, n)$ code for this discrete memoryless multicast network (DM-MN) consists of

- a message set $[1 : 2^{nR}]$,

- a source encoder that assigns a symbol $x_{1i}(m, y_1^{i-1})$ to each message $m \in [1 : 2^{nR}]$ and received sequence y_1^{i-1} for $i \in [1 : n]$,

- a set of relay encoders, where encoder $k \in [2:N]$ assigns a symbol $x_{ki}(y_k^{i-1})$ to every received sequence y_k^{i-1} for $i \in [1 : n]$, and

- a set of decoders, where decoder $k \in \mathcal{D}$ assigns an estimate $\hat{m}_k$ or an error message e to each y_k^n.

We assume that the message M is uniformly distributed over the message set. The average probability of error is defined as $P_e^{(n)} = \mathsf{P}\{\hat{M}_k \neq M \text{ for some } k \in \mathcal{D}\}$. A rate R is said to be achievable if there exists a sequence of $(2^{nR}, n)$ codes such that $\lim_{n \to \infty} P_e^{(n)} = 0$. The capacity of the DM-MN is the supremum of the set of achievable rates.

Consider the following special cases:

- If $N = 2$, $Y_1 = Y_2$, $X_2 = \emptyset$, and $\mathcal{D} = \{2\}$, then the DM-MN reduces to the DMC with feedback.

- If $N = 3$, $X_3 = Y_1 = \emptyset$, and $\mathcal{D} = \{3\}$, the DM-MN reduces to the DM-RC.

- If $X_2 = \cdots = X_N = \emptyset$ and $\mathcal{D} = [2:N]$, then the DM-MN reduces to the DM-BC with common message (with feedback if $Y_1 = (Y_2, \ldots, Y_N)$ or without feedback if $Y_1 = \emptyset$).

- If $\mathcal{D} = \{N\}$, then the DM-MN reduces to a discrete memoryless *unicast* network.

As we already know from the special case of the DM-RC, the capacity of the DM-MN is not known in general.

Cutset bound. The cutset bounds for the graphical multicast network in Section 15.1 and the DM-RC in Section 16.2 can be generalized to the DM-MN.

Theorem 18.1 (Cutset Bound for the DM-MN). The capacity of the DM-MN with destination set $\mathcal{D}$ is upper bounded as

$$C \le \max_{p(x^N)} \min_{k \in \mathcal{D}} \min_{\mathcal{S}:1\in\mathcal{S},k\in\mathcal{S}^c} I(X(\mathcal{S}); Y(\mathcal{S}^c)|X(\mathcal{S}^c)).$$

Proof. To establish this cutset upper bound, let $k \in \mathcal{D}$ and consider a cut $(\mathcal{S}, \mathcal{S}^c)$ such that $1 \in \mathcal{S}$ and $k \in \mathcal{S}^c$. Then by Fano's inequality, $H(M|Y^n(\mathcal{S}^c)) \le H(M|Y_k^n) \le n\epsilon_n$, where ϵ_n tends to zero as $n \to \infty$. Now consider

$$
\begin{aligned}
nR &= H(M) \\
&\le I(M; Y^n(\mathcal{S}^c)) + n\epsilon_n \\
&= \sum_{i=1}^{n} I(M; Y_i(\mathcal{S}^c)|Y^{i-1}(\mathcal{S}^c)) + n\epsilon_n \\
&= \sum_{i=1}^{n} I(M; Y_i(\mathcal{S}^c)|Y^{i-1}(\mathcal{S}^c), X_i(\mathcal{S}^c)) + n\epsilon_n \\
&\le \sum_{i=1}^{n} I(M, Y^{i-1}(\mathcal{S}^c); Y_i(\mathcal{S}^c)|X_i(\mathcal{S}^c)) + n\epsilon_n \\
&\le \sum_{i=1}^{n} I(X_i(\mathcal{S}), M, Y^{i-1}(\mathcal{S}^c); Y_i(\mathcal{S}^c)|X_i(\mathcal{S}^c)) + n\epsilon_n \\
&= \sum_{i=1}^{n} I(X_i(\mathcal{S}); Y_i(\mathcal{S}^c)|X_i(\mathcal{S}^c)) + n\epsilon_n.
\end{aligned}
$$

Introducing a time-sharing random variable $Q \sim \text{Unif}[1:n]$ independent of all other random variables and defining $X(\mathcal{S}) = X_Q(\mathcal{S})$, $Y(\mathcal{S}^c) = Y_Q(\mathcal{S}^c)$, and $X(\mathcal{S}^c) = X_Q(\mathcal{S}^c)$ so that $Q \to X^N \to Y^N$, we obtain

$$
\begin{aligned}
nR &\le nI(X_Q(\mathcal{S}); Y_Q(\mathcal{S}^c)|X_Q(\mathcal{S}^c), Q) + n\epsilon_n \\
&\le nI(X_Q(\mathcal{S}); Y_Q(\mathcal{S}^c)|X_Q(\mathcal{S}^c)) + n\epsilon_n \\
&= nI(X(\mathcal{S}); Y(\mathcal{S}^c)|X(\mathcal{S}^c)) + n\epsilon_n,
\end{aligned}
$$

which completes the proof of Theorem 18.1.

As discussed in Example 16.2, this cutset bound is not tight in general.

18.2 NETWORK DECODE–FORWARD

We now consider coding schemes and corresponding lower bounds on the capacity of the DM-MN. First we generalize the decode–forward scheme for the relay channel to DM multicast networks.

Theorem 18.2 (Network Decode–Forward Lower Bound). The capacity of the DM-MN $p(y^N|x^N)$ with destination set $\mathcal{D}$ is lower bounded as

$$C \geq \max_{p(x^N)} \min_{k \in [1:N-1]} I(X^k; Y_{k+1}|X_{k+1}^N).$$

Remark 18.1. For $N = 3$ and $X_3 = \emptyset$, the network decode–forward lower bound reduces to the decode–forward lower bound for the relay channel in Section 16.4.

Remark 18.2. The network decode–forward lower bound is tight when the DM-MN is *degraded*, i.e.,

$$p(y_{k+2}^N|x^N, y^{k+1}) = p(y_{k+2}^N|x_{k+1}^N, y_{k+1})$$

for $k \in [1:N-2]$. The converse follows by evaluating the cutset bound in Theorem 18.1 only for cuts of the form $\mathcal{S} = [1:k]$, $k \in [1:N-1]$, instead of all possible cuts. Hence

$$C \leq \max_{p(x^N)} \min_{k \in [1:N-1]} I(X^k; Y_{k+1}^N|X_{k+1}^N) = \max_{p(x^N)} \min_{k \in [1:N-1]} I(X^k; Y_{k+1}|X_{k+1}^N),$$

where the equality follows by the degradedness condition $X^k \to (Y_{k+1}, X_{k+1}^N) \to Y_{k+2}^N$.

Remark 18.3. The network decode–forward lower bound is also tight for broadcasting over the relay channel $p(y_2, y_3|x_1, x_2)$, i.e., $\mathcal{D} = \{2, 3\}$ (see Problem 16.18). For this case, the capacity is

$$C = \max_{p(x_1, x_2)} \min\{I(X_1; Y_2|X_2), I(X_1, X_2; Y_3)\}.$$

Remark 18.4. The decode–forward lower bound holds for any set of destination nodes $\mathcal{D} \subseteq [2:N]$. The bound can be further improved by removing some relay nodes and relabeling the nodes in the best order.

18.2.1 Sliding Window Decoding for the DM-RC

The achievability proof of the network decode–forward lower bound in Theorem 18.2 involves the new idea of *sliding window decoding*. For clarity of presentation, we first use this decoding scheme to provide an alternative proof of the decode–forward lower bound for the relay channel

$$C \geq \max_{p(x_1, x_2)} \min\{I(X_1, X_2; Y_3), I(X_1; Y_2|X_2)\}. \tag{18.1}$$

The codebook generation, encoding, and relay encoding are the same as in the decode–forward scheme for the DM-RC in Section 16.4. However, instead of using backward decoding, the receiver decodes for the message sequence in the forward direction using a different procedure.

As before, we use a block Markov coding scheme to send a sequence of $(b-1)$ i.i.d. messages, $m_j \in [1 : 2^{nR}]$, $j \in [1 : b-1]$, in b transmission blocks, each consisting of n transmissions.

Codebook generation. We use the same codebook generation as in cooperative multihop relaying discussed in Section 16.4.2. Fix the pmf $p(x_1, x_2)$ that attains the decode–forward lower bound in (18.1). For each block $j \in [1 : b]$, randomly and independently generate 2^{nR} sequences $x_2^n(m_{j-1})$, $m_{j-1} \in [1 : 2^{nR}]$, each according to $\prod_{i=1}^n p_{X_2}(x_{2i})$. For each $m_{j-1} \in [1 : 2^{nR}]$, randomly and conditionally independently generate 2^{nR} sequences $x_1^n(m_j|m_{j-1})$, $m_j \in [1 : 2^{nR}]$, each according to $\prod_{i=1}^n p_{X_1|X_2}(x_{1i}|x_{2i}(m_{j-1}))$. This defines the codebook $C_j = \{(x_1^n(m_j|m_{j-1}), x_2^n(m_{j-1})): m_{j-1}, m_j \in [1 : 2^{nR}]\}$ for $j \in [1 : b]$. The codebooks are revealed to all parties.

Encoding and decoding are explained with the help of Table 18.1.

Block	1	2	3	$\cdots$	$b-1$	b
X_1	$x_1^n(m_1|1)$	$x_1^n(m_2|m_1)$	$x_1^n(m_3|m_2)$	$\cdots$	$x_1^n(m_{b-1}|m_{b-2})$	$x_1^n(1|m_{b-1})$
Y_2	$\tilde{m}_1$	$\tilde{m}_2$	$\tilde{m}_3$	$\cdots$	$\tilde{m}_{b-1}$	$\emptyset$
X_2	$x_2^n(1)$	$x_2^n(\tilde{m}_1)$	$x_2^n(\tilde{m}_2)$	$\cdots$	$x_2^n(\tilde{m}_{b-2})$	$x_2^n(\tilde{m}_{b-1})$
Y_3	$\emptyset$	$\hat{m}_1$	$\hat{m}_2$	$\cdots$	$\hat{m}_{b-2}$	$\hat{m}_{b-1}$

Table 18.1. Encoding and sliding window decoding for the relay channel.

Encoding. Encoding is again the same as in the coherent multihop coding scheme. To send m_j in block j, the sender transmits $x_1^n(m_j|m_{j-1})$ from codebook C_j, where $m_0 = m_b = 1$ by convention.

Relay encoding. Relay encoding is also the same as in the coherent multihop coding scheme. By convention, let $\tilde{m}_0 = 1$. At the end of block j, the relay finds the unique message $\tilde{m}_j$ such that $(x_1^n(\tilde{m}_j|\tilde{m}_{j-1}), x_2^n(\tilde{m}_{j-1}), y_2^n(j)) \in T_\epsilon^{(n)}$. In block $j+1$, it transmits $x_2^n(\tilde{m}_j)$ from codebook C_{j+1}.

Sliding window decoding. At the end of block $j+1$, the receiver finds the unique message $\hat{m}_j$ such that $(x_1^n(\hat{m}_j|\hat{m}_{j-1}), x_2^n(\hat{m}_{j-1}), y_3^n(j)) \in T_\epsilon^{(n)}$ and $(x_2^n(\hat{m}_j), y_3^n(j+1)) \in T_\epsilon^{(n)}$ simultaneously.

Analysis of the probability of error. We analyze the probability of decoding error for M_j

averaged over codebooks. Assume without loss of generality that $M_{j-1} = M_j = 1$. Then the decoder makes an error only if one or more of the following events occur:

$$\tilde{\mathcal{E}}(j-1) = \{\tilde{M}_{j-1} \neq 1\},$$

$$\tilde{\mathcal{E}}(j) = \{\tilde{M}_j \neq 1\},$$

$$\mathcal{E}(j-1) = \{\hat{M}_{j-1} \neq 1\},$$

$$\mathcal{E}_1(j) = \{(X_1^n(\tilde{M}_j|\hat{M}_{j-1}), X_2^n(\hat{M}_{j-1}), Y_3^n(j)) \notin \mathcal{T}_\epsilon^{(n)} \text{ or } (X_2^n(\tilde{M}_j), Y_3^n(j+1)) \notin \mathcal{T}_\epsilon^{(n)}\},$$

$$\mathcal{E}_2(j) = \{(X_1^n(m_j|\hat{M}_{j-1}), X_2^n(\hat{M}_{j-1}), Y_3^n(j)) \in \mathcal{T}_\epsilon^{(n)} \text{ and } (X_2^n(m_j), Y_3^n(j+1)) \in \mathcal{T}_\epsilon^{(n)}$$

$$\text{for some } m_j \neq \tilde{M}_j\}.$$

Thus the probability of error is upper bounded as

$$P(\mathcal{E}(j)) = P\{\hat{M}_j \neq 1\}$$

$$\leq P(\tilde{\mathcal{E}}(j-1) \cup \tilde{\mathcal{E}}(j) \cup \mathcal{E}(j-1) \cup \mathcal{E}_1(j) \cup \mathcal{E}_2(j))$$

$$\leq P(\tilde{\mathcal{E}}(j-1)) + P(\tilde{\mathcal{E}}(j)) + P(\mathcal{E}(j-1))$$

$$+ P(\mathcal{E}_1(j) \cap \tilde{\mathcal{E}}^c(j-1) \cap \tilde{\mathcal{E}}^c(j) \cap \mathcal{E}^c(j-1)) + P(\mathcal{E}_2(j) \cap \tilde{\mathcal{E}}^c(j)).$$

By the independence of the codebooks, the LLN, and the packing lemma, the first, second, and fourth terms tend to zero as $n \to \infty$ if $R < I(X_1; Y_2|X_2) - \delta(\epsilon)$, and by induction, the third term tends to zero as $n \to \infty$. For the last term, consider

$$P(\mathcal{E}_2(j) \cap \tilde{\mathcal{E}}^c(j)) = P\{(X_1^n(m_j|\hat{M}_{j-1}), X_2^n(\hat{M}_{j-1}), Y_3^n(j)) \in \mathcal{T}_\epsilon^{(n)},$$

$$(X_2^n(m_j), Y_3^n(j+1)) \in \mathcal{T}_\epsilon^{(n)} \text{ for some } m_j \neq 1, \text{ and } \tilde{M}_j = 1\}$$

$$\leq \sum_{m_j \neq 1} P\{(X_1^n(m_j|\hat{M}_{j-1}), X_2^n(\hat{M}_{j-1}), Y_3^n(j)) \in \mathcal{T}_\epsilon^{(n)},$$

$$(X_2^n(m_j), Y_3^n(j+1)) \in \mathcal{T}_\epsilon^{(n)}, \text{ and } \tilde{M}_j = 1\}$$

$$\overset{(a)}{=} \sum_{m_j \neq 1} P\{(X_1^n(m_j|\hat{M}_{j-1}), X_2^n(\hat{M}_{j-1}), Y_3^n(j)) \in \mathcal{T}_\epsilon^{(n)} \text{ and } \tilde{M}_j = 1\}$$

$$\cdot P\{(X_2^n(m_j), Y_3^n(j+1)) \in \mathcal{T}_\epsilon^{(n)} | \tilde{M}_j = 1\}$$

$$\leq \sum_{m_j \neq 1} P\{(X_1^n(m_j|\hat{M}_{j-1}), X_2^n(\hat{M}_{j-1}), Y_3^n(j)) \in \mathcal{T}_\epsilon^{(n)}\}$$

$$\cdot P\{(X_2^n(m_j), Y_3^n(j+1)) \in \mathcal{T}_\epsilon^{(n)} | \tilde{M}_j = 1\}$$

$$\overset{(b)}{\leq} 2^{nR} 2^{-n(I(X_1; Y_3|X_2)-\delta(\epsilon))} 2^{-n(I(X_2; Y_3)-\delta(\epsilon))},$$

which tends to zero as $n \to \infty$ if $R < I(X_1; Y_3|X_2) + I(X_2; Y_3) - 2\delta(\epsilon) = I(X_1, X_2; Y_3) - 2\delta(\epsilon)$. Here step (a) follows since, by the independence of the codebooks, the events

$$\{(X_1^n(m_j|\hat{M}_{j-1}), X_2^n(\hat{M}_{j-1}), Y_3^n(j)) \in \mathcal{T}_\epsilon^{(n)}\}$$

and

$$\{(X_2^n(m_j), Y_3^n(j+1)) \in \mathcal{T}_\epsilon^{(n)}\}$$

are conditionally independent given $\tilde{M}_j = 1$ for $m_j \neq 1$, and (b) follows by the independence of the codebooks and the joint typicality lemma. This completes the proof of the decode–forward lower bound for the DM-RC using sliding block decoding.

18.2.2   Proof of the Network Decode–Forward Lower Bound

We now extend decode–forward using sliding window decoding for the relay channel to the DM-MN. Each relay node $k \in [2:N-1]$ recovers the message and forwards it to the next node. A sequence of $(b - N + 2)$ i.i.d. messages $m_j \in [1:2^{nR}]$, $j \in [1:b-N+2]$, is to be sent over the channel in b blocks, each consisting of n transmissions. Note that the average rate over the b blocks is $R(b - N + 2)/b$, which can be made as close to R as desired.

Codebook generation. Fix $p(x^N)$. For every $j \in [1:b]$, randomly and independently generate a sequence $x_N^n(j)$ according to $\prod_{i=1}^n p_{X_N}(x_{Ni})$ (this is similar to coded time sharing). For every relay node $k = N - 1, N - 2, \ldots, 1$ and every $(m_{j-N+2}, \ldots, m_{j-k})$, generate 2^{nR} conditionally independent sequences $x_k^n(m_{j-k+1}|m_{j-N+2}^{j-k})$, $m_{j-k+1} \in [1:2^{nR}]$, each according to $\prod_{i=1}^n p_{X_k|X_{k+1}^N}(x_{ki}|x_{k+1,i}(m_{j-k}|m_{j-N+2}^{j-k-1}), \ldots, x_{N-1,i}(m_{j-N+2}), x_{Ni}(j))$. This defines the codebook

$$\mathcal{C}_j = \{(x_1^n(m_j|m_{j-N+2}^{j-1}), x_2^n(m_{j-1}|m_{j-N+2}^{j-2}), \ldots, x_{N-1}^n(m_{j-N+2}), x_N^n(j)):$$

$$m_{j-N+2}, \ldots, m_j \in [1:2^{nR}]\}$$

for each block $j \in [1:b]$. The codebooks are revealed to all parties.

Encoding and decoding for $N = 4$ are explained with the help of Table 18.2.

Encoding. Let $m_j \in [1:2^{nR}]$ be the new message to be sent in block j. Source node 1 transmits $x_1^n(m_j|m_{j-N+2}^{j-1})$ from codebook $\mathcal{C}_j$. At the end of block $j + k - 2$, relay node $k \in [2:N-1]$ has an estimate $\hat{m}_{kj}$ of the message m_j. In block $j + k - 1$, it transmits $x_k^n(\hat{m}_{kj}|\hat{m}_{k,j+k-N+1}^{j-1})$ from codebook $\mathcal{C}_{j+k-1}$. Node N transmits $x_N^n(j)$ in block $j \in [1:b]$.

Decoding and analysis of the probability of error. The decoding procedures for message m_j are as follows. At the end of block $j + k - 2$, node k finds the unique message $\hat{m}_{kj}$ such that

$$(x_1^n(\hat{m}_{kj}), x_2^n, \ldots, x_N^n(j), y_k^n(j)) \in \mathcal{T}_\epsilon^{(n)},$$

$$(x_2^n(\hat{m}_{kj}), x_3^n, \ldots, x_N^n(j+1), y_k^n(j+1)) \in \mathcal{T}_\epsilon^{(n)},$$

$$\vdots$$

$$(x_{k-1}^n(\hat{m}_{kj}), x_k^n, \ldots, x_N^n(j+k-2), y_k^n(j+k-2)) \in \mathcal{T}_\epsilon^{(n)},$$

where the dependence of codewords on previous message indices $\hat{m}_{k,j-N+2}^{j-1}$ is suppressed for brevity. For example, for the case $N = 4$, node 4 at the end of block $j + 2$ finds the unique message $\hat{m}_{4j}$ such that

$$(x_1^n(\hat{m}_{4j}|\hat{m}_{4,j-2}, \hat{m}_{4,j-1}), x_2^n(\hat{m}_{4,j-1}|\hat{m}_{4,j-2}), x_3^n(\hat{m}_{4,j-2}), x_4^n(j), y_4^n(j)) \in \mathcal{T}_\epsilon^{(n)},$$

$$(x_2^n(\hat{m}_{4j}|\hat{m}_{4,j-1}), x_3^n(\hat{m}_{4,j-1}), x_4^n(j+1), y_4^n(j+1)) \in \mathcal{T}_\epsilon^{(n)},$$

$$(x_3^n(\hat{m}_{4j}), x_4^n(j+2), y_4^n(j+2)) \in \mathcal{T}_\epsilon^{(n)}.$$

Block	1	2	3	$\cdots$	j
X_1	$x_1^n(m_1\|1,1)$	$x_1^n(m_2\|1,m_1)$	$x_1^n(m_3\|m_1^2)$	$\cdots$	$x_1^n(m_j\|m_{j-2}^{j-1})$
Y_2	$\hat{m}_{21}$	$\hat{m}_{22}$	$\hat{m}_{23}$	$\cdots$	$\hat{m}_{2j}$
X_2	$x_2^n(1\|1)$	$x_2^n(\hat{m}_{21}\|1)$	$x_2^n(\hat{m}_{22}\|\hat{m}_{21})$	$\cdots$	$x_2^n(\hat{m}_{2,j-1}\|\hat{m}_{2,j-2})$
Y_3	$\emptyset$	$\hat{m}_{31}$	$\hat{m}_{32}$	$\cdots$	$\hat{m}_{3,j-1}$
X_3	1	1	$x_3^n(\hat{m}_{31})$	$\cdots$	$x_3^n(\hat{m}_{3,j-2})$
Y_4	$\emptyset$	$\emptyset$	$\hat{m}_{41}$	$\cdots$	$\hat{m}_{4,j-2}$

Block	$j+1$	$\cdots$	$b-2$	$b-1$	b
X_1	$x_1^n(m_{j+1}\|m_{j-1}^j)$	$\cdots$	$x_1^n(m_{b-2}\|m_{b-4}^{b-3})$	$x_1^n(1\|m_{b-3}^{b-2})$	$x_1^n(1\|m_{b-2},1)$
Y_2	$\hat{m}_{2,j+1}$	$\cdots$	$\hat{m}_{2,b-2}$	$\emptyset$	$\emptyset$
X_2	$x_2^n(\hat{m}_{2j}\|\hat{m}_{2,j-1})$	$\cdots$	$x_2^n(\hat{m}_{2,b-3}\|\hat{m}_{2,b-4})$	$x_2^n(\hat{m}_{2,b-2}\|\hat{m}_{2,b-3})$	$x_2^n(1\|\hat{m}_{2,b-2})$
Y_3	$\hat{m}_{3,j}$	$\cdots$	$\hat{m}_{3,b-3}$	$\hat{m}_{3,b-2}$	$\emptyset$
X_3	$x_3^n(\hat{m}_{3,j-1})$	$\cdots$	$x_3^n(\hat{m}_{3,b-4})$	$x_3^n(\hat{m}_{3,b-3})$	$x_3^n(\hat{m}_{3,b-2})$
Y_4	$\hat{m}_{4,j-1}$	$\cdots$	$\hat{m}_{4,b-4}$	$\hat{m}_{4,b-3}$	$\hat{m}_{4,b-2}$

Table 18.2. Encoding and decoding for network decode–forward for $N = 4$.

Following similar steps to the proof of the sliding window decoding for the relay channel, by the independence of the codebooks, the LLN, the joint typicality lemma, and induction, it can be shown that the probability of error tends to zero as $n \to \infty$ if

$$R < \sum_{k'=1}^{k-1} I(X_{k'}; Y_k | X_{k'+1}^N) - \delta(\epsilon) = I(X^{k-1}; Y_k | X_k^N) - \delta(\epsilon).$$

This completes the proof of Theorem 18.2.

18.3   NOISY NETWORK CODING

The compress–forward coding scheme for the relay channel and network coding for the graphical multicast network can be extended to DM multicast networks.

> **Theorem 18.3 (Noisy Network Coding Lower Bound).** The capacity of the DM-MN $p(y^N|x^N)$ with destination set $\mathcal{D}$ is lower bounded as
>
> $$C \ge \max \min_{k \in \mathcal{D}} \min_{\mathcal{S}:1 \in \mathcal{S}, k \in \mathcal{S}^c} \left(I(X(\mathcal{S}); \hat{Y}(\mathcal{S}^c), Y_k | X(\mathcal{S}^c)) - I(Y(\mathcal{S}); \hat{Y}(\mathcal{S}) | X^N, \hat{Y}(\mathcal{S}^c), Y_k) \right),$$
>
> where the maximum is over all conditional pmfs $\prod_{k=1}^N p(x_k) p(\hat{y}_k | y_k, x_k)$ and $\hat{Y}_1 = \emptyset$ by convention.

Note that this lower bound differs from the cutset bound in Theorem 18.1 in the following three aspects:

- The first term is similar to the cutset bound with $Y(\mathcal{S}^c)$ replaced by the "compressed version" $\hat{Y}(\mathcal{S}^c)$.

- There is a subtracted term that captures the rate penalty for describing the compressed version.

- The maximum is over independent X^N (no coherent cooperation).

Remark 18.5. For $N = 3$, the noisy network coding lower bound reduces to the compress–forward lower bound for the DM-RC in Section 16.7.

Remark 18.6. As in compress–forward for the relay channel, the noisy network coding scheme can be improved by coded time sharing (i.e., by incorporating a time-sharing random variable Q into the characterization of the lower bound).

Before proving Theorem 18.3, we discuss several special cases.

18.3.1   Deterministic Multicast Network

Consider the deterministic multicast network depicted in Figure 18.2, where the received symbol Y_k at each node $k \in [1:N]$ is a function of the transmitted symbols, i.e., $Y_k = y_k(X^N)$. Note that this model generalizes the graphical multicast network. It also captures the interference and broadcasting aspects of the DMN without the noise; hence it is a good model for networks with "high SNR."

The cutset bound for this deterministic multicast network simplifies to

$$C \le \max_{p(x^N)} \min_{k \in \mathcal{D}} \min_{\mathcal{S}:1 \in \mathcal{S}, k \in \mathcal{S}^c} H(Y(\mathcal{S}^c) | X(\mathcal{S}^c)). \tag{18.2}$$

In the other direction, by setting $\hat{Y}_k = Y_k$ for all $k \in [2:N]$, the noisy network coding lower bound simplifies to

$$C \ge \max_{\prod_{k=1}^N p(x_k)} \min_{k \in \mathcal{D}} \min_{\mathcal{S}:1 \in \mathcal{S}, k \in \mathcal{S}^c} H(Y(\mathcal{S}^c) | X(\mathcal{S}^c)). \tag{18.3}$$

Note that the maximum here is taken over all product pmfs instead of all joint pmfs as in

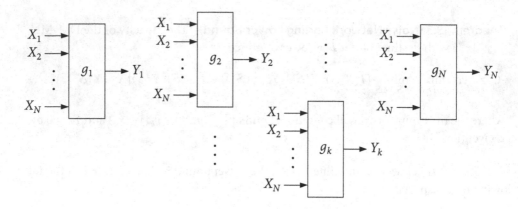

Figure 18.2. Deterministic network.

the cutset bound in (18.2). As such, these two bounds do not coincide in general. There are several interesting special cases where the bounds coincide, however.

Graphical multicast network with cycles. Consider a multicast network modeled by a directed cyclic graph $(\mathcal{N}, \mathcal{E}, \mathcal{C})$ with destination set $\mathcal{D}$. Each node $k \in [1:N]$ transmits $X_k = (X_{kl}: (k,l) \in \mathcal{E})$ and receives $Y_k = (X_{jk}: (j,k) \in \mathcal{E})$. Thus, each link $(j,k) \in \mathcal{E}$ carries a symbol $X_{jk} \in \mathcal{X}_{jk}$ noiselessly from node j to node k with link capacity $C_{jk} = \log|\mathcal{X}_{jk}|$. Now for each cut $(\mathcal{S}, \mathcal{S}^c)$ separating source node 1 and some destination node,

$$H(Y(\mathcal{S}^c)|X(\mathcal{S}^c)) \le \sum_{k \in \mathcal{S}^c} H(X_{jk}: j \in \mathcal{S}, (j,k) \in \mathcal{E})$$

$$\le \sum_{k \in \mathcal{S}^c} \sum_{j \in \mathcal{S}, (j,k) \in \mathcal{E}} H(X_{jk})$$

$$\le \sum_{\substack{(j,k) \in \mathcal{E} \\ j \in \mathcal{S}, k \in \mathcal{S}^c}} C_{jk}$$

$$= C(\mathcal{S})$$

with equality if $(X_{jk}: (j,k) \in \mathcal{E})$ has a uniform product pmf. Hence, the cutset bound in (18.2) and the lower bound in (18.3) coincide, and the capacity is

$$C = \min_{k \in \mathcal{D}} \min_{\mathcal{S}: 1 \in \mathcal{S}, k \in \mathcal{S}^c} C(\mathcal{S}).$$

Note that this result extends the network coding theorem in Section 15.3 to graphical networks with cycles.

Deterministic multicast network with no interference. Suppose that each node in the deterministic network receives a collection of single-variable functions of input symbols, i.e., $Y_k = (y_{k1}(X_1), \ldots, y_{kN}(X_N))$, $k \in [1:N]$. Then it can be shown that the cutset bound in (18.2) is attained by a product input pmf and thus the capacity is

$$C = \max_{\prod_{k=1}^N p(x_k)} \min_{k \in \mathcal{D}} \min_{\mathcal{S}: 1 \in \mathcal{S}, k \in \mathcal{S}^c} \sum_{j \in \mathcal{S}} H(Y_{1j}: l \in \mathcal{S}^c). \qquad (18.4)$$

For example, if $N = 3$, $\mathcal{D} = \{2, 3\}$, $Y_2 = (y_{21}(X_1), y_{23}(X_3))$, and $Y_3 = (y_{31}(X_1), y_{32}(X_2))$, then

$$C = \max_{p(x_1)p(x_2)p(x_3)} \min\{H(Y_{21}, Y_{31}), H(Y_{21}) + H(Y_{23}), H(Y_{31}) + H(Y_{32})\}.$$

Finite-field deterministic multicast network (FFD-MN). Suppose that each node receives a linear function

$$Y_j = \sum_{k=1}^{N} g_{jk} X_k,$$

where g_{jk}, $j, k \in [1 : N]$, and X_k, $k \in [1 : N]$, take values in the finite field $\mathbb{F}_q$. Then it can be easily shown that the cutset bound is attained by the uniform product input pmf. Hence the capacity is

$$C = \min_{k \in \mathcal{D}} \; \min_{\mathcal{S}: 1 \in \mathcal{S}, k \in \mathcal{S}^c} \mathrm{rank}(G(\mathcal{S})) \log q,$$

where $G(\mathcal{S})$ for each cut $(\mathcal{S}, \mathcal{S}^c)$ is defined such that

$$\begin{bmatrix} Y(\mathcal{S}) \\ Y(\mathcal{S}^c) \end{bmatrix} = \begin{bmatrix} G'(\mathcal{S}) & G(\mathcal{S}^c) \\ G(\mathcal{S}) & G'(\mathcal{S}^c) \end{bmatrix} \begin{bmatrix} X(\mathcal{S}) \\ X(\mathcal{S}^c) \end{bmatrix} + \begin{bmatrix} Z(\mathcal{S}) \\ Z(\mathcal{S}^c) \end{bmatrix},$$

for some submatrices $G'(\mathcal{S})$ and $G'(\mathcal{S}^c)$.

Remark 18.7. The lower bound in (18.3) does not always coincide with the cutset bound. For example, consider the deterministic relay channel with $X_3 = Y_1 = \emptyset$, $Y_2 = y_2(X_1, X_2)$, $Y_3 = y_3(X_1, X_2)$, and $\mathcal{D} = \{3\}$. Then the capacity is

$$C = \max_{p(x_1, x_2)} \min\{H(Y_3), H(Y_2, Y_3 | X_2)\},$$

and is achieved by partial decode–forward as discussed in Section 16.6. In the other direction, the noisy network coding lower bound in (18.3) simplifies to

$$C \geq \max_{p(x_1)p(x_2)} \min\{H(Y_3), H(Y_2, Y_3 | X_2)\},$$

which can be strictly lower than the capacity.

18.3.2 Wireless Erasure Multicast Network

Consider a wireless data network with packet loss modeled by a hypergraph $\mathcal{H} = (\mathcal{N}, \mathcal{E}, \mathcal{C})$ with random input erasures as depicted in Figure 18.3. Each node $k \in [1 : N]$ broadcasts a symbol X_k to a subset of nodes $\mathcal{N}_k$ over a hyperedge $(k, \mathcal{N}_k)$ and receives $Y_k = (Y_{kj} : k \in \mathcal{N}_j)$ from nodes j for $k \in \mathcal{N}_j$, where

$$Y_{kj} = \begin{cases} e & \text{with probability } p_{kj}, \\ X_j & \text{with probability } 1 - p_{kj}. \end{cases}$$

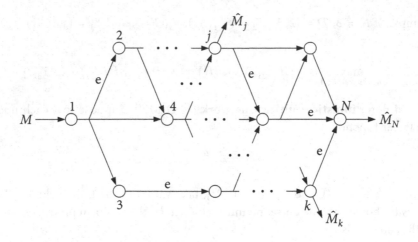

Figure 18.3. Wireless erasure multicast network.

Note that the capacity of each hyperedge $(k, \mathcal{N}_k)$ (with no erasure) is $C_k = \log |\mathcal{X}_k|$. We assume that the erasures are independent of each other. Assume further that the erasure pattern of the entire network is known at each destination node.

The capacity of this network is

$$C = \min_{j \in \mathcal{D}} \; \min_{\mathcal{S}: 1 \in \mathcal{S}, j \in \mathcal{S}^c} \; \sum_{k \in \mathcal{S}: \mathcal{N}_k \cap \mathcal{S}^c \neq \emptyset} \left(1 - \prod_{l \in \mathcal{N}_k \cap \mathcal{S}^c} p_{lk} \right) C_k . \tag{18.5}$$

The proof follows by evaluating the noisy network coding lower bound using the uniform product pmf on X^N and $\hat{Y}_k = Y_k$, $k \in [2:N]$, and showing that this choice attains the maximum in the cutset bound in Theorem 18.1.

18.3.3   Noisy Network Coding for the DM-RC

While the noisy network coding scheme achieves the same compress–forward lower bound for the relay channel, it involves several new ideas not encountered in the compress–forward scheme. First, unlike other block Markov coding techniques we have seen so far, in noisy network coding the same message $m \in [1 : 2^{nbR}]$ is sent over b blocks (each with n transmissions), while the relays send compressed versions of the received sequences in the previous block. Second, instead of using Wyner–Ziv coding to send the compression index, the relay sends the compression index without binning. Third, the receiver performs simultaneous joint typicality decoding on the received sequences from all b blocks without explicitly decoding for the compression indices.

For clarity of presentation, we first use the noisy network coding scheme to obtain an alternative proof of the compress–forward lower bound for the relay channel

$$C \geq \max_{p(x_1)p(x_2)p(\hat{y}_2|y_2,x_2)} \min\{I(X_1, X_2; Y_3) - I(Y_2; \hat{Y}_2|X_1, X_2, Y_3), I(X_1; \hat{Y}_2, Y_3|X_2)\}. \tag{18.6}$$

Codebook generation. Fix the conditional pmf $p(x_1)p(x_2)p(\hat{y}_2|y_2, x_2)$ that attains the lower bound. For each $j \in [1 : b]$, we generate an independent codebook as follows. Randomly and independently generate 2^{nbR} sequences $x_1^n(j, m)$, $m \in [1 : 2^{nbR}]$, each according to $\prod_{i=1}^{n} p_{X_1}(x_{1i})$. Randomly and independently generate 2^{nR_2} sequences $x_2^n(l_{j-1})$, $l_{j-1} \in [1 : 2^{nR_2}]$, each according to $\prod_{i=1}^{n} p_{X_2}(x_{2i})$. For each $l_{j-1} \in [1 : 2^{nR_2}]$, randomly and conditionally independently generate 2^{nR_2} sequences $\hat{y}_2^n(l_j|l_{j-1})$, $l_j \in [1 : 2^{nR_2}]$, each according to $\prod_{i=1}^{n} p_{\hat{Y}_2|X_2}(\hat{y}_{2i}|x_{2i}(l_{j-1}))$. This defines the codebook

$$C_j = \{(x_1^n(j, m), x_2^n(l_{j-1}), \hat{y}_2^n(l_j|l_{j-1})) : m \in [1 : 2^{nbR}], l_j, l_{j-1} \in [1 : 2^{nR_2}]\}$$

for $j \in [1 : b]$. The codebooks are revealed to all parties.

Encoding and decoding are explained with the help of Table 18.3.

Block	1	2	3	$\cdots$	$b - 1$	b
X_1	$x_1^n(1, m)$	$x_1^n(2, m)$	$x_1^n(3, m)$	$\cdots$	$x_1^n(b - 1, m)$	$x_1^n(b, m)$
Y_2	$\hat{y}_2^n(l_1|1), l_1$	$\hat{y}_2^n(l_2|l_1), l_2$	$\hat{y}_2^n(l_3|l_2), l_3$	$\cdots$	$\hat{y}_2^n(l_{b-1}|l_{b-2}), l_{b-1}$	$\hat{y}_2^n(l_b|l_{b-1}), l_b$
X_2	$x_2^n(1)$	$x_2^n(l_1)$	$x_2^n(l_2)$	$\cdots$	$x_2^n(l_{b-2})$	$x_2^n(l_{b-1})$
Y_3	$\emptyset$	$\emptyset$	$\emptyset$	$\cdots$	$\emptyset$	$\hat{m}$

Table 18.3. Noisy network coding for the relay channel.

Encoding. To send $m \in [1 : 2^{nbR}]$, the sender transmits $x_1^n(j, m)$ from C_j in block j.

Relay encoding. By convention, let $l_0 = 1$. At the end of block j, the relay finds an index l_j such that $(y_2^n(j), \hat{y}_2^n(l_j|l_{j-1}), x_2^n(l_{j-1})) \in T_{\epsilon'}^{(n)}$. If there is more than one such index, it selects one of them uniformly at random. If there is no such index, it selects an index from $[1 : 2^{nR_2}]$ uniformly at random. The relay then transmits $x_2^n(l_j)$ from C_{j+1} in block $j + 1$.

Decoding. Let $\epsilon > \epsilon'$. At the end of block b, the receiver finds the unique index $\hat{m}$ such that $(x_1^n(j, \hat{m}), x_2^n(l_{j-1}), \hat{y}_2^n(l_j|l_{j-1}), y_3^n(j)) \in T_{\epsilon}^{(n)}$ for all $j \in [1 : b]$ for some $l_1, l_2, \ldots, l_b$.

Analysis of the probability of error. To bound the probability of error, assume without loss of generality that $M = 1$ and $L_1 = L_2 = \cdots = L_b = 1$. Then the decoder makes an error only if one or more of the following events occur:

$$\mathcal{E}_1 = \{(Y_2^n(j), \hat{Y}_2^n(l_j|1), X_2^n(1)) \notin T_{\epsilon'}^{(n)} \text{ for all } l_j \text{ for some } j \in [1 : b]\},$$

$$\mathcal{E}_2 = \{(X_1^n(j, 1), X_2^n(1), \hat{Y}_2^n(1|1), Y_3^n(j)) \notin T_{\epsilon}^{(n)} \text{ for some } j \in [1 : b]\},$$

$$\mathcal{E}_3 = \{(X_1^n(j, m), X_2^n(l_{j-1}), \hat{Y}_2^n(l_j|l_{j-1}), Y_3^n(j)) \in T_{\epsilon}^{(n)} \text{ for all } j \text{ for some } l^b, m \neq 1\}.$$

Thus, the probability of error is upper bounded as

$$P(\mathcal{E}) \le P(\mathcal{E}_1) + P(\mathcal{E}_2 \cap \mathcal{E}_1^c) + P(\mathcal{E}_3).$$

By the covering lemma and the union of events bound (over b blocks), $P(\mathcal{E}_1)$ tends to zero as $n \to \infty$ if $R_2 > I(\hat{Y}_2; Y_2|X_2) + \delta(\epsilon')$. By the conditional typicality lemma and the union of events bound, the second term $P(\mathcal{E}_2 \cap \mathcal{E}_1^c)$ tends to zero as $n \to \infty$. For the third term, define the events

$$\tilde{\mathcal{E}}_j(m, l_{j-1}, l_j) = \{(X_1^n(j, m), X_2^n(l_{j-1}), \hat{Y}_2^n(l_j|l_{j-1}), Y_3^n(j)) \in \mathcal{T}_\epsilon^{(n)}\}.$$

Now consider

$$P(\mathcal{E}_3) = P\left(\bigcup_{m \ne 1} \bigcup_{l^b} \bigcap_{j=1}^{b} \tilde{\mathcal{E}}_j(m, l_{j-1}, l_j) \right)$$

$$\le \sum_{m \ne 1} \sum_{l^b} P\left(\bigcap_{j=1}^{b} \tilde{\mathcal{E}}_j(m, l_{j-1}, l_j) \right)$$

$$\overset{(a)}{=} \sum_{m \ne 1} \sum_{l^b} \prod_{j=1}^{b} P(\tilde{\mathcal{E}}_j(m, l_{j-1}, l_j))$$

$$\le \sum_{m \ne 1} \sum_{l^b} \prod_{j=2}^{b} P(\tilde{\mathcal{E}}_j(m, l_{j-1}, l_j)), \tag{18.7}$$

where (a) follows since the codebooks are generated independently for each block $j \in [1:b]$ and the channel is memoryless. Note that if $m \ne 1$ and $l_{j-1} = 1$, then $X_1^n(j, m) \sim \prod_{i=1}^{n} p_{X_1}(x_{1i})$ is independent of $(\hat{Y}_2^n(l_j|l_{j-1}), X_2^n(l_{j-1}), Y_3^n(j))$. Hence, by the joint typicality lemma,

$$P(\tilde{\mathcal{E}}_j(m, l_{j-1}, l_j)) = P\{(X_1^n(j, m), X_2^n(l_{j-1}), \hat{Y}_2^n(l_j|l_{j-1}), Y_3^n(j)) \in \mathcal{T}_\epsilon^{(n)}\}$$
$$\le 2^{-n(I_1 - \delta(\epsilon))},$$

where $I_1 = I(X_1; \hat{Y}_2, Y_3|X_2)$. Similarly, if $m \ne 1$ and $l_{j-1} \ne 1$, then $(X_1^n(j, m), X_2^n(l_{j-1}), \hat{Y}_2^n(l_j|l_{j-1})) \sim \prod_{i=1}^{n} p_{X_1}(x_{1i}) p_{X_2, \hat{Y}_2}(x_{2i}, \hat{y}_{2i})$ is independent of $Y_3^n(j)$. Hence

$$P(\tilde{\mathcal{E}}_j(m, l_{j-1}, l_j)) \le 2^{-n(I_2 - \delta(\epsilon))},$$

where $I_2 = I(X_1, X_2; Y_3) + I(\hat{Y}_2; X_1, Y_3|X_2)$. If $\pi(1|l^{b-1}) = k/(b-1)$, i.e., l^{b-1} has k ones, then

$$\prod_{j=2}^{b} P(\tilde{\mathcal{E}}_j(m, l_{j-1}, l_j)) \le 2^{-n(kI_1 + (b-1-k)I_2 - (b-1)\delta(\epsilon))}.$$

Continuing with the bound in (18.7), we have

$$\sum_{m\neq 1}\sum_{1^b}\prod_{j=2}^{b} P(\tilde{\mathcal{E}}_j(m, l_{j-1}, l_j)) = \sum_{m\neq 1}\sum_{l_b}\sum_{1^{b-1}}\prod_{j=2}^{b} P(\tilde{\mathcal{E}}_j(m, l_{j-1}, l_j))$$

$$\leq \sum_{m\neq 1}\sum_{l_b}\sum_{j=0}^{b-1} \binom{b-1}{j} 2^{n(b-1-j)R_2} \cdot 2^{-n(jI_1+(b-1-j)I_2-(b-1)\delta(\epsilon))}$$

$$= \sum_{m\neq 1}\sum_{l_b}\sum_{j=0}^{b-1} \binom{b-1}{j} 2^{-n(jI_1+(b-1-j)(I_2-R_2)-(b-1)\delta(\epsilon))}$$

$$\leq \sum_{m\neq 1}\sum_{l_b}\sum_{j=0}^{b-1} \binom{b-1}{j} 2^{-n(b-1)(\min\{I_1, I_2-R_2\}-\delta(\epsilon))}$$

$$\leq 2^{nbR} \cdot 2^{nR_2} \cdot 2^b \cdot 2^{-n(b-1)(\min\{I_1, I_2-R_2\}-\delta(\epsilon))},$$

which tends to zero as $n \to \infty$ if

$$R < \frac{b-1}{b}(\min\{I_1, I_2 - R_2\} - \delta'(\epsilon)) - \frac{R_2}{b}.$$

Finally, by eliminating $R_2 > I(\hat{Y}_2; Y_2|X_2) + \delta(\epsilon')$, substituting $I_1 = I(X_1; \hat{Y}_2, Y_3|X_2)$ and $I_2 = I(X_1, X_2; Y_3) + I(\hat{Y}_2; X_1, Y_3|X_2)$, and taking $b \to \infty$, we have shown that the probability of error tends to zero as $n \to \infty$ if

$$R < \min\{I(X_1; \hat{Y}_2, Y_3|X_2), I(X_1, X_2; Y_3) - I(\hat{Y}_2; Y_2|X_1, X_2, Y_3)\} - \delta'(\epsilon) - \delta(\epsilon').$$

This completes the proof of the compress–forward lower bound for the relay channel in (18.6) using noisy network coding.

18.3.4 Proof of the Noisy Network Coding Lower Bound

We now establish the noisy network coding lower bound for the DM-MN. First consider the unicast special case with $(N - 2)$ relay nodes $k \in [2 : N - 1]$ and a single destination node N. Assume without loss of generality that $X_N = \emptyset$ (otherwise, we can consider a new destination node $N + 1$ with $Y_{N+1} = Y_N$ and $X_{N+1} = \emptyset$). To simplify notation, let $\hat{Y}_1 = \emptyset$ and $\hat{Y}_N = Y_N$.

Codebook generation. Fix the conditional pmf $\prod_{k=1}^{N-1} p(x_k)p(\hat{y}_k|y_k, x_k)$ that attains the lower bound. For each $j \in [1 : b]$, randomly and independently generate 2^{nbR} sequences $x_1^n(j, m)$, $m \in [1 : 2^{nbR}]$, each according to $\prod_{i=1}^{n} p_{X_1}(x_{1i})$. For each $k \in [2 : N - 1]$, randomly and independently generate 2^{nR_k} sequences $x_k^n(l_{k,j-1})$, $l_{k,j-1} \in [1 : 2^{nR_k}]$, each according to $\prod_{i=1}^{n} p_{X_k}(x_{ki})$. For each $k \in [2 : N - 1]$ and $l_{k,j-1} \in [1 : 2^{nR_k}]$, randomly and conditionally independently generate 2^{nR_k} sequences $\hat{y}_k^n(l_{kj}|l_{k,j-1})$, $l_{kj} \in [1 : 2^{nR_k}]$, each according to $\prod_{i=1}^{n} p_{\hat{Y}_k|X_k}(\hat{y}_{ki}|x_{ki}(l_{k,j-1}))$. This defines the codebook

$$C_j = \{(x_1^n(m, j), x_2^n(l_{2,j-1}), \dots, x_{N-1}^n(l_{N-1,j-1}), \hat{y}_2^n(l_{2j}|l_{2,j-1}), \dots, \hat{y}_{N-1}^n(l_{N-1,j}|l_{N-1,j-1})):$$

$$m \in [1 : 2^{nbR}], \ l_{k,j-1}, l_{kj} \in [1 : 2^{nR_k}], \ k \in [2 : N - 1]\}$$

for $j \in [1 : b]$. The codebooks are revealed to all parties.

Encoding and decoding are explained with the help of Table 18.4.

Block	1	2	$\cdots$	$b-1$	b
X_1	$x_1^n(1, m)$	$x_1^n(2, m)$	$\cdots$	$x_1^n(b-1, m)$	$x_1^n(b, m)$
Y_k	$\hat{y}_k^n(l_{k1}\lvert 1), l_{k1}$	$\hat{y}_k^n(l_{k2}\lvert l_{k1}), l_{k2}$	$\cdots$	$\hat{y}_k^n(l_{k,b-1}\lvert l_{k,b-2}), l_{k,b-1}$	$\hat{y}_k^n(l_{kb}\lvert l_{k,b-1}), l_{kb}$
X_k	$x_k^n(1)$	$x_k^n(l_{k1})$	$\cdots$	$x_k^n(l_{k,b-2})$	$x_k^n(l_{k,b-1})$
Y_N	$\emptyset$	$\emptyset$	$\cdots$	$\emptyset$	$\hat{m}$

Table 18.4. Encoding and decoding for the noisy network coding lower bound.

Encoding. To send $m \in [1 : 2^{nbR}]$, source node 1 transmits $x^n(j, m)$ in block j.

Relay encoding. Relay node k, upon receiving $y_k^n(j)$, finds an index l_{kj} such that $(y_k^n(j),$ $\hat{y}_k^n(l_{kj}\lvert l_{k,j-1}), x_k^n(l_{k,j-1})) \in \mathcal{T}_{\epsilon'}^{(n)}$. If there is more than one such index, it selects one of them uniformly at random. If there is no such index, it selects an index from $[1 : 2^{nR_k}]$ uniformly at random. Relay node k then transmits $x_k^n(l_{kj})$ in block $j + 1$.

Decoding and analysis of the probability of error. Let $\epsilon > \epsilon'$. At the end of block b, the receiver finds the unique index $\hat{m}$ such that

$$(x_1^n(j, \hat{m}), x_2^n(l_{2,j-1}), \dots, x_{N-1}^n(l_{N-1,j-1}),$$
$$\hat{y}_2^n(l_{2j}\lvert l_{2,j-1}), \dots, \hat{y}_{N-1}^n(l_{N-1,j}\lvert l_{N-1,j-1}), y_N^n(j)) \in \mathcal{T}_\epsilon^{(n)} \text{ for every } j \in [1 : b]$$

for some $l^b = (l_1, \dots, l_b)$, where $l_j = (l_{2j}, \dots, l_{N-1,j})$, $j \in [1 : b]$, and $l_{k0} = 1$, $k \in [2 : N - 1]$, by convention.

Analysis of the probability of error. To bound the probability of error, assume without loss of generality that $M = 1$ and $L_{kj} = 1$, $k \in [2 : N - 1]$, $j \in [1 : b]$. The decoder makes an error only if one or more of the following events occur:

$$\mathcal{E}_1 = \{(Y_k^n(j), \hat{Y}_k^n(l_{kj}\lvert 1), X_k^n(1)) \notin \mathcal{T}_{\epsilon'}^{(n)}$$
$$\text{for all } l_{kj} \text{ for some } j \in [1 : b], k \in [2 : N - 1]\},$$
$$\mathcal{E}_2 = \{(X_1^n(j, 1), X_2^n(1), \dots, X_{N-1}^n(1), \hat{Y}_2^n(1\lvert 1), \dots, \hat{Y}_{N-1}^n(1\lvert 1), Y_N^n(j)) \notin \mathcal{T}_\epsilon^{(n)}$$
$$\text{for some } j \in [1 : b]\},$$
$$\mathcal{E}_3 = \{(X_1^n(j, m), X_2^n(l_{2j}), \dots, X_{N-1}^n(l_{N-1,j}), \hat{Y}_2^n(l_{2j}\lvert l_{2,j-1}), \dots, \hat{Y}_{N-1}^n(l_{N-1,j}\lvert l_{N-1,j-1}),$$
$$Y_N^n(j)) \in \mathcal{T}_\epsilon^{(n)} \text{ for all } j \in [1 : b] \text{ for some } l^b, m \neq 1\}.$$

Thus, the probability of error is upper bounded as

$$P(\mathcal{E}) \le P(\mathcal{E}_1) + P(\mathcal{E}_2 \cap \mathcal{E}_1^c) + P(\mathcal{E}_3).$$

By the covering lemma and the union of events bound, $P(\mathcal{E}_1)$ tends to zero as $n \to \infty$ if $R_k > I(\hat{Y}_k; Y_k | X_k) + \delta(\epsilon')$, $k \in [2:N-1]$. By the Markov lemma and the union of events bound, the second term $P(\mathcal{E}_2 \cap \mathcal{E}_1^c)$ tends to zero as $n \to \infty$. For the third term, define the events

$$\tilde{\mathcal{E}}_j(m, l_{j-1}, l_j) = \{(X_1^n(j, m), X_2^n(l_{2,j-1}), \ldots, X_{N-1}^n(l_{N-1,j-1}),$$
$$\hat{Y}_2^n(l_{2j}|l_{2,j-1}), \ldots \hat{Y}_{N-1}^n(l_{N-1,j}|l_{N-1,j-1}), Y_N^n(j)) \in \mathcal{T}_\epsilon^{(n)}\}.$$

Then

$$P(\mathcal{E}_3) = P\left(\bigcup_{m \neq 1} \bigcup_{l^b} \bigcap_{j=1}^{b} \tilde{\mathcal{E}}_j(m, l_{j-1}, l_j)\right)$$

$$\leq \sum_{m \neq 1} \sum_{l^b} P\left(\bigcap_{j=1}^{b} \tilde{\mathcal{E}}_j(m, l_{j-1}, l_j)\right)$$

$$\stackrel{(a)}{=} \sum_{m \neq 1} \sum_{l^b} \prod_{j=1}^{b} P(\tilde{\mathcal{E}}_j(m, l_{j-1}, l_j))$$

$$\leq \sum_{m \neq 1} \sum_{l^b} \prod_{j=2}^{b} P(\tilde{\mathcal{E}}_j(m, l_{j-1}, l_j)), \tag{18.8}$$

where (a) follows since the codebook is generated independently for each block.

For each l^b and $j \in [2:b]$, define the subset of nodes

$$\mathcal{S}_j(l^b) = \{1\} \cup \{k \in [2:N-1]: l_{k,j-1} \neq 1\}.$$

Note that $\mathcal{S}_j(l^b)$ depends only on l_{j-1} and hence can be written as $\mathcal{S}_j(l_{j-1})$. Now by the joint typicality lemma,

$$P(\tilde{\mathcal{E}}_j(m, l_{j-1}, l_j)) \leq 2^{-n(I_1(\mathcal{S}_j(l_{j-1})) + I_2(\mathcal{S}_j(l_{j-1})) - \delta(\epsilon))}, \tag{18.9}$$

where

$$I_1(\mathcal{S}) = I(X(\mathcal{S}); \hat{Y}(\mathcal{S}^c) | X(\mathcal{S}^c)),$$
$$I_2(\mathcal{S}) = \sum_{k \in \mathcal{S}} I(\hat{Y}_k; \hat{Y}(\mathcal{S}^c \cup \{k' \in \mathcal{S}: k' < k\}), X^{N-1} | X_k).$$

Furthermore

$$\sum_{l_{j-1}} 2^{-n(I_1(\mathcal{S}_j(l_{j-1})) + I_2(\mathcal{S}_j(l_{j-1})) - \delta(\epsilon))} \leq \sum_{\mathcal{S}:1 \in \mathcal{S}, N \in \mathcal{S}^c} \sum_{l_{j-1}:\mathcal{S}_j(l_{j-1})=\mathcal{S}} 2^{-n(I_1(\mathcal{S}_j(l_{j-1})) + I_2(\mathcal{S}_j(l_{j-1})) - \delta(\epsilon))}$$

$$\leq \sum_{\mathcal{S}:1 \in \mathcal{S}, N \in \mathcal{S}^c} 2^{-n(I_1(\mathcal{S}) + I_2(\mathcal{S}) - \sum_{k \in \mathcal{S}} R_k - \delta(\epsilon))}$$

$$\leq 2^{N-2} \cdot 2^{-n[\min_{\mathcal{S}}(I_1(\mathcal{S}) + I_2(\mathcal{S}) - \sum_{k \in \mathcal{S}} R_k - \delta(\epsilon))]}.$$

Continuing with the bound in (18.8), we have

$$
\sum_{m \neq 1} \sum_{1^b} \prod_{j=2}^{b} \mathsf{P}(\tilde{\mathcal{E}}_j(m, l_{j-1}, l_j)) \leq \sum_{m \neq 1} \sum_{l_b} \prod_{j=2}^{b} \left(\sum_{l_{j-1}} 2^{-n(I_1(\mathcal{S}_j(l_{j-1})) + I_2(\mathcal{S}_j(l_{j-1})) - \delta(\epsilon))} \right)
$$

$$
\leq 2^{(N-2)(b-1)} \cdot 2^{n[bR + \sum_{k=2}^{N-1} R_k - (b-1)\min_{\mathcal{S}}(I_1(\mathcal{S}) + I_2(\mathcal{S}) - \sum_{k \in \mathcal{S}} R_k - \delta(\epsilon))]},
$$

which tends to zero as $n \to \infty$ if

$$
R < \frac{b-1}{b} \left(\min_{\mathcal{S}: 1 \in \mathcal{S}, N \in \mathcal{S}^c} \left(I_1(\mathcal{S}) + I_2(\mathcal{S}) - \sum_{k \in \mathcal{S}} R_k \right) - \delta(\epsilon) \right) - \frac{1}{b} \sum_{k=2}^{N-1} R_k.
$$

By eliminating $R_k > I(\hat{Y}_k; Y_k | X_k) + \delta(\epsilon')$, noting that

$$
I_2(\mathcal{S}) - \sum_{k \in \mathcal{S}} I(\hat{Y}_k; Y_k | X_k) = - \sum_{k \in \mathcal{S}} I(\hat{Y}_k; Y_k | X^{N-1}, \hat{Y}(\mathcal{S}^c), \hat{Y}(\{k' \in \mathcal{S}: k' < k\}))
$$

$$
= - \sum_{k \in \mathcal{S}} I(\hat{Y}_k; Y(\mathcal{S}) | X^{N-1}, \hat{Y}(\mathcal{S}^c), \hat{Y}(\{k' \in \mathcal{S}: k' < k\}))
$$

$$
= -I(\hat{Y}(\mathcal{S}); Y(\mathcal{S}) | X^{N-1}, \hat{Y}(\mathcal{S}^c)),
$$

and taking $b \to \infty$, the probability of error tends to zero as $n \to \infty$ if

$$
R < \min_{\mathcal{S}: 1 \in \mathcal{S}, N \in \mathcal{S}^c} \left(I_1(\mathcal{S}) + I_2(\mathcal{S}) - \sum_{k \in \mathcal{S}} I(\hat{Y}_k; Y_k | X_k) \right) - (N-2)\delta(\epsilon') - \delta(\epsilon)
$$

$$
= \min_{\mathcal{S}: 1 \in \mathcal{S}, N \in \mathcal{S}^c} (I(X(\mathcal{S}); \hat{Y}(\mathcal{S}^c) | X(\mathcal{S}^c)) - I(\hat{Y}(\mathcal{S}); Y(\mathcal{S}) | X^{N-1}, \hat{Y}(\mathcal{S}^c))) - \delta'(\epsilon).
$$

This completes the proof of the noisy network coding lower bound in Theorem 18.3 for a single destination node N.

In general when $X_N \neq \emptyset$ and $\hat{Y}_N \neq Y_N$, it can be easily seen by relabeling the nodes that the above condition becomes

$$
R < \min_{\mathcal{S}: 1 \in \mathcal{S}, N \in \mathcal{S}^c} (I(X(\mathcal{S}); \hat{Y}(\mathcal{S}^c), Y_N | X(\mathcal{S}^c)) - I(\hat{Y}(\mathcal{S}); Y(\mathcal{S}) | X^N, \hat{Y}(\mathcal{S}^c), Y_N)) - \delta'(\epsilon).
$$

Now to prove achievability for the general multicast case, note that each relay encoder operates in the same manner at the same rate regardless of which node is the destination. Therefore, if each destination node performs the same multiblock decoding procedure as in the single-destination case described above, the probability of decoding error for destination node $k \in \mathcal{D}$ tends to zero as $n \to \infty$ if

$$
R < \min_{\mathcal{S}: 1 \in \mathcal{S}, k \in \mathcal{S}^c} (I(X(\mathcal{S}); \hat{Y}(\mathcal{S}^c), Y_k | X(\mathcal{S}^c)) - I(\hat{Y}(\mathcal{S}); Y(\mathcal{S}) | X^N, \hat{Y}(\mathcal{S}^c), Y_k)).
$$

This completes the proof of Theorem 18.3.

18.4   DISCRETE MEMORYLESS MULTIMESSAGE NETWORK

We consider extensions of the upper and lower bounds on the multicast capacity presented in the previous sections to multimessage networks. Assume that node $j \in [1:N]$ wishes to send a message M_j to a set of destination nodes $\mathcal{D}_j \subseteq [1:N]$. Note that a node may not be a destination for any message or a destination for one or more messages, and may serve as a relay for messages from other nodes.

A $(2^{nR_1}, \ldots, 2^{nR_N}, n)$ code for the DMN consists of

- N message sets $[1 : 2^{nR_1}], \ldots, [1 : 2^{nR_N}]$,

- a set of encoders, where encoder $j \in [1:N]$ assigns a symbol $x_{ji}(m_j, y_j^{i-1})$ to each pair (m_j, y_j^{i-1}) for $i \in [1:n]$, and

- a set of decoders, where decoder $k \in \bigcup_j \mathcal{D}_j$ assigns an estimate $(\hat{m}_{jk} : j \in [1:N], k \in \mathcal{D}_j)$ or an error message e to each pair (m_k, y_k^n).

Assume that $(M_1, \ldots, M_N)$ is uniformly distributed over $[1 : 2^{nR_1}] \times \cdots \times [1 : 2^{nR_N}]$. The average probability of error is defined as

$$P_e^{(n)} = \mathrm{P}\{\hat{M}_{jk} \neq M_j \text{ for some } j \in [1:N], k \in \mathcal{D}_j\}.$$

A rate tuple $(R_1, \ldots, R_N)$ is said to be achievable if there exists a sequence of codes such that $\lim_{n \to \infty} P_e^{(n)} = 0$. The capacity region of the DMN is the closure of the set of achievable rates.

Note that this model includes the DM-MAC, DM-IC, DM-RC, and DM-TWC (with and without feedback) as special cases. In particular, when $X_N = \emptyset$ and $\mathcal{D}_1 = \cdots = \mathcal{D}_{N-1} = \{N\}$, the corresponding network is referred to as the DM-MAC with *generalized feedback*. However, our DMN model does *not* include broadcast networks with multiple messages communicated by a single source to different destination nodes.

The cutset bound can be easily extended to multimessage multicast networks.

Theorem 18.4 (Cutset Bound for the DMN). If a rate tuple $(R_1, \ldots, R_N)$ is achievable for the DMN $p(y^N | x^N)$ with destination sets $(\mathcal{D}_1, \ldots, \mathcal{D}_N)$, then it must satisfy the inequality

$$\sum_{j \in \mathcal{S} : \mathcal{D}_j \cap \mathcal{S}^c \neq \emptyset} R_j \leq I(X(\mathcal{S}); Y(\mathcal{S}^c) | X(\mathcal{S}^c))$$

for all $\mathcal{S} \subset [1:N]$ such that $\mathcal{S}^c \cap \mathcal{D}(\mathcal{S}) \neq \emptyset$ for some pmf $p(x^N)$, where $\mathcal{D}(\mathcal{S}) = \bigcup_{j \in \mathcal{S}} \mathcal{D}_j$.

This cutset outer bound can be improved for some special network models. For example, when no cooperation between the nodes is possible (or allowed), that is, the encoder of each node is a function only of its own message, the bound can be tightened by conditioning on a time-sharing random variable Q and replacing $p(x^N)$ with $p(q) \prod_{j=1}^{N} p(x_j | q)$. This modification makes the bound tight for the N-sender DM-MAC. It also yields a tighter outer bound on the capacity region of the N sender–receiver pair interference channel.

18.4.1 Noisy Network Coding for the Multimessage Multicast Network

We extend noisy network coding to multimessage multicast networks, where the sets of destination nodes are the same, i.e., $\mathcal{D}_j = \mathcal{D}$ for all $j \in [1:N]$.

Theorem 18.5 (Multimessage Multicast Noisy Network Coding Inner Bound).
A rate tuple $(R_1, \ldots, R_N)$ is achievable for the DM multimessage multicast network $p(y^N|x^N)$ with destinations $\mathcal{D}$ if

$$\sum_{j \in \mathcal{S}} R_j < \min_{k \in \mathcal{S}^c \cap \mathcal{D}} I(X(\mathcal{S}); \hat{Y}(\mathcal{S}^c), Y_k | X(\mathcal{S}^c), Q) - I(Y(\mathcal{S}); \hat{Y}(\mathcal{S}) | X^N, \hat{Y}(\mathcal{S}^c), Y_k, Q)$$

for all $\mathcal{S} \subset [1:N]$ such that $\mathcal{S}^c \cap \mathcal{D} \neq \emptyset$ for some pmf $p(q) \prod_{j=1}^N p(x_j|q) p(\hat{y}_j|y_j, x_j, q)$.

This inner bound is tight for several classes of networks. For the deterministic network $Y^N = y^N(X^N)$, the cutset bound in Theorem 18.4 simplifies to the set of rate tuples $(R_1, \ldots, R_N)$ such that

$$\sum_{j \in \mathcal{S}} R_j \le H(Y(\mathcal{S}^c) | X(\mathcal{S}^c)) \tag{18.10}$$

for all $\mathcal{S} \subset [1:N]$ with $\mathcal{D} \cap \mathcal{S}^c \neq \emptyset$ for some $p(x^N)$. In the other direction, by setting $\hat{Y}_k = Y_k$, $k \in [1:N]$, the inner bound in Theorem 18.5 simplifies to the set of rate tuples $(R_1, \ldots, R_N)$ such that

$$\sum_{j \in \mathcal{S}} R_j < H(Y(\mathcal{S}^c) | X(\mathcal{S}^c)) \tag{18.11}$$

for all $\mathcal{S} \subset [1:N]$ with $\mathcal{D} \cap \mathcal{S}^c \neq \emptyset$ for some $\prod_{j=1}^N p(x_j)$. It can be easily shown that the two bounds coincide for the graphical multimessage multicast network, the deterministic network with no interference, and the deterministic finite-field network, extending the single-message results in Section 15.4 and Section 18.3.1. Similarly, it can be shown that the capacity region of the wireless erasure multimessage multicast network is the set of all rate tuples $(R_1, \ldots, R_N)$ such that

$$\sum_{j \in \mathcal{S}} R_j \le \sum_{j \in \mathcal{S}: \mathcal{N}_j \cap \mathcal{S}^c \neq \emptyset} \left(1 - \prod_{l \in \mathcal{N}_j \cap \mathcal{S}^c} p_{lj} \right) C_j \tag{18.12}$$

for all $\mathcal{S} \subset [1:N]$ with $\mathcal{D} \cap \mathcal{S}^c \neq \emptyset$. This extends the results for the wireless erasure multicast network in Section 18.3.2.

Proof of Theorem 18.5 (outline). Randomly generate an independent codebook $\mathcal{C}_j = \{x_k^n(m_k, l_{k,j-1}), \hat{y}_k^n(l_{kj}|m_k, l_{k,j-1}): m_k \in [1:2^{nbR_k}], l_{kj}, l_{k,j-1} \in [1:2^{n\tilde{R}_k}], k \in [1:N]\}$ for each block $j \in [1:b]$, where $l_{k0} = 1$, $k \in [1:N]$, by convention. At the end of block j, node k finds an index l_{kj} such that $(y_k^n(j), \hat{y}_k^n(l_{kj}|m_k, l_{k,j-1}), x_k^n(m_k, l_{k,j-1})) \in \mathcal{T}_{\epsilon'}^{(n)}$ and transmits $x_k^n(m_k, l_{kj})$ from $\mathcal{C}_{j+1}$ in block $j + 1$. The probability of error for this joint typicality encoding step tends to zero as $n \to \infty$ if $\tilde{R}_k > I(\hat{Y}_k; Y_k|X_k) + \delta(\epsilon')$ as in the proof of Theorem 18.3. Let $\hat{l}_{kj} = l_{kj}$ and $\hat{m}_{kk} = m_k$ by convention. At the end of block b, decoder $k \in \mathcal{D}$ finds the unique message tuple $(\hat{m}_{1k}, \ldots, \hat{m}_{Nk})$, such that

$$(x_1^n(\hat{m}_{1k}, \hat{l}_{1,j-1}), \ldots, x_N^n(\hat{m}_{Nk}, \hat{l}_{N,j-1}),$$
$$\hat{y}_1^n(\hat{l}_{1j}|\hat{m}_{1k}, \hat{l}_{1,j-1}), \ldots, \hat{y}_N^n(\hat{l}_{Nj}|\hat{m}_{Nk}, \hat{l}_{N,j-1}), y_k^n(j)) \in \mathcal{T}_{\epsilon}^{(n)} \tag{18.13}$$

for all $j \in [1:b]$ for some $(\hat{l}_1, \ldots, \hat{l}_b)$, where $\hat{l}_j = (\hat{l}_{1j}, \ldots, \hat{l}_{Nj})$. Following similar steps to the proof of Theorem 18.3, we define $\mathcal{S}_j(\mathbf{m}, \mathbf{l}^b) = \{k' \in [1:N]: m_{k'} \neq 1 \text{ or } l_{k',j-1} \neq 1\}$ and $\mathcal{T}(\mathbf{m}) = \{k' \in [1:N]: m_{k'} \neq 1\} \subseteq \mathcal{S}_j(\mathbf{m}, \mathbf{l}^b)$, where $\mathbf{m} = (m_1, \ldots, m_N)$. Then, it can be shown that the probability of error corresponding to $P(\mathcal{E}_3)$ in the proof of Theorem 18.3 tends to zero as $n \to \infty$ if

$$\sum_{k' \in \mathcal{T}} R_{k'} < \frac{b-1}{b}\left(I_1(\mathcal{S}) + I_2(\mathcal{S}) - \sum_{k' \in \mathcal{S}} \tilde{R}_{k'} - \delta(\epsilon)\right) - \frac{1}{b}\left(\sum_{k' \neq k} \tilde{R}_{k'}\right)$$

for every $\mathcal{S}, \mathcal{T} \subset [1:N]$ such that $\emptyset \neq \mathcal{T} \subseteq \mathcal{S}$ and $k \in \mathcal{S}^c$, where $I_1(\mathcal{S})$ and $I_2(\mathcal{S})$ are defined as before. By eliminating $\tilde{R}_{k'}$, $k' \in [1:N]$, taking $b \to \infty$, and observing that each proper subset $\mathcal{T}$ of $\mathcal{S}$ corresponds to an inactive inequality, we obtain the condition in Theorem 18.5.

18.4.2*   Noisy Network Coding for General Multimessage Networks

We now extend noisy network coding to general multimessage networks. As a first step, we note that Theorem 18.5 continues to hold for general networks with *multicast completion* of destination nodes, that is, when every message is decoded by all the destination nodes in $\mathcal{D} = \bigcup_{j=1}^N \mathcal{D}_j$. Thus, we can obtain an inner bound on the capacity region of the DMN in the same form as the inner bound in Theorem 18.5 with $\mathcal{D} = \bigcup_{j=1}^N \mathcal{D}_j$.

This multicast-completion inner bound can be improved by noting that noisy network coding transforms a multihop relay network $p(y^N|x^N)$ into a single-hop network $p(\tilde{y}^N|x^N)$, where the effective output at destination node k is $\tilde{Y}_k = (Y_k, \hat{Y}^N)$ and the compressed channel outputs $\hat{Y}^N$ are described to the destination nodes with some rate penalty. This observation leads to improved coding schemes that combine noisy network coding with decoding techniques for interference channels.

Simultaneous nonunique decoding. Each receiver decodes for all the messages and compression indices without requiring the correct recovery of unintended messages and the compression indices; see the simultaneous-nonunique-decoding inner bound in Section 6.2. This approach yields the following inner bound on the capacity region that consists of all rate tuples $(R_1, \ldots, R_N)$ such that

$$\sum_{k' \in \mathcal{S}} R_{k'} < \min_{k \in \mathcal{S}^c \cap \mathcal{D}(\mathcal{S})} (I(X(\mathcal{S}); \hat{Y}(\mathcal{S}^c), Y_k|X(\mathcal{S}^c), Q) - I(Y(\mathcal{S}); \hat{Y}(\mathcal{S})|X^N, \hat{Y}(\mathcal{S}^c), Y_k, Q))$$

$$\tag{18.14}$$

for all $S \subset [1:N]$ with $S^c \cap \mathcal{D}(S) \neq \emptyset$ for some conditional pmf $p(q) \prod_{k=1}^{N} p(x_k|q) \cdot p(\hat{y}_k|y_k, x_k, q)$. The proof of this inner bound is similar to that of Theorem 18.5, except that decoder k finds the unique index tuple $(\hat{m}_{k'k}: k' \in \mathcal{N}_k)$, where $\mathcal{N}_k = \{k' \in [1:N]: k \in \mathcal{D}_{k'}\}$, such that (18.13) holds for all $j \in [1:b]$ for some $(\hat{l}_1, \ldots, \hat{l}_b)$ and $(\hat{m}_{k'k}: k' \notin \mathcal{N}_k)$.

Treating interference as noise. As an alternative to decoding for all the messages, each destination node can simply treat interference as noise as in the interference channel; see Section 6.2. This can be combined with superposition coding of the message (not intended for every destination) and the compression index at each node to show that a rate tuple $(R_1, \ldots, R_N)$ is achievable for the DMN if

$$\sum_{k' \in \mathcal{T}} R_{k'} < I(X(\mathcal{T}), U(S); \hat{Y}(S^c), Y_k | X(\mathcal{N}_k \cap \mathcal{T}^c), U(S^c), Q)$$
$$- I(Y(S); \hat{Y}(S) | X(\mathcal{N}_k), U^N, \hat{Y}(S^c), Y_k, Q)$$

for all $S, \mathcal{T} \subset [1:N]$, $k \in \mathcal{D}(S)$ such that $S^c \cap \mathcal{D}(S) \neq \emptyset$ and $S \cap \mathcal{N}_k \subseteq \mathcal{T} \subseteq \mathcal{N}_k$ for some conditional pmf $p(q) \prod_{k=1}^{N} p(u_k, x_k|q) p(\hat{y}_k|y_k, u_k, q)$, where $\mathcal{N}_k = \{k' \in [1:N]: k \in \mathcal{D}_{k'}\}$ is the set of sources for node k. To prove achievability, we use a randomly and independently generated codebook $C_j = \{u_k^n(l_{k,j-1}), x_k^n(m_k|l_{k,j-1}), \hat{y}_k^n(l_{kj}|l_{k,j-1}): m_k \in [1:2^{nbR_k}],$ $l_{kj}, l_{k,j-1} \in [1:2^{n\tilde{R}_k}], k \in [1:N]\}$ for each block $j \in [1:b]$. Upon receiving $y_k^n(j)$ at the end of block j, node $k \in [1:N]$ finds an index $l_{kj} \in [1:2^{n\tilde{R}_k}]$ such that $(u_k^n(l_{k,j-1}),$ $\hat{y}_k^n(l_{kj}|l_{k,j-1}), y_k^n(j)) \in \mathcal{T}_{\epsilon'}^{(n)}$, and transmits the codeword $x_k^n(m_k|l_{kj})$ in block $j + 1$. The probability of error for this joint typicality encoding step tends to zero as $n \to \infty$ if $\tilde{R}_k > I(\hat{Y}_k; Y_k|U_k) + \delta(\epsilon')$. At the end of block b, decoder k finds the unique index tuple $(\hat{m}_{k'k}:$ $k' \in \mathcal{N}_k)$ such that $((x_{k'}^n(\hat{m}_{k'k}|\hat{l}_{k',j-1}): k' \in \mathcal{N}_k), u_1^n(\hat{l}_{1,j-1}), \ldots, u_N^n(\hat{l}_{N,j-1}), \hat{y}_1^n(\hat{l}_{1j}|\hat{l}_{1,j-1}),$ $\ldots, \hat{y}_N^n(\hat{l}_{Nj}|\hat{l}_{N,j-1}), y_k^n(j)) \in \mathcal{T}_{\epsilon}^{(n)}$ for all $j \in [1:b]$ for some $(l_1, \ldots, l_b)$. Defining $S_j(l^b) = \{k \in [1:N]: l_{k,j-1} \neq 1\}$ and $\mathcal{T}(\mathbf{m}) = \{k' \in \mathcal{N}_k: m_{k'} \neq 1\}$, and following similar steps to the proof of Theorems 18.3 and 18.5, it can be shown that the probability of error corresponding to $P(\mathcal{E}_3)$ in the proof of Theorem 18.3 tends to zero as $n \to \infty$ if

$$\sum_{k' \in \mathcal{T}} R_{k'} < \frac{b-1}{b}\left(I_1(S, \mathcal{T}) + I_2(S, \mathcal{T}) - \sum_{k' \in S} \tilde{R}_{k'} - \delta(\epsilon)\right) - \frac{1}{b}\left(\sum_{k' \neq k} \tilde{R}_{k'}\right)$$

for all $S \subset [1:N]$ and $\mathcal{T} \subseteq \mathcal{N}_k$ such that $k \in S^c \cap \mathcal{T}^c$, where

$$I_1(S, \mathcal{T}) = I(X((S \cup \mathcal{T}) \cap \mathcal{N}_k), U(S); \hat{Y}(S^c), Y_k | X((S^c \cap \mathcal{T}^c) \cap \mathcal{N}_k), U(S^c)),$$
$$I_2(S, \mathcal{T}) = \sum_{k' \in S} I(\hat{Y}_{k'}'; \hat{Y}(S^c \cup \{k'' \in S: k'' < k'\}), Y_k, X(\mathcal{N}_k), U^N | U_{k'}').$$

Eliminating $\tilde{R}_k$, $k \in [1:N]$, taking $b \to \infty$, and removing inactive inequalities completes the proof of the inner bound.

Remark 18.8. As for the interference channel, the rates achieved by the above coding schemes can be improved by using more sophisticated techniques such as superposition coding and rate splitting.

SUMMARY

- Discrete memoryless network (DMN): Generalizes graphical networks, DM-RC, and single-hop networks with and without feedback

- Cutset bounds for the DMN

- Sliding window decoding for network decode–forward

- Noisy network coding:

  - Same message is sent multiple times using independent codebooks

  - No Wyner–Ziv binning

  - Simultaneous nonunique decoding without requiring the recovery of compression bin indices

  - Includes compress–forward for the relay channel and network coding for graphical networks as special cases

  - Extensions to multimessage networks via interference channel coding schemes

- **Open problems:**

  18.1. What is the common-message capacity of a general DM network (that is, multicast with $\mathcal{D} = [2:N]$)?

  18.2. What is the capacity of a general deterministic multicast network?

  18.3. How should partial decode–forward be extended to noisy networks?

  18.4. How should noisy network coding and interference alignment be combined?

BIBLIOGRAPHIC NOTES

The cutset bound in Theorem 18.1 is due to El Gamal (1981). Aref (1980) established the capacity of physically degraded unicast networks by extending decode–forward via binning for the relay channel. He also established the capacity of deterministic unicast networks by extending partial decode–forward. The sliding window decoding scheme was developed by Carleial (1982). The network decode–forward lower bound for the general network in Theorem 18.2 is due to Xie and Kumar (2005) and Kramer, Gastpar, and Gupta (2005). This improves upon a previous lower bound based on an extension of decode–forward via binning by Gupta and Kumar (2003). Kramer et al. (2005) also provided an extension of compress–forward for the relay channel. As in the original compress–forward scheme, this extension involves Wyner–Ziv coding and sequential decoding. Decode–forward of compression indices is used to enhance the performance of the scheme.

In an independent line of work, Ratnakar and Kramer (2006) extended network coding to establish the capacity of deterministic multicast networks with no interference in (18.4). Subsequently Avestimehr, Diggavi, and Tse (2011) further extended this result to obtain the lower bound on the capacity of general deterministic multicast networks in (18.3), and showed that it is tight for finite-field deterministic networks, which they used to approximate the capacity of Gaussian multicast networks. As in the original proof of the network coding theorem (see the Bibliographic Notes for Chapter 15), the proof by Avestimehr et al. (2011) is divided into two steps. In the first step, *layered* networks as exemplified in Figure 18.4 are considered and it is shown that if the rate R satisfies the lower bound, then the end-to-end mapping is one-to-one with high probability. The proof is then extended to *nonlayered* networks by considering a time-expanded layered network with b blocks, and it is shown that if the rate bR is less than b times the lower bound in (18.3) for sufficiently large b, then the end-to-end mapping is again one-to-one with high probability.

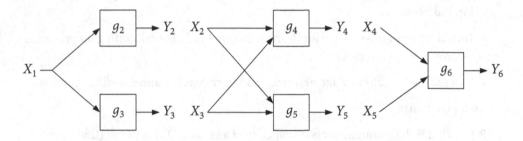

Figure 18.4. A layered deterministic network example.

The noisy network coding scheme in Section 18.3 generalizes compress–forward for the relay channel and network coding and its extensions to deterministic networks to noisy networks. The extensions of noisy network coding to multimessage networks in Section 18.4 are due to Lim, Kim, El Gamal, and Chung (2011). Other extensions of noisy network coding can be found in Lim, Kim, El Gamal, and Chung (2010). The inner bound on the capacity region of deterministic multimessage multicast networks was established by Perron (2009). The capacity region of wireless erasure networks in (18.12) is due to Dana, Gowaikar, Palanki, Hassibi, and Effros (2006), who extended linear network coding to packet erasures.

In the discrete memoryless network model in Section 18.4, each node wishes to communicate at most one message. A more general model of multiple messages at each node was considered by Kramer (2007), where a cutset bound similar to Theorem 18.4 was established. The simplest example of such multihop broadcast networks is the relay broadcast channel studied by Kramer, Gastpar, and Gupta (2005) and Liang and Kramer (2007).

PROBLEMS

18.1. Provide the details of the probability of error analysis for the network decode–forward lower bound in Section 18.2.2.

18.2. Consider the cutset bound on the multicast capacity for deterministic networks in (18.2). Show that the product input pmf attains the maximum in the bound for the deterministic network with no interference and that the uniform input pmf attains the maximum for the finite-field deterministic network.

18.3. Consider the wireless erasure multicast network in Section 18.3.2. Show that the cutset bound in Theorem 18.1 and the noisy network coding lower bound in Theorem 18.3 coincide and simplify to the capacity expression in (18.5).

18.4. Derive the inequality in (18.9) using the fact that $m \neq 1$ and $l_{k,j-1} \neq 1$ for $k \in S_j(l_{j-1})$.

18.5. Prove the cutset bound for noisy multimessage networks in Theorem 18.4.

18.6. Complete the details of the achievability proof of Theorem 18.5.

18.7. *Broadcasting over a diamond network.* Consider a DMN $p(y_2, y_3, y_4 | x_1, x_2, x_3) = p(y_2|x_1)p(y_3|x_1)p(y_4|x_2, x_3)$. Node 1 wishes to communicate a common message M to all other nodes. Find the capacity.

18.8. *Two-way relay channel.* Consider a DMN $p(y_1, y_2, y_3 | x_1, x_2, x_3)$. Node 1 wishes to communicate a message $M_1 \in [1 : 2^{nR_1}]$ to node 2 and node 2 wishes to communicate a message $M_2 \in [1 : 2^{nR_2}]$ to node 1 with the help of relay node 3.

(a) Characterize the cutset bound for this channel.

(b) Noting that the channel is a multimessage multicast network with $\mathcal{D} = \{1, 2\}$, simplify the noisy network coding inner bound in Theorem 18.5.

18.9. *Interference relay channel.* Consider a DMN $p(y_3, y_4, y_5 | x_1, x_2, x_3)$. Node 1 wishes to communicate a message $M_1 \in [1 : 2^{nR_1}]$ to node 4 and node 2 wishes to communicate a message $M_2 \in [1 : 2^{nR_2}]$ to node 5 with the help of relay node 3.

(a) Characterize the cutset bound for this channel.

(b) Simplify the noisy network coding inner bound with simultaneous nonunique decoding in (18.14) for this channel.

CHAPTER 19

Gaussian Networks

In this chapter, we discuss models for wireless multihop networks that generalize the Gaussian channel models we studied earlier. We extend the cutset bound and the noisy network coding inner bound on the capacity region of the multimessage DMN presented in Chapter 18 to Gaussian networks. We show through a Gaussian two-way relay channel example that noisy network coding can outperform decode–forward and amplify–forward, achieving rates within a constant gap of the cutset bound while the inner bounds achieved by these other schemes can have an arbitrarily large gap to the cutset bound. More generally, we show that noisy network coding for the Gaussian multimessage multicast network achieves rates within a constant gap of the capacity region independent of network topology and channel gains. For Gaussian networks with other messaging demands, e.g., general multiple-unicast networks, however, no such constant gap results exist in general. Can we still obtain some guarantees on the capacity of these networks?

To address this question, we introduce the scaling-law approach to capacity, where we seek to find the order of capacity scaling as the number of nodes in the network becomes large. In addition to providing some guarantees on network capacity, the study of capacity scaling sheds light on the role of cooperation through relaying in combating interference and path loss in large wireless networks. We first illustrate the scaling-law approach via a simple unicast network example that shows how relaying can dramatically increase the capacity by reducing the effect of high path loss. We then present the Gupta–Kumar random network model in which the nodes are randomly distributed over a geographical area and the goal is to determine the capacity scaling law that holds for most such networks. We establish lower and upper bounds on the capacity scaling law for the multiple-unicast case. The lower bound is achieved via a cellular time-division scheme in which the messages are sent simultaneously using a simple multihop scheme with nodes in cells along the lines from each source to its destination acting as relays. We show that this scheme achieves much higher rates than direct transmission with time division, which demonstrates the role of relaying in mitigating interference in large networks. This cellular time-division scheme also outperforms noncellular multihop through spatial reuse of time enabled by high path loss. Finally, we derive an upper bound on the capacity scaling law using the cutset bound and a network augmentation technique. This upper bound becomes tighter as the path loss exponent increases and has essentially the same order as the cellular time-division lower bound under the absorption path loss model.

19.1 GAUSSIAN MULTIMESSAGE NETWORK

Consider an N-node Gaussian network

$$Y_k = \sum_{j=1}^{N} g_{kj} X_j + Z_k, \quad k \in [1:N],$$

where g_{kj} is the gain from the transmitter of node j to the receiver of node k, and the noise components Z_k, $k \in [1:N]$, are i.i.d. N(0, 1). We assume expected average power constraint P on each X_j, i.e., $\sum_{i=1}^{n} \mathsf{E}(x_{ji}^2(m_j, Y_j^{i-1})) \le nP$, $m_j \in [1 : 2^{nR_j}]$, $j \in [1:N]$. We consider a general multimessage demand where each node j wishes to send a message M_j to a set of destination nodes $\mathcal{D}_j$. The definitions of a code, probability of error, achievability, and capacity region follow those for the multimessage DMN in Section 18.4.

Consider the following special cases:

- If $X_N = \emptyset$ and $\mathcal{D}_j = \{N\}$ for $j \in [1:N-1]$, then the network reduces to the $(N-1)$-sender Gaussian MAC with *generalized* feedback.

- If $N = 2k$, $X_{k+1} = \cdots = X_N = Y_1 = \cdots = Y_k = \emptyset$, $\mathcal{D}_j = \{j + k\}$ for $j \in [1 : k]$, then the network reduces to the k-user-pair Gaussian IC.

- If $N = 3$, $X_3 = Y_1 = \emptyset$, $\mathcal{D}_1 = \{3\}$, and $R_2 = 0$, then the network reduces to the Gaussian RC.

The Gaussian network can be equivalently written in a vector form

$$Y^N = GX^N + Z^N, \tag{19.1}$$

where X^N is the channel input vector, $G \in \mathbb{R}^{N \times N}$ is the channel gain matrix, and Z^N is a vector of i.i.d. N(0, 1) noise components. Using this vector form, the cutset bound in Theorem 18.4 can be easily adapted to the Gaussian network model.

Theorem 19.1 (Cutset Bound for the Gaussian Multimessage Network). If a rate tuple $(R_1, \ldots, R_N)$ is achievable for the Gaussian multimessage network with destination sets $(\mathcal{D}_1, \ldots, \mathcal{D}_N)$, then it must satisfy the inequality

$$\sum_{j \in \mathcal{S}: \mathcal{D}_j \cap \mathcal{S}^c \ne \emptyset} R_j \le \frac{1}{2} \log |I + G(\mathcal{S}) K(\mathcal{S} | \mathcal{S}^c) G^T(\mathcal{S})|$$

for all $\mathcal{S}$ such that $\mathcal{S}^c \cap \mathcal{D}(\mathcal{S}) \ne \emptyset$ for some covariance matrix $K \succeq 0$ with $K_{jj} \le P$, $j \in [1:N]$. Here $\mathcal{D}(\mathcal{S}) = \bigcup_{j \in \mathcal{S}} \mathcal{D}_j$, $K(\mathcal{S} | \mathcal{S}^c)$ is the conditional covariance matrix of $X(\mathcal{S})$ given $X(\mathcal{S}^c)$ for $X^N \sim$ N(0, K), and $G(\mathcal{S})$ is defined such that

$$\begin{bmatrix} Y(\mathcal{S}) \\ Y(\mathcal{S}^c) \end{bmatrix} = \begin{bmatrix} G'(\mathcal{S}) & G(\mathcal{S}^c) \\ G(\mathcal{S}) & G'(\mathcal{S}^c) \end{bmatrix} \begin{bmatrix} X(\mathcal{S}) \\ X(\mathcal{S}^c) \end{bmatrix} + \begin{bmatrix} Z(\mathcal{S}) \\ Z(\mathcal{S}^c) \end{bmatrix},$$

for some gain submatrices $G'(\mathcal{S})$ and $G'(\mathcal{S}^c)$.

When no cooperation between the nodes is possible, the cutset bound can be tightened as for the DMN by conditioning on a time-sharing random variable Q and considering $X^N|\{Q = q\} \sim N(0, K(q))$, where $K(q)$ is diagonal and $E_Q(K_{jj}(Q)) \leq P$. This yields the improved bound with conditions

$$\sum_{j \in S: \mathcal{D}_j \cap S^c \neq \emptyset} R_j \leq \frac{1}{2} E_Q(\log |I + G(S)K(S|Q)G^T(S)|)$$

for all S such that $S^c \cap \mathcal{D}(S) \neq \emptyset$, where $K(S|Q)$ is the (random) covariance matrix of $X(S)$ given Q.

19.1.1 Noisy Network Coding Inner Bound

The inner bound on the capacity region of the DM multimessage multicast network in Theorem 18.5 can be also adapted to Gaussian networks. By adding the power constraints, we can readily obtain the noisy network coding inner bound that consists of all rate tuples $(R_1, \ldots, R_N)$ such that

$$\sum_{j \in S} R_j < \min_{k \in S^c \cap \mathcal{D}} I(X(S); \hat{Y}(S^c), Y_k | X(S^c), Q) - I(Y(S); \hat{Y}(S) | X^N, \hat{Y}(S^c), Y_k, Q) \quad (19.2)$$

for all S satisfying $S^c \cap \mathcal{D} \neq \emptyset$ for some conditional distribution $p(q) \prod_{j=1}^{N} F(x_j|q) \cdot F(\hat{y}_j|y_j, x_j, q)$ such that $E(X_j^2) \leq P$ for $j \in [1:N]$. The optimizing conditional distribution of the inner bound in (19.2) is not known in general. To compare this noisy network coding inner bound to the cutset bound in Theorem 19.1 and to other inner bounds, we set $Q = \emptyset$, $X_j, j \in [1:N]$, i.i.d. $N(0, P)$, and

$$\hat{Y}_k = Y_k + \hat{Z}_k, \quad k \in [1:N],$$

where $\hat{Z}_k \sim N(0, 1)$, $k \in [1:N]$, are independent of each other and of (X^N, Y^N). Substituting in (19.2), we have

$$I(Y(S); \hat{Y}(S)|X^N, \hat{Y}(S^c), Y_k) \stackrel{(a)}{\leq} I(\hat{Y}(S); Y(S)|X^N)$$
$$= h(\hat{Y}(S)|X^N) - h(\hat{Y}(S)|Y(S), X^N)$$
$$= \frac{|S|}{2} \log(4\pi e) - \frac{|S|}{2} \log(2\pi e)$$
$$= \frac{|S|}{2}$$

for each $k \in \mathcal{D}$ and S such that $S^c \cap \mathcal{D} \neq \emptyset$. Here step (a) follows since $(\hat{Y}(S^c), Y_k) \rightarrow (X^N, Y(S)) \rightarrow \hat{Y}(S)$ form a Markov chain. Furthermore

$$I(X(S); \hat{Y}(S^c), Y_k | X(S^c)) \geq I(X(S); \hat{Y}(S^c) | X(S^c))$$
$$= h(\hat{Y}(S^c)|X(S^c)) - h(\hat{Y}(S^c)|X^N)$$
$$= \frac{1}{2} \log\left((2\pi e)^{|S^c|} |2I + PG(S)G^T(S)|\right) - \frac{|S^c|}{2} \log(4\pi e)$$
$$= \frac{1}{2} \log\left|I + \frac{P}{2}G(S)G^T(S)\right|.$$

Hence, we obtain the inner bound characterized by the set of inequalities

$$\sum_{j \in S} R_j < \frac{1}{2} \log \left| I + \frac{P}{2} G(S) G^T(S) \right| - \frac{|S|}{2} \qquad (19.3)$$

for all S with $S^c \cap D \neq \emptyset$.

Remark 19.1. As in the compress–forward lower bound for the Gaussian RC in Section 16.7, the choice of $\hat{Y}_k = Y_k + \hat{Z}_k$ with $\hat{Z}_k \sim N(0, 1)$ can be improved upon by optimizing over the average powers of $\hat{Z}_k$, $k \in [1 : N]$, for the given channel gain matrix. The bound can be improved also by time sharing. It is not known, however, if Gaussian test channels are optimal.

In the following, we compare this noisy network coding inner bound to the cutset bound and other inner bounds on the capacity region.

19.1.2 Gaussian Two-Way Relay Channel

Consider the 3-node Gaussian two-way relay channel with no direct links depicted in Figure 19.1 with outputs

$$Y_1 = g_{13} X_3 + Z_1,$$
$$Y_2 = g_{23} X_3 + Z_2,$$
$$Y_3 = g_{31} X_1 + g_{32} X_2 + Z_3,$$

where the noise components Z_k, $k = 1, 2, 3$, are i.i.d. $N(0, 1)$. We assume expected average power constraint P on each of X_1, X_2, and X_3. Denote the received SNR for the signal from node j to node k as $S_{kj} = g_{kj}^2 P$. Node 1 wishes to communicate a message M_1 to node 2 and node 2 wishes to communicate a message M_2 to node 1 with the help of relay node 3, i.e., $D = \{1, 2\}$; see Problem 18.8 for a more general DM counterpart.

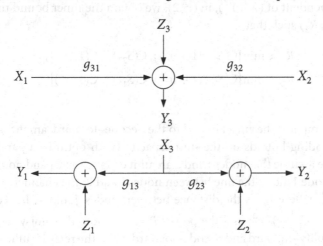

Figure 19.1. Gaussian two-way relay channel with no direct links.

The capacity region of this multimessage multicast network is not known in general. We compare the following outer and inner bounds on the capacity region.

Cutset bound. The cutset bound in Theorem 19.1 can be readily specialized to this Gaussian two-way channel. If a rate pair (R_1, R_2) is achievable, then it must satisfy the inequalities

$$R_1 \le \min\{C(S_{31}), C(S_{23})\},$$
$$R_2 \le \min\{C(S_{32}), C(S_{13})\}. \tag{19.4}$$

Decode–forward inner bound. The decode–forward coding scheme for the DM-RC in Section 16.4 can be extended to this two-way relay channel. Node 3 recovers both M_1 and M_2 over the multiple access channel $Y_3 = g_{31}X_1 + g_{32}X_2 + Z_3$ and broadcasts them. It can be easily shown that a rate pair (R_1, R_2) is achievable if

$$R_1 < \min\{C(S_{31}), C(S_{23})\},$$
$$R_2 < \min\{C(S_{32}), C(S_{13})\}, \tag{19.5}$$
$$R_1 + R_2 < C(S_{31} + S_{32}).$$

Amplify–forward inner bound. The amplify–forward relaying scheme for the RFD Gaussian RC in Section 16.8 can be easily extended to this setting by having node 3 send a scaled version of its received symbol. The corresponding inner bound consists of all rate pairs (R_1, R_2) such that

$$R_1 < C\left(\frac{S_{23}S_{31}}{1 + S_{23} + S_{31} + S_{32}}\right),$$
$$R_2 < C\left(\frac{S_{13}S_{32}}{1 + S_{13} + S_{31} + S_{32}}\right). \tag{19.6}$$

Noisy network coding inner bound. By setting $Q = \emptyset$ and $\hat{Y}_3 = Y_3 + \hat{Z}_3$, where $\hat{Z}_3 \sim N(0, \sigma^2)$ is independent of (X^3, Y^3), in (19.2), we obtain the inner bound that consists of all rate pairs (R_1, R_2) such that

$$R_1 < \min\{C(S_{31}/(1 + \sigma^2)), C(S_{23}) - C(1/\sigma^2)\},$$
$$R_2 < \min\{C(S_{32}/(1 + \sigma^2)), C(S_{13}) - C(1/\sigma^2)\} \tag{19.7}$$

for some $\sigma^2 > 0$.

Figure 19.2 compares the cutset bound to the decode–forward, amplify–forward, and noisy network coding bounds on the sum-capacity (with optimized parameters). The plots in the figure assume that nodes 1 and 2 are unit distance apart and node 3 is distance $r \in [0, 1]$ from node 1 along the line between nodes 1 and 2; the channel gains are of the form $g_{kj} = r_{kj}^{-3/2}$, where r_{kj} is the distance between nodes j and k, hence $g_{13} = g_{31} = r^{-3/2}$, $g_{23} = g_{32} = (1 - r)^{-3/2}$; and the power $P = 10$. Note that noisy network coding outperforms amplify–forward and decode–forward when the relay is sufficiently far from both destination nodes.

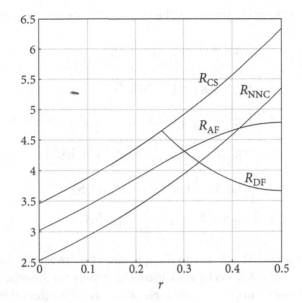

Figure 19.2. Comparison of the cutset bound R_{CS}, decode–forward lower bound R_{DF}, amplify–forward lower bound R_{AF}, and noisy network coding lower bound R_{NNC} on the sum-capacity of the Gaussian two-way relay channel as the function of the distance r between nodes 1 and 3.

In general, it can be shown that noisy network coding achieves the capacity region within $1/2$ bit per dimension, while the other schemes have an *unbounded* gap to the cutset bound as $P \to \infty$ (see Problem 19.2).

Remark 19.2. Unlike the case of the RFD Gaussian relay channel studied in Section 16.8, noisy network coding does not always outperform amplify–forward. The reason is that both destination nodes are required to recover the compression index and hence its rate is limited by the worse channel. This limitation can be overcome by sending *layered* descriptions of Y_3^n such that the weaker receiver recovers the coarser description while the stronger receiver recovers both descriptions.

19.1.3   Multimessage Multicast Capacity Region within a Constant Gap

We show that noisy network coding achieves the capacity region of the Gaussian multimessage network $Y^N = GX^N + Z^N$ within a constant gap uniformly for any channel gain matrix G.

Theorem 19.2 (Constant Gap for Gaussian Multimessage Multicast Network).
For the Gaussian multimessage multicast network, if a rate tuple $(R_1, \ldots, R_N)$ is in the cutset bound in Theorem 19.1, then the rate tuple $(R_1 - \Delta, \ldots, R_N - \Delta)$ is achievable, where $\Delta = (N/2) \log 6$.

Proof. Note that the cutset bound in Theorem 19.1 can be loosened as

$$\sum_{j \in S} R_j \leq \frac{1}{2} \log \left| I + G(S) K_{X(S)} G^T(S) \right|$$

$$= \frac{1}{2} \log \left| I + K_{X(S)} G^T(S) G(S) \right|$$

$$\leq \frac{1}{2} \log \left| I + K_{X(S)} G^T(S) G(S) + \frac{2}{P} K_{X(S)} + \frac{P}{2} G^T(S) G(S) \right|$$

$$= \frac{1}{2} \log \left(\left| I + \frac{2}{P} K_{X(S)} \right| \left| I + \frac{P}{2} G^T(S) G(S) \right| \right)$$

$$\overset{(a)}{\leq} \frac{|S|}{2} \log 3 + \frac{1}{2} \log \left| I + \frac{P}{2} G(S) G^T(S) \right|$$

$$\leq \frac{N}{2} \log 3 + \frac{1}{2} \log \left| I + \frac{P}{2} G(S) G^T(S) \right| \tag{19.8}$$

for all S such that $\mathcal{D} \cap S^c \neq \emptyset$, where $K_{X(S)}$ denotes the covariance matrix of $X(S)$ when $X^N \sim N(0, K)$, and (a) follows by Hadamard's inequality. In the other direction, by loosening the inner bound in (19.3), a rate tuple $(R_1, \ldots, R_N)$ is achievable if

$$\sum_{j \in S} R_j < \frac{1}{2} \log \left| I + \frac{P}{2} G(S) G^T(S) \right| - \frac{N}{2} \tag{19.9}$$

for all S such that $\mathcal{D} \cap S^c \neq \emptyset$. Comparing (19.8) and (19.9) completes the proof of Theorem 19.2.

19.2 CAPACITY SCALING LAWS

As we have seen, the capacity of Gaussian networks is known only in very few special cases. For the multimessage multicast case, we are able to show that the capacity region for any Gaussian network is within a constant gap of the cutset bound. No such constant gap results exist, however, for other multimessage demands. The scaling laws approach to capacity provides another means for obtaining guarantees on the capacity of a Gaussian network. It aims to establish the optimal scaling order of the capacity as the number of nodes grows.

In this section, we focus on Gaussian multiple-unicast networks in which each node in a source-node set S wishes to communicate a message to a distinct node in a disjoint destination-node set $\mathcal{D}$. The rest of the nodes as well as the source and destination nodes themselves can also act as relays. We define the *symmetric network capacity* $C(N)$ as the supremum of the set of symmetric rates R such that the rate tuple $(R, \ldots, R)$ is achievable. We seek to establish the scaling law for $C(N)$, that is, to find a function $g(N)$ such that $C(N) = \Theta(g(N))$; see Notation.

We illustrate this approach through the following simple example. Consider the N-node Gaussian unicast network depicted in Figure 19.3. Assume the power law path loss (channel gain) $g(r) = r^{-\nu/2}$, where r is the distance and $\nu > 2$ is the path loss exponent.

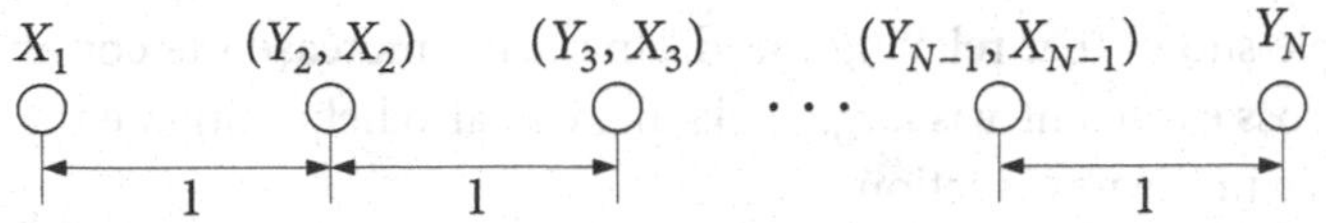

Figure 19.3. Gaussian unicast network.

Hence the received signal at node k is

$$Y_k = \sum_{j=1, j \neq k}^{N-1} |j-k|^{-v/2} X_j + Z_k, \quad k \in [2:N].$$

We assume expected average power constraint P on each X_j, $j \in [1:N-1]$.

Source node 1 wishes to communicate a message to destination node N with the other nodes acting as relays to help the communication; thus the source and relay encoders are specified by $x_1^n(m)$ and $x_{ji}(y_j^{i-1})$, $i \in [1:n]$, for $j \in [2:N-1]$. As we discussed in Chapter 16, the capacity of this network is not known for $N = 3$ for any nonzero channel parameter values. How does $C(N)$ scale with N? To answer this question consider the following bounds on $C(N)$.

Lower bounds. Consider a simple multihop relaying scheme, where signals are Gaussian and interference is treated as noise. In each transmission block, the source transmits a new message to the first relay and node $j \in [2:N-1]$ transmits its most recently received message to node $j+1$. Then $C(N) \geq \min_j C(P/(I_j + 1))$. Now the interference power at node $j \in [2:N]$ is $I_j = \sum_{k=1, k \neq j-1, j}^{N-1} |j-k|^{-v} P$. Since $v > 2$, $I_j = O(1)$ for all j. Hence $C(N) = \Omega(1)$.

Upper bound. Consider the cooperative broadcast upper bound on the capacity

$$C(N) \leq \sup_{F(x^{N-1}): E(X_j^2) \leq P, j \in [1:N-1]} I(X_1; Y_2, \ldots, Y_N | X_2, \ldots, X_{N-1})$$

$$\leq \frac{1}{2} \log |I + APA^T|$$

$$= \frac{1}{2} \log |I + PA^T A|,$$

where $A = \begin{bmatrix} 1 & 2^{-v/2} & 3^{-v/2} & \cdots & (N-1)^{-v/2} \end{bmatrix}^T$. Hence

$$C(N) \leq C\left(\left(\sum_{j=1}^{N-1} \frac{1}{j^v} \right) P \right).$$

Since $v > 2$, $C(N) = O(1)$, which is the same scaling as achieved by the simple multihop scheme. Thus, we have shown that $C(N) = \Theta(1)$.

Remark 19.3. The maximum rate achievable by direct transmission from the source to the destination using the same total system power NP is $C(PN(N-1)^{-v}) = \Theta(N^{1-v})$. Since $v > 2$, this rate tends to zero as $N \to \infty$.

This example shows that relaying can dramatically increase the communication rate when the path loss exponent v is large. Relaying can also help mitigate the effect of interference as we see in the next section.

19.3   GUPTA–KUMAR RANDOM NETWORK

The Gupta–Kumar random network approach aims to establish capacity scaling laws that apply to *most* large ad-hoc wireless networks. The results can help our understanding of the role of cooperation in large networks, which in turn can guide network architecture design and coding scheme development.

We assume a "constant density" network with $2N$ nodes, each randomly and independently placed according to a uniform pdf over a square of area N as illustrated in Figure 19.4. The nodes are randomly partitioned into N source–destination (S-D) pairs. Label the source nodes as $1, 2, \ldots, N$ and the destination nodes as $N + 1, N + 2, \ldots, 2N$.

Once generated, the node locations and the S-D assignments are assumed to be fixed and known to the network architect (code designer). We allow each node, in addition to being either a source or a destination, to act as a relay to help other nodes communicate their messages.

We assume the Gaussian network model in (19.1) with power law path loss, that is, if the distance between nodes j and k is r_{jk}, then the channel gain $g_{jk} = r_{jk}^{-v/2}$ for $v > 2$. Hence, the output signal at each node $k \in [1 : 2N]$ is

$$Y_k = \sum_{j=1, j \neq k}^{2N} r_{kj}^{-v/2} X_j + Z_k.$$

We consider a multiple-unicast setting in which source node $j \in [1 : N]$ wishes to

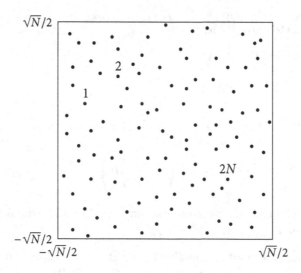

Figure 19.4. Gupta–Kumar random network.

communicate a message $M_j \in [1 : 2^{nR_j}]$ reliably to destination node $k = j + N$. The messages are assumed to be independent and uniformly distributed. We wish to determine the scaling law for the symmetric capacity $C(N)$ that holds *with high probability* (w.h.p.), that is, with probability $\geq (1 - \epsilon_N)$, where ϵ_N tends to zero as $N \to \infty$. In other words, the scaling law holds for most large networks generated in this random manner.

We establish the following bounds on the symmetric capacity.

Theorem 19.3. The symmetric capacity of the random network model with path loss exponent $v > 2$ has the following order bounds:

1. Lower bound: $C(N) = \Omega(N^{-1/2}(\log N)^{-(v+1)/2})$ w.h.p.

2. Upper bound: $C(N) = O(N^{-1/2+1/v} \log N)$ w.h.p.

In other words, there exist constants $a_1, a_2 > 0$ such that

$$\lim_{N\to\infty} P\{a_1 N^{-1/2}(\log N)^{-(v+1)/2} \leq C(N) \leq a_2 N^{-1/2+1/v} \log N\} = 1.$$

Before proving the upper and lower order bounds on the symmetric capacity scaling, consider the following simple transmission schemes.

Direct transmission. Suppose that there is only a single randomly chosen S-D pair. Then it can be readily checked that the S-D pair is $\Omega(N^{-1/2})$ apart w.h.p. and thus direct transmission achieves the rate $\Omega(N^{-v/2})$ w.h.p. Hence, for N S-D pairs, using time division with power control achieves the symmetric rate $\Omega(N^{-v/2})$ w.h.p.

Multihop relaying. Consider a single randomly chosen S-D pair. As we mentioned above, the S-D pair is $\Omega(N^{-1/2})$ apart. Furthermore, it can be shown that with high probability, there are roughly $\Omega((N/\log N)^{1/2})$ relays placed close to the straight line from the source to the destination with distance $O((\log N)^{1/2})$ between every two consecutive relays. Using the multihop scheme in Section 19.2 with these relays, we can show that $\Omega((\log N)^{-v/2})$ is achievable w.h.p. Hence, using time division and multihop relaying (without power control), we can achieve the lower bound on the symmetric capacity $C(N) = \Omega((\log N)^{-v/2}/N)$ w.h.p., which is a huge improvement over direct transmission when the path loss exponent $v > 2$ is large.

Remark 19.4. Using relaying, each node transmits at a much lower power than using direct transmission. This has the added benefit of reducing interference between the nodes, which can be exploited through *spatial reuse* of time/frequency to achieve higher rates.

19.3.1   Proof of the Lower Bound

To prove the lower bound in Theorem 19.3, consider the cellular time-division scheme illustrated in Figure 19.5 with cells of area $\log N$ (to guarantee that no cell is empty w.h.p.). As shown in the figure, the cells are divided into nine groups. We assume equal transmission rates for all S-D pairs. A block Markov transmission scheme is used, where each

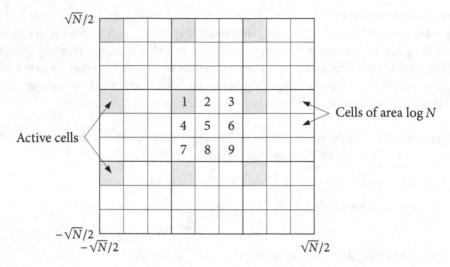

Figure 19.5. Cellular time-division scheme.

source node sends messages over several transmission blocks. Each transmission block is divided into nine *cell-blocks*. A cell is said to be *active* if its nodes are allowed to transmit. Each cell is active only during one out of the nine cell-blocks. Nodes in inactive cells act as receivers. As shown in Figure 19.6, each message is sent from its source node to the destination node using other nodes in cells along the straight line joining them (referred to as an *S-D line*) as relays.

Transmission from each node in an active cell to nodes in its four neighboring cells is performed using Gaussian random codes with power P and each receiver treats interference from other senders as noise. Let $S(N)$ be the maximum number of sources in a cell and $L(N)$ be the maximum number of S-D lines passing through a cell, over all cells.

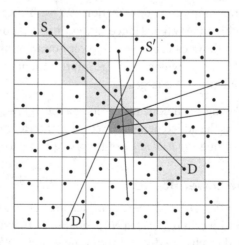

Figure 19.6. Messages transmitted via relays along S-D lines.

Each cell-block is divided into $S(N) + L(N)$ node-blocks for time-division transmission by nodes inside each active cell as illustrated in Figure 19.7. Each source node in an active cell broadcasts a *new* message during its node-block using a Gaussian random code with power P. One of the nodes in the active cell acts as a relay for the S-D pairs that communicate their messages through this cell. It relays the messages during the allotted $L(N)$ node-blocks using a Gaussian random code with power P.

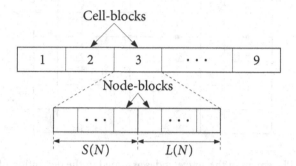

Figure 19.7. Time-division scheme.

Analysis of the probability of failure. The cellular time-division scheme fails if one or both of the following events occur:

$\mathcal{E}_1 = \{$there is a cell with no nodes in it$\}$,

$\mathcal{E}_2 = \{$transmission from a node in a cell to a node in a neighboring cell fails$\}$.

Then the probability that the scheme fails is upper bounded as

$$P(\mathcal{E}) \leq P(\mathcal{E}_1) + P(\mathcal{E}_2 \cap \mathcal{E}_1^c).$$

It is straightforward to show that $P(\mathcal{E}_1)$ tends to zero as $N \to \infty$. Consider the second term $P(\mathcal{E}_2 \cap \mathcal{E}_1^c)$. From the cell geometry, the distance between each transmitting node in a cell and each receiving node in its neighboring cells is always less than or equal to $(5 \log N)^{1/2}$. Since each transmitting node uses power P, the received power at a node in a neighboring cell is always greater than or equal to $(5 \log N)^{-\nu/2}P$. Under worst-case placement of the sender, the receiver, and the interfering transmitters during a cell-block (see Figure 19.8), it can be shown that the total average interference power at a receiver from all other transmitting nodes is

$$I \leq \sum_{j=1}^{\infty} \frac{2P}{((3j-2)^2 \log N)^{\nu/2}} + \sum_{j=1}^{\infty} \sum_{k=1}^{\infty} \frac{4P}{(((3j-2)^2 + (3k-1)^2) \log N)^{\nu/2}}. \tag{19.10}$$

Hence, if $\nu > 2$, $I \leq a_3 (\log N)^{-\nu/2}$ for some constant $a_3 > 0$.

Since we are using Gaussian random codes, the probability of error tends to zero as the node-block length $n \to \infty$ if the transmission rate for each node block is less than

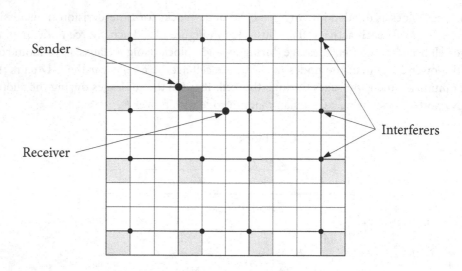

Figure 19.8. Placement of the active nodes assumed in the derivation of the bound on interference power.

$C((5 \log N)^{-v/2}P/(1 + a_3(\log N)^{-v/2}))$. Thus, for a fixed network, $P(\mathcal{E}_2 \cap \mathcal{E}_1^c)$ tends to zero as $n \to \infty$ if the symmetric rate

$$R(N) < \frac{1}{9(S(N) + L(N))}\, C\left(\frac{(5 \log N)^{-v/2}P}{1 + a_3(\log N)^{-v/2}}\right). \tag{19.11}$$

In the following, we bound $S(N) + L(N)$ over a random network.

Lemma 19.1. $S(N) + L(N) = O((N \log N)^{1/2})$ w.h.p.

The proof of this lemma is given in Appendix 19A. Combining Lemma 19.1 and the bound on $R(N)$ in (19.11), we have shown that $C(N) = \Omega(N^{-1/2}(\log N)^{-(v+1)/2})$ w.h.p., which completes the proof of the lower bound in Theorem 19.3.

Remark 19.5. The lower bound achieved by the cellular time-division scheme represents a vast improvement over time division with multihop, which, by comparison, achieves $C(N) = \Omega((\log N)^{-v/2}/N)$ w.h.p. This improvement is the result of spatial reuse of time (or frequency), which enables simultaneous transmission with relatively low interference due to the high path loss.

19.3.2 Proof of the Upper Bound

We prove the upper bound in Theorem 19.3, i.e., $C(N) = O(N^{-1/2+1/v} \log N)$ w.h.p. For a given random network, divide the square area of the network into two halves. Assume the case where there are at least $N/3$ sources on the left half and at least a third of them transmit to destinations on the right half. Since the locations of sources and destinations are chosen independently, it can be easily shown that the probability of this event tends to one as $N \to \infty$. We relabel the nodes so that these sources are $1, \ldots, N'$, where $N' \geq N/9$, and the corresponding destinations are $N + 1, \ldots, N + N'$.

By the cutset bound in Theorem 18.4, the symmetric capacity for these source nodes, and hence the symmetric capacity for all source nodes, are upper bounded by

$$\max_{F(x^N)} \frac{1}{N'} I(X^{N'}; Y^{N+N'}_{N+1} \mid X^N_{N'+1}) \leq \max_{F(x^{N'})} \frac{1}{N'} I(X^{N'}; \tilde{Y}^{N+N'}_{N+1}),$$

where $\tilde{Y}_k = \sum_{j=1}^{N'} g_{kj} X_j + Z_k$ for $k \in [N + 1 : N + N']$. Since the symmetric capacity of the original network is upper bounded by the symmetric capacity of these N' source–destination pairs, from this point on, we consider the subnetwork consisting only of these source–destination pairs and ignore the reception at the source nodes and the transmission at the destination nodes. To simplify the notation, we relabel N' as N and $\tilde{Y}_k$ as Y_k for $k \in [N + 1 : 2N]$, which does not affect the order of the upper bound since $N' = \Theta(N)$. Thus, each source node $j \in [1 : N]$ transmits X_j with the same average power constraint P and each destination node $k \in [N + 1 : 2N]$ receives

$$Y_k = \sum_{j=1}^{N} g_{kj} X_j + Z_k.$$

We upper bound $(1/N)I(X^N; Y^{2N}_{N+1})$ for this $2N$-user interference channel.

Let node j (source or destination) be at random location (U_j, V_j). We create an *augmented network* by adding $2N$ mirror nodes as depicted in Figure 19.9. For every destination node Y_j, $j \in [N + 1 : 2N]$, we add a sender node X_j at location $(-U_{N+j}, V_{N+j})$, and for every source node X_j, $j \in [1 : N]$, we add a receiver node Y_j at location $(-U_j, V_j)$.

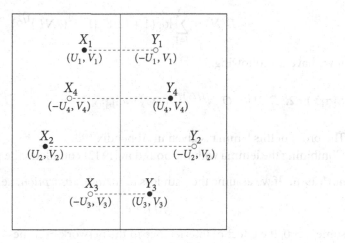

Figure 19.9. Augmented network.

The received vector of this augmented network is

$$Y^{2N} = GX^{2N} + Z^{2N},$$

where Z^{2N} is a vector of i.i.d. $N(0, 1)$ noise components. The gain matrix G is symmetric and $G_{jj} = (2|U_j|)^{-\nu/2}$. Furthermore, it can be shown that $G \succeq 0$ (for $\nu > 0$).

Now consider

$$NC(N) \leq \sup_{F(x^{2N}): \mathbb{E}|X_j^2| \leq P, j\in[1:2N]} I(X^N; Y_{N+1}^{2N}|X_{N+1}^{2N})^{\bullet}$$

$$\leq \sup_{F(x^{2N}): \mathbb{E}|X_j^2| \leq P, j\in[1:2N]} I(X^{2N}; Y^{2N})$$

$$\leq \max_{K_{\mathbf{X}} \succeq 0 : \text{tr}(K_{\mathbf{X}}) \leq 2NP} \frac{1}{2} \log |I + GK_{\mathbf{X}}G^T|$$

$$= \max_{P_j : \sum_{j=1}^{2N} P_j \leq 2NP} \frac{1}{2} \sum_{j=1}^{2N} \log(1 + P_j \lambda_j^2)$$

$$\leq \frac{1}{2} \sum_{j=1}^{2N} \log(1 + 2NP\lambda_j^2)$$

$$\leq \sum_{j=1}^{2N} \log(1 + (2NP)^{1/2}\lambda_j)$$

$$= \log |I_{2N} + (2NP)^{1/2}G|$$

$$\leq \sum_{j=1}^{2N} \log(1 + (2NP)^{1/2}G_{jj}), \tag{19.12}$$

where P_j and λ_j, $j \in [1:2N]$, are the eigenvalues of the positive semidefinite matrices $K_{\mathbf{X}}$ and G, respectively, and $G_{jj} = (2|U_j|)^{-\nu/2}$. Define

$$D(N) = \sum_{j=1}^{2N} \log(1 + (2|U_j|)^{-\nu/2}(2NP)^{1/2}).$$

Then we have the following.

Lemma 19.2. $D(N) = O(N^{1/2+1/\nu} \log N)$ w.h.p.

The proof of this lemma is given in Appendix 19B.

Combining the lemma with the bound in (19.12) completes the proof of Theorem 19.3.

Remark 19.6. If we assume the path loss to include *absorption*, i.e.,

$$g(r) = e^{-\gamma r/2} r^{-\nu/2}$$

for some $\gamma > 0$, the effect of interference in the network becomes more localized and the upper bound on $C(N)$ reduces to $O(N^{-1/2}(\log N)^2)$ w.h.p., which has roughly the same order as the lower bound.

SUMMARY

- Cutset bound for the Gaussian network

- Noisy network coding achieves within a constant gap of the cutset bound for Gaussian multimessage multicast networks

- Relaying plays a key role in combating high path loss and interference in large wireless networks

- Scaling laws for network capacity

- Random network model

- Cellular time-division scheme:

  - Outperforms time division with relaying via spatial reuse of time/frequency enabled by high path loss

  - Achieves close to the symmetric capacity order for most networks as the path loss exponent becomes large, and is order-optimal under the absorption model

- Use of network augmentation in the proof of the symmetric capacity upper bound

- **Open problems:**

  19.1. What is the capacity region of the Gaussian two-way relay channel with no direct links?

  19.2. What is the symmetric capacity scaling law for the random network model?

BIBLIOGRAPHIC NOTES

The noisy network coding inner bound on the capacity region of the Gaussian multimessage multicast network in (19.3) and the constant gap result in Theorem 19.2 were established by Lim, Kim, El Gamal, and Chung (2011). The Gaussian two-way relay channel with and without direct links was studied by Rankov and Wittneben (2006), Katti, Maric, Goldsmith, Katabi, and Médard (2007), Nam, Chung, and Lee (2010), and Lim, Kim, El Gamal, and Chung (2011, 2010). The layered noisy network coding scheme mentioned in Remark 19.2 was proposed by Lim, Kim, El Gamal, and Chung (2010), who showed that it can significantly improve the achievable rates over nonlayered noisy network coding.

The random network model was first introduced by Gupta and Kumar (2000). They analyzed the network under two *network theoretic* models for successful transmission, the signal-to-interference ratio (SIR) model and the *protocol* model. They roughly showed that the symmetric capacity under these models scales as $\Theta(N^{-1/2})$. Subsequent work under these network theoretic models include Grossglauser and Tse (2002) and El Gamal, Mammen, Prabhakar, and Shah (2006a,b).

The capacity scaling of a random network was first studied by Xie and Kumar (2004). Subsequent work includes Gastpar and Vetterli (2005) and Özgür, Lévêque, and Preissmann (2007). The lower bound in Theorem 19.3 is due to El Gamal, Mammen, Prabhakar, and Shah (2006a). This lower bound was improved to $C(N) = \Omega(N^{-1/2})$ w.h.p. by Franceschetti, Dousse, Tse, and Thiran (2007). The upper bound in Theorem 19.3 was established by Lévêque and Telatar (2005). Analysis of scaling laws based on physical limitations on electromagnetic wave propagation was studied in Franceschetti, Migliore, and Minero (2009) and Özgür, Lévêque, and Tse (2010).

PROBLEMS

19.1. Prove the cutset bound in Theorem 19.1.

19.2. Consider the Gaussian two-way relay channel in Section 19.1.2.

(a) Derive the cutset bound in (19.4), the decode–forward inner bound in (19.5), the amplify–forward inner bound in (19.6), and the noisy network coding inner bound in (19.7).

(b) Suppose that in the decode–forward coding scheme, node 3 uses network coding and broadcasts the modulo-2 sum of the binary sequence representations of M_1 and M_2, instead of (M_1, M_2), and nodes 1 and 2 find each other's message by first recovering the modulo-2 sum. Show that this modified coding scheme yields the lower bound

$$R_1 < \min\{C(S_{31}), C(S_{13}), C(S_{23})\},$$
$$R_2 < \min\{C(S_{32}), C(S_{13}), C(S_{23})\},$$
$$R_1 + R_2 < C(S_{31} + S_{32}).$$

Note that this bound is worse than the decode–forward lower bound when node 3 broadcasts (M_1, M_2)! Explain this surprising result.

(c) Let $g_{31} = g_{32} = 1$ and $g_{13} = g_{23} = 2$. Show that the gap between the decode–forward inner bound and the cutset bound is unbounded.

(d) Let $g_{31} = g_{13} = g_{23} = 1$ and $g_{32} = \sqrt{P}$. Show that the gap between the amplify–forward inner bound and the cutset bound is unbounded.

19.3. Consider the cellular time-division scheme in Section 19.3.1.

(a) Show that $P(\mathcal{E}_1)$ tends to zero as $n \to \infty$.

(b) Verify the upper bound on the total average interference power in (19.10).

19.4. *Capacity scaling of the N-user-pair Gaussian IC.* Consider the N-user-pair symmetric Gaussian interference channel

$$Y^N = GX^N + Z^N,$$

where the channel gain matrix is

$$G = \begin{bmatrix} 1 & a & \cdots & a \\ a & 1 & \cdots & a \\ \vdots & \vdots & \ddots & \vdots \\ a & a & \cdots & 1 \end{bmatrix}.$$

Assume average power constraint P on each sender. Denote the symmetric capacity by $C(N)$.

(a) Using time division with power control, show that the symmetric capacity is lower bounded as

$$C(N) \geq \frac{1}{N} C(NP).$$

(b) Show that the symmetric capacity is upper bounded as $C(N) \leq C(P)$. (Hint: Consider the case $a = 0$.)

(c) Tighten the bound in part (b) and show that

$$C(N) \leq \frac{1}{2N} \log |I + GG^T P|$$

$$= \frac{1}{2N} \log \left((1 + (a-1)^2 P)^{k-1} (1 + (a(N-1)+1)^2 P) \right).$$

(d) Show that when $a = 1$, the symmetric capacity is $C(N) = (1/N) C(P)$.

APPENDIX 19A   PROOF OF LEMMA 19.1

It is straightforward to show that the number of sources in each cell $S(N) = O(\log N)$ w.h.p. We now bound $L(N)$. Consider a *torus* with the same area and the same square cell division as the square area. For each S-D pair on the torus, send each packet along the four possible lines connecting them. Clearly for every configuration of nodes, each cell in the torus has at least as many S-D lines crossing it as in the original square. The reason we consider the torus is that the pmf of the number of lines in each cell becomes the same, which greatly simplifies the proof.

Let H_j be the total number of hops taken by packets traveling along one of the four lines between S-D pair j, $j \in [1:N]$. It is not difficult to see that the expected length of each path is $\Theta(N^{1/2})$. Since the hops are along cells having side-length $(\log N)^{1/2}$,

$$E(H_j) = \Theta((N/\log N)^{1/2}).$$

Fix a cell $c \in [1:N/\log N]$ and define E_{jc} to be the indicator of the event that a line between S-D pair $j \in [1:N]$ passes through cell $c \in [1:N/\log N]$, i.e.,

$$E_{jc} = \begin{cases} 1 & \text{if a hop of S-D pair } j \text{ is in cell } c, \\ 0 & \text{otherwise.} \end{cases}$$

Summing up the total number of hops in the cells in two different ways, we obtain

$$\sum_{j=1}^{N} \sum_{c=1}^{N/\log N} E_{jc} = \sum_{j=1}^{N} H_j.$$

Taking expectations on both sides and noting that the probabilities $P\{E_{jc} = 1\}$ are equal for all $j \in [1 : N]$ because of the symmetry on the torus, we obtain

$$\frac{N^2}{\log N} P\{E_{jc} = 1\} = N \, E(H_j),$$

or equivalently,

$$P\{E_{jc} = 1\} = \Theta((\log N/N)^{1/2})$$

for $j \in [1 : N]$ and $c \in [1 : N/\log N]$. Now for a fixed cell c, the total number of lines passing through it is $L_c = \sum_{j=1}^{N} E_{jc}$. This is the sum of N i.i.d. Bernoulli random variables since the positions of the nodes are independent and E_{jc} depends only on the positions of the source and destination nodes of S-D pair j. Moreover

$$E(L_c) = \sum_{j=1}^{N} P\{E_{jc} = 1\} = \Theta((N \log N)^{1/2})$$

for every cell c. Hence, by the Chernoff bound,

$$P\{L_c > (1 + \delta) \, E(L_c)\} \le \exp(- E(L_c)\delta^2/3).$$

Choosing $\delta = 2\sqrt{2 \log N / E(L_c)}$ yields

$$P\{L_c > (1 + \delta) \, E(L_c)\} \le 1/N^2.$$

Since $\delta = o(1)$, $L_c = O(E(L_c))$ with probability $\ge 1 - 1/N^2$. Finally using the union of events bound over $N/\log N$ cells shows that $L(N) = \max_{c \in [1:N/\log N]} L_c = O((N \log N)^{1/2})$ with probability $\ge 1 - 1/(N \log N)$ for sufficiently large N.

APPENDIX 19B   PROOF OF LEMMA 19.2

Define

$$W(N) = \log \left(1 + (2U)^{-\nu/2}(2NP)^{1/2}\right),$$

where $U \sim \text{Unif}[0, N^{1/2}]$. Since $D(N)$ is the sum of i.i.d. random variables, we have

$$E(D(N)) = N \, E(W(N)),$$
$$\text{Var}(D(N)) = N \, \text{Var}(W(N)).$$

We find upper and lower bounds on $E(W(N))$ and an upper bound on $E(W^2(N))$. For simplicity, we assume the natural logarithm here since we are only interested in order results. Let $a = e^{-\frac{v-1}{2}} P^{1/2}$, $v' = v/2$, $k = N^{1/2}$, and $u_0 = (ak)^{1/v'}$. Consider

$$k\,E(W(N)) = \int_0^k \log(1 + au^{-v'} k)\,du$$

$$= \int_0^1 \log(1 + au^{-v'} k)\,du + \int_1^{u_0} \log(1 + au^{-v'} k)\,du$$

$$+ \int_{u_0}^k \log(1 + au^{-v'} k)\,du \tag{19.13}$$

$$\leq \int_0^1 \log((1 + ak)u^{-v'})\,du + \int_1^{u_0} \log(1 + ak)\,du + \int_{u_0}^k au^{-v'} k\,du$$

$$= \log(1 + ak) + v' \int_0^1 \log(1/u)\,du + (u_0 - 1)\log(1 + ak)$$

$$+ \frac{ak}{v' - 1}(u_0^{-(v'-1)} - k^{-(v'-1)}).$$

Thus, there exists a constant $b_1 > 0$ such that for N sufficiently large,

$$E(W(N)) \leq b_1 N^{-1/2 + 1/v} \log N. \tag{19.14}$$

Now we establish a lower bound on $E(W(N))$. From (19.13), we have

$$k\,E(W(N)) \geq \int_1^{u_0} \log(1 + au^{-v'} k)\,du \geq \int_1^{u_0} \log(1 + a)\,du = (u_0 - 1)\log(1 + a).$$

Thus there exists a constant $b_2 > 0$ such that for N sufficiently large,

$$E(W(N)) \geq b_2 N^{-1/2 + 1/v}. \tag{19.15}$$

Next we find an upper bound on $E(W^2(N))$. Consider

$$k\,E(W^2(N)) = \int_0^1 (\log(1 + au^{-v'} k))^2\,du + \int_1^{u_0} (\log(1 + au^{-v'} k))^2\,du$$

$$+ \int_{u_0}^k (\log(1 + au^{-v'} k))^2\,du$$

$$\leq \int_0^1 (\log((1 + ak)u^{-v'}))^2\,du + \int_1^{u_0} (\log(1 + ak))^2\,du + \int_{u_0}^k a^2 u^{-2v'} k^2\,du$$

$$\leq (\log(1 + ak))^2 + (v')^2 \int_0^1 (\log(1/u))^2\,du$$

$$+ 2v' \log(1 + ak) \int_0^1 \log(1/u)\,du + (u_0 - 1)(\log(1 + ak))^2$$

$$+ \frac{a^2 k^2}{2v' - 1}\left(u_0^{-(2v'-1)} - k^{-(2v'-1)}\right).$$

Thus there exists a constant $b_3 > 0$ such that for N sufficiently large,

$$E(W^2(N)) \leq b_3 N^{-1/2+1/\nu}(\log N)^2. \tag{19.16}$$

Finally, using the Chebyshev lemma in Appendix B and substituting from (19.14), (19.15), and (19.16), then for N sufficiently large, we have

$$
\begin{aligned}
P\{D(N) \geq 2b_1 N^{1/2+1/\nu} \log N\} &\leq P\{D(N) \geq 2 E(D(N))\} \\
&\leq \frac{\text{Var}(D(N))}{(E[D(N)])^2} \\
&\leq \frac{N E(W^2(N))}{N^2(E[W(N)])^2} \\
&\leq \frac{b_3 N^{-1/2+1/\nu}(\log N)^2}{b_2^2 N^{2/\nu}} \\
&= \left(\frac{b_3}{b_2^2}\right) N^{-1/2-1/\nu}(\log N)^2,
\end{aligned}
$$

which tends to zero as $N \to \infty$. This completes the proof of the lemma.

CHAPTER 20

Compression over Graphical Networks

In this chapter, we study communication of correlated sources over networks represented by graphs, which include as special cases communication of independent messages over graphical networks studied in Chapter 15 and lossy and lossless source coding over noiseless links studied in Part II of the book. We first consider networks modeled by a directed acyclic graph. We show that the optimal coding scheme for communicating correlated sources losslessly to the same set of destination nodes (multisource multicast) is to perform separate Slepian–Wolf coding and linear network coding. For the lossy case, we consider the multiple description network in which a node observes a source and wishes to communicate it with prescribed distortions to other nodes in the network. We establish the optimal rate–distortion region for special classes of this network.

We then consider interactive communication over noiseless public broadcast channels. We establish the optimal rate region for multiway lossless source coding (coding for omniscience/CFO problem) in which each node has a source and the nodes communicate in several rounds over a noiseless public broadcast channel so that a subset of them can losslessly recover all the sources. Achievability is proved via independent rounds of Slepian–Wolf coding; hence interaction does not help in this case. Finally, we discuss the two-way lossy source coding problem. We establish the rate–distortion region for a given number of communication rounds, and show through an example that two rounds of communication can strictly outperform a single round. Hence, interaction can help reduce the rates in lossy source coding over graphical networks.

20.1 DISTRIBUTED LOSSLESS SOURCE–NETWORK CODING

In Chapter 15, we showed that linear network coding can achieve the cutset bound for graphical multimessage multicast networks. We extend this result to lossless source coding of correlated sources over graphical networks. We model an N-node network by a directed acyclic graph $\mathcal{G} = (\mathcal{N}, \mathcal{E})$, and assume that the nodes are ordered so that there is no path from node k to node j if $j < k$. Let $(X_1, \ldots, X_N)$ be an N-DMS. Node $j \in [1 : N - 1]$ observes the source X_j and wishes to communicate it losslessly to a set of destination

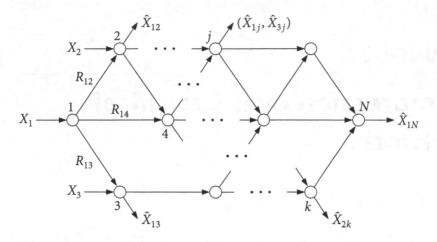

Figure 20.1. Graphical multisource network.

nodes $\mathcal{D}_j \subseteq [j+1:N]$ as depicted in Figure 20.1. The problem is to determine the set of achievable link rate tuples $(R_{jk} : (j,k) \in \mathcal{E})$.

A $((2^{nR_{jk}} : (j,k) \in \mathcal{E}), n)$ lossless source code for the multisource network $\mathcal{G} = (\mathcal{N}, \mathcal{E})$ is defined in a similar manner to the multimessage case. The probability of error is defined as

$$P_e^{(n)} = \mathsf{P}\{\hat{X}_{jk}^n \neq X_j^n \text{ for some } j \in [1:N-1], k \in \mathcal{D}_j\}.$$

A rate tuple $(R_{jk} : (j,k) \in \mathcal{E})$ is said to be achievable if there exists a sequence of $((2^{nR_{jk}} : (j,k) \in \mathcal{E}), n)$ codes such that $\lim_{n\to\infty} P_e^{(n)} = 0$. The optimal rate region $\mathscr{R}^*$ is the closure of the set of achievable rates tuples.

Since this problem is a generalization of the multiple independent message case, the optimal rate region is not known in general. The cutset bound in Theorem 15.4 can be extended to the following.

Theorem 20.1 (Cutset Bound for the Graphical Multisource Network). If the rate tuple $(R_{jk} : (j,k) \in \mathcal{E})$ is achievable, then it must satisfy the inequality

$$R(\mathcal{S}) \geq H(X(\mathcal{J}(\mathcal{S})) \,|\, X(\mathcal{S}^c))$$

for all $\mathcal{S} \subset [1:N]$ such that $\mathcal{S}^c \cap \mathcal{D}_j \neq \emptyset$ for some $j \in [1:N-1]$, where

$$R(\mathcal{S}) = \sum_{(j,k):j\in\mathcal{S},k\in\mathcal{S}^c} R_{jk}$$

and $\mathcal{J}(\mathcal{S}) = \{j \in \mathcal{S} : \mathcal{S}^c \cap \mathcal{D}_j \neq \emptyset\}$.

The cutset outer bound is tight when the sources are to be sent to the same set of destination nodes. Suppose that $X_{k+1} = \cdots = X_N = \emptyset$ and $\mathcal{D}_1 = \cdots = \mathcal{D}_k = \mathcal{D}$.

> **Theorem 20.2.** The optimal rate region $\mathscr{R}^*$ for lossless multisource multicast coding of a k-DMS $(X_1, \ldots, X_k)$ is the set of rate tuples $(R_{jl} : (j, l) \in \mathcal{E})$ such that
>
> $$R(\mathcal{S}) \geq H(X(\mathcal{S} \cap [1 : k]) \mid X(\mathcal{S}^c \cap [1 : k]))$$
>
> for all $\mathcal{S} \subset [1 : N]$ such that $\mathcal{S}^c \cap \mathcal{D} \neq \emptyset$.

The converse of this theorem follows by the cutset bound. The proof of achievability follows by separate Slepian–Wolf coding of the sources at rates $R_1, \ldots, R_k$ and linear network coding to send the bin indices.

Theorems 20.1 and 20.2 can be easily extended to wireless networks with cycles modeled by a directed cyclic hypergraph. This is illustrated in the following.

Example 20.1 (Source coding with a two-way relay). Consider the noiseless two-way relay channel depicted in Figure 20.2, which is the source coding counterpart of the DM and Gaussian two-way relay channels in Problem 18.8 and Section 19.1.2, respectively.

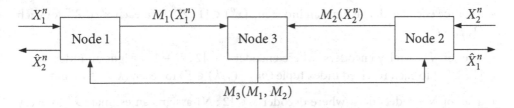

Figure 20.2. Source coding over a noiseless two-way relay channel.

Let (X_1, X_2) be a 2-DMS. Node 1 observes X_1 and node 2 observes X_2. Communication is performed in two rounds. In the first round, node $j = 1, 2$ encodes X_j^n into an index M_j at rate R_j and sends it to the relay node 3. Node 3 then encodes the index pair (M_1, M_2) into an index M_3 at rate R_3 and broadcasts it to both source nodes. Upon receiving M_3, nodes 1 and 2 find the estimates $\hat{X}_2^n$ and $\hat{X}_1^n$, respectively, based on their respective sources and the index M_3. By the extension of Theorem 20.2 to noiseless wireless networks, the optimal rate region is the set of rate triples (R_1, R_2, R_3) such that

$$R_1 \geq H(X_1 \mid X_2),$$
$$R_2 \geq H(X_2 \mid X_1),$$
$$R_3 \geq \max\{H(X_1 \mid X_2), H(X_2 \mid X_1)\}.$$

The converse follows by a simple cutset argument. To prove achievability, we use Slepian–Wolf coding and linear network coding. Node 1 sends the nR_1-bit bin index M_1 of X_1^n and node 2 sends the nR_2-bit bin index M_2 of X_2^n. Assume without loss of generality that $R_1 \geq R_2$. The relay pads the binary sequence representation of M_2 with zeros to obtain the nR_1-bit index M_2'. It then broadcasts the modulo-2 sum $M_3 = M_1 \oplus M_2'$. Upon receiving M_3, each node decodes the other node's index to recover its source. Following the achievability proof of the Slepian–Wolf theorem, the probability of error tends to zero as $n \to \infty$ if $R_1 > H(X_1|X_2) + \delta(\epsilon)$, $R_2 > H(X_2|X_1) + \delta(\epsilon)$ and $R_3 \geq \max\{R_1, R_2\}$.

20.2   MULTIPLE DESCRIPTION NETWORK CODING

The optimal rate–distortion region for lossy source coding over graphical networks is not known even for the single-hop case discussed in Chapter 12. Hence, we focus our discussion on the multiple description network, which extends multiple description coding to graphical networks and is well motivated by multimedia distribution over communication networks as mentioned in the introduction of Chapter 13. Consider an N-node communication network modeled by a directed acyclic graph $\mathcal{G} = (\mathcal{N}, \mathcal{E})$. Let X be a DMS and $d_j(x, \hat{x}_j)$, $j \in [2:N]$, be a set of distortion measures. Source node 1 observes the source X and node $j \in [2:N]$ wishes to reconstruct it with a prescribed distortion D_j. We wish to find the set of achievable link rates.

A $((2^{nR_{jk}} : (j, k) \in \mathcal{E}), n)$ multiple description code for the graphical network $\mathcal{G} = (\mathcal{N}, \mathcal{E})$ consists of

- a source encoder that assigns an index $m_{1j}(x^n) \in [1 : 2^{nR_{1j}}]$ to each $x^n \in \mathcal{X}^n$ for each $(1, j) \in \mathcal{E}$,

- a set of $N - 2$ relay encoders, where encoder $k \in [2 : N - 1]$ assigns an index $m_{kl} \in [1 : 2^{nR_{kl}}]$ to each received index tuple $(m_{jk} : (j, k) \in \mathcal{E})$ for each $(k, l) \in \mathcal{E}$, and

- a set of $N - 1$ decoders, where decoder $k \in [2 : N]$ assigns an estimate $\hat{x}_j^n$ to every received index tuple $(m_{jk} : (j, k) \in \mathcal{E})$.

A rate–distortion tuple $((R_{jk} : (j, k) \in \mathcal{E}), D_2, \ldots, D_N)$ is said to be achievable if there exists a sequence of $((2^{nR_{jk}} : (j, k) \in \mathcal{E}), n)$ codes such that

$$\limsup_{n \to \infty} \mathsf{E}(d_j(X^n, \hat{X}_j^n)) \leq D_j, \quad j \in [2:N].$$

The rate–distortion region $\mathcal{R}(D_2, \ldots, D_N)$ for the multiple description network is the closure of the set of rate tuples $(R_{jk} : (j, k) \in \mathcal{E})$ such that $((R_{jk} : (j, k) \in \mathcal{E}), D_2, \ldots, D_N)$ is achievable.

It is easy to establish the following cutset outer bound on the rate–distortion region.

Theorem 20.3 (Cutset Bound for the Multiple Description Network). If the rate–distortion tuple $((R_{jk} : (j, k) \in \mathcal{E}), D_2, \ldots, D_N)$ is achievable, then it must satisfy the inequality

$$R(\mathcal{S}) \geq I(X; \hat{X}(\mathcal{S}^c))$$

for all $\mathcal{S} \subset [1 : N]$ such that $1 \in \mathcal{S}$ for some conditional pmf $p(\hat{x}_2^N | x)$ that satisfies the constraints $\mathsf{E}(d_j(X, \hat{X}_j)) \leq D_j$, $j \in [2:N]$.

If d_j, $j \in [2 : N]$, are Hamming distortion measures and $D_j = 0$ or 1, then the problem reduces to communication of a single message at rate $H(X)$ over the graphical multicast network in Chapter 15 with the set of destination nodes $\mathcal{D} = \{j \in [2:N]: D_j = 0\}$. The rate–distortion region is the set of edge rate tuples such that the resulting min-cut capacity C (minimized over all destination nodes) satisfies the condition $H(X) \leq C$.

The rate–distortion region for the multiple description network is not known in general. In the following, we present a few nontrivial special cases for which the optimal rate–distortion tradeoff is known.

20.2.1 Cascade Multiple Description Network

The *cascade* multiple description network is a special case of the multiple description network, where the graph has a line topology, i.e., $\mathcal{E} = \{(j-1, j): j \in [2:N]\}$. The rate–distortion region for this network is known.

Theorem 20.4. The rate–distortion region for a cascade multiple description network with distortion tuple $(D_2, \ldots, D_N)$ is the set of rate tuples $(R_{j-1,j}: j \in [2:N])$ such that

$$R_{j-1,j} \geq I(X; \hat{X}([j:N])), \quad j \in [2:N],$$

for some conditional pmf $p(\hat{x}_2^N | x)$ that satisfies the constraints $\mathsf{E}(d_j(X, \hat{X}_j)) \leq D_j, j \in [2:N]$.

For example, consider the 3-node cascade multiple description network depicted in Figure 20.3. By Theorem 20.4, the rate–distortion region with distortion pair (D_2, D_3) is the set of rate pairs (R_{12}, R_{23}) such that

$$R_{12} \geq I(X; \hat{X}_2, \hat{X}_3),$$
$$R_{23} \geq I(X; \hat{X}_3)$$

for some conditional pmf $p(\hat{x}_2, \hat{x}_3 | x)$ that satisfies the constraints $\mathsf{E}(d_j(X, \hat{X}_j)) \leq D_j, j = 2, 3$.

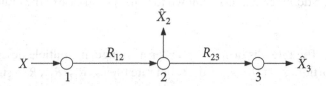

Figure 20.3. Three-node cascade multiple description network.

Proof of Theorem 20.4. The proof of the converse follows immediately by relaxing the cutset bound in Theorem 20.3 only to the cuts $(\mathcal{S}, \mathcal{S}^c) = ([1:j-1], [j:N]), j \in [2:N]$.

To prove achievability, fix the conditional pmf $p(\hat{x}_2^N | x)$ that satisfies the distortion constraints (with a factor $1/(1 + \epsilon)$) as in the achievability proofs for single-hop lossy source coding problems). We generate the codebook sequentially for $j = N, N - 1, \dots, 2$. For each $(\tilde{m}_{j+1}, \dots, \tilde{m}_N)$, randomly and independently generate $2^{n\tilde{R}_j}$ sequences $\hat{x}_j^n(\tilde{m}_j, \dots, \tilde{m}_N)$, $\tilde{m}_j \in [1 : 2^{n\tilde{R}_j}]$, each according to $\prod_{i=1}^{n} p_{\hat{X}_j | \hat{X}_{j+1}^N}(\hat{x}_{ji} | \hat{x}_{j+1,i}(\tilde{m}_{j+1}), \dots, \hat{x}_{Ni}(\tilde{m}_N))$. By the covering lemma and the typical average lemma, if

$$\tilde{R}_j > I(X; \hat{X}_j | \hat{X}_{j+1}^N) + \delta(\epsilon) \quad \text{for all } j \in [2 : N],$$

then there exists a tuple $(\hat{x}_2^n(\tilde{m}_2^N), \hat{x}_3^n(\tilde{m}_3^N), \dots, \hat{x}_N^n(\tilde{m}_N))$ that satisfies the distortion constraints asymptotically. Letting

$$R_{j-1,j} = \tilde{R}_j + \tilde{R}_{j+1} + \dots + \tilde{R}_N, \quad j \in [2 : N],$$

and using Fourier–Motzkin elimination, we obtain the desired inequalities

$$R_{j-1,j} > I(X; \hat{X}_j^N) + \delta(\epsilon), \quad j \in [2 : N].$$

This completes the proof of Theorem 20.4.

20.2.2 Triangular Multiple Description Network

Consider the 3-node triangular multiple description network depicted in Figure 20.4.

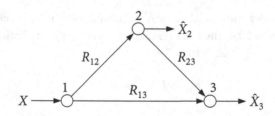

Figure 20.4. Triangular multiple description network.

The rate–distortion region is again known for this special case of the multiple description network.

Theorem 20.5. The rate–distortion region for a triangular multiple description network with distortion pair (D_2, D_3) is the set of rate tuples (R_{12}, R_{13}, R_{23}) such that

$$R_{23} \geq I(U; X),$$
$$R_{12} \geq I(U, \hat{X}_2; X),$$
$$R_{13} \geq I(\hat{X}_3; X | U)$$

for some conditional pmf $p(u|x)p(\hat{x}_2|u, x)p(\hat{x}_3|u, x)$ that satisfies the constraints $E(d_j(X, \hat{X}_j)) \leq D_j, j = 2, 3$.

To illustrate this theorem, consider the following.

Example 20.2. Let X be a WGN(P) source and d_2, d_3 be squared error distortion measures. The rate–distortion region for distortion pair (D_2, D_3) is the set of rate tuples (R_{12}, R_{13}, R_{23}) such that

$$R_{12} \geq \mathsf{R}\left(\frac{P}{D_2}\right),$$

$$R_{12} + R_{13} \geq \mathsf{R}\left(\frac{P}{D_3}\right),$$

$$R_{13} + R_{23} \geq \mathsf{R}\left(\frac{P}{D_3}\right).$$

The proof of the converse follows immediately by applying the cutset bound in Theorem 20.3 to the quadratic Gaussian case. The proof of achievability follows by the Gaussian successive refinement coding in Example 13.3. This rate–distortion region can be equivalently represented as the set of rate tuples (R_{12}, R_{13}, R_{23}) such that

$$R_{12} \geq \mathsf{R}\left(\frac{P}{D_2}\right),$$

$$R_{13} \geq \mathsf{R}\left(\frac{D'}{D_3}\right),$$

$$R_{23} \geq \mathsf{R}\left(\frac{P}{D'}\right)$$

for some $D' \geq \max\{D_2, D_3\}$. This equivalent characterization follows by evaluating Theorem 20.5 directly using the Gaussian test channels for the successive refinement scheme in Section 13.5.

Proof of Theorem 20.5. To prove achievability, we first generate a description U of X at rate $\tilde{R}_3 > I(U; X)$. Conditioned on each description U, we generate reconstructions $\hat{X}_2$ and $\hat{X}_3$ of X at rates $\tilde{R}_2 > I(X; \hat{X}_2|U)$ and $R_{13} > I(X; \hat{X}_3|U)$, respectively. By substituting $R_{12} = \tilde{R}_2 + \tilde{R}_3$ and $R_{23} = \tilde{R}_3$, and using the Fourier–Motzkin elimination procedure in Appendix D, the inequalities in the theorem guarantee the achievability of the desired distortions.

To prove the converse, we use the auxiliary random variable identification $U_i = (M_{23}, X^{i-1})$. First consider

$$nR_{23} \geq H(M_{23})$$

$$= \sum_{i=1}^{n} I(M_{23}, X^{i-1}; X_i)$$

$$= \sum_{i=1}^{n} I(U_i; X_i)$$

$$= nI(U_Q, Q; X_Q)$$

$$= nI(U; X),$$

where Q is a time-sharing random variable, $U = (U_Q, Q)$, and $X = X_Q$. Next, since M_{23} is a function of M_{12}, we have

$$
\begin{aligned}
nR_{12} &\geq H(M_{12}) \\
&\geq H(M_{12}|M_{23}) \\
&\geq I(M_{12}; X^n|M_{23}) \\
&\geq \sum_{i=1}^{n} I(\hat{X}_{2i}; X_i|M_{23}, X^{i-1}) \\
&= nI(\hat{X}_2; X|U),
\end{aligned}
$$

which depends only on $p(x)p(u|x)p(\hat{x}_2|u, x)$. Similarly,

$$
nR_{13} \geq nI(\hat{X}_3; X|U),
$$

which depends only on $p(x)p(u|x)p(\hat{x}_3|u, x)$. Even though as defined, $\hat{X}_2$ and $\hat{X}_3$ are not in general conditionally independent given (U, X), we can assume without loss of generality that they are, since none of the mutual information and distortion terms in the theorem is a function of the joint conditional pmf of $(\hat{X}_2, \hat{X}_3)$ given (U, X). This completes the proof of the theorem.

Remark 20.1. Theorems 20.4 and 20.5 can be extended to the network with edge set

$$
\mathcal{E} = \{(1, 2), (2, 3), (3, 4), \ldots, (N - 1, N), (1, N)\}.
$$

In this case, the rate–distortion region for distortion tuple $(D_2, D_3, \ldots, D_N)$ is the set of rate tuples $(R_{12}, R_{23}, R_{34}, \ldots, R_{N-1,N}, R_{1N})$ such that

$$
\begin{aligned}
R_{N-1,N} &\geq I(X; U), \\
R_{j-1,j} &\geq I(X; \hat{X}_j^{N-1}, U), \quad j \in [2 : N - 1], \\
R_{1N} &\geq I(X; \hat{X}_N|U)
\end{aligned} \tag{20.1}
$$

for some conditional pmf $p(u|x)p(\hat{x}_2^{N-1}|u, x)p(\hat{x}_N|u, x)$ such that $\mathsf{E}(d_j(X, \hat{X}_j)) \leq D_j$, $j \in [2 : N]$.

Remark 20.2. The multiple description network can be generalized by considering wireless networks modeled by a hypergraph or by adding correlated side information at various nodes; see Problems 20.9 and 20.11.

20.3   INTERACTIVE SOURCE CODING

As we showed in Chapter 17, interaction among nodes can increase the capacity of multi-user networks. Does interactive communication also help reduce the rates needed for encoding correlated sources? We investigate this question by considering source coding over noiseless public broadcast channels.

20.3.1 Two-Way Lossless Source Coding

Consider the two-way lossless source coding setting depicted in Figure 20.5. Let (X_1, X_2) be a 2-DMS and assume a 2-node communication network in which node 1 observes X_1 and node 2 observes X_2. The nodes interactively communicate over a noiseless bi-directional link so that each node can losslessly recover the source observed by the other node. We wish to determine the number of communication rounds needed and the set of achievable rates for this number of rounds.

Assume without loss of generality that node 1 sends the first index and that the number of rounds of communication q is even. A $(2^{nr_1}, \ldots, 2^{nr_q}, n)$ code for two-way lossless source coding consists of

- two encoders, one for each node, where in round $l_j \in \{j, j+2, \ldots, q-2+j\}$, encoder $j = 1, 2$ sends an index $m_{l_j}(x_j^n, m^{l_j-1}) \in [1 : 2^{nr_{l_j}}]$, that is, a function of its source vector and all previously transmitted indices, and

- two decoders, one for each node, where decoder 1 assigns an estimate $\hat{x}_2^n$ to each received index and source sequence pair (m^q, x_1^n) and decoder 2 assigns an estimate $\hat{x}_1^n$ to each index and source sequence pair (m^q, x_2^n).

The probability of error is defined as $P_e^{(n)} = P\{(\hat{X}_1^n, \hat{X}_2^n) \neq (X_1^n, X_2^n)\}$. The total transmission rate for node $j = 1, 2$ is defined as $R_j = \sum_{l_j=j,j+2,\ldots,q-2+j} r_{l_j}$. A rate pair (R_1, R_2) is said to be achievable if there exists a sequence of $(2^{nr_1}, \ldots, 2^{nr_q}, n)$ codes for some q such that $\lim_{n\to\infty} P_e^{(n)} = 0$ and $R_j = \sum_{l_j=j,j+2,\ldots,q-2+j} r_{l_j}$. The optimal rate region $\mathcal{R}^*$ for two-way lossless source coding is the closure of the set of achievable rate pairs (R_1, R_2).

The optimal rate region for two-way lossless source coding is easy to establish. By the Slepian–Wolf theorem, we know that $R_1 > H(X_1|X_2)$ and $R_2 > H(X_2|X_1)$ are sufficient and can be achieved in two independent communication rounds. It is also straightforward to show that this set of rate pairs is necessary. Hence, interaction does not help in this lossless source coding setting.

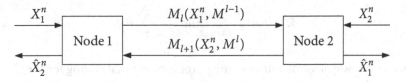

Figure 20.5. Two-way lossless source coding.

Remark 20.3. For error-free compression, it can be shown that

$$R_1 \geq H(X_1),$$
$$R_2 \geq H(X_2|X_1)$$

is necessary in some cases, for example, when $p(x, y) > 0$ for all $(x, y) \in \mathcal{X} \times \mathcal{Y}$. By contrast, without interaction, $R_1 \geq H(X_1)$ and $R_2 \geq H(X_2)$ are sometimes necessary. Hence interaction can help in error-free compression.

20.3.2 CFO Problem

We now consider the more general problem of multiway lossless source coding of an N-DMS $(X_1, \ldots, X_N)$ over an N-node wireless network, where node $j \in [1 : N]$ observes the source X_j. The nodes interactively communicate over a noiseless public broadcast channel so that each node in some subset $\mathcal{A} \subseteq [1 : N]$ can losslessly recover all the sources. This problem is referred to as *communication for omniscience* (CFO).

As for the 2-node special case in the previous subsection, we assume that the nodes communicate in a round-robin fashion in q rounds, where q is divisible by N and can depend on the code block length. Node $j \in [1 : N]$ broadcasts in rounds $j, N + j, \ldots, q - N + j$. Note that this "protocol" can achieve any other node communication order by having some of the nodes send a null message in some of their allotted rounds.

A $(2^{nr_1}, \ldots, 2^{nr_q}, n)$ code for the CFO problem consists of

- a set of encoders, where in round $l_j \in \{j, N + j, \ldots, q - N + j\}$, encoder $j \in [1 : N]$ sends an index $m_{l_j}(x_j^n, m^{l_j - 1}) \in [1 : 2^{nr_{l_j}}]$, that is, a function of its source vector and all previously transmitted indices, and

- a set of decoders, where decoder $k \in \mathcal{A}$ assigns an estimate $(\hat{x}_{1k}^n, \ldots, \hat{x}_{Nk}^n)$ to each index sequence m^q and source sequence $x_k^n \in \mathcal{X}_k^n$.

The probability of error is defined as

$$P_e^{(n)} = \mathsf{P}\{(\hat{X}_{1k}^n, \ldots, \hat{X}_{Nk}^n) \neq (X_1^n, \ldots, X_N^n) \text{ for some } k \in \mathcal{A}\}.$$

The total transmission rate for node j is defined as $R_j = \sum_{l_j = j, j+N, \ldots, q-N+j} r_{l_j}$. A rate tuple $(R_1, \ldots, R_N)$ is said to be achievable if there exists a sequence of $(2^{nr_1}, \ldots, 2^{nr_q}, n)$ codes for some q such that $\lim_{n \to \infty} P_e^{(n)} = 0$ and $R_j = \sum_{l_j = j, j+N, \ldots, q-N+j} r_{l_j}$. The optimal rate region $\mathscr{R}^*(\mathcal{A})$ for the CFO problem is the closure of the set of achievable rate tuples $(R_1, \ldots, R_N)$. In some cases, we are interested in the minimum sum-rate

$$R_{\text{CFO}}(\mathcal{A}) = \min_{(R_1, \ldots, R_N) \in \mathscr{R}^*(\mathcal{A})} \sum_{j=1}^N R_j.$$

As in the two-way lossless source coding problem, optimal coding for the CFO problem involves N rounds of Slepian–Wolf coding.

Theorem 20.6. The optimal rate region $\mathscr{R}^*(\mathcal{A})$ for the CFO problem is the set of rate tuples $(R_1, R_2, \ldots, R_N)$ such that

$$\sum_{j \in \mathcal{S}} R_j \geq H(X(\mathcal{S}) | X(\mathcal{S}^c))$$

for every subset $\mathcal{S} \subseteq [1 : N] \setminus \{k\}$ and every $k \in \mathcal{A}$.

For $N = 2$ and $\mathcal{A} = \{1, 2\}$, the optimal rate region is the set of rate pairs (R_1, R_2) such that

$$R_1 \geq H(X_1|X_2),$$
$$R_2 \geq H(X_2|X_1),$$

and the minimum sum-rate is $R_{\mathrm{CFO}}(\{1, 2\}) = H(X_1|X_2) + H(X_2|X_1)$, which recovers the two-way case in the previous subsection. Note that the optimal rate region for the set $\mathcal{A}$ is the intersection of the optimal rates for the sets $\{k\}$, $k \in \mathcal{A}$, i.e., $\mathscr{R}^*(\mathcal{A}) = \bigcap_{k \in \mathcal{A}} \mathscr{R}^*(\{k\})$.

Example 20.3. Let $N = 3$ and $\mathcal{A} = \{1, 2\}$. The optimal rate region is the set of rate triples (R_1, R_2, R_3) such that

$$R_1 \geq H(X_1|X_2, X_3),$$
$$R_2 \geq H(X_2|X_1, X_3),$$
$$R_3 \geq H(X_3|X_1, X_2),$$
$$R_1 + R_3 \geq H(X_1, X_3|X_2),$$
$$R_2 + R_3 \geq H(X_2, X_3|X_1).$$

Using Fourier–Motzkin elimination in Appendix D, we can show that

$$R_{\mathrm{CFO}}(\{1, 2\}) = H(X_1|X_2) + H(X_2|X_1) + \max\{H(X_3|X_1), H(X_3|X_2)\}.$$

Remark 20.4. The minimum sum-rate $R_{\mathrm{CFO}}(\mathcal{A})$ can be computed efficiently using duality results from linear programming. In particular, a maximum of N constraints, as compared to 2^N for naive computation, are required.

20.3.3 Two-Way Lossy Source Coding

We now turn our attention to interactive lossy source coding; again see Figure 20.5. Let (X_1, X_2) be a 2-DMS and d_1, d_2 be two distortion measures. Node 1 observes X_1 and node 2 observes X_2. The nodes communicate in rounds over a noiseless bidirectional link so that each node reconstruct the source at the other node with a prescribed distortion. We wish to find the optimal number of communication rounds and the rate–distortion region for this number of rounds.

We again assume that node 1 sends the first index and the number of rounds q is even. A $(2^{nr_1}, \ldots, 2^{nr_q}, n)$ code is defined as for the lossless case. A rate–distortion quadruple (R_1, R_2, D_1, D_2) is said to be achievable if there exists a sequence of $(2^{nr_1}, \ldots, 2^{nr_q}, n)$ codes such that $\limsup_{n \to \infty} \mathsf{E}(d_j(X_j^n, \hat{X}_j^n)) \leq D_j$ and $R_j = \sum_{l_j = j, j+2, \ldots, q-2+j} r_{l_j}$, $j = 1, 2$. The q-round rate–distortion region $\mathscr{R}^{(q)}(D_1, D_2)$ is the closure of the set of rate pairs (R_1, R_2) such that (R_1, R_2, D_1, D_2) is achievable in q rounds. The rate–distortion region $\mathscr{R}(D_1, D_2)$ is the closure of the set of rate pairs (R_1, R_2) such that (R_1, R_2, D_1, D_2) is achievable for some q.

The rate–distortion region $\mathscr{R}(D_1, D_2)$ is not known in general. We first consider simple inner and outer bounds.

Simple inner bound. Consider the inner bound on the rate–distortion region for two-way lossy source coding obtained by having each node perform an independent round of Wyner–Ziv coding considering the source at the other node as side information. This simple scheme, which requires only two rounds of one-way communication, yields the inner bound consisting of rate pairs (R_1, R_2) such that

$$R_1 > I(X_1; U_1 | X_2),$$
$$R_2 > I(X_2; U_2 | X_1) \tag{20.2}$$

for some conditional pmf $p(u_1 | x_1) p(u_2 | x_2)$ and functions $\hat{x}_1(u_1, x_2)$ and $\hat{x}_2(u_2, x_1)$ that satisfy the constraints $E(d_j(X_j, \hat{X}_j)) \leq D_j$, $j = 1, 2$.

Simple outer bound. Even if each encoder (but not the decoders) has access to the source observed by the other node, the rate pair (R_1, R_2) must satisfy the inequalities

$$R_1 \geq I(X_1; \hat{X}_1 | X_2),$$
$$R_2 \geq I(X_2; \hat{X}_2 | X_1) \tag{20.3}$$

for some conditional pmf $p(\hat{x}_1 | x_1, x_2) p(\hat{x}_2 | x_1, x_2)$ that satisfies $E(d_j(X_j, \hat{X}_j)) \leq D_j$, $j = 1, 2$.

These bounds can be sometimes tight.

Example 20.4. Let (X_1, X_2) be a 2-WGN(P, ρ) source and assume squared error distortion measures. The inner and outer bounds in (20.2) and (20.3) coincide, since in this case the Wyner–Ziv rate is the same as the rate when the side information is available at both the encoder and the decoder. Hence, the rate–distortion region $\mathscr{R}(D_1, D_2)$ is the set of rate pairs (R_1, R_2) such that $R_1 \geq R((1 - \rho^2)P/D_1)$ and $R_2 \geq R((1 - \rho^2)P/D_2)$, and is achieved by two independent rounds of Wyner–Ziv coding.

The inner and outer bounds in (20.2) and (20.3), however, are not tight in general, and interactive communication is needed to achieve the rate–distortion region. To reduce the rates, the two nodes interactively build additional correlation between their respective knowledge of the sources instead of using two independent rounds of Wyner–Ziv coding.

Theorem 20.7. The q-round rate–distortion region $\mathscr{R}^{(q)}(D_1, D_2)$ for two-way lossy source coding is the set of rate pairs (R_1, R_2) such that

$$R_1 \geq I(X_1; U^q | X_2),$$
$$R_2 \geq I(X_2; U^q | X_1)$$

for some conditional pmf $\prod_{l=1}^{q} p(u_l | u^{l-1}, x_{j_l})$ with $|\mathcal{U}_l| \leq |\mathcal{X}_{j_l}| \cdot (\prod_{j=1}^{l-1} |\mathcal{U}_j|) + 1$ and functions $\hat{x}_1(u^q, x_2)$ and $\hat{x}_2(u^q, x_1)$ that satisfy the constraints $E(d_j(X_j, \hat{X}_j)) \leq D_j$, $j = 1, 2$, where $j_l = 1$ if l is odd and $j_l = 2$ if l is even.

Note that the rate–distortion region in this theorem is computable for any given number of rounds q. However, there is no bound on q that holds in general; hence this theorem does not yield a computable characterization of the rate–distortion region $\mathcal{R}(D_1, D_2)$. The following example shows that interactive communication can achieve a strictly lower sum-rate than one round of Wyner–Ziv coding.

Example 20.5. Let (X_1, X_2) be a DSBS(p), $d_2 \equiv 0$ (that is, there is no distortion constraint on reconstructing X_2), and

$$d_1(x_1, \hat{x}_1) = \begin{cases} 0 & \text{if } \hat{x}_1 = x_1, \\ 1 & \text{if } \hat{x}_1 = e, \\ \infty & \text{if } \hat{x}_1 = 1 - x_1. \end{cases}$$

With one-way communication from node 1 to node 2, it can be easily shown that the optimal Wyner–Ziv rate is $R^{(1)} = (1 - D_1)H(p)$.

Now consider the rate–distortion region in Theorem 20.7 for two rounds of communication, i.e., $q = 2$. We set the conditional pmf $p(u_1|x_2)$ as a BSC(α), the conditional pmf $p(u_2|x_1, u_1)$ as in Table 20.1, and the function $\hat{x}_1(u_2, u_1, x_2) = u_2$. Evaluating the sum-rate for $p = 0.03$, $\alpha = 0.3$, and $\beta = 0.3$, we obtain

$$R^{(2)}_{\text{sum}} = I(X_2; U_1|X_1) + I(U_2; X_1|X_2, U_1) = 0.0858,$$

and the corresponding distortion is $E(d_1(X_1, \hat{X}_1)) = 0.5184$. By comparison, the one-way rate for this distortion is $R^{(1)} = 0.0936$. Hence, interactive communication can reduce the rate.

(x_1, u_1) \ u_2	0	e	1
$(0, 0)$	$1 - \beta$	β	0
$(0, 1)$	0	1	0
$(1, 0)$	0	1	0
$(1, 1)$	0	β	$1 - \beta$

Table 20.1. Test channel $p(u_2|x_1, u_1)$ for Example 20.5.

Proof of Theorem 20.7. Achievability is established by performing Wyner–Ziv coding in each round. Fix q and a joint pmf $p(u^q|x_1, x_2)$ and reconstruction functions of the forms specified in the theorem. In odd rounds $l \in [1 : q]$, node 1 sends the bin index of the description U_l of X_1 given U^{l-1} to node 2 at rate $r_{1l} > I(X_1; U_l|U^{l-1}, X_2)$, and in even rounds node 2 sends the index of the description U_l of X_2 at rate $r_{2l} > I(X_2; U_l|U^{l-1}, X_1)$. Summing up the rates for each node establishes the required bounds on $R_1 = \sum_{l \text{ odd}} r_{1l}$

and $R_2 = \sum_{l \text{ even}} r_{2l}$. At the end of the q rounds, node 1 forms the estimate $\hat{x}_2(U^q, X_1)$ of X_2 and node 2 forms the estimate $\hat{x}_1(U^q, X_2)$ of X_1. The details follow the achievability proof of the Wyner–Ziv theorem in Section 11.3.

The proof of the converse involves careful identification of the auxiliary random variables. Consider

$$nR_1 \geq \sum_{l \text{ odd}}^{q-1} H(M_l)$$

$$\geq H(M_1, M_3, \ldots, M_{q-1})$$

$$\geq I(X_1^n; M_1, M_3, \ldots, M_{q-1} | X_2^n)$$

$$= H(X_1^n | X_2^n) - H(X_1^n | X_2^n, M_1, M_3, \ldots, M_{q-1})$$

$$\overset{(a)}{=} H(X_1^n | X_2^n) - H(X_1^n | X_2^n, M^q)$$

$$\geq \sum_{i=1}^{n} \left(H(X_{1i} | X_{2i}) - H(X_{1i} | X_{2i}, X_{2,i+1}^n, X_1^{i-1}, M^q) \right)$$

$$= \sum_{i=1}^{n} I(X_{1i}; X_{2,i+1}^n, X_1^{i-1}, M^q | X_{2i})$$

$$\overset{(b)}{=} \sum_{i=1}^{n} I(X_{1i}; U_{1i}, \ldots, U_{qi} | X_{2i}),$$

where (a) follows since $(M_2, M_4, \ldots, M_q)$ is a function of $(M_1, M_3, \ldots, M_{q-1}, X_2^n)$ and (b) follows by the auxiliary random variable identifications $U_{1i} = (M_1, X_1^{i-1}, X_{2,i+1}^n)$ and $U_{li} = M_l$ for $l \in [2:q]$. The bound on nR_2 can be similarly established with the same identifications of U_{li}. Next, we establish the following properties of $U^q(i) = (U_{1i}, \ldots, U_{qi})$.

Lemma 20.1. For every $i \in [1:n]$, the following Markov relationships hold:

1. $X_{2i} \to (X_{1i}, U_{1i}, \ldots, U_{l-1,i}) \to U_{li}$ for l odd.

2. $X_{1i} \to (X_{2i}, U_{1i}, \ldots, U_{l-1,i}) \to U_{li}$ for l even.

3. $X_2^{i-1} \to (X_{2i}, U^q(i)) \to X_{1i}$.

4. $X_{1,i+1}^n \to (X_{1i}, U^q(i)) \to X_{2i}$.

The proof of this lemma is given in Appendix 20A. We also need the following simple lemma.

Lemma 20.2. Suppose $Y \to Z \to W$ form a Markov chain and $d(y, \hat{y})$ is a distortion measure. Then for every reconstruction function $\hat{y}(z, w)$, there exists a reconstruction function $\hat{y}^*(z)$ such that

$$E[d(Y, \hat{y}^*(Z))] \leq E[d(Y, \hat{y}(Z, W))].$$

To prove Lemma 20.2, consider

$$E[d(Y, \hat{y}(Z, W))] = \sum_{y,z,w} p(y, z, w) d(y, \hat{y}(z, w))$$

$$= \sum_{y,z} p(y, z) \sum_{w} p(w|z) d(y, \hat{y}(z, w))$$

$$\overset{(a)}{\geq} \sum_{y,z} p(y, z) \min_{w(z)} d(y, \hat{y}(z, w(z)))$$

$$= E[d(Y, \hat{y}^*(Z))],$$

where (a) follows since the minimum is less than or equal to the average.

Continuing with the converse proof of Theorem 20.7, we introduce a time-sharing random variable T and set $U_1 = (T, U_{1T})$ and $U_l = U_{lT}$ for $l \in [2:q]$. Noting that (X_{1T}, X_{2T}) has the same pmf as (X_1, X_2), we have

$$R_1 \geq \frac{1}{n} \sum_{i=1}^{n} I(X_{1i}; U^q(i)|X_{2i}) = I(X_{1T}; U^q(T)|X_{2T}, T) = I(X_1; U^q|X_2)$$

and

$$R_2 \geq I(X_2; U^q|X_1).$$

Using the first two statements in Lemma 20.1 and the identification of U^q, it can be easily verified that $p(u^q|x_1, x_2) = \prod_{l=1}^{q} p(u_l|u^{l-1}, x_{j_l})$ as desired. Finally, to check that the distortion constraints are satisfied, note from Lemma 20.2 and the third statement of Lemma 20.1 that for every $i \in [1:n]$, there exists a reconstruction function $\hat{x}_1^*$ such that

$$E[d(X_{1i}, \hat{x}_1^*(i, U^q(i), X_{2i}))] \leq E[d(X_{1i}, \hat{x}_{1i}(M^q, X_2^n))].$$

Hence

$$E[d(X_1, \hat{x}_1^*(U^q, X_2))] = \frac{1}{n} \sum_{i=1}^{n} E[d(X_{1i}, \hat{x}_1^*(i, U^q(i), X_{2i})) \mid T = i]$$

$$\leq \frac{1}{n} \sum_{i=1}^{n} E[d(X_{1i}, \hat{x}_{1i}(M^q, X_2^n))].$$

Letting $n \to \infty$ shows that $E[d(X_1, \hat{x}_1^*(U^q, X_2))] \leq D_1$. The other distortion constraint can be verified similarly. This completes the proof of Theorem 20.7.

SUMMARY

- Cutset bounds for source coding networks

- Distributed lossless source–network coding: Separate Slepian–Wolf coding and network coding suffices for multicasting correlated sources over graphical networks

- Multiple description network: Rate–distortion regions for tree and triangular networks

- CFO problem:

  - Noninteractive rounds of Slepian–Wolf coding achieve the cutset bound

  - Interaction does not reduce the rates in lossless source coding over noiseless public broadcast channels

- Two-way lossy source coding:

  - q-round rate–distortion region

  - Identification of auxiliary random variables satisfying Markovity in the proof of the converse

  - Interaction can reduce the rates in lossy source coding over noiseless bidirectional links

BIBLIOGRAPHIC NOTES

The optimal rate region for distributed lossless source-network coding in Theorem 20.2 was established by Effros, Médard, Ho, Ray, Karger, Koetter, and Hassibi (2004). The lossless source coding problem over a noiseless two-way relay channel in Example 20.1 is a special case of the three-source setup investigated by Wyner, Wolf, and Willems (2002). The cascade multiple description network was formulated by Yamamoto (1981), who established the rate–distortion region. The rate–distortion region for the triangular network in Theorem 20.5 was also established by Yamamoto (1996).

El Gamal and Orlitsky (1984) showed that $R_1 \geq H(X_1)$, $R_2 \geq H(X_2|X_1)$ is sometimes necessary for interactive error-free source coding. The CFO problem was formulated by Csiszár and Narayan (2004), who established the optimal rate region. The q-round rate–distortion region for two-way lossy source coding was established by Kaspi (1985). Example 20.5, which shows that two rounds of communication can strictly outperform a single round, is due to Ma and Ishwar (2010).

PROBLEMS

20.1. Prove Theorem 20.1.

20.2. Provide the details of the proof of Theorem 20.2.

20.3. Show that the rate–distortion region of the generalized triangular multiple description network in Remark 20.1 is given by (20.1).

20.4. Prove Theorem 20.6.

20.5. Provide the details of the proof of the rate–distortion region for two-way quadratic Gaussian source coding in Example 20.4.

20.6. *Dual-cascade multiple description network.* Consider the multiple description network depicted in Figure 20.6. Find the rate–distortion region $\mathscr{R}(D_2, D_3, D_4, D_5)$. Remark: This result can be easily extended to networks with multiple cascades.

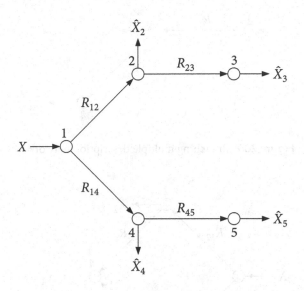

Figure 20.6. Dual-cascade multiple description network.

20.7. *Branching multiple description network.* Consider the 4-node branching multiple description network depicted in Figure 20.7. Show that a rate triple (R_{12}, R_{23}, R_{24}) is achievable for distortion triple (D_2, D_3, D_4) if

$$R_{12} > I(X; \hat{X}_2, \hat{X}_3, \hat{X}_4 | U) + 2I(X; U) + I(\hat{X}_1; \hat{X}_2 | U),$$
$$R_{23} > I(X; \hat{X}_3, U),$$
$$R_{24} > I(X; \hat{X}_4, U),$$
$$R_{23} + R_{24} > I(X; \hat{X}_3, \hat{X}_4 | U) + 2I(X; U) + I(\hat{X}_1; \hat{X}_2 | U)$$

for some conditional pmf $p(u, \hat{x}_2, \hat{x}_3, \hat{x}_4 | x)$ such that $\mathsf{E}(d_j(X, \hat{X}_j)) \le D_j, j = 2, 3, 4$.

20.8. *Diamond multiple description network.* Consider the 4-node diamond multiple network depicted in Figure 20.8. Show that a rate quadruple $(R_{12}, R_{13}, R_{24}, R_{34})$ is achievable for distortion triple (D_2, D_3, D_4) if

$$R_{12} > I(X; \hat{X}_2, U_2, U_4),$$
$$R_{13} > I(X; \hat{X}_3, U_3, U_4),$$
$$R_{24} > I(X; U_2, U_4),$$
$$R_{34} > I(X; U_3, U_4),$$
$$R_{24} + R_{34} > I(X; \hat{X}_4, U_2, U_3 | U_4) + 2I(U_4; X) + I(U_2; U_3 | U_4)$$

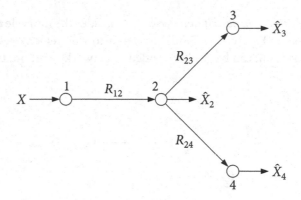

Figure 20.7. Branching multiple description network.

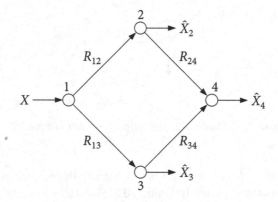

Figure 20.8. Diamond multiple description network.

for some conditional pmf $p(u_2, u_3, u_4, \hat{x}_2, \hat{x}_3, \hat{x}_4|x)$ such that $E(d_j(X, \hat{X}_j)) \le D_j$, $j = 2, 3, 4$.

20.9. *Cascade multiple description network with side information.* The cascade multiple description network can be extended to various side information scenarios. Here we consider two cases. In the first case, the rate–distortion region is known. In the second case, the rate–distortion region is known only for the quadratic Gaussian case. In the following assume the 2-DMS (X, Y) and distortion measures d_2 and d_3. The definitions of a code, achievability and rate–distortion region for each scenario are straightforward extensions of the case with no side information.

(a) Consider the case where the side information Y is available at both node 1 and node 2 as depicted in Figure 20.9. Show that the rate–distortion region $\mathcal{R}(D_2, D_3)$ is the set of rate pairs (R_{12}, R_{23}) such that

$$R_{12} \ge I(X; \hat{X}_2, \hat{X}_3|Y),$$
$$R_{23} \ge I(X, Y; \hat{X}_3)$$

for some conditional pmf $p(\hat{x}_2, \hat{x}_3 | x, y)$ that satisfies $E(d_j(X, \hat{X}_j)) \le D_j$, $j = 2, 3$.

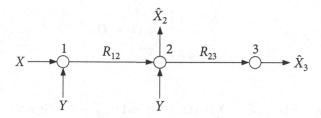

Figure 20.9. Cascade multiple description network with side information at node 1 and node 2.

(b) Consider the case where the side information Y is available only at node 3 as depicted in Figure 20.10. Establish the inner bound on the rate–distortion region for distortion pair (D_2, D_3) that consists of the set of rate pairs (R_{12}, R_{23}) such that

$$R_{23} > I(X; U, V | Y),$$

$$R_{12} > I(X; \hat{X}_2, U) + I(X; V | U, Y)$$

for some conditional pmf $p(u, v | x)$ and function $\hat{x}_3(u, v, y)$ that satisfy the constraints $E(d_j(X, \hat{X}_j)) \le D_j$, $j = 2, 3$.

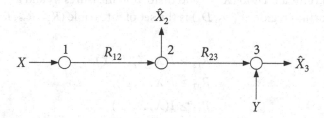

Figure 20.10. Cascade multiple description network with side information at node 3.

Remark: The first part is due to Permuter and Weissman (2010) and the second part is due to Vasudevan, Tian, and Diggavi (2006).

20.10. *Quadratic Gaussian extension.* Consider the cascade multiple description network with side information depicted in Figure 20.10. Let (X, Y) be a 2-WGN with $X \sim N(0, P)$ and $Y = X + Z$, where $Z \sim N(0, N)$ is independent of X, and let d_2 and d_3 be squared error distortion measures. Show that the rate–distortion region $\mathcal{R}(D_2, D_3)$ is the set of rate pairs (R_{12}, R_{23}) such that

$$R_{12} \ge R\left(\frac{P}{D_2}\right),$$

$$R_{23} \ge R\left(\frac{P_{X|Y}}{D_3}\right)$$

for $D_3 \geq D_2^*$, and

$$R_{12} \geq R\left(\frac{PN}{D_3(D_2 + N)}\right),$$

$$R_{23} \geq R\left(\frac{P_{X|Y}}{D_3}\right)$$

for $D_3 < D_2^*$, where $D_2^* = D_2 N/(D_2 + N)$ and $P_{X|Y} = PN/(P + N)$.

(Hint: Evaluate the inner bound in Problem 20.9 for this case. For $D_3 \geq D_2^*$, set $X = \hat{X}_2 + Z_2$, where $\hat{X}_2 \sim N(0, P - D_2)$ and $Z_2 \sim N(0, D_2)$. Next, let $\hat{X}_2 = U + Z_3$, where $Z_3 \sim N(0, D_3 N/(N - D_3) - D_2)$, $U \sim N(0, P_U)$, and $\hat{X}_3 = E(X|U, Y)$. For $D_3 < D_2^*$, set $U = X + Z_2$ and $V = X + Z_3$, where $Z_2 \sim N(0, P_2)$ and $Z_3 \sim N(0, P_3)$ are independent, and show that P_2 and P_3 can be chosen such that $1/P + 1/P_2 = 1/D_2$ and $1/P + 1/P_2 + 1/P_3 + 1/N = 1/D_3$. To prove the inequality on R_{12}, assume that node 3 receives the message M_{12} in addition to M_{23}. Argue that the minimum R_{12} for this modified setup is equal to the minimum rate (at the same distortions) for lossy source coding when side information may be absent discussed in Section 11.4.)

20.11. *Triangular multiple description network with side information.* Consider the 3-node multiple description network with side information at nodes 1 and 2 as depicted in Figure 20.11 for a 2-DMS (X, Y) and distortion measures d_2 and d_3. Show that the rate–distortion region $\mathcal{R}(D_2, D_3)$ is the set of rate triple (R_{12}, R_{23}, R_{13}) such that

$$R_{12} \geq I(X; \hat{X}_2, U | Y),$$
$$R_{23} \geq I(X, Y; U),$$
$$R_{13} \geq I(X; \hat{X}_3 | U)$$

for some conditional pmf $p(\hat{x}_2, u|x, y)p(\hat{x}_3|x, u)$ that satisfies $E(d_j(X, \hat{X}_j)) \leq D_j$, $j = 2, 3$.

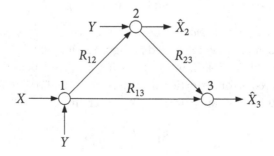

Figure 20.11. Triangular multiple description network with side information.

20.12. *Conditional CFO problem.* Let $(X_1, \ldots, X_N, Y)$ be an $(N+1)$-DMS. Consider a variation on the CFO problem in Section 20.3.2, where node $j \in [1:N]$ observes the common side information Y as well as its own source X_j. Find the optimal rate region $\mathscr{R}^*(\mathcal{A})$.

20.13. *Lossy source coding over a noiseless two-way relay channel.* Consider the lossy version of the source coding problem over a noiseless two-way relay channel in Example 20.1. Let d_1 and d_2 be two distortion measures. Suppose that source node 1 wishes to reconstruct the source X_2 with distortion D_2 and source node 2 wishes to reconstruct X_1 with distortion D_1.

(a) Show that if the rate–distortion quintuple $(R_1, R_2, R_3, D_1, D_2)$ is achievable, then it must satisfy the inequalities

$$R_1 \geq I(X_1; U_1 | X_2),$$
$$R_2 \geq 0,$$
$$R_3 \geq I(X_2; U_2 | X_1, U_1)$$

for some conditional pmf $p(u_1|x_1)p(u_2|x_2, u_1)$ and functions $\hat{x}_1(u_1, u_2, x_2)$ and $\hat{x}_2(u_1, u_2, x_1)$ such that $\mathsf{E}(d_j(X_j, \hat{X}_j)) \leq D_j$, $j = 1, 2$. (Hint: Consider the two-round communication between node 1 and supernode $\{2, 3\}$.)

(b) Show that the rate–distortion quintuple $(R_1, R_2, R_3, D_1, D_2)$ is achievable if

$$R_1 > R_{\text{SI-D2}}(D_1),$$
$$R_1 > R_{\text{SI-D1}}(D_2),$$
$$R_3 > \max\{R_{\text{SI-D2}}(D_1), R_{\text{SI-D1}}(D_2)\},$$

where

$$R_{\text{SI-D2}}(D_1) = \min_{p(u|x_1), \hat{x}_1(u, x_2): \mathsf{E}(d_1(X_1, \hat{X}_1)) \leq D_1} I(X_1; U | X_2),$$
$$R_{\text{SI-D1}}(D_2) = \min_{p(u|x_2), \hat{x}_2(u, x_1): \mathsf{E}(d_2(X_2, \hat{X}_2)) \leq D_2} I(X_2; U | X_1).$$

(c) Find the rate–distortion region when the sources are independent.

(d) Find the rate–distortion region when (X_1, X_2) is a 2-WGN(P, ρ) source and d_1 and d_2 are squared error distortion measures.

Remark: This problem was studied by Su and El Gamal (2010).

APPENDIX 20A   PROOF OF LEMMA 20.1

We first prove the following.

Lemma 20.3. Let (Y_1, Y_2, Z_1, Z_2) be a quadruple of random variables with joint pmf $p(y_1, y_2, z_1, z_2) = p(y_1, z_1)p(y_2, z_2)$ and M_l be a function of (Y_1, Y_2, M^{l-1}) for l odd and of (Z_1, Z_2, M^{l-1}) for l even. Then

$$I(Y_2; Z_1 | M^q, Y_1, Z_2) = 0,$$

$$I(Z_1; M_l | M^{l-1}, Y_1, Z_2) = 0 \quad \text{for } l \text{ odd,}$$

$$I(Y_2; M_l | M^{l-1}, Y_1, Z_2) = 0 \quad \text{for } l \text{ even.}$$

To prove this lemma, consider

$$\begin{aligned}
I(Y_2; Z_1 | M^q, Y_1, Z_2) &= H(Y_2 | M^q, Y_1, Z_2) - H(Y_2 | M^q, Y_1, Z_2, Z_1) \\
&= H(Y_2 | M^{q-1}, Y_1, Z_2) - I(Y_2; M_q | M^{q-1}, Y_1, Z_2) \\
&\quad - H(Y_2 | M^q, Y_1, Z_2, Z_1) \\
&\le I(Y_2; Z_1 | M^{q-1}, Y_1, Z_2) \\
&= H(Z_1 | M^{q-1}, Y_1, Z_2) - H(Z_1 | M^{q-1}, Y_1, Y_2, Z_2) \\
&= H(Z_1 | M^{q-2}, Y_1, Z_2) - I(Z_1; M_{q-1} | M^{q-2}, Y_1, Z_2) \\
&\quad - H(Z_1 | M^{q-1}, Y_1, Y_2, Z_2) \\
&\le I(Y_2; Z_1 | M^{q-2}, Y_1, Z_2).
\end{aligned}$$

Continuing this process gives $I(Y_2; Z_1 | Y_1, Z_2) = 0$. Hence, all the inequalities hold with equality, which completes the proof of Lemma 20.3.

We are now ready to prove Lemma 20.1. First consider the first statement $X_{2i} \to (X_{1i}, U^{l-1}(i)) \to U_l(i)$ for $l > 1$. Since $U_{li} = M_l$,

$$I(U_l(i); X_{2i} | X_{1i}, U_i^{l-1}) \le I(M_l; X_{2i}, X_2^{i-1} | X_{1i}, M^{l-1}, X_1^{i-1}, X_{2,i+1}^n).$$

Setting $Y_1 = (X_1^{i-1}, X_{1i})$, $Y_2 = X_{1,i+1}^n$, $Z_1 = (X_2^{i-1}, X_{2i})$, and $Z_2 = X_{2,i+1}^n$ in Lemma 20.3 yields $I(M_l; X_{2i}, X_2^{i-1} | X_{1,i}, M^{l-1}, X_1^{i-1}, X_{2,i+1}^n) = 0$ as desired. For $l = 1$, note that

$$\begin{aligned}
I(X_{2i}; M_1, X_1^{i-1}, X_{2,i+1}^n | X_{1i}) &\le I(X_{2i}; M_1, X_1^{i-1}, X_{2,i+1}^n, X_{1,i+1}^n | X_{1i}) \\
&= I(X_{2i}; X_1^{i-1}, X_{2,i+1}^n, X_{1,i+1}^n | X_{1i}) \\
&= 0.
\end{aligned}$$

The second statement can be similarly proved. To prove the third statement, note that

$$I(X_{1i}; X_2^{i-1} | M^q, X_{2i}, X_{2,i+1}^n, X_1^{i-1}) \le I(X_{1i}, X_{1,i+1}^n; X_2^{i-1} | M^q, X_{2i}, X_{2,i+1}^n, X_1^{i-1}).$$

Setting $Y_1 = X_1^{i-1}$, $Y_2 = (X_{1,i+1}^n, X_{1i})$, $Z_1 = X_2^{i-1}$, and $Z_2 = (X_{2,i+1}^n, X_{2i})$ in Lemma 20.3 yields $I(X_{1i}, X_1^{i-1}; X_2^{i-1} | M^q, X_{2i}, X_{2,i+1}^n, X_1^{i-1}) = 0$ as desired. The last statement can be similarly proved.

PART IV

EXTENSIONS

CHAPTER 21

Communication for Computing

In the first three parts of the book we investigated the limits on information flow in networks whose task is to communicate (or store) distributed information. In many real-world distributed systems, such as multiprocessors, peer-to-peer networks, networked mobile agents, and sensor networks, the task of the network is to compute a function, make a decision, or coordinate an action based on distributed information. Can the communication rate needed to perform such a task at some node be reduced relative to communicating all the sources to this node?

This question has been formulated and studied in computer science under communication complexity and gossip algorithms, in control and optimization under distributed consensus, and in information theory under coding for computing and the μ-sum problem, among other topics. In this chapter, we study information theoretic models for distributed computing over networks. In some cases, we find that the total communication rate can be significantly reduced when the task of the network is to compute a function of the sources rather than to communicate the sources themselves, while in other cases, no such reduction is possible.

We first show that the Wyner–Ziv theorem in Chapter 11 extends naturally to the case when the decoder wishes to compute a function of the source and the side information. We provide a refined characterization of the lossless special case of this result in terms of conditional graph entropy. We then discuss distributed coding for computing. Although the rate–distortion region for this case is not known in general (even when the goal is to reconstruct the sources themselves), we show through examples that the total communication rate needed for computing can be significantly lower than for communicating the sources themselves. The first example we discuss is the μ-sum problem, where the decoder wishes to reconstruct a weighted sum of two separately encoded Gaussian sources with a prescribed quadratic distortion. We establish the rate–distortion region for this setting by reducing the problem to the CEO problem discussed in Chapter 12. The second example is lossless computing of the modulo-2 sum of a DSBS. Surprisingly, we find that using the *same* linear code at both encoders can outperform Slepian–Wolf coding.

Next, we consider coding for computing over multihop networks.

- We extend the result on two-way lossy source coding in Chapter 20 to computing a function of the sources and show that interaction can help reduce the total communication rate even for lossless computing.

- We establish the optimal rate region for cascade lossless coding for computing and inner and outer bounds on the rate–distortion region for the lossy case.

- We present an information theoretic formulation of the distributed averaging problem and establish upper and lower bounds on the network rate–distortion function that differ by less than a factor of 2 for some large networks.

Finally, we show through an example that even when the sources are independent, source–channel separation may fail when the goal is to compute a function of several sources over a noisy multiuser channel.

21.1 CODING FOR COMPUTING WITH SIDE INFORMATION

Consider the communication system for computing with side information depicted in Figure 21.1. Let (X, Y) be a 2-DMS and $d(z, \hat{z})$ be a distortion measure. Suppose that the receiver wishes to reconstruct a function $Z = g(X, Y)$ of the 2-DMS with distortion D. What is the rate–distortion function $R_g(D)$, that is, the minimum rate needed to achieve distortion D for computing the function $g(X, Y)$?

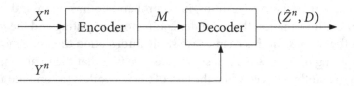

Figure 21.1. Coding for computing with side information at the decoder.

We define a $(2^{nR}, n)$ code, achievability, and the rate–distortion function $R_g(D)$ for this setup as in Chapter 11. Following the proof of the Wyner–Ziv theorem, we can readily determine the rate–distortion function.

Theorem 21.1. The rate–distortion function for computing $Z = g(X, Y)$ with side information Y is

$$R_g(D) = \min I(X; U|Y),$$

where the minimum is over all conditional pmfs $p(u|x)$ with $|\mathcal{U}| \le |\mathcal{X}| + 1$ and functions $\hat{z}(u, y)$ such that $E(d(Z, \hat{Z})) \le D$.

To illustrate this result, consider the following.

Example 21.1. Let (X, Y) be a 2-WGN(P, ρ) source and d be a squared error distortion measure. If the function to be computed is $g(X, Y) = (X + Y)/2$, that is, the average of the source and the side information, then $R_g(D) = R((1 - \rho^2)P/4D)$, and is achieved by Wyner–Ziv coding of the source $X/2$ with side information Y.

21.1.1   Lossless Coding for Computing

Now suppose that the receiver wishes to recover the function $Z = g(X, Y)$ losslessly. What is the optimal rate R_g^*? Clearly

$$H(Z|Y) \le R_g^* \le H(X|Y). \tag{21.1}$$

These bounds sometimes coincide, for example, if $Z = X$, or if (X, Y) is a DSBS(p) and $g(X, Y) = X \oplus Y$. The bounds do not coincide in general, however, as illustrated in the following.

Example 21.2. Let $X = (V_1, V_2, \dots, V_{10})$, where the V_j, $j \in [1:10]$, are i.i.d. Bern(1/2), and $Y \sim$ Unif$[1:10]$. Suppose $g(X, Y) = V_Y$. The lower bound gives $H(V_Y|Y) = 1$ bit and the upper bound gives $H(X|Y) = 10$ bits. It can be shown that $R_g^* = 10$ bits, that is, the decoder must be able to recover X losslessly.

Optimal rate. As we showed in Sections 3.6 and 11.2, lossless source coding can be viewed as a special case of lossy source coding. Similarly, the optimal rate for lossless computing with side information R_g^* can be obtained from the rate–distortion function in Theorem 21.1 by assuming a Hamming distortion measure and setting $D = 0$. This yields

$$R_g^* = R_g(0) = \min I(X; U|Y),$$

where the minimum is over all conditional pmfs $p(u|x)$ such that $H(Z|U, Y) = 0$.

The characterization of the optimal rate R_g^* can be further refined. We first introduce some needed definitions. Let $\mathcal{G} = (\mathcal{N}, \mathcal{E})$ be an undirected simple graph (i.e., a graph with no self-loops). A set of nodes in $\mathcal{N}$ is said to be *independent* if no two nodes are connected to each other through an edge. An independent set is said to be *maximally independent* if it is not a subset of any other independent set. Define $\Gamma(\mathcal{G})$ to be the collection of maximally independent sets of $\mathcal{G}$. For example, the collection of maximally independent sets for the graph in Figure 21.2 is $\Gamma(\mathcal{G}) = \{\{1, 3, 5\}, \{1, 3, 6\}, \{1, 4, 6\}, \{2, 5\}, \{2, 6\}\}$.

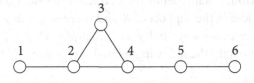

Figure 21.2. Example graph.

Graph entropy. Let X be a random variable over the nodes of the graph $\mathcal{G}$. Define a random variable $W \in \Gamma(\mathcal{G})$ with conditional pmf $p(w|x)$ such that $p(w|x) = 0$ if $x \notin w$, and hence for every x, $\sum_{w:x\in w} p(w|x) = 1$. The *graph entropy* of X is defined as

$$H_{\mathcal{G}}(X) = \min_{p(w|x)} I(X; W).$$

To better understand this definition, consider the following.

Example 21.3. Let $\mathcal{G}$ be a graph with no edges, i.e., $\mathcal{E} = \emptyset$. Then, the graph entropy $H_{\mathcal{G}}(X) = 0$, since $\Gamma(\mathcal{G}) = \{\mathcal{N}\}$ has a single element.

Example 21.4. Let $\mathcal{G}$ be a complete graph. Then, the graph entropy is $H_{\mathcal{G}}(X) = H(X)$ and is attained by $W = \{X\}$.

Example 21.5. Let X be uniformly distributed over $\mathcal{N} = \{1, 2, 3\}$ and assume that the graph $\mathcal{G}$ has only a single edge $(1, 3)$. Then, $\Gamma(\mathcal{G}) = \{\{1, 2\}, \{2, 3\}\}$. By convexity, $I(X; W)$ is minimized when $p(\{1, 2\}|2) = p(\{2, 3\}|2) = 1/2$. Thus

$$H_{\mathcal{G}}(X) = H(W) - H(W|X) = 1 - \frac{1}{3} = \frac{2}{3}.$$

Conditional graph entropy. Now let (X, Y) be a pair of random variables and let the nodes of the graph $\mathcal{G}$ be the support set of X. The conditional graph entropy of X given Y is defined as

$$H_{\mathcal{G}}(X|Y) = \min I(X; W|Y),$$

where the minimum is over all conditional pmfs $p(w|x)$ such that $W \in \Gamma(\mathcal{G})$ and $p(w|x) = 0$ if $x \notin w$. To explain this definition, consider the following.

Example 21.6. Let $\mathcal{E} = \emptyset$. Then $H_{\mathcal{G}}(X|Y) = 0$.

Example 21.7. Let $\mathcal{G}$ be a complete graph. Then $H_{\mathcal{G}}(X|Y) = H(X|Y)$.

Example 21.8. Let (X, Y) be uniformly distributed over $\{(x, y) \in \{1, 2, 3\}^2 : x \neq y\}$ and assume that $\mathcal{G}$ has a single edge $(1, 3)$. By convexity, $I(X; W|Y)$ is minimized by setting $p(\{1, 2\}|2) = p(\{2, 3\}|2) = 1/2$. Thus

$$H_{\mathcal{G}}(X|Y) = H(W|Y) - H(W|X, Y) = \frac{1}{3} + \frac{2}{3} H\left(\frac{1}{4}\right) - \frac{1}{3} = \frac{2}{3} H\left(\frac{1}{4}\right).$$

Refined characterization. Finally, define the *characteristic graph* $\mathcal{G}$ of the (X, Y, g) triple such that the set of nodes is the support of X and there is an edge between two distinct nodes x and x' iff there exists a symbol y such that $p(x, y), p(x', y) > 0$ and $g(x, y) \neq g(x', y)$. Then, we can obtain the following refined expression for the optimal rate.

Theorem 21.2. The optimal rate for losslessly computing the function $g(X, Y)$ with side information Y is

$$R_g^* = H_{\mathcal{G}}(X|Y).$$

To illustrate this result, consider the following.

Example 21.9 (Online card game). Recall the online card game discussed in Chapter 1, where Alice and Bob each select one card without replacement from a virtual hat with three cards labeled 1,2,3. The one with the larger number wins. Let the 2-DMS (X, Y)

represent the numbers on Alice and Bob's cards. Bob wishes to find who won the game, i.e., $g(X, Y) = \max\{X, Y\}$. Since $g(x, y) \neq g(x', y)$ iff $(x, y, x') = (1, 2, 3)$ or $(3, 2, 1)$, the characteristic graph $\mathcal{G}$ has one edge $(1, 3)$. Hence the triple $(X, Y, \mathcal{G})$ is the same as in Example 21.8 and the optimal rate is $R_g^* = (2/3)H(1/4)$.

Proof of Theorem 21.2. To prove the theorem, we need to show that

$$\min_{p(u|x)} I(U; X|Y) = \min_{p(w|x)} I(W; X|Y),$$

where the first minimum is over all conditional pmfs $p(u|x)$ such that $H(Z|U, Y) = 0$ and the second minimum is over all conditional pmfs $p(w|x)$ such that $W \in \Gamma(\mathcal{G})$ and $X \in W$. Consider $p(w|x)$ that achieves the second minimum. Since w is a maximally independent set with respect to the characteristic graph of (X, Y, g), for each $y \in \mathcal{Y}$, $g(x, y)$ is constant for all $x \in w$ with $p(x, y) > 0$. Hence, $g(x, y)$ can be uniquely determined from (w, y) whenever $p(w, x, y) > 0$, or equivalently, $H(g(X, Y)|W, Y) = 0$. By identifying $U = W$, it follows that $\min_{p(u|x)} I(U; X|Y) \leq \min_{p(w|x)} I(W; X|Y)$.

To establish the other direction of the inequality, let $p(u|x)$ be the conditional pmf that attains the first minimum and define $W = w(U)$, where $w(u) = \{x: p(u, x) > 0\}$. If $p(w, x) > 0$, then there exists a u such that $w(u) = w$ and $p(u, x) > 0$, which implies that $x \in w$. Thus, $p(w|x) = 0$ if $x \notin w$. Now suppose that $x \in w$ and $p(w) > 0$. Then there exists a u such that $w(u) = w$ and $p(u, x) > 0$. Hence by the Markovity of $U \rightarrow X \rightarrow Y$, $p(x, y) > 0$ implies that $p(u, x, y) > 0$. But since $H(g(X, Y)|U, Y) = 0$, the pair (u, y) must determine $g(x, y)$ uniquely. Therefore, if $x, x' \in w$ and $p(x, y), p(x', y) > 0$, then $g(x, y) = g(x', y)$, which implies that W is maximally independent. Finally, since W is a function of U, $I(U; X|Y) \geq I(W; X|Y) \geq \min_{p(w|x)} I(W; X|Y)$. This completes the proof of Theorem 21.2.

21.2 DISTRIBUTED CODING FOR COMPUTING

Consider the distributed communication system for computing depicted in Figure 21.3 for a 2-DMS (X_1, X_2) and distortion measure $d(z, \hat{z})$. Suppose that the decoder wishes to compute a function $Z = g(X_1, X_2)$ of the 2-DMS with distortion D. The goal is to find the rate–distortion region $\mathcal{R}_g(D)$ for computing g.

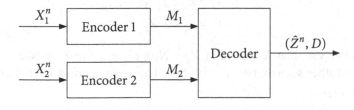

Figure 21.3. Distributed coding for computing.

A $(2^{nR_1}, 2^{nR_2}, n)$ code, achievability, and the rate–distortion region $\mathcal{R}_g(D)$ can be defined as in Chapter 12. The rate–distortion region for this problem is not known in general. In the following we discuss special cases.

21.2.1 μ-Sum Problem

Let (Y_1, Y_2) be a 2-WGN$(1, \rho)$ source with $\rho \in (0, 1)$ and $\boldsymbol{\mu} = [\mu_1 \; \mu_2]^T$. The sources are separately encoded with the goal of computing $Z = \boldsymbol{\mu}^T \mathbf{Y}$ at the decoder with a prescribed mean squared error distortion D. The μ-sum rate–distortion region $\mathcal{R}_\mu(D)$ is characterized in the following.

Theorem 21.3. Let $\mu_1 \mu_2 = \delta^2 s_1 s_2 / \rho > 0$, where $s_1 = \rho(\rho + \mu_1/\mu_2)/(1 - \rho^2)$, $s_2 = \rho(\rho + \mu_2/\mu_1)/(1 - \rho^2)$, and $\delta = 1/(1 + s_1 + s_2)$. The μ-sum rate–distortion region $\mathcal{R}_\mu(D)$ is the set of rate pairs (R_1, R_2) such that

$$R_1 \geq r_1 + \mathsf{R}\left(\frac{\left(1 + s_2(1 - 2^{-2r_2})\right)^{-1}}{D + \delta}\right),$$

$$R_2 \geq r_2 + \mathsf{R}\left(\frac{\left(1 + s_1(1 - 2^{-2r_1})\right)^{-1}}{D + \delta}\right),$$

$$R_1 + R_2 \geq r_1 + r_2 + \mathsf{R}\left(\frac{1}{D + \delta}\right)$$

for some $r_1, r_2 \geq 0$ that satisfy the condition $D + \delta \geq (1 + s_1(1 - 2^{-2r_1}) + s_2(1 - 2^{-2r_2}))^{-1}$.

Note that if (R_1, R_2) is achievable with distortion D for $\boldsymbol{\mu}$, then for any $b > 0$, (R_1, R_2) is also achievable with distortion $b^2 D$ for $b\boldsymbol{\mu}$ since

$$\mathsf{E}\left[(b\mu_1 Y_1 + b\mu_2 Y_2 - \tilde{Z})^2\right] \leq b^2 D$$

if $\mathsf{E}[(\mu_1 Y_1 + \mu_2 Y_2 - \hat{Z})^2] \leq D$, where $\tilde{Z} = b\hat{Z}$ is the estimate of bZ. Thus, this theorem characterizes the rate–distortion region for any $\boldsymbol{\mu}$ with $\mu_1 \mu_2 > 0$.

Proof of Theorem 21.3. We show that the μ-sum problem is equivalent to the quadratic Gaussian CEO problem discussed in Section 12.4 and use this equivalence to find $\mathcal{R}_\mu(D)$. Since Y_1 and Y_2 are jointly Gaussian, they can be expressed as

$$Y_1 = a_1 X + Z_1,$$
$$Y_2 = a_2 X + Z_2$$

for some random variables $X \sim \mathsf{N}(0, 1)$, $Z_1 \sim \mathsf{N}(0, 1 - a_1^2)$, and $Z_2 \sim \mathsf{N}(0, 1 - a_2^2)$, independent of each other, such that $a_1, a_2 \in (0, 1)$ and $a_1 a_2 = \rho$. Consider the MMSE estimate of X given (Y_1, Y_2)

$$\tilde{X} = \mathsf{E}(X | Y_1, Y_2) = [a_1 \; a_2] K_Y^{-1} \mathbf{Y} = \left[\frac{a_1 - \rho a_2}{1 - \rho^2} \; \frac{a_2 - \rho a_1}{1 - \rho^2}\right] \mathbf{Y}.$$

We now choose a_j, $j = 1, 2$, such that $\tilde{X} = \boldsymbol{\mu}^T \mathbf{Y}$, i.e.,

$$\frac{a_1 - \rho a_2}{1 - \rho^2} = \mu_1,$$

$$\frac{a_2 - \rho a_1}{1 - \rho^2} = \mu_2.$$

Solving for a_1, a_2, we obtain

$$a_1^2 = \rho \frac{\rho + \mu_1/\mu_2}{1 + \rho\mu_1/\mu_2} = \frac{s_1}{1 + s_1},$$

$$a_2^2 = \rho \frac{\rho + \mu_2/\mu_1}{1 + \rho\mu_2/\mu_1} = \frac{s_2}{1 + s_2}.$$

It can be readily checked that the constraint $\rho = a_1 a_2$ is equivalent to the normalization

$$\mu_1 \mu_2 = \frac{\rho}{(\rho + \mu_1/\mu_2)(\rho + \mu_2/\mu_1)}$$

and that the corresponding mean squared error is

$$E((X - \tilde{X})^2) = 1 - \left(\frac{a_1 - \rho a_2}{1 - \rho^2} a_1 + \frac{a_2 - \rho a_1}{1 - \rho^2} a_2 \right) = 1 - \frac{a_1^2 + a_2^2 - 2\rho^2}{1 - \rho^2} = \delta.$$

Now let $\tilde{Z} = X - \tilde{X}$, then for every U such that $X \to (Y_1, Y_2) \to U$ form a Markov chain,

$$E[(X - E(X|U))^2] = E[(\boldsymbol{\mu}^T \mathbf{Y} + \tilde{Z} - E(\boldsymbol{\mu}^T \mathbf{Y} + \tilde{Z}|U))^2]$$
$$= \delta + E[(\boldsymbol{\mu}^T \mathbf{Y} - E(\boldsymbol{\mu}^T \mathbf{Y}|U))^2].$$

Therefore, every code that achieves distortion D in the μ-sum problem can be used to achieve distortion $D + \delta$ in the CEO problem and vice versa.

The observations for the corresponding CEO problem are $Y_j/a_j = X + \tilde{Z}_j$, $j = 1, 2$, where $\tilde{Z}_j = Z_j/a_j \sim N(0, 1/s_j)$ since $s_j = a_j^2/(1 - a_j^2)$. Hence by Theorem 12.4, the rate–distortion region for the CEO problem with distortion $D + \delta$ is the set of rate pairs (R_1, R_2) satisfying the inequalities in the theorem. Finally, by equivalence, this region is also the rate–distortion region for the μ-sum problem, which completes the proof of Theorem 21.3.

Remark 21.1. When $\mu_1 \mu_2 = 0$, the μ-sum problem is equivalent to the quadratic Gaussian distributed source coding discussed in Section 12.3 with $D_1 \geq 1$ or $D_2 \geq 1$, and the rate–distortion region is given by Theorem 12.3 with proper normalization. Thus, for $\mu_1 \mu_2 \geq 0$, the Berger–Tung inner bound is tight. When $\mu_1 \mu_2 < 0$, however, it can be shown that the Berger–Tung inner bound is not tight in general.

21.2.2   Distributed Lossless Computing

Consider the lossless special case of distributed coding for computing. As for the general lossy case, the optimal rate region is not known in general. Consider the following simple inner and outer bounds on $\mathcal{R}_g^*$.

Inner bound. The Slepian–Wolf region consisting of all rate pairs (R_1, R_2) such that

$$
\begin{aligned}
R_1 &\geq H(X_1 | X_2), \\
R_2 &\geq H(X_2 | X_1), \\
R_1 + R_2 &\geq H(X_1, X_2)
\end{aligned}
\tag{21.2}
$$

constitutes an inner bound on the optimal rate region for any function $g(X_1, X_2)$.

Outer bound. Even when the receiver knows X_1, $R_2 \geq H(Z|X_1)$ is still necessary. Similarly, we must have $R_1 \geq H(Z|X_2)$. These inequalities constitute an outer bound on the optimal rate region for any function $Z = g(X_1, X_2)$ that consists of all rate pairs (R_1, R_2) such that

$$
\begin{aligned}
R_1 &\geq H(Z|X_2), \\
R_2 &\geq H(Z|X_1).
\end{aligned}
\tag{21.3}
$$

The above simple inner bound is sometimes tight, e.g., when $Z = (X_1, X_2)$. The following is an interesting example in which the outer bound is tight.

Example 21.10 (Distributed computing of the modulo-2 sum of a DSBS). Let (X_1, X_2) be a DSBS(p), that is, $X_1 \sim \text{Bern}(1/2)$ and $Z \sim \text{Bern}(p)$ are independent and $X_2 = X_1 \oplus Z$. The decoder wishes to losslessly compute the modulo-2 sum of the two sources $Z = X_1 \oplus X_2$. The inner bound in (21.2) on the optimal rate region reduces to $R_1 \geq H(p)$, $R_2 \geq H(p)$, $R_1 + R_2 \geq 1 + H(p)$, while the outer bound in (21.3) reduces to $R_1 \geq H(p)$, $R_2 \geq H(p)$.

We now show that the outer bound can be achieved using random linear codes! Randomly generate $k \times n$ binary parity-check matrix H with i.i.d. Bern($1/2$) entries. Suppose that encoder 1 sends the binary k-vector $H X_1^n$ and encoder 2 sends the binary k-vector $H X_2^n$. Then the receiver adds the two binary k-vectors to obtain $H X_1^n \oplus H X_2^n = H Z^n$. Following similar steps to the random linear binning achievability proof of the lossless source coding theorem in Remark 10.2, it can be readily shown that the probability of decoding error averaged over the random encoding matrix H tends to zero as $n \to \infty$ if $k/n > H(p) + \delta(\epsilon)$.

Remark 21.2. The outer bound in (21.3) can be tightened by using Theorem 21.2 twice, once when the receiver knows X_1 and a second time when it knows X_2. Defining the characteristic graphs $\mathcal{G}_1$ for (X_1, X_2, g) and $\mathcal{G}_2$ for (X_2, X_1, g) as before, we obtain the tighter outer bound consisting of all rate pairs (R_1, R_2) such that

$$
\begin{aligned}
R_1 &\geq H_{\mathcal{G}_1}(X_1 | X_2), \\
R_2 &\geq H_{\mathcal{G}_2}(X_2 | X_1).
\end{aligned}
$$

21.3 INTERACTIVE CODING FOR COMPUTING

Consider the two-way communication system for computing depicted in Figure 21.4 for a 2-DMS (X_1, X_2) and two distortion measures d_1 and d_2. Node 1 wishes to reconstruct a function $Z_1 = g_1(X_1, X_2)$ and node 2 wishes to reconstruct a function $Z_2 = g_2(X_1, X_2)$ with distortions D_1 and D_2, respectively. The goal is to find the rate–distortion region $\mathscr{R}(D_1, D_2)$ defined as in Section 20.3.3.

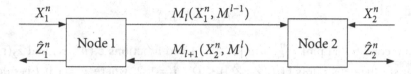

Figure 21.4. Two-way coding for computing.

We have the following simple bounds on the rate–distortion region.

Inner bound. Using two independent rounds of Wyner–Ziv coding, we obtain the inner bound consisting of all rate pairs (R_1, R_2) such that

$$R_1 > I(X_1; U_1 | X_2),$$
$$R_2 > I(X_2; U_2 | X_1) \tag{21.4}$$

for some conditional pmf $p(u_1 | x_1) p(u_2 | x_2)$ and functions $\hat{z}_1(u_1, x_2)$ and $\hat{z}_2(u_2, x_1)$ that satisfy the constraints $\mathsf{E}(d_j(Z_j, \hat{Z}_j)) \le D_j$, $j = 1, 2$.

Outer bound. Even if each encoder knows the source at the other node, every achievable rate pair must satisfy the inequalities

$$R_1 \ge I(Z_2; \hat{Z}_2 | X_2),$$
$$R_2 \ge I(Z_1; \hat{Z}_1 | X_1) \tag{21.5}$$

for some conditional pmf $p(\hat{z}_1 | x_1, x_2) p(\hat{z}_2 | x_1, x_2)$ such that $\mathsf{E}(d_j(Z_j, \hat{Z}_j)) \le D_j$, $j = 1, 2$.
These bounds are sometimes tight.

Example 21.11. Let (X_1, X_2) be a 2-WGN(P, ρ) source and $Z_1 = Z_2 = Z = (X_1 + X_2)/2$. The rate–distortion region for mean squared error distortion $D_1 = D_2 = D$ is the set of rate pairs (R_1, R_2) such that

$$R_1 \ge \mathsf{R}\left(\frac{(1 - \rho^2)P}{4D}\right),$$
$$R_2 \ge \mathsf{R}\left(\frac{(1 - \rho^2)P}{4D}\right).$$

It is easy to see that this region coincides with both the inner bound in (21.4) and the outer bound in (21.5). Note that this problem is equivalent (up to scaling) to the two-way lossy source coding problem for a 2-WGN source in Example 20.4.

The bounds in (21.4) and (21.5) are not tight in general. Following Theorem 20.7 in Section 20.3.3, we can readily establish the rate–distortion region for q rounds of communication.

Theorem 21.4. The q-round rate–distortion region $\mathcal{R}_q(D_1, D_2)$ for computing functions Z_1 and Z_2 with distortion pair (D_1, D_2) is the set of rate pairs (R_1, R_2) such that

$$R_1 \geq I(X_1; U^q | X_2),$$
$$R_2 \geq I(X_2; U^q | X_1)$$

for some conditional pmf $\prod_{l=1}^{q} p(u_l | u^{l-1}, x_{j_l})$ and functions $\hat{z}_1(u^q, x_1)$ and $\hat{z}_2(u^q, x_2)$ that satisfy the constraints $\mathsf{E}(d_j(Z_j, \hat{Z}_j)) \leq D_j$, $j = 1, 2$, where $j_l = 1$ if l is odd and $j_l = 2$ if l is even.

Note that the above region is computable for every given q. However, there is no bound on q in general.

21.3.1 Interactive Coding for Lossless Computing

Now consider the two-way coding for computing setting in which node 1 wishes to recover $Z_1(X_1, X_2)$ and node 2 wishes to recover $Z_2(X_1, X_2)$ losslessly. As we have seen in Section 20.3, when $Z_1 = X_2$ and $Z_2 = X_1$, the optimal rate region is achieved by two independent rounds of Slepian–Wolf coding. For arbitrary functions, Theorem 21.4 for the lossy setting can be specialized to yield the following characterization of the optimal rate region for q rounds.

Theorem 21.5. The optimal q-round rate region $\mathcal{R}_q^*$ for lossless computing of the functions Z_1 and Z_2 is the set of rate pairs (R_1, R_2) such that

$$R_1 \geq I(X_1; U^q | X_2),$$
$$R_2 \geq I(X_2; U^q | X_1)$$

for some conditional pmf $\prod_{l=1} p(u_l | u^{l-1}, x_{j_l})$ that satisfy $H(g_1(X_1, X_2) | X_1, U^q) = 0$ and $H(g_2(X_1, X_2) | X_2, U^q) = 0$, where $j_l = 1$ if l is odd and $j_l = 2$ if l is even.

In some cases, two independent rounds are sufficient, for example, if $Z_1 = X_2$ and $Z_2 = X_1$, or if (X_1, X_2) is a DSBS and $Z_1 = Z_2 = X_1 \oplus X_2$. In general, however, interactivity can help reduce the transmission rate.

Assume that $Z_2 = \emptyset$ and relabel Z_1 as Z. Consider two rounds of communication. In the first round, node 1 sends a message to node 2 that depends on X_1 and in the second round, node 2 sends a second message to node 1 that depends on X_2 and the first message. By Theorem 21.5, the optimal rate region for two rounds of communication is the set of rate pairs (R_1, R_2) such that

$$R_1 \geq I(U_1; X_1|X_2),$$
$$R_2 \geq I(U_2; X_2|U_1, X_1)$$

for some conditional pmf $p(u_1|x_1)p(u_2|u_1, x_2)$ that satisfies $H(Z|U_1, U_2, X_1) = 0$.

The following example shows that two rounds can achieve strictly lower rates than one round.

Example 21.12. Let $X_1 \sim \text{Bern}(p)$, $p \ll 1/2$, and $X_2 \sim \text{Bern}(1/2)$ be two independent sources, and $Z = X_1 \cdot X_2$. The minimum rate for one round of communication from node 2 to node 1 is $R^{(1)} = 1$ bit/symbol.

Consider the following two-round coding scheme. In the first round, node 1 uses Slepian–Wolf coding to communicate X_1 losslessly to node 2 at rate $R_1 > H(p)$. In the second round, node 2 sets $Y = X_2$ if $X_1 = 1$ and $Y = e$ if $X_1 = 0$. It then uses Slepian–Wolf coding to send Y losslessly to node 1 at rate $R_2 > H(Y|X_1) = p$. Thus the sum-rate for the two rounds is

$$R_{\text{sum}}^{(2)} = H(p) + p < 1$$

for p sufficiently small.

The sum-rate can be reduced further by using more than two rounds.

Example 21.13. Let (X_1, X_2) be a DSBS(p), $Z_1 = Z_2 = X_1 \cdot X_2$. For two rounds of communication a rate pair (R_1, R_2) is achievable if $R_1 > H_G(X_1|X_2) = H(X_1|X_2) = H(p)$ and $R_2 > H(Z_1|X_1) = (1/2)H(p)$. Optimality follows by Theorem 21.2 for the bound on R_1 and by the outer bound for the lossy case in (21.5) with Hamming distortion $D = 0$ for the bound on R_2.

Consider the following 3-round coding scheme. Set the auxiliary random variables in Theorem 21.5 as $U_1 = (1 - X_1) \cdot W$, where $W \sim \text{Bern}(1/2)$ is independent of X_1, $U_2 = X_2 \cdot (1 - U_1)$, and $U_3 = X_1 \cdot U_2$. Since $U_3 = X_1 \cdot X_2$, both nodes can compute the product losslessly. This gives the upper bound on the sum-rate

$$R_{\text{sum}}^{(3)} < \frac{5}{4}H(p) + \frac{1}{2}H\left(\frac{1-p}{2}\right) - \frac{1-p}{2} < \frac{3}{2}H(p).$$

Note that the sum-rate can be reduced further by using more rounds.

21.4 CASCADE CODING FOR COMPUTING

Consider the cascade communication system for computing depicted in Figure 21.5 for a 2-DMS (X_1, X_2) and distortion measure $d(z, \hat{z})$. The decoder wishes to reconstruct the function $Z = g(X_1, X_2)$ with prescribed distortion D. The definitions of a $(2^{nR_1}, 2^{nR_2}, n)$ code, achievability, and rate–distortion region are similar to those for the cascade multiple description network in Section 20.2. Again we wish to find the rate–distortion region $\mathscr{R}_g(D)$.

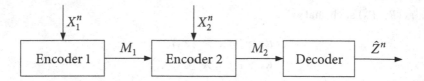

Figure 21.5. Cascade coding for computing.

The rate–distortion region for this problem is not known in general, even when X_1 and X_2 are independent! In the following, we discuss inner and outer bounds on the rate–distortion region.

Cutset outer bound. If a rate pair (R_1, R_2) is achievable with distortion D for cascade coding for computing, then it must satisfy the inequalities

$$R_1 \geq I(X_1; U | X_2),$$
$$R_2 \geq I(Z; \hat{Z}) \tag{21.6}$$

for some conditional pmf $p(u|x_1)p(\hat{z}|x_2, u)$ such that $E(d(Z, \hat{Z})) \leq D$.

Local computing inner bound. Encoder 1 uses Wyner–Ziv coding to send a description U of its source X_1 to encoder 2 at rate $R_1 > I(X_1; U | X_2)$. Encoder 2 sends the estimate $\hat{Z}$ of Z based on (U, X_2) at rate $R_2 > I(X_2, U; \hat{Z})$ to the decoder. This gives the *local computing* inner bound on the rate distortion region $\mathcal{R}_g(D)$ that consists of all rate pairs (R_1, R_2) such that

$$R_1 > I(X_1; U | X_2),$$
$$R_2 > I(X_2, U; \hat{Z}) \tag{21.7}$$

for some conditional pmf $p(u|x_1)p(\hat{z}|x_2, u)$ that satisfies the constraint $E(d(Z, \hat{Z})) \leq D$.

The local computing inner bound coincides with the cutset bound in (21.6) for the special case of lossless computing of Z. To show this, let d be a Hamming distortion measure and consider the case of $D = 0$. Then the inner bound in (21.7) simplifies to the set of rate pairs (R_1, R_2) such that

$$R_1 > I(X_1; U | X_2),$$
$$R_2 > H(Z)$$

for some conditional pmf $p(u|x_1)$ that satisfies the condition $H(Z|U, X_2) = 0$. Now, consider the cutset bound in 21.6 and set $\hat{Z} = Z$. Since $\hat{Z} \to (X_2, U) \to X_1$ form a Markov chain and Z is a function of (X_1, X_2), we must have $H(Z|U, X_2) = 0$. As in Theorem 21.2, the inequality $R_1 > I(X_1; U | X_2)$ can be further refined to the conditional graph entropy $H_G(X_1 | X_2)$.

Forwarding inner bound. Consider the following alternative coding scheme for cascade coding for computing. Encoder 1 again sends a description U_1 of its source X_1 at rate

$R_1 > I(X_1; U_1|X_2)$. Encoder 2 forwards the description from encoder 1 together with a description U_2 of its source X_2 given U_1 at total rate $R_2 > I(X_1; U_1) + I(X_2; U_2|U_1)$. The decoder then computes the estimate $\hat{Z}(U_1, U_2)$. This yields the *forwarding* inner bound on the rate–distortion region $\mathscr{R}_g(D)$ that consists of all rate pairs (R_1, R_2) such that

$$R_1 > I(X_1; U_1|X_2),$$
$$R_2 > I(U_1; X_1) + I(U_2; X_2|U_1) \tag{21.8}$$

for some conditional pmf $p(u_1|x_1)p(u_2|x_2, u_1)$ and function $\hat{z}(u_1, u_2)$ that satisfy the constraint $E(d(Z, \hat{Z})) \le D$.

It is straightforward to show that forwarding is optimal when X_1 and X_2 are independent and $Z = (X_1, X_2)$. The lossless computing example shows that local computing can outperform forwarding. The following is another example where the same conclusion holds.

Example 21.14. Let X_1 and X_2 be independent WGN(1) sources, $Z = X_1 + X_2$, and d be a squared error distortion measure. Consider the forwarding inner bound in (21.8). It can be shown that the minimum sum-rate for this region is attained by setting $U_1 = \hat{X}_1$, $U_2 = \hat{X}_2$, and independent Gaussian test channels. This yields the forwarding sum-rate

$$R_F = R_1 + R_2 = 2R\left(\frac{1}{D_1}\right) + R\left(\frac{1}{D_2}\right) = R\left(\frac{1}{D_1^2 D_2}\right)$$

such that $D_1 + D_2 \le D$. Optimizing over D_1, D_2 yields the minimum forwarding sum-rate

$$R_F^* = R\left(\frac{27}{4D^3}\right),$$

which is achieved for $D_1 = 2D_2 = 2D/3$.

Next consider the local computing inner bound in (21.7). Let $U = X_1 + W_1$, $V = E(X_1|U) + X_2 + W_2$, and $\hat{Z} = E(Z|V)$, where $W_1 \sim N(0, D_1/(1 - D_1))$ and $W_2 \sim N(0, (2 - D_1)D_2/(2 - D_1 - D_2))$ are independent of each other and of (X_1, X_2). The local computing sum-rate for this choice of test channels is

$$R_{LC} = R_1 + R_2 = R\left(\frac{1}{D_1}\right) + R\left(\frac{2 - D_1}{D_2}\right)$$

subject to the distortion constraint $D_1 + D_2 \le D$. Optimizing over D_1, D_2, we obtain

$$R_{LC}^* = R\left(\frac{1}{2\left(1 - \sqrt{1 - D/2}\right)^2}\right) < R_F^*$$

for all $D \in (0, 1]$. Thus local computing can outperform forwarding.

Combined inner bound. Local computing and forwarding can be combined as follows. Encoder 1 uses Wyner–Ziv coding to send the description pair (U, V) of X_1 to encoder 2.

Encoder 2 forwards the description V and the estimate $\hat{Z}$ based on (U, V, X_2) to the decoder. This yields the inner bound on the rate–distortion region $\mathcal{R}_g(D)$ that consists of all rate pairs (R_1, R_2) such that

$$R_1 > I(X_1; U, V | X_2),$$
$$R_2 > I(X_1; V) + I(X_2, U; \hat{Z} | V)$$

for some conditional pmf $p(u, v | x_1) p(\hat{z} | x_2, u, v)$ that satisfies $\mathsf{E}(d(Z, \hat{Z})) \le D$. This inner bound, however, does not coincide with the cutset bound in (21.6) in general.

Remark 21.3 (Cascade versus distributed coding for computing). The optimal rate region for lossless computing is known for the cascade coding case but is not known in general for the distributed coding case. This suggests that the cascade source coding problem is more tractable than distributed source coding for computing. This is not always the case, however. For example, while the rate–distortion region for Gaussian sources is known for the μ-sum problem, it is not known for the cascade case even when X_1 and X_2 are independent.

21.5 DISTRIBUTED LOSSY AVERAGING

Distributed averaging is a canonical example of distributed consensus, which has been extensively studied in control and computer science. The problem arises, for example, in distributed coordination of autonomous agents, distributed computing in sensor networks, and statistics collection in peer-to-peer networks. We discuss a lossy source coding formulation of this problem.

Consider a network modeled by a graph with N nodes and a set of undirected edges representing noiseless bidirectional links. Node $j \in [1 : N]$ observes an independent WGN(1) source X_j. Each node wishes to estimate the average $Z = (1/N) \sum_{j=1}^{N} X_j$ to the same prescribed mean squared error distortion D. Communication is performed in q rounds. In each round an edge (node-pair) is selected and the two nodes communicate over a noiseless bidirectional link in several subrounds using block codes as in two-way coding for lossy computing in Section 21.3. Let r_l be the total communication rate in round $l \in [1 : q]$ and define the *network sum-rate* as $R = \sum_{l=1}^{q} r_l$. A rate–distortion pair (R, D) is said to be achievable if there exist a number of rounds q and associated sequence of edge selections and codes such that

$$\limsup_{n \to \infty} \frac{1}{n} \sum_{i=1}^{n} \mathsf{E}((\hat{Z}_{ji} - Z_i)^2) \le D \quad \text{for all } j \in [1 : N],$$

where $Z_i = (1/N) \sum_{j=1}^{N} X_{ji}$ and $\hat{Z}_{ji}$ is the estimate of Z_i at node j. The *network rate–distortion function* $R(D)$ is the infimum of network sum-rates R such that (R, D) is achievable.

Note that $R(D) = 0$ if $D \ge (N-1)/N^2$, and is achieved by the estimates $\hat{Z}_{ji} = X_{ji}/N$ for $j \in [1 : N]$ and $i \in [1 : n]$. Also, $R(D)$ is known completely for $N = 2$; see Example 21.11.

The network rate–distortion function is not known in general, however. Consider the following upper and lower bounds.

Theorem 21.6 (Cutset Bound for Distributed Lossy Averaging). The network rate–distortion function for the distributed lossy averaging problem is lower bounded as

$$R(D) \geq N\,\mathsf{R}\left(\frac{N-1}{N^2 D}\right).$$

This cutset bound is established by considering the information flow into each node from the rest of the nodes when they are combined into a single "supernode." The bound turns out to be achievable within a factor of two for sufficiently large N and $D < 1/N^2$ for every connected network that contains a *star* subnetwork, that is, every network having a node with $N - 1$ edges.

Theorem 21.7. The network rate–distortion function for a network with star subnetwork is upper bounded as

$$R(D) \leq 2(N-1)\,\mathsf{R}\left(\frac{2N-3}{N^2 D}\right)$$

for $D < (N-1)/N^2$.

To prove this upper bound, suppose that there are $(N-1)$ edges between node 1 and nodes $2, \ldots, N$. Node $j \in [2:N]$ sends the description of the Gaussian reconstruction $\hat{X}_j$ of its source X_j to node 1 at a rate higher than $I(X_j; \hat{X}_j) = \mathsf{R}(1/D')$, where $D' \in (0,1)$ is determined later. Node 1 computes the estimates

$$\hat{Z}_1 = \frac{1}{N}X_1 + \frac{1}{N}\sum_{j=2}^{N}\hat{X}_j,$$

$$U_j = \hat{Z}_1 - \frac{1}{N}\hat{X}_j,$$

and sends the description of the Gaussian reconstruction $\hat{U}_j$ of U_j to node $j \in [2:N]$ at a rate higher than

$$I(U_j; \hat{U}_j) = \mathsf{R}\left(\frac{1}{D'}\right).$$

Node $j \in [2:N]$ then computes the estimate $\hat{Z}_j = (1/N)X_j + \hat{U}_j$. The corresponding distortion is

$$D_1 = \mathsf{E}\left[(\hat{Z}_1 - Z)^2\right] = \mathsf{E}\left[\left(\frac{1}{N}\sum_{j=2}^{N}(X_j - \hat{X}_j)\right)^2\right] = \frac{N-1}{N^2}D',$$

and for $j \in [2:N]$

$$D_j = \mathsf{E}\left[(\hat{Z}_j - Z)^2\right]$$

$$= \mathsf{E}\left[\left((\hat{U}_j - U_j) + \frac{1}{N}\sum_{k \neq 1, j}(\hat{X}_k - X_k)\right)^2\right]$$

$$= \left(\frac{1}{N^2} + \frac{N-2}{N^2}(1 - D')\right)D' + \frac{N-2}{N^2}D'$$

$$\leq \frac{2N-3}{N^2}D'.$$

Now choose $D' = N^2 D/(2N - 3)$ to satisfy the distortion constraint. Then every sum-rate higher than

$$2(N-1)\mathsf{R}\left(\frac{1}{D'}\right) = 2(N-1)\mathsf{R}\left(\frac{2N-3}{N^2 D}\right)$$

is achievable. This completes the proof of Theorem 21.7.

21.6   COMPUTING OVER A MAC

As we discussed in Chapter 14, source–channel separation holds for communicating independent sources over a DM-MAC, but does not necessarily hold when the sources are correlated. We show through an example that even when the sources are independent, separation does not necessarily hold when the goal is to compute a function of the sources over a DM-MAC.

Let (U_1, U_2) be a pair of independent Bern$(1/2)$ sources and suppose that the function $Z = U_1 \oplus U_2$ is to be computed over a modulo-2 sum DM-MAC followed by a BSC(p) as depicted in Figure 21.6. Sender $j = 1, 2$ encodes the source sequence U_j^k into a channel input sequence X_j^n. Hence the transmission rate is $r = k/n$ bits per transmission. We wish to find the necessary and sufficient condition for computing Z losslessly.

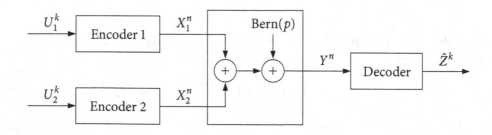

Figure 21.6. Computing over a MAC example.

If we use separate source and channel coding, $Z = U_1 \oplus U_2$ can be computed losslessly over the DM-MAC at rate r if

$$rH(U_1) < I(X_1; Y|X_2, Q),$$
$$rH(U_2) < I(X_2; Y|X_1, Q),$$
$$r(H(U_1) + H(U_2)) < I(X_1, X_2; Y|Q)$$

for some pmf $p(q)p(x_1|q)p(x_2|q)$. This reduces to $r(H(U_1) + H(U_2)) = 2r < 1 - H(p)$. Thus, a rate r is achievable via separate source and channel coding if $r < (1 - H(p))/2$. In the other direction, considering the cut at the input of the BSC part of the channel yields the upper bound

$$rH(U_1 \oplus U_2) \le H(X_1 \oplus X_2) \le 1 - H(p),$$

or equivalently, $r \le 1 - H(p)$.

We now show that the upper bound is achievable using random linear codes as in Example 21.10.

Codebook generation. Randomly and independently generate an $l \times k$ binary matrix A and an $n \times l$ binary matrix B, each with i.i.d. Bern(1/2) entries.

Encoding. Encoder 1 sends $X_1^n = BAU_1^k$ and encoder 2 sends $X_2^n = BAU_2^k$.

Decoding. The receiver first decodes for $A(U_1^k \oplus U_2^k)$ and then decodes for $Z^k = U_1^k \oplus U_2^k$. If $l < n(1 - H(p) - \delta(\epsilon))$, the receiver can recover $A(U_1^k \oplus U_2^k)$ with probability of error (averaged over the random matrix B) that tends to zero as $n \to \infty$; see Section 3.1.3. In addition, if $l > k(1 + \delta(\epsilon))$, then the receiver can recover Z^k with probability of error (averaged over the random matrix A) that tends to zero as $n \to \infty$. Therefore, $r < 1 - H(p)$ is achievable using joint source–channel coding.

SUMMARY

- Information theoretic framework for distributed computing
- Optimal rate for lossless computing with side information via conditional graph entropy
- μ-sum problem is equivalent to the quadratic Gaussian CEO problem
- Using the same linear code can outperform random codes for distributed lossless computing
- Interaction can reduce the rates in lossless coding for computing
- Cascade coding for computing:
  - Local computing
  - Forwarding
  - Neither scheme outperforms the other in general

- Information theoretic formulation of the distributed averaging problem

- Source–channel separation is suboptimal for computing even for independent sources

- **Open problems:**

  21.1. What is the optimal rate region for distributed lossless computing of $X_1 \cdot X_2$, where (X_1, X_2) is a DSBS?

  21.2. What is the rate–distortion region of cascade coding for computing the sum of independent WGN sources?

BIBLIOGRAPHIC NOTES

The rate–distortion function for lossy computing with side information in Theorem 21.1 was established by Yamamoto (1982). The refined characterization for the lossless case in Theorem 21.2 using conditional graph entropy and Example 21.2 are due to Orlitsky and Roche (2001). The rate–distortion region for the μ-sum in Theorem 21.3 was established by Wagner, Tavildar, and Viswanath (2008). Krithivasan and Pradhan (2009) showed that lattice coding can outperform Berger–Tung coding when $\mu_1\mu_2 < 0$. Han and Kobayashi (1987) established necessary and sufficient conditions on the function $g(X_1, X_2)$ such that the Slepian–Wolf region in (21.2) is optimal for distributed lossless computing. The lossless computing problem of the modulo-2 sum of a DSBS in Example 21.10 is due to Körner and Marton (1979). Their work has motivated the example of distributed computing over a MAC in Section 21.6 by Nazer and Gastpar (2007a) as well as coding schemes for Gaussian networks that use structured, instead of random, coding (Nazer and Gastpar 2007b, Wilson, Narayanan, Pfister, and Sprintson 2010).

Two-way coding for computing is closely related to communication complexity introduced by Yao (1979). In this setup each node has a value drawn from a finite set and both nodes wish to compute the same function with no errors. Communication complexity is measured by the minimum number of bits that need to be exchanged. Work in this area is described in Kushilevitz and Nisan (1997). Example 21.11 is due to Su and El Gamal (2009). The optimal two-round rate region for lossless computing in Theorem 21.5 was established by Orlitsky and Roche (2001), who also demonstrated the benefit of interaction via Example 21.12. The optimal rate region for an arbitrary number of rounds was established by Ma and Ishwar (2008), who showed that three rounds can strictly outperform two rounds via Example 21.13. They also characterized the optimal rate region $\mathscr{R}^* = \bigcup_q \mathscr{R}_q^*$ for this example. Cascade coding for computing in Section 21.4 was studied by Cuff, Su, and El Gamal (2009), who established the local computing, forwarding, and combined inner bounds.

The distributed averaging problem has been extensively studied in control and computer science, where each node is assumed to have a nonrandom real number and wishes to estimate the average by communicating synchronously, that is, in each round the nodes communicate with their neighbors and update their estimates simultaneously, or asynchronously, that is, in each round a randomly selected subset of the nodes communicate and update their estimates. The cost of communication is measured by the number of communication rounds needed. This setup was further extended to quantized averaging, where the nodes exchange quantized versions of their observations. Results on these formulations can be found, for example, in Tsitsiklis (1984), Xiao and Boyd (2004), Boyd, Ghosh, Prabhakar, and Shah (2006), Xiao, Boyd, and Kim (2007), Kashyap, Başar, and Srikant (2007), Nedić, Olshevsky, Ozdaglar, and Tsitsiklis (2009), and survey papers by Olfati-Saber, Fax, and Murray (2007) and Shah (2009). The information theoretic treatment in Section 21.5 is due to Su and El Gamal (2009). Another information theoretic formulation can be found in Ayaso, Shah, and Dahleh (2010).

PROBLEMS

21.1. Evaluate Theorem 21.1 to show that the rate–distortion function in Example 21.1 is
$R_g(D) = R((1 - \rho^2)P/4D)$.

21.2. Establish the upper and lower bounds on the optimal rate for lossless computing with side information in (21.1).

21.3. Evaluate Theorem 21.2 to show that the optimal rate in Example 21.2 is $R_g^* = 10$.

21.4. Prove achievability of the combined inner bound in Section 21.4.

21.5. Prove the cutset bound for distributed lossy averaging in Theorem 21.6.

21.6. *Cutset bound for distributed lossy averaging over tree networks.* Show that the cutset bound in Theorem 21.6 can be improved for a tree network to yield

$$R(D) \geq \frac{N-1}{2N} \log\left(\frac{1}{2N^3 D^2}\right).$$

(Hint: In a tree, removing an edge partitions the network into two disconnected subnetworks. Hence, we can use the cutset bound for two-way quadratic Gaussian lossy coding.)

21.7. *Compute–compress inner bound.* Consider the lossy source coding over a noiseless two-way relay channel in Problem 20.13, where (X_1, X_2) is a DSBS(p), and d_1 and d_2 are Hamming distortion measures. Consider the following compute–compress coding scheme. Node $j = 1, 2$ sends the index M_j to node 3 at rate R_j so that node 3 can losslessly recover their mod-2 sum $Z^n = X_1^n \oplus X_2^n$. The relay then performs lossy source coding on Z^n and sends the index M_3 at rate R_3 to nodes 1 and 2. Node 1 first reconstructs an estimate $\hat{Z}_1^n$ of Z^n and then computes the estimate $\hat{X}_2^n = X_1^n \oplus \hat{Z}_1^n$ of X_2^n. Node 2 computes the reconstruction $\hat{X}_1^n$ similarly.

(a) Find the inner bound on the rate–distortion region achieved by this coding scheme. (Hint: For lossless computing of Z^n at node 3, use random linear binning as in Example 21.10. For lossy source coding of Z^n at nodes 1 and 2, use the fact that Z^n is independent of X_1^n and of X_2^n.)

(b) Show that the inner bound in part (a) is strictly larger than the inner bound in part (b) of Problem 20.13 that uses two rounds of Wyner–Ziv coding.

(c) Use this compute–compress idea to establish the inner bound on the rate–distortion region for a general 2-DMS (X_1, X_2) and distortion measures d_1, d_2 that consists of all rate triples (R_1, R_2, R_3) such that

$$R_1 \geq I(X_1; U_1 | X_2, Q),$$
$$R_2 \geq I(X_2; U_2 | X_1, Q),$$
$$R_1 + R_2 \geq I(X_1, X_2; U_1, U_2 | Q),$$
$$R_3 \geq I(Z; W | X_1, Q),$$
$$R_3 \geq I(Z; W | X_2, Q)$$

for some conditional pmf $p(q)p(u_1|x_1, q)p(u_2|x_2, q)p(w|z, q)$ and functions $z(x_1, x_2)$, $\hat{x}_1(w, x_2)$, and $\hat{x}_2(w, x_1)$ that satisfy the constraints $H(V|U_1, U_2) = 0$ and

$$E[d_1(X_1, \hat{X}_1(W, X_2))] \leq D_1,$$
$$E[d_2(X_2, \hat{X}_2(W, X_1))] \leq D_2.$$

CHAPTER 22

Information Theoretic Secrecy

Confidentiality of information is a key consideration in many networking applications, including e-commerce, online banking, and intelligence operations. How can information be communicated reliably to the legitimate users, while keeping it secret from eavesdroppers? How does such a secrecy constraint on communication affect the limits on information flow in the network?

In this chapter, we study these questions under the information theoretic notion of secrecy, which requires each eavesdropper to obtain essentially no information about the messages sent from knowledge of its received sequence, the channel statistics, and the codebooks used. We investigate two approaches to achieve secure communication. The first is to exploit the statistics of the channel from the sender to the legitimate receivers and the eavesdroppers. We introduce the wiretap channel as a 2-receiver broadcast channel with a legitimate receiver and an eavesdropper, and establish its secrecy capacity, which is the highest achievable secret communication rate. The idea is to design the encoder so that the channel from the sender to the receiver becomes effectively stronger than the channel to the eavesdropper; hence the receiver can recover the message but the eavesdropper cannot. This wiretap coding scheme involves multicoding and randomized encoding.

If the channel from the sender to the receiver is weaker than that to the eavesdropper, however, secret communication at a positive rate is not possible. This brings us to the second approach to achieve secret communication, which is to use a secret key shared between the sender and the receiver but unknown to the eavesdropper. We show that the rate of such secret key must be at least as high as the rate of the confidential message. This raises the question of how the sender and the receiver can agree on such a long secret key in the first place. After all, if they had a confidential channel with sufficiently high capacity to communicate the key, then why not use it to communicate the message itself!

We show that if the sender and the receiver have access to correlated sources (e.g., through a satellite beaming common randomness to them), then they can still agree on a secret key even when the channel has zero secrecy capacity. We first consider the source model for key agreement, where the sender communicates with the receiver over a noiseless public broadcast channel to generate a secret key from their correlated sources. We establish the secret key capacity from one-way public communication. The coding scheme involves the use of double random binning to generate the key while keeping it secret from the eavesdropper. We then obtain upper and lower bounds on the secret key capacity for multiple rounds of communication and show that interaction can increase the key rate.

As a more general model for key agreement, we consider the channel model in which the sender broadcasts a random sequence over a noisy channel to generate a correlated output sequence at the receiver (and unavoidably another correlated sequence at the eavesdropper). The sender and the receiver then communicate over a noiseless public broadcast channel to generate the key from these correlated sequences as in the source model. We illustrate this setting via Maurer's example in which the channel to the receiver is a degraded version of the channel to the eavesdropper and hence no secret communication is possible without a key. We show that the availability of correlated sources through channel transmissions, however, makes it possible to generate a secret key at a positive rate.

22.1 WIRETAP CHANNEL

Consider the point-to-point communication system with an eavesdropper depicted in Figure 22.1. We assume a *discrete memoryless wiretap channel* (DM-WTC) $(\mathcal{X}, p(y, z|x), \mathcal{Y} \times \mathcal{Z})$ with sender X, legitimate receiver Y, and eavesdropper Z. The sender X (Alice) wishes to communicate a message M to the receiver Y (Bob) while keeping it secret from the eavesdropper Z (Eve).

A $(2^{nR}, n)$ secrecy code for the DM-WTC consists of

- a message set $[1 : 2^{nR}]$,

- a *randomized encoder* that generates a codeword $X^n(m)$, $m \in [1 : 2^{nR}]$, according to a conditional pmf $p(x^n|m)$, and

- a decoder that assigns an estimate $\hat{m} \in [1 : 2^{nR}]$ or an error message e to each received sequence $y^n \in \mathcal{Y}^n$.

The message M is assumed to be uniformly distributed over the message set. The *information leakage rate* associated with the $(2^{nR}, n)$ secrecy code is defined as

$$R_{\mathrm{L}}^{(n)} = \frac{1}{n} I(M; Z^n).$$

The average probability of error for the secrecy code is defined as $P_e^{(n)} = \mathrm{P}\{\hat{M} \neq M\}$. A rate–leakage pair (R, R_{L}) is said to be achievable if there exists a sequence of $(2^{nR}, n)$ codes

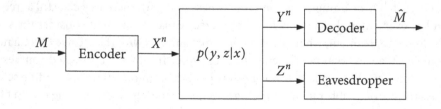

Figure 22.1. Point-to-point communication system with an eavesdropper.

such that

$$\lim_{n \to \infty} P_e^{(n)} = 0,$$

$$\limsup_{n \to \infty} R_{\mathrm{L}}^{(n)} \le R_{\mathrm{L}}.$$

The rate–leakage region $\mathscr{R}^*$ is the closure of the set of achievable rate–leakage pairs (R, R_{L}). As in the DM-BC (see Lemma 5.1), the rate–leakage region depends on the channel conditional pmf $p(y, z|x)$ only through the conditional marginals $p(y|x)$ and $p(z|x)$. We focus mainly on the *secrecy capacity* $C_{\mathrm{S}} = \max\{R \colon (R, 0) \in \mathscr{R}^*\}$.

Remark 22.1. As we already know, using a randomized encoder does not help communication when there is no secrecy constraint; see Problem 3.6. We show that under a secrecy constraint, however, a randomized encoder can help the sender hide the message from the eavesdropper; hence it can increase the secret communication rate.

The secrecy capacity has a simple characterization.

Theorem 22.1. The secrecy capacity of the DM-WTC is

$$C_{\mathrm{S}} = \max_{p(u,x)} \big(I(U; Y) - I(U; Z) \big),$$

where $|\mathcal{U}| \le |\mathcal{X}|$.

If the channel to the eavesdropper is a degraded version of the channel to the receiver, i.e., $p(y, z|x) = p(y|x)p(z|y)$,

$$I(U; Y) - I(U; Z) = I(U; Y|Z) \le I(X; Y|Z) = I(X; Y) - I(X; Z). \qquad (22.1)$$

Hence, the secrecy capacity simplifies to

$$C_{\mathrm{S}} = \max_{p(x)} \big(I(X; Y) - I(X; Z) \big). \qquad (22.2)$$

More generally, this capacity expression holds if Y is more capable than Z, i.e., $I(X; Y) \ge I(X; Z)$ for all $p(x)$. To show this, note that

$$I(U; Y) - I(U; Z) = I(X; Y) - I(X; Z) - \big(I(X; Y|U) - I(X; Z|U) \big)$$
$$\le I(X; Y) - I(X; Z),$$

since the more capable condition implies that $I(X; Y|U) - I(X; Z|U) \ge 0$ for all $p(u, x)$.

To illustrate Theorem 22.1 and the special case in (22.2), consider the following.

Example 22.1 (Binary symmetric wiretap channel). Consider the BS-WTC with channel outputs $Y = X \oplus Z_1$ and $Z = X \oplus Z_2$, where $Z_1 \sim \text{Bern}(p_1)$ and $Z_2 \sim \text{Bern}(p_2)$. The secrecy capacity of this BS-WTC is

$$C_{\mathrm{S}} = [H(p_2) - H(p_1)]^+,$$

where $[a]^+ = \max\{0, a\}$. If $p_1 \geq p_2$, the eavesdropper can recover any message intended for the receiver; thus $C_S = 0$. To prove the converse, we set $X \sim \text{Bern}(p)$ in (22.2) and consider

$$I(X; Y) - I(X; Z) = H(p_2) - H(p_1) - (H(p * p_2) - H(p * p_1)). \tag{22.3}$$

For $p < 1/2$, the function $p * u$ is monotonically increasing in $u \in [0, 1/2]$. This implies that $H(p * p_2) - H(p * p_1) \geq 0$. Hence, setting $p = 1/2$ optimizes the expression in (22.3) and C_S can be achieved using binary symmetric random codes.

Example 22.2 (Gaussian wiretap channel). Consider the Gaussian WTC with outputs $Y = X + Z_1$ and $Z = X + Z_2$, where $Z_1 \sim N(0, N_1)$ and $Z_2 \sim N(0, N_2)$. Assume an almost-sure average power constraint

$$P\left\{\sum_{i=1}^{n} X_i^2(m) \leq nP\right\} = 1, \quad m \in [1 : 2^{nR}].$$

The secrecy capacity of the Gaussian WTC is

$$C_S = \left[C\left(\frac{P}{N_1}\right) - C\left(\frac{P}{N_2}\right) \right]^+.$$

To prove the converse, assume without loss of generality that the channel is physically degraded and $Z_2 = Z_1 + Z_2'$, where $Z_2' \sim N(0, N_2 - N_1)$. Consider

$$I(X; Y) - I(X; Z) = \frac{1}{2} \log\left(\frac{N_2}{N_1}\right) - (h(Z) - h(Y)).$$

By the entropy power inequality (EPI),

$$h(Z) - h(Y) = h(Y + Z_2') - h(Y)$$

$$\geq \frac{1}{2} \log(2^{2h(Z_2')} + 2^{2h(Y)}) - h(Y)$$

$$= \frac{1}{2} \log(2\pi e(N_2 - N_1) + 2^{2h(Y)}) - h(Y).$$

Now since the function $(1/2) \log(2\pi e(N_2 - N_1) + 2^{2u}) - u$ is monotonically decreasing in u and $h(Y) \leq (1/2) \log(2\pi e(P + N_1))$,

$$h(Z) - h(Y) \geq \frac{1}{2} \log(2\pi e(N_2 - N_1) + 2\pi e(P + N_1)) - \frac{1}{2} \log(2\pi e(P + N_1))$$

$$= \frac{1}{2} \log\left(\frac{P + N_2}{P + N_1}\right).$$

Hence

$$C_S \leq \frac{1}{2} \log\left(\frac{N_2}{N_1}\right) - \frac{1}{2} \log\left(\frac{P + N_2}{P + N_1}\right) = C\left(\frac{P}{N_1}\right) - C\left(\frac{P}{N_2}\right),$$

which is attained by $X \sim N(0, P)$.

Example 22.2 can be further extended to the vector case, which is not degraded in general.

Example 22.3 (Gaussian vector wiretap channel). Consider the Gaussian vector WTC

$$\mathbf{Y} = G_1\mathbf{X} + \mathbf{Z}_1,$$
$$\mathbf{Z} = G_2\mathbf{X} + \mathbf{Z}_2,$$

where $\mathbf{Z}_1 \sim N(0, I)$ and $\mathbf{Z}_2 \sim N(0, I)$, and assume almost-sure average power constraint P on $\mathbf{X}$. It can be shown that the secrecy capacity is

$$C_S = \max_{K_{\mathbf{X}} \geq 0 : \mathrm{tr}(K_{\mathbf{X}}) \leq P} \left(\frac{1}{2} \log |I + G_1 K_{\mathbf{X}} G_1^T| - \frac{1}{2} \log |I + G_2 K_{\mathbf{X}} G_2^T| \right).$$

Note that the addition of the spatial dimension through multiple antennas allows for beamforming the signal away from the eavesdropper.

22.1.1 Achievability Proof of Theorem 22.1

The coding scheme for the DM-WTC is illustrated in Figure 22.2. We assume that $C_S > 0$ and fix the pmf $p(u, x)$ that attains it. Thus $I(U; Y) - I(U; Z) > 0$. We use multicoding and a two-step randomized encoding scheme to help hide the message from the eavesdropper. For each message m, we generate a subcodebook $C(m)$ consisting of $2^{n(\tilde{R}-R)}$ $u^n(l)$ sequences. To send m, the encoder randomly chooses one of the $u^n(l)$ sequences in its subcodebook. It then randomly generates the codeword $X^n \sim \prod_{i=1}^n p_{X|U}(x_i|u_i(l))$ and transmits it. The receiver can recover the index l if $\tilde{R} < I(U; Y)$. For each subcodebook $C(m)$, the eavesdropper has roughly $2^{n(\tilde{R}-R-I(U;Z))}$ $u^n(l)$ sequences such that $(u^n(l), z^n) \in \mathcal{T}_\epsilon^{(n)}$. Thus, if $\tilde{R} - R > I(U; Z)$, the eavesdropper has a roughly equal number (in the exponent) of jointly typical sequences from each subcodebook and hence has almost no information about the actual message sent.

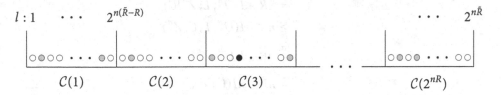

Figure 22.2. Coding scheme for the wiretap channel. The black circle corresponds to $(u^n(l), y^n) \in \mathcal{T}_\epsilon^{(n)}$ and the gray circles correspond to $(u^n(l), z^n) \in \mathcal{T}_\epsilon^{(n)}$.

In the following we provide details of this scheme.

Codebook generation. For each message $m \in [1 : 2^{nR}]$, generate a subcodebook $C(m)$ consisting of $2^{n(\tilde{R}-R)}$ randomly and independently generated sequences $u^n(l)$, $l \in [(m - 1)2^{n(\tilde{R}-R)} + 1 : m2^{n(\tilde{R}-R)}]$, each according to $\prod_{i=1}^n p_U(u_i)$. The codebook $C = \{C(m): m \in [1 : 2^{nR}]\}$ is revealed to all parties (including the eavesdropper).

Encoding. To send message $m \in [1 : 2^{nR}]$, the encoder chooses an index L uniformly at random from $[(m-1)2^{n(\tilde{R}-R)} + 1 : m2^{n(\tilde{R}-R)}]$, generates $X^n(m) \sim \prod_{i=1}^{n} p_{X|U}(x_i|u_i(L))$, and transmits it.

Decoding. The decoder finds the unique message $\hat{m}$ such that $(u^n(l), y^n) \in \mathcal{T}_{\epsilon}^{(n)}$ for some $u^n(l) \in \mathcal{C}(\hat{m})$.

Analysis of the probability of error. By the LLN and the packing lemma, the probability of error $\mathsf{P}(\mathcal{E})$ averaged over the random codebook and encoding tends to zero as $n \to \infty$ if $\tilde{R} < I(U; Y) - \delta(\epsilon)$.

Analysis of the information leakage rate. Let M be the message sent and L be the randomly selected index by the encoder. Every codebook $\mathcal{C}$ induces a conditional pmf on (M, L, U^n, Z^n) of the form

$$p(m, l, u^n, z^n \,|\, \mathcal{C}) = 2^{-nR} 2^{-n(\tilde{R}-R)} p(u^n \,|\, l, \mathcal{C}) \prod_{i=1}^{n} p_{Z|U}(z_i|u_i).$$

In particular, $p(u^n, z^n) = \prod_{i=1}^{n} p_{U,Z}(u_i, z_i)$. Now we bound the amount of information leakage averaged over codebooks. Consider

$$
\begin{aligned}
I(M; Z^n | \mathcal{C}) &= H(M|\mathcal{C}) - H(M|Z^n, \mathcal{C}) \\
&= nR - H(M, L|Z^n, \mathcal{C}) + H(L|Z^n, M, \mathcal{C}) \\
&= nR - H(L|Z^n, \mathcal{C}) + H(L|Z^n, M, \mathcal{C}), \quad (22.4)
\end{aligned}
$$

where the last step follows since the message M is a function of the index L. We first establish a lower bound on the equivocation term $H(L|Z^n, \mathcal{C})$ in (22.4), which is tantamount to lower bounding the number of U^n sequences that are jointly typical with Z^n. Consider

$$
\begin{aligned}
H(L|Z^n, \mathcal{C}) &= H(L|\mathcal{C}) - I(L; Z^n|\mathcal{C}) \\
&= n\tilde{R} - I(L; Z^n|\mathcal{C}) \\
&= n\tilde{R} - I(U^n, L; Z^n|\mathcal{C}) \\
&\geq n\tilde{R} - I(U^n, L, \mathcal{C}; Z^n) \\
&\overset{(a)}{=} n\tilde{R} - I(U^n; Z^n) \\
&\overset{(b)}{=} n\tilde{R} - nI(U; Z), \quad (22.5)
\end{aligned}
$$

where (a) follows since $(L, \mathcal{C}) \to U^n \to Z^n$ form a Markov chain and (b) follows since by construction $p(u^n, z^n) = \prod_{i=1}^{n} p_{U,Z}(u_i, z_i)$. Next, we establish an upper bound on the second equivocation term in (22.4), which is tantamount to upper bounding the number of U^n sequences in the subcodebook $\mathcal{C}(M)$ that are jointly typical with Z^n.

Lemma 22.1. If $\tilde{R} - R \geq I(U; Z)$, then

$$\limsup_{n \to \infty} \frac{1}{n} H(L|Z^n, M, \mathcal{C}) \leq \tilde{R} - R - I(U; Z) + \delta(\epsilon).$$

The proof of this lemma is given in Appendix 22A. Substituting from (22.5) and this lemma into (22.4) and using the condition that $\tilde{R} < I(U;Y) - \delta(\epsilon)$, we have shown that $\limsup_{n\to\infty}(1/n)I(M;Z^n|\mathcal{C}) \le \delta(\epsilon)$, if $R < I(U;Y) - I(U;Z) - \delta(\epsilon)$. Therefore, there must exist a sequence of codes such that $P_e^{(n)}$ tends to zero and $R_L^{(n)} \le \delta(\epsilon)$ as $n \to \infty$, which completes the proof of achievability.

Remark 22.2. Instead of sending a random X^n, we can randomly generate a *satellite codebook* consisting of $2^{nR'}$ codewords $x^n(l, l')$ for each l. The encoder then randomly transmits one of the codewords in the chosen satellite codebook. To achieve secrecy, we must have $R' > I(X;Z|U) + \delta(\epsilon)$ in addition to $\tilde{R} - R \ge I(U;Z)$. As we discuss in Section 22.1.3, this approach is useful for establishing lower bounds on the secrecy capacity of wiretap channels with more than one legitimate receiver.

22.1.2   Converse Proof of Theorem 22.1

Consider any sequence of $(2^{nR}, n)$ codes such that $P_e^{(n)}$ tends to zero and $R_L^{(n)} \le \epsilon$ as $n \to \infty$. Then, by Fano's inequality, $H(M|Y^n) \le n\epsilon_n$ for some ϵ_n that tends to zero as $n \to \infty$. Now consider

$$nR \le I(M;Y^n) + n\epsilon_n$$

$$= I(M;Y^n) - I(M;Z^n) + nR_L^{(n)} + n\epsilon_n$$

$$\le \sum_{i=1}^{n}(I(M;Y_i|Y^{i-1}) - I(M;Z_i|Z_{i+1}^n)) + n(\epsilon + \epsilon_n)$$

$$\overset{(a)}{=} \sum_{i=1}^{n}(I(M,Z_{i+1}^n;Y_i|Y^{i-1}) - I(M,Y^{i-1};Z_i|Z_{i+1}^n)) + n(\epsilon + \epsilon_n)$$

$$\overset{(b)}{=} \sum_{i=1}^{n}(I(M;Y_i|Y^{i-1},Z_{i+1}^n) - I(M;Z_i|Y^{i-1},Z_{i+1}^n)) + n(\epsilon + \epsilon_n)$$

$$\overset{(c)}{=} \sum_{i=1}^{n}(I(U_i;Y_i|V_i) - I(U_i;Z_i|V_i)) + n(\epsilon + \epsilon_n)$$

$$\overset{(d)}{=} n(I(U;Y|V) - I(U;Z|V)) + n(\epsilon + \epsilon_n)$$

$$\le n \max_v(I(U;Y|V = v) - I(U;Z|V = v)) + n(\epsilon + \epsilon_n)$$

$$\overset{(e)}{\le} nC_S + n(\epsilon + \epsilon_n),$$

where (a) and (b) follow by the Csiszár sum identity in Section 2.3, (c) follows by identifying the auxiliary random variables $V_i = (Y^{i-1}, Z_{i+1}^n)$ and $U_i = (M, V_i)$, and (d) follows by introducing a time-sharing random variable Q and defining $U = (U_Q, Q)$, $V = (V_Q, Q)$, $Y = Y_Q$, and $Z = Z_Q$. Step (e) follows since $U \to X \to (Y, Z)$ form a Markov chain given $\{V = v\}$. The bound on the cardinality of U can be proved using the convex cover method in Appendix C. This completes the proof of the converse.

22.1.3 Extensions

The wiretap channel model and its secrecy capacity can be extended in several directions.

Secrecy capacity for multiple receivers or eavesdroppers. The secrecy capacity for more than one receiver and/or eavesdropper is not known. We discuss a lower bound on the secrecy capacity of a 2-receiver 1-eavesdropper wiretap channel that involves several ideas beyond wiretap channel coding. Consider a DM-WTC $p(y_1, y_2, z|x)$ with sender X, two legitimate receivers Y_1 and Y_2, and a single eavesdropper Z. The sender wishes to communicate a common message $M \in [1:2^{nR}]$ reliably to both receivers Y_1 and Y_2 while keeping it asymptotically secret from the eavesdropper Z, i.e., $\lim_{n\to\infty}(1/n)I(M; Z^n) = 0$.

A straightforward extension of Theorem 22.1 to this case yields the lower bound

$$C_S \geq \max_{p(u,x)} \min\{I(U; Y_1) - I(U; Z), \ I(U; Y_2) - I(U; Z)\}. \tag{22.6}$$

Now suppose that Z is a degraded version of Y_1. Then from (22.1), $I(U; Y_1) - I(U; Z) \leq I(X; Y_1) - I(X; Z)$ for all $p(u, x)$, with equality if $U = X$. However, no such inequality holds in general for the second term; hence taking $U = X$ does not attain the maximum in (22.6) in general. Using the satellite codebook idea in Remark 22.2 and indirect decoding in Section 8.2, we can obtain the following tighter lower bound.

Proposition 22.1. The secrecy capacity of the 2-receiver 1-eavesdropper DM-WTC $p(y_1, y_2, z|x)$ is lower bounded as

$$C_S \geq \max_{p(u,x)} \min\{I(X; Y_1) - I(X; Z), \ I(U; Y_2) - I(U; Z)\}.$$

To prove achievability of this lower bound, for each message $m \in [1:2^{nR}]$, generate a subcodebook $\mathcal{C}(m)$ consisting of $2^{n(\tilde{R}-R)}$ randomly and independently generated sequences $u^n(l_0)$, $l_0 \in [(m-1)2^{n(\tilde{R}-R)} + 1 : m2^{n(\tilde{R}-R)}]$, each according to $\prod_{i=1}^{n} p_U(u_i)$. For each l_0, conditionally independently generate 2^{nR_1} sequences $x^n(l_0, l_1)$, $l_1 \in [1:2^{nR_1}]$, each according to $\prod_{i=1}^{n} p_{X|U}(x_i|u_i(l_0))$. To send $m \in [1:2^{nR}]$, the encoder chooses an index pair (L_0, L_1) uniformly at random from $[(m-1)2^{n(\tilde{R}-R)} + 1 : m2^{n(\tilde{R}-R)}] \times [1:2^{nR_1}]$, and transmits $x^n(L_0, L_1)$. Receiver Y_2 decodes for the message directly through U and receiver Y_1 decodes for the message *indirectly* through (U, X) as detailed in Section 8.1. Then the probability of error tends to zero as $n \to \infty$ if

$$\tilde{R} < I(U; Y_2) - \delta(\epsilon),$$
$$\tilde{R} + R_1 < I(X; Y_1) - \delta(\epsilon).$$

It can be further shown that M is kept asymptotically secret from the eavesdropper Z if

$$\tilde{R} - R > I(U; Z) + \delta(\epsilon),$$
$$R_1 > I(X; Z|U) + \delta(\epsilon).$$

Eliminating $\tilde{R}$ and R_1 by the Fourier–Motzkin procedure in Appendix D, we obtain the lower bound in Proposition 22.1.

Broadcast channel with common and confidential messages. Consider a DM-BC $p(y, z|x)$, where the sender wishes to communicate a common message M_0 reliably to both receivers Y and Z in addition to communicating a confidential message M_1 to Y and keeping it partially secret from Z. A $(2^{nR_0}, 2^{nR_1}, n)$ code for the DM-BC with common and confidential messages consists of

- two message sets $[1 : 2^{nR_0}]$ and $[1 : 2^{nR_1}]$,

- an encoder that randomly assigns a codeword $X^n(m_0, m_1)$ to each (m_0, m_1) according to a conditional pmf $p(x^n|m_0, m_1)$, and

- two decoders, where decoder 1 assigns an estimate $(\hat{m}_{01}, \hat{m}_{11}) \in [1 : 2^{nR_0}] \times [1 : 2^{nR_1}]$ or an error message e to each sequence y^n, and decoder 2 assigns an estimate $\hat{m}_{02} \in [1 : 2^{nR_0}]$ or an error message e to each sequence z^n.

Assume that the message pair (M_0, M_1) is uniformly distributed over $[1 : 2^{nR_0}] \times [1 : 2^{nR_1}]$. The average probability of error is defined as $P_e^{(n)} = \mathsf{P}\{(\hat{M}_{01}, \hat{M}_1) \neq (M_0, M_1) \text{ or } \hat{M}_{02} \neq M_0\}$. The information leakage rate at receiver Z is defined as $R_L^{(n)} = (1/n)I(M_1; Z^n)$. A rate–leakage triple (R_0, R_1, R_L) is said to be achievable if there exists a sequence of codes such that $\lim_{n\to\infty} P_e^{(n)} = 0$ and $\limsup_{n\to\infty} R_L^{(n)} \leq R_L$. The rate–leakage region is the closure of the set of achievable rate triples (R_0, R_1, R_L).

Theorem 22.2. The rate–leakage region of the 2-receiver DM-BC $p(y, z|x)$ is the set of rate triples (R_0, R_1, R_L) such that

$$R_0 \leq \min\{I(U; Z), I(U; Y)\},$$
$$R_1 \leq [I(V; Y|U) - I(V; Z|U)]^+ + R_L,$$
$$R_0 + R_1 \leq I(U; Z) + I(V; Y|U),$$
$$R_0 + R_1 \leq I(V; Y)$$

for some pmf $p(u)p(v|u)p(x|v)$ with $|\mathcal{U}| \leq |\mathcal{X}| + 3$, $|\mathcal{V}| \leq (|\mathcal{X}| + 1)(|\mathcal{X}| + 3)$, where $[a]^+ = \max\{a, 0\}$.

The proof of achievability uses superposition coding and randomized encoding for M_1 as in the achievability proof of Theorem 22.1. The proof of the converse follows similar steps to the converse proofs for the DM-BC with degraded message sets in Section 8.1 and for the DM-WTC in Section 22.1.2.

22.2   CONFIDENTIAL COMMUNICATION VIA SHARED KEY

If the channel to the eavesdropper is less noisy than the channel to the receiver, no secret communication can take place at a positive rate without a secret key shared between the sender and the receiver. How long must this secret key be to ensure secrecy?

To answer this question, we consider the secure communication system depicted in Figure 22.3, where the sender wishes to communicate a message M to the receiver over a public DMC $p(y|x)$ in the presence of an eavesdropper who observes the channel output. The sender and the receiver share a secret key K, which is unknown to the eavesdropper, and use the key to keep the message secret from the eavesdropper. We wish to find the optimal tradeoff between the secrecy capacity and the rate of the secret key.

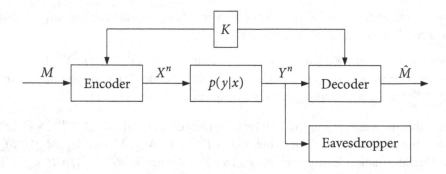

Figure 22.3. Secure communication over a public channel.

A $(2^{nR}, 2^{nR_K}, n)$ secrecy code for the DMC consists of

- a message set $[1 : 2^{nR}]$ and a key set $[1 : 2^{nR_K}]$,

- a *randomized* encoder that generates a codeword $X^n(m, k) \in \mathcal{X}^n$ according to a conditional pmf $p(x^n|m, k)$ for each message–key pair $(m, k) \in [1 : 2^{nR}] \times [1 : 2^{nR_K}]$, and

- a decoder that assigns an estimate $\hat{m} \in [1 : 2^{nR}]$ or an error message e to each received sequence $y^n \in \mathcal{Y}^n$ and key $k \in [1 : 2^{nR_K}]$.

We assume that the message–key pair (M, K) is uniformly distributed over $[1 : 2^{nR}] \times [1 : 2^{nR_K}]$. The *information leakage rate* associated with the $(2^{nR}, 2^{nR_K}, n)$ secrecy code is $R_L^{(n)} = (1/n)I(M; Y^n)$. The average probability of error and achievability are defined as for the DM-WTC. The rate–leakage region $\mathcal{R}^*$ is the set of achievable rate triples (R, R_K, R_L). The secrecy capacity with key rate R_K is defined as $C_S(R_K) = \max\{R: (R, R_K, 0) \in \mathcal{R}^*\}$.

The secrecy capacity is upper bounded by the key rate until saturated by the channel capacity.

Theorem 22.3. The secrecy capacity of the DMC $p(y|x)$ with key rate R_K is

$$C_S(R_K) = \min\left\{R_K, \max_{p(x)} I(X; Y)\right\}.$$

To prove achievability for this theorem, assume without loss of generality that $R = R_K$. The sender first encrypts the message M by taking the bitwise modulo-2 sum L of the binary expansions of M and K (referred to as a *one-time pad*), and transmits the encrypted message L using an optimal channel code for the DMC. If $R < C_S$, the receiver can recover L and decrypt it to recover M. Since M and L are independent, M and Y^n are also independent. Thus, $I(M; Y^n) = 0$ and we can achieve *perfect secrecy*.

To prove the converse, consider any sequence of $(2^{nR}, 2^{nR_K}, n)$ codes such that $P_e^{(n)}$ tends to zero and $R_L^{(n)} \le \epsilon$ as $n \to \infty$. Then we have

$$nR \le I(M; Y^n, K) + n\epsilon_n$$
$$= I(M; Y^n | K) + n\epsilon_n$$
$$\le I(M, K; Y^n) + n\epsilon_n$$
$$\le I(X^n; Y^n) + n\epsilon_n$$
$$\le n \max_{p(x)} I(X; Y) + n\epsilon_n.$$

Furthermore,

$$nR \le H(M|Y^n) + I(M; Y^n)$$
$$\overset{(a)}{\le} H(M|Y^n) + n\epsilon$$
$$\le H(M, K|Y^n) + n\epsilon$$
$$= H(K|Y^n) + H(M|Y^n, K) + n\epsilon$$
$$\overset{(b)}{\le} nR_K + n(\epsilon + \epsilon_n),$$

where (a) follows by the secrecy constraint and (b) follows by Fano's inequality. This completes the proof of Theorem 22.3.

Wiretap channel with secret key. A secret key between the sender and the receiver can be combined with wiretap channel coding to further increase the secrecy capacity. If R_K is the rate of the secret key, then the secrecy capacity of a more capable DM-WTC $p(y, z|x)$ is

$$C_S(R_K) = \max_{p(x)} \min\{I(X; Y) - I(X; Z) + R_K, \, I(X; Y)\}. \tag{22.7}$$

Such a secret key can be generated via interactive communication. For example, if noiseless causal feedback is available from the receiver to the sender but not to the eavesdropper, then it can be distilled into a secret key; see Problem 22.10. In the following, we show that even nonsecure interaction can be used to generate a secret key.

22.3   SECRET KEY AGREEMENT: SOURCE MODEL

Suppose that the sender and the receiver observe correlated sources. Then it turns out that they can agree on a secret key through interactive communication over a public channel that the eavesdropper has complete access to. We first discuss this key agreement scheme under the *source model*.

Consider a network with two sender–receiver nodes, an eavesdropper, and a 3-DMS (X_1, X_2, Z) as depicted in Figure 22.4. Node 1 observes the DMS X_1, node 2 observes the DMS X_2, and the eavesdropper observes the DMS Z. Nodes 1 and 2 communicate over a noiseless public broadcast channel that the eavesdropper has complete access to with the goal of agreeing on a key that the eavesdropper has almost no information about. We wish to find the maximum achievable secret key rate. We assume that the nodes communicate in a round robin fashion over q rounds as in the interactive source coding setup studied in Section 20.3. Assume without loss of generality that q is even and that node 1 transmits during the odd rounds $l = 1, 3, \ldots, q - 1$ and node 2 transmits during the even rounds.

A $(2^{nr_1}, \ldots, 2^{nr_q}, n)$ key agreement code consists of

- two randomized encoders, where in odd rounds $l = 1, 3, \ldots, q - 1$, encoder 1 generates an index $M_l \in [1 : 2^{nr_l}]$ according to a conditional pmf $p(m_l | x_1^n, m^{l-1})$ (that is, a random mapping given its source vector and all previously transmitted indices), and in even rounds $l = 2, 4, \ldots, q$, encoder 2 generates an index $M_l \in [1 : 2^{nr_l}]$ according to $p(m_l | x_2^n, m^{l-1})$, and

- two decoders, where decoder $j = 1, 2$ generates a key K_j according to a conditional pmf $p(k_j | m^q, x_j^n)$ (that is, a random mapping given its source sequence and all received indices).

The probability of error for the key agreement code is defined as $P_e^{(n)} = \mathsf{P}\{K_1 \neq K_2\}$. The *key leakage rate* is defined as $R_{\mathrm{L}}^{(n)} = \max_{j \in \{1,2\}} (1/n) I(K_j; Z^n, M^q)$. A key rate–leakage pair (R, R_{L}) is said to be achievable if there exists a sequence of $(2^{nr_1}, 2^{nr_2}, \ldots, 2^{nr_q}, n)$ codes such that $\lim_{n \to \infty} P_e^{(n)} = 0$, $\limsup_{n \to \infty} R_{\mathrm{L}}^{(n)} \leq R_{\mathrm{L}}$, and $\liminf_{n \to \infty} (1/n) H(K_j) \geq R$ for $j = 1, 2$. The key rate–leakage region $\mathscr{R}^*$ is the closure of the set of achievable rate-leakage pairs (R, R_{L}). As in the wiretap channel, we focus on the *secret key capacity* $C_{\mathrm{K}} = \max\{R : (R, 0) \in \mathscr{R}^*\}$. The secret key capacity is not known in general.

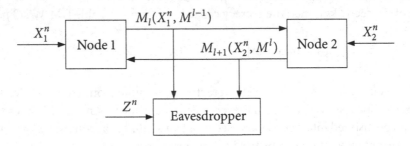

Figure 22.4. Source model for secret key agreement.

22.3.1 Secret Key Agreement from One-Way Communication

We first consider the case where the public communication is from node 1 to node 2 only, i.e., $q = 1$. We establish the following single-letter characterization of the one-round secret key capacity.

Theorem 22.4. The one-round secret key capacity for the 3-DMS (X_1, X_2, Z) is

$$C_{\mathrm{K}}^{(1)} = \max_{p(u,v|x_1)} \left(I(V; X_2|U) - I(V; Z|U) \right),$$

where $|\mathcal{U}| \le |\mathcal{X}_1|$ and $|\mathcal{V}| \le |\mathcal{X}_1|$.

Note that the auxiliary random variable U can be taken as a function of V with cardinality bounds $|\mathcal{U}| \le |\mathcal{X}_1|$ and $|\mathcal{V}| \le |\mathcal{X}_1|^2$. Also note that if $X_1 \to X_2 \to Z$ form a Markov chain, then the one-round secret key capacity simplifies to $C_{\mathrm{K}}^{(1)} = I(X_1; X_2|Z)$.

Remark 22.3 (One-way secret key agreement). Suppose that the public communication is either from node 1 to node 2 or vice versa, that is, $M^q = (M_1, \emptyset)$ or $(\emptyset, M_2)$. By changing the roles of nodes 1 and 2 in Theorem 22.4, it can be readily shown that the *one-way secret key capacity* is

$$\begin{aligned}
C_{\mathrm{K\text{-}OW}} = \max\{ & \max_{p(u,v|x_1)} \left(I(V; X_2|U) - I(V; Z|U) \right), \\
& \max_{p(u,v|x_2)} \left(I(V; X_1|U) - I(V; Z|U) \right) \}.
\end{aligned} \tag{22.8}$$

Clearly, $C_{\mathrm{K}}^{(1)} \le C_{\mathrm{K\text{-}OW}} \le C_{\mathrm{K}}$. As we show in Example 22.4, these inequalities can be strict.

Converse proof of Theorem 22.4. Consider

$$nR \le H(K_1)$$
$$\overset{(a)}{\le} H(K_1) - H(K_1|X_2^n, M) + n\epsilon_n$$
$$= I(K_1; X_2^n, M) + n\epsilon_n$$
$$\overset{(b)}{\le} I(K_1; X_2^n, M) - I(K_1; Z^n, M) + n(\epsilon + \epsilon_n)$$
$$= I(K_1; X_2^n|M) - I(K_1; Z^n|M) + n(\epsilon + \epsilon_n)$$
$$\overset{(c)}{=} \sum_{i=1}^{n} \left(I(K_1; X_{2i}|M, X_2^{i-1}, Z_{i+1}^n) - I(K_1; Z_i|M, X_2^{i-1}, Z_{i+1}^n) \right) + n(\epsilon + \epsilon_n)$$
$$\overset{(d)}{=} \sum_{i=1}^{n} \left(I(V_i; X_{2i}|U_i) - I(V_i; Z_i|U_i) \right) + n(\epsilon + \epsilon_n)$$
$$\overset{(e)}{=} n \left(I(V; X_2|U) - I(V; Z|U) \right) + n(\epsilon + \epsilon_n),$$

where (a) follows by Fano's inequality, (b) follows by the secrecy constraint, (c) follows by the converse proof of Theorem 22.1, (d) follows by identifying $U_i = (M, X_2^{i-1}, Z_{i+1}^n)$ and $V_i = (K_1, U_i)$, and (e) follows by introducing a time-sharing random variable Q and defining $U = (U_Q, Q)$, $V = (V_Q, Q)$, $X_{1Q} = X_1$, $X_{2Q} = X_2$, and $Z_Q = Z$. Finally, observing that the pmf $p(u, v|x_1, x_2, z) = p(v|x_1)p(u|v)$ completes the converse proof of Theorem 22.4.

22.3.2 Achievability Proof of Theorem 22.4

We first consider the special case of $U = \emptyset$ and $V = X_1$, and then generalize the result. Assume that $I(X_1; X_2) - I(X_1; Z) = H(X_1|Z) - H(X_1|X_2) > 0$. To generate the key, we use the *double random binning* scheme illustrated in Figure 22.5.

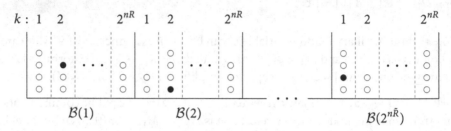

Figure 22.5. Key generation scheme. The black circles correspond to $(x_1^n, x_2^n) \in T_\epsilon^{(n)}$ and the white circles correspond to $(x_1^n, z^n) \in T_\epsilon^{(n)}$.

Codebook generation. Randomly and independently partition the set of sequences $\mathcal{X}_1^n$ into $2^{n\tilde{R}}$ bins $\mathcal{B}(m)$, $m \in [1 : 2^{n\tilde{R}}]$. Randomly and independently partition the sequences in each nonempty bin $\mathcal{B}(m)$ into 2^{nR} sub-bins $\mathcal{B}(m, k)$, $k \in [1 : 2^{nR}]$. The bin assignments are revealed to all parties.

Encoding. Given a sequence x_1^n, node 1 finds the index pair (m, k) such that $x_1^n \in \mathcal{B}(m, k)$. It then sends the index m to both node 2 and the eavesdropper.

Decoding and key generation. Node 1 sets its key $K_1 = k$. Upon receiving m, node 2 finds the unique sequence $\hat{x}_1^n \in \mathcal{B}(m)$ such that $(\hat{x}_1^n, x_2^n) \in T_\epsilon^{(n)}$. If such a unique sequence exists, then it sets $K_2 = \hat{k}$ such that $\hat{x}_1^n \in \mathcal{B}(m, \hat{k})$. If there is no such sequence or there is more than one, it sets $K_2 = 1$.

Analysis of the probability of error. We analyze the probability of error averaged over (X_1^n, X_2^n) and the random bin assignments. Since node 2 makes an error only if $\hat{X}_1^n \neq X_1^n$,

$$P(\mathcal{E}) = P\{K_1 \neq K_2\} \leq P\{\hat{X}_1^n \neq X_1^n\},$$

which, by the achievability proof of the Slepian–Wolf theorem, tends to zero as $n \to \infty$ if $\tilde{R} > H(X_1|X_2) + \delta(\epsilon)$.

Analysis of the key rate. We first establish the following bound on the key rate averaged over the random bin assignments $\mathcal{C}$.

Lemma 22.2. If $R < H(X_1) - 4\delta(\epsilon)$, then $\liminf_{n \to \infty}(1/n)H(K_1|\mathcal{C}) \geq R - \delta(\epsilon)$.

The proof of this lemma is in Appendix 22B.
Next we show that $H(K_2|\mathcal{C})$ is close to $H(K_1|\mathcal{C})$. By Fano's inequality,

$$H(K_2|\mathcal{C}) = H(K_1|\mathcal{C}) + H(K_2|K_1, \mathcal{C}) - H(K_1|K_2, \mathcal{C}) \leq H(K_1|\mathcal{C}) + 1 + nR\,P(\mathcal{E})$$

and similarly $H(K_2|\mathcal{C}) \geq H(K_1|\mathcal{C}) - 1 - nR\,\mathsf{P}(\mathcal{E})$. Thus, if $\tilde{R} > H(X_1|X_2) + \delta(\epsilon)$,

$$\lim_{n\to\infty} \frac{1}{n} \left| H(K_2|\mathcal{C}) - H(K_1|\mathcal{C}) \right| = 0, \qquad (22.9)$$

hence by Lemma 22.2, $\liminf_{n\to\infty}(1/n)H(K_2|\mathcal{C}) \geq R - \delta(\epsilon)$.

Analysis of the key leakage rate. Consider the key leakage rate averaged over $\mathcal{C}$. Following a similar argument to (22.9), we only need to consider the K_1 term

$$
\begin{aligned}
I(K_1; Z^n, M|\mathcal{C}) &= I(K_1, X_1^n; Z^n, M|\mathcal{C}) - I(X_1^n; Z^n, M|K_1, \mathcal{C}) \\
&= I(X_1^n; Z^n, M|\mathcal{C}) - H(X_1^n|K_1, \mathcal{C}) + H(X_1^n|Z^n, M, K_1, \mathcal{C}) \\
&= H(X_1^n) - H(X_1^n|Z^n, M, \mathcal{C}) - H(X_1^n|K_1, \mathcal{C}) + H(X_1^n|Z^n, M, K_1, \mathcal{C}).
\end{aligned}
$$

Now

$$
\begin{aligned}
H(X_1^n|Z^n, M, \mathcal{C}) &= H(X_1^n|Z^n) + H(M|Z^n, X_1^n, \mathcal{C}) - H(M|Z^n, \mathcal{C}) \\
&= H(X_1^n|Z^n) - H(M|Z^n, \mathcal{C}) \\
&\geq H(X_1^n|Z^n) - H(M) \\
&= n(H(X_1|Z) - \tilde{R})
\end{aligned}
$$

and

$$
\begin{aligned}
H(X_1^n|K_1, \mathcal{C}) &= H(X_1^n) + H(K_1|X_1^n, \mathcal{C}) - H(K_1|\mathcal{C}) \\
&= H(X_1^n) - H(K_1|\mathcal{C}) \\
&\geq n(H(X_1) - R).
\end{aligned}
$$

Substituting, we have

$$I(K_1; Z^n, M|\mathcal{C}) \leq n(\tilde{R} + R - H(X_1|Z)) + H(X_1^n|Z^n, M, K_1, \mathcal{C}). \qquad (22.10)$$

We bound the remaining term in the following.

Lemma 22.3. If $R < H(X_1|Z) - H(X_1|X_2) - 2\delta(\epsilon)$, then

$$\limsup_{n\to\infty} \frac{1}{n} H(X_1^n|Z^n, M, K_1, \mathcal{C}) \leq H(X_1|Z) - \tilde{R} - R + \delta(\epsilon).$$

The proof of this lemma is in Appendix 22C.

Substituting from this lemma into the bound in (22.10) and taking limits shows that $(1/n)I(K_1; Z^n, M|\mathcal{C}) \leq \delta(\epsilon)$ as $n \to \infty$ if $R < H(X_1|Z) - H(X_1|X_2) - 2\delta(\epsilon)$.

To summarize, we have shown that if $R < H(X_1|Z) - H(X_1|X_2) - 2\delta(\epsilon)$, then $\mathsf{P}(\mathcal{E})$ tends to zero as $n \to \infty$ and for $j = 1, 2$,

$$\liminf_{n\to\infty} \frac{1}{n} H(K_j|\mathcal{C}) \geq R - \delta(\epsilon),$$

$$\limsup_{n\to\infty} \frac{1}{n} I(K_j; Z^n, M|\mathcal{C}) \leq \delta(\epsilon).$$

Therefore, there exists a sequence of key agreement codes such that $P_e^{(n)}$ tends to zero, $H(K_j) \geq R - \delta(\epsilon)$, $j = 1, 2$, and $R_L^{(n)} \leq \delta(\epsilon)$ as $n \to \infty$.

Finally, to complete the proof of achievability for a general (U, V), we modify codebook generation as follows. Node 1 generates (U^n, V^n) according to $\prod_{i=1}^n p(u_i, v_i | x_{1i})$. Node 1 then sends U^n over the public channel. Following the above proof with V in place of X_1 at node 1, (U, X_2) at node 2, and (U, Z) at the eavesdropper proves the achievability of $R < I(V; X_2, U) - I(V; Z, U) = I(V; X_2 | U) - I(V; Z | U)$.

Remark 22.4 (One-round secret key agreement with rate constraint). As in the wiretap channel case (see Remark 22.2), instead of generating (U^n, V^n) randomly, we can generate codewords $u^n(l)$, $v^n(l, l')$ and use Wyner–Ziv coding to send the indices. Thus the one-round secret key capacity under a rate constraint $r_1 \leq R_1$ on the public discussion link is lower bounded as

$$C_K^{(1)}(R_1) \geq \max \left(I(V; X_2 | U) - I(V; Z | U) \right), \tag{22.11}$$

where the maximum is over all conditional pmfs $p(u, v | x_1)$ such that $R_1 \geq I(U, V; X_1) - I(U, V; X_2)$. It can be shown that this lower bound is tight, establishing the secret key capacity for this case.

22.3.3 Lower Bound on the Secret Key Capacity

The secret key capacity for general interactive communication is not known. We establish the following lower bound on the q-round secrecy capacity.

Theorem 22.5. The q-round secret key capacity for the 3-DMS (X_1, X_2, Z) is lower bounded as

$$C_K^{(q)} \geq \max \sum_{l=1}^q (I(U_l; X_{j_{l+1}} | U^{l-1}) - I(U_l; Z | U^{l-1}))$$

$$= \max \left(H(U^q | Z) - \sum_{l=1}^q H(U_l | U^{l-1}, X_{j_{l+1}}) \right),$$

where the maximum is over all conditional pmfs $\prod_{l=1}^q p(u_l | u^{l-1}, x_{j_l})$ and $j_l = 1$ if l is odd and $j_l = 2$ if l is even.

This lower bound can be strictly tighter than the one-way secret key capacity in (22.8), as demonstrated in the following.

Example 22.4. Let $X_1 = (X_{11}, X_{12})$, where X_{11} and X_{12} are independent Bern(1/2) random variables. The joint conditional pmf for $X_2 = (X_{21}, X_{22})$ and $Z = (Z_1, Z_2)$ is defined in Figure 22.6.

By setting $U_1 = X_{11}$ and $U_2 = X_{21}$ in Theorem 22.5, interactive communication can achieve the secret key rate

$$I(X_1; X_2 | Z) = I(X_{11}; X_{21} | Z_1) + I(X_{12}; X_{22} | Z_2), \tag{22.12}$$

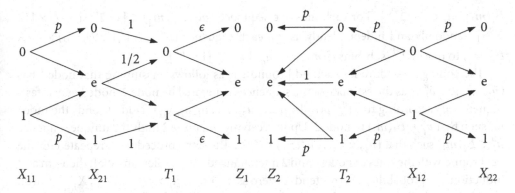

Figure 22.6. DMS for key agreement in Example 22.4; $p, \epsilon \in (0, 1)$.

which will be shown to be optimal in Theorem 22.6. This, however, is strictly larger than the one-round secrecy key capacity $C_K^{(1)}$.

To show this, first note that the one-round secret key capacity from X_1 to X_2 in Theorem 22.4 depends only on the marginal pmfs $p(x_1, x_2)$ and $p(x_1, z)$. As shown in Figure 22.6, $(X_1, X_2) = (X_{11}, X_{12}, X_{21}, X_{22})$ has the same joint pmf as $(X_1, X_2') = (X_{11}, X_{12}, X_{21}, T_2)$. Furthermore, $X_1 \to X_2' \to Z$ form a Markov chain. Thus, the one-round secret key capacity is

$$C_K^{(1)} = I(X_1; X_2'|Z) = I(X_{11}; X_{21}|Z_1) + I(X_{12}; T_2|Z_2).$$

But this rate is strictly less than the interactive communication rate in (22.12), since

$$
\begin{aligned}
I(X_{12}; T_2|Z_2) &= I(X_{12}; T_2) - I(X_{12}; Z_2) \\
&< I(X_{12}; T_2) - I(X_{22}; Z_2) \\
&= I(X_{12}; X_{22}) - I(X_{22}; Z_2) \\
&= I(X_{12}; X_{22}|Z_2).
\end{aligned}
$$

Similarly, note that $(X_1, X_2) = (X_{11}, X_{12}, X_{21}, X_{22})$ has the same joint pmf as $(X_1', X_2) = (T_1, X_{12}, X_{21}, X_{22})$, $X_2 \to X_1' \to Z$ form a Markov chain, and $I(X_{11}; Z_1) < I(X_{21}; Z_1)$. Hence, the one-round secret key capacity from X_2 to X_1 is strictly less than the interactive communication rate in (22.12).

Thus, as we have shown for channel coding in Chapter 17, source coding in Chapter 20, and coding for computing in Chapter 21, interaction can also improve the performance of coding for secrecy.

Proof of Theorem 22.5 (outline). Fix the pmf $\prod_{l=1}^q p(u_l|u^{l-1}, x_{j_l})$, $j_l = 1$ if l is odd and $j_l = 2$ if l is even, that attains the maximum in the lower bound in Theorem 22.5. For each $l \in [1:q]$, randomly and independently assign each sequence $u_l^n \in \mathcal{U}_l^n$ to one of $2^{n\tilde{R}_l}$ bins

$B_l(m_l)$, $m_l \in [1 : 2^{n\tilde{R}_l}]$. For each message sequence $(m_1, \ldots, m_q) \in [1 : 2^{n\tilde{R}_1}] \times \cdots \times [1 : 2^{n\tilde{R}_q}]$, randomly and independently assign each sequence $(u_1^n, \ldots, u_q^n) \in B_1(m_1) \times \cdots \times B_q(m_q)$ to one of 2^{nR} sub-bins $\mathcal{B}(m_1, \ldots, m_q, k)$, $k \in [1 : 2^{nR}]$.

The coding procedure for each odd round l is as follows. Assuming that node 1 has $(\hat{u}_2^n, \hat{u}_4^n, \ldots, \hat{u}_{l-1}^n)$ as the estimate of the sequences generated by node 2, node 1 generates a sequence u_l^n according to $\prod_{i=1}^n p(u_{li} | u_{1i}, \hat{u}_{2i}, u_{3i}, \ldots, \hat{u}_{l-1,i}, x_{1i})$. Node 1 sends the index m_l such that $u_l^n \in B_l(m_l)$ to node 2. Upon receiving m_l, node 2 finds the unique sequence $\hat{u}_l^n \in B_l(m_l)$ such that $(\hat{u}_1^n, u_2^n, \ldots, \hat{u}_l^n, x_2^n) \in \mathcal{T}_\epsilon^{(n)}$. The same procedure is repeated for the next round with the roles of nodes 1 and 2 interchanged. By the Slepian–Wolf theorem and induction, the probability of error tends to zero as $n \to \infty$ if $\tilde{R}_l > H(U_l | U^{l-1}, X_{j_{l+1}}) + \delta(\epsilon)$.

At the end of round q, if the unique set of sequences $(u_1^n, \hat{u}_2^n, \ldots, u_{l-1}^n, \hat{u}_l^n)$ is available at node 1, it finds the sub-bin index $\hat{k}$ to which the set of sequences belongs and sets $K_1 = \hat{k}$. Otherwise it sets K_1 to a random index in $[1 : 2^{nR}]$. Node 2 generates K_2 similarly. The key rates and the eavesdropper's key leakage rate can be analyzed by following similar steps to the one-way case in Section 22.3.1. If $\tilde{R}_l > H(U_l | U^{l-1}, X_{j_{l+1}}) + \delta(\epsilon)$ and $R + \sum_{l=1}^q \tilde{R}_l < H(U^q | Z)$, then $H(K_1 | \mathcal{C})$, $H(K_2 | \mathcal{C}) \geq n(R - \epsilon_n)$ and $(1/n)I(K_1; Z^n, M^q | \mathcal{C}) \leq \delta(\epsilon)$ as $n \to \infty$. This completes the proof of achievability.

Remark 22.5. As in the one-way communication case, the key rate may improve by first exchanging $U_1, U_2, \ldots, U_{q'-1}$ for some $q' \leq q$ and then conditionally double binning $U_{q'}, U_{q'+1}, \ldots, U_q$ to generate the key. This yields the lower bound

$$C_K^{(q)} \geq \max \sum_{l=q'}^q \left(I(U_l; X_{j_{l+1}} | U^{l-1}) - I(U_l; Z | U^{l-1}) \right),$$

where the maximum is over all conditional pmfs $\prod_{l=1}^q p(u_l | u^{l-1}, x_{j_l})$ and $q' \in [1 : q]$.

22.3.4 Upper Bound on the Secret Key Capacity

We establish the following general upper bound on the secret key capacity.

Theorem 22.6. The secret key capacity for the 3-DMS (X_1, X_2, Z) is upper bounded as

$$C_K \leq \min_{p(\tilde{z}|z)} I(X_1; X_2 | \tilde{Z}),$$

where $|\tilde{Z}| \leq |Z|$.

This bound is tight when $X_1 \to X_2 \to Z$ or $X_2 \to X_1 \to Z$, in which case $C_K = I(X_1; X_2 | Z)$. The proof follows immediately from the lower bound on the one-round secret key capacity in Theorem 22.4 and the Markov condition. In particular, consider the following interesting special cases.

- Z is known everywhere: Suppose that Z is a common part between X_1 and X_2. Then nodes 1 and 2 know the eavesdropper's sequence Z^n and the secrecy capacity is $C_K = I(X_1; X_2 | Z)$.

- $Z = \emptyset$: Suppose that the eavesdropper has access to the public communication but does not have correlated prior information. The secrecy capacity for this case is $C_K = I(X_1; X_2)$. This is in general strictly larger than the Gács–Körner–Witsenhausen common information $K(X_1; X_2)$, which is the maximum amount of common randomness nodes 1 and 2 can agree on *without* public information; see Section 14.2.1 and Problem 22.9.

The theorem also shows that the interactive secret key rate in (22.12) for Example 22.4 is optimal.

In all the cases above, the upper bound in Theorem 22.6 is tight with $\bar{Z} = Z$. In the following, we show that the upper bound can be strictly tighter with $\bar{Z} \neq Z$.

Example 22.5. Let X_1 and X_2 be independent Bern(1/2) sources and $Z = X_1 \oplus X_2$. Then $I(X_1; X_2|Z) = 1$. However, $I(X_1; X_2) = 0$. By taking $\bar{Z} = \emptyset$, the upper bound in Theorem 22.6 is tight and yields $C_K = 0$.

Proof of Theorem 22.6. We first consider the case where the encoding and key generation mappings are *deterministic* and establish the upper bound on the key rate $R \leq I(X_1; X_2|Z)$. We then generalize the proof for arbitrary $\bar{Z}$ and randomized codes. Given a sequence of deterministic codes such that $\lim_{n\to\infty} P_e^{(n)} = 0$ and

$$\max_{j\in\{1,2\}} \limsup_{n\to\infty} \frac{1}{n} I(K_j; Z^n, M^q) = 0,$$

we establish an upper bound on the key rate $\min\{(1/n)H(K_1), (1/n)H(K_2)\}$. By Fano's inequality, it suffices to provide a bound only on $(1/n)H(K_1)$. Consider

$$H(K_1) \leq H(K_1|M^q, Z^n) + I(K_1; M^q, Z^n)$$
$$\stackrel{(a)}{\leq} H(K_1|M^q, Z^n) + n\epsilon$$
$$= H(K_1, X_1^n|M^q, Z^n) - H(X_1^n|M^q, K_1, Z^n) + n\epsilon$$
$$= H(X_1^n|M^q, Z^n) + H(K_1|X_1^n, M^q, Z^n) - H(X_1^n|M^q, K_1, Z^n) + n\epsilon$$
$$\stackrel{(b)}{\leq} H(X_1^n|M^q, Z^n) - H(X_1^n|M^q, K_1, X_2^n, Z^n) + n\epsilon$$
$$= H(X_1^n|M^q, Z^n) - H(X_1^n, K_1|M^q, X_2^n, Z^n) + H(K_1|M^q, X_2^n, Z^n) + n\epsilon$$
$$\stackrel{(c)}{=} H(X_1^n|M^q, Z^n) - H(X_1^n|M^q, X_2^n, Z^n) + H(K_1|M^q, X_2^n, Z^n) + n\epsilon$$
$$\stackrel{(d)}{\leq} I(X_1^n; X_2^n|M^q, Z^n) + H(K_1|K_2) + n\epsilon$$
$$\stackrel{(e)}{\leq} I(X_1^n; X_2^n|M^q, Z^n) + n(\epsilon + \epsilon_n),$$

where (a) follows by the secrecy constraint, (b) and (c) follow since $H(K_1|M^q, X_1^n) = 0$, (d) follows since $H(K_1|M^q, X_2^n, Z^n) = H(K_1|M^q, K_2, X_2^n, Z^n) \leq H(K_1|K_2)$, and finally (e) follows by the condition that $P\{K_1 \neq K_2\} \leq \epsilon_n$ and Fano's inequality.

Next, since q is even by convention and encoding is deterministic by assumption,

$H(M_q|X_2^n, M^{q-1}) = 0.$ Hence

$$\begin{aligned}
I(X_1^n; X_2^n|M^q, Z^n) &= H(X_1^n|M^q, Z^n) - H(X_1^n|M^q, X_2^n, Z^n) \\
&= H(X_1^n|M^q, Z^n) - H(X_1^n|M^{q-1}, X_2^n, Z^n) \\
&\le H(X_1^n|M^{q-1}, Z^n) - H(X_1^n|M^{q-1}, X_2^n, Z^n) \\
&= I(X_1^n; X_2^n|Z^n, M^{q-1}).
\end{aligned}$$

Using the same procedure, we expand $I(X_1^n; X_2^n|M^{q-1}, Z^n)$ in the other direction to obtain

$$\begin{aligned}
I(X_1^n; X_2^n|M^{q-1}, Z^n) &= H(X_2^n|M^{q-1}, Z^n) - H(X_2^n|M^{q-1}, X_1^n, Z^n) \\
&\le I(X_1^n; X_2^n|M^{q-2}, Z^n).
\end{aligned}$$

Repeating this procedure q times, we obtain $I(X_1^n; X_2^n|M^q, Z^n) \le I(X_1^n; X_2^n|Z^n) = nI(X_1; X_2|Z)$. Substituting and taking $n \to \infty$ shows that $R = \limsup_{n \to \infty}(1/n)H(K_1) \le I(X_1; X_2|Z) + \epsilon$ for every $\epsilon > 0$ and hence $R \le I(X_1; X_2|Z)$.

We can readily improve the bound by allowing the eavesdropper to pass Z through an arbitrary channel $p(\bar{z}|z)$ to generate a new random variable $\bar{Z}$. However, any secrecy code for the original system with (X_1, X_2, Z) is also a secrecy code for $(X_1, X_2, \bar{Z})$, since

$$\begin{aligned}
I(K_1; \bar{Z}, M^q) &= H(K_1) - H(K_1|\bar{Z}, M^q) \\
&\le H(K_1) - H(K_1|\bar{Z}, Z, M^q) \\
&= I(K_1; M^q, Z).
\end{aligned}$$

This establishes the upper bound $R \le I(X_1; X_2|\bar{Z})$ for every $p(\bar{z}|z)$.

Finally, we consider the case with randomized encoding and key generation. Note that randomization is equivalent to having random variables W_1 at node 1 and W_2 at node 2 that are independent of each other and of (X_1^n, X_2^n, Z^n) and performing encoding and key generation using deterministic functions of the source sequence, messages, and the random variable at each node. Using this equivalence and the upper bound on the key rate for deterministic encoding obtained above, the key rate for randomized encoding is upper bounded as

$$\begin{aligned}
R &\le \min_{p(\bar{z}|z)} \sup_{p(w_1)p(w_2)} I(X_1, W_1; X_2, W_2|\bar{Z}) \\
&= \min_{p(\bar{z}|z)} \sup_{p(w_1)p(w_2)} \left(I(X_1, W_1; W_2|\bar{Z}) + I(X_1, W_1; X_2|\bar{Z}, W_2) \right) \\
&= \min_{p(\bar{z}|z)} \sup_{p(w_1)p(w_2)} I(X_1, W_1; X_2|\bar{Z}) \\
&= \min_{p(\bar{z}|z)} \sup_{p(w_1)p(w_2)} I(X_1; X_2|\bar{Z}) \\
&= \min_{p(\bar{z}|z)} I(X_1; X_2|\bar{Z}).
\end{aligned}$$

This completes the proof of Theorem 22.6.

22.3.5   Extensions to Multiple Nodes

Consider a network with N nodes, an eavesdropper, and an $(N + 1)$-DMS $(X_1, \dots, X_N, Z)$. Node $j \in [1 : N]$ observes source X_j and the eavesdropper observes Z. The nodes communicate over a noiseless public broadcast channel in a round robin fashion over q rounds, where q is divisible by N, such that node $j \in [1 : N]$ communicates in rounds $j, N + j, \dots,$ $q - N + j$. A $(2^{nr_1}, \dots, 2^{nr_q}, n)$ code for key agreement consists of

- a set of encoders, where in round $l_j \in \{j, N + j, \dots, q - N + j\}$, encoder $j \in [1 : N]$ generates an index $M_{l_j} \in [1 : 2^{nr_{l_j}}]$ according to a conditional pmf $p(m_{l_j}|x_j^n, m^{l_j-1})$, and

- a set of decoders, where decoder $j \in [1 : N]$ generates a key K_j according to a conditional pmf $p(k_j|m^q, x_j^n)$.

The probability of error for the key agreement code is defined as $P_e^{(n)} = 1 - \mathsf{P}\{K_1 = K_2 = \dots = K_N\}$. The key leakage rate, achievability, key rate–leakage region, and secret key capacity are defined as for the 2-node case in the beginning of this section. Consider the following bounds on the secret key capacity.

Lower bound. The lower bound on the secret key capacity for 2 nodes in Theorem 22.5 can be extended to N nodes.

Proposition 22.2. The q-round secret key capacity for the $(N + 1)$-DMS $(X_1, \dots, X_N, Z)$ is lower bounded as

$$C_K^{(q)} \geq \max \sum_{l=q'}^{q} \left(\min_{j \in [1:N]} I(U_l; X_j | U^{l-1}) - I(U_l; Z | U^{l-1}) \right),$$

where the maximum is over all conditional pmfs $\prod_{j=1}^{N} \prod_{l_j} p(u_{l_j}|u^{l_j-1}, x_j)$ and $q' \in [1 : q]$.

Upper bound. The upper bound on the secret key capacity for 2 nodes in Theorem 22.6 can also be extended to N nodes. The upper bound for N nodes has a compact representation in terms of the minimum sum-rate of the CFO problem in Section 20.3.2. Let $R_{\mathrm{CFO}}([1 : N]|Z)$ be the minimum sum-rate for the CFO problem when all the nodes know Z, that is, $R_{\mathrm{CFO}}([1 : N]|Z)$ for the $(N + 1)$-DMS $(X_1, \dots, X_N, Z)$ is equivalent to $R_{\mathrm{CFO}}([1 : N])$ for the N-DMS $((X_1, Z), \dots, (X_N, Z))$.

Proposition 22.3. The secret key capacity for the $(N + 1)$-DMS $(X_1, \dots, X_N, Z)$ is upper bounded as

$$C_K \leq \min_{p(\bar{z}|z)} (H(X^N|\bar{Z}) - R_{\mathrm{CFO}}([1 : N]|\bar{Z})),$$

where $|\bar{\mathcal{Z}}| \leq |\mathcal{Z}|$.

To establish this upper bound, it suffices to bound $H(K_1)$, which results in an upper bound on the achievable key rate. As for the 2-node case, we first establish the bound for $\bar{Z} = Z$ and no randomization. Consider

$$H(K_1) = H(K_1 | M^q, Z^n) + I(K_1; M^q, Z^n)$$

$$\overset{(a)}{\leq} H(K_1 | M^q, Z^n) + n\epsilon$$

$$= H(K_1, X^n([1:N]) | M^q, Z^n) - H(X^n([1:N]) | K_1, M^q, Z^n) + n\epsilon$$

$$\overset{(b)}{=} H(X^n([1:N]) | M^q, Z^n) - H(X^n([1:N]) | K_1, M^q, Z^n) + n\epsilon$$

$$= H(X^n([1:N]), M^q | Z^n) - H(M^q | Z^n) - H(X^n([1:N]) | K_1, M^q, Z^n) + n\epsilon$$

$$\overset{(c)}{=} H(X^n([1:N]) | Z^n) - H(M^q | Z^n) - H(X^n([1:N]) | K_1, M^q, Z^n) + n\epsilon$$

$$= nH(X^N | Z) - H(M^q | Z^n) - H(X^n([1:N]) | K_1, M^q, Z^n) + n\epsilon$$

$$\overset{(d)}{=} nH(X^N | Z) - \sum_{j=1}^{N} nR'_j + n\epsilon, \tag{22.13}$$

where (a) follows by the secrecy constraint, (b) and (c) follow since both K_1 and M^q are functions of $X^n([1:N])$, and (d) follows by defining

$$nR'_j = \sum_{l_j} H(M_{l_j} | M^{l_j-1}, Z^n) + H(X^n_j | K_1, M^q, X^n([1:j-1]), Z^n), \quad j \in [1:N].$$

Now for every proper subset $\mathcal{J}$ of $[1:N]$,

$$nH(X(\mathcal{J}) | X(\mathcal{J}^c), Z)$$

$$= H(X^n(\mathcal{J}) | X^n(\mathcal{J}^c), Z^n)$$

$$= H(M^q, K_1, X^n(\mathcal{J}) | X^n(\mathcal{J}^c), Z^n)$$

$$= \sum_{j=1}^{N} \sum_{l_j} H(M_{l_j} | M^{l_j-1}, X^n(\mathcal{J}^c), Z^n) + H(K_1 | M^q, X^n(\mathcal{J}^c), Z^n)$$

$$\quad + \sum_{j\in\mathcal{J}} H(X^n_j | K_1, M^q, X^n([1:j-1]), X^n(\mathcal{J}^c \cap [j+1:N]), Z^n)$$

$$\overset{(a)}{\leq} \sum_{j=1}^{N} \sum_{l_j} H(M_{l_j} | M^{l_j-1}, X^n(\mathcal{J}^c), Z^n) + \sum_{j\in\mathcal{J}} H(X^n_j | K_1, M^q, X^n([1:j-1]), Z^n) + n\epsilon_n$$

$$\overset{(b)}{=} \sum_{j\in\mathcal{J}} \left(\sum_{l_j} H(M_{l_j} | M^{l_j-1}, Z^n) + H(X^n_j | K_1, M^q, X^n([1:j-1]), Z^n) \right) + n\epsilon_n$$

$$= \sum_{j\in\mathcal{J}} nR'_j + n\epsilon_n,$$

where (a) follows by Fano's inequality (since $\mathcal{J}^c$ is nonempty) and (b) follows since M_{l_j} is a function of (M^{l_j-1}, X^n_j) and thus $H(M_{l_j} | M^{l_j-1}, X^n(\mathcal{J}^c), Z^n) = 0$ if $j \notin \mathcal{J}$. Hence, by

Theorem 20.6, the rate tuple $(R'_1 + \epsilon_n, \ldots, R'_N + \epsilon_n)$ is in the optimal rate region for the CFO problem when all the nodes know Z. In particular,

$$\sum_{j=1}^{N} R'_j \geq R_{\text{CFO}}([1:N]|Z) - N\epsilon_n.$$

Substituting in (22.13), we have $(1/n)H(K_1) \leq H(X([1:N])|Z) - R_{\text{CFO}}([1:N]|Z) + N\epsilon_n + \epsilon$, which completes the proof of the upper bound without randomization. The proof of the bound with $\bar{Z}$ and randomization follows the same steps as for the 2-node case.

Secret key capacity when all nodes know Z. Suppose that the eavesdropper observes Z and node $j \in [1:N]$ observes (X_j, Z).

Theorem 22.7. The secret key capacity when all nodes know Z is

$$C_K = H(X^N|Z) - R_{\text{CFO}}([1:N]|Z).$$

The converse follows by noting that $\bar{Z} = Z$ minimizes the upper bound in Proposition 22.3. Achievability follows by setting $U_j = (X_j, Z)$, $j \in [1:N]$, in Proposition 22.2.

We give an alternative proof of achievability that uses the coding scheme for the CFO problem explicitly. The codebook generation for the public communications between the nodes is the same as that for the CFO problem. To generate the secret key codebook, we randomly and independently partition the set of $x^n([1:N])$ sequences into 2^{nR} bins. The index of the bin is the key to be agreed upon. The nodes communicate to achieve omniscience, that is, each node $j \in [1:N]$ broadcasts the bin index M_j of its source sequence. The decoder for each node first recovers $\hat{x}^n([1:N])$ and then sets the secret key equal to the bin index of $\hat{x}^n([1:N])$. The analysis of the probability of error follows that for the CFO problem with side information Z (see Problem 20.12).

The analysis of the key leakage rate is similar to that for the 2-node case. We consider the leakage only for K_1. The other cases follow by Fano's inequality as in the 2-node case. Denote $\mathbf{X}^n = X^n([1:N])$ and consider

$$I(K_1; M^N, Z^n|\mathcal{C}) = I(K_1, \mathbf{X}^n; M^N, Z^n|\mathcal{C}) - I(\mathbf{X}^n; M^N, Z^n|K_1\mathcal{C})$$

$$\overset{(a)}{=} I(\mathbf{X}^n; M^N, Z^n|\mathcal{C}) - I(\mathbf{X}^n; M^N, Z^n|K_1, \mathcal{C})$$

$$= I(\mathbf{X}^n; M^N|Z^n, \mathcal{C}) + I(\mathbf{X}^n; Z^n|\mathcal{C}) - I(\mathbf{X}^n; M^N, Z^n|K_1, \mathcal{C})$$

$$\leq H(M^N) + I(\mathbf{X}^n; Z^n) - H(\mathbf{X}^n|K_1, \mathcal{C}) + H(\mathbf{X}^n|M^N, K_1, Z^n, \mathcal{C})$$

$$\overset{(b)}{\leq} n(R_{\text{CFO}}([1:N]|Z) + \delta(\epsilon)) + (H(\mathbf{X}^n) - H(\mathbf{X}^n|Z^n))$$

$$- (H(\mathbf{X}^n) + H(K_1|\mathbf{X}^n, \mathcal{C}) - H(K_1|\mathcal{C})) + H(\mathbf{X}^n|M^N, K_1, Z^n, \mathcal{C})$$

$$\overset{(c)}{\leq} n(R_{\text{CFO}}([1:N]|Z) - H(\mathbf{X}^n|Z) + R + \delta(\epsilon)) + H(\mathbf{X}^n|M^N, K_1, Z^n, \mathcal{C}),$$

$$(22.14)$$

where (a) and (c) follow since K_1 is a function of $(\mathbf{X}^n, \mathcal{C})$ and (b) follows since $H(M^N) \leq$

$n(R_{\mathrm{CFO}}([1:N]|Z) + \delta(\epsilon))$ in the CFO coding scheme. Using a similar approach to the 2-node case, it can be shown that if $R < H(X^N|Z) - R_{\mathrm{CFO}}([1:N]|Z) - \delta(\epsilon)$, then

$$\limsup_{n\to\infty} \frac{1}{n} H(\mathbf{X}^n \mid M^N, K_1, Z^n, \mathcal{C}) \le H(\mathbf{X}^n|Z) - R - R_{\mathrm{CFO}}([1:N]|Z) + \delta(\epsilon).$$

Substituting in bound (22.14) yields $I(K_1; M^N, Z^n) \le n\delta(\epsilon)$, which completes the proof of achievability.

Remark 22.6. Consider the case where only a subset of the nodes $\mathcal{A} \subseteq [1:N]$ wish to agree on a secret key and the rest of the nodes act as "helpers" to achieve a higher secret key rate. As before, all nodes observe the source Z at the eavesdropper. In this case, it can be shown that

$$C_K(\mathcal{A}) = H(X^N|Z) - R_{\mathrm{CFO}}(\mathcal{A}|Z). \tag{22.15}$$

Note that this secret key capacity can be larger than the capacity without helpers in Theorem 22.7.

Remark 22.7. Since $R_{\mathrm{CFO}}(\mathcal{A}|Z)$ can be computed efficiently (see Remark 20.4), the secret key capacity with and without helpers in Theorem 22.7 and in (22.15), respectively, can be also computed efficiently.

22.4 SECRET KEY AGREEMENT: CHANNEL MODEL

In the source model for key agreement, we assumed that the nodes have access to correlated sources. The channel model for key agreement provides a natural way for generating such sources. It generalizes the source model by assuming a DM-WTC $p(y, z|x)$ from one of the nodes to the other node and the eavesdropper, in addition to the noiseless public broadcast channel. The input to the DM-WTC is not limited to a DMS X and communication is performed in multiple rounds. Each round consists of a number of transmissions over the DM-WTC followed by rounds of interactive communication over the public channel. The definitions of the secret key rate and achievability are similar to those for the source model. In the following, we illustrate this channel model through the following ingenious example.

22.4.1 Maurer's Example

Consider the BS-WTC in Example 22.1. Assume that the receiver is allowed to feed some information back to the sender through a noiseless public broadcast channel of unlimited capacity. This information is also available to the eavesdropper. The sender (Alice) and the receiver (Bob) wish to agree on a key for secure communication that the eavesdropper (Eve) is ignorant of. What is the secret key capacity?

Maurer established the secret key capacity using a wiretap channel coding scheme over a "virtual" degraded WTC from Bob to Alice and Eve as depicted in Figure 22.7. Alice

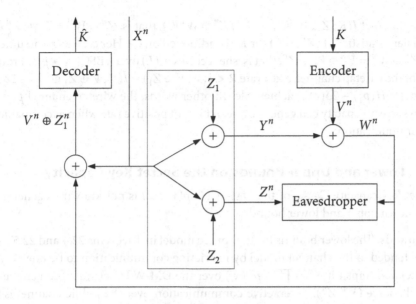

Figure 22.7. Maurer's key agreement scheme.

first sends a random sequence X^n. Bob then randomly picks a key $K \in [1 : 2^{nR}]$, adds its codeword V^n to the received sequence Y^n, and sends the sum $W^n = V^n \oplus Y^n$ over the public channel. Alice uses her knowledge of X^n to obtain a better version of V^n than Eve's and proceeds to decode it to find $\hat{K}$.

Now we present the details of this scheme. Let X and V be independent Bern(1/2) random variables. Randomly and independently generate $2^{n\tilde{R}}$ sequences $v^n(l)$, $l \in [1 : 2^{n\tilde{R}}]$, each according to $\prod_{i=1}^{n} p_V(v_i)$, and partition them into 2^{nR} equal-size bins $\mathcal{B}(k)$, $k \in [1 : 2^{nR}]$. This codebook is revealed to all parties.

Alice sends the randomly generated sequence $X^n \sim \prod_{i=1}^{n} p_X(x_i)$ over the BS-WTC. Bob picks an index L uniformly at random from the set $[1 : 2^{n\tilde{R}}]$. The secret key is the index K such that $v^n(L) \in \mathcal{B}(K)$. Bob then sends $W^n = v^n(L) \oplus Y^n$ over the public channel.

Upon receiving W^n, Alice adds it to X^n to obtain $v^n(L) \oplus Z_1^n$. Thus in effect Alice receives $v^n(L)$ over a BSC(p_1). Alice declares $\hat{l}$ is sent by Bob if it is the unique index such that $(v^n(\hat{l}), v^n(L) \oplus Z_1^n) \in \mathcal{T}_\epsilon^{(n)}$ and then declares the bin index $\hat{k}$ of $v^n(\hat{l})$ as the key estimate. By the packing lemma, the probability of decoding error tends to zero if $\tilde{R} < I(V; V \oplus Z_1) - \delta(\epsilon) = 1 - H(p_1) - \delta(\epsilon)$.

We now show that the key K satisfies the secrecy constraint. Let $\mathcal{C}$ denote the codebook (random bin assignments). Then

$$
\begin{aligned}
H(K|Z^n, W^n, \mathcal{C}) &= H(K|Z^n \oplus W^n, W^n, \mathcal{C}) \\
&= H(K, W^n|Z^n \oplus W^n, \mathcal{C}) - H(W^n|Z^n \oplus W^n, \mathcal{C}) \\
&= H(K|Z^n \oplus W^n, \mathcal{C}) + H(W^n|K, Z^n \oplus W^n, \mathcal{C}) - H(W^n|Z^n \oplus W^n, \mathcal{C}) \\
&= H(K|Z^n \oplus W^n, \mathcal{C}),
\end{aligned}
$$

which implies that $I(K; Z^n, W^n | \mathcal{C}) = I(K; Z^n \oplus W^n | \mathcal{C})$, that is, $Z^n \oplus W^n = Z_1^n \oplus Z_2^n \oplus V^n$ is a "sufficient statistic" of (Z^n, W^n) for K (conditioned on $\mathcal{C}$). Hence, we can assume that Eve has $Z^n \oplus W^n = V^n \oplus Z_1^n \oplus Z_2^n$, that is, she receives $v^n(L)$ over a BSC($p_1 * p_2$). From the analysis of the wiretap channel, a key rate $R < I(V; V \oplus Z_1) - I(V; V \oplus Z_1 \oplus Z_2) - 2\delta(\epsilon) = H(p_1 * p_2) - H(p_1) - 2\delta(\epsilon)$ is achievable. In other words, the wiretap channel $p(y, z | x)$ with zero secrecy capacity can generate a secret key of positive rate when accompanied by public communication.

22.4.2 Lower and Upper Bounds on the Secret Key Capacity

The secret key capacity $C_{\text{K-CM}}$ of the DM-WTC $p(y, z | x)$ is not known in general. We discuss known upper and lower bounds.

Lower bounds. The lower bounds for the source model in Theorems 22.4 and 22.5 can be readily extended to the channel model by restricting communication to be one-way. Fix a pmf $p(x)$ and transmit $X^n \sim \prod_{i=1}^{n} p_X(x_i)$ over the DM-WTC $p(y, z | x)$ to generate the correlated source (Y^n, Z^n). Interactive communication over the public channel is then used to generate the key. This restriction to one-round communication turns the problem into an *equivalent* source model with the 3-DMS $(X, Y, Z) \sim p(x)p(y, z | x)$. Thus, if $C_K(X; Y | Z)$ denotes the secret key capacity for the 3-DMS (X, Y, Z), then

$$C_{\text{K-CM}} \geq \max_{p(x)} C_K(X; Y | Z). \tag{22.16}$$

In particular, using the lower bound in Theorem 22.4 with $U = \emptyset$ yields

$$C_{\text{K-CM}} \geq \max_{p(v,x)} \big(I(V; Y) - I(V; Z) \big),$$

which is the *secrecy capacity* of the DM-WTC (without any public communication) and corresponds to the simple key agreement scheme of sending a secret key from node 1 to node 2 via wiretap channel coding.

For the general N-receiver 1-eavesdropper DM-WTC $p(y_1, \ldots, y_N, z | x)$, the secrecy capacity is lower bounded as

$$C_{\text{K-CM}} \geq \max_{p(x)} C_K(X; Y_1; \ldots; Y_N | Z), \tag{22.17}$$

where $C_K(X; Y_1; \ldots; Y_N | Z)$ is the secret key capacity for the $(N + 2)$-DMS (X, Y^N, Z).

Upper bounds. The secret key capacity for the DM-WTC $p(y, z | x)$ is upper bounded as

$$C_{\text{K-CM}} \leq \max_{p(x)} \min_{p(\bar{z}|z)} I(X; Y | \bar{Z}). \tag{22.18}$$

It can be readily verified that this upper bound is tight for Maurer's example. The proof of this upper bound follows similar steps to the upper bound on the secret key capacity for the source model in Section 22.3.4.

For the general N-receiver 1-eavesdropper DM-WTC $p(y_1, \ldots, y_N, z|x)$, the secrecy capacity is upper bounded as

$$C_{\text{K-CM}} \leq \max_{p(x)} \min_{p(\bar{z}|z)} (H(X, Y^N|\bar{Z}) - R_{\text{CFO}}([1:N+1]|\bar{Z})), \qquad (22.19)$$

where $R_{\text{CFO}}([1:N+1]|\bar{Z})$ is the minimum CFO sum-rate for the $(N+1)$-DMS $((X, \bar{Z}),$ $(Y_1, \bar{Z}), \ldots, (Y_N, \bar{Z}))$. This upper bound coincides with the lower bound in (22.17) when all nodes know the eavesdropper's source Z, establishing the secret key capacity. Thus in this case, the channel is used solely to provide correlated sources to the nodes.

22.4.3 Connection between Key Agreement and the Wiretap Channel

Note that in Maurer's example we generated a key by constructing a *virtual* degraded wiretap channel from Y to (X, Z). This observation can be generalized to obtain the one-round secret key capacity for the source model from the wiretap channel result. We provide an alternative achievability proof of Theorem 22.4 by converting the source model for key agreement into a wiretap channel. For simplicity, we consider achievability of the rate $R < I(X; Y) - I(X; Z)$. The general case can be proved as discussed in the achievability proof in Section 22.3.1.

Fix $V \sim \text{Unif}(\mathcal{X})$ independent of (X, Y, Z). For $i \in [1:n]$, node 1 turns the public discussion channel into a wiretap channel by sending $V_i \oplus X_i$ over it. The equivalent wiretap channel is then $V \to ((Y, V \oplus X), (Z, V \oplus X))$. By Theorem 22.1, we can transmit a confidential message at a rate

$$R < I(V; Y, V \oplus X) - I(V; Z, V \oplus X)$$
$$= I(V; Y|V \oplus X) - I(V; Z|V \oplus X)$$
$$= H(Y|V \oplus X) - H(Y|V \oplus X, V) - H(Z|V \oplus X) + H(Z|V \oplus X, V)$$
$$\stackrel{(a)}{=} H(Y) - H(Y|X, V) - H(Z) + H(Z|X, V)$$
$$= H(Y) - H(Y|X) - H(Z) + H(Z|X)$$
$$= I(X; Y) - I(X; Z),$$

where (a) follows since $V \oplus X$ is independent of X and hence of (Y, Z) by the Markov relation $V \oplus X \to X \to (Y, Z)$.

SUMMARY

- Information theoretic notion of secrecy

- Discrete memoryless wiretap channel (DM-WTC):

  - Multicoding

  - Random selection of the codeword

  - Random mapping of the codeword to the channel input sequence

- Upper bound on the average leakage rate via the analysis of the number of jointly typical sequences

- Secrecy capacity of public communication is upper bounded by the key rate

- Source model for secret key agreement via public communication:

  - Double random binning

  - Lower bound on the average key rate via the analysis of the random pmf

  - Interaction over the public channel can increase the secret key rate

  - CFO rate characterizes the secret key capacity when all nodes know the eavesdropper's source

- Randomized encoding achieves higher rates for confidential communication and key agreement than deterministic encoding

- Channel model for secret key agreement:

  - Generation of correlated sources via a noisy channel

  - Wiretap channel coding for key generation

- **Open problems:**

  22.1. What is the secrecy capacity of the 2-receiver 1-eavesdropper DM-WTC?

  22.2. What is the secrecy capacity of the 1-receiver 2-eavesdropper DM-WTC?

  22.3. What is the 2-round secret key capacity for the 3-DMS (X_1, X_2, Z)?

BIBLIOGRAPHIC NOTES

Information theoretic secrecy was pioneered by Shannon (1949b). He considered communication of a message (plaintext) M from Alice to Bob over a noiseless public broadcast channel in the presence of an eavesdropper (Eve) who observes the channel output (ciphertext) L. Alice and Bob share a key, which is unknown to Eve, and they can use it to encrypt and decrypt M into L and vice versa. Shannon showed that to achieve perfect secrecy, that is, information leakage $I(M; L) = 0$, the size of the key must be at least as large as the size of the message, i.e., $H(K) \geq H(M)$.

This negative result has motivated subsequent work on the wiretap channel and secret key agreement. The wiretap channel was first introduced by Wyner (1975c), who established the secrecy capacity of the degraded wiretap channel. The secrecy capacity of the scalar Gaussian wiretap channel in Example 22.2 is due to Leung and Hellman (1978). The secrecy capacity for the vector case in Example 22.3 is due to Oggier and Hassibi (2008) and Khisti and Wornell (2010a,b). Csiszár and Körner (1978) established the rate–leakage region of a general DM-BC with common and confidential messages in Theorem 22.2, which includes Theorem 22.1 as a special case. Chia and El Gamal (2009) extended wire-tap channel coding to the cases of two receivers and one eavesdropper and of one receiver and two eavesdroppers, and established inner bounds on the rate–leakage region that are tight in some special cases. For the 2-receiver 1-eavesdropper case, they showed that the secrecy capacity is lower bounded as

$$C_S \geq \max \min\{I(U_0, U_1; Y_1|Q) - I(U_0, U_1; Z|Q), I(U_0, U_2; Y_2|Q) - I(U_0, U_2; Z|Q)\},$$

where the maximum is over all pmfs $p(q, u_0, u_1, u_2, x) = p(q, u_0)p(u_1|u_0)p(u_2, x|u_1) = p(q, u_0)p(u_2|u_0)p(u_1, x|u_0)$ such that

$$I(U_1, U_2; Z|U_0) \leq I(U_1; Z|U_0) + I(U_2; Z|U_0) - I(U_1; U_2|U_0).$$

The proof involved Marton coding in addition to wiretap channel coding, satellite code-book generation, and indirect decoding. Proposition 22.1 is a special case of this result. Wiretap channel coding has been extended to other multiuser channels; see the survey by Liang, Poor, and Shamai (2009).

The secrecy capacity of the DM-WTC with a secret key in (22.7) is due to Yamamoto (1997). Ahlswede and Cai (2007) and Ardestanizadeh, Franceschetti, Javidi, and Kim (2009) studied secure feedback coding schemes to generate such a shared key, which are optimal when the wiretap channel is physically degraded. Chia and El Gamal (2010) stud-ied wiretap channels with state and proposed a coding scheme that combines wiretap channel coding with a shared key extracted from the channel state information to achieve a lower bound on the secrecy capacity.

The source model for secret key agreement was introduced by Ahlswede and Csiszár (1993), who established the one-way secret key capacity in Section 22.3.1. The one-round secret key capacity with rate constraint in Remark 22.4 is due to Csiszár and Narayan (2000). Gohari and Anantharam (2010a) established the lower bound on the secret key capacity for interactive key agreement between 2 nodes in Section 22.3.3 and its extension to multiple nodes in Section 22.3.5. Example 22.4 is also due to them. The upper bound for the 2-node case in Theorem 22.6 is due to Maurer and Wolf (1999) and Example 22.5 is due to A. A. Gohari. The upper bound for multiple nodes in Proposition 22.3 is due to Csiszár and Narayan (2004), who also established the secret key capacity in Theorem 22.7 when all the nodes know Z and the extension in (22.15) when some nodes are helpers. A tighter upper bound was established by Gohari and Anantharam (2010a).

The example in Section 22.4.1 is due to Maurer (1993), who also established the upper bound on the secret key capacity in (22.18). Csiszár and Narayan (2008) established the upper bound in (22.19) and showed that it is tight when every node knows the eavesdrop-per's source. Tighter lower and upper bounds were given by Gohari and Anantharam (2010b).

Most of the above results can be extended to a stronger notion of secrecy (Maurer and Wolf 2000). For example, Csiszár and Narayan (2004) established the secret key capacity in Theorem 22.7 under the secrecy constraint

$$\limsup_{n\to\infty} I(K_j; Z^n, M^q) = 0, \quad j \in [1:N].$$

Information theoretic secrecy concerns confidentiality at the physical layer. Most secrecy systems are built at higher layers and use the notion of computational secrecy, which exploits the computational hardness of recovering the message from the ciphertext without knowing the key (assuming $P \neq NP$); see, for example, Menezes, van Oorschot, and Vanstone (1997) and Katz and Lindell (2007).

PROBLEMS

22.1. Verify that $(U, V) \to X \to (Y, Z)$ form a Markov chain in the proof of the converse for Theorem 22.1 in Section 22.1.2.

22.2. Provide the details of the proof of the lower bound on the 2-receiver DM-WTC secrecy capacity in Proposition 22.1.

22.3. Prove Theorem 22.2.

22.4. Establish the secrecy capacity of the DM-WTC $p(y, z|x)$ with secret key rate R_K in (22.7). (Hint: Randomly and independently generate wiretap channel codebooks C_{m_K} for each key $m_K \in [1 : 2^{nR_K}]$.)

22.5. Provide the details of the proof of the lower bound on the secret key capacity in Theorem 22.5.

22.6. Establish the upper bound in (22.18) on the secret key capacity for the DM-WTC $p(x_2, z|x_1)$.

22.7. *Secrecy capacity of a BSC-BEC WTC.* Consider the DM-WTC in Example 5.4.

(a) Suppose that the channel to the receiver is a BEC(ϵ) and the channel to the eavesdropper is a BSC(p). Find the secrecy capacity for each of the four regimes of ϵ and p.

(b) Now suppose that the channels to the receiver and the eavesdropper in part (a) are exchanged. Find the secrecy capacity.

22.8. *Secret key capacity.* Consider the sources $X_1, X_2, Z \in \{0, 1, 2, 3\}$ with the joint pmf

$$p_{X_1, X_2, Z}(0, 0, 0) = p_{X_1, X_2, Z}(0, 1, 1) = p_{X_1, X_2, Z}(1, 0, 1) = p_{X_1, X_2, Z}(1, 1, 0) = \frac{1}{8},$$

$$p_{X_1, X_2, Z}(2, 2, 2) = p_{X_1, X_2, Z}(3, 3, 3) = \frac{1}{4}.$$

Find the secret key capacity for the 3-DMS (X_1, X_2, Z).

22.9. *Key generation without communication.* Consider a secret key agreement setup, where node 1 observes X_1, node 2 observes X_2, and the eavesdropper observes Z. Show that the secret key capacity without (public) communication is

$$C_K = H(X_0|Z),$$

where X_0 is the common part of X_1 and X_2 as defined in Section 14.1.

22.10. *Secrecy capacity of the DMC with feedback.* Consider the physically degraded DM-WTC $p(y|x)p(z|y)$. Assume that there is noiseless causal secure feedback from the receiver to the sender. Show that the secrecy capacity is

$$C_S = \max_{p(x)} \max\{I(X;Y|Z) + H(Y|X,Z), \; I(X;Y)\}.$$

(Hint: Due to the secure noiseless feedback, the sender and the receiver can agree on a secret key of rate $H(Y|X,Z)$ *independent* of the message and the transmitted random codeword; see Problem 22.9. Using block Markov coding, this key can be used to increase the secrecy capacity for the next transmission block as in (22.7).) Remark: This result is due to Ahlswede and Cai (2007).

APPENDIX 22A   PROOF OF LEMMA 22.1

We bound $H(L|Z^n, M = m, C)$ for every m. We first bound the probabilities of the following events. Fix $L = l$ and a sequence $z^n \in \mathcal{T}_{\epsilon}^{(n)}$. Let

$$N(m, l, z^n) = \left|\{U^n(\tilde{l}) \in C(m) \colon (U^n(\tilde{l}), z^n) \in \mathcal{T}_{\epsilon}^{(n)}, \; \tilde{l} \neq l\}\right|.$$

It is not difficult to show that

$$2^{n(\tilde{R}-R-I(U;Z)-\delta(\epsilon)-\epsilon_n)} \leq \mathsf{E}(N(m, l, z^n)) \leq 2^{n(\tilde{R}-R-I(U;Z)+\delta(\epsilon)-\epsilon_n)}$$

and

$$\mathrm{Var}(N(m, l, z^n)) \leq 2^{n(\tilde{R}-R-I(U;Z)+\delta(\epsilon)-\epsilon_n)},$$

where ϵ_n tends to zero as $n \to \infty$.

Define the random event $\mathcal{E}_2(m, l, z^n) = \{N(m, l, z^n) \geq 2^{n(\tilde{R}-R-I(U;Z)+\delta(\epsilon)-\epsilon_n/2)+1}\}$. By the Chebyshev inequality,

$$\begin{aligned}
\mathsf{P}(\mathcal{E}_2(m, l, z^n)) &= \mathsf{P}\{N(m, l, z^n) \geq 2^{n(\tilde{R}-R-I(U;Z)+\delta(\epsilon)-\epsilon_n/2)+1}\} \\
&\leq \mathsf{P}\{N(m, l, z^n) \geq \mathsf{E}(N(m, l, z^n)) + 2^{n(\tilde{R}-R-I(U;Z)+\delta(\epsilon)-\epsilon_n/2)}\} \\
&\leq \mathsf{P}\{|N(m, l, z^n) - \mathsf{E}(N(m, l, z^n))| \geq 2^{n(\tilde{R}-R-I(U;Z)+\delta(\epsilon)-\epsilon_n/2)}\} \\
&\leq \frac{\mathrm{Var}(N(m, l, z^n))}{2^{2n(\tilde{R}-R-I(U;Z)+\delta(\epsilon)-\epsilon_n/2)}}. \qquad (22.20)
\end{aligned}$$

Thus if $\tilde{R} - R - I(U;Z) \geq 0$, then $\mathsf{P}(\mathcal{E}_2(m, l, z^n))$ tends to zero as $n \to \infty$ for every m.

Next, for each message m, define $N(m) = |\{U^n(\tilde{l}) \in C(m) \colon (U^n(\tilde{l}), Z^n) \in \mathcal{T}_{\epsilon}^{(n)}, \; \tilde{l} \neq L\}|$ and $\mathcal{E}_2(m) = \{N(m) \geq 2^{n(\tilde{R}-R-I(U;Z)+\delta(\epsilon)-\epsilon_n/2)+1}\}$. Finally, define the indicator random variable $E(m) = 1$ if $(U^n(L), Z^n) \notin \mathcal{T}_{\epsilon}^{(n)}$ or the event $\mathcal{E}_2(m)$ occurs, and $E(m) = 0$ otherwise. By the union of events bound,

$$\mathsf{P}\{E(m) = 1\} \leq \mathsf{P}\{(U^n(L), Z^n) \notin \mathcal{T}_{\epsilon}^{(n)}\} + \mathsf{P}(\mathcal{E}_2(m)).$$

We bound each term. By the LLN, the first term tends to zero as $n \to \infty$. For $P(\mathcal{E}_2(m))$,

$$P(\mathcal{E}_2(m)) \leq \sum_{z^n \in T_\epsilon^{(n)}} p(z^n) \, P\{\mathcal{E}_2(m)|Z^n = z^n\} + P\{Z^n \notin T_\epsilon^{(n)}\}$$

$$= \sum_{z^n \in T_\epsilon^{(n)}} \sum_l p(z^n) p(l|z^n) \, P\{\mathcal{E}_2(m)|Z^n = z^n, L = l\} + P\{Z^n \notin T_\epsilon^{(n)}\}$$

$$= \sum_{z^n \in T_\epsilon^{(n)}} \sum_l p(z^n) p(l|z^n) \, P\{\mathcal{E}_2(m, z^n, l)\} + P\{Z^n \notin T_\epsilon^{(n)}\}.$$

By (22.20), the first term tends to zero as $n \to \infty$ if $\tilde{R} - R \geq I(U; Z)$, and by the LLN, the second term tends to zero as $n \to \infty$.

We are now ready to upper bound the equivocation. Consider

$$H(L|Z^n, M = m, C) \leq 1 + P\{E(m) = 1\} H(L|Z^n, M = m, E(m) = 1, C)$$
$$+ H(L|Z^n, M = m, E(m) = 0, C)$$
$$\leq 1 + n(\tilde{R} - R) \, P\{E(m) = 1\} + H(L|Z^n, M = m, E(m) = 0, C)$$
$$\leq 1 + n(\tilde{R} - R) \, P\{E(m) = 1\} + \log\left(2^{n(\tilde{R}-R-I(U;Z)+\delta(\epsilon)-\epsilon_n/2)+1}\right).$$

Now since $P\{E(m) = 1\}$ tends to zero as $n \to \infty$ if $\tilde{R} - R \geq I(U; Z)$, then for every m,

$$\limsup_{n\to\infty} \frac{1}{n} H(L|Z^n, M = m, C) \leq \tilde{R} - R - I(U; Z) + \delta'(\epsilon).$$

This completes the proof of the lemma.

APPENDIX 22B   PROOF OF LEMMA 22.2

Consider

$$H(K_1|C) \geq P\{X^n \in T_\epsilon^{(n)}\} H(K_1|C, X^n \in T_\epsilon^{(n)}) \geq (1 - \epsilon_n') H(K_1|C, X^n \in T_\epsilon^{(n)})$$

for some ϵ_n' that tends to zero as $n \to \infty$. Let $P(k_1)$ be the *random* pmf of K_1 given $\{X^n \in T_\epsilon^{(n)}\}$, where the randomness is induced by the random bin assignments C. By symmetry, $P(k_1)$, $k_1 \in [1 : 2^{nR}]$, are identically distributed. Let $B'(1) = \bigcup_m B(m, 1)$ and express $P(1)$ in terms of a weighted sum of indicator functions as

$$P(1) = \sum_{x^n \in T_\epsilon^{(n)}} \frac{p(x^n)}{P\{X^n \in T_\epsilon^{(n)}\}} \cdot E(x^n),$$

where

$$E(x^n) = \begin{cases} 1 & \text{if } x^n \in B'(1), \\ 0 & \text{otherwise.} \end{cases}$$

Then, it can be easily shown that

$$E_{\mathcal{C}}(P(1)) = 2^{-nR},$$

$$\text{Var}(P(1)) = 2^{-nR}(1 - 2^{-nR}) \sum_{x^n \in T_\epsilon^{(n)}} \left(\frac{p(x^n)}{P\{X^n \in T_\epsilon^{(n)}\}} \right)^2$$

$$\leq 2^{-nR} 2^{n(H(X)+\delta(\epsilon))} \frac{2^{-2n(H(X)-\delta(\epsilon))}}{(1 - \epsilon_n')^2}$$

$$\leq 2^{-n(R+H(X)-4\delta(\epsilon))}$$

for n sufficiently large. Thus, by the Chebyshev lemma in Appendix B,

$$P\{|P(1) - E(P(1))| \geq \epsilon\, E(P(1))\} \leq \frac{\text{Var}(P(1))}{(\epsilon\, E[P(1)])^2} \leq \frac{2^{-n(H(X)-R-4\delta(\epsilon))}}{\epsilon^2},$$

which tends to zero as $n \to \infty$ if $R < H(X) - 4\delta(\epsilon)$. Now, by symmetry,

$$H(K_1 | \mathcal{C}, X^n \in T_\epsilon^{(n)})$$
$$= 2^{nR}\, E[P(1)\log(1/P(1))]$$
$$\geq 2^{nR}\, P\{|P(1) - E(P(1))| < \epsilon 2^{-nR}\} \cdot E[P(1)\log(1/P(1)) \,|\, |P(1) - E(P(1))| < \epsilon 2^{-nR}]$$
$$\geq \left(1 - \frac{2^{-n(H(X)-R-4\delta(\epsilon))}}{\epsilon^2} \right) \cdot \left(nR(1 - \epsilon) - (1 - \epsilon)\log(1 + \epsilon) \right)$$
$$\geq n(R - \delta(\epsilon))$$

for n sufficiently large and $R < H(X) - 4\delta(\epsilon)$. Thus, we have shown that if $R < H(X) - 4\delta(\epsilon)$, $\lim\inf_{n\to\infty} H(K_1 | \mathcal{C}) \geq R - \delta(\epsilon)$, which completes the proof of the lemma.

APPENDIX 22C   PROOF OF LEMMA 22.3

Let $E_1 = 1$ if $(X_1^n, Z^n) \notin T_\epsilon^{(n)}$ and $E_1 = 0$ otherwise. Note that by the LLN, $P\{E_1 = 1\}$ tends to zero as $n \to \infty$. Consider

$$H(X_1^n | Z^n, M, K_1, \mathcal{C}) \leq H(X_1^n, E_1 | Z^n, M, K_1, \mathcal{C})$$
$$\leq 1 + n\, P\{E_1 = 1\}\log|\mathcal{X}_1| + \sum_{(z^n, m, k_1)} p(z^n, m, k_1 | E_1 = 0)$$
$$\cdot H(X_1^n | Z^n = z^n, M = m, K_1 = k_1, E_1 = 0, \mathcal{C}).$$

Now, for a codebook $\mathcal{C}$ and $z^n \in T_\epsilon^{(n)}$, let $N(z^n, \mathcal{C})$ be the number of sequences $x_1^n \in \mathcal{B}(m, k_1) \cap T_\epsilon^{(n)}(X_1|z^n)$, and define

$$E_2(z^n, \mathcal{C}) = \begin{cases} 1 & \text{if } N(z^n, \mathcal{C}) \geq 2\, E(N(z^n, \mathcal{C})), \\ 0 & \text{otherwise.} \end{cases}$$

It is easy to show that

$$E(N(z^n, \mathcal{C})) = 2^{-n(\tilde{R}+R)} |\mathcal{T}_\epsilon^{(n)}(X_1|z^n)|,$$

$$\text{Var}(N(z^n, \mathcal{C})) \le 2^{-n(\tilde{R}+R)} |\mathcal{T}_\epsilon^{(n)}(X_1|z^n)|.$$

Then, by the Chebyshev lemma in Appendix B,

$$P\{E_2(z^n, \mathcal{C}) = 1\} \le \frac{\text{Var}(N(z^n, \mathcal{C}))}{(E[N(z^n, \mathcal{C})])^2} \le 2^{-n(H(X_1|Z) - \tilde{R} - R - \delta(\epsilon))},$$

which tends to zero as $n \to \infty$ if $\tilde{R} + R < H(X_1|Z) - \delta(\epsilon)$, i.e., $R < H(X_1|Z) - H(X_1|X_2) - 2\delta(\epsilon)$. Now consider

$$H(X_1^n \mid Z^n = z^n, M = m, K_1 = k_1, E_1 = 0, \mathcal{C})$$
$$\le H(X_1^n, E_2 \mid Z^n = z^n, M = m, K_1 = k_1, E_1 = 0, \mathcal{C})$$
$$\le 1 + n P\{E_2 = 1\} \log |\mathcal{X}_1| + H(X_1^n \mid Z^n = z^n, M = m, K_1 = k_1, E_1 = 0, E_2 = 0, \mathcal{C})$$
$$\le 1 + n P\{E_2 = 1\} \log |\mathcal{X}_1| + n(H(X_1|Z) - \tilde{R} - R + \delta(\epsilon)),$$

which implies that

$$H(X_1^n | Z^n, M, K_1, \mathcal{C})$$
$$\le 2 + n P\{E_1 = 1\} \log |\mathcal{X}_1| + n P\{E_2 = 1\} \log |\mathcal{X}_1| + n(H(X_1|Z) - \tilde{R} - R + \delta(\epsilon)).$$

Taking $n \to \infty$ completes the proof of the lemma.

CHAPTER 23

Wireless Fading Channels

So far we have modeled wireless channels as time-invariant Gaussian channels. Real-world wireless channels, however, are time-varying due to multiple signal paths and user mobility. The wireless fading channel models capture these effects by allowing the gains in the Gaussian channels to change randomly over time. Since the channel gain information is typically available at each receiver (through training sequences) and may be available at each sender (via feedback from the receivers), wireless fading channels can be viewed as channels with random state, where the state (channel gain) information is available at the decoders and fully or partially available at the encoders. Depending on the fading model and coding delay, wireless fading channel capacity may or may not be well defined. Moreover, even when capacity is well defined, it may not be a good measure of performance in practice because it is overly pessimistic, or because achieving it requires very long coding delay.

In this chapter we study several canonical wireless fading channels under the block fading model with stationary ergodic channel gains. We introduce several coding approaches under the fast and slow fading assumptions, including water-filling in time, compound channel coding, outage coding, broadcasting, and adaptive coding. We compare these approaches using performance metrics motivated by practice. Finally, we show that significant increase in rates can be achieved by exploiting fading through interference alignment as we demonstrated for the k-user-pair symmetric QED-IC in Section 6.9.

23.1 GAUSSIAN FADING CHANNEL

Consider the Gaussian fading channel

$$Y_i = G_i X_i + Z_i, \quad i \in [1:n],$$

where $\{G_i\}$ is a *channel gain* process that models fading in wireless communication and $\{Z_i\}$ is a WGN(1) process independent of $\{G_i\}$.

In practice, the channel gain typically varies at a much longer time scale than symbol transmission time. This motivates the simplified *block fading model* depicted in Figure 23.1, where the gain G_i is assumed to be constant over each *coherence time interval* $[(l-1)k+1:lk]$ of length k for $l = 1, 2, \ldots$, and the block gain process $\{\overline{G}_l\}_{l=1}^{\infty} = \{G_{lk}\}_{l=1}^{\infty}$ is stationary ergodic. Assuming this model, we consider two coding paradigms.

- **Fast fading.** Here the code block length spans a large number of coherence time intervals and thus the channel is ergodic with a well-defined Shannon capacity (sometimes referred to as the *ergodic capacity*). However, coding over a large number of coherence time intervals results in excessive delay.

- **Slow fading.** Here the code block length is in the order of the coherence time interval and therefore the channel is not ergodic and does not have a Shannon capacity in general. We discuss alternative coding approaches and corresponding performance metrics for this case.

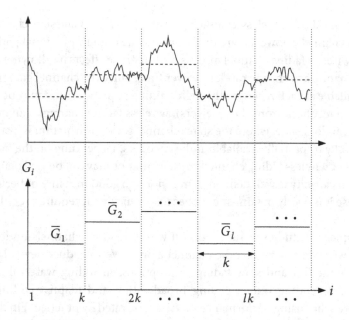

Figure 23.1. Wireless channel fading process and its block fading model.

In the following two sections, we consider coding under fast and slow fading with channel gain availability only at the decoder or at both the encoder and the decoder. When the channel gain is available only at the decoder, we assume average power constraint P on X, i.e., $\sum_{i=1}^{n} x_i^2(m) \le nP$, $m \in [1 : 2^{nR}]$, and when the channel gain is also available at the encoder, we assume *expected* average power constraint P on X, i.e.,

$$\sum_{i=1}^{n} \mathsf{E}(x_i^2(m, G_i)) \le nP, \quad m \in [1 : 2^{nR}].$$

23.2 CODING UNDER FAST FADING

In fast fading, we code over many coherence time intervals (i.e., $n \gg k$) and the block gain process $\{\overline{G}_l\}$ is stationary ergodic, for example, an i.i.d. process.

Channel gain available only at the decoder. First we show that the ergodic capacity for this case is

$$C_{\text{GI-D}} = E_G[C(G^2 P)].$$

Recall from Remark 7.4 that for a DMC with stationary ergodic state $p(y|x, s)$, the capacity when the state information is available only at the decoder is

$$C_{\text{SI-D}} = \max_{p(x)} I(X; Y|S).$$

This result can be readily extended to the Gaussian fading channel with a stationary ergodic block gain process $\{\overline{G}_l\}$ having the marginal distribution $F_G(g_l)$ and with power constraint P to obtain the capacity

$$
\begin{aligned}
C_{\text{SI-D}}(P) &= \sup_{F(x):E(X^2)\leq P} I(X; Y|G) \\
&= \sup_{F(x):E(X^2)\leq P} (h(Y|G) - h(Y|G, X)) \\
&= \sup_{F(x):E(X^2)\leq P} h(GX + Z|G) - h(Z) \\
&= E_G[C(G^2 P)],
\end{aligned}
$$

where the supremum is attained by $X \sim N(0, P)$.

Now the Gaussian fading channel under fast fading can be decomposed in time into k parallel Gaussian fading channels, the first corresponding to transmission times $1, k + 1, 2k + 1, \ldots$, the second corresponding to transmission times $2, k + 2, \ldots$, and so on, with the same stationary ergodic channel gain process and average power constraint P. Hence, $C_{\text{GI-D}} \geq C_{\text{SI-D}}(P)$. The converse can be readily proved using standard arguments.

Channel gain available at the encoder and decoder. The ergodic capacity in this case is

$$
\begin{aligned}
C_{\text{GI-ED}} &= \max_{F(x|g):E(X^2)\leq P} I(X; Y|G) \\
&= \max_{F(x|g):E(X^2)\leq P} (h(Y|G) - h(Y|G, X)) \\
&= \max_{F(x|g):E(X^2)\leq P} h(GX + Z|G) - h(Z) \\
&\stackrel{(a)}{=} \max_{\phi(g):E(\phi(G))\leq P} E_G[C(G^2 \phi(G))],
\end{aligned}
$$

where $F(x|g)$ is the conditional cdf of X given $\{G = g\}$ and (a) follows since the maximum is attained by $X|\{G = g\} \sim N(0, \phi(g))$. This can be proved by Remark 7.4 and using similar arguments to the case with channel gain available only at the decoder.

The above capacity expression can be simplified further. Using a Lagrange multiplier λ, we can show that the capacity is attained by

$$\phi^*(g) = \left[\lambda - \frac{1}{g^2}\right]^+,$$

where λ is chosen to satisfy

$$E_G(\phi^*(G)) = E_G\left(\left[\lambda - \frac{1}{G^2}\right]^+\right) = P.$$

Note that this power allocation corresponds to water-filling *over time*; see Figure 3.9 in Section 3.4.3. At high SNR, however, the capacity gain from water-filling vanishes and $C_{\text{GI-ED}} - C_{\text{GI-D}}$ tends to zero as $P \to \infty$.

23.3 CODING UNDER SLOW FADING

Under the assumption of slow fading, we code over a single coherence time interval (i.e., $n = k$) and the notion of channel capacity is not well defined in general. As before, we consider the cases where the channel gain is available only at the decoder and where it is available at both the encoder and the decoder.

23.3.1 Channel Gain Available Only at the Decoder

When the encoder does not know the gain, we have several coding options.

Compound channel approach. We code against the worst channel to guarantee reliable communication. The (Shannon) capacity under this coding approach can be readily found by extending the capacity of the compound channel in Section 7.2 to the Gaussian case, which yields

$$C_{\text{CC}} = \inf_{g \in \mathcal{G}} C(g^2 P).$$

This compound channel approach is impractical when fading results in very low channel gain. Hence, we consider the following alternative coding approaches that are more useful in practice.

Outage capacity approach. In this approach we transmit at a rate higher than the compound channel capacity C_{CC} and tolerate some information loss when the channel gain is too low for the message to be recovered. If the probability of such an *outage* event is small, then we can reliably communicate the message most of the time. More precisely, if we can tolerate an outage probability p_{out}, that is, a loss of a fraction p_{out} of the messages on average, then we can communicate at any rate lower than the outage capacity

$$C_{\text{out}} = \max_{g : \text{P}\{G < g\} \le p_{\text{out}}} C(g^2 P).$$

Broadcast channel approach. For simplicity of presentation, assume two fading states g_1 and g_2 with $g_1 > g_2$. We view the channel as a Gaussian BC with gains g_1 and g_2, and use superposition coding to send a common message to both receivers at rate $\tilde{R}_0 < C(g_2^2 \bar{\alpha} P/(1 + \alpha g_2^2 P))$, $\alpha \in [0,1]$, and a private message to the stronger receiver at rate $\tilde{R}_1 < C(g_1^2 \alpha P)$. If the gain is g_2, the receiver of the fading channel can recover the common message at rate $R_2 = \tilde{R}_0$, and if the gain is g_1, it can recover both messages at total rate

$R_1 = \tilde{R}_0 + \tilde{R}_1$. Assuming $P\{G = g_1\} = p$ and $P\{G = g_2\} = \bar{p}$, we can compute the *broadcast capacity* as

$$C_{BC} = \max_{\alpha \in [0,1]} \left(p\, C(g_1^2 \alpha P) + C\left(\frac{g_2^2 \bar{\alpha} P}{1 + \alpha g_2^2 P} \right) \right).$$

This approach is best suited for sending multimedia (video or music) over a fading channel using successive refinement as discussed in Chapter 13. If the channel gain is low, the receiver recovers only the low-fidelity description of the source and if the gain is high, it also recovers the refinement and obtains the high-fidelity description.

23.3.2 Channel Gain Available at Both the Encoder and the Decoder

Compound channel approach. If the channel gain is available at the encoder, the compound channel capacity is

$$C_{CC\text{-}E} = \inf_{g \in \mathcal{G}} C(g^2 P) = C_{CC}.$$

Thus, the capacity is the same as when the encoder does not know the state.

Adaptive coding. Instead of communicating at the capacity of the channel with the worst gain, we adapt the transmission rate to the channel gain and communicate at the maximum rate $C_g = C(g^2 P)$ when the gain is g. We define the *adaptive capacity* as

$$C_A = E_G[C(G^2 P)].$$

Note that this is the same as the ergodic capacity when the channel gain is available only at the decoder. The adaptive capacity, however, is just a convenient performance metric and *not* a capacity in the Shannon sense.

Adaptive coding with power control. Since the encoder knows the channel gain, it can adapt the power as well as the transmission rate. In this case, we define the *power-control adaptive capacity* as

$$C_{PA} = \max_{\phi(g):E_G(\phi(G)) \le P} E_G[C(G^2 \phi(G))],$$

where the maximum is attained by the water-filling power allocation that satisfies the constraint $E_G(\phi(G)) \le P$. Note that the power-control adaptive capacity is identical to the ergodic capacity when the channel gain is available at both the encoder and the decoder. Again C_{PA} is not a capacity in the Shannon sense.

Remark 23.1. Although using power control achieves a higher rate on average, in some practical situations, such as under the Federal Communications Commission (FCC) regulation, the power constraint must be satisfied in each coding block.

In the following, we compare the performance of the above coding schemes.

Example 23.1. Assume two fading states g_1 and g_2 with $g_1 > g_2$ and $P\{G = g_1\} = p$. In Figure 23.2, we compare the performance metrics C_{CC}, C_{out}, C_{BC}, C_A, and C_{PA} for different values of $p \in [0, 1]$. The broadcast channel approach is effective when the better channel occurs more often ($p \approx 1$) and power control is particularly effective when the channel varies frequently ($p \approx 1/2$).

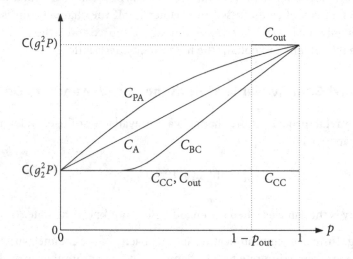

Figure 23.2. Comparison of performance metrics.

23.4   GAUSSIAN VECTOR FADING CHANNEL

As mentioned in Chapter 9, in the presence of fading, using multiple antennas can increase the capacity via spatial diversity. To demonstrate this improvement, we model a MIMO fading channel by the Gaussian vector fading channel

$$\mathbf{Y}(i) = G(i)\mathbf{X}(i) + \mathbf{Z}(i), \quad i \in [1:n],$$

where $\mathbf{Y}(i)$ is an r-dimensional vector, the input $\mathbf{X}(i)$ is a t-dimensional vector, $\{G(i)\}$ is the channel gain matrix process that models random multipath fading, and $\{\mathbf{Z}(i)\}$ is a WGN(I_r) process independent of $\{G(i)\}$. We assume power constraint P on $\mathbf{X}$.

As in the scalar case, we assume the block fading model, that is, $G(i)$ is constant over each coherence time interval but varies according to a stationary ergodic process over consecutive coherence time intervals. As before, we can investigate various channel gain availability scenarios and coding strategies under fast and slow fading assumptions. In the following, we assume fast fading and study the effect of the number of antennas on the capacity when the channel gain matrices are available at the decoder.

If the random channel gain matrices $G(1), G(2), \ldots, G(n)$ are identically distributed, then the ergodic capacity when the gain matrices are available only at the decoder is

$$C = \max_{\mathrm{tr}(K_\mathbf{X}) \leq P} \mathrm{E}\left[\frac{1}{2}\log |I_r + GK_\mathbf{X}G^T|\right].$$

The maximization in the capacity expression can be further simplified if G is *isotropic*, that is, the joint pdf of the matrix elements $(G_{jk}: j \in [1:r], k \in [1:t])$ is invariant under orthogonal transformations. For example, if Rayleigh fading is assumed, that is, if $G_{jk} \sim N(0, 1)$, and $G_{jk}, j \in [1:r], k \in [1:t]$, are mutually independent, then G is isotropic.

Theorem 23.1. If the channel gain matrix G is isotropic, then the ergodic capacity when the channel gain is available only at the decoder is

$$C = \mathrm{E}\left[\frac{1}{2}\log\left|I_r + \frac{P}{t}GG^T\right|\right] = \frac{1}{2}\sum_{j=1}^{\min\{t,r\}} \mathrm{E}\left[\log\left(1 + \frac{P}{t}\Lambda_j\right)\right],$$

which is attained by $K_\mathbf{X}^* = (P/t)I_t$, where Λ_j are random nonzero eigenvalues of GG^T.

At small P (low SNR), assuming Rayleigh fading with $\mathrm{E}(|G_{jk}|^2) = 1$, the capacity expression simplifies to

$$
\begin{aligned}
C &= \frac{1}{2}\sum_{j=1}^{\min\{t,r\}} \mathrm{E}\left[\log\left(1 + \frac{P}{t}\Lambda_j\right)\right] \\
&\approx \frac{P}{2t\log 2}\sum_{j=1}^{\min\{t,r\}} \mathrm{E}(\Lambda_j) \\
&= \frac{P}{2t\log 2}\, \mathrm{E}[\mathrm{tr}(GG^T)] \\
&= \frac{P}{2t\log 2}\, \mathrm{E}\left[\sum_{j,k}|G_{jk}|^2\right] \\
&= \frac{rP}{2\log 2}.
\end{aligned}
$$

This is an r-fold SNR (and capacity) gain compared to the single-antenna case. By contrast, at large P (high SNR),

$$
\begin{aligned}
C &= \frac{1}{2}\sum_{j=1}^{\min\{t,r\}} \mathrm{E}\left[\log\left(1 + \frac{P}{t}\Lambda_j\right)\right] \\
&\approx \frac{1}{2}\sum_{j=1}^{\min\{t,r\}} \mathrm{E}\left[\log\left(\frac{P}{t}\Lambda_j\right)\right] \\
&= \frac{\min\{t, r\}}{2}\log\left(\frac{P}{t}\right) + \frac{1}{2}\sum_{j=1}^{\min\{t,r\}} \mathrm{E}(\log \Lambda_j).
\end{aligned}
$$

This is a $\min\{t, r\}$-fold increase in capacity over the single-antenna case.

23.5 GAUSSIAN FADING MAC

Consider the Gaussian fading MAC

$$Y_i = G_{1i}X_{1i} + G_{2i}X_{2i} + Z_i, \quad i \in [1:n],$$

where $\{G_{1i} \in \mathcal{G}_1\}$ and $\{G_{2i} \in \mathcal{G}_2\}$ are channel gain processes and $\{Z_i\}$ is a WGN(1) process, independent of $\{G_{1i}\}$ and $\{G_{2i}\}$. Assume power constraint P on each of X_1 and X_2. As before, we assume the block fading model, where in block $l = 1, 2, \ldots, G_{ji} = \overline{G}_{jl}$ for $i \in [(l-1)n + 1 : ln]$ and $j = 1, 2$, and consider fast and slow fading scenarios.

23.5.1 Fast Fading

Channel gains available only at the decoder. The ergodic capacity region in this case is the set of rate pairs (R_1, R_2) such that

$$R_1 \le \mathsf{E}_{G_1}[\mathsf{C}(G_1^2 P)],$$
$$R_2 \le \mathsf{E}_{G_2}[\mathsf{C}(G_2^2 P)],$$
$$R_1 + R_2 \le \mathsf{E}_{G_1,G_2}[\mathsf{C}((G_1^2 + G_2^2)P)].$$

This can be shown by evaluating the capacity region characterization for the DM-MAC with DM state when the state information is available only at the decoder (see Section 7.4.2) under average power constraint P on each sender. In particular, the ergodic sum-capacity is

$$C_{\text{GI-D}} = \mathsf{E}_{G_1,G_2}[\mathsf{C}((G_1^2 + G_2^2)P)].$$

Channel gains available at both encoders and the decoder. The ergodic capacity region in this case is the set of rate pairs (R_1, R_2) such that

$$R_1 \le \mathsf{E}[\mathsf{C}(G_1^2 \phi_1(G_1, G_2))],$$
$$R_2 \le \mathsf{E}[\mathsf{C}(G_2^2 \phi_2(G_1, G_2))],$$
$$R_1 + R_2 \le \mathsf{E}[\mathsf{C}(G_1^2 \phi_1(G_1, G_2) + G_2^2 \phi_2(G_1, G_2))]$$

for some ϕ_1 and ϕ_2 that satisfy the constraints $\mathsf{E}(\phi_j(G_1, G_2)) \le P$, $j = 1, 2$. This can be shown by evaluating the capacity region characterization for the DM-MAC with DM state when the state information is available at both encoders and the decoder (see Section 7.4.2) under expected average power constraint P on each sender. In particular, the ergodic sum-capacity $C_{\text{GI-ED}}$ for this case can be computed by solving the optimization problem

$$\text{maximize } \mathsf{E}[\mathsf{C}(G_1^2 \phi_1(G_1, G_2) + G_2^2 \phi_2(G_1, G_2))]$$
$$\text{subject to } \mathsf{E}[\phi_j(G_1, G_2)] \le P, \quad j = 1, 2.$$

Channel gains available at their respective encoders and at the decoder. Consider the case where both channel gains are available at the receiver but each sender knows only the

gain of its channel to the receiver. This scenario is motivated by practical wireless communication systems in which each sender can estimate its channel gain via electromagnetic
reciprocity from a training sequence globally transmitted by the receiver (access point or
base station). Complete gain availability at the receiver can be achieved by feeding back
these gains from the senders to the receiver. Complete channel gain availability at the
senders, however, is not always practical as it requires either communication between the
senders or an additional round of feedback from the access point.

The ergodic capacity region in this case is the set of rate pairs (R_1, R_2) such that

$$R_1 \le \mathsf{E}[\mathsf{C}(G_1^2 \phi_1(G_1))],$$
$$R_2 \le \mathsf{E}[\mathsf{C}(G_2^2 \phi_2(G_2))],$$
$$R_1 + R_2 \le \mathsf{E}[\mathsf{C}(G_1^2 \phi_1(G_1) + G_2^2 \phi_2(G_2))]$$

for some ϕ_1 and ϕ_2 that satisfy the constraints $\mathsf{E}(\phi_j(G_j)) \le P$, $j = 1, 2$. This again can
be established by evaluating the capacity region characterization for the DM-MAC with
DM state when the state information is available at the decoder and each state component
is available at its respective encoder under expected average power constraint P on each
sender. In particular, the sum-capacity $C_{\mathrm{DGI\text{-}ED}}$ can be computed by solving the optimization problem

$$\text{maximize } \mathsf{E}[\mathsf{C}(G_1^2 \phi_1(G_1) + G_2^2 \phi_2(G_2))]$$
$$\text{subject to } \mathsf{E}[\phi_j(G_j)] \le P, \quad j = 1, 2.$$

23.5.2 Slow Fading

If the channel gains are available only at the decoder or at both encoders and the decoder,
the compound channel, outage, and adaptive coding approaches and corresponding performance metrics can be analyzed as in the point-to-point Gaussian fading channel case.
Hence, we discuss these coding approaches only for the case when each channel gains are
available at their respective encoders and at the decoder.

Compound channel approach. The capacity region using the compound channel approach is the set of rate pairs (R_1, R_2) such that

$$R_1 \le \min_{g_1} \mathsf{C}(g_1^2 P),$$
$$R_2 \le \min_{g_2} \mathsf{C}(g_2^2 P),$$
$$R_1 + R_2 \le \min_{g_1, g_2} \mathsf{C}((g_1^2 + g_2^2)P).$$

This is the same as the case where the channel gains are available only at the decoder.

Adaptive coding. Each sender can adapt its rate to the channel gain so that its message
can be reliably communicated regardless of the other sender's channel gain. This coding
approach makes the fading MAC equivalent to a channel that comprises multiple MACs

(one for each channel gain pair) with shared inputs and separate outputs as illustrated in Figure 23.3 for $\mathcal{G}_j = \{g_{j1}, g_{j2}\}$, $j = 1, 2$. Note that the capacity region of this equivalent multiple MACs is the set of rate quadruples $(R_{11}, R_{12}, R_{21}, R_{22})$ such that

$$
\begin{aligned}
R_{11} &\leq C(g_{11}^2 P), \\
R_{12} &\leq C(g_{12}^2 P), \\
R_{21} &\leq C(g_{21}^2 P), \\
R_{22} &\leq C(g_{22}^2 P), \\
R_{11} + R_{21} &\leq C((g_{11}^2 + g_{21}^2)P), \\
R_{11} + R_{22} &\leq C((g_{11}^2 + g_{22}^2)P), \\
R_{12} + R_{21} &\leq C((g_{12}^2 + g_{21}^2)P), \\
R_{12} + R_{22} &\leq C((g_{12}^2 + g_{22}^2)P).
\end{aligned}
\tag{23.1}
$$

Hence, the adaptive capacity region of the original fading MAC is the set of rate pairs (R_1, R_2) such that

$$
\begin{aligned}
R_1 &= P\{G_1 = g_{11}\}R_{11} + P\{G_1 = g_{12}\}R_{12}, \\
R_2 &= P\{G_2 = g_{21}\}R_{21} + P\{G_2 = g_{22}\}R_{22},
\end{aligned}
$$

where the rate quadruple $(R_{11}, R_{12}, R_{21}, R_{22})$ satisfies the inequalities in (23.1). A more

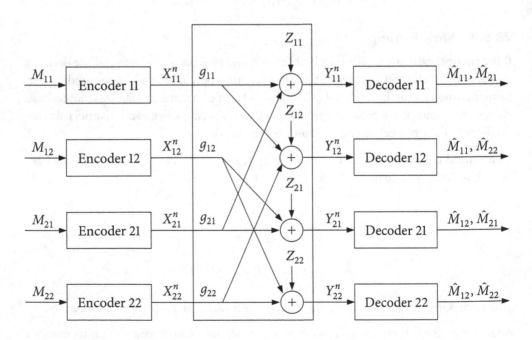

Figure 23.3. Equivalent channel with multiple MACs for adaptive coding over the Gaussian fading MAC.

explicit characterization of the adaptive capacity region can be obtained by the Fourier–Motzkin elimination procedure.

In general, the adaptive capacity region is the set of rate pairs $(R_1, R_2) = (E[R_1(G_1)], E[R_2(G_2)])$ such that the adaptive rate pair $(R_1(g_1), R_2(g_2))$ satisfies the inequalities

$$R_1(g_1) \le C(g_1^2 P), \quad g_1 \in \mathcal{G}_1,$$
$$R_2(g_2) \le C(g_2^2 P), \quad g_2 \in \mathcal{G}_2, \tag{23.2}$$
$$R_1(g_1) + R_2(g_2) \le C((g_1^2 + g_2^2) P), \quad (g_1, g_2) \in \mathcal{G}_1 \times \mathcal{G}_2.$$

In particular, the adaptive sum-capacity C_A can be computed by solving the optimization problem

$$\begin{aligned}
\text{maximize} \quad & E(R_1(G_1)) + E(R_2(G_2)) \\
\text{subject to} \quad & R_1(g_1) \le C(g_1^2 P), \quad g_1 \in \mathcal{G}_1, \\
& R_2(g_2) \le C(g_2^2 P), \quad g_2 \in \mathcal{G}_2, \\
& R_1(g_1) + R_2(g_2) \le C((g_1^2 + g_2^2) P), \quad (g_1, g_2) \in \mathcal{G}_1 \times \mathcal{G}_2.
\end{aligned}$$

Adaptive coding with power control. Each sender can further adapt its power to the channel gain according to the power allocation function $\phi_j(g_j)$ that satisfies the constraint $E(\phi(G_j)) \le P, j = 1, 2$. The power-control adaptive capacity region can be characterized as in (23.2) with P replaced by $\phi(g_j)$. In particular, the power-control adaptive sum-capacity C_{PA} can be computed by solving the optimization problem

$$\begin{aligned}
\text{maximize} \quad & E(R_1(G_1)) + E(R_2(G_2)) \\
\text{subject to} \quad & E(\phi_j(G_j)) \le P, \quad j = 1, 2, \\
& R_1(g_1) \le C(g_1^2 \phi_1(g_1)), \quad g_1 \in \mathcal{G}_1, \\
& R_2(g_2) \le C(g_2^2 \phi_2(g_2)), \quad g_2 \in \mathcal{G}_2, \\
& R_1(g_1) + R_2(g_2) \le C(g_1^2 \phi_1(g_1) + g_2^2 \phi_2(g_2)), \quad (g_1, g_2) \in \mathcal{G}_1 \times \mathcal{G}_2.
\end{aligned}$$

In the following, we compare the sum-capacities achieved by the above approaches.

Example 23.2. Assume that the gains $G_1, G_2 \in \{g^{(1)}, g^{(2)}\}$, where states $g^{(1)} > g^{(2)}$ and $P\{G_j = g^{(1)}\} = p, j = 1, 2$. In Figure 23.4, we compare the sum-capacities C_{CC} (compound channel), C_A (adaptive), C_{PA} (power-control adaptive), $C_{\mathrm{GI-D}}$ (ergodic with gains available only at the decoder), $C_{\mathrm{DGI-ED}}$ (ergodic with gains available at their respective encoder and the decoder), and $C_{\mathrm{GI-ED}}$ (ergodic with gains available at both encoders and the decoder) for different values of $p \in [0, 1]$. At low SNR (Figure 23.4a), power control is useful for adaptive coding; see C_{PA} versus C_A. At high SNR (Figure 23.4b), centralized knowledge of channel gains significantly improves upon distributed knowledge of channel gains.

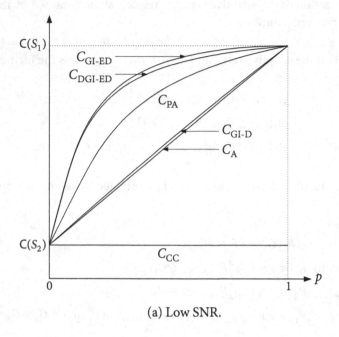

(a) Low SNR.

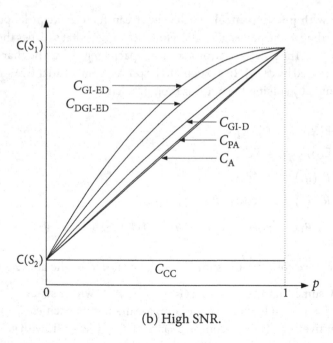

(b) High SNR.

Figure 23.4. Comparison of performance metrics for the fading MAC. Here $S_j = (g^{(j)})^2 P$, $j = 1, 2$.

23.6   GAUSSIAN FADING BC

Consider the Gaussian fading BC

$$Y_{1i} = G_{1i}X_i + Z_{1i},$$
$$Y_{2i} = G_{2i}X_i + Z_{2i}, \quad i \in [1:n],$$

where $\{G_{1i} \in \mathcal{G}_1\}$ and $\{G_{2i} \in \mathcal{G}_2\}$ are channel gain processes and $\{Z_{1i}\}$ and $\{Z_{2i}\}$ are WGN(1) processes, independent of $\{G_{1i}\}$ and $\{G_{2i}\}$. Assume power constraint P on X. As before, we assume the block fading model, where the channel gains are fixed in each coherence time interval but vary according to a stationary ergodic process over different intervals.

Fast fading. When the channel gain is available only at the decoders, the ergodic capacity region is not known in general. The difficulty is that the encoder does not know in which direction the channel is degraded during each coherence time interval. When the channel gain is available also at the encoder, however, the direction of channel degradedness is known and the capacity region can be easily established. Let $E(g_1, g_2) = 1$ if $g_2 \geq g_1$ and $E(g_1, g_2) = 0$ otherwise, and $\bar{E} = 1 - E$. Then, the ergodic capacity region is the set of rate pairs (R_1, R_2) such that

$$R_1 \leq \mathsf{E}\left(\mathsf{C}\left(\frac{\alpha G_1^2 P}{1 + \bar{\alpha} G_1^2 PE(G_1, G_2)}\right)\right),$$
$$R_2 \leq \mathsf{E}\left(\mathsf{C}\left(\frac{\bar{\alpha} G_2^2 P}{1 + \alpha G_2^2 P\bar{E}(G_1, G_2)}\right)\right)$$

for some $\alpha \in [0, 1]$.

Slow fading. The interesting case is when the channel gains are available only at the decoders. Again we can investigate different approaches to coding. For example, using the compound channel approach, we code for the worst channel. Let g_1^* and g_2^* be the lowest gains for the channel to Y_1 and the channel to Y_2, respectively, and assume without loss of generality that $g_1^* \geq g_2^*$. Then, the compound capacity region is the set of rate pairs (R_1, R_2) such that

$$R_1 \leq \mathsf{C}(g_1^* \alpha P),$$
$$R_2 \leq \mathsf{C}\left(\frac{g_2^* \bar{\alpha} P}{1 + g_2^* \alpha P}\right)$$

for some $\alpha \in [0, 1]$. As before, the adaptive coding capacity regions are the same as the corresponding ergodic capacity regions.

23.7   GAUSSIAN FADING IC

Consider a k-user-pair Gaussian fading IC

$$\mathbf{Y}(i) = G(i)\mathbf{X}(i) + \mathbf{Z}(i), \quad i \in [1:n],$$

where $\{G_{jj'}(i)\}_{i=1}^{\infty}$, $j, j' \in [1:k]$, are the random channel gain processes from sender j' to receiver j, and $\{Z_{ji}\}_{i=1}^{\infty}$, $j \in [1:k]$, are WGN(1) processes, independent of the channel gain processes. Assume power constraint P on X_j, $j \in [1:k]$. As before, assume the block fading model, where the channel gains are fixed in each coherence time interval but varies according to a stationary ergodic processes over different intervals.

We consider fast fading where the channel gains $\{G(i)\}$ are available at all the encoders and the decoders. Assuming that the marginal distribution of each channel gain process $\{G_{jj'}(i)\}_{i=1}^{n}$, $j, j' \in [1:k]$, is symmetric, that is, $\mathsf{P}\{G_{jj'}(i) \le g\} = \mathsf{P}\{G_{jj'}(i) \ge -g\}$, we can apply the interference alignment technique introduced in Section 6.9 to achieve higher rates than time division or treating interference as Gaussian noise.

We illustrate this result through the following simple example. Suppose that for each $i \in [1:n]$, $G_{jj'}(i)$, $j, j' \in [1:k]$, are i.i.d. Unif$\{-g, +g\}$.

Time division. Using time division with power control, we can easily show that the ergodic sum-capacity is lower bounded as

$$C_{\text{GI-ED}} \ge \mathsf{C}(g^2 kP).$$

Thus the rate per user tends to zero as $k \to \infty$.

Treating interference as Gaussian noise. Using Gaussian point-to-point channel codes and treating interference as *Gaussian* noise, the ergodic sum-capacity is lower bounded as

$$C_{\text{GI-ED}} \ge k\,\mathsf{C}\left(\frac{g^2 P}{g^2(k-1)P + 1}\right).$$

Thus again the rate per user tends to zero as $k \to \infty$.

Ergodic interference alignment. Using interference alignment over time, we show that the ergodic sum-capacity is

$$C_{\text{GI-ED}} = \frac{k}{2}\,\mathsf{C}(2g^2 P).$$

The proof of the converse follows by noting that the pairwise sum-rate $R_j + R_{j'}$, $j \ne j'$, is upper bounded by the sum-capacity of the two 2-user-pair Gaussian ICs with strong interference:

$$Y_j = gX_j + gX_{j'} + Z_j,$$
$$Y_{j'} = gX_j - gX_{j'} + Z_{j'},$$

and

$$Y_j = gX_j + gX_{j'} + Z_j,$$
$$Y_{j'} = gX_j + gX_{j'} + Z_{j'}.$$

Note that these two Gaussian ICs have the same capacity region.

Achievability can be proved by treating the channel gain matrices G^n as a time-sharing sequence as in Section 7.4.1 and repeating each codeword twice such that the interfering signals are aligned. For a channel gain matrix G, let G^* be its *conjugate* channel gain matrix, that is, $G_{jj'} + G^*_{jj'} = 0$ for $j \neq j'$ and $G_{jj} = G^*_{jj}$. Under our channel model, there are $|\mathcal{G}| = 2^{k^2}$ channel gain matrices and $|\mathcal{G}|/2$ pairs of $\{G, G^*\}$, and $p(G) = 1/|\mathcal{G}|$ for all G.

Following the coding procedure for the DMC with DM state available at both the encoder and the decoder, associate with each channel gain matrix G a FIFO buffer of length n. Divide the message M_j into $|\mathcal{G}|/2$ equal-rate messages $M_{j,G}$ of rate $2R_j/|\mathcal{G}|$, and generate codewords $x^n_j(m_{j,G}, G)$ for each pair $\{G, G^*\}$. Store each codeword twice into FIFO buffers corresponding to each G and its conjugate. Then transmit the codewords by multiplexing over the buffers based on the channel gain matrix sequence.

Now since the same codeword is repeated twice, the demultiplexed channel outputs corresponding to each gain pair $\{G, G^*\}$ are

$$Y(G) = GX + Z(G),$$
$$Y(G^*) = G^*X + Z(G^*)$$

with the same X. Here the corresponding noise vectors $Z(G)$ and $Z(G^*)$ are independent because they have different transmission times. Thus, for each gain pair $\{G, G^*\}$, the effective channel is

$$\tilde{Y} = (G + G^*)X + Z(G) + Z(G^*),$$

which has no interference since $G + G^*$ is diagonal! Hence, the probability of error tends to zero as $n \to \infty$ if the rate

$$R_{j,G} < (1 - \epsilon)p(G) C(2g^2P),$$

or equivalently, if

$$R_j < \frac{1 - \epsilon}{2} C(2g^2P).$$

This completes the proof of achievability, establishing the sum-capacity of the channel.

Remark 23.2. This ergodic interference alignment technique can be extended to arbitrary symmetric fading distributions by quantizing the channel gain matrices.

SUMMARY

- Gaussian fading channel models

- Coding under fast fading:

  - Ergodic capacity

  - Water-filling in time

- The ergodic capacity of a MIMO fading channel increases linearly with the number of antennas

- Coding under slow fading:

  - Compound channel approach

  - Outage capacity approach

  - Broadcast channel approach

  - Adaptive coding with or without power control

- Ergodic interference alignment

- **Open problem 23.1.** What is the ergodic capacity region of the Gaussian fading BC under fast fading when the channel gain information is available only at the decoders?

BIBLIOGRAPHIC NOTES

The block fading model was first studied by Ozarow, Shamai, and Wyner (1994), who also introduced the notion of outage capacity. The ergodic capacity of the Gaussian fading channel was established by Goldsmith and Varaiya (1997). Cover (1972) introduced the broadcast channel approach to compound channels, which was later applied to Gaussian fading channels with slow fading by Shamai (1997) and Shamai and Steiner (2003). The Gaussian vector fading channel was first considered by Telatar (1999) and Foschini (1996), who established the ergodic capacity in Theorem 23.1.

The ergodic sum-capacity of the Gaussian fading MAC when channel gains are available at both encoders and the decoder was established by Knopp and Humblet (1995). This result was generalized to the capacity region by Tse and Hanly (1998). The ergodic capacity region when the channel gains are available at the decoder and each respective encoder was established by Cemal and Steinberg (2005) and Jafar (2006). The adaptive coding approach under the same channel gain availability assumption was studied by Shamai and Telatar (1999) and Hwang, Malkin, El Gamal, and Cioffi (2007).

The ergodic capacity region of the Gaussian fading BC when the channel gains are available at the encoder and both decoders was established by Li and Goldsmith (2001). Ergodic interference alignment was introduced by Cadambe and Jafar (2008). The example in Section 23.7 is due to Nazer, Gastpar, Jafar, and Vishwanath (2009), who studied both the k-user QED-IC and Gaussian IC. A survey of the literature on Gaussian fading channels can be found in Biglieri, Proakis, and Shamai (1998). More recent developments as well as applications to wireless systems are discussed in Tse and Viswanath (2005) and Goldsmith (2005).

The capacity scaling for random wireless fading networks was studied in Jovičić, Viswanath, and Kulkarni (2004), Xue, Xie, and Kumar (2005), Xie and Kumar (2006), and Ahmad, Jovičić, and Viswanath (2006). Özgür, Lévêque, and Tse (2007) showed that under Rayleigh fading with channel gains available at all the nodes, the order of the symmetric ergodic capacity is lower bounded as $C(N) = \Omega(N^{-\epsilon})$ for every $\epsilon > 0$; see also Ghaderi, Xie, and Shen (2009). Özgür et al. (2007) used a 3-phase cellular hierarchical cooperation scheme, where in phase 1 source nodes in every cell exchange their messages, in phase 2 the source node and other nodes in the same cell cooperatively transmit the intended message for each source–destination pair as for the point-to-point Gaussian vector fading channel, and in phase 3 receiver nodes in every cell exchange the quantized versions of their received outputs. The communication in the first and third phases can be performed using the same scheme recursively by further dividing the cells into subcells. This result demonstrates the potential improvement in capacity by transmission cooperation and opportunistic coding using the channel fading information. This result was further extended to networks with varying node densities by Özgür, Johari, Tse, and Lévêque (2010). The hierarchical cooperation approach was also extended to scaling laws on the *capacity region* of wireless networks by Niesen, Gupta, and Shah (2009).

PROBLEMS

23.1. Find the compound channel capacity region of the Gaussian fading MAC under slow fading (a) when the channel gains are available only at the decoder and (b) when the channel gains are available at their respective encoders and at the decoder.

23.2. Establish the ergodic capacity region of the Gaussian fading BC under fast fading when the channel gains are available at the encoder and both decoders.

23.3. *Outage capacity.* Consider the slow-fading channel $Y = GX + Z$, where $G \sim$ Unif$[0, 1]$, $Z \sim N(0, 1)$, and assume average power constraint P. Find the outage capacity (a) when $P = 1$ and $p_{\text{out}} = 0.2$, and (b) when $P = 2$ and $p_{\text{out}} = 0.1$.

23.4. *On-off fading MAC.* Consider the Gaussian fading MAC under slow fading when the channel gains are available at their respective encoders and at the decoder. Suppose that G_1 and G_2 are i.i.d. Bern(p).

(a) Find the adaptive coding capacity region.

(b) Find the adaptive sum capacity.

23.5. *DM-MAC with distributed state information.* Consider the DM-MAC with DM state $p(y|x_1, x_2, s_1, s_2)p(s_1, s_2)$, where sender $j = 1, 2$ wishes to communicate a message $M_j \in [1 : 2^{nR_j}]$ to the receiver. Suppose that both states are available at the decoder and each state is causally available at its corresponding encoder. Find the adaptive coding capacity region of this channel, which is the capacity region of multiple MACs, one for each channel gain pair $(s_1, s_2) \in \mathcal{S}_1 \times \mathcal{S}_2$, defined similarly to the Gaussian case.

CHAPTER 24

Networking and Information Theory

The source and network models we discussed so far capture many essential ingredients of real-world communication networks, including

- noise,

- multiple access,

- broadcast,

- interference,

- time variation and uncertainty about channel statistics,

- distributed compression and computing,

- joint source–channel coding,

- multihop relaying,

- node cooperation,

- interaction and feedback, and

- secure communication.

Although a general theory for information flow under these models remains elusive, we have seen that there are several coding techniques—some of which are optimal or close to optimal—that promise significant performance improvements over today's practice. Still, the models we discussed do not capture other key aspects of real-world networks.

- We assumed that data is always available at the communication nodes. In real-world networks, data is bursty and the nodes have finite buffer sizes.

- We assumed that the network has a known and fixed number of users. In real-world networks, users can enter and leave the network at will.

- We assumed that the network operation is centralized and communication over the network is synchronous. Many real-world networks are decentralized and communication is asynchronous.

- We analyzed performance assuming arbitrarily long delays. In many networking applications, delay is a primary concern.

- We ignored the overhead (protocol) needed to set up the communication as well as the cost of feedback and channel state information.

While these key aspects of real-world networks have been at the heart of the field of computer networks, they have not been satisfactorily addressed by network information theory, either because of their incompatibility with the basic asymptotic approach of information theory or because the resulting models are messy and intractable. There have been several success stories at the intersection of networking and network information theory, however. In this chapter we discuss three representative examples.

We first consider the channel coding problem for a DMC with random data arrival. We show that reliable communication is feasible provided that the data arrival rate is less than the channel capacity. Similar results can be established for multiuser channels and multiple data streams. A key new ingredient in this study is the notion of queue stability.

The second example we discuss is motivated by the random medium access control scheme for sharing a channel among multiple senders such as in the ALOHA network. We model a 2-sender 1-receiver random access system by a modulo-2 sum MAC with multiplicative binary state available partially at each sender and completely at the receiver. We apply various coding approaches introduced in Chapter 23 to this model and compare the corresponding performance metrics.

Finally, we investigate the effect of asynchrony on the capacity region of the DM-MAC. We extend the synchronous multiple access communication system setup in Chapter 4 to multiple transmission blocks in order to incorporate unknown transmission delays. When the delay is small relative to the transmission block length, the capacity region does not change. However, when we allow arbitrary delay, time sharing cannot be used and hence the capacity region can be smaller than for the synchronous case.

24.1   RANDOM DATA ARRIVALS

In the point-to-point communication system setup in Section 3.1 and subsequent extensions to multiuser channels, we assumed that data is always available at the encoder. In many networking applications, however, data is bursty and it may or may not be available at the senders when the channel is free. Moreover, the amount of data at a sender may exceed its finite buffer size, which results in data loss even before transmission takes place. It turns out that under fairly general data arrival models, if the data rate λ bits/transmission is below the capacity C of the channel, then the data can be reliably communicated to the receiver, while if $\lambda > C$, data cannot be reliably communicated either because the incoming data exceeds the sender's queue size or because transmission rate exceeds the channel capacity. We illustrate this general result using a simple random data arrival process.

Consider the point-to-point communication system with random data arrival at its input depicted in Figure 24.1. Suppose that data packets arrive at the encoder at the "end"

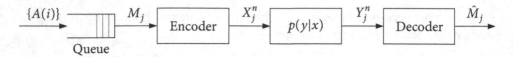

Figure 24.1. Communication system with random data arrival.

of transmission time $i = 1, 2, \ldots$ according to an i.i.d. process $\{A(i)\}$, where

$$A(i) = \begin{cases} k & \text{with probability } p, \\ 0 & \text{with probability } \bar{p}. \end{cases}$$

Thus, a packet randomly and uniformly chosen from the set of k-bit sequences arrives at the encoder with probability p and no packet arrives with probability $\bar{p}$. Assume that the packets arriving at different transmission times are independent of each other.

A $(2^{nR}, n)$ *augmented* block code for the DMC consists of

- an augmented message set $[1 : 2^{nR}] \cup \{0\}$,

- an encoder that assigns a codeword $x^n(m)$ to each $m \in [1 : 2^{nR}] \cup \{0\}$, and

- a decoder that assigns a message $\hat{m} \in [1 : 2^{nR}] \cup \{0\}$ or an error message e to each received sequence y^n.

The code is used in consecutive transmission blocks as follows. Let $Q(i)$ be the number of bits (backlog) in the sender's queue at the "beginning" of transmission time $i = 1, 2, \ldots$. At the beginning of time jn, $j = 1, 2, \ldots$, that is, at the beginning of transmission block j, nR bits are taken out of the queue if $Q(jn) \geq nR$. The bits are represented by a message $M_j \in [1 : 2^{nR}]$ and the codeword $x^n(m_j)$ is sent over the DMC. If $Q(jn) < nR$, no bits are taken out of the queue and the "0-message" codeword $x^n(0)$ is sent. Thus, the backlog $Q(i)$ is a time-varying Markov process with transition law

$$Q(i + 1) = \begin{cases} Q(i) - nR + A(i) & \text{if } i = jn \text{ and } Q(i) \geq nR, \\ Q(i) + A(i) & \text{otherwise.} \end{cases} \tag{24.1}$$

The queue is said to be *stable* if $\sup_i \mathsf{E}(Q(i)) \leq B$ for some constant $B < \infty$. By the Markov inequality, queue stability implies that the probability of data loss can be made as small as desired with a finite buffer size. Define the *arrival rate* $\lambda = kp$ as the product of the packet arrival rate $p \in (0, 1]$ and packet size k bits. We have the following sufficient and necessary conditions on the stability of the queue in terms of the transmission rate R and the arrival rate λ.

Lemma 24.1. If $\lambda < R$, then the queue is stable. Conversely, if the queue is stable, then $\lambda \leq R$.

The proof of this lemma is given in Appendix 24.1.

Let $p_j = \mathsf{P}\{M_j = 0\}$ be the probability that the sender queue has less than nR bits at the beginning of transmission block j. By the definition of the arrival time process, $M_j | \{M_j \neq 0\} \sim \text{Unif}[1 : 2^{nR}]$. Define the probability of error in transmission block j as

$$P_{ej}^{(n)} = P\{\hat{M}_j \neq M_j\} = p_j \, P\{\hat{M}_j \neq 0 \,|\, M_j = 0\} + \frac{(1 - p_j)}{2^{nR}} \sum_{m=1}^{2^{nR}} P\{\hat{M}_j \neq m \,|\, M_j = m\}.$$

The data arriving at the encoder according to the process $\{A(i)\}$ is said to be reliably communicated at rate R over the DMC if the queue is stable and there exists a sequence of $(2^{nR}, n)$ augmented codes such that $\lim_{n \to \infty} \sup_j P_{ej}^{(n)} = 0$. We wish to find the necessary and sufficient condition for reliable communication of the data over the DMC.

Theorem 24.1. The random data arrival process $\{A(i)\}$ with arrival rate λ can be reliably communicated at rate R over a DMC $p(y|x)$ with capacity C if $\lambda < R < C$. Conversely, if the process $\{A(i)\}$ can be reliably communicated at rate R over this DMC, then $\lambda \leq R \leq C$.

Proof. To prove achievability, let $\lambda < R < C$. Then the queue is stable by Lemma 24.1 and there exists a sequence of $(2^{nR} + 1, n)$ (regular) channel codes such that both the average probability of error $P_e^{(n)}$ and $P\{\hat{M} \neq M | M = m'\}$ for some m' tend to zero as $n \to \infty$. By relabeling $m' = 0$, we have shown that there exists a sequence of $(2^{nR}, n)$ augmented codes such that $P_{ej}^{(n)}$ tends to zero as $n \to \infty$ for every j.

To prove the converse, note first that $\lambda \leq R$ from Lemma 24.1. Now, for each j, following similar steps to the converse proof of the channel coding theorem in Section 3.1.4, we obtain

$$\begin{aligned}
nR &= H(M_j \,|\, M_j \neq 0) \\
&\leq I(M_j; Y^n \,|\, M_j \neq 0) + n\epsilon_n \\
&\leq \sum_{i=1}^{n} I(X_i; Y_i \,|\, M_j \neq 0) + n\epsilon_n \\
&\leq n(C + \epsilon_n).
\end{aligned} \tag{24.2}$$

This completes the proof of Theorem 24.1.

Remark 24.1. Theorem 24.1 continues to hold for arrival processes for which Lemma 24.1 holds. It can be also extended to multiuser channels with random data arrivals at each sender. For example, consider the case of a DM-MAC with two independent i.i.d. arrival processes $\{A_1(i)\}$ and $\{A_2(i)\}$ of arrival rates λ_1 and λ_2, respectively. The *stability region* $\mathcal{S}$ for the two sender queues is the closure of the set of arrival rates (λ_1, λ_2) such that both queues are stable. We define the augmented code $(2^{nR_1}, 2^{nR_2}, n)$, the average probability of error, and achievability as for the point-to-point case. Let $\mathcal{C}$ be the capacity region of the DM-MAC. Then it can be readily shown that $\mathcal{S} = \mathcal{C}$. Note that the same result holds when the packet arrivals (but not the packet contents) are correlated.

Remark 24.2. The conclusion that randomly arriving data can be communicated reliably over a channel when the arrival rate is less than the capacity trivializes the effect of randomness in data arrival. In real-world applications, packet delay constraints are as important as queue stability. However, the above result, and the asymptotic approach of information theory in general, does not capture such constraints well.

24.2 RANDOM ACCESS CHANNEL

The previous section dealt with random data arrivals at the senders. In this section, we consider random data arrivals at the receivers. We discuss random access, which is a popular scheme for medium access control in local area networks. In these networks, the number of senders is not fixed a priori and hence using time division can be inefficient. The random access scheme improves upon time division by having each active sender transmit its packets in randomly selected transmission blocks. In practical random access control systems, however, the packets are encoded at a fixed rate and if more than one sender transmits in the same block, the packets are lost. It turns out that we can do better by using more sophisticated coding schemes.

We model a random access channel by a modulo-2 sum MAC with multiplicative binary state components as depicted in Figure 24.2. The output of the channel at time i is

$$Y_i = S_{1i} \cdot X_{1i} \oplus S_{2i} \cdot X_{2i},$$

where the states S_{1i} and S_{2i} are constant over each *access time interval* $[(l-1)k + 1 : lk]$ of length k for $l = 1, 2, \ldots$, and the processes $\{\bar{S}_{1l}\}_{l=1}^{\infty} = \{S_{1,(l-1)k+1}\}_{l=1}^{\infty}$ and $\{\bar{S}_{2l}\}_{l=1}^{\infty} = \{S_{2,(l-1)k+1}\}_{l=1}^{\infty}$ are independent Bern(p) processes. Sender $j = 1, 2$ is active (has a packet to transmit) when $S_j = 1$ and is inactive when $S_j = 0$. We assume that the receiver knows which senders are active in each access time interval, but each sender knows only its own activity.

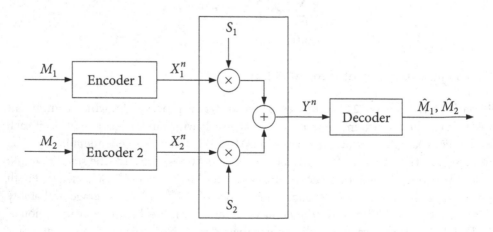

Figure 24.2. Random access channel.

Note that this model is analogous to the Gaussian fading MAC in Section 23.5, where the channel gains are available at the receiver but each sender knows only the gain of its own channel. Since each sender becomes active at random in each block, the communication model corresponds to the slow fading scenario. Using the analogy to the fading MAC, we consider different coding approaches and corresponding performance metrics for the random access channel. Unlike the fading MAC, however, no coordination is allowed between the senders in the random access channel.

Compound channel approach. In this approach, we code for the worst case in which no packets are to be transmitted (i.e., $S_{1i} = S_{2i} = 0$ for all $i \in [1:n]$). Hence, the capacity region is $\{(0, 0)\}$.

ALOHA. In this approach, sender $j = 1, 2$ transmits at rate $R_j = 1$ when it is active and at rate $R_j = 0$ when it is not. When there is collision (that is, both senders are active), decoding simply fails. The *ALOHA sum-capacity* (that is, the average total throughput) is

$$C_{\text{ALOHA}} = p(1 - p) + p(1 - p) = 2p(1 - p).$$

Adaptive coding. By reducing the rates in the ALOHA approach to $\tilde{R}_j \le 1$ when sender $j = 1, 2$ is active (so that the messages can be recovered even under collision), we can increase the average throughput.

To analyze the achievable rates for this approach, consider the 2-sender 3-receiver channel depicted in Figure 24.3. It can be easily shown that the capacity region of this channel is the set of rate pairs $(\tilde{R}_1, \tilde{R}_2)$ such that $\tilde{R}_1 + \tilde{R}_2 \le 1$ and is achieved using simultaneous decoding without time sharing; see Problem 24.4. Hence, any rate pair $(\tilde{R}_1, \tilde{R}_2)$ in the capacity region of the 2-sender 3-receiver channel is achievable for the random access channel, even though each sender is aware only of its own activity. In particular, the *adaptive coding sum-capacity* is

$$C_{\text{A}} = \max_{(\tilde{R}_1, \tilde{R}_2): \tilde{R}_1 + \tilde{R}_2 \le 1} (P\{S_1 = 1\}\tilde{R}_1 + P\{S_2 = 1\}\tilde{R}_2) = p.$$

Broadcast channel approach. In the ALOHA approach, the messages cannot be recovered at all when there is a collision. In the adaptive coding approach, both messages must

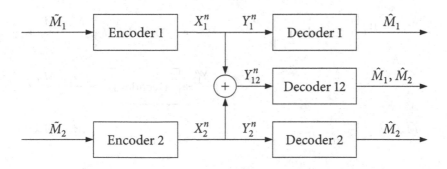

Figure 24.3. Adaptive coding for the random access channel.

be recovered even when there is a collision. The broadcast approach combines these two approaches by requiring only part of each message to be recovered when there is a collision and the rest of the message to be also recovered when there is no collision. This is achieved using superposition coding. To analyze the achievable rates for this strategy, consider the 2-sender 3-receiver channel depicted in Figure 24.4. Here the message pair $(\tilde{M}_{j0}, \tilde{M}_{jj})$ from the active sender j is to be recovered when there is no collision, while the message pair $(\tilde{M}_{10}, \tilde{M}_{20})$, one from each sender is to be recovered when there is collision. It can be shown (see Problem 24.5) that the capacity region of this 2-sender 3-receiver channel is the set of rate quadruples $(\tilde{R}_{10}, \tilde{R}_{11}, \tilde{R}_{20}, \tilde{R}_{22})$ such that

$$\tilde{R}_{10} + \tilde{R}_{20} + \tilde{R}_{11} \le 1,$$
$$\tilde{R}_{10} + \tilde{R}_{20} + \tilde{R}_{22} \le 1. \tag{24.3}$$

As for the adaptive coding case, this region can be achieved using simultaneous decoding without time sharing. Note that taking $(\tilde{R}_{11}, \tilde{R}_{22}) = (0, 0)$ reduces to the adaptive coding case. The average throughput of sender $j \in \{1, 2\}$ is

$$R_j = p(1 - p)(\tilde{R}_{j0} + \tilde{R}_{jj}) + p^2 \tilde{R}_{j0} = p\tilde{R}_{j0} + p(1 - p)\tilde{R}_{jj}.$$

Thus, the *broadcast sum-capacity* is

$$C_{BC} = \max \left(p(\tilde{R}_{10} + \tilde{R}_{20}) + p(1 - p)(\tilde{R}_{11} + \tilde{R}_{22}) \right),$$

where the maximum is over all rate quadruples in the capacity region in (24.3). By symmetry, it can be readily checked that

$$C_{BC} = \max\{2p(1 - p), p\}.$$

Note that this sum-capacity is achieved by setting $\tilde{R}_{11} = \tilde{R}_{22} = 1$, $\tilde{R}_{10} = \tilde{R}_{20} = 0$ for $p \le 1/2$, and $\tilde{R}_{10} = \tilde{R}_{20} = 1/2$, $\tilde{R}_{11} = \tilde{R}_{22} = 0$ for $p \ge 1/2$. Hence, ignoring collision (ALOHA) is throughput-optimal when $p \le 1/2$, while the broadcast channel approach reduces to adaptive coding when $p \ge 1/2$.

Figure 24.5 compares the sum-capacities C_{CC} (compound channel approach), C_{ALOHA} (ALOHA), C_A (adaptive coding), and C_{BC} (broadcast channel approach). Note that the broadcast channel approach performs better than adaptive coding when the senders are active less often ($p \le 1/2$).

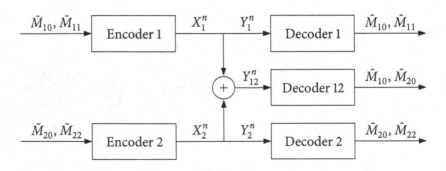

Figure 24.4. Broadcast coding for the random access channel.

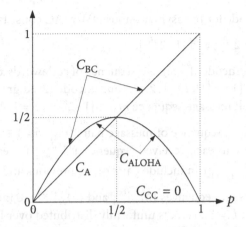

Figure 24.5. Comparison of the sum-capacities of the random access channel—C_{CC} for the compound approach, C_{ALOHA} for ALOHA, C_{A} for adaptive coding, and C_{BC} for the broadcast channel approach.

24.3 ASYNCHRONOUS MAC

In the single-hop channel models we discussed in Part II of the book, we assumed that the transmissions from the senders to the receivers are synchronized (at both the symbol and block levels). In practice, such complete synchronization is often not feasible. How does the lack of synchronization affect the capacity region of the channel? We answer this question for the asynchronous multiple access communication system depicted in Figure 24.6.

Suppose that sender $j = 1, 2$ wishes to communicate an i.i.d. message sequence $(M_{j1}, M_{j2}, \ldots)$. Assume that the same codebook is used in each transmission block. Further assume that symbols are synchronized, but that the blocks sent by the two encoders incur arbitrary delays $d_1, d_2 \in [0 : d]$, respectively, for some $d \le n - 1$. Assume that the encoders and the decoder do not know the delays a priori. The received sequence Y^n is distributed according to

$$p(y^n | x_{1,1-d_1}^n, x_{2,1-d_2}^n) = \prod_{i=1}^{n} p_{Y|X_1,X_2}(y_i | x_{1,i-d_1}, x_{2,i-d_2}),$$

where the symbols with negative indices are from the previous transmission block.

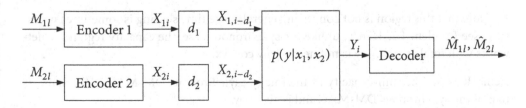

Figure 24.6. Asynchronous multiple access communication system.

A $(2^{nR_1}, 2^{nR_2}, n, d)$ code for the asynchronous DM-MAC consists of

- two message sets $[1 : 2^{nR_1}]$ and $[1 : 2^{nR_2}]$,

- two encoders, where encoder 1 assigns a sequence of codewords $x_1^n(m_{1l})$ to each message sequence $m_{1l} \in [1 : 2^{nR_1}]$, $l = 1, 2, \ldots$, and encoder 2 assigns a sequence of codewords $x_2^n(m_{2l})$ to each message sequence $m_{2l} \in [1 : 2^{nR_2}]$, $l = 1, 2, \ldots$, and

- a decoder that assigns a sequence of message pairs $(\hat{m}_{1l}, \hat{m}_{2l}) \in [1 : 2^{nR_1}] \times [1 : 2^{nR_2}]$ or an error message e to each received sequence $y_{(l-1)n+1}^{ln+d}$ for each $l = 1, 2, \ldots$ (the received sequence $y_{(l-1)n+1}^{ln+d}$ can include parts of the previous and next blocks).

We assume that the message sequences $\{M_{1l}\}_{l=1}^{\infty}$ and $\{M_{2l}\}_{l=1}^{\infty}$ are independent and each message pair (M_{1l}, M_{2l}), $l = 1, 2, \ldots$, is uniformly distributed over $[1 : 2^{nR_1}] \times [1 : 2^{nR_2}]$. The average probability of error is defined as

$$P_e^{(n)} = \max_{d_1, d_2 \in [0:d]} \sup_l P_{el}^{(n)}(d_1, d_2),$$

where $P_{el}^{(n)}(d_1, d_2) = P\{(\hat{M}_{1l}, \hat{M}_{2l}) \neq (M_{1l}, M_{2l}) | d_1, d_2\}$. Note that by the memoryless property of the channel and the definition of the code, $\sup_l P_{el}^{(n)}(d_1, d_2) = P_{el}^{(n)}(d_1, d_2)$ for all l. Thus in the following, we drop the subscript l. Achievability and the capacity region are defined as for the synchronous DM-MAC.

We consider two degrees of asynchrony.

Mild asynchrony. Suppose that d/n tends to zero as $n \to \infty$. Then, it can be shown that the capacity region is the same as for the synchronous case.

Total asynchrony. Suppose that d_1 and d_2 can vary from 0 to $(n-1)$, i.e., $d = n - 1$. In this case, time sharing is no longer feasible and the capacity region reduces to the following.

Theorem 24.2. The capacity region of the totally asynchronous DM-MAC is the set of all rate pairs (R_1, R_2) such that

$$R_1 \leq I(X_1; Y|X_2),$$
$$R_2 \leq I(X_2; Y|X_1),$$
$$R_1 + R_2 \leq I(X_1, X_2; Y)$$

for some pmf $p(x_1)p(x_2)$.

Note that this region is not convex in general, since time sharing is sometimes necessary; see Problem 4.2. Hence, unlike the synchronous case, the capacity region for networks with total asynchrony is not necessarily convex.

Remark 24.3. The sum-capacity of the totally asynchronous DM-MAC is the same as that of the synchronous DM-MAC and is given by

$$C_{\text{sum}} = \max_{p(x_1)p(x_2)} I(X_1, X_2; Y).$$

Remark 24.4. The capacity region of the Gaussian MAC does not change with asynchrony because time sharing is not required. However, under total asynchrony, simultaneous decoding is needed to achieve all the points in the capacity region.

Remark 24.5. Theorem 24.2 also shows that the capacity of a point-to-point channel does not change with asynchrony.

We prove Theorem 24.2 in the next two subsections.

24.3.1 Proof of Achievability

Divide each n-transmission block into $(b + 1)$ subblocks each consisting of k symbols as illustrated in Figure 24.7; thus $n = (b + 1)k$ and the delays range from 0 to $(b + 1)k - 1$. The first subblock labeled $j = 0$ is the *preamble* subblock. Also divide the message pair (M_1, M_2) into b independent submessage pairs $(M_{1j}, M_{2j}) \in [1 : 2^{kR_1}] \times [1 : 2^{kR_2}]$, $j \in [1 : b]$, and send them in the following b subblocks. Note that the resulting rate pair for this code, $(bR_1/(b + 1), bR_2/(b + 1))$, can be made arbitrarily close to (R_1, R_2) as $b \to \infty$.

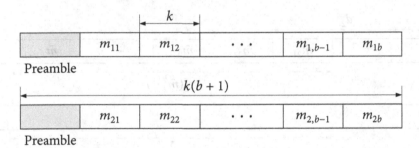

Figure 24.7. Transmission block divided into subblocks.

Codebook generation. Fix a product pmf $p(x_1)p(x_2)$. Randomly and independently generate a codebook for each subblock. Randomly generate a preamble codeword $x_1^k(0)$ according to $\prod_{i=1}^{k} p_{X_1}(x_{1i})$. For each $j \in [1 : b]$, randomly and independently generate 2^{kR_1} codewords $x_1^k(m_{1j})$, $m_{1j} \in [1 : 2^{kR_1}]$, each according to $\prod_{i=1}^{k} p_{X_1}(x_{i1})$. Similarly generate a preamble codeword $x_2^k(0)$ and codewords $x_2^k(m_{2j})$, $m_{2j} \in [1 : 2^{kR_2}]$, $j \in [1 : b]$, each according to $\prod_{i=1}^{k} p_{X_2}(x_{i2})$.

Encoding. To send the submessages m_1^b, encoder 1 first transmits its preamble codeword $x_1^k(0)$ followed by $x_1^k(m_{1j})$ for each $j \in [1 : b]$. Similarly, encoder 2 transmits its preamble codeword $x_2^k(0)$ followed by $x_2^k(m_{2j})$ for each $j \in [1 : b]$.

Decoding. The decoding procedure consists of two steps—preamble decoding and message decoding. The decoder declares $\hat{d}_1$ to be the estimate for d_1 if it is the unique number in $[0 : (b + 1)k - 1]$ such that $(x_1^k(0), y_{\hat{d}_1+1}^{\hat{d}_1+k}) \in \mathcal{T}_\epsilon^{(n)}$. Similarly, the decoder declares

$\hat{d}_2$ to be the estimate for d_2 if it is the unique number in $[0 : (b + 1)k - 1]$ such that $\left(x_2^k(0), y_{\hat{d}_2+1}^{\hat{d}_2+k} \right) \in T_\epsilon^{(n)}$.

Assume without loss of generality that $\hat{d}_1 \le \hat{d}_2$. Referring to Figure 24.8, define the sequences

$$\mathbf{x}_1(m_1^b) = \left(x_{1,\delta+1}^k(0), x_1^k(m_{11}), x_1^k(m_{12}), \ldots, x_1^k(m_{1b}), x_1^\delta(0) \right),$$
$$\mathbf{x}_2(\tilde{m}_2^{b+1}) = \left(x_2^k(\tilde{m}_{21}), x_2^k(\tilde{m}_{22}), \ldots, x_2^k(\tilde{m}_{2,b+1}) \right),$$
$$\mathbf{y} = y_{\hat{d}_1+\delta+1}^{(b+1)k+\hat{d}_1+\delta},$$

where $\delta = \hat{d}_2 - \hat{d}_1 \pmod k$. The receiver declares that $\hat{m}_1^b$ is the sequence of submessages sent by sender 1 if it is the unique submessage sequence such that $(\mathbf{x}_1(\hat{m}_1^b), \mathbf{x}_2(\tilde{m}_2^{b+1}), \mathbf{y}) \in T_\epsilon^{(n)}$ for some $\tilde{m}_2^{b+1}$.

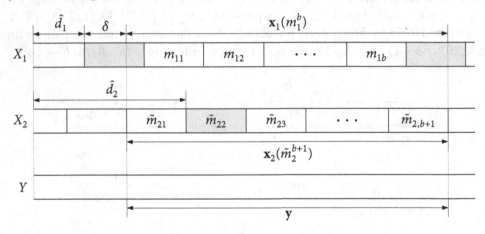

Figure 24.8. Asynchronous transmission and received sequence.

To recover the message sequence m_2^b, the same procedure is repeated beginning with the preamble of sender 2.

Analysis of the probability of error. We bound the probability of decoding error for the submessages M_1^b from sender 1 averaged over the codes. Assume without loss of generality that $M_1^b = \mathbf{1} = (1, \ldots, 1)$ and $d_1 \le d_2$. Let $\tilde{M}_2^{b+1}$, $\mathbf{X}_1(M_1^b)$, $\mathbf{X}_2(\tilde{M}_2^{b+1})$, and $\mathbf{Y}$ be defined as before (see Figure 24.8) with (d_1, d_2) in place of $(\hat{d}_1, \hat{d}_2)$. The decoder makes an error only if one or more of the following events occur:

$$\mathcal{E}_0 = \left\{ (\hat{d}_1(Y^{2n-1}), \hat{d}_2(Y^{2n-1})) \ne (d_1, d_2) \right\},$$
$$\mathcal{E}_{11} = \left\{ (\mathbf{X}_1(\mathbf{1}), \mathbf{X}_2(\tilde{M}_2^{b+1}), \mathbf{Y}) \notin T_\epsilon^{(n)} \right\},$$
$$\mathcal{E}_{12} = \left\{ (\mathbf{X}_1(m_1^b), \mathbf{X}_2(\tilde{m}_2^{b+1}), \mathbf{Y}) \in T_\epsilon^{(n)} \text{ for some } m_1^b \ne \mathbf{1}, \tilde{m}_2^{b+1} \ne \tilde{M}_2^{b+1} \right\}.$$

Thus, the probability of decoding error for M_1^b is upper bounded as

$$\mathsf{P}(\mathcal{E}_1) \le \mathsf{P}(\mathcal{E}_0) + \mathsf{P}(\mathcal{E}_{11} \cap \mathcal{E}_0^c) + \mathsf{P}(\mathcal{E}_{12} \cap \mathcal{E}_0^c). \tag{24.4}$$

To bound the first term, the probability of preamble decoding error, define the events

$$\mathcal{E}_{01} = \{(X_1^k(0), Y_{d_1+1}^{d_1+k}) \notin \mathcal{T}_\epsilon^{(n)}\},$$

$$\mathcal{E}_{02} = \{(X_2^k(0), Y_{d_2+1}^{d_2+k}) \notin \mathcal{T}_\epsilon^{(n)}\},$$

$$\mathcal{E}_{03} = \{(X_1^k(0), Y_{\tilde{d}_1+1}^{\tilde{d}_1+k}) \in \mathcal{T}_\epsilon^{(n)} \text{ for some } \tilde{d}_1 \neq d_1, \tilde{d}_1 \in [0:(b+1)k-1]\},$$

$$\mathcal{E}_{04} = \{(X_2^k(0), Y_{\tilde{d}_2+1}^{\tilde{d}_2+k}) \in \mathcal{T}_\epsilon^{(n)} \text{ for some } \tilde{d}_2 \neq d_2, \tilde{d}_2 \in [0:(b+1)k-1]\}.$$

Then

$$\mathsf{P}(\mathcal{E}_0) \leq \mathsf{P}(\mathcal{E}_{01}) + \mathsf{P}(\mathcal{E}_{02}) + \mathsf{P}(\mathcal{E}_{03}) + \mathsf{P}(\mathcal{E}_{04}).$$

By the LLN, the first two terms tend to zero as $k \to \infty$. To bound the other two terms, we use the following.

Lemma 24.2. Let $(X, Y) \sim p(x, y) \neq p(x)p(y)$ and $(X^n, Y^n) \sim \prod_{i=1}^n p_{X,Y}(x_i, y_i)$. If $\epsilon > 0$ is sufficiently small, then there exists $\gamma(\epsilon) > 0$ that depends only on $p(x, y)$ such that

$$\mathsf{P}\{(X^k, Y_{d+1}^{d+k}) \in \mathcal{T}_\epsilon^{(k)}\} \leq 2^{-k\gamma(\epsilon)}$$

for every $d \neq 0$.

The proof of this lemma is given in Appendix 24B.

Now using this lemma with $X^k \leftarrow X_1^k(0)$ and $Y_{d+1}^{d+k} \leftarrow Y_{\tilde{d}_1+1}^{\tilde{d}_1+k}$, we have

$$\mathsf{P}\{(X_1^k(0), Y_{\tilde{d}_1+1}^{\tilde{d}_1+k}) \in \mathcal{T}_\epsilon^{(k)}\} \leq 2^{-k\gamma(\epsilon)}$$

for $\tilde{d}_1 < d_1$, and the same bound holds also for $\tilde{d}_1 > d_1$ by changing the role of X and Y in the lemma. Thus, by the union of events bound,

$$\mathsf{P}(\mathcal{E}_{03}) \leq (b+1)k2^{-k\gamma(\epsilon)},$$

which tends to zero as $k \to \infty$. Similarly, $\mathsf{P}(\mathcal{E}_{04})$ tends to zero as $k \to \infty$.

We continue with bounding the last two terms in (24.4). By the LLN, $\mathsf{P}(\mathcal{E}_{11} \cap \mathcal{E}_0^c)$ tends to zero as $n \to \infty$. To upper bound $\mathsf{P}(\mathcal{E}_{12} \cap \mathcal{E}_0^c)$, define the events

$$\mathcal{E}(\mathcal{J}_1, \mathcal{J}_2) = \{(\mathbf{X}_1(m_1^b), \mathbf{X}_2(\bar{m}_2^{b+1}), \mathbf{Y}) \in \mathcal{T}_\epsilon^{(n)} \text{ for } m_{1j_1} = 1, j_1 \notin \mathcal{J}_1, \bar{m}_{2j_2} = \tilde{M}_{2j_2}, j_2 \notin \mathcal{J}_2$$
$$\text{and some } m_{1j_1} \neq 1, j_1 \in \mathcal{J}_1, \bar{m}_{2j_2} \neq \tilde{M}_{2j_2}, j_2 \in \mathcal{J}_2\}$$

for each $\mathcal{J}_1 \subseteq [1:b]$ and $\mathcal{J}_2 \subseteq [1:b+1]$. Then

$$\mathsf{P}(\mathcal{E}_{12} \cap \mathcal{E}_0^c) \leq \sum_{\emptyset \neq \mathcal{J}_1 \subseteq [1:b], \mathcal{J}_2 \subseteq [1:b+1]} \mathsf{P}(\mathcal{E}(\mathcal{J}_1, \mathcal{J}_2)).$$

We bound each term. Consider the event $\mathcal{E}(\mathcal{J}_1, \mathcal{J}_2)$ illustrated in Figure 24.9 for $b = 5$, $\mathcal{J}_1 = \{1, 3\}$, $\mathcal{J}_2 = \{3, 4\}$. The $(b+1)k$ transmissions are divided into the following four groups:

- Transmissions where both m_{1j_1} and $\bar{m}_{2j_2}$ are correct: Each symbol in this group is generated according to $p(x_1)p(x_2)p(y|x_1, x_2)$. Assume that there are k_1 such symbols.

- Transmissions where m_{1j_1} is in error but $\bar{m}_{2j_2}$ is correct: Each symbol in this group is generated according to $p(x_1)p(x_2)p(y|x_2)$. Assume that there are k_2 such symbols.

- Transmissions where $\bar{m}_{2j_2}$ is in error but m_{1j_1} is correct: Each symbol in this group is generated according to $p(x_1)p(x_2)p(y|x_1)$. Assume that there are k_3 such symbols.

- Transmissions where both m_{1j_1} and $\bar{m}_{2j_2}$ are in error: Each symbol in this group is generated according to $p(x_1)p(x_2)p(y)$. Assume that there are k_4 such symbols.

Note that $k_1 + k_2 + k_3 + k_4 = (b+1)k$, $k_2 + k_4 = k|\mathcal{J}_1|$, and $k_3 + k_4 = k|\mathcal{J}_2|$.

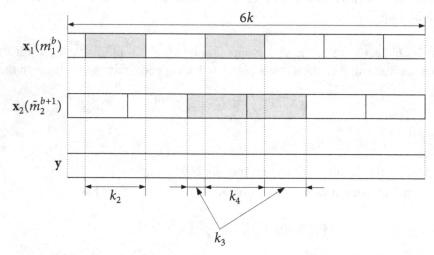

Figure 24.9. Illustration of error event $\mathcal{E}(\mathcal{J}_1, \mathcal{J}_2)$ partitioning into four groups. The shaded subblocks denote the messages in error.

Now, by the independence of the subblock codebooks and the joint typicality lemma,

$$P\{(\mathbf{X}_1(m_1^b), \mathbf{X}_2(\bar{m}_2^{b+1}), \mathbf{Y}) \in \mathcal{T}_\epsilon^{(n)}\}$$
$$\leq 2^{-k_2(I(X_1;Y|X_2)-\delta(\epsilon))} \cdot 2^{-k_3(I(X_2;Y|X_1)-\delta(\epsilon))} \cdot 2^{-k_4(I(X_1,X_2;Y)-\delta(\epsilon))} \qquad (24.5)$$

for each submessage sequence pair $(m_1^b, \bar{m}_2^{b+2})$ with the given error location. Furthermore, the total number of such submessage sequence pairs is upper bounded by $2^{k(|\mathcal{J}_1|R_1+|\mathcal{J}_2|R_2)}$. Thus, by the union of events bound and (24.5), we have

$$P(\mathcal{E}(\mathcal{J}_1, \mathcal{J}_2)) \leq 2^{k(|\mathcal{J}_1|R_1+|\mathcal{J}_2|R_2)} \cdot 2^{-k_2(I(X_1;Y|X_2)-\delta(\epsilon))-k_3(I(X_2;Y|X_1)-\delta(\epsilon))-k_4(I(X_1,X_2;Y)-\delta(\epsilon))}$$
$$= 2^{-k_2(I(X_1;Y|X_2)-R_1-\delta(\epsilon))} \cdot 2^{-k_3(I(X_2;Y|X_1)-R_2-\delta(\epsilon))} \cdot 2^{-k_4(I(X_1,X_2;Y)-R_1-R_2-\delta(\epsilon))},$$

which tends to zero as $k \to \infty$ if $R_1 < I(X_1; Y|X_2) - \delta(\epsilon)$, $R_2 < I(X_2; Y|X_1) - \delta(\epsilon)$, and $R_1 + R_2 < I(X_1, X_2; Y) - \delta(\epsilon)$.

The probability of decoding error for M_2^b can be bounded similarly. This completes the achievability proof of Theorem 24.2.

24.3.2 Proof of the Converse

Given a sequence of $(2^{nR_1}, 2^{nR_2}, n, d = n - 1)$ codes such that $\lim_{n \to \infty} P_e^{(n)} = 0$, we wish to show that the rate pair (R_1, R_2) must satisfy the inequalities in Theorem 24.2 for some product pmf $p(x_1)p(x_2)$. Recall that the codebook is used independently in consecutive blocks. Assume that $d_1 = 0$ and the receiver can synchronize the decoding with the transmitted sequence from sender 1. The probability of error in this case is

$$\max_{d_2 \in [0:n-1]} \sup_l P_{el}^{(n)}(0, d_2) \le \max_{d_1, d_2 \in [0:n-1]} \sup_l P_{el}^{(n)}(d_1, d_2) = P_e^{(n)}.$$

Further assume that $D_2 \sim \text{Unif}[0:n-1]$. Then the expected probability of error is upper bounded as $\mathsf{E}_{D_2}(\sup_l P_{el}^{(n)}(0, D_2)) \le P_e^{(n)}$. We now prove the converse under these more relaxed assumptions.

To simplify the notation and ignore the edge effect, we assume that the communication started in the distant past, so (X_1^n, X_2^n, Y^n) has the same distribution as $(X_{1,n+1}^{2n}, X_{2,n+1}^{2n}, Y_{n+1}^{2n})$. Consider decoding the received sequence $Y^{(\kappa+1)n-1}$ to recover the sequence of κ message pairs $(M_{1l}, M_{2l}) \in [1:2^{nR_1}] \times [1:2^{nR_2}]$, $l \in [1:\kappa]$.

By Fano's inequality,

$$H(M_{1l}, M_{2l} | Y^{(\kappa+1)n}, D_2) \le H(M_{1l}, M_{2l} | Y^{(\kappa+1)n-1}) \le n\epsilon_n$$

for $l \in [1:\kappa]$, where ϵ_n tends to zero as $n \to \infty$.

Following the converse proof for the synchronous DM-MAC in Section 4.5, it is easy to show that

$$\kappa nR_1 \le \sum_{i=1}^{(\kappa+1)n} I(X_{1i}; Y_i | X_{2,i-D_2}, D_2) + \kappa n\epsilon_n,$$

$$\kappa nR_2 \le \sum_{i=1}^{(\kappa+1)n} I(X_{2,i-D_2}; Y_i | X_{1i}, D_2) + \kappa n\epsilon_n,$$

$$\kappa n(R_1 + R_2) \le \sum_{i=1}^{(\kappa+1)n} I(X_{1i}, X_{2,i-D_2}; Y_i | D_2) + \kappa n\epsilon_n.$$

Now let $Q \sim \text{Unif}[1:n]$ (not over $[1:(\kappa+1)n-1]$) be the time-sharing random variable independent of $(X_1^{\kappa n}, X_2^{\kappa n}, Y^{(\kappa+1)n}, D_2)$. Then

$$\kappa nR_1 \le \sum_{l=1}^{\kappa+1} nI(X_{1,Q+(l-1)n}; Y_{Q+(l-1)n} | X_{2,Q+(l-1)n-D_2}, D_2, Q) + \kappa n\epsilon_n$$

$$\overset{(a)}{=} (\kappa + 1)nI(X_{1Q}; Y_Q | X_{2,Q-D_2}, Q, D_2) + \kappa n\epsilon_n$$

$$= (\kappa + 1)nI(X_1; Y | X_2, Q, D_2) + \kappa n\epsilon_n$$

$$\overset{(b)}{\le} (\kappa + 1)nI(X_1; Y | X_2) + \kappa n\epsilon_n,$$

where $X_1 = X_{1Q}$, $X_2 = X_{2,Q-D_2}$, $Y = Y_Q$, (a) follows since the same codebook is used over blocks, and (b) follows since $(Q, D_2) \rightarrow (X_1, X_2) \rightarrow Y$ form a Markov chain. Similarly

$$\kappa n R_2 \le (\kappa + 1) n I(X_2; Y \mid X_1) + \kappa n \epsilon_n,$$
$$\kappa n (R_1 + R_2) \le (\kappa + 1) n I(X_1, X_2; Y) + \kappa n \epsilon_n.$$

Note that since $D_2 \sim \text{Unif}[0 : n - 1]$ is independent of Q, X_2 is independent of Q and thus of X_1. Combining the above inequalities, and letting $n \rightarrow \infty$ and then $\kappa \rightarrow \infty$ completes the proof of Theorem 24.2.

SUMMARY

- DMC with random arrival model:

  - Queue stability

  - Channel capacity is the limit on the arrival rate for reliable communication

  - Extensions to multiuser channels

- Random access channel as a MAC with state:

  - Compound channel approach

  - ALOHA

  - Adaptive coding

  - Broadcast channel approach

- Asynchronous MAC:

  - Capacity region does not change under mild asynchrony

  - Capacity region under total asynchrony reduces to the synchronous capacity region without time sharing

  - Subblock coding and synchronization via preamble decoding

  - Simultaneous decoding increases the rates under asynchrony

- **Open problem 24.1.** What is the capacity region of the asynchronous MAC when $d = \alpha n$ for $\alpha \in (0, 1)$?

BIBLIOGRAPHIC NOTES

The "unconsummated union" between information theory and networking was surveyed by Ephremides and Hajek (1998). This survey includes several topics at the intersection of the two fields, including multiple access protocols, timing channels, effective bandwidth of bursty data sources, deterministic constraints on data streams, queuing theory, and switching networks. The result on the stability region of a DM-MAC mentioned in Remark 24.1 can be found, for example, in Kalyanarama Sesha Sayee and Mukherji (2006). The random access (collision) channel is motivated by the ALOHA System first described in Abramson (1970). A comparative study of information theoretic and collision resolution approaches to the random access channel is given by Gallager (1985). The adaptive coding approach in Section 24.2 is an example of the DM-MAC with distributed state information studied in Hwang, Malkin, El Gamal, and Cioffi (2007). The broadcast channel approach to the random access channel is due to Minero, Franceschetti, and Tse (2009). They analyzed the broadcast channel approach for the N-sender random access channel and demonstrated that simultaneous decoding can greatly improve the average throughput over simple collision resolution approaches as sketched in Figure 24.10.

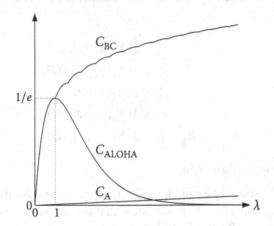

Figure 24.10. Comparison of the sum-capacities (average throughputs) of ALOHA (C_{ALOHA}), adaptive coding (C_A), and broadcast channel approach (C_{BC}) versus the load (average number of active senders) λ.

Cover, McEliece, and Posner (1981) showed that mild asynchrony does not affect the capacity region of the DM-MAC. Massey and Mathys (1985) studied total asynchrony in the collision channel without feedback and showed that time sharing cannot be used. The capacity region of the totally asynchronous DM-MAC in Theorem 24.2 is due to Poltyrev (1983) and Hui and Humblet (1985). Verdú (1989) extended this result to multiple access channels with memory and showed that unlike the memoryless case, asynchrony can in general reduce the sum-capacity.

PROBLEMS

24.1. Provide the details of the converse proof of Theorem 24.1 by justifying the second inequality in (24.2).

24.2. Consider the DM-MAC $p(y_1, y_2|x)$ with two i.i.d. arrival processes $\{A_1(i)\}$ and $\{A_2(i)\}$ of arrival rates λ_1 and λ_2, respectively. Show that the stability region $\mathscr{S}$ is equal to the capacity region $\mathscr{C}$.

24.3. *Nonslotted DMC with random data arrivals.* Consider the DMC with random data arrival process $\{A(i)\}$ as defined in Section 24.1. Suppose that the sender transmits a codeword if there are more than nR bits in the queue and transmits a fixed symbol, otherwise. Find the necessary and sufficient conditions for reliable communication (that is, the queue is stable and the message is recovered).

24.4. *Two-sender three-receiver channel with 2 messages.* Consider a DM 2-sender 3-receiver channel $p(y_1|x_1)p(y_2|x_2)p(y_{12}|x_1, x_2)$, where the message demands are specified as in Figure 24.3.

(a) Show that the capacity region of this channel is the set of rate pairs $(\tilde{R}_1, \tilde{R}_2)$ such that

$$\tilde{R}_1 \le I(X_1; Y_1|Q),$$
$$\tilde{R}_1 \le I(X_1; Y_{12}|X_2, Q),$$
$$\tilde{R}_2 \le I(X_2; Y_2|Q),$$
$$\tilde{R}_2 \le I(X_2; Y_{12}|X_1, Q),$$
$$\tilde{R}_1 + \tilde{R}_2 \le I(X_1, X_2; Y_{12}|Q)$$

for some $p(q)p(x_1|q)p(x_2|q)$.

(b) Consider the special case in Figure 24.4, where X_1 and X_2 are binary, and $Y_1 = X_1$, $Y_2 = X_2$, and $Y_{12} = X_1 \oplus X_2$. Show that the capacity region reduces to the set of rate pairs $(\tilde{R}_1, \tilde{R}_2)$ such that $\tilde{R}_1 + \tilde{R}_2 \le 1$ and can be achieved without time sharing.

24.5. *Two-sender three-receiver channel with 4 messages.* Consider a DM 2-sender 3-receiver channel $p(y_1|x_1)p(y_2|x_2)p(y_{12}|x_1, x_2)$, where the message demands are specified as in Figure 24.4.

(a) Show that a rate quadruple $(\tilde{R}_{10}, \tilde{R}_{11}, \tilde{R}_{20}, \tilde{R}_{22})$ is achievable if

$$\tilde{R}_{11} \le I(X_1; Y_1|U_1, Q),$$
$$\tilde{R}_{10} + \tilde{R}_{11} \le I(X_1; Y_1|Q),$$
$$\tilde{R}_{22} \le I(X_2; Y_2|U_2, Q),$$
$$\tilde{R}_{20} + \tilde{R}_{22} \le I(X_2; Y_2|Q),$$
$$\tilde{R}_{10} + \tilde{R}_{20} \le I(U_1, U_2; Y_{12}, Q),$$
$$\tilde{R}_{10} \le I(U_1; Y_{12}|U_2, Q),$$
$$\tilde{R}_{20} \le I(U_2; Y_{12}|U_1, Q)$$

for some pmf $p(q)p(u_1, x_1|q)p(u_2, x_2|q)$.

(b) Consider the special case in Figure 24.4, where X_1 and X_2 are binary, and $Y_1 = X_1$, $Y_2 = X_2$, and $Y_{12} = X_1 \oplus X_2$. Show that the above inner bound simplifies to (24.3). (Hint: Show that both regions have the same five extreme points $(1, 0, 0, 0)$, $(0, 1, 0, 0)$, $(0, 0, 1, 0)$, $(0, 0, 0, 1)$, and $(0, 1, 0, 1)$.)

(c) Prove the converse for the capacity region in (24.3).

24.6. *MAC and BC with known delays.* Consider the DM-MAC and the DM-BC with constant delays d_1 and d_2 known at the senders and the receivers. Show that the capacity regions for these channels coincide with those without any delays.

24.7. *Mild asynchrony.* Consider the DM-MAC with delays $d_1, d_2 \in [0:d]$ such that d/n tends to zero as $n \to \infty$. Show that the capacity region is equal to that without any delays. (Hint: Consider all contiguous codewords of length $n - d$ and perform joint typicality decoding using y_{d+1}^n for each delay pair.)

APPENDIX 24A   PROOF OF LEMMA 24.1

We first prove the converse, that is, the necessity of $\lambda \le R$. By the transition law for $Q(i)$ in (24.1),

$$Q(i + 1) \ge \begin{cases} Q(i) - nR + A(i) & \text{if } i = jn, \\ Q(i) + A(i) & \text{otherwise.} \end{cases}$$

Hence, by summing over i and telescoping, we have $Q(jn + 1) \ge \sum_{i=1}^{jn} A(i) - jnR$. By taking expectation on both sides and using the stability condition, we have $\infty > B \ge E(Q(jn + 1)) \ge jn(\lambda - R)$ for $j = 1, 2, \ldots$. This implies that $R \ge \lambda$.

Next we prove the sufficiency of $\lambda < R$ using an elementary form of Foster–Lyapunov techniques (Meyn and Tweedie 2009). Let $\tilde{Q}_j = Q((j - 1)n + 1)$ for $j = 1, 2, \ldots$ and $\tilde{A}_j = \sum_{i=(j-1)n+1}^{jn} A(i)$. Then, by the queue transition law,

$$\tilde{Q}_{j+1} = \begin{cases} \tilde{Q}_j - nR + \tilde{A}_j & \text{if } \tilde{Q}_j \ge nR, \\ \tilde{Q}_j + \tilde{A}_j & \text{otherwise} \end{cases}$$

$$\le \max\{\tilde{Q}_j - nR, \ nR\} + \tilde{A}_j$$

$$= \max\{\tilde{Q}_j - 2nR, \ 0\} + \tilde{A}_j + nR.$$

Since $(\max\{\tilde{Q}_j - 2nR, \ 0\})^2 \le (\tilde{Q}_j - 2nR)^2$,

$$\tilde{Q}_{j+1}^2 \le \tilde{Q}_j^2 + (2nR)^2 + (\tilde{A}_j + nR)^2 - 2\tilde{Q}_j(nR - \tilde{A}_j).$$

By taking expectation on both sides and using the independence of $\tilde{Q}_j$ and $\tilde{A}_j$ and the fact that $E(\tilde{A}_j) = n\lambda$ and $E((\tilde{A}_j + nR)^2) \le n^2(k + R)^2$, we obtain

$$E(\tilde{Q}_{j+1}^2) \le E(\tilde{Q}_j^2) + n^2((k + R)^2 + 4R^2)) - 2n(R - \lambda) E(\tilde{Q}_j),$$

or equivalently,

$$E(\tilde{Q}_j) \leq \frac{n((k+R)^2 + 4R^2)}{2(R-\lambda)} + \frac{E(\tilde{Q}_j^2) - E(\tilde{Q}_{j+1}^2)}{2n(R-\lambda)}.$$

Since $\tilde{Q}_1 = 0$, summing over j and telescoping, we have

$$\frac{1}{b} \sum_{j=1}^{b} E(\tilde{Q}_j) \leq \frac{n((k+R)^2 + 4R^2)}{2(R-\lambda)} + \frac{E(\tilde{Q}_1^2) - E(\tilde{Q}_{b+1}^2)}{2nb(R-\lambda)} \leq \frac{n((k+R)^2 + 4R^2)}{2(R-\lambda)}.$$

Recall the definition of $\tilde{Q}_j = Q((j-1)n + 1)$ and note that $Q(i) \leq Q((j-1)n+1) + kn$ for $i \in [(j-1)n+1 : jn]$. Therefore, we have *stability in the mean*, that is,

$$\sup_l \frac{1}{l} \sum_{i=1}^{l} E(Q(i)) \leq B < \infty. \tag{24.6}$$

To prove stability, i.e., $\sup_i E(Q(i)) < \infty$, which is a stronger notion than stability in the mean in (24.6), we note that the Markov chain $\{\tilde{Q}_j\}$ is positively recurrent; otherwise, the stability in the mean would not hold. Furthermore, it can be readily checked that the Markov chain is aperiodic. Hence, the chain has a unique limiting distribution and $E(\tilde{Q}_j)$ converges to a limit (Meyn and Tweedie 2009). But by the Cesàro mean lemma (Hardy 1992, Theorem 46), $(1/b)\sum_{j=1}^{b} E(\tilde{Q}_j) < \infty$ for all b implies that $\lim_j E(\tilde{Q}_j) < \infty$. Thus, $\sup_j E(\tilde{Q}_j) < \infty$ (since $E(\tilde{Q}_j) < \infty$ for all j). Finally, using the same argument as before, we can conclude that $\sup_i E(Q(i)) < \infty$, which completes the proof of stability.

APPENDIX 24B   PROOF OF LEMMA 24.2

First consider the case $d \geq k$ (indices for the underlying k-sequences do not overlap). Then $(X_1, Y_{d+1}), (X_2, Y_{d+2}), \ldots$ are i.i.d. with $(X_i, Y_{d+i}) \sim p_X(x_i)p_Y(y_{d+i})$. Hence, by the joint typicality lemma,

$$P\{(X^k, Y_{d+1}^{d+k}) \in \mathcal{T}_\epsilon^{(n)}(X, Y)\} \leq 2^{-k(I(X;Y)-\delta(\epsilon))}.$$

Next consider the case $d \in [1 : k-1]$. Then X^k and Y_{d+1}^{d+k} have overlapping indices and are no longer independent of each other. Suppose that $\epsilon > 0$ is sufficiently small that $(1-\epsilon)p_{X,Y}(x^*, y^*) \geq (1+\epsilon)p_X(x^*)p_Y(y^*)$ for some (x^*, y^*). Let $p = p_X(x^*)p_Y(y^*)$ and $q = p_{X,Y}(x^*, y^*)$. For $i \in [1 : k]$, define $\tilde{Y}_i = Y_{d+i}$ and

$$E_i = \begin{cases} 1 & \text{if } (X_i, \tilde{Y}_i) = (x^*, y^*), \\ 0 & \text{otherwise.} \end{cases}$$

Now consider

$$\pi(x, y | X^k, \tilde{Y}^k) = \frac{|\{i : (X_i, \tilde{Y}_i) = (x^*, y^*)\}|}{k} = \frac{1}{k} \sum_{i=1}^{k} E_i.$$

Since $\{(X_i, \tilde{Y}_i)\}$ is stationary ergodic with $p_{X_i, \tilde{Y}_i}(x^*, y^*) = p$ for all $i \in [1:k]$,

$$P\{(X^k, Y_{d+1}^{d+k}) \in \mathcal{T}_\epsilon^{(n)}(X, Y)\} \leq P\{\pi(x^*, y^* | X^k, \tilde{Y}^k) \geq (1-\epsilon)q\}$$
$$\leq P\{\pi(x^*, y^* | X^k, \tilde{Y}^k) \geq (1+\epsilon)p\},$$

which, by Birkhoff's ergodic theorem (Petersen 1983, Section 2.2), tends to zero as $n \to \infty$. To show the exponential tail, however, we should bound $P\{\pi(x^*, y^* | X^k, \tilde{Y}_{d+1}^{d+k}) \geq (1+\epsilon)p\}$ more carefully.

Assuming that k is even, consider the following two cases. First suppose that d is odd. Let $U^{k/2} = ((X_{2i-1}, \tilde{Y}_{d+2i-1}): i \in [1:k/2])$ be the subsequence of odd indices. Then $U^{k/2}$ is i.i.d. with $p_{U_i}(x^*, y^*) = p$ and by the Chernoff bound in Appendix B,

$$P\{\pi(x^*, y^* | U^{k/2}) \geq (1+\epsilon)p\} \leq e^{-kp\epsilon^2/6}.$$

Similarly, let $V^{k/2} = \{(X_{2i}, \tilde{Y}_{2i})\}_{i=1}^{k/2}$ be the subsequence of even indices. Then

$$P\{\pi(x^*, y^* | V^{k/2}) \geq (1+\epsilon)p\} \leq e^{-kp\epsilon^2/6}.$$

Thus, by the union of events bound,

$$P\{\pi(x^*, y^* | X^k, \tilde{Y}^k) \geq (1+\epsilon)p\}$$
$$= P\{\pi(x^*, y^* | U^{k/2}, V^{k/2}) \geq (1+\epsilon)p\}$$
$$\leq P\{\pi(x^*, y^* | U^{k/2}) \geq (1+\epsilon)p \text{ or } \pi(x^*, y^* | V^{k/2}) \geq (1+\epsilon)p\}$$
$$\leq 2e^{-kp\epsilon^2/6}.$$

Next suppose that d is even. We can construct two i.i.d. subsequences by alternating even and odd indices for every d indices, we have

$$U^{k/2} = ((X_i, \tilde{Y}_i): i \text{ odd} \in [(2l-1)d + 1 : 2ld], i \text{ even} \in [2ld + 1 : 2(l+1)d]).$$

For example, if $d = 2$, then $U^{k/2} = ((X_i, Y_{d+i}): i = 1, 4, 5, 8, 9, 12, \ldots)$. The rest of the analysis is the same as before. This completes the proof of the lemma.

APPENDICES

APPENDIX A

Convex Sets and Functions

We review elementary results on convex sets and functions. The readers are referred to classical texts such as Eggleston (1958) and Rockafellar (1970) for proofs of these results and more in depth treatment of convexity.

Recall that a set $\mathcal{R} \subseteq \mathbb{R}^d$ is said to be *convex* if $\mathbf{x}, \mathbf{y} \in \mathcal{R}$ implies that $\alpha \mathbf{x} + \bar{\alpha} \mathbf{y} \in \mathcal{R}$ for all $\alpha \in [0, 1]$. Thus, a convex set includes a line segment between any pair of points in the set. Consequently, a convex set includes every *convex combination* $\mathbf{x}$ of any finite number of elements $\mathbf{x}_1, \dots, \mathbf{x}_k$, i.e.,

$$\mathbf{x} = \sum_{j=1}^{k} \alpha_j \mathbf{x}_j$$

for some $(\alpha_1, \dots, \alpha_k)$ such that $\alpha_j \geq 0$, $j \in [1 : k]$, and $\sum_{j=1}^{k} \alpha_j = 1$, that is, a convex set includes every possible "average" of any tuple of points in the set.

The *convex hull* of a set $\mathcal{R} \subseteq \mathbb{R}^d$, denoted as $\mathrm{co}(\mathcal{R})$, is the union of all finite convex combinations of elements in $\mathcal{R}$. Equivalently, the convex hull of $\mathcal{R}$ is the smallest convex set containing $\mathcal{R}$. The *convex closure* of $\mathcal{R}$ is the closure of the convex hull of $\mathcal{R}$, or equivalently, the smallest closed convex set containing $\mathcal{R}$.

> **Fenchel–Eggleston–Carathéodory theorem.** Any point in the convex closure of a connected compact set $\mathcal{R} \in \mathbb{R}^d$ can be represented as a convex combination of at most d points in $\mathcal{R}$.

Let $\mathcal{R}$ be a convex set and $\mathbf{x}_0$ be a point on its boundary. Suppose $\mathbf{a}^T \mathbf{x} \leq \mathbf{a}^T \mathbf{x}_0$ for all $\mathbf{x} \in \mathcal{R}$ and some $\mathbf{a} \neq 0$. Then the hyperplane $\{\mathbf{y} \colon \mathbf{a}^T (\mathbf{x} - \mathbf{x}_0) = 0\}$ is referred to as a *supporting hyperplane* to $\mathcal{R}$ at the point $\mathbf{x}_0$. The *supporting hyperplane theorem* (Eggleston 1958) states that at least one such hyperplane exists.

Any closed bounded convex set $\mathcal{R}$ can be described as the intersection of closed half spaces characterized by supporting hyperplanes. This provides a condition to check if two convex sets are identical.

> **Lemma A.1.** Let $\mathcal{R} \subseteq \mathbb{R}^d$ be convex. Let $\mathcal{R}_1 \subseteq \mathcal{R}_2$ be two bounded convex subsets of $\mathcal{R}$, closed relative to $\mathcal{R}$. If every supporting hyperplane of $\mathcal{R}_2$ intersects with $\mathcal{R}_1$, then $\mathcal{R}_1 = \mathcal{R}_2$.

Sometimes it is easier to consider supporting hyperplanes of the smaller set.

Lemma A.2. Let $\mathscr{R} \subseteq \mathbb{R}^d$ be convex. Let $\mathscr{R}_1 \subseteq \mathscr{R}_2$ be two bounded convex subsets of $\mathscr{R}$, closed relative to $\mathscr{R}$. Let $\mathscr{A}$ be a subset of the boundary points of $\mathscr{R}_1$ such that its convex hull includes $\mathscr{R}_1$. If each $\mathbf{x}_0 \in \mathscr{A}$ has a unique supporting hyperplane and lies on the boundary of $\mathscr{R}_2$, then $\mathscr{R}_1 = \mathscr{R}_2$.

A real-valued function $g(\mathbf{x})$ is said to be *convex* if its *epigraph* $\{(\mathbf{x}, a): g(\mathbf{x}) \le a\}$ is convex for all a. If $g(\mathbf{x})$ is twice differentiable, it is convex iff

$$\nabla^2 g(\mathbf{x}) = \left(\frac{\partial^2 g(\mathbf{x})}{\partial x_i \partial x_j} \right)_{ij}$$

is positive semidefinite for all $\mathbf{x}$. If $-g(\mathbf{x})$ is convex, then $g(\mathbf{x})$ is *concave*. The following are a few examples of convex and concave functions:

- Entropy: $g(\mathbf{x}) = -\sum_j x_j \log x_j$ is concave on the probability simplex.

- Log determinant: $g(X) = \log|X|$ is concave on the set of positive definite matrices.

- Maximum: The function $\sup_\theta g_\theta(\mathbf{x})$ of any collection of convex functions $\{g_\theta(\mathbf{x})\}$ is convex.

APPENDIX B

Probability and Estimation

We review some basic probability and mean squared error estimation results. More detailed discussion including proofs of some of these results can be found in textbooks on the subject, such as Durrett (2010), Mitzenmacher and Upfal (2005), and Kailath, Sayed, and Hassibi (2000).

Probability Bounds and Limits

Union of events bound. Let $\mathcal{E}_1, \mathcal{E}_2, \ldots, \mathcal{E}_k$ be events. Then

$$P\left(\bigcup_{j=1}^{k} \mathcal{E}_j\right) \le \sum_{j=1}^{k} P(\mathcal{E}_j).$$

Jensen's inequality. Let $X \in \mathcal{X}$ (or $\mathbb{R}$) be a random variable with finite mean $E(X)$ and g be a real-valued convex function over $\mathcal{X}$ (or $\mathbb{R}$) with finite expectation $E(g(X))$. Then

$$E(g(X)) \ge g(E(X)).$$

Consider the following variant of the standard Chebyshev inequality.

Chebyshev lemma. Let X be a random variable with finite mean $E(X)$ and variance $\text{Var}(X)$, and let $\delta > 0$. Then

$$P\{|X - E(X)| \ge \delta E(X)\} \le \frac{\text{Var}(X)}{(\delta E(X))^2}.$$

In particular,

$$P\{X \le (1 - \delta) E(X)\} \le \frac{\text{Var}(X)}{(\delta E(X))^2},$$

$$P\{X \ge (1 + \delta) E(X)\} \le \frac{\text{Var}(X)}{(\delta E(X))^2}.$$

Chernoff bound. Let $X_1, X_2, \ldots$ be a sequence of independent identically distributed (i.i.d.) Bern(p) random variables, and let $\delta \in (0, 1)$. Then

$$P\left\{\sum_{i=1}^{n} X_i \ge n(1 + \delta)p\right\} \le e^{-np\delta^2/3},$$

$$P\left\{\sum_{i=1}^{n} X_i \le n(1 - \delta)p\right\} \le e^{-np\delta^2/2}.$$

Weak law of large numbers (LLN). Let $X_1, X_2, \ldots$ be a sequence of i.i.d. random variables with finite mean $E(X)$ and variance, then for every $\epsilon > 0$,

$$\lim_{n \to \infty} P\left\{ \left| \frac{1}{n} \sum_{i=1}^{n} X_i - E(X) \right| \geq \epsilon \right\} = 0.$$

In other words, $(1/n) \sum_{i=1}^{n} X_i$ converges to $E(X)$ in probability.

Dominated convergence theorem. Let $\{g_n(x)\}$ be a sequence of real-valued functions such that $|g_n(x)| \leq \phi(x)$ for some integrable function $\phi(x)$ (i.e., $\int \phi(x)dx < \infty$). Then

$$\int \lim_{n \to \infty} g_n(x) \, dx = \lim_{n \to \infty} \int g_n(x) \, dx.$$

Functional Representation Lemma

The following lemma shows that any conditional pmf can be represented as a deterministic function of the conditioning variable and an independent random variable. This lemma has been used to establish identities between information quantities optimized over different sets of distributions (Hajek and Pursley 1979, Willems and van der Meulen 1985).

> **Functional Representation Lemma.** Let $(X, Y, Z) \sim p(x, y, z)$. Then, we can represent Z as a function of (Y, W) for some random variable W of cardinality $|\mathcal{W}| \leq |\mathcal{Y}|(|\mathcal{Z}| - 1) + 1$ such that W is independent of Y and $X \to (Y, Z) \to W$ form a Markov chain.

To prove this lemma, it suffices to show that Z can be represented as a function of (Y, W) for some W independent of Y with $|\mathcal{W}| \leq |\mathcal{Y}|(|\mathcal{Z}| - 1) + 1$. The Markov relation $X \to (Y, Z) \to W$ is guaranteed by generating X according to the conditional pmf $p(x|y, z)$, which results in the desired joint pmf $p(x, y, z)$ on (X, Y, Z). We first illustrate the proof through the following.

Example B.1. Suppose that (Y, Z) is a DSBS(p) for some $p \in (0, 1/2)$, i.e., $p_{Y,Z}(0, 0) = p_{Y,Z}(1, 1) = (1 - p)/2$ and $p_{Y,Z}(0, 1) = p_{Y,Z}(1, 0) = p/2$. Let $\mathcal{P} = \{F(z|y): (y, z) \in \mathcal{Y} \times \mathcal{Z}\}$ be the set of values that the conditional cdf $F(z|y)$ takes and let W be a random variable independent of Y with cdf $F(w)$ such that $\{F(w): w \in \mathcal{W}\} = \mathcal{P}$ as shown in Figure B.1. Then $\mathcal{P} = \{0, p, 1 - p, 1\}$ and W is ternary with pmf

$$p(w) = \begin{cases} p & w = 1, \\ 1 - 2p & w = 2, \\ p & w = 3. \end{cases}$$

Now we can write $Z = g(Y, W)$, where

$$g(y, w) = \begin{cases} 0 & (y, w) = (0, 1) \text{ or } (0, 2) \text{ or } (1, 1), \\ 1 & (y, w) = (0, 3) \text{ or } (1, 2) \text{ or } (1, 3). \end{cases}$$

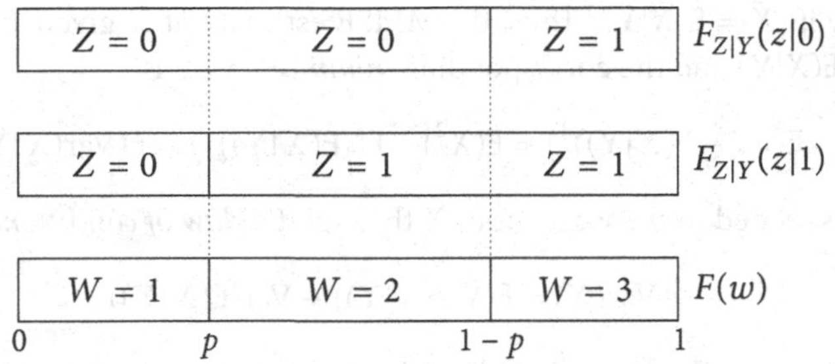

$Z = 0$	$Z = 0$	$Z = 1$	$F_{Z|Y}(z|0)$

$Z = 0$	$Z = 1$	$Z = 1$	$F_{Z|Y}(z|1)$

$W = 1$	$W = 2$	$W = 3$	$F(w)$
0	p	$1 - p$	1

Figure B.1. Conditional pmf for the DSBS(p) and its functional representation.

It is straightforward to see that for each $y \in \mathcal{Y}$, $Z = g(y, W) \sim F(z|y)$.

We can easily generalize the above example to an arbitrary pmf $p(y, z)$. Assume without loss of generality that $\mathcal{Y} = \{1, 2, \ldots, |\mathcal{Y}|\}$ and $\mathcal{Z} = \{1, 2, \ldots, |\mathcal{Z}|\}$. Let $\mathcal{P} = \{F(z|y) : (y, z) \in \mathcal{Y} \times \mathcal{Z}\}$ be the set of values that $F(z|y)$ takes and define W to be a random variable independent of Y with cdf $F(w)$ that takes values in $\mathcal{P}$. It is easy to see that $|\mathcal{W}| = |\mathcal{P}| - 1 \le |\mathcal{Y}|(|\mathcal{Z}| - 1) + 1$. Now consider the function

$$g(y, w) = \min\{z \in \mathcal{Z} : F(w) \le F(z|y)\}.$$

Then

$$P\{g(Y, W) \le z \mid Y = y\} = P\{g(y, W) \le z \mid Y = y\}$$
$$\overset{(a)}{=} P\{F(W) \le F(z|y) \mid Y = y\}$$
$$\overset{(b)}{=} P\{F(W) \le F(z|y)\}$$
$$\overset{(c)}{=} F(z|y),$$

where (a) follows since $g(y, w) \le z$ iff $F(w) \le F(z|y)$, (b) follows by the independence of Y and W, and (c) follows since $F(w)$ takes values in $\mathcal{P} = \{F(z|y)\}$. Hence we can write $Z = g(Y, W)$, which completes the proof of the functional representation lemma.

Mean Squared Error Estimation

Let $(X, Y) \sim F(x, y)$ and $\text{Var}(X) < \infty$. The minimum mean squared error (MMSE) estimate of X given an observation vector $\mathbf{Y}$ is a (measurable) function $\hat{x}(\mathbf{Y})$ of $\mathbf{Y}$ that minimizes the mean squared error (MSE) $E((X - \hat{X})^2)$.

To find the MMSE estimate, note that $E(Xg(\mathbf{Y})) = E(E(X|\mathbf{Y})g(\mathbf{Y}))$ for every function $g(\mathbf{Y})$ of the observation vector. Hence, the MSE is lower bounded by

$$E((X - \hat{X})^2) = E((X - E(X|\mathbf{Y}) + E(X|\mathbf{Y}) - \hat{X})^2)$$
$$= E((X - E(X|\mathbf{Y}))^2) + E((E(X|\mathbf{Y}) - \hat{X})^2)$$
$$\ge E((X - E(X|\mathbf{Y}))^2)$$

with equality iff $\hat{X} = E(X|\mathbf{Y})$. Thus, the MMSE estimate of X given $\mathbf{Y}$ is the conditional expectation $E(X|\mathbf{Y})$ and the corresponding *minimum* MSE is

$$E((X - E(X|\mathbf{Y}))^2) = E(X^2) - E((E(X|\mathbf{Y}))^2) = E(\text{Var}(X|\mathbf{Y})).$$

The MMSE is related to the variance of X through the *law of conditional variances*

$$\text{Var}(X) = E(\text{Var}(X|\mathbf{Y})) + \text{Var}(E(X|\mathbf{Y})).$$

Linear estimation. The linear MMSE estimate of X given $\mathbf{Y} = (Y_1, \ldots, Y_n)$ is an affine function $\hat{X} = \mathbf{a}^T\mathbf{Y} + b$ that minimizes the MSE. For simplicity we first consider the case $E(X) = 0$ and $E(\mathbf{Y}) = 0$. The linear MMSE estimate of X given $\mathbf{Y}$ is

$$\hat{X} = \mathbf{a}^T\mathbf{Y},$$

where $\mathbf{a}$ is such that the estimation error $(X - \hat{X})$ is orthogonal to $\mathbf{Y}$, i.e.,

$$E((X - \mathbf{a}^T\mathbf{Y})Y_i) = 0 \quad \text{for every } i \in [1:n],$$

or equivalently, $\mathbf{a}^T K_{\mathbf{Y}} = K_{X\mathbf{Y}}$. If $K_{\mathbf{Y}}$ is nonsingular, then the linear MMSE estimate is

$$\hat{X} = K_{X\mathbf{Y}}K_{\mathbf{Y}}^{-1}\mathbf{Y}$$

with corresponding *minimum* MSE

$$
\begin{aligned}
E((X - \hat{X})^2) &\overset{(a)}{=} E((X - \hat{X})X) \\
&= E(X^2) - E(K_{X\mathbf{Y}}K_{\mathbf{Y}}^{-1}\mathbf{Y}X) \\
&= K_X - K_{X\mathbf{Y}}K_{\mathbf{Y}}^{-1}K_{\mathbf{Y}X},
\end{aligned}
$$

where (a) follows since $E((X - \hat{X})\hat{X}) = 0$ by the orthogonality principle.

When X or $\mathbf{Y}$ has a nonzero mean, the linear MMSE estimate is determined by finding the MMSE estimate of $X' = X - E(X)$ given $\mathbf{Y}' = \mathbf{Y} - E(\mathbf{Y})$ and setting $\hat{X} = \hat{X}' + E(X)$. Thus, if $K_{\mathbf{Y}}$ is nonsingular, the linear MMSE estimate of X given $\mathbf{Y}$ is

$$\hat{X} = K_{X\mathbf{Y}}K_{\mathbf{Y}}^{-1}(\mathbf{Y} - E(\mathbf{Y})) + E(X)$$

with corresponding MSE

$$K_X - K_{X\mathbf{Y}}K_{\mathbf{Y}}^{-1}K_{\mathbf{Y}X}.$$

Thus, unlike the (nonlinear) MMSE estimate, the linear MMSE estimate is a function only of $E(X)$, $E(\mathbf{Y})$, $\text{Var}(X)$, $K_{\mathbf{Y}}$, and $K_{X\mathbf{Y}}$.

Since the linear MMSE estimate is not optimal in general, its MSE is higher than that of the MMSE in general, i.e.,

$$K_X - K_{X\mathbf{Y}}K_{\mathbf{Y}}^{-1}K_{\mathbf{Y}X} \geq E(\text{Var}(X|\mathbf{Y})).$$

The following fact on matrix inversion is often useful in calculating the linear MMSE estimate.

Matrix Inversion Lemma. If a square block matrix

$$K = \begin{bmatrix} K_{11} & K_{12} \\ K_{21} & K_{22} \end{bmatrix}$$

is invertible, then

$$(K_{11} - K_{12}K_{22}^{-1}K_{21})^{-1} = K_{11}^{-1} + K_{11}^{-1}K_{12}(K_{22} - K_{21}K_{11}^{-1}K_{12})^{-1}K_{21}K_{11}^{-1}.$$

Example B.2. Let X be the signal with mean μ and variance P, and the observations be $Y_i = X + Z_i$ for $i = [1:n]$, where each noise component Z_i has zero mean and variance N. Assume that Z_i and Z_j are uncorrelated for every $i \neq j \in [1:n]$, and X and Z_i are uncorrelated for every $i \in [1:n]$.

We find the linear MMSE estimate $\hat{X}$ of X given $\mathbf{Y} = (Y_1, \ldots, Y_n)$. Since $K_{\mathbf{Y}} = P\mathbf{11}^T + NI$, by the matrix inversion lemma (with $K_{11} = NI$, $K_{12} = \mathbf{1}$, $K_{21} = \mathbf{1}^T$, $K_{22} = -1/P$),

$$K_{\mathbf{Y}}^{-1} = \frac{1}{N}I - \frac{P}{N(nP + N)}\mathbf{11}^T.$$

Also note that $\mathsf{E}(\mathbf{Y}) = \mu\mathbf{1}$ and $K_{XY} = P\mathbf{1}^T$. Hence, the linear MMSE estimate is

$$\hat{X} = K_{XY}K_{\mathbf{Y}}^{-1}(\mathbf{Y} - \mathsf{E}(\mathbf{Y})) + \mathsf{E}(X)$$

$$= \frac{P}{nP + N}\sum_{i=1}^{n}(Y_i - \mu) + \mu$$

$$= \frac{P}{nP + N}\sum_{i=1}^{n}Y_i + \frac{N}{nP + N}\mu$$

and its MSE is

$$\mathsf{E}((X - \hat{X})^2) = P - K_{XY}K_{\mathbf{Y}}^{-1}K_{YX} = \frac{PN}{nP + N}.$$

Gaussian Random Vectors

Let $\mathbf{X} = (X_1, \ldots, X_n)$ be a random vector with mean μ and covariance matrix $K \succeq 0$. We say that $(X_1, \ldots, X_n)$ is *jointly Gaussian* and $\mathbf{X}$ is a *Gaussian random vector* if every linear combination $\mathbf{a}^T\mathbf{X}$ is a Gaussian random variable. If $K \succ 0$, then the joint pdf is well-defined as

$$f(\mathbf{x}) = \frac{1}{\sqrt{(2\pi)^n|K|}} e^{-\frac{1}{2}(\mathbf{x}-\mu)^T K^{-1}(\mathbf{x}-\mu)}.$$

A Gaussian random vector $\mathbf{X} = (X_1, \ldots, X_n)$ satisfies the following properties:

1. If $X_1, \ldots, X_n$ are uncorrelated, that is, K is diagonal, then $X_1, \ldots, X_n$ are independent. This can be verified by substituting $K_{ij} = 0$ for all $i \neq j$ in the joint pdf.

2. A linear transformation of **X** is a Gaussian random vector, that is, for every real $m \times n$ matrix A,

$$\mathbf{Y} = A\mathbf{X} \sim \mathrm{N}(A\boldsymbol{\mu},\, AKA^T).$$

This can be verified from the characteristic function of **Y**.

3. The marginals of **X** are Gaussian, that is, $X(\mathcal{S})$ is jointly Gaussian for any subset $\mathcal{S} \subseteq [1:n]$. This follows from the second property.

4. The conditionals of **X** are Gaussian; more specifically, if

$$\mathbf{X} = \begin{bmatrix} \mathbf{X}_1 \\ \mathbf{X}_2 \end{bmatrix} \sim \mathrm{N}\left(\begin{bmatrix} \mu_1 \\ \mu_2 \end{bmatrix}, \begin{bmatrix} K_{11} & K_{12} \\ K_{21} & K_{22} \end{bmatrix} \right),$$

where $\mathbf{X}_1 = (X_1, \ldots, X_k)$ and $\mathbf{X}_2 = (X_{k+1}, \ldots, X_n)$, then

$$\mathbf{X}_2 \mid \{\mathbf{X}_1 = \mathbf{x}\} \sim \mathrm{N}\left(K_{21}K_{11}^{-1}(\mathbf{x} - \mu_1) + \mu_2,\; K_{22} - K_{21}K_{11}^{-1}K_{12} \right).$$

This follows from the above properties of Gaussian random vectors.

Note that the last property implies that if $(X, \mathbf{Y})$ is jointly Gaussian, then the MMSE estimate of X given $\mathbf{Y}$ is linear. Since uncorrelation implies independence for Gaussian random vectors, the error $(X - \hat{X})$ of the MMSE estimate and the observation $\mathbf{Y}$ are independent.

APPENDIX C

Cardinality Bounding Techniques

We introduce techniques for bounding the cardinalities of auxiliary random variables.

Convex Cover Method

The convex cover method is based on the following lemma (Ahlswede and Körner 1975, Wyner and Ziv 1976), which is a direct consequence of the Fenchel–Eggleston–Carathéodory theorem in Appendix A.

> **Support Lemma.** Let $\mathcal{X}$ be a finite set and $\mathcal{U}$ be an arbitrary set. Let $\mathcal{P}$ be a connected compact subset of pmfs on $\mathcal{X}$ and $p(x|u) \in \mathcal{P}$, indexed by $u \in \mathcal{U}$, be a collection of (conditional) pmfs on $\mathcal{X}$. Suppose that $g_j(\pi)$, $j = 1, \ldots, d$, are real-valued continuous functions of $\pi \in \mathcal{P}$. Then for every $U \sim F(u)$ defined on $\mathcal{U}$, there exist a random variable $U' \sim p(u')$ with $|\mathcal{U}'| \le d$ and a collection of conditional pmfs $p(x|u') \in \mathcal{P}$, indexed by $u' \in \mathcal{U}'$, such that for $j = 1, \ldots, d$,
>
> $$\int_{\mathcal{U}} g_j(p(x|u)) \, dF(u) = \sum_{u' \in \mathcal{U}'} g_j(p(x|u'))p(u').$$

We now show how this lemma is used to bound the cardinality of auxiliary random variables. For concreteness, we focus on bounding the cardinality of the auxiliary random variable U in the characterization of the capacity region of the (physically) degraded DM-BC $p(y_1|x)p(y_2|y_1)$ in Theorem 5.2.

Let $U \sim F(u)$ and $X | \{U = u\} \sim p(x|u)$, where U takes values in an arbitrary set $\mathcal{U}$. Let $\mathcal{R}(U, X)$ be the set of rate pairs (R_1, R_2) such that

$$R_1 \le I(X; Y_1|U),$$
$$R_2 \le I(U; Y_2).$$

We prove that given any (U, X), there exists (U', X) with $|\mathcal{U}'| \le \min\{|\mathcal{X}|, |\mathcal{Y}_1|, |\mathcal{Y}_2|\} + 1$ such that $\mathcal{R}(U, X) = \mathcal{R}(U', X)$, which implies that it suffices to consider auxiliary random variables with $|\mathcal{U}| \le \min\{|\mathcal{X}|, |\mathcal{Y}_1|, |\mathcal{Y}_2|\} + 1$.

We first show that it suffices to take $|\mathcal{U}| \le |\mathcal{X}| + 1$. Assume without loss of generality that $\mathcal{X} = \{1, 2, \ldots, |\mathcal{X}|\}$. Given (U, X), consider the set $\mathcal{P}$ of all pmfs on $\mathcal{X}$ (which is connected and compact) and the following $|\mathcal{X}| + 1$ continuous functions on $\mathscr{P}$:

$$g_j(\pi) = \begin{cases} \pi(j) & j = 1, \ldots, |\mathcal{X}| - 1, \\ H(Y_1) & j = |\mathcal{X}|, \\ H(Y_2) & j = |\mathcal{X}| + 1. \end{cases}$$

Clearly, the first $|\mathcal{X}| - 1$ functions are continuous. The last two functions are also continuous in π by continuity of the entropy function and linearity of $p(y_1) = \sum_x p(y_1|x)\pi(x)$ and $p(y_2) = \sum_x p(y_2|x)\pi(x)$ in π. Now by the support lemma, we can find a random variable U' taking at most $|\mathcal{X}| + 1$ values such that

$$H(Y_1|U) = \int_{\mathscr{U}} H(Y_1|U = u)\, dF(u) = \sum_{u'} H(Y_1|U' = u')p(u') = H(Y_1|U'),$$

$$H(Y_2|U) = \int_{\mathscr{U}} H(Y_2|U = u)\, dF(u) = \sum_{u'} H(Y_2|U' = u')p(u') = H(Y_2|U'),$$

$$\int_{\mathscr{U}} p(x|u)\, dF(u) = p(x) = \sum_{u'} p_{X|U}(x|u')p(u')$$

for $x = 1, \ldots, |\mathcal{X}| - 1$ and hence for $x = |\mathcal{X}|$. Since $p(x)$ determines $p(x, y_1) = p(x) \cdot p(y_1|x)$ and $p(y_2) = \sum_x p(x)p(y_2|x)$, $H(Y_1|X)$ and $H(Y_2)$ are also preserved, and

$$I(X; Y_1|U) = H(Y_1|U) - H(Y_1|X) = H(Y_1|U') - H(Y_1|X) = I(X; Y_1|U'),$$
$$I(U; Y_2) = H(Y_2) - H(Y_2|U) = H(Y_2) - H(Y_2|U') = I(U'; Y_2).$$

Thus we have shown that $\mathscr{R}(U, X) = \mathscr{R}(U', X)$ for some U' with $|\mathcal{U}'| \leq |\mathcal{X}| + 1$.

Now we derive the bound $|\mathcal{U}'| \leq |\mathcal{Y}_2| + 1$. Again assume that $\mathcal{Y}_2 = \{1, 2, \ldots, |\mathcal{Y}_2|\}$. Consider the following $|\mathcal{Y}_2| + 1$ continuous functions on $\mathscr{P}$:

$$g_j(\pi) = \begin{cases} p_{Y_2}(j) = \sum_x p_{Y_2|X}(j|x)\pi(x) & j = 1, \ldots, |\mathcal{Y}_2| - 1, \\ H(Y_2) & j = |\mathcal{Y}_2|, \\ I(X; Y_1) & j = |\mathcal{Y}_2| + 1. \end{cases}$$

Again by the support lemma, there exists U' (not necessarily the same as the one above) with $|\mathcal{U}'| \leq |\mathcal{Y}_2| + 1$ such that $p(y_2) = \sum_{u',x} p(y_2|x)p(x|u')p(u')$, $H(Y_2)$, $H(Y_2|U')$, and $I(X; Y_1|U')$ stay the same as with the original U. Hence $\mathscr{R}(U, X) = \mathscr{R}(U', X)$. Note that the pmf $p(x)$ of X itself is not necessarily preserved under U', which, however, does not affect the quantities of interest.

In the same manner, we can show that $\mathscr{R}(U, X) = \mathscr{R}(U', X)$ for some U' with $|\mathcal{U}'| \leq |\mathcal{Y}_1| + 1$ by preserving $p(y_1)$ instead of $p(y_2)$. In this case, the physical degradedness of the channel guarantees that preserving $p(y_1)$ also preserves $H(Y_2)$.

Combining the above three steps, we have shown that for every U there exists a random variable $U' \sim p(u')$ with $|\mathcal{U}'| \leq \min\{|\mathcal{X}|, |\mathcal{Y}_1|, |\mathcal{Y}_2|\} + 1$ such that $\mathscr{R}(U, X) = \mathscr{R}(U', X)$. This establishes the desired cardinality bound on $\mathcal{U}$ in the capacity region characterization of the degraded broadcast channel in Theorem 5.2.

Remark C.1. We can apply the same technique to bound the cardinality of the time-sharing random variable Q appearing in the characterization of the DM-MAC capacity region. Recall that the capacity region of the DM-MAC is the set of rate pairs (R_1, R_2) such that

$$R_1 \le I(X_1; Y|X_2, Q),$$
$$R_2 \le I(X_2; Y|X_1, Q),$$
$$R_1 + R_2 \le I(X_1, X_2; Y|Q)$$

for some pmf $p(q)p(x_1|q)p(x_2|q)$. By considering the set $\mathscr{P}$ of all *product* pmfs on $\mathcal{X}_1 \times \mathcal{X}_2$ (which is a connected compact subset of all pmfs on $\mathcal{X}_1 \times \mathcal{X}_2$) and the following three continuous functions:

$$g_1(p(x_1|q)p(x_2|q)) = I(X_1; Y|X_2, Q = q),$$
$$g_2(p(x_1|q)p(x_2|q)) = I(X_2; Y|X_1, Q = q),$$
$$g_3(p(x_1|q)p(x_2|q)) = I(X_1, X_2; Y|Q = q),$$

we can easily establish the cardinality bound $|\mathcal{Q}| \le 3$. Although this is weaker than the bound $|\mathcal{Q}| \le 2$ derived in Chapter 4, the technique described here is more general and can be easily extended to other scenarios encountered in the book.

Extension to Multiple Random Variables

The above technique can be extended to give cardinality bounds for capacity regions with two or more auxiliary random variables (Csiszár and Körner 1978, Nair and El Gamal 2009).

As an example, consider the 3-receiver degraded DM-BC $p(y_1, y_2, y_3|x)$. The capacity region is the set of rate triples (R_1, R_2, R_3) such that

$$R_1 \le I(X; Y_1|V),$$
$$R_2 \le I(V; Y_2|U),$$
$$R_3 \le I(U; Y_3)$$

for some pmf $p(u)p(v|u)p(x|v)$. We show that it suffices to take $|\mathcal{U}| \le |\mathcal{X}| + 2$ and $|\mathcal{V}| \le (|\mathcal{X}| + 1)(|\mathcal{X}| + 2)$.

First fix $p(x|v)$ and consider the following $|\mathcal{X}| + 2$ continuous functions of $p(v|u)$:

$$p(x|u) = \sum_v p(x|v)p(v|u), \quad x = 1, \ldots, |\mathcal{X}| - 1,$$
$$I(X; Y_1|V, U = u) = I(X; Y_1|U = u) - I(V; Y_1|U = u),$$
$$I(V; Y_2|U = u),$$
$$H(Y_3|U = u).$$

As in the 2-receiver DM-BC example (with the bound $|\mathcal{U}| \le |\mathcal{Y}_2| + 1$), there exists U' with

$|\mathcal{U}'| \leq |\mathcal{X}| + 2$ such that $p(x)$, $I(X; Y_1|V) = I(X; Y_1|V, U)$, $I(V; Y_2|U)$, and $I(U; Y_3)$ are preserved. Let V' denote the corresponding random variable.

Now for each $u' \in \mathcal{U}'$, consider the following $|\mathcal{X}| + 1$ continuous functions of $p(x|v', u')$: $p(x|v', u')$, $H(Y_1|V' = v', U' = u')$, and $H(Y_2|V' = v', U' = u')$. Then as in the 2-receiver DM-BC example (with the bound $|\mathcal{U}| \leq |\mathcal{X}| + 1$), for each u', there exists $V''|\{U' = u'\} \sim p(v''|u')$ such that the cardinality of the support of $p(v''|u')$ is upper bounded by $|\mathcal{X}| + 1$ and the quantities $p(x|u')$, $I(X; Y_1|V', U' = u')$, and $I(V'; Y_2|U' = u')$ are preserved. By relabeling the support of $p(v''|u')$ for each $u' \in \mathcal{U}'$ and redefining $p(x|v'', u')$ accordingly, we can construct V'' with $|\mathcal{V}''| \leq |\mathcal{X}| + 1$. However, this choice does not satisfy the Markov chain condition $U' \to V'' \to X$ in general!

Instead, consider $V''' = (U', V'')$. Then $|\mathcal{V}'''| \leq |\mathcal{U}'| \cdot |\mathcal{V}''| \leq (|\mathcal{X}| + 1)(|\mathcal{X}| + 2)$ and $U' \to V''' \to X \to (Y_1, Y_2, Y_3)$ form a Markov chain. Furthermore,

$$
\begin{aligned}
I(X; Y_1|V''') &= I(X; Y_1|U') - I(V'''; Y_1|U') \\
&= I(X; Y_1|U') - I(V''; Y_1|U') \\
&= I(X; Y_1|U') - I(V'; Y_1|U') \\
&= I(X; Y_1|V') \\
&= I(X; Y_1|V).
\end{aligned}
$$

Similarly

$$
I(V'''; Y_2|U') = I(V''; Y_2|U') = I(V'; Y_2|U') = I(V; Y_2|U),
$$
$$
I(U'; Y_3) = I(U; Y_3).
$$

This completes the proof of the cardinality bound.

Perturbation Method

The technique described above does not provide cardinality bounds for all capacity/rate–distortion regions. Most notably, the cardinality bounds on U_0, U_1, and U_2 in Marton's inner bound for the DM-BC in Theorem 8.4 require a different technique based on a perturbation method introduced by Gohari and Anantharam (2009) and further simplified by Jog and Nair (2010).

To be concrete, we consider the maximum sum-rate for Marton's inner bound in Theorem 8.3,

$$
\max_{p(u_1, u_2),\, x(u_1, u_2)} \left(I(U_1; Y_1) + I(U_2; Y_2) - I(U_1; U_2) \right), \tag{C.1}
$$

and show that it suffices to take $|\mathcal{U}_1|, |\mathcal{U}_2| \leq |\mathcal{X}|$.

Let (U_1, U_2, X) be the random variables that attain the maximum in (C.1) and let $(U_1', U_2', X') \sim p_\epsilon(u_1', u_2', x') = p_{U_1, U_2, X}(u_1', u_2', x')(1 + \epsilon\phi(u_1'))$ be its *perturbed* version. We assume that $1 + \epsilon\phi(u_1) \geq 0$ for all u_1 and $E(\phi(U_1)) = \sum_{u_1} p(u_1)\phi(u_1) = 0$, so that $p_\epsilon(u_1', u_2', x')$ is a valid pmf. We further assume that

$$
E(\phi(U_1)|X = x) = \sum_{u_1, u_2} p(u_1, u_2|x)\phi(u_1) = 0
$$

for all $x \in \mathcal{X}$, which implies that $p_\epsilon(x') = p_X(x')$. In other words, the perturbation preserves the pmf of X and hence the pmf of (Y_1, Y_2). Note that there exists a nonzero perturbation that satisfies this condition (along with $E(\phi(U_1)) = 0$) as long as $|\mathcal{U}_1| \geq |\mathcal{X}| + 1$, since there are $|\mathcal{X}| + 1$ linear constraints. Note also that X' continues to be a function of (U_1', U_2'), since $p_{U_1, U_2, X}(u_1', u_2', x') = 0$ implies that $p_\epsilon(u_1', u_2', x') = 0$. Now consider

$$I(U_1'; Y_1') + I(U_2'; Y_2') - I(U_1'; U_2')$$
$$= H(Y_1') + H(Y_2') + H(U_1', U_2') - H(U_1', Y_1') - H(U_2', Y_2')$$
$$= H(Y_1) + H(Y_2) + H(U_1', U_2') - H(U_1', Y_1') - H(U_2', Y_2')$$
$$= H(Y_1) + H(Y_2) + H(U_1, U_2) - H(U_1, Y_1) + \epsilon H_\phi(U_1, U_2) - \epsilon H_\phi(U_1, Y_1) - H(U_2', Y_2'),$$

where

$$H_\phi(U_1, U_2) = - \sum_{u_1, u_2} p(u_1, u_2)\phi(u_1) \log p(u_1, u_2),$$

$$H_\phi(U_1, Y_1) = - \sum_{u_1, y_1} p(u_1, y_1)\phi(u_1) \log p(u_1, y_1).$$

Since $p(u_1, u_2, x)$ attains the maximum in (C.1),

$$\left. \frac{\partial^2}{\partial \epsilon^2} (I(U_1'; Y_1') + I(U_2'; Y_2') - I(U_1'; U_2')) \right|_{\epsilon=0} = - \left. \frac{\partial^2}{\partial \epsilon^2} H(U_2', Y_2') \right|_{\epsilon=0} \leq 0.$$

It can be shown by simple algebra that this is equivalent to $E[(E(\phi(U_1)|U_2, Y_2))^2] \leq 0$. In particular, $E(\phi(U_1)|U_2 = u_2, Y_2 = y_2) = 0$ whenever $p(u_2, y_2) > 0$. Hence $p_\epsilon(u_2', y_2') = p_{U_2, Y_2}(u_2', y_2')$ and $H(U_2', Y_2') = H(U_2, Y_2)$. Therefore

$$I(U_1'; Y_1') + I(U_2'; Y_2') - I(U_1'; U_2')$$
$$= H(Y_1) + H(Y_2) + H(U_1, U_2) - H(U_1, Y_1) - H(U_2, Y_2) + \epsilon H_\phi(U_1, U_2) - \epsilon H_\phi(U_1, Y_1).$$

Once again by the optimality of $p(u_1, u_2)$, we have

$$\frac{\partial}{\partial \epsilon} I(U_1'; Y_1') + I(U_2'; Y_2') - I(U_1'; U_2') = H_\phi(U_1, U_2) - H_\phi(U_1, Y_1) = 0,$$

which, in turn, implies that

$$I(U_1'; Y_1') + I(U_2'; Y_2') - I(U_1'; U_2') = I(U_1; Y_1) + I(U_2; Y_2) - I(U_1; U_2),$$

that is, $(U_1', U_2', X') \sim p_\epsilon(u_1', u_2', x')$ also attains the maximum in (C.1). Finally, we choose the largest $\epsilon > 0$ such that $1 + \epsilon \phi(u_1) \geq 0$, that is, $1 + \epsilon \phi(u_1^*) = 0$ for some $u_1^* \in \mathcal{U}_1$. Then $p_\epsilon(u_1^*) = 0$, i.e., $|\mathcal{U}_1'| \leq |\mathcal{U}_1| - 1$, and the maximum is still attained.

We can repeat the same argument as long as $|\mathcal{U}_1'| \geq |\mathcal{X}| + 1$. Hence by induction, we can take $|\mathcal{U}_1'| = |\mathcal{X}|$ while preserving $I(U_1; Y_1) + I(U_2; Y_2) - I(U_1; U_2)$. Using the same argument for U_2, we can take $|\mathcal{U}_2'| = |\mathcal{X}|$ as well.

APPENDIX D

Fourier–Motzkin Elimination

Suppose that $\mathcal{R} \subseteq \mathbb{R}^d$ is the set of tuples $(r_1, r_2, \ldots, r_d)$ that satisfy a finite system of linear inequalities $Ar^d \leq b^k$, i.e.,

$$
\begin{aligned}
a_{11}r_1 + a_{12}r_2 + \cdots + a_{1d}r_d &\leq b_1 \\
a_{21}r_1 + a_{22}r_2 + \cdots + a_{2d}r_d &\leq b_2 \\
&\vdots \qquad \leq \vdots \\
a_{k1}r_1 + a_{k2}r_2 + \cdots + a_{kd}r_d &\leq b_k.
\end{aligned}
\tag{D.1}
$$

Such $\mathcal{R}$ is referred to as a *polyhedron* (or a *polytope* if it is bounded).

Let $\mathcal{R}' \subseteq \mathbb{R}^{d-1}$ be the projection of $\mathcal{R}$ onto the hyperplane $\{r_1 = 0\}$ (or any hyperplane parallel to it). In other words,

1. if $(r_1, r_2, \ldots, r_d) \in \mathcal{R}$, then $(r_2, \ldots, r_d) \in \mathcal{R}'$; and

2. if $(r_2, \ldots, r_d) \in \mathcal{R}'$, then there exists an r_1 such that $(r_1, r_2, \ldots, r_d) \in \mathcal{R}$.

Thus $\mathcal{R}'$ captures the exact collection of inequalities satisfied by $(r_2, \ldots, r_d)$. How can we find the system of linear inequalities that characterize $\mathcal{R}'$ from the original system of k inequalities for $\mathcal{R}$?

The Fourier–Motzkin elimination procedure (see, for example, Ziegler 1995) is a systematic method for finding the system of linear inequalities in $(r_2, \ldots, r_d)$. We explain this procedure through the following.

Example D.1. Consider the system of inequalities

$$
\begin{aligned}
-r_1 \quad &\leq -1, \\
-2r_2 &\leq -1, \\
-r_1 + r_2 &\leq 0, \\
-r_1 - r_2 &\leq -2, \\
+r_1 + 4r_2 &\leq 15, \\
+2r_1 - r_2 &\leq 3.
\end{aligned}
\tag{D.2}
$$

These inequalities characterize the set $\mathcal{R}$ shown in Figure D.1.

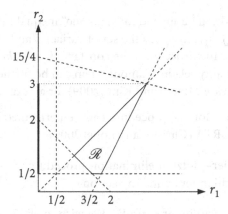

Figure D.1. The set $\mathcal{R}$ of (r_1, r_2) satisfying the inequalities in (D.2).

Note that $r_2 \in \mathcal{R}'$ iff there exists an r_1 (for given r_2) that satisfies all the inequalities, which in turn happens iff every lower bound on r_1 is less than or equal to every upper bound on r_1. Hence, the sufficient and necessary condition for $r_2 \in \mathcal{R}'$ becomes

$$\max\{1, r_2, 2 - r_2\} \le \min\{(r_2 + 3)/2, 15 - 4r_2\}$$

(6 inequalities total) in addition to the inequality $-2r_2 \le -1$ in the original system, which does not involve r_1. Solving for r_2 and removing redundant inequalities, it can be easily checked that $\mathcal{R}' = \{r_2 : 1/2 \le r_2 \le 3\}$.

The extension of this method to d dimensions is as follows.

1. Partition the system of inequalities in (D.1) into three sets,

    (a) inequalities without r_1, i.e., $a_{j1} = 0$,

    (b) upper bounds on r_1, i.e., $a_{j1} > 0$, and

    (c) lower bounds on r_1, i.e., $a_{j1} < 0$.

    Note that each upper or lower bound is an affine equation in $(r_2, \dots, r_d)$.

2. The first set of inequalities is copied verbatim for $\mathcal{R}'$ as inequalities in $(r_2, \dots, r_d)$.

3. We then generate a new system of inequalities by writing each lower bound as less than or equal to each upper bound.

In this manner, we obtain a new system of inequalities in $(r_2, \dots, r_d)$.

Remark D.1. The number of inequalities can increase rapidly—in the worst case, there can be $k^2/4$ inequalities generated from the original k inequalities. Some inequalities can be inactive and thus redundant.

Remark D.2. We can project onto an arbitrary hyperplane by taking an affine transformation of the variables. We can also eliminate multiple variables by applying the method successively.

Remark D.3. This method can be applied to "symbolic" inequalities as well. In this case, we can consider $(r_1, \ldots, r_d, b_1, \ldots, b_k)$ as the set of variables and eliminate unnecessary variables among $r_1, \ldots, r_d$. In particular, if the constants $b_1, \ldots, b_k$ are information theoretic quantities, the inequality relations among them can be further incorporated to remove inactive inequalities; see Minero and Kim (2009) for an example.

Remark D.4. The Fourier–Motzkin procedure can be performed easily by using computer programs such as PORTA (Christof and Löbel 2009).

We illustrate the Fourier–Motzkin elimination procedure with the final step in the proof of the Han–Kobayashi inner bound in Section 6.5.

Example D.2 (Han–Kobayashi inner bound). We wish to eliminate R_{10} and R_{20} in the system of inequalities

$$R_1 - R_{10} < I_1,$$
$$R_1 < I_2,$$
$$R_1 - R_{10} + R_{20} < I_3,$$
$$R_1 + R_{20} < I_4,$$
$$R_2 - R_{20} < I_5,$$
$$R_2 < I_6,$$
$$R_2 - R_{20} + R_{10} < I_7,$$
$$R_2 + R_{10} < I_8,$$
$$R_{10} \geq 0,$$
$$R_1 - R_{10} \geq 0,$$
$$R_{20} \geq 0,$$
$$R_2 - R_{20} \geq 0.$$

Step 1 (elimination of R_{10}). We have three upper bounds on R_{10}:

$$R_{10} < I_7 - R_2 + R_{20},$$
$$R_{10} < I_8 - R_2,$$
$$R_{10} \leq R_1,$$

and three lower bounds on R_{10}:

$$R_{10} > R_1 - I_1,$$
$$R_{10} > R_1 + R_{20} - I_3,$$
$$R_{10} \geq 0.$$

Comparing the upper and lower bounds, copying the six inequalities in the original system that do not involve R_{10}, and removing trivial or redundant inequalities $I_1 > 0$, $R_1 \geq 0$,

$R_2 < I_8$ (since $I_6 \leq I_8$), and $R_2 - R_{20} < I_7$ (since $I_5 \leq I_7$), we obtain the new system of inequalities in (R_{20}, R_1, R_2):

$$R_1 < I_2,$$
$$R_1 + R_{20} < I_4,$$
$$R_2 - R_{20} < I_5,$$
$$R_2 < I_6,$$
$$R_1 + R_2 - R_{20} < I_1 + I_7,$$
$$R_1 + R_2 < I_1 + I_8,$$
$$R_1 + R_2 < I_3 + I_7,$$
$$R_1 + R_2 + R_{20} < I_3 + I_8,$$
$$R_{20} < I_3,$$
$$R_{20} \geq 0,$$
$$R_2 - R_{20} \geq 0.$$

Step 2 (elimination of R_{20}). Comparing the upper bounds on R_{20}

$$R_{20} < I_4 - R_1,$$
$$R_{20} < I_3 + I_8 - R_1 - R_2,$$
$$R_{20} < I_3,$$
$$R_{20} \leq R_2$$

with the lower bounds

$$R_{20} > R_2 - I_5,$$
$$R_{20} > R_1 + R_2 - I_1 - I_7,$$
$$R_{20} \geq 0,$$

copying inequalities that do not involve R_{20}, and removing trivial or redundant inequalities $I_3 > 0$, $I_5 > 0$, $R_2 \geq 0$, $R_1 + R_2 < I_1 + I_3 + I_7$, $R_1 < I_4$ (since $I_2 \leq I_4$), $R_1 + R_2 < I_3 + I_8$ (since $I_1 \leq I_3$), and $2R_1 + 2R_2 < I_1 + I_3 + I_7 + I_8$ (since it is the sum of $R_1 + R_2 < I_1 + I_8$ and $R_1 + R_2 < I_3 + I_7$), we obtain nine inequalities in (R_1, R_2):

$$R_1 < I_2,$$
$$R_1 < I_1 + I_7,$$
$$R_2 < I_6,$$
$$R_2 < I_3 + I_5,$$
$$R_1 + R_2 < I_1 + I_8,$$
$$R_1 + R_2 < I_3 + I_7,$$
$$R_1 + R_2 < I_4 + I_5,$$
$$2R_1 + R_2 < I_1 + I_4 + I_7,$$
$$R_1 + 2R_2 < I_3 + I_5 + I_8.$$

APPENDIX E

Convex Optimization

We review basic results in convex optimization. For details, readers are referred to Boyd and Vandenberghe (2004).

An optimization problem

$$\text{minimize } g_0(\mathbf{x})$$
$$\text{subject to } g_j(\mathbf{x}) \leq 0, \quad j \in [1:k],$$
$$A\mathbf{x} = \mathbf{b}$$

with variables $\mathbf{x}$ is *convex* if the functions g_j, $j \in [0:k]$, are convex. We denote by

$$\mathscr{D} = \{\mathbf{x}: g_j(\mathbf{x}) \leq 0, j \in [1:k], A\mathbf{x} = \mathbf{b}\}$$

the set of *feasible points* (the *domain* of the optimization problem). The convex optimization problem is said to be *feasible* if $\mathscr{D} \neq \emptyset$. The *optimal value* of the problem is denoted by $p^* = \inf\{g_0(\mathbf{x}): \mathbf{x} \in \mathscr{D}\}$ (or $-\infty$ if the problem is infeasible). Any $\mathbf{x}$ that attains the infimum is said to be *optimal* and is denoted by $\mathbf{x}^*$.

Example E.1. Linear program (LP):

$$\text{minimize } \mathbf{c}^T \mathbf{x}$$
$$\text{subject to } A\mathbf{x} = \mathbf{b},$$
$$x_j \geq 0 \text{ for all } j.$$

Example E.2. Differential entropy maximization under correlation constraint:

$$\text{maximize } \log |K|$$
$$\text{subject to } K_{j,j+k} = a_k, \quad k \in [0:l].$$

Note that this problem is a special case of matrix determinant maximization (max-det) with linear matrix inequalities (Vandenberghe, Boyd, and Wu 1998).

We define the *Lagrangian* associated with a feasible optimization problem

$$\text{minimize } g_0(\mathbf{x})$$
$$\text{subject to } g_j(\mathbf{x}) \leq 0, \quad j \in [1:k],$$
$$A\mathbf{x} = \mathbf{b}$$

as

$$L(\mathbf{x}, \boldsymbol{\lambda}, \boldsymbol{v}) = g_0(\mathbf{x}) + \sum_{j=1}^{k} \lambda_j g_j(\mathbf{x}) + \boldsymbol{v}^T (A\mathbf{x} - \mathbf{b}).$$

We further define the *Lagrangian dual function* (or *dual function* in short) as

$$\phi(\boldsymbol{\lambda}, \boldsymbol{v}) = \inf_{\mathbf{x}} L(\mathbf{x}, \boldsymbol{\lambda}, \boldsymbol{v}).$$

It can be easily seen that for any $(\boldsymbol{\lambda}, \boldsymbol{v})$ with $\lambda_j \geq 0$ for all j and any feasible $\mathbf{x}$, $\phi(\boldsymbol{\lambda}, \boldsymbol{v}) \leq g_0(\mathbf{x})$. This leads to the (Lagrange) *dual problem*

$$\begin{aligned} & \text{maximize} \quad \phi(\boldsymbol{\lambda}, \boldsymbol{v}) \\ & \text{subject to} \quad \lambda_j \geq 0, \quad j \in [1:k]. \end{aligned}$$

Note that $\phi(\boldsymbol{\lambda}, \boldsymbol{v})$ is concave and hence the dual problem is convex (regardless of whether the primal problem is convex or not). The optimal value of the dual problem is denoted by d^* and the dual optimal point is denoted by $(\boldsymbol{\lambda}^*, \boldsymbol{v}^*)$.

The original optimization problem, its feasible set, and optimal value are sometimes referred to as the primal problem, primal feasible set, and primal optimal value, respectively.

Example E.3. Consider the LP discussed above with $\mathbf{x} = x^n$. The Lagrangian is

$$L(\mathbf{x}, \boldsymbol{\lambda}, \boldsymbol{v}) = \mathbf{c}^T \mathbf{x} - \sum_{j=1}^{n} \lambda_j x_j + \boldsymbol{v}^T (A\mathbf{x} - \mathbf{b}) = -\mathbf{b}^T \boldsymbol{v} + (\mathbf{c} + A^T \boldsymbol{v} - \boldsymbol{\lambda})^T \mathbf{x}$$

and the dual function is

$$\begin{aligned} \phi(\boldsymbol{\lambda}, \boldsymbol{v}) &= -\mathbf{b}^T \boldsymbol{v} + \inf_{\mathbf{x}} (\mathbf{c} + A^T \boldsymbol{v} - \boldsymbol{\lambda})^T \mathbf{x} \\ &= \begin{cases} -\mathbf{b}^T \boldsymbol{v} & \text{if } A^T \boldsymbol{v} - \boldsymbol{\lambda} + \mathbf{c} = 0, \\ -\infty & \text{otherwise.} \end{cases} \end{aligned}$$

Hence, the dual problem is

$$\begin{aligned} & \text{maximize} \quad -\mathbf{b}^T \boldsymbol{v} \\ & \text{subject to} \quad A^T \boldsymbol{v} - \boldsymbol{\lambda} + \mathbf{c} = 0, \\ & \qquad\qquad\quad \lambda_j \geq 0 \quad \text{for all } j, \end{aligned}$$

which is another LP.

By the definition of the dual function and dual problem, we already know that the lower bound on the primal optimal value is $d^* \leq p^*$. When this bound is tight (i.e., $d^* = p^*$), we say that *strong duality* holds. One simple sufficient condition for strong duality to hold is the following.

Slater's Condition. If the primal problem is convex and there exists a feasible $\mathbf{x}$ in the relative interior of $\mathcal{D}$, i.e., $g_j(\mathbf{x}) < 0$, $j \in [1:k]$, and $A\mathbf{x} = \mathbf{b}$, then strong duality holds.

If strong duality holds (say, Slater's condition holds), then

$$g_0(\mathbf{x}^*) = \phi(\boldsymbol{\lambda}^*, \mathbf{v}^*)$$

$$\leq \inf_{\mathbf{x}} g_0(\mathbf{x}) + \sum_{j=1}^{k} \lambda_j^* g_j(\mathbf{x}) + (\mathbf{v}^*)^T(A\mathbf{x} - \mathbf{b})$$

$$\leq g_0(\mathbf{x}^*) + \sum_{j=1}^{k} \lambda_j^* g_j(\mathbf{x}^*) + (\mathbf{v}^*)^T(A\mathbf{x}^* - \mathbf{b})$$

$$\leq g_0(\mathbf{x}^*).$$

Following the equality conditions, we obtain the following sufficient and necessary condition, commonly referred to as the *Karush–Kuhn–Tucker (KKT) condition*, for the primal optimal point $\mathbf{x}^*$ and dual optimal point $(\boldsymbol{\lambda}^*, \mathbf{v}^*)$:

1. $\mathbf{x}^*$ minimizes $L(\mathbf{x}, \boldsymbol{\lambda}^*, \mathbf{v}^*)$. This condition can be easily checked if $L(\mathbf{x}, \boldsymbol{\lambda}^*, \mathbf{v}^*)$ is differentiable.

2. Complementary slackness: $\lambda_j^* g_j(\mathbf{x}^*) = 0$, $j \in [1:k]$.

Example E.4. Consider the determinant maximization problem

$$\text{maximize } \log|X + K|$$
$$\text{subject to } X \succeq 0,$$
$$\text{tr}(X) \leq P,$$

where K is a given positive definite matrix.

Noting that $X \succeq 0$ iff $\text{tr}(YX) \geq 0$ for every $Y \succeq 0$, we form the Lagrangian

$$L(X, Y, \lambda) = \log|X + K| + \text{tr}(YX) - \lambda(\text{tr}(X) - P).$$

Since the domain $\mathcal{D}$ has a nonempty interior (for example, $X = (P/2)I$ is feasible), Slater's condition is satisfied and strong duality holds. Hence the KKT condition characterizes the optimal X^* and (Y^*, λ^*):

1. X^* maximizes the Lagrangian. Since the derivative of $\log|Y|$ is Y^{-1} for $Y \succ 0$ and the derivative of $\text{tr}(AY)$ is A (Boyd and Vandenberghe 2004),

$$\frac{\partial}{\partial X}L(X, Y^*, \lambda^*) = (X + K)^{-1} + Y^* - \lambda I = 0.$$

2. Complementary slackness:

$$\text{tr}(Y^*X^*) = 0,$$
$$\lambda^*(\text{tr}(X^*) - P) = 0.$$

Bibliography

Abramson, N. (1970). The ALOHA system: Another alternative for computer communications. In *Proc. AFIPS Joint Comput. Conf.*, Houston, TX, pp. 281–285. [615]

Ahlswede, R. (1971). Multiway communication channels. In *Proc. 2nd Int. Symp. Inf. Theory*, Tsahkadsor, Armenian SSR, pp. 23–52. [xvii, 99]

Ahlswede, R. (1973). A constructive proof of the coding theorem for discrete memoryless channels in case of complete feedback. In *Trans. 6th Prague Conf. Inf. Theory, Statist. Decision Functions, Random Processes (Tech Univ., Prague, 1971)*, pp. 1–22. Academia, Prague. [454]

Ahlswede, R. (1974). The capacity region of a channel with two senders and two receivers. *Ann. Probability*, 2(5), 805–814. [xvii, 99, 159, 191]

Ahlswede, R. (1978). Elimination of correlation in random codes for arbitrarily varying channels. *Probab. Theory Related Fields*, 44(2), 159–175. [125, 191]

Ahlswede, R. (1979). Coloring hypergraphs: A new approach to multi-user source coding—I. *J. Combin. Inf. Syst. Sci.*, 4(1), 76–115. [270]

Ahlswede, R. (1985). The rate–distortion region for multiple descriptions without excess rate. *IEEE Trans. Inf. Theory*, 31(6), 721–726. [335]

Ahlswede, R. and Cai, N. (2007). Transmission, identification, and common randomness capacities for wire-tape channels with secure feedback from the decoder. In R. Ahlswede, L. Bäumer, N. Cai, H. Aydinian, V. Blinovsky, C. Deppe, and H. Mashurian (eds.) *General Theory of Information Transfer and Combinatorics*, pp. 258–275. Springer, Berlin. [577, 579]

Ahlswede, R., Cai, N., Li, S.-Y. R., and Yeung, R. W. (2000). Network information flow. *IEEE Trans. Inf. Theory*, 46(4), 1204–1216. [377, 378]

Ahlswede, R. and Csiszár, I. (1993). Common randomness in information theory and cryptography—I: Secret sharing. *IEEE Trans. Inf. Theory*, 39(4), 1121–1132. [577]

Ahlswede, R., Gács, P., and Körner, J. (1976). Bounds on conditional probabilities with applications in multi-user communication. *Probab. Theory Related Fields*, 34(2), 157–177. Correction (1977). ibid, 39(4), 353–354. [271]

Ahlswede, R. and Körner, J. (1975). Source coding with side information and a converse for degraded broadcast channels. *IEEE Trans. Inf. Theory*, 21(6), 629–637. [125, 271, 631]

Ahlswede, R. and Wolfowitz, J. (1969). Correlated decoding for channels with arbitrarily varying channel probability functions. *Inf. Control*, 14(5), 457–473. [191]

Ahlswede, R. and Wolfowitz, J. (1970). The capacity of a channel with arbitrarily varying channel probability functions and binary output alphabet. *Z. Wahrsch. verw. Gebiete*, 15(3), 186–194. [191]

Ahmad, S. H. A., Jovičić, A., and Viswanath, P. (2006). On outer bounds to the capacity region of wireless networks. *IEEE Trans. Inf. Theory*, 52(6), 2770–2776. [598]

Aleksic, M., Razaghi, P., and Yu, W. (2009). Capacity of a class of modulo-sum relay channels. *IEEE Trans. Inf. Theory*, 55(3), 921–930. [418]

Annapureddy, V. S. and Veeravalli, V. V. (2009). Gaussian interference networks: Sum capacity in the low interference regime and new outer bounds on the capacity region. *IEEE Trans. Inf. Theory*, 55(7), 3032–3050. [159]

Ardestanizadeh, E. (2010). *Feedback communication systems: Fundamental limits and control-theoretic approach*. Ph.D. thesis, University of California, San Diego, La Jolla, CA. [455]

Ardestanizadeh, E., Franceschetti, M., Javidi, T., and Kim, Y.-H. (2009). Wiretap channel with secure rate-limited feedback. *IEEE Trans. Inf. Theory*, 55(12), 5353–5361. [577]

Aref, M. R. (1980). *Information flow in relay networks*. Ph.D. thesis, Stanford University, Stanford, CA. [481]

Arıkan, E. (2009). Channel polarization: A method for constructing capacity-achieving codes for symmetric binary-input memoryless channels. *IEEE Trans. Inf. Theory*, 55(7), 3051–3073. [69]

Artstein, S., Ball, K. M., Barthe, F., and Naor, A. (2004). Solution of Shannon's problem on the monotonicity of entropy. *J. Amer. Math. Soc.*, 17(4), 975–982. [32]

Avestimehr, A. S., Diggavi, S. N., and Tse, D. N. C. (2011). Wireless network information flow: A deterministic approach. *IEEE Trans. Inf. Theory*, 57(4), 1872–1905. [159, 482]

Ayaso, O., Shah, D., and Dahleh, M. A. (2010). Information theoretic bounds for distributed computation over networks of point-to-point channels. *IEEE Trans. Inf. Theory*, 56(12), 6020–6039. [547]

Bandemer, B. (2009). Capacity region of the 2-user-pair symmetric deterministic IC. Applet available at http://www.stanford.edu/~bandemer/detic/detic2/. [159]

Bandemer, B. and El Gamal, A. (2011). Interference decoding for deterministic channels. *IEEE Trans. Inf. Theory*, 57(5), 2966–2975. [160]

Bandemer, B., Vazquez-Vilar, G., and El Gamal, A. (2009). On the sum capacity of a class of cyclically symmetric deterministic interference channels. In *Proc. IEEE Int. Symp. Inf. Theory*, Seoul, Korea, pp. 2622–2626. [160]

Barron, A. R. (1985). The strong ergodic theorem for densities: Generalized Shannon–McMillan–Breiman theorem. *Ann. Probab.*, 13(4), 1292–1303. [32]

Beckner, W. (1975). Inequalities in Fourier analysis. *Ann. Math.*, 102(1), 159–182. [31]

Berger, T. (1968). Rate distortion theory for sources with abstract alphabets and memory. *Inf. Control*, 13(3), 254–273. [70]

Berger, T. (1978). Multiterminal source coding. In G. Longo (ed.) *The Information Theory Approach to Communications*, pp. 171–231. Springer-Verlag, New York. [32, 313]

Berger, T. and Yeung, R. W. (1989). Multiterminal source encoding with one distortion criterion. *IEEE Trans. Inf. Theory*, 35(2), 228–236. [313, 315]

Berger, T. and Zhang, Z. (1983). Minimum breakdown degradation in binary source encoding. *IEEE Trans. Inf. Theory*, 29(6), 807–814. [335]

Berger, T., Zhang, Z., and Viswanathan, H. (1996). The CEO problem. *IEEE Trans. Inf. Theory*, 42(3), 887–902. [313]

Bergmans, P. P. (1973). Random coding theorem for broadcast channels with degraded components. *IEEE Trans. Inf. Theory*, 19(2), 197–207. [125]

Bergmans, P. P. (1974). A simple converse for broadcast channels with additive white Gaussian noise. *IEEE Trans. Inf. Theory*, 20(2), 279–280. [125]

Bierbaum, M. and Wallmeier, H. (1979). A note on the capacity region of the multiple-access channel. *IEEE Trans. Inf. Theory*, 25(4), 484. [100]

Biglieri, E., Calderbank, R., Constantinides, A., Goldsmith, A. J., Paulraj, A., and Poor, H. V. (2007). *MIMO Wireless Communications*. Cambridge University Press, Cambridge. [254]

Biglieri, E., Proakis, J., and Shamai, S. (1998). Fading channels: Information-theoretic and communications aspects. *IEEE Trans. Inf. Theory*, 44(6), 2619–2692. [598]

Blachman, N. M. (1965). The convolution inequality for entropy powers. *IEEE Trans. Inf. Theory*, 11(2), 267–271. [31]

Blackwell, D., Breiman, L., and Thomasian, A. J. (1959). The capacity of a class of channels. *Ann. Math. Statist.*, 30(4), 1229–1241. [191]

Blackwell, D., Breiman, L., and Thomasian, A. J. (1960). The capacity of a certain channel classes under random coding. *Ann. Math. Statist.*, 31(3), 558–567. [191]

Borade, S., Zheng, L., and Trott, M. (2007). Multilevel broadcast networks. In *Proc. IEEE Int. Symp. Inf. Theory*, Nice, France, pp. 1151–1155. [220]

Boyd, S., Ghosh, A., Prabhakar, B., and Shah, D. (2006). Randomized gossip algorithms. *IEEE Trans. Inf. Theory*, 52(6), 2508–2530. [547]

Boyd, S. and Vandenberghe, L. (2004). *Convex Optimization*. Cambridge University Press, Cambridge. [257, 640, 642]

Brascamp, H. J. and Lieb, E. H. (1976). Best constants in Young's inequality, its converse, and its generalization to more than three functions. *Adv. Math.*, 20(2), 151–173. [31]

Breiman, L. (1957). The individual ergodic theorem of information theory. *Ann. Math. Statist.*, 28(3), 809–811. Correction (1960). 31(3), 809–810. [32]

Bresler, G., Parekh, A., and Tse, D. N. C. (2010). The approximate capacity of the many-to-one and one-to-many Gaussian interference channel. *IEEE Trans. Inf. Theory*, 56(9), 4566–4592. [159, 160]

Bresler, G. and Tse, D. N. C. (2008). The two-user Gaussian interference channel: A deterministic view. *Euro. Trans. Telecomm.*, 19(4), 333–354. [159]

Bross, S. I., Lapidoth, A., and Wigger, M. A. (2008). The Gaussian MAC with conferencing encoders. In *Proc. IEEE Int. Symp. Inf. Theory*, Toronto, Canada, pp. 2702–2706. [130]

Bucklew, J. A. (1987). The source coding theorem via Sanov's theorem. *IEEE Trans. Inf. Theory*, 33(6), 907–909. [70]

Butman, S. (1976). Linear feedback rate bounds for regressive channels. *IEEE Trans. Inf. Theory*, 22(3), 363–366. [454]

Cadambe, V. and Jafar, S. A. (2008). Interference alignment and degrees of freedom of the K-user interference channel. *IEEE Trans. Inf. Theory*, 54(8), 3425–3441. [159, 160, 598]

Cadambe, V., Jafar, S. A., and Shamai, S. (2009). Interference alignment on the deterministic channel and application to fully connected Gaussian interference channel. *IEEE Trans. Inf. Theory*, 55(1), 269–274. [160]

Caire, G. and Shamai, S. (1999). On the capacity of some channels with channel state information. *IEEE Trans. Inf. Theory*, 45(6), 2007–2019. [196]

Caire, G. and Shamai, S. (2003). On the achievable throughput of a multiantenna Gaussian broadcast channel. *IEEE Trans. Inf. Theory*, 49(7), 1691–1706. [253]

Cannons, J., Dougherty, R., Freiling, C., and Zeger, K. (2006). Network routing capacity. *IEEE Trans. Inf. Theory*, 52(3), 777–788. [378]

Carleial, A. B. (1975). A case where interference does not reduce capacity. *IEEE Trans. Inf. Theory*, 21(5), 569–570. [159]

Carleial, A. B. (1978). Interference channels. *IEEE Trans. Inf. Theory*, 24(1), 60–70. [159]

Carleial, A. B. (1982). Multiple-access channels with different generalized feedback signals. *IEEE Trans. Inf. Theory*, 28(6), 841–850. [454, 481]

Cemal, Y. and Steinberg, Y. (2005). The multiple-access channel with partial state information at the encoders. *IEEE Trans. Inf. Theory*, 51(11), 3992–4003. [192, 598]

Chen, J. (2009). Rate region of Gaussian multiple description coding with individual and central distortion constraints. *IEEE Trans. Inf. Theory*, 55(9), 3991–4005. [335]

Chen, J., Tian, C., Berger, T., and Hemami, S. S. (2006). Multiple description quantization via Gram–Schmidt orthogonalization. *IEEE Trans. Inf. Theory*, 52, 5197–5217. [335]

Cheng, R. S. and Verdú, S. (1993). Gaussian multiaccess channels with ISI: Capacity region and multiuser water-filling. *IEEE Trans. Inf. Theory*, 39(3), 773–785. [253]

Chia, Y.-K. and El Gamal, A. (2009). 3-receiver broadcast channels with common and confidential messages. In *Proc. IEEE Int. Symp. Inf. Theory*, Seoul, Korea, pp. 1849–1853. [577]

Chia, Y.-K. and El Gamal, A. (2010). Wiretap channel with causal state information. In *Proc. IEEE Int. Symp. Inf. Theory*, Austin, TX, pp. 2548–2552. [577]

Chong, H.-F., Motani, M., Garg, H. K., and El Gamal, H. (2008). On the Han–Kobayashi region for the interference channel. *IEEE Trans. Inf. Theory*, 54(7), 3188–3195. [159]

Chou, P. A., Wu, Y., and Jain, K. (2003). Practical network coding. In *Proc. 41st Ann. Allerton Conf. Comm. Control Comput.*, Monticello, IL. [378]

Christof, T. and Löbel, A. (2009). PORTA: Polyhedron representation transformation algorithm. Software available at http://typo.zib.de/opt-long_projects/Software/Porta/. [638]

Cohen, A. S. and Lapidoth, A. (2002). The Gaussian watermarking game. *IEEE Trans. Inf. Theory*, 48(6), 1639–1667. [192]

Costa, M. H. M. (1983). Writing on dirty paper. *IEEE Trans. Inf. Theory*, 29(3), 439–441. [192]

Costa, M. H. M. (1985). A new entropy power inequality. *IEEE Trans. Inf. Theory*, 31(6), 751–760. [32]

Costa, M. H. M. and Cover, T. M. (1984). On the similarity of the entropy power inequality and the Brunn–Minkowski inequality. *IEEE Trans. Inf. Theory*, 30(6), 837–839. [31]

Costa, M. H. M. and El Gamal, A. (1987). The capacity region of the discrete memoryless interference channel with strong interference. *IEEE Trans. Inf. Theory*, 33(5), 710–711. [159]

Cover, T. M. (1972). Broadcast channels. *IEEE Trans. Inf. Theory*, 18(1), 2–14. [xvii, 8, 125, 598]

Cover, T. M. (1975a). A proof of the data compression theorem of Slepian and Wolf for ergodic sources. *IEEE Trans. Inf. Theory*, 21(2), 226–228. [270]

Cover, T. M. (1975b). An achievable rate region for the broadcast channel. *IEEE Trans. Inf. Theory*, 21(4), 399–404. [69, 220]

Cover, T. M. (1975c). Some advances in broadcast channels. In A. J. Viterbi (ed.) *Advances in Communication Systems*, vol. 4, pp. 229–260. Academic Press, San Francisco. [99]

Cover, T. M. (1998). Comments on broadcast channels. *IEEE Trans. Inf. Theory*, 44(6), 2524–2530. [125]

Cover, T. M. and Chiang, M. (2002). Duality between channel capacity and rate distortion with two-sided state information. *IEEE Trans. Inf. Theory*, 48(6), 1629–1638. [288]

Cover, T. M. and El Gamal, A. (1979). Capacity theorems for the relay channel. *IEEE Trans. Inf. Theory*, 25(5), 572–584. [418, 454]

Cover, T. M., El Gamal, A., and Salehi, M. (1980). Multiple access channels with arbitrarily correlated sources. *IEEE Trans. Inf. Theory*, 26(6), 648–657. [355]

Cover, T. M. and Kim, Y.-H. (2007). Capacity of a class of deterministic relay channels. In *Proc. IEEE Int. Symp. Inf. Theory*, Nice, France, pp. 591–595. [418]

Cover, T. M. and Leung, C. S. K. (1981). An achievable rate region for the multiple-access channel with feedback. *IEEE Trans. Inf. Theory*, 27(3), 292–298. [454]

Cover, T. M., McEliece, R. J., and Posner, E. C. (1981). Asynchronous multiple-access channel capacity. *IEEE Trans. Inf. Theory*, 27(4), 409–413. [615]

Cover, T. M. and Thomas, J. A. (2006). *Elements of Information Theory*. 2nd ed. Wiley, New York. [20, 31, 32, 70]

Csiszár, I. and Körner, J. (1978). Broadcast channels with confidential messages. *IEEE Trans. Inf. Theory*, 24(3), 339–348. [576, 633]

Csiszár, I. and Körner, J. (1981a). On the capacity of the arbitrarily varying channel for maximum probability of error. *Z. Wahrsch. Verw. Gebiete*, 57(1), 87–101. [191]

Csiszár, I. and Körner, J. (1981b). *Information Theory: Coding Theorems for Discrete Memoryless Systems*. Akadémiai Kiadó, Budapest. [xvii, 32, 70, 71, 99, 100, 125, 191]

Csiszár, I. and Narayan, P. (1988). The capacity of the arbitrarily varying channel revisited: Positivity, constraints. *IEEE Trans. Inf. Theory*, 34(2), 181–193. [191]

Csiszár, I. and Narayan, P. (2000). Common randomness and secret key generation with a helper. *IEEE Trans. Inf. Theory*, 46(2), 344–366. [577]

Csiszár, I. and Narayan, P. (2004). Secrecy capacities for multiple terminals. *IEEE Trans. Inf. Theory*, 50(12), 3047–3061. [520, 577]

Csiszár, I. and Narayan, P. (2008). Secrecy capacities for multiterminal channel models. *IEEE Trans. Inf. Theory*, 54(6), 2437–2452. [577]

Cuff, P., Su, H.-I., and El Gamal, A. (2009). Cascade multiterminal source coding. In *Proc. IEEE Int. Symp. Inf. Theory*, Seoul, Korea, pp. 1199–1203. [546]

Dana, A. F., Gowaikar, R., Palanki, R., Hassibi, B., and Effros, M. (2006). Capacity of wireless erasure networks. *IEEE Trans. Inf. Theory*, 52(3), 789–804. [482]

De Bruyn, K., Prelov, V. V., and van der Meulen, E. C. (1987). Reliable transmission of two correlated sources over an asymmetric multiple-access channel. *IEEE Trans. Inf. Theory*, 33(5), 716–718. [357]

Devroye, N., Mitran, P., and Tarokh, V. (2006). Achievable rates in cognitive radio channels. *IEEE Trans. Inf. Theory*, 52(5), 1813–1827. [196]

Diaconis, P. and Freedman, D. (1999). Iterated random functions. *SIAM Rev.*, 41(1), 45–76. [433]

Dobrushin, R. L. (1959a). General formulation of Shannon's main theorem in information theory. *Uspkhi Mat. Nauk*, 14(6), 3–104. English translation (1963). *Amer. Math. Soc. Transl.*, 33(2), 323–438. [31]

Dobrushin, R. L. (1959b). Optimum information transmission through a channel with unknown parameters. *Radio Eng. Electron.*, 4(12), 1–8. [191]

Dougherty, R., Freiling, C., and Zeger, K. (2005). Insufficiency of linear coding in network information flow. *IEEE Trans. Inf. Theory*, 51(8), 2745–2759. [378]

Dougherty, R., Freiling, C., and Zeger, K. (2011). Network coding and matroid theory. *Proc. IEEE*, 99(3), 388–405. [378]

Dueck, G. (1978). Maximal error capacity regions are smaller than average error capacity regions for multi-user channels. *Probl. Control Inf. Theory*, 7(1), 11–19. [100]

Dueck, G. (1979). The capacity region of the two-way channel can exceed the inner bound. *Inf. Control*, 40(3), 258–266. [454]

Dueck, G. (1980). Partial feedback for two-way and broadcast channels. *Inf. Control*, 46(1), 1–15. [454]

Dueck, G. (1981a). A note on the multiple access channel with correlated sources. *IEEE Trans. Inf. Theory*, 27(2), 232–235. [355]

Dueck, G. (1981b). The strong converse of the coding theorem for the multiple-access channel. *J. Combin. Inf. Syst. Sci.*, 6(3), 187–196. [99]

Dunham, J. G. (1978). A note on the abstract-alphabet block source coding with a fidelity criterion theorem. *IEEE Trans. Inf. Theory*, 24(6), 760. [70]

Durrett, R. (2010). *Probability: Theory and Examples*. 4th ed. Cambridge University Press, Cambridge. [430, 625]

Effros, M., Médard, M., Ho, T., Ray, S., Karger, D. R., Koetter, R., and Hassibi, B. (2004). Linear network codes: A unified framework for source, channel, and network coding. In *Advances in Network Information Theory*, pp. 197–216. American Mathematical Society, Providence, RI. [520]

Eggleston, H. G. (1958). *Convexity*. Cambridge University Press, Cambridge. [623]

El Gamal, A. (1978). The feedback capacity of degraded broadcast channels. *IEEE Trans. Inf. Theory*, 24(3), 379–381. [454]

El Gamal, A. (1979). The capacity of a class of broadcast channels. *IEEE Trans. Inf. Theory*, 25(2), 166–169. [125, 220]

El Gamal, A. (1980). Capacity of the product and sum of two unmatched broadcast channels. *Probl. Inf. Transm.*, 16(1), 3–23. [253]

El Gamal, A. (1981). On information flow in relay networks. In *Proc. IEEE National Telecomm. Conf.*, vol. 2, pp. D4.1.1–D4.1.4. New Orleans, LA. [481]

El Gamal, A. and Aref, M. R. (1982). The capacity of the semideterministic relay channel. *IEEE Trans. Inf. Theory*, 28(3), 536. [418]

El Gamal, A. and Costa, M. H. M. (1982). The capacity region of a class of deterministic interference channels. *IEEE Trans. Inf. Theory*, 28(2), 343–346. [159]

El Gamal, A. and Cover, T. M. (1980). Multiple user information theory. *Proc. IEEE*, 68(12), 1466–1483. [xvii, 99]

El Gamal, A. and Cover, T. M. (1982). Achievable rates for multiple descriptions. *IEEE Trans. Inf. Theory*, 28(6), 851–857. [334]

El Gamal, A., Hassanpour, N., and Mammen, J. (2007). Relay networks with delays. *IEEE Trans. Inf. Theory*, 53(10), 3413–3431. [419]

El Gamal, A., Mammen, J., Prabhakar, B., and Shah, D. (2006a). Optimal throughput–delay scaling in wireless networks—I: The fluid model. *IEEE Trans. Inf. Theory*, 52(6), 2568–2592. [499, 500]

El Gamal, A., Mammen, J., Prabhakar, B., and Shah, D. (2006b). Optimal throughput–delay scaling in wireless networks—II: Constant-size packets. *IEEE Trans. Inf. Theory*, 52(11), 5111–5116. [499]

El Gamal, A., Mohseni, M., and Zahedi, S. (2006). Bounds on capacity and minimum energy-per-bit for AWGN relay channels. *IEEE Trans. Inf. Theory*, 52(4), 1545–1561. [418]

El Gamal, A. and Orlitsky, A. (1984). Interactive data compression. In *Proc. 25th Ann. Symp. Found. Comput. Sci.*, Washington DC, pp. 100–108. [270, 520]

El Gamal, A. and van der Meulen, E. C. (1981). A proof of Marton's coding theorem for the discrete memoryless broadcast channel. *IEEE Trans. Inf. Theory*, 27(1), 120–122. [220]

El Gamal, A. and Zahedi, S. (2005). Capacity of a class of relay channels with orthogonal components. *IEEE Trans. Inf. Theory*, 51(5), 1815–1817. [418]

Elia, N. (2004). When Bode meets Shannon: Control-oriented feedback communication schemes. *IEEE Trans. Automat. Control*, 49(9), 1477–1488. [454]

Elias, P. (1955). Coding for noisy channels. In *IRE Int. Conv. Rec.*, vol. 3, part 4, pp. 37–46. [69]

Elias, P., Feinstein, A., and Shannon, C. E. (1956). A note on the maximum flow through a network. *IRE Trans. Inf. Theory*, 2(4), 117–119. [2, 377]

Ephremides, A. and Hajek, B. E. (1998). Information theory and communication networks: An unconsummated union. *IEEE Trans. Inf. Theory*, 44(6), 2416–2434. [614]

Equitz, W. H. R. and Cover, T. M. (1991). Successive refinement of information. *IEEE Trans. Inf. Theory*, 37(2), 269–275. Addendum (1993). ibid, 39(4), 1465–1466. [335]

Erez, U., Shamai, S., and Zamir, R. (2005). Capacity and lattice strategies for canceling known interference. *IEEE Trans. Inf. Theory*, 51(11), 3820–3833. [192]

Erez, U. and Zamir, R. (2004). Achieving $\frac{1}{2}\log(1 + \text{SNR})$ on the AWGN channel with lattice encoding and decoding. *IEEE Trans. Inf. Theory*, 50(10), 2293–2314. [69]

Etkin, R., Tse, D. N. C., and Wang, H. (2008). Gaussian interference channel capacity to within one bit. *IEEE Trans. Inf. Theory*, 54(12), 5534–5562. [159, 164]

Fano, R. M. (1952). Transmission of information. Unpublished course notes, Massachusetts Institute of Technology, Cambridge, MA. [31]

Feinstein, A. (1954). A new basic theorem of information theory. *IRE Trans. Inf. Theory*, 4(4), 2–22. [69]

Ford, L. R., Jr. and Fulkerson, D. R. (1956). Maximal flow through a network. *Canad. J. Math.*, 8(3), 399–404. [2, 377]

Forney, G. D., Jr. (1972). Information theory. Unpublished course notes, Stanford University, Stanford, CA. [69]

Foschini, G. J. (1996). Layered space-time architecture for wireless communication in a fading environment when using multi-element antennas. *Bell Labs Tech. J.*, 1(2), 41–59. [253, 598]

Fragouli, C. and Soljanin, E. (2007a). Network coding fundamentals. *Found. Trends Netw.*, 2(1), 1–133. [378]

Fragouli, C. and Soljanin, E. (2007b). Network coding applications. *Found. Trends Netw.*, 2(2), 135–269. [378]

Franceschetti, M., Dousse, O., Tse, D. N. C., and Thiran, P. (2007). Closing the gap in the capacity of wireless networks via percolation theory. *IEEE Trans. Inf. Theory*, 53(3), 1009–1018. [500]

Franceschetti, M., Migliore, M. D., and Minero, P. (2009). The capacity of wireless networks: Information-theoretic and physical limits. *IEEE Trans. Inf. Theory*, 55(8), 3413–3424. [500]

Fu, F.-W. and Yeung, R. W. (2002). On the rate–distortion region for multiple descriptions. *IEEE Trans. Inf. Theory*, 48(7), 2012–2021. [335]

Gaarder, N. T. and Wolf, J. K. (1975). The capacity region of a multiple-access discrete memoryless channel can increase with feedback. *IEEE Trans. Inf. Theory*, 21(1), 100–102. [454]

Gács, P. and Körner, J. (1973). Common information is far less than mutual information. *Probl. Control Inf. Theory*, 2(2), 149–162. [355]

Gallager, R. G. (1963). *Low-Density Parity-Check Codes*. MIT Press, Cambridge, MA. [69]

Gallager, R. G. (1965). A simple derivation of the coding theorem and some applications. *IEEE Trans. Inf. Theory*, 11(1), 3–18. [69]

Gallager, R. G. (1968). *Information Theory and Reliable Communication*. Wiley, New York. [69–72]

Gallager, R. G. (1974). Capacity and coding for degraded broadcast channels. *Probl. Inf. Transm.*, 10(3), 3–14. [125]

Gallager, R. G. (1985). A perspective on multiaccess channels. *IEEE Trans. Inf. Theory*, 31(2), 124–142. [615]

Gardner, R. J. (2002). The Brunn–Minkowski inequality. *Bull. Amer. Math. Soc. (N.S.)*, 39(3), 355–405. [31]

Gastpar, M., Rimoldi, B., and Vetterli, M. (2003). To code, or not to code: Lossy source-channel communication revisited. *IEEE Trans. Inf. Theory*, 49(5), 1147–1158. [71]

Gastpar, M. and Vetterli, M. (2005). On the capacity of large Gaussian relay networks. *IEEE Trans. Inf. Theory*, 51(3), 765–779. [500]

Gelfand, S. I. (1977). Capacity of one broadcast channel. *Probl. Inf. Transm.*, 13(3), 106–108. [220]

Gelfand, S. I. and Pinsker, M. S. (1980a). Capacity of a broadcast channel with one deterministic component. *Probl. Inf. Transm.*, 16(1), 24–34. [220]

Gelfand, S. I. and Pinsker, M. S. (1980b). Coding for channel with random parameters. *Probl. Control Inf. Theory*, 9(1), 19–31. [192]

Gelfand, S. I. and Pinsker, M. S. (1984). On Gaussian channels with random parameters. In *Proc. 6th Int. Symp. Inf. Theory*, Tashkent, USSR, part 1, pp. 247–250. [192]

Geng, Y., Gohari, A. A., Nair, C., and Yu, Y. (2011). The capacity region for two classes of product broadcast channels. In *Proc. IEEE Int. Symp. Inf. Theory*, Saint Petersburg, Russia. [221]

Ghaderi, J., Xie, L.-L., and Shen, X. (2009). Hierarchical cooperation in ad hoc networks: optimal clustering and achievable throughput. *IEEE Trans. Inf. Theory*, 55(8), 3425–3436. [598]

Ghasemi, A., Motahari, A. S., and Khandani, A. K. (2010). Interference alignment for the K user MIMO interference channel. In *Proc. IEEE Int. Symp. Inf. Theory*, Austin, TX, pp. 360–364. [159]

Gohari, A. A. and Anantharam, V. (2008). An outer bound to the admissible source region of broadcast channels with arbitrarily correlated sources and channel variations. In *Proc. 46th Ann. Allerton Conf. Comm. Control Comput.*, Monticello, IL, pp. 301–308. [220]

Gohari, A. A. and Anantharam, V. (2009). Evaluation of Marton's inner bound for the general broadcast channel. In *Proc. IEEE Int. Symp. Inf. Theory*, Seoul, Korea, pp. 2462–2466. [221, 634]

Gohari, A. A. and Anantharam, V. (2010a). Information-theoretic key agreement of multiple terminals—I: Source model. *IEEE Trans. Inf. Theory*, 56(8), 3973–3996. [577]

Gohari, A. A. and Anantharam, V. (2010b). Information-theoretic key agreement of multiple terminals—II: Channel model. *IEEE Trans. Inf. Theory*, 56(8), 3997–4010. [577]

Gohari, A. A., El Gamal, A., and Anantharam, V. (2010). On an outer bound and an inner bound for the general broadcast channel. In *Proc. IEEE Int. Symp. Inf. Theory*, Austin, TX, pp. 540–544. [220]

Golay, M. J. E. (1949). Note on the theoretical efficiency of information reception with PPM. *Proc. IRE*, 37(9), 1031. [70]

Goldsmith, A. J. (2005). *Wireless Communications*. Cambridge University Press, Cambridge. [598]

Goldsmith, A. J. and Varaiya, P. P. (1997). Capacity of fading channels with channel side information. *IEEE Trans. Inf. Theory*, 43(6), 1986–1992. [191, 598]

Gou, T. and Jafar, S. A. (2009). Capacity of a class of symmetric SIMO Gaussian interference channels within $O(1)$. In *Proc. IEEE Int. Symp. Inf. Theory*, Seoul, Korea, pp. 1924–1928. [160]

Gou, T. and Jafar, S. A. (2010). Degrees of freedom of the K user $M \times N$ MIMO interference channel. *IEEE Trans. Inf. Theory*, 56(12), 6040–6057. [159]

Gray, R. M. (1990). *Entropy and Information Theory*. Springer, New York. [23, 71]

Gray, R. M. and Wyner, A. D. (1974). Source coding for a simple network. *Bell Syst. Tech. J.*, 53(9), 1681–1721. [355]

Grossglauser, M. and Tse, D. N. C. (2002). Mobility increases the capacity of ad hoc wireless networks. *IEEE/ACM Trans. Netw.*, 10(4), 477–486. [499]

Gupta, A. and Verdú, S. (2011). Operational duality between lossy compression and channel coding. *IEEE Trans. Inf. Theory*, 57(6), 3171–3179. [288]

Gupta, P. and Kumar, P. R. (2000). The capacity of wireless networks. *IEEE Trans. Inf. Theory*, 46(2), 388–404. [499]

Gupta, P. and Kumar, P. R. (2003). Toward an information theory of large networks: An achievable rate region. *IEEE Trans. Inf. Theory*, 49(8), 1877–1894. [481]

Hajek, B. E. and Pursley, M. B. (1979). Evaluation of an achievable rate region for the broadcast channel. *IEEE Trans. Inf. Theory*, 25(1), 36–46. [221, 626]

Han, T. S. (1979). The capacity region of general multiple-access channel with certain correlated sources. *Inf. Control*, 40(1), 37–60. [99, 130]

Han, T. S. (1998). An information-spectrum approach to capacity theorems for the general multiple-access channel. *IEEE Trans. Inf. Theory*, 44(7), 2773–2795. [191]

Han, T. S. and Costa, M. H. M. (1987). Broadcast channels with arbitrarily correlated sources. *IEEE Trans. Inf. Theory*, 33(5), 641–650. [355]

Han, T. S. and Kobayashi, K. (1981). A new achievable rate region for the interference channel. *IEEE Trans. Inf. Theory*, 27(1), 49–60. [99, 159]

Han, T. S. and Kobayashi, K. (1987). A dichotomy of functions $F(X, Y)$ of correlated sources (X, Y) from the viewpoint of the achievable rate region. *IEEE Trans. Inf. Theory*, 33(1), 69–76. [546]

Hardy, G. H. (1992). *Divergent Series*. 2nd ed. American Mathematical Society, Providence, RI. [618]

Heegard, C. and Berger, T. (1985). Rate distortion when side information may be absent. *IEEE Trans. Inf. Theory*, 31(6), 727–734. [288, 291]

Heegard, C. and El Gamal, A. (1983). On the capacity of computer memories with defects. *IEEE Trans. Inf. Theory*, 29(5), 731–739. [192, 195]

Hekstra, A. P. and Willems, F. M. J. (1989). Dependence balance bounds for single-output two-way channels. *IEEE Trans. Inf. Theory*, 35(1), 44–53. [454]

Ho, T. and Lun, D. (2008). *Network Coding: An Introduction*. Cambridge University Press, Cambridge. [378]

Ho, T., Médard, M., Koetter, R., Karger, D. R., Effros, M., Shi, J., and Leong, B. (2006). A random linear network coding approach to multicast. *IEEE Trans. Inf. Theory*, 52(10), 4413–4430. [378]

Hoeffding, W. (1963). Probability inequalities for sums of bounded random variables. *J. Amer. Statist. Assoc.*, 58, 13–30. [73]

Horstein, M. (1963). Sequential transmission using noiseless feedback. *IEEE Trans. Inf. Theory*, 9(3), 136–143. [454]

Høst-Madsen, A. and Zhang, J. (2005). Capacity bounds and power allocation for wireless relay channels. *IEEE Trans. Inf. Theory*, 51(6), 2020–2040. [418]

Hu, T. C. (1963). Multi-commodity network flows. *Oper. Res.*, 11(3), 344–360. [378]

Hughes-Hartogs, D. (1975). *The capacity of the degraded spectral Gaussian broadcast channel*. Ph.D. thesis, Stanford University, Stanford, CA. [127, 253]

Hui, J. Y. N. and Humblet, P. A. (1985). The capacity region of the totally asynchronous multiple-access channel. *IEEE Trans. Inf. Theory*, 31(2), 207–216. [615]

Hwang, C.-S., Malkin, M., El Gamal, A., and Cioffi, J. M. (2007). Multiple-access channels with distributed channel state information. In *Proc. IEEE Int. Symp. Inf. Theory*, Nice, France, pp. 1561–1565. [598, 615]

Jafar, S. A. (2006). Capacity with causal and noncausal side information: A unified view. *IEEE Trans. Inf. Theory*, 52(12), 5468–5474. [191, 598]

Jafar, S. A. and Vishwanath, S. (2010). Generalized degrees of freedom of the symmetric Gaussian K user interference channel. *IEEE Trans. Inf. Theory*, 56(7), 3297–3303. [159, 160]

Jaggi, S., Sanders, P., Chou, P. A., Effros, M., Egner, S., Jain, K., and Tolhuizen, L. M. G. M. (2005). Polynomial time algorithms for multicast network code construction. *IEEE Trans. Inf. Theory*, 51(6), 1973–1982. [378]

Jog, V. and Nair, C. (2010). An information inequality for the BSSC channel. In *Proc. UCSD Inf. Theory Appl. Workshop*, La Jolla, CA. [221, 634]

Jovičić, A. and Viswanath, P. (2009). Cognitive radio: An information-theoretic perspective. *IEEE Trans. Inf. Theory*, 55(9), 3945–3958. [196]

Jovičić, A., Viswanath, P., and Kulkarni, S. R. (2004). Upper bounds to transport capacity of wireless networks. *IEEE Trans. Inf. Theory*, 50(11), 2555–2565. [598]

Kailath, T., Sayed, A. H., and Hassibi, B. (2000). *Linear Estimation*. Prentice-Hall, Englewood Cliffs, NJ. [442, 625]

Kalyanarama Sesha Sayee, K. C. V. and Mukherji, U. (2006). A multiclass discrete-time processor-sharing queueing model for scheduled message communication over multiaccess channels with joint maximum-likelihood decoding. In *Proc. 44th Ann. Allerton Conf. Comm. Control Comput.*, Monticello, IL, pp. 615–622. [615]

Kashyap, A., Başar, T., and Srikant, R. (2007). Quantized consensus. *Automatica*, 43(7), 1192–1203. [547]

Kaspi, A. H. (1985). Two-way source coding with a fidelity criterion. *IEEE Trans. Inf. Theory*, 31(6), 735–740. [520]

Kaspi, A. H. (1994). Rate–distortion function when side-information may be present at the decoder. *IEEE Trans. Inf. Theory*, 40(6), 2031–2034. [288]

Kaspi, A. H. and Berger, T. (1982). Rate–distortion for correlated sources with partially separated encoders. *IEEE Trans. Inf. Theory*, 28(6), 828–840. [314]

Katti, S., Maric, I., Goldsmith, A. J., Katabi, D., and Médard, M. (2007). Joint relaying and network coding in wireless networks. In *Proc. IEEE Int. Symp. Inf. Theory*, Nice, France, pp. 1101–1105. [499]

Katz, J. and Lindell, Y. (2007). *Introduction to Modern Cryptography*. Chapman & Hall/CRC, Boca Raton, FL. [578]

Keshet, G., Steinberg, Y., and Merhav, N. (2008). Channel coding in the presence of side information. *Found. Trends Comm. Inf. Theory*, 4(6), 445–586. [192]

Khisti, A. and Wornell, G. W. (2010a). Secure transmission with multiple antennas—I: The MISOME wiretap channel. *IEEE Trans. Inf. Theory*, 56(7), 3088–3104. [576]

Khisti, A. and Wornell, G. W. (2010b). Secure transmission with multiple antennas—II: The MIMOME wiretap channel. *IEEE Trans. Inf. Theory*, 56(11), 5515–5532. [576]

Kim, Y.-H. (2008a). Capacity of a class of deterministic relay channels. *IEEE Trans. Inf. Theory*, 54(3), 1328–1329. [418]

Kim, Y.-H. (2008b). A coding theorem for a class of stationary channels with feedback. *IEEE Trans. Inf. Theory*, 54(4), 1488–1499. [71, 455]

Kim, Y.-H. (2010). Feedback capacity of stationary Gaussian channels. *IEEE Trans. Inf. Theory*, 56(1), 57–85. [454]

Kimura, A. and Uyematsu, T. (2006). Multiterminal source coding with complementary delivery. In *Proc. IEEE Int. Symp. Inf. Theory Appl.*, Seoul, Korea, pp. 189–194. [292]

Knopp, R. and Humblet, P. A. (1995). Information capacity and power control in single-cell multi-user communications. In *Proc. IEEE Int. Conf. Comm.*, Seattle, WA, vol. 1, pp. 331–335. [598]

Koetter, R. and Médard, M. (2003). An algebraic approach to network coding. *IEEE/ACM Trans. Netw.*, 11(5), 782–795. [378]

Kolmogorov, A. N. (1956). On the Shannon theory of information transmission in the case of continuous signals. *IRE Trans. Inf. Theory*, 2(4), 102–108. [31]

Korada, S. and Urbanke, R. (2010). Polar codes are optimal for lossy source coding. *IEEE Trans. Inf. Theory*, 56(4), 1751–1768. [69]

Körner, J. and Marton, K. (1977a). Comparison of two noisy channels. In I. Csiszár and P. Elias (eds.) *Topics in Information Theory (Colloquia Mathematica Societatis János Bolyai, Keszthely, Hungary, 1975)*, pp. 411–423. North-Holland, Amsterdam. [125]

Körner, J. and Marton, K. (1977b). General broadcast channels with degraded message sets. *IEEE Trans. Inf. Theory*, 23(1), 60–64. [220]

Körner, J. and Marton, K. (1977c). Images of a set via two channels and their role in multi-user communication. *IEEE Trans. Inf. Theory*, 23(6), 751–761. [220]

Körner, J. and Marton, K. (1979). How to encode the modulo-two sum of binary sources. *IEEE Trans. Inf. Theory*, 25(2), 219–221. [546]

Kramer, G. (1999). Feedback strategies for a class of two-user multiple-access channels. *IEEE Trans. Inf. Theory*, 45(6), 2054–2059. [455]

Kramer, G. (2003). Capacity results for the discrete memoryless network. *IEEE Trans. Inf. Theory*, 49(1), 4–21. [455]

Kramer, G. (2004). Outer bounds on the capacity of Gaussian interference channels. *IEEE Trans. Inf. Theory*, 50(3), 581–586. [159]

Kramer, G. (2007). Topics in multi-user information theory. *Found. Trends Comm. Inf. Theory*, 4(4/5), 265–444. [454, 482]

Kramer, G., Gastpar, M., and Gupta, P. (2005). Cooperative strategies and capacity theorems for relay networks. *IEEE Trans. Inf. Theory*, 51(9), 3037–3063. [481, 482]

Kramer, G. and Nair, C. (2009). Comments on "Broadcast channels with arbitrarily correlated sources". In *Proc. IEEE Int. Symp. Inf. Theory*, Seoul, Korea, pp. 2777–2779. [355]

Kramer, G. and Savari, S. A. (2006). Edge-cut bounds on network coding rates. *J. Network Syst. Manage.*, 14(1), 49–67. [380]

Krithivasan, D. and Pradhan, S. S. (2009). Lattices for distributed source coding: Jointly Gaussian sources and reconstruction of a linear function. *IEEE Trans. Inf. Theory*, 55(12), 5628–5651. [546]

Kushilevitz, E. and Nisan, N. (1997). *Communication Complexity*. Cambridge University Press, Cambridge. [546]

Kuznetsov, A. V. and Tsybakov, B. S. (1974). Coding in a memory with defective cells. *Probl. Inf. Transm.*, 10(2), 52–60. [192]

Lancaster, P. and Rodman, L. (1995). *Algebraic Riccati Equations*. Oxford University Press, New York. [442]

Laneman, J. N., Tse, D. N. C., and Wornell, G. W. (2004). Cooperative diversity in wireless networks: efficient protocols and outage behavior. *IEEE Trans. Inf. Theory*, 50(12), 3062–3080. [418]

Lapidoth, A. and Narayan, P. (1998). Reliable communication under channel uncertainty. *IEEE Trans. Inf. Theory*, 44(6), 2148–2177. [191, 192]

Leung, C. S. K. and Hellman, M. E. (1978). The Gaussian wire-tap channel. *IEEE Trans. Inf. Theory*, 24(4), 451–456. [576]

Lévêque, O. and Telatar, İ. E. (2005). Information-theoretic upper bounds on the capacity of large extended ad hoc wireless networks. *IEEE Trans. Inf. Theory*, 51(3), 858–865. [500]

Li, L. and Goldsmith, A. J. (2001). Capacity and optimal resource allocation for fading broadcast channels—I: Ergodic capacity. *IEEE Trans. Inf. Theory*, 47(3), 1083–1102. [598]

Li, S.-Y. R., Yeung, R. W., and Cai, N. (2003). Linear network coding. *IEEE Trans. Inf. Theory*, 49(2), 371–381. [378]

Liang, Y. (2005). *Multiuser communications with relaying and user cooperation*. Ph.D. thesis, University of Illinois, Urbana-Champaign, IL. [220]

Liang, Y. and Kramer, G. (2007). Rate regions for relay broadcast channels. *IEEE Trans. Inf. Theory*, 53(10), 3517–3535. [482]

Liang, Y., Kramer, G., and Poor, H. V. (2011). On the equivalence of two achievable regions for the broadcast channel. *IEEE Trans. Inf. Theory*, 57(1), 95–100. [220]

Liang, Y., Kramer, G., and Shamai, S. (2008). Capacity outer bounds for broadcast channels. In *Proc. IEEE Inf. Theory Workshop*, Porto, Portugal, pp. 2–4. [220]

Liang, Y., Poor, H. V., and Shamai, S. (2009). Information theoretic security. *Found. Trends Comm. Inf. Theory*, 5(4/5), 355–580. [577]

Liang, Y. and Veeravalli, V. V. (2005). Gaussian orthogonal relay channels: Optimal resource allocation and capacity. *IEEE Trans. Inf. Theory*, 51(9), 3284–3289. [418]

Liao, H. H. J. (1972). *Multiple access channels*. Ph.D. thesis, University of Hawaii, Honolulu, HI. [xvii, 99]

Lidl, R. and Niederreiter, H. (1997). *Finite Fields*. 2nd ed. Cambridge University Press, Cambridge. [372, 381]

Lieb, E. H. (1978). Proof of an entropy conjecture of Wehrl. *Comm. Math. Phys.*, 62(1), 35–41. [31]

Lim, S. H., Kim, Y.-H., El Gamal, A., and Chung, S.-Y. (2010). Layered noisy network coding. In *Proc. IEEE Wireless Netw. Coding Workshop*, Boston, MA. [482, 499]

Lim, S. H., Kim, Y.-H., El Gamal, A., and Chung, S.-Y. (2011). Noisy network coding. *IEEE Trans. Inf. Theory*, 57(5), 3132–3152. [482, 499]

Lim, S. H., Minero, P., and Kim, Y.-H. (2010). Lossy communication of correlated sources over multiple access channels. In *Proc. 48th Ann. Allerton Conf. Comm. Control Comput.*, Monticello, IL, pp. 851–858. [355]

Linder, T., Zamir, R., and Zeger, K. (2000). On source coding with side-information-dependent distortion measures. *IEEE Trans. Inf. Theory*, 46(7), 2697–2704. [289]

Ma, N. and Ishwar, P. (2008). Two-terminal distributed source coding with alternating messages for function computation. In *Proc. IEEE Int. Symp. Inf. Theory*, Toronto, Canada, pp. 51–55. [546]

Ma, N. and Ishwar, P. (2010). Interaction strictly improves the Wyner–Ziv rate distortion function. In *Proc. IEEE Int. Symp. Inf. Theory*, Austin, TX, pp. 61–65. [520]

Maddah-Ali, M. A., Motahari, A. S., and Khandani, A. K. (2008). Communication over MIMO X channels: Interference alignment, decomposition, and performance analysis. *IEEE Trans. Inf. Theory*, 54(8), 3457–3470. [159, 160]

Madiman, M. and Barron, A. R. (2007). Generalized entropy power inequalities and monotonicity properties of information. *IEEE Trans. Inf. Theory*, 53(7), 2317–2329. [32]

Marko, H. (1973). The bidirectional communication theory: A generalization of information theory. *IEEE Trans. Comm.*, 21(12), 1345–1351. [455]

Marton, K. (1979). A coding theorem for the discrete memoryless broadcast channel. *IEEE Trans. Inf. Theory*, 25(3), 306–311. [220]

Massey, J. L. (1990). Causality, feedback, and directed information. In *Proc. IEEE Int. Symp. Inf. Theory Appl.*, Honolulu, HI, pp. 303–305. [455]

Massey, J. L. and Mathys, P. (1985). The collision channel without feedback. *IEEE Trans. Inf. Theory*, 31(2), 192–204. [615]

Maurer, U. M. (1993). Secret key agreement by public discussion from common information. *IEEE Trans. Inf. Theory*, 39(3), 733–742. [577]

Maurer, U. M. and Wolf, S. (1999). Unconditionally secure key agreement and the intrinsic conditional information. *IEEE Trans. Inf. Theory*, 45(2), 499–514. [577]

Maurer, U. M. and Wolf, S. (2000). Information-theoretic key agreement: From weak to strong secrecy for free. In *Advances in Cryptology—EUROCRYPT 2000 (Bruges, Belgium)*, pp. 351–368. Springer, Berlin. [577]

McEliece, R. J. (1977). *The Theory of Information and Coding*. Addison-Wesley, Reading, MA. [70]

McMillan, B. (1953). The basic theorems of information theory. *Ann. Math. Statist.*, 24(2), 196–219. [32]

Menezes, A. J., van Oorschot, P. C., and Vanstone, S. A. (1997). *Handbook of Applied Cryptography*. CRC Press, Boca Raton, FL. [578]

Meyn, S. and Tweedie, R. L. (2009). *Markov Chains and Stochastic Stability*. 2nd ed. Cambridge University Press, Cambridge. [617, 618]

Minero, P., Franceschetti, M., and Tse, D. N. C. (2009). Random access: An information-theoretic perspective. Preprint available at http://arxiv.org/abs/0912.3264/. [615]

Minero, P. and Kim, Y.-H. (2009). Correlated sources over broadcast channels. In *Proc. IEEE Int. Symp. Inf. Theory*, Seoul, Korea, pp. 2780–2784. [355, 638]

Mitzenmacher, M. and Upfal, E. (2005). *Probability and Computing*. Cambridge University Press, Cambridge. [625]

Mohseni, M. (2006). *Capacity of Gaussian vector broadcast channels*. Ph.D. thesis, Stanford University, Stanford, CA. [254]

Motahari, A. S., Gharan, S. O., Maddah-Ali, M. A., and Khandani, A. K. (2009). Real interference alignment: Exploring the potential of single antenna systems. Preprint available at http://arxiv.org/abs/0908.2282/. [159, 160]

Motahari, A. S. and Khandani, A. K. (2009). Capacity bounds for the Gaussian interference channel. *IEEE Trans. Inf. Theory*, 55(2), 620–643. [159]

Moulin, P. and O'Sullivan, J. A. (2003). Information-theoretic analysis of information hiding. *IEEE Trans. Inf. Theory*, 49(3), 563–593. [192]

Nair, C. (2008). An outer bound for 2-receiver discrete memoryless broadcast channels. Preprint available at http://arxiv.org/abs/0807.3593/. [220]

Nair, C. (2010). Capacity regions of two new classes of two-receiver broadcast channels. *IEEE Trans. Inf. Theory*, 56(9), 4207–4214. [125, 221]

Nair, C. and El Gamal, A. (2007). An outer bound to the capacity region of the broadcast channel. *IEEE Trans. Inf. Theory*, 53(1), 350–355. [220]

Nair, C. and El Gamal, A. (2009). The capacity region of a class of three-receiver broadcast channels with degraded message sets. *IEEE Trans. Inf. Theory*, 55(10), 4479–4493. [220, 633]

Nair, C. and Wang, Z. V. (2008). On the inner and outer bounds for 2-receiver discrete memoryless broadcast channels. In *Proc. UCSD Inf. Theory Appl. Workshop*, La Jolla, CA, pp. 226–229. [221]

Nam, W., Chung, S.-Y., and Lee, Y. H. (2010). Capacity of the Gaussian two-way relay channel to within $\frac{1}{2}$ bit. *IEEE Trans. Inf. Theory*, 56(11), 5488–5494. [499]

Nazer, B. and Gastpar, M. (2007a). Computation over multiple-access channels. *IEEE Trans. Inf. Theory*, 53(10), 3498–3516. [546]

Nazer, B. and Gastpar, M. (2007b). Lattice coding increases multicast rates for Gaussian multiple-access networks. In *Proc. 45th Ann. Allerton Conf. Comm. Control Comput.*, Monticello, IL, pp. 1089–1096. [546]

Nazer, B., Gastpar, M., Jafar, S. A., and Vishwanath, S. (2009). Ergodic interference alignment. In *Proc. IEEE Int. Symp. Inf. Theory*, Seoul, Korea, pp. 1769–1773. [160, 598]

Nedić, A., Olshevsky, A., Ozdaglar, A., and Tsitsiklis, J. N. (2009). On distributed averaging algorithms and quantization effects. *IEEE Trans. Automat. Control*, 54(11), 2506–2517. [547]

Niesen, U., Gupta, P., and Shah, D. (2009). On capacity scaling in arbitrary wireless networks. *IEEE Trans. Inf. Theory*, 55(9), 3959–3982. [599]

Oggier, F. and Hassibi, B. (2008). The MIMO wiretap channel. In *Proc. 3rd Int. Symp. Comm. Control Signal Processing*, St. Julians, Malta, pp. 213–218. [576]

Olfati-Saber, R., Fax, J. A., and Murray, R. M. (2007). Consensus and cooperation in networked multi-agent systems. *Proc. IEEE*, 95(1), 215–233. [547]

Oohama, Y. (1997). Gaussian multiterminal source coding. *IEEE Trans. Inf. Theory*, 43(6), 1912–1923. [313]

Oohama, Y. (1998). The rate–distortion function for the quadratic Gaussian CEO problem. *IEEE Trans. Inf. Theory*, 44(3), 1057–1070. [313]

Oohama, Y. (2005). Rate–distortion theory for Gaussian multiterminal source coding systems with several side informations at the decoder. *IEEE Trans. Inf. Theory*, 51(7), 2577–2593. [313, 314]

Ooi, J. M. and Wornell, G. W. (1998). Fast iterative coding techniques for feedback channels. *IEEE Trans. Inf. Theory*, 44(7), 2960–2976. [454]

Orlitsky, A. and El Gamal, A. (1990). Average and randomized communication complexity. *IEEE Trans. Inf. Theory*, 36(1), 3–16. [316]

Orlitsky, A. and Roche, J. R. (2001). Coding for computing. *IEEE Trans. Inf. Theory*, 47(3), 903–917. [32, 546]

Ozarow, L. H. (1980). On a source-coding problem with two channels and three receivers. *Bell Syst. Tech. J.*, 59(10), 1909–1921. [335]

Ozarow, L. H. (1984). The capacity of the white Gaussian multiple access channel with feedback. *IEEE Trans. Inf. Theory*, 30(4), 623–629. [454]

Ozarow, L. H. and Leung, C. S. K. (1984). An achievable region and outer bound for the Gaussian broadcast channel with feedback. *IEEE Trans. Inf. Theory*, 30(4), 667–671. [454]

Ozarow, L. H., Shamai, S., and Wyner, A. D. (1994). Information theoretic considerations for cellular mobile radio. *IEEE Trans. Veh. Tech.*, 43(2), 359–378. [598]

Özgür, A., Johari, R., Tse, D. N. C., and Lévêque, O. (2010). Information-theoretic operating regimes of large wireless networks. *IEEE Trans. Inf. Theory*, 56(1), 427–437. [599]

Özgür, A., Lévêque, O., and Preissmann, E. (2007). Scaling laws for one- and two-dimensional random wireless networks in the low-attenuation regimes. *IEEE Trans. Inf. Theory*, 53(10), 3573–3585. [500]

Özgür, A., Lévêque, O., and Tse, D. N. C. (2007). Hierarchical cooperation achieves optimal capacity scaling in ad hoc networks. *IEEE Trans. Inf. Theory*, 53(10), 3549–3572. [598]

Özgür, A., Lévêque, O., and Tse, D. N. C. (2010). Linear capacity scaling in wireless networks: Beyond physical limits? In *Proc. UCSD Inf. Theory Appl. Workshop*, La Jolla, CA. [500]

Permuter, H. H., Kim, Y.-H., and Weissman, T. (2011). Interpretations of directed information in portfolio theory, data compression, and hypothesis testing. *IEEE Trans. Inf. Theory*, 57(3), 3248–3259. [455]

Permuter, H. H. and Weissman, T. (2010). Cascade and triangular source coding with side information at the first two nodes. In *Proc. IEEE Int. Symp. Inf. Theory*, Austin, TX, pp. 31–35. [523]

Permuter, H. H., Weissman, T., and Goldsmith, A. J. (2009). Finite state channels with time-invariant deterministic feedback. *IEEE Trans. Inf. Theory*, 55(2), 644–662. [455]

Perron, E. (2009). *Information-theoretic secrecy for wireless networks*. Ph.D. thesis, École Polytechnique Fédérale de Lausanne, Lausanne, Switzerland. [482]

Petersen, K. (1983). *Ergodic Theory*. Cambridge University Press, Cambridge. [619]

Pinsker, M. S. (1964). *Information and Information Stability of Random Variables and Processes*. Holden-Day, San Francisco. [23]

Pinsker, M. S. (1968). The probability of error in block transmission in a memoryless Gaussian channel with feedback. *Probl. Inf. Transm.*, 4(4), 3–19. [454]

Pinsker, M. S. (1978). Capacity of noiseless broadcast channels. *Probl. Inf. Transm.*, 14(2), 28–34. [220]

Poltyrev, G. S. (1977). The capacity of parallel broadcast channels with degraded components. *Probl. Inf. Transm.*, 13(2), 23–35. [127, 253]

Poltyrev, G. S. (1983). Coding in an asynchronous multiple-access channel. *Probl. Inf. Transm.*, 19(3), 12–21. [615]

Prabhakaran, V., Tse, D. N. C., and Ramchandran, K. (2004). Rate region of the quadratic Gaussian CEO problem. In *Proc. IEEE Int. Symp. Inf. Theory*, Chicago, IL, pp. 117. [313, 314]

Pradhan, S. S., Chou, J., and Ramchandran, K. (2003). Duality between source and channel coding and its extension to the side information case. *IEEE Trans. Inf. Theory*, 49(5), 1181–1203. [288]

Puri, R., Pradhan, S. S., and Ramchandran, K. (2005). *n*-channel symmetric multiple descriptions—II: An achievable rate–distortion region. *IEEE Trans. Inf. Theory*, 51(4), 1377–1392. [335]

Rankov, B. and Wittneben, A. (2006). Achievable rate region for the two-way relay channel. In *Proc. IEEE Int. Symp. Inf. Theory*, Seattle, WA, pp. 1668–1672. [499]

Ratnakar, N. and Kramer, G. (2006). The multicast capacity of deterministic relay networks with no interference. *IEEE Trans. Inf. Theory*, 52(6), 2425–2432. [481]

Richardson, T. and Urbanke, R. (2008). *Modern Coding Theory*. Cambridge University Press, Cambridge. [69]

Rimoldi, B. (1994). Successive refinement of information: Characterization of the achievable rates. *IEEE Trans. Inf. Theory*, 40(1), 253–259. [335]

Rockafellar, R. T. (1970). *Convex Analysis*. Princeton University Press, Princeton, NJ. [623]

Rosenzweig, A., Steinberg, Y., and Shamai, S. (2005). On channels with partial channel state information at the transmitter. *IEEE Trans. Inf. Theory*, 51(5), 1817–1830. [192]

Royden, H. L. (1988). *Real Analysis*. 3rd ed. Macmillan, New York. [23]

Sato, H. (1976). Information transmission through a channel with relay. Technical Report B76-7, The Aloha Systems, University of Hawaii, Honolulu, HI. [418, 419]

Sato, H. (1977). Two-user communication channels. *IEEE Trans. Inf. Theory*, 23(3), 295–304. [159]

Sato, H. (1978a). An outer bound to the capacity region of broadcast channels. *IEEE Trans. Inf. Theory*, 24(3), 374–377. [220]

Sato, H. (1978b). On the capacity region of a discrete two-user channel for strong interference. *IEEE Trans. Inf. Theory*, 24(3), 377–379. [159]

Schalkwijk, J. P. M. (1966). A coding scheme for additive noise channels with feedback—II: Band-limited signals. *IEEE Trans. Inf. Theory*, 12(2), 183–189. [454]

Schalkwijk, J. P. M. (1982). The binary multiplying channel: A coding scheme that operates beyond Shannon's inner bound region. *IEEE Trans. Inf. Theory*, 28(1), 107–110. [455]

Schalkwijk, J. P. M. (1983). On an extension of an achievable rate region for the binary multiplying channel. *IEEE Trans. Inf. Theory*, 29(3), 445–448. [455]

Schalkwijk, J. P. M. and Kailath, T. (1966). A coding scheme for additive noise channels with feedback—I: No bandwidth constraint. *IEEE Trans. Inf. Theory*, 12(2), 172–182. [454]

Schein, B. and Gallager, R. G. (2000). The Gaussian parallel relay channel. In *Proc. IEEE Int. Symp. Inf. Theory*, Sorrento, Italy, pp. 22. [418]

Schrijver, A. (2003). *Combinatorial Optimization*. 3 vols. Springer-Verlag, Berlin. [378]

Shah, D. (2009). Gossip algorithms. *Found. Trends Netw.*, 3(1), 1–125. [547]

Shamai, S. (1997). A broadcast strategy for the Gaussian slowly fading channel. In *Proc. IEEE Int. Symp. Inf. Theory*, Ulm, Germany, pp. 150. [598]

Shamai, S. and Steiner, A. (2003). A broadcast approach for a single-user slowly fading MIMO channel. *IEEE Trans. Inf. Theory*, 49(10), 2617–2635. [598]

Shamai, S. and Telatar, İ. E. (1999). Some information theoretic aspects of decentralized power control in multiple access fading channels. In *Proc. IEEE Inf. Theory Netw. Workshop*, Metsovo, Greece, pp. 23. [598]

Shamai, S. and Wyner, A. D. (1990). A binary analog to the entropy-power inequality. *IEEE Trans. Inf. Theory*, 36(6), 1428–1430. [31]

Shang, X., Kramer, G., and Chen, B. (2009). A new outer bound and the noisy-interference sum-rate capacity for Gaussian interference channels. *IEEE Trans. Inf. Theory*, 55(2), 689–699. [159]

Shannon, C. E. (1948). A mathematical theory of communication. *Bell Syst. Tech. J.*, 27(3), 379–423, 27(4), 623–656. [2, 3, 31, 32, 69–71]

Shannon, C. E. (1949a). Communication in the presence of noise. *Proc. IRE*, 37(1), 10–21. [70]

Shannon, C. E. (1949b). Communication theory of secrecy systems. *Bell Syst. Tech. J.*, 28(4), 656–715. [576]

Shannon, C. E. (1956). The zero error capacity of a noisy channel. *IRE Trans. Inf. Theory*, 2(3), 8–19. [12]

Shannon, C. E. (1958). Channels with side information at the transmitter. *IBM J. Res. Develop.*, 2(4), 289–293. [191]

Shannon, C. E. (1959). Coding theorems for a discrete source with a fidelity criterion. In *IRE Int. Conv. Rec.*, vol. 7, part 4, pp. 142–163. Reprint with changes (1960). In R. E. Machol (ed.) *Information and Decision Processes*, pp. 93–126. McGraw-Hill, New York. [2, 4, 70, 71, 288]

Shannon, C. E. (1961). Two-way communication channels. In *Proc. 4th Berkeley Symp. Math. Statist. Probab.*, vol. I, pp. 611–644. University of California Press, Berkeley. [xvii, 99, 454]

Shayevitz, O. and Feder, M. (2011). Optimal feedback communication via posterior matching. *IEEE Trans. Inf. Theory*, 57(3), 1186–1222. [454]

Shayevitz, O. and Wigger, M. A. (2010). An achievable region for the discrete memoryless broadcast channel with feedback. In *Proc. IEEE Int. Symp. Inf. Theory*, Austin, TX, pp. 450–454. [457]

Shepp, L. A., Wolf, J. K., Wyner, A. D., and Ziv, J. (1969). Binary communication over the Gaussian channel using feedback with a peak energy constraint. *IEEE Trans. Inf. Theory*, 15(4), 476–478. [454]

Slepian, D. (1976). On bandwidth. *Proc. IEEE*, 64(3), 292–300. [70]

Slepian, D. and Wolf, J. K. (1973a). Noiseless coding of correlated information sources. *IEEE Trans. Inf. Theory*, 19(4), 471–480. [xvii, 6, 270]

Slepian, D. and Wolf, J. K. (1973b). A coding theorem for multiple access channels with correlated sources. *Bell Syst. Tech. J.*, 52(7), 1037–1076. [99, 130]

Stam, A. J. (1959). Some inequalities satisfied by the quantities of information of Fisher and Shannon. *Inf. Control*, 2(2), 101–112. [31]

Steinberg, Y. (2005). Coding for the degraded broadcast channel with random parameters, with causal and noncausal side information. *IEEE Trans. Inf. Theory*, 51(8), 2867–2877. [192]

Steinsaltz, D. (1999). Locally contractive iterated function systems. *Ann. Probab.*, 27(4), 1952–1979. [433]

Su, H.-I. and El Gamal, A. (2009). Distributed lossy averaging. In *Proc. IEEE Int. Symp. Inf. Theory*, Seoul, Korea, pp. 1453–1457. [546, 547]

Su, H.-I. and El Gamal, A. (2010). Two-way source coding through a relay. In *Proc. IEEE Int. Symp. Inf. Theory*, Austin, TX, pp. 176–180. [525]

Tatikonda, S. and Mitter, S. (2009). The capacity of channels with feedback. *IEEE Trans. Inf. Theory*, 55(1), 323–349. [455]

Tavildar, S., Viswanath, P., and Wagner, A. B. (2006). The Gaussian many-help-one distributed source coding problem. In *Proc. IEEE Inf. Theory Workshop*, Chengdu, China, pp. 596–600. [313]

Telatar, İ. E. (1999). Capacity of multi-antenna Gaussian channels. *Euro. Trans. Telecomm.*, 10(8), 585–595. [253, 598]

Telatar, İ. E. and Tse, D. N. C. (2007). Bounds on the capacity region of a class of interference channels. In *Proc. IEEE Int. Symp. Inf. Theory*, Nice, France, pp. 2871–2874. [159]

Tse, D. N. C. and Hanly, S. V. (1998). Multiaccess fading channels—I: Polymatroid structure, optimal resource allocation and throughput capacities. *IEEE Trans. Inf. Theory*, 44(7), 2796–2815. [99, 598]

Tse, D. N. C. and Viswanath, P. (2005). *Fundamentals of Wireless Communication*. Cambridge University Press, Cambridge. [598]

Tsitsiklis, J. N. (1984). *Problems in decentralized decision making and computation*. Ph.D. thesis, Massachusetts Institute of Technology, Cambridge, MA. [547]

Tung, S.-Y. (1978). *Multiterminal source coding*. Ph.D. thesis, Cornell University, Ithaca, NY. [313]

Uhlmann, W. (1966). Vergleich der hypergeometrischen mit der Binomial-Verteilung. *Metrika*, 10(1), 145–158. [316]

van der Meulen, E. C. (1971a). Three-terminal communication channels. *Adv. Appl. Probab.*, 3(1), 120–154. [418]

van der Meulen, E. C. (1971b). The discrete memoryless channel with two senders and one receiver. In *Proc. 2nd Int. Symp. Inf. Theory*, Tsahkadsor, Armenian SSR, pp. 103–135. [99]

van der Meulen, E. C. (1975). Random coding theorems for the general discrete memoryless broadcast channel. *IEEE Trans. Inf. Theory*, 21(2), 180–190. [220]

van der Meulen, E. C. (1977). A survey of multi-way channels in information theory: 1961–1976. *IEEE Trans. Inf. Theory*, 23(1), 1–37. [xvii, 99, 125]

van der Meulen, E. C. (1981). Recent coding theorems and converses for multi-way channels—I: The broadcast channel (1976–1980). In J. K. Skwyrzinsky (ed.) *New Concepts in Multi-User Communication*, pp. 15–51. Sijthoff & Noordhoff, Alphen aan den Rijn. [125]

van der Meulen, E. C. (1985). Recent coding theorems and converses for multi-way channels— II: The multiple access channel (1976–1985). Department Wiskunde, Katholieke Universiteit Leuven, Leuven, Belgium. [99]

van der Meulen, E. C. (2007). A survey of the relay channel. In E. Biglieri and L. Györfi (eds.) *Multiple Access Channels: Theory and Practice*, pp. 73–96. IOS Press, Boston. [419]

Vandenberghe, L., Boyd, S., and Wu, S.-P. (1998). Determinant maximization with linear matrix inequality constraints. *SIAM J. Matrix Anal. Appl.*, 19(2), 499–533. [640]

Vasudevan, D., Tian, C., and Diggavi, S. N. (2006). Lossy source coding for a cascade communication system with side-informations. In *Proc. 44th Ann. Allerton Conf. Comm. Control Comput.*, Monticello, IL, pp. 561–568. [523]

Venkataramani, R., Kramer, G., and Goyal, V. K. (2003). Multiple description coding with many channels. *IEEE Trans. Inf. Theory*, 49(9), 2106–2114. [335]

Verdú, S. (1989). Multiple-access channels with memory with and without frame synchronism. *IEEE Trans. Inf. Theory*, 35(3), 605–619. [615]

Verdú, S. (1990). On channel capacity per unit cost. *IEEE Trans. Inf. Theory*, 36(5), 1019–1030. [70]

Verdú, S. and Guo, D. (2006). A simple proof of the entropy-power inequality. *IEEE Trans. Inf. Theory*, 52(5), 2165–2166. [31]

Verdú, S. and Han, T. S. (1994). A general formula for channel capacity. *IEEE Trans. Inf. Theory*, 40(4), 1147–1157. [71]

Vishwanath, S., Jindal, N., and Goldsmith, A. J. (2003). Duality, achievable rates, and sum-rate capacity of Gaussian MIMO broadcast channels. *IEEE Trans. Inf. Theory*, 49(10), 2658–2668. [253]

Viswanath, P. and Tse, D. N. C. (2003). Sum capacity of the vector Gaussian broadcast channel and uplink-downlink duality. *IEEE Trans. Inf. Theory*, 49(8), 1912–1921. [253]

Viswanathan, H. and Berger, T. (1997). The quadratic Gaussian CEO problem. *IEEE Trans. Inf. Theory*, 43(5), 1549–1559. [313]

Wagner, A. B. and Anantharam, V. (2008). An improved outer bound for multiterminal source coding. *IEEE Trans. Inf. Theory*, 54(5), 1919–1937. [313]

Wagner, A. B., Kelly, B. G., and Altuğ, Y. (2011). Distributed rate–distortion with common components. *IEEE Trans. Inf. Theory*, 57(7), 4035–4057. [314]

Wagner, A. B., Tavildar, S., and Viswanath, P. (2008). Rate region of the quadratic Gaussian two-encoder source-coding problem. *IEEE Trans. Inf. Theory*, 54(5), 1938–1961. [313, 546]

Wang, J., Chen, J., and Wu, X. (2010). On the sum rate of Gaussian multiterminal source coding: New proofs and results. *IEEE Trans. Inf. Theory*, 56(8), 3946–3960. [313]

Wang, J., Chen, J., Zhao, L., Cuff, P., and Permuter, H. H. (2011). On the role of the refinement layer in multiple description coding and scalable coding. *IEEE Trans. Inf. Theory*, 57(3), 1443–1456. [335]

Wang, Z. V. and Nair, C. (2010). The capacity region of a class of broadcast channels with a sequence of less noisy receivers. In *Proc. IEEE Int. Symp. Inf. Theory*, Austin, TX, pp. 595–598. [125, 129]

Weingarten, H., Steinberg, Y., and Shamai, S. (2006). The capacity region of the Gaussian multiple-input multiple-output broadcast channel. *IEEE Trans. Inf. Theory*, 52(9), 3936–3964. [254]

Weissman, T. and El Gamal, A. (2006). Source coding with limited-look-ahead side information at the decoder. *IEEE Trans. Inf. Theory*, 52(12), 5218–5239. [288, 289]

Weldon, E. J., Jr. (1963). *Asymptotic error coding bounds for the binary symmetric channel with feedback*. Ph.D. thesis, University of Florida, Gainesville, FL. [454]

Willems, F. M. J. (1982). The feedback capacity region of a class of discrete memoryless multiple access channels. *IEEE Trans. Inf. Theory*, 28(1), 93–95. [455]

Willems, F. M. J. (1990). The maximal-error and average-error capacity region of the broadcast channel are identical: A direct proof. *Probl. Control Inf. Theory*, 19(4), 339–347. [125, 223]

Willems, F. M. J. and van der Meulen, E. C. (1985). The discrete memoryless multiple-access channel with cribbing encoders. *IEEE Trans. Inf. Theory*, 31(3), 313–327. [418, 422, 626]

Wilson, M. P., Narayanan, K., Pfister, H. D., and Sprintson, A. (2010). Joint physical layer coding and network coding for bidirectional relaying. *IEEE Trans. Inf. Theory*, 56(11), 5641–5654. [546]

Witsenhausen, H. S. (1974). Entropy inequalities for discrete channels. *IEEE Trans. Inf. Theory*, 20(5), 610–616. [31]

Witsenhausen, H. S. (1975). On sequences of pairs of dependent random variables. *SIAM J. Appl. Math.*, 28(1), 100–113. [355]

Witsenhausen, H. S. (1976a). Values and bounds for the common information of two discrete random variables. *SIAM J. Appl. Math.*, 31(2), 313–333. [355]

Witsenhausen, H. S. (1976b). The zero-error side information problem and chromatic numbers. *IEEE Trans. Inf. Theory*, 22(5), 592–593. [270]

Witsenhausen, H. S. (1980). Indirect rate distortion problems. *IEEE Trans. Inf. Theory*, 26(5), 518–521. [334]

Witsenhausen, H. S. and Wyner, A. D. (1975). A conditional entropy bound for a pair of discrete random variables. *IEEE Trans. Inf. Theory*, 21(5), 493–501. [31]

Witsenhausen, H. S. and Wyner, A. D. (1981). Source coding for multiple descriptions—II: A binary source. *Bell Syst. Tech. J.*, 60(10), 2281–2292. [334]

Wolf, J. K., Wyner, A. D., and Ziv, J. (1980). Source coding for multiple descriptions. *Bell Syst. Tech. J.*, 59(8), 1417–1426. [334]

Wolf, J. K., Wyner, A. D., Ziv, J., and Körner, J. (1984). Coding for write-once memory. *Bell Syst. Tech. J.*, 63(6), 1089–1112. [222]

Wolfowitz, J. (1957). The coding of messages subject to chance errors. *Illinois J. Math.*, 1(4), 591–606. [69]

Wolfowitz, J. (1960). Simultaneous channels. *Arch. Rational Mech. Anal.*, 4(1), 371–386. [191]

Wolfowitz, J. (1978). *Coding Theorems of Information Theory*. 3rd ed. Springer-Verlag, Berlin. [191]

Wu, W., Vishwanath, S., and Arapostathis, A. (2007). Capacity of a class of cognitive radio channels: Interference channels with degraded message sets. *IEEE Trans. Inf. Theory*, 53(11), 4391–4399. [196]

Wyner, A. D. (1966). The capacity of the band-limited Gaussian channel. *Bell Syst. Tech. J.*, 45(3), 359–395. [70]

Wyner, A. D. (1968). On the Schalkwijk–Kailath coding scheme with a peak energy constraint. *IEEE Trans. Inf. Theory*, 14(1), 129–134. [454]

Wyner, A. D. (1973). A theorem on the entropy of certain binary sequences and applications—II. *IEEE Trans. Inf. Theory*, 19(6), 772–777. [125, 270]

Wyner, A. D. (1974). Recent results in the Shannon theory. *IEEE Trans. Inf. Theory*, 20(1), 2–10. [99, 270]

Wyner, A. D. (1975a). The common information of two dependent random variables. *IEEE Trans. Inf. Theory*, 21(2), 163–179. [355]

Wyner, A. D. (1975b). On source coding with side information at the decoder. *IEEE Trans. Inf. Theory*, 21(3), 294–300. [271]

Wyner, A. D. (1975c). The wire-tap channel. *Bell Syst. Tech. J.*, 54(8), 1355–1387. [576]

Wyner, A. D., Wolf, J. K., and Willems, F. M. J. (2002). Communicating via a processing broadcast satellite. *IEEE Trans. Inf. Theory*, 48(6), 1243–1249. [520]

Wyner, A. D. and Ziv, J. (1973). A theorem on the entropy of certain binary sequences and applications—I. *IEEE Trans. Inf. Theory*, 19(6), 769–772. [31]

Wyner, A. D. and Ziv, J. (1976). The rate–distortion function for source coding with side information at the decoder. *IEEE Trans. Inf. Theory*, 22(1), 1–10. [288, 631]

Xiao, L. and Boyd, S. (2004). Fast linear iterations for distributed averaging. *Syst. Control Lett.*, 53(1), 65–78. [547]

Xiao, L., Boyd, S., and Kim, S.-J. (2007). Distributed average consensus with least-mean-square deviation. *J. Parallel Distrib. Comput.*, 67(1), 33–46. [547]

Xie, L.-L. and Kumar, P. R. (2004). A network information theory for wireless communication: scaling laws and optimal operation. *IEEE Trans. Inf. Theory*, 50(5), 748–767. [500]

Xie, L.-L. and Kumar, P. R. (2005). An achievable rate for the multiple-level relay channel. *IEEE Trans. Inf. Theory*, 51(4), 1348–1358. [481]

Xie, L.-L. and Kumar, P. R. (2006). On the path-loss attenuation regime for positive cost and linear scaling of transport capacity in wireless networks. *IEEE Trans. Inf. Theory*, 52(6), 2313–2328. [598]

Xue, F., Xie, L.-L., and Kumar, P. R. (2005). The transport capacity of wireless networks over fading channels. *IEEE Trans. Inf. Theory*, 51(3), 834–847. [598]

Yamamoto, H. (1981). Source coding theory for cascade and branching communication systems. *IEEE Trans. Inf. Theory*, 27(3), 299–308. [520]

Yamamoto, H. (1982). Wyner–Ziv theory for a general function of the correlated sources. *IEEE Trans. Inf. Theory*, 28(5), 803–807. [546]

Yamamoto, H. (1996). Source coding theory for a triangular communication systems. *IEEE Trans. Inf. Theory*, 42(3), 848–853. [520]

Yamamoto, H. (1997). Rate–distortion theory for the Shannon cipher system. *IEEE Trans. Inf. Theory*, 43(3), 827–835. [577]

Yang, Y., Zhang, Y., and Xiong, Z. (2010). A new sufficient condition for sum-rate tightness in quadratic Gaussian MT source coding. In *Proc. UCSD Inf. Theory Appl. Workshop*, La Jolla, CA. [313]

Yao, A. C.-C. (1979). Some complexity questions related to distributive computing. In *Proc. 11th Ann. ACM Symp. Theory Comput.*, Atlanta, GA, pp. 209–213. [546]

Yeung, R. W. (2008). *Information Theory and Network Coding*. Springer, New York. [378, 380]

Yeung, R. W., Li, S.-Y. R., Cai, N., and Zhang, Z. (2005a). Network coding theory—I: Single source. *Found. Trends Comm. Inf. Theory*, 2(4), 241–329. [378]

Yeung, R. W., Li, S.-Y. R., Cai, N., and Zhang, Z. (2005b). Network coding theory—II: Multiple source. *Found. Trends Comm. Inf. Theory*, 2(5), 330–381. [378]

Yu, W. and Cioffi, J. M. (2004). Sum capacity of Gaussian vector broadcast channels. *IEEE Trans. Inf. Theory*, 50(9), 1875–1892. [254]

Yu, W., Rhee, W., Boyd, S., and Cioffi, J. M. (2004). Iterative water-filling for Gaussian vector multiple-access channels. *IEEE Trans. Inf. Theory*, 50(1), 145–152. [253]

Yu, W., Sutivong, A., Julian, D. J., Cover, T. M., and Chiang, M. (2001). Writing on colored paper. In *Proc. IEEE Int. Symp. Inf. Theory*, Washington DC, pp. 302. [192]

Zamir, R. (2009). Lattices are everywhere. In *Proc. UCSD Inf. Theory Appl. Workshop*, La Jolla, CA, pp. 392–421. [69]

Zamir, R. and Feder, M. (1993). A generalization of the entropy power inequality with applications. *IEEE Trans. Inf. Theory*, 39(5), 1723–1728. [32]

Zeng, C.-M., Kuhlmann, F., and Buzo, A. (1989). Achievability proof of some multiuser channel coding theorems using backward decoding. *IEEE Trans. Inf. Theory*, 35(6), 1160–1165. [418, 454]

Zhang, Z. and Berger, T. (1987). New results in binary multiple descriptions. *IEEE Trans. Inf. Theory*, 33(4), 502–521. [335]

Ziegler, G. M. (1995). *Lectures on Polytopes*. Springer-Verlag, New York. [636]

Common Symbols

α, β	numbers in $[0, 1]$
δ, ϵ	small positive numbers
ρ	correlation coefficient
B	input cost constraint
b	input cost/number of transmission blocks
C	capacity
C	Gaussian capacity function
$\mathscr{C}$	capacity region
D	distortion constraint/relative entropy
d	distortion measure/dimension
E	indicator variable
E	expectation
$\mathcal{E}$	error events/set of edges
e	error message
F	cumulative distribution function (cdf)
$\mathbb{F}$	field
f	probability density function (pdf)
G	channel gain matrix/generator matrix
$\mathcal{G}$	graph
g	channel gain/generic function
H	entropy/parity check matrix
h	differential entropy
I	mutual information/interference-to-noise ratio (INR)
i	transmission time
j, k, l	generic indices
K	covariance matrix/crosscovariance matrix/secret key

M, m	message/index
N	number of nodes in a network/noise power
$\mathcal{N}$	set of nodes
n	transmission block length
P	power constraint
$P_e^{(n)}$	probability of error
P	probability
p	probability mass function (pmf)
Q	time-sharing random variable/Q-function
q	number of rounds
R	rate
R	Gaussian rate function
$\mathscr{R}$	rate region
$\mathbb{R}$	set of real numbers
S	signal-to-noise ratio (SNR)/random channel state
s	channel state
$\mathcal{T}_\epsilon^{(n)}$	typical set
U, V, W, X, Y, Z	source/channel/auxiliary random variables
$\mathcal{U}, \mathcal{V}, \mathcal{W}, \mathcal{X}, \mathcal{Y}, \mathcal{Z}$	source/channel/auxiliary random variable alphabets

Author Index

Subject Index

Printed in the United States
By Bookmasters